D0087075

To Martha, Niki, George, and Sophia

CONTENTS

PREFACE

All men by nature desire knowledge

ARISTOTLE
Metaphysics, Book I

Knowledge, however meager, is usable
if we know the amount of
uncertainty in it

CR Rao

Statistics has become an integral part of scientific investigations in virtually all disciplines and is used extensively in industry and government organizations. *Probability & Statistics with R for Engineers and Scientists* offers a comprehensive introduction to the most commonly used statistical ideas and methods.

This book evolved from lecture notes for a one-semester course aimed mainly at undergraduate students in engineering and the natural sciences, as well as mathematics education majors and graduate students from various disciplines. The choice of examples, exercises, and data sets reflects the diversity of this audience.

The mathematical level has been kept relatively modest. Students who have completed one semester of differential and integral calculus should find almost all the exposition accessible. In particular, substantial use of calculus is made only in Chapters 3 and 4 and the third section of Chapter 6. Matrix algebra is used only in Chapter 12, which is usually not taught in a one-semester course.

THE R SOFTWARE PACKAGE

The widespread use of statistics is supported by a number of statistical software packages. Thus, modern courses on statistical methodology familiarize students with reading and interpreting software output. In sharp contrast to other books with the same intended audience, this book emphasizes not only the interpretation of software output, but also the *generation* of this output.

I decided to emphasize the software R (launched in 1984), which is sponsored by the Free Software Foundation. R is now used by the vast majority of statistics graduate students for thesis research, is a leader in new software development,[1] and is increasingly accepted in industry.[2] Moreover, R can be downloaded for free so students do not have to go to computer labs for their assignments. (To download R, go to the site http://www.R-project.org/ and follow the instructions.)

---

[1] See, e.g., http://www.r-bloggers.com/r-and-the-journal-of-computational-and-graphical-statistics.

[2] See the *New York Times* article "Data Analysts Captivated by R's Power," by Ashlee Vance, January 6, 2009.

TEACHING INNOVATIONS AND CHAPTER CONTENT

In addition to the use of a software package as an integral part of teaching probability and statistics, this book contains a number of other innovative approaches, reflecting the teaching philosophy that: (a) students should be intellectually challenged and (b) major concepts should be introduced as early as possible.

This text's major innovations occur in Chapters 1 and 4. Chapter 1 covers most of the important statistical concepts including sampling concepts, random variables, the population mean and variance for finite populations, the corresponding sample statistics, and basic graphics (histograms, stem and leaf plots, scatterplots, matrix scatterplots, pie charts and bar graphs). It goes on to introduce the notions of statistical experiments, comparative studies, and corresponding comparative graphics. The concepts and ideas underlying comparative studies, including main effects and interactions, are interesting in themselves, and their early introduction helps engage students in "statistical thinking."

Chapter 4, which deals with joint (mainly bivariate) distributions, covers the standard topics (marginal and conditional distributions, and independence of random variables), but also introduces the important concepts of covariance and correlation, along with the notion of a regression function. The simple linear regression model is discussed extensively, as it arises in the hierarchical model approach for defining the bivariate normal distribution.

Additional innovations are scattered throughout the rest of the chapters. Chapter 2 is devoted to the definition and basic calculus of probability. Except for the use of R to illustrate some concepts and the early introduction of probability mass function, this material is fairly standard. Chapter 3 gives a more general definition of the mean value and variance of a random variable and connects it to the simple definition given in Chapter 1. The common probability models for discrete and continuous random variables are discussed. Additional models commonly used in reliability studies are presented in the exercises. Chapter 5 discusses the distribution of sums and the Central Limit Theorem. The method of least squares, method of moments, and method of maximum likelihood are discussed in Chapter 6. Chapters 7 and 8 cover interval estimation and hypothesis testing, respectively, for the mean, median, and variance as well as the parameters of the simple linear regression model. Chapters 9 and 10 cover inference procedures for two and $k > 2$ samples, respectively, including paired data and randomized block designs. Nonparametric, or rank-based, inference is discussed alongside traditional methods of inference in Chapters 7 through 10. Chapter 11 is devoted to the analysis of two-factor, three-factor, and fractional factorial designs. Polynomial and multiple regression, and related topics such as weighted least squares, variable selection, multicollinearity, and logistic regression are presented in Chapter 12. The final chapter, Chapter 13, develops procedures used in statistical process control.

DATA SETS

This book contains both real life data sets, with identified sources, and simulated data sets. They can all be found at

www.pearsonhighered.com/akritas

Clicking on the name of a particular data set links to the corresponding data file. Importing data sets into R from the URL is easy when using the *read.table* command. As an example, you can import the data set BearsData.txt into the R data frame *br* by copying and pasting its URL into a read.table command:

```
br=read.table("http://media.pearsoncmg.com/cmg/pmmg_mml_shared/
  mathstatsresources/Akritas/BearsData.txt", header=T)
```

The data sets can also be downloaded to your computer and then imported into R from there.

Throughout the book, the *read.table* command will include only the name of the particular data set to be imported into R. For example, the command for importing the bear data into R will be given as

```
br=read.table("BearsData.txt", header=T)
```

SUGGESTED COVERAGE

This book has enough material for a year-long course, but can also be adapted for courses of one semester or two quarters. In a one-semester course, meeting three times a week, I cover selected topics from Chapters 1 through 10 and, recalling briefly the concepts of main effects and interaction (first introduced in Chapter 1), I finish the course by explaining the R commands and output for two-way analysis of variance. I typically deemphasize joint continuous distributions in Chapter 4 and may skip one or more of the following topics: multinomial distribution (Section 4.6.4), the method of maximum likelihood (Section 6.3.2), sign confidence intervals for the median (Section 7.3.4), the comparison of two variances (Section 9.4), the paired T test for proportions (Section 9.5.3), the Wilcoxon signed-rank test (Section 9.5.4), and the chi-square test for proportions (Section 10.2.3). It is possible to include material from Chapter 13 on statistical process control (for example after Chapter 8) by omitting additional material. One suggestion is to omit the section on comparing estimators (Section 6.4), confidence intervals and tests for a normal variance (Sections 7.3.5 and 8.3.6), and randomized block designs (Section 10.4).

ACKNOWLEDGMENTS

I greatly appreciate the support of the Department of Statistics at Penn State University and express my sincere thanks to colleagues, instructors, and graduate students who used various editions of the lecture notes and provided many suggestions for improvement over the years. I also thank all the people at Pearson for a highly professional and cordial collaboration through the various stages of production of the book. Special thanks go to Mary Sanger who supervised the last stages of production with an exceptionally high level of care and professionalism.

I am very grateful for numerous edits and substantive suggestions I received by the following reviewers: Keith Friedman, University of Texas at Austin; Steven T. Garren, James Madison University; Songfeng Zheng, Missouri State University; Roger Johnson, South Dakota School of Mines & Technology; Subhash Kochar, Portland State University; Michael Levine, Purdue University; Karin Reinhold, SUNY at Albany; Kingsley A. Reeves, Jr., University of South Florida; Katarina Jegdic, University of Houston Downtown; Lianming Wang, University of South Carolina; Lynne Butler, Haverford College; John Callister, Cornell University.

Chapter 1

Basic Statistical Concepts

1.1 Why Statistics?

Statistics deals with collecting, processing, summarizing, analyzing, and interpreting data. On the other hand, scientists and engineers deal with such diverse issues as the development of new products, effective use of materials and labor, solving production problems, quality improvement and reliability, and, of course, basic research. The usefulness of statistics as a tool for dealing with the above problems is best seen through some specific case studies mentioned in the following example.

Example 1.1-1

Examples of specific case studies arising in the sciences and engineering include

1. estimating the coefficient of thermal expansion of a metal;
2. comparing two methods of cloud seeding for hail and fog suppression at international airports;
3. comparing two or more methods of cement preparation in terms of compressive strength;
4. comparing the effectiveness of three cleaning products in removing four different types of stains;
5. predicting the failure time of a beam on the basis of stress applied;
6. assessing the effectiveness of a new traffic regulatory measure in reducing the weekly rate of accidents;
7. testing a manufacturer's claim regarding the quality of its product;
8. studying the relation between salary increases and employee productivity in a large corporation;
9. estimating the proportion of US citizens age 18 and over who are in favor of expanding solar energy sources; and
10. determining whether the content of lead in the water of a certain lake is within the safety limit. ■

The reason why tasks like the above require statistics is **variability**. Thus, if all cement prepared according to the same method had the same compressive strength, the task of comparing the different methods in case study 3 would not require statistics; it would suffice to compare the compressive strength of one cement specimen prepared from each method. However, the strength of different cement

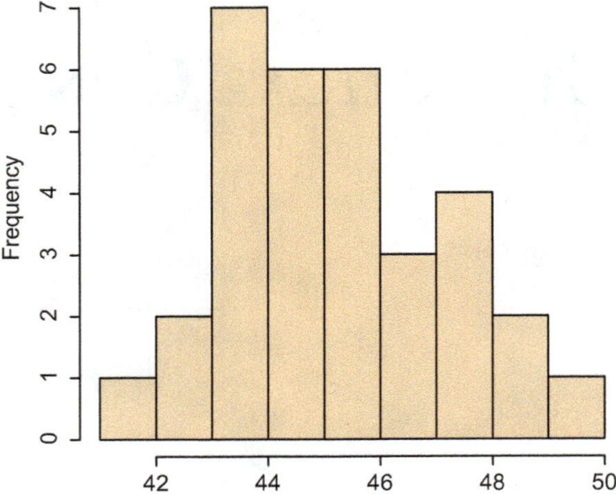

Figure 1-1 Histogram of 32 compressive strength measurements.

specimens prepared by the same method will, in general, differ. Figure 1-1 shows the histogram for 32 compressive strength measurements.[1] (See Section 1.5 for a discussion about histograms.) Similarly, if all beams fail at the same time under a given stress level, the prediction problem in case study 5 would not require statistics. A similar comment applies to all the case studies mentioned in Example 1.1-1.

An appreciation of the complications caused by variability begins by realizing that the problem of case study 3, as stated, is ambiguous. Indeed, if the hardness differs among preparations of the same cement mixture, then what does it mean to compare the hardness of different cement mixtures? A more precise statement of the problem would be to compare the *average* (or *mean*) hardness of the different cement mixtures. Similarly, the estimation problem in case study 1 is stated more precisely by referring to the average (or mean) thermal expansion.

It should also be mentioned that, due to variability, the familiar words *average* and *mean* have a technical meaning in statistics that can be made clear through the concepts of *population* and *sample*. These concepts are discussed in the next section.

1.2 Populations and Samples

As the examples of case studies mentioned in Example 1.1-1 indicate, statistics becomes relevant whenever the study involves the investigation of certain characteristic(s) of members (objects or subjects) in a certain **population** or populations. In statistics the word population is used to denote the set of all objects or subjects relevant to the particular study that are exposed to the same treatment or method. The members of a population are called **population units**.

Example 1.2-1

(a) In Example 1.1-1, case study 1, the characteristic under investigation is the thermal expansion of a metal in the population of all specimens of the particular metal.

(b) In Example 1.1-1, case study 3, we have two or more populations, one for each type of cement mixture, and the characteristic under investigation is compressive strength. Population units are the cement preparations.

(c) In Example 1.1-1, case study 5, the characteristic of interest is time to failure of a beam under a given stress level. Each stress level used in the study

[1] Compressive strength, in MPa (megapascal units), of test cylinders 6 in. in diameter by 12 in. high, using water/cement ratio of 0.4, measured on the 28th day after they were made.

corresponds to a separate population that consists of all beams that will be exposed to that stress level.

(d) In Example 1.1-1, case study 8, we have two characteristics, salary increase and productivity, for each subject in the population of employees of a large corporation. ■

In Example 1.2-1, part (c), we see that all populations consist of the same type of beams but are distinguished by the fact that beams of different populations will be exposed to different stress levels. Similarly, in Example 1.1-1, case study 2, the two populations consist of the same type of clouds distinguished by the fact that they will be seeded by different methods.

As mentioned in the previous section, the characteristic of interest varies among members of the same population. This is called the **inherent** or **intrinsic variability** of a population. A consequence of intrinsic variability is that complete, or *population-level*, understanding of characteristic(s) of interest requires a **census**, that is, examination of all members of the population. For example, full understanding of the relation between salary and productivity, as it applies to the population of employees of a large corporation (Example 1.1-1, case study 8), requires obtaining information on these two characteristics for all employees of the particular large corporation. Typically, however, census is not conducted due to cost and time considerations.

Example 1.2-2

(a) Cost and time considerations make it impractical to conduct a census of all US citizens age 18 and over in order to determine the proportion of these citizens who are in favor of expanding solar energy sources.

(b) Cost and time considerations make it impractical to analyze all the water in a lake in order to determine the lake's content of lead. ■

Moreover, census is often not feasible because the population is **hypothetical** or **conceptual**, in the sense that not all members of the population are available for examination.

Example 1.2-3

(a) If the objective is to study the quality of a product (as in Example 1.1-1, case studies 7 and 4), the relevant population consists not only of the available supply of this product, but also that which will be produced in the future. Thus, the relevant population is hypothetical.

(b) In a study aimed at reducing the weekly rate of accidents (Example 1.1-1, case study 6) the relevant population consists not only of the one-week time periods on which records have been kept, but also of future one-week periods. Thus, the relevant population is hypothetical. ■

In studies where it is either impractical or infeasible to conduct a census (which is the vast majority of cases), answers to questions regarding population-level properties/attributes of characteristic(s) under investigation are obtained by **sampling** the population. Sampling refers to the process of selecting a number of population units and recording their characteristic(s). For example, determination of the proportion of US citizens age 18 and over who are in favor of expanding solar energy sources is based on a sample of such citizens. Similarly, the determination of whether or not the content of lead in the water of a certain lake is within the safety limit must be based on water samples. The good news is that if the sample is suitably drawn from the population, then the **sample properties/attributes** of the characteristic of interest resemble (though they are not identical to) the **population properties/attributes**.

Example 1.2-4

(a) A *sample proportion* (i.e., the proportion in a chosen sample) of US citizens who favor expanding the use of solar energy approximates (but is, in general, different from) the *population proportion*. (Precise definitions of sample proportion and population proportion are given in Section 1.6.1.)

(b) The average concentration of lead in water samples (*sample average*) approximates (but is, in general, different from) the average concentration in the entire lake (*population average*). (Precise definitions of sample average and population average are given in Section 1.6.2.)

(c) The relation between salary and productivity manifested in a sample of employees approximates (but is, in general, different from) the relation in the entire population of employees of a large corporation.

Example 1.2-5

The easier-to-measure chest girth of bears is often used to estimate the harder-to-measure weight. Chest girth and weight measurements for 50 bears residing in a given forested area are marked with "x" in Figure 1-2. The colored circles indicate the chest girth and weight measurements of the bears in a sample of size 10.[2] The black line captures the roughly linear relationship between chest girth and weight in the population of 50 black bears, while the colored line does the same for the sample.[3] It is seen that the relationship between chest girth and weight suggested by the sample is similar but not identical to that of the population.

Sample properties of the characteristic of interest also differ from sample to sample. This is another consequence of the intrinsic variability of the population from which samples are drawn. For example, the number of US citizens, in a sample of size 20, who favor expanding solar energy will (most likely) be different from the corresponding number in a different sample of 20 US citizens. (See also the examples in Section 1.6.2.) The term **sampling variability** is used to describe such differences in the characteristic of interest from sample to sample.

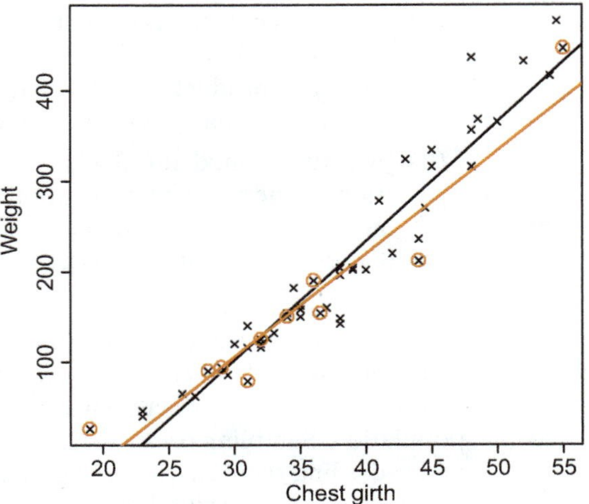

Figure 1-2 Population and sample relationships between chest girth (in) and weight (lb) of black bears.

[2] The sample was obtained by the method of *simple random sampling* described in Section 1.3.
[3] The lines were fitted by the method of *least squares* described in Chapter 6.

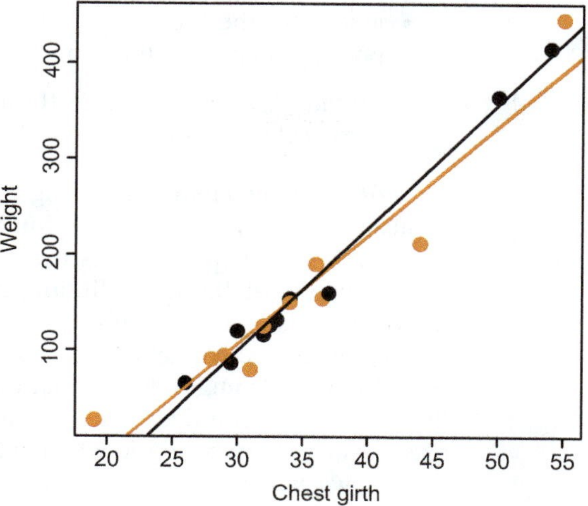

Figure 1-3 Variability in the relationships between chest girth and weight of black bears suggested by two different samples of size 10.

Example 1.2-6

As an illustration of sampling variability, a second sample of size 10 was taken from the population of 50 black bears described in Example 1.2-5. Figure 1-3 shows the chest girth and weight measurements for the original sample in colored dots while those for the second sample are shown in black dots. The sampling variability is demonstrated by the colored and black lines, which suggest somewhat different relationships between chest girth and weight, although both lines approximate the population relationship. ■

One must never lose sight of the fact that all scientific investigations aim at discovering the population-level properties/attributes of the characteristic(s) of interest. In particular, the problems in all the case studies mentioned in Example 1.1-1 refer to population-level properties. Thus, the technical meaning of the familiar word *average* (or *mean*), which was alluded to at the end of Section 1.1, is that of the population average (or mean); see Section 1.6.2 for a precise definition.

Population-level properties/attributes of characteristic(s) are called **population parameters**. Examples include the population mean (or average) and the population proportion that were referred to in Example 1.2-4. These and some additional examples of population parameters are defined in Sections 1.6 and 1.7. Further examples of population parameters, to be discussed in later chapters, include the *correlation coefficient* between two characteristics, e.g., between salary increase and productivity or between chest girth and weight. The corresponding sample properties/attributes of characteristics are called **statistics**, which is a familiar term because of its use in *sports statistics*. The sample mean (or average), sample proportion, and some additional statistics are defined in Sections 1.6 and 1.7, while further statistics are introduced in later chapters.

A sample can be thought of as a window that provides a glimpse into the population. However, due to sampling variability, a sample cannot yield accurate information regarding the population properties/attributes of interest. Using the new terminology introduced in the previous paragraph, this can be restated as: statistics approximate corresponding population parameters but are, in general, not equal to them.

Because only sample information is available, population parameters remain unknown. **Statistical inference** is the branch of statistics dealing with the uncertainty issues that arise in extrapolating to the population the information contained in the sample. Statistical inference helps decision makers choose actions in the absence of accurate knowledge about the population by

- assessing the accuracy with which statistics approximate corresponding population parameters; and
- providing an appraisal of the probability of making the wrong decision, or incorrect prediction.

For example, city officials might want to know whether a new industrial plant is pushing the average air pollution beyond the acceptable limits. Air samples are taken and the air pollution is measured in each sample. The sample average, or sample mean, of the air pollution measurements must then be used to decide if the overall (i.e., population-level) average air pollution is elevated enough to justify taking corrective action. In the absence of accurate knowledge, there is a risk that city officials might decide that the average air pollution exceeds the acceptable limit, when in fact it does not, or, conversely, that the average air pollution does not exceed the acceptable limit, when in fact it does.

As we will see in later chapters, statistical inference mainly takes the form of **estimation** (both **point** and, the more useful, **interval** estimation) of the population parameter(s) of interest, and of **testing** various **hypotheses** regarding the value of the population parameter(s) of interest. For example, estimation would be used in the task of estimating the average coefficient of thermal expansion of a metal (Example 1.1-1, case study 1), while the task of testing a manufacturer's claim regarding the quality of its product (Example 1.1-1, case study 7) involves hypothesis testing. Finally, the principles of statistical inference are also used in the problem of **prediction**, which arises, for example, if we would like to predict the failure time of a particular beam on the basis of the stress to which it will be exposed (Example 1.1-1, case study 5). The majority of the statistical methods presented in this book fall under the umbrella of statistical inference.

Exercises

1. A car manufacturer wants to assess customer satisfaction for cars sold during the previous year.

(a) Describe the population involved.

(b) Is the population involved hypothetical or not?

2. A field experiment is conducted to compare the yield of three varieties of corn used for biofuel. Each variety will be planted on 10 randomly selected plots and the yield will be measured at the time of harvest.

(a) Describe the population(s) involved.

(b) What is the characteristic of interest?

(c) Describe the sample(s).

3. An automobile assembly line is manned by two shifts a day. The first shift accounts for two-thirds of the overall production. Quality control engineers want to compare the average number of nonconformances per car in each of the two shifts.

(a) Describe the population(s) involved.

(b) Is (are) the population(s) involved hypothetical or not?

(c) What is the characteristic of interest?

4. A consumer magazine article titled "How Safe Is the Air in Airplanes" reports that the air quality, as quantified by the degree of staleness, was measured on 175 domestic flights.

(a) Identify the population of interest.

(b) Identify the sample.

(c) What is the characteristic of interest?

5. In an effort to determine the didactic benefits of computer activities when used as an integral part of a statistics course for engineers, one section is taught using the traditional method, while another is taught with computer activities. At the end of the semester, each student's score on the same test is recorded. To eliminate unnecessary variability, both sections were taught by the same professor.

(a) Is there one or two populations involved in the study?

(b) Describe the population(s) involved.

(c) Is (are) the population(s) involved hypothetical or not?

(d) What is (are) the sample(s) in this study?

1.3 Some Sampling Concepts

1.3.1 REPRESENTATIVE SAMPLES

Proper extrapolation of sample information to the population, that is, valid statistical inference, requires that the sample be **representative** of the population. For example, extrapolation of the information from a sample that consists of those who work in the oil industry to the population of US citizens will unavoidably lead to wrong conclusions about the prevailing public opinion regarding the use of solar energy.

A famous (or infamous) example that demonstrates what can go wrong when a non-representative sample is used is the *Literary Digest* poll of 1936. The magazine *Literary Digest* had been extremely successful in predicting the results in US presidential elections, but in 1936 it predicted a 3-to-2 victory for Republican Alf Landon over the Democratic incumbent Franklin Delano Roosevelt. The blunder was due to the use of a non-representative sample, which is discussed further in Section 1.3.4. It is worth mentioning that the prediction of the *Literary Digest* magazine was wrong even though it was based on 2.3 million responses (out of 10 million questionnaires sent). On the other hand, Gallup correctly predicted the outcome of that election by surveying only 50,000 people.

The notion of representativeness of a sample, though intuitive, is hard to pin down because there is no way to tell just by looking at a sample whether or not it is representative. Thus we adopt an indirect definition and say that a sample is representative if it leads to valid statistical inference. The only assurance that the sample will be representative comes from the method used to select the sample. Some of these sampling methods are discussed below.

1.3.2 SIMPLE RANDOM SAMPLING AND STRATIFIED SAMPLING

The most straightforward method for obtaining a representative sample is called **simple random sampling**. A sample of size n, selected from some population, is a simple random sample if the selection process ensures that every sample of size n has an equal chance of being selected. In particular, every member of the population has the same chance of being included in the sample.

A common way to select a simple random sample of size n from a finite population consisting of N units is to number the population units from $1, \ldots, N$, use a **random number generator** to randomly select n of these numbers, and form the sample from the units that correspond to the n selected numbers. A random number generator for selecting a simple random sample simulates the process of writing each number from $1, \ldots, N$ on slips of paper, putting the slips in a box, mixing them thoroughly, selecting one slip at random, and recording the number on the slip. The process is repeated (without replacing the selected slips in the box) until n distinct numbers from $1, \ldots, N$ have been selected.

Example 1.3-1

Sixty KitchenAid professional grade mixers are manufactured per day. Prior to shipping, a simple random sample of 12 must be selected from each day's production and carefully rechecked for possible defects.

(a) Describe a procedure for obtaining a simple random sample of 12 mixers from a day's production of 60 mixers.

(b) Use R to implement the procedure described in part (a).

Solution

As a first step we identify each mixer with a number from 1 to 60. Next, we write each number from 1 to 60 on separate, identical slips of paper, put all 60 slips of paper in a box, and mix them thoroughly. Finally, we select 12 slips from the box, one at a time and without replacement. The 12 numbers selected specify the desired sample of size $n = 12$ mixers from a day's production of 60. This process can be implemented in R with the command

Simple Random Sampling in R

```
y=sample(seq(1, 60), size=12)
```

(1.3.1)

The command without the $y=$, that is, *sample(seq(1, 60), size=12)*, will result in the 12 random numbers being typed in the R console; with the command as stated the random numbers are stored in the object *y* and can be seen by typing the letter "*y*." A set of 12 numbers thus obtained is 6, 8, 57, 53, 31, 35, 2, 4, 16, 7, 49, 41. ■

Clearly, the above technique cannot be used with hypothetical/infinite populations. However, measurements taken according to a set of well-defined instructions can assure that the essential properties of simple random sampling hold. For example, in comparing the compressive strength of cement mixtures, guidelines can be established for the mixture preparations and the measurement process to assure that the sample of measurements taken is representative.

As already mentioned, simple random sampling guarantees that every population unit has the same chance of being included in the sample. However, the mere fact that every population unit has the same chance of being included in the sample does not guarantee that the sampling process is simple random. This is illustrated in the following example.

Example 1.3-2

In order to select a representative sample of 10 from a group of 100 undergraduate students consisting of 50 male and 50 female students, the following sampling method is implemented: (a) assign numbers 1–50 to the male students and use a random number generator to select five of them; (b) repeat the same for the female students. Does this method yield a simple random sample of 10 students?

Solution

First note that the sampling method described guarantees that every student has the same chance (1 out of 10) of being selected. However, this sampling excludes all samples with unequal numbers of male and female students. For example, samples consisting of 4 male and 6 female students are excluded, that is, have zero chance of being selected. Hence, the condition for simple random sampling, namely, that each sample of size 10 has equal chance of being selected, is violated. It follows that the method described does not yield a simple random sample. ■

The sampling method of Example 1.3-2 is an example of what is called **stratified sampling**. Stratified sampling can be used whenever the population of interest consists of well-defined subgroups, or sub-populations, which are called **strata**. Examples of strata are ethnic groups, types of cars, age of equipment, different labs where water samples are sent for analysis, and so forth. Essentially, a stratified sample consists of simple random samples from each of the strata. A common method of choosing the within-strata sample sizes is to make the sample representation of each stratum equal to its population representation. This method of *proportionate allocation* is used in Example 1.3-2. Stratified samples are also representative, that is, they allow for valid statistical inference. In fact, if population units belonging to the same stratum tend to be more homogenous (i.e., similar) than population units belonging in different strata, then stratified sampling provides more accurate information regarding the entire population, and thus it is preferable.

1.3.3 SAMPLING WITH AND WITHOUT REPLACEMENT

In sampling from a finite population, one can choose to do the sampling **with replacement** or **without replacement**. Sampling with replacement means that after a unit is selected and its characteristic is recorded, it is replaced back into the population and may therefore be selected again. Tossing a fair coin can be thought of as sampling with replacement from the population {*Heads, Tails*}. In sampling without replacement, each unit can be included only once in the sample. Hence, simple random sampling is sampling without replacement.

It is easier to analyze the properties of a sample drawn with replacement because each selected unit is drawn from the same (the original) population of N units. (Whereas, in sampling without replacement, the second selection is drawn from a reduced population of $N-1$ units, the third is drawn from a further reduced population of $N-2$ units, and so forth.) On the other hand, including population unit(s) more than once (which is possible when sampling with replacement) clearly does not enhance the representativeness of the sample. Hence, the conceptual convenience of sampling with replacement comes with a cost, and, for this reason, it is typically avoided (but see the next paragraph). However, the cost is negligible when the population size is much larger than the sample size. This is because the likelihood of a unit being included twice in the sample is negligible, so that sampling with and without replacement are essentially equivalent. In such cases, we can pretend that a sample obtained by simple random sampling (i.e., without replacement) has the same properties as a sample obtained with replacement.

A major application of sampling with replacement occurs in the statistical method known by the name of **bootstrap**. Typically, however, this useful and widely used tool for statistical inference is not included in introductory textbooks.

1.3.4 NON-REPRESENTATIVE SAMPLING

Non-representative samples arise whenever the sampling plan is such that a part, or parts, of the population of interest are either excluded from, or systematically under-represented in, the sample.

Typical non-representative samples are the so-called **self-selected** and **convenience** samples. As an example of a self-selected sample, consider a magazine that conducts a reply-card survey of its readers, then uses information from cards that were returned to make statements like "80% of readers have purchased cellphones with digital camera capabilities." In this case, readers who like to update and try new technology are more likely to respond indicating their purchases. Thus, the proportion of purchasers of cellphones with digital camera capabilities in the sample of returned cards will likely be much higher than it is amongst all readers. As an example of a convenience sample, consider using the students in your statistics class as a sample of students at your university. Note that this sampling plan excludes students from majors that do not require a statistics course. Moreover, most students take statistics in their sophomore or junior year and thus freshmen and seniors will be under-represented.

Perhaps the most famous historical example of a sampling blunder is the 1936 pre-election poll by the *Literary Digest* magazine. For its poll, the *Literary Digest* used a sample of 10 million people selected mainly from magazine subscribers, car owners, and telephone directories. In 1936, those who owned telephones or cars, or subscribed to magazines, were more likely to be wealthy individuals who were not happy with the Democratic incumbent. Thus, it was a convenience sample that excluded (or severely under-represented) parts of the population. Moveover, only 2.3 million responses were returned from the 10 million questionnaires that were sent. Obviously, those who felt strongly about the election were more likely to respond, and a majority of them wanted change. Thus, the *Literary Digest* sample was self-selected, in addition to being a sample of convenience. (The *Literary Digest* went bankrupt, while Gallup survived to make another blunder another day [in the 1948 Dewey-Truman contest].)

The term *selection bias* refers to the systematic exclusion or under-representation of some part(s) of the population of interest. Selection bias, which is inherent in self-selected and convenience samples, is the typical cause of non-representative samples. Simple random sampling and stratified sampling avoid selection bias. Other sampling methods that avoid selection bias do exist, and in some situations they can be less costly or easier to implement. But in this book we will mainly assume that the samples are simple random samples, with occasional passing reference to stratified sampling.

Exercises

1. The person designing the study of Exercise 5 in Section 1.2, aimed at determining the didactic benefits of computer activities, can make one of the two choices: (i) make sure that the students know which of the two sections will be taught with computer activities, so they can make an informed choice, or (ii) not make available any information regarding the teaching method of the two sections. Which of these two choices provides a closer approximation to simple random sampling?

2. A type of universal remote for home theater systems is manufactured in three distinct locations. Twenty percent of the remotes are manufactured in location A, 50% in location B, and 30% in location C. The quality control team (QCT) wants to inspect a simple random sample (SRS) of 100 remotes to see if a recently reported problem with the menu feature has been corrected. The QCT requests that each location send to the QC Inspection Facility a SRS of remotes from their recent production as follows: 20 from location A, 50 from B, and 30 from C.

(a) Does the sampling scheme described produce a simple random sample of size 100 from the recent production of remotes?

(b) Justify your answer in part (a). If you answer no, then what kind of sampling is it?

3. A civil engineering student, working on his thesis, plans a survey to determine the proportion of all current drivers in his university town that regularly use their seat belt. He decides to interview his classmates in the three classes he is currently enrolled.

(a) What is the population of interest?

(b) Do the student's classmates constitute a simple random sample from the population of interest?

(c) What name have we given to the sample that the student collected?

(d) Do you think that this sample proportion is likely to overestimate or underestimate the true proportion of all drivers who regularly use their seat belt?

4. In the Macworld Conference Expo Keynote Address on January 9, 2007, Steve Jobs announced a new product, the iPhone. A technology consultant for a consumer magazine wants to select 15 devices from the pilot lot of 70 iPhones to inspect feature coordination. Describe a method for obtaining a simple random sample of 15 from the lot of 70 iPhones. Use R to select a sample of 15. Give the R commands and the sample you obtained.

5. A distributor has just received a shipment of 90 drain pipes from a major manufacturer of such pipes. The distributor wishes to select a sample of size 5 to carefully inspect for defects. Describe a method for obtaining a simple random sample of 5 pipes from the shipment of 90 pipes. Use R to implement the method. Give the R commands and the sample you obtained.

6. A service agency wishes to assess its clients' views on quality of service over the past year. Computer records identify 1000 clients over the past 12 months, and a decision is made to select 100 clients to survey.

(a) Describe a procedure for selecting a simple random sample of 100 clients from last year's population of 1000 clients.

(b) The population of 1000 clients consists of 800 Caucasian-Americans, 150 African-Americans and 50 Hispanic-Americans. Describe an alternative procedure for selecting a representative random sample of size 100 from the population of 1000 clients.

(c) Give the R commands for implementing the sampling procedures described in parts (a) and (b).

7. A car manufacturer wants information about customer satisfaction for cars sold during the previous year. The particular manufacturer makes three different types of cars. Describe and discuss two different random sampling methods that might be employed.

8. A particular product is manufactured in two facilities, A and B. Facility B is more modern and accounts for 70% of the total production. A quality control engineer wishes to obtain a simple random sample of 50 from the entire production during the past hour. A coin is flipped and each time the flip results in heads, the engineer selects an item at random from those produced in facility A, and each time the flip results in tails, the engineer selects an item at random from those produced in facility B. Does this sampling scheme result in simple random sampling? Explain your answer.

9. An automobile assembly line operates for two shifts a day. The first shift accounts for two-thirds of the overall production. The task of quality control engineers is to monitor the number of nonconformances per car. Each day a simple random sample of 6 cars from the first shift, and a simple random sample of 3 cars from the second shift is taken, and the number of nonconformances per car is recorded. Does this sampling scheme produce a simple random sample of size 9 from the day's production? Justify your answer.

1.4 Random Variables and Statistical Populations

The characteristics of interest in all study examples given in Section 1.1 can be **quantitative** in the sense that they can be measured and thus can be expressed as numbers. Though quantitative characteristics are more common, **categorical**, including **qualitative**, characteristics also arise. Two examples of qualitative characteristics are gender and type of car, while strength of opinion is (ordinal) categorical. Since statistical procedures are applied on numerical data sets, numbers are assigned for expressing categorical characteristics. For example, −1 can be used to denote that a subject is male, and +1 to denote a female subject.

A characteristic of any type expressed as a number is called a **variable**. Categorical variables are a particular kind of **discrete** variables. Quantitative variables can also be discrete. For example, all variables expressing counts, such as the number in favor of a certain proposition, are discrete. Quantitative variables expressing measurements on a continuous scale, such as measurements of length, strength, weight, or time to failure, are examples of **continuous** variables. Finally, variables can be **univariate**, **bivariate**, or **multivariate** depending on whether one or two or more characteristics are measured, or recorded, on each population unit.

Example 1.4-1

(a) In a study aimed at determining the relation between productivity and salary increase, two characteristics are recorded on each population unit (productivity and salary increase), resulting in a bivariate variable.

(b) Consider the study that surveys US citizens age 18 and over regarding their opinion on solar energy. If an additional objective of the study is to determine how this opinion varies among different age groups, then the age of each individual in the sample is also recorded, resulting in a bivariate variable. If, in addition, the study aims to investigate how this opinion varies between genders, then the gender of each individual in the sample is also recorded, resulting in a multivariate variable.

(c) Consider the environmental study that measures the content of lead in water samples from a lake in order to determine if the concentration of lead exceeds the safe limits. If other contaminants are also of concern, then the content of these other contaminants is also measured in each water sample, resulting in a multivariate variable. ■

Due to the intrinsic variability, the value of the (possibly multivariate) variable varies among population units. It follows that when a population unit is randomly sampled from a population, its value is not known a priori. The value of the variable of a population unit that will be randomly sampled will be denoted by a capital letter, such as X. The fact that X is not known a priori justifies the term **random variable** for X.

> A random variable, X, denotes the value of the variable of a population unit that will be sampled.

The population from which a random variable was drawn will be called the **underlying population** of the random variable. Such terminology is particularly helpful in studies involving several populations, as are all studies that compare the performance of two or more methods or products; see, for example, case study 3 of Example 1.1-1.

Finally, we need a term for the entire collection of values that the variable under investigation takes among the units in the population. Stated differently, suppose that each unit in the population is labeled by the value of the variable under investigation, and the values in all labels are collected. This collection of values is called the **statistical population**. Note that if two (or more) population units have the same value of the variable, then this value appears two (or more) times in the statistical population.

Example 1.4-2 Consider the study that surveys US citizens age 18 and over regarding their opinion on solar energy. Suppose that the opinion is rated on the scale $0, 1, \ldots, 10$, and imagine each member of the population labeled by the value of their opinion. The statistical population contains as many 0's as there are people with opinion rated 0, as many 1's as there are people whose opinion is rated 1, and so forth. ■

The word "population" will be used to refer either to the population of units or to the statistical population. The context, or an explanation, will make clear which is the case.

In the above discussion, a random variable was introduced as the numerical outcome of random sampling from a (statistical) population. More generally, the concept of a random variable applies to the outcome of any action or process that generates a random numerical outcome. For example, the process of taking the arithmetic average of a simple random sample (see Section 1.6 for details) generates a random numerical outcome which, therefore, is a random variable.

Exercises

1. In a population of 500 tin plates, the number of plates with 0, 1, and 2 scratches is $N_0 = 190$, $N_1 = 160$, and $N_2 = 150$.

(a) Identify the variable of interest and the statistical population.

(b) Is the variable of interest quantitative or qualitative?

(c) Is the variable of interest univariate, bivariate, or multivariate?

2. Consider the following examples of populations, together with the variable/characteristic measured on each population unit.

(a) All undergraduate students currently enrolled at PSU. Variable: major.

(b) All campus restaurants. Variable: seating capacity.

(c) All books in Penn State libraries. Variable: frequency of check-out.

(d) All steel cylinders made in a given month. Variable: diameter.

For each of the above examples, describe the statistical population, state whether the variable of interest is quantitative or qualitative, and specify another variable that could be measured on the population units.

3. At the final assembly point of BMW cars in Graz, Austria, the car's engine and transmission arrive from Germany and France, respectively. A quality control inspector, visiting for the day, selects a simple random sample of n cars from the N cars available for inspection, and records the total number of engine and transmission nonconformances for each of the n cars.

(a) Is the variable of interest univariate, bivariate or multivariate?

(b) Is the variable of interest quantitative or qualitative?

(c) Describe the statistical population.

(d) Suppose the number of nonconformances in the engine and transmission are recorded separately for each car. Is the new variable univariate, bivariate, or multivariate?

4. In Exercise 4 in Section 1.2, a consumer magazine article reports that the air quality, as quantified by the degree of staleness, was measured on 175 domestic flights.

(a) Identify the variable of interest and the statistical population.

(b) Is the variable of interest quantitative or qualitative?

(c) Is the variable of interest univariate or multivariate?

5. A car manufacturing company that makes three different types of cars wants information about customer satisfaction for cars sold during the previous year. Each customer is asked for the type of car he or she bought last year and to rate his or her level of satisfaction on a scale from 1–6.

(a) Identify the variable recorded and the statistical population.

(b) Is the variable of interest univariate?

(c) Is the variable of interest quantitative or categorical?

1.5 Basic Graphics for Data Visualization

This section describes some of the most common graphics for data presentation and visualization. Additional graphics are introduced throughout this book.

1.5.1 HISTOGRAMS AND STEM AND LEAF PLOTS

Histograms and stem and leaf plots offer ways of organizing and displaying data. Histograms consist of dividing the range of the data into consecutive intervals, or *bins*, and constructing a box, or vertical bar, above each bin. The height of each box represents the bin's *frequency*, which is the number of observations that fall in the bin. Alternatively, the heights can be adjusted so the histogram's area (i.e., the total area defined by the boxes) equals one.

R will automatically choose the number of bins but it also allows user-specified intervals. Moreover, R offers the option of constructing a *smooth histogram*. Figure 1-4 shows a histogram (with area adjusted to one) of the Old Faithful geyser's eruption durations with a smooth histogram superimposed. (The data are from the R data frame *faithful*.)

Stem and leaf plots offer a somewhat different way for organizing and displaying data. They retain more information about the original data than histograms but do not offer as much flexibility in selecting the bins. The basic idea is to think of each observation as the *stem*, which consists of the beginning digit(s), and the *leaf*, which consists of the first of the remaining digits. In spite of different grouping of the observations, the stem and leaf display of the Old Faithful geyser's eruption durations shown in Figure 1-5 reveals a similar bimodal (i.e., having two modes or peaks) shape.

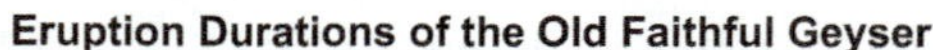

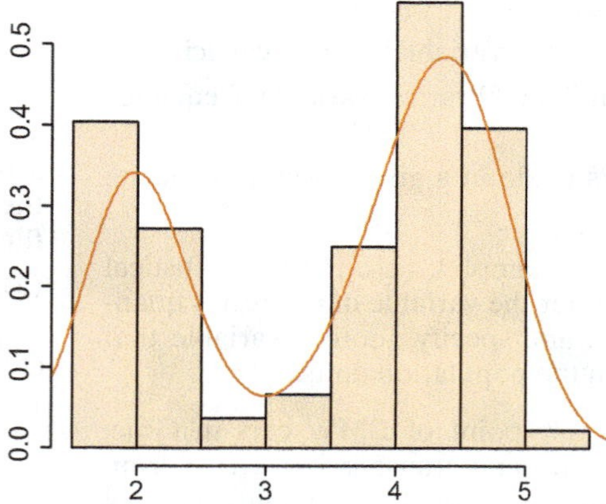

Figure 1-4 Histogram and smooth histogram for 272 eruption durations (min).

```
16 | 070355555588
18 | 000022233333335577777777888822335777888
20 | 00002223378800035778
22 | 0002335578023578
24 | 00228
26 | 23
28 | 080
30 | 7
32 | 2337
34 | 250077
36 | 0000823577
38 | 2333335582225577
40 | 0000003357788888002233555577778
42 | 03335555778800233333555577778
44 | 02222335557780000000023333357778888
46 | 0000233357700000023578
48 | 00000022335800333
50 | 0370
```

Figure 1-5 Stem and leaf plot for the 272 eruption durations.

With the R object x containing the data (e.g., $x = faithful\$eruptions$), the R commands for histograms and the stem and leaf plot are [# is the comment character]

R Commands for Histograms, Smooth Histograms, and Stem and Leaf Plots

```
hist(x) # basic frequency histogram

hist(x, freq=FALSE) # histogram area = 1

plot(density(x)) # basic smooth histogram

hist(x, freq=F); lines(density(x)) # superimposes
  the two

stem(x) # basic stem and leaf plot

stem(x, scale=1) # equivalent to the above
```

(1.5.1)

REMARK 1.5-1

1. The main label of a figure and the labels for the axes are controlled by *main = " ", xlab = " ", ylab = " "*, respectively; leaving a blank space between the quotes results in no labels. The color can also be specified. For example, the commands used for Figure 1-4 are *x = faithful$eruptions; hist(x, freq = F, main = "Eruption Durations of the Old Faithful Geyser", xlab = " ", col = "grey"); lines(density(x), col = "red")*.
2. To override the automatic selection of bins one can either specify the number of bins (for example *breaks = 6*), or specify explicitly the break points of the bins. Try *hist(faithful$eruptions, breaks = seq(1.2, 5.3, 0.41))*.
3. For additional control parameters type *?hist*, *?density*, or *?stem* on the R console. ◁

As an illustration of the role of the *scale* parameter in the stem command (whose default value is 1), consider the data on US beer production (in millions of barrels)

```
3 | 566699
4 | 11122444444
4 | 6678899
5 | 022334
5 | 5
```

for different quarters during the period 1975–1982. Entering the data in the R object *x* through *x = c(35, 36, 36, 36, 39, 39, 41, 41, 41, 42, 42, 44, 44, 44, 44, 44, 44, 46, 46, 47, 48, 48, 49, 49, 50, 52, 52, 53, 53, 54, 55)*, the command *stem(x, scale = 0.5)* results in the above stem and leaf display. Note that leaves within each stem have been split into the low half (integers from 0 through 4) and the upper half (integers from 5 through 9).

1.5.2 SCATTERPLOTS

Scatterplots are useful for exploring the relationship between two and three variables. For example, Figures 1-2 and 1-3 show such scatterplots for the variables bear chest girth and bear weight for a population of black bears and a sample drawn from that population. These scatterplots suggested a fairly strong *positive association* between chest girth and weight (i.e., bigger chest girth suggests a heavier bear), so that chest girth can be used for predicting a bear's weight. In this section we will see some enhanced versions of the basic scatterplot and a three-dimensional (3D) scatterplot.

Scatterplots with Subclass Identification The scatterplot in Figure 1-6 is similar to the scatterplot of Figure 1-2 but uses colors to distinguish between male and female bears. The additional insight gained from Figure 1-6 is that the relationship between the variables chest girth and weight is similar for both genders in that population of black bears.

Scatterplot Matrix As the name suggests, a scatterplot matrix is a matrix of scatterplots for all pairs of variables in a data set. In fact, two scatterplots are produced for every pair of variables, with each variable being plotted once on the *x*-axis and once on the *y*-axis. Figure 1-7 gives the matrix of all pairwise scatterplots between the

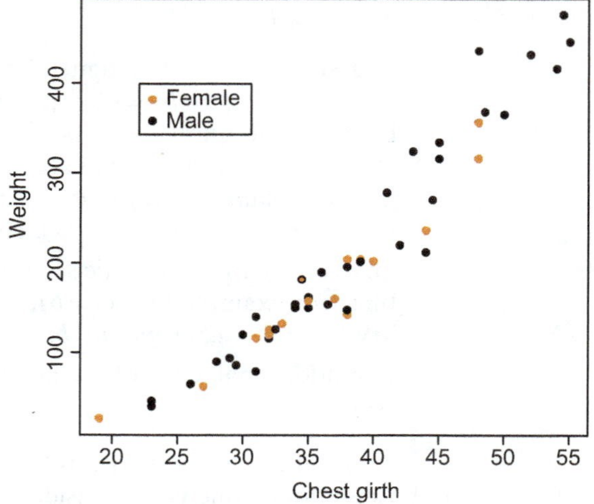

Figure 1-6 Bear weight vs chest girth scatterplot.

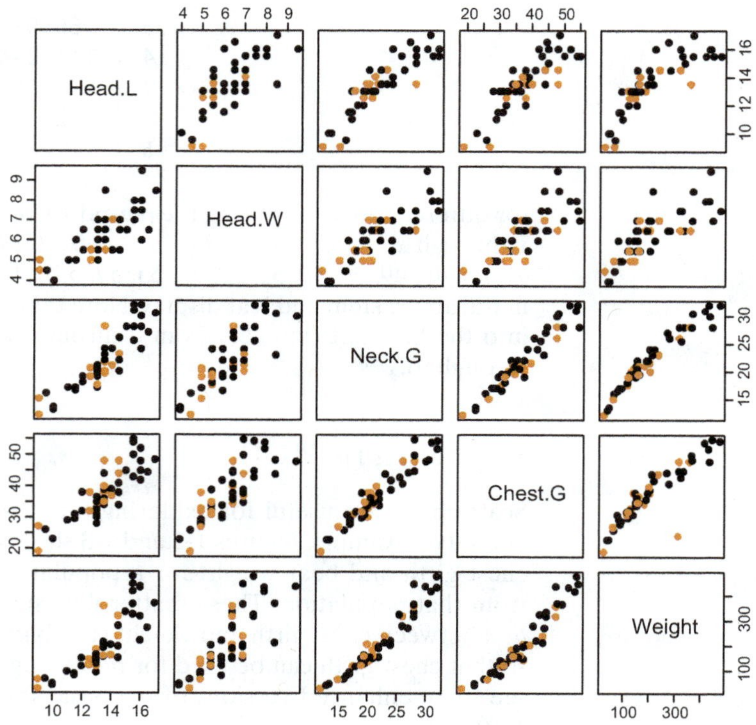

Figure 1-7 Scatterplot matrix for bear measurements.

different measurements taken on the black bears. The scatterplot in location (2,1), that is, in row 2 and column 1, has Head.L (head length) on the *x*-axis and Head.W (head width) on the *y*-axis, while the scatterplot in location (1,2) has Head.W on the *x*-axis and Head.L on the *y*-axis.

Scatterplot matrices are useful for identifying which variable serves as the best predictor for another variable. For example, Figure 1-7 suggests that a bear's chest girth and neck girth are the two best single predictors for a bear's weight.

With the data read into data frame *br* (for example by *br = read.table("BearsData.txt", header = T)*), the R commands that generated Figures 1-6 and 1-7 are:[4]

R Commands for Figures 1-6 and 1-7

```
attach(br) # so variables can be referred to by name

plot(Chest.G, Weight, pch=21, bg=c("red",
  "green")[unclass(Sex)]) # Figure 1-6

legend( x=22, y=400, pch=c(21, 21), col=c("red",
  "green"), legend=c("Female", "Male")) # add legend in
  Figure 1-6

pairs(br[4:8], pch=21,bg=c("red", "green")[unclass(Sex)]) #
  Figure 1-7
```

Scatterplots with Marginal Histograms This enhancement of the basic scatterplot shows individual histograms for the two variables used in the scatterplot. Figure 1-8 shows such an enhancement for the scatterplot of Figure 1-6.[5] The term *marginal*, which is justified by the fact the histograms appear on the margins of the scatterplot, is commonly used to refer to the statistical population of individual variables in a multivariate data set; see also Chapter 4.

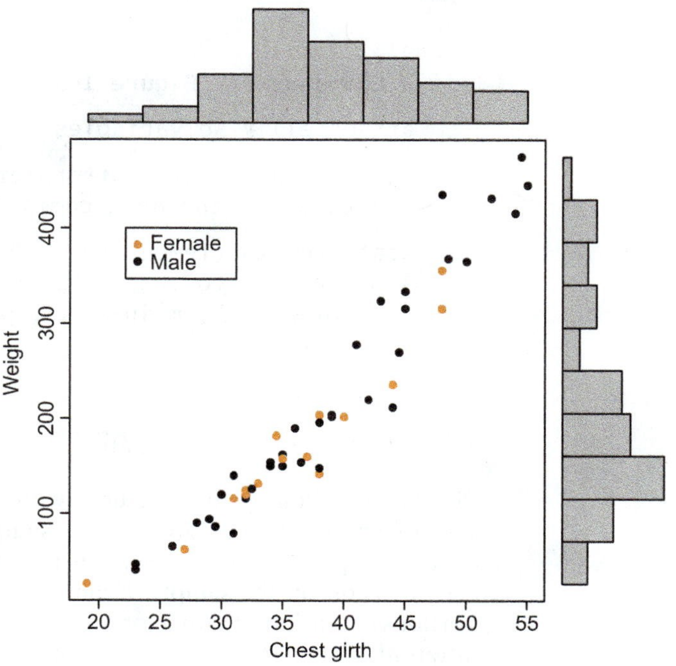

Figure 1-8 Scatterplot of bear weight vs chest girth showing the marginal histograms.

[4] Attempts to estimate a bear's weight from its chest girth measurements go back to Neil F. Payne (1976). Estimating live weight of black bears from chest girth measurements, *The Journal of Wildlife Management*, 40(1): 167–169. The data used in Figure 1-7 is a subset of a data set contributed to Minitab by Dr. Gary Alt.

[5] The R commands that generated Figure 1-8 are given at http://www.stat.psu.edu/~mga/401/fig/ScatterHist.txt; they are a variation of the commands given in an example on http://www.r-bloggers.com/example-8-41-scatterplot-with-marginal-histograms.

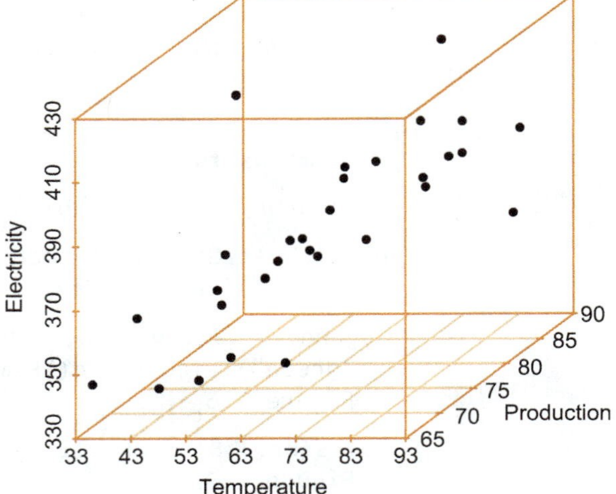

Figure 1-9 3D scatterplot for temperature, production, and electricity.

3D Scatterplot Figure 1-9 pertains to data on electricity consumed in an industrial plant in 30 consecutive 30-day periods, together with the average temperature and amount (in tons) of production. This figure gives a three dimensional view of the joint effect of temperature and production volume on electricity consumed.

With the data read into data frame *el*, for example, by *el=read.table("ElectrProdTemp.txt", header=T)*, the R commands used to generate this figure are:

R Commands for Figure 1-9

```
attach(el) # so variables can be referred to by name

install.packages("scatterplot3d"); library(scatterplot3d) #
  needed for the next command

scatterplot3d(Temperature, Production, Electricity,
  angle=35, col.axis="blue", col.grid="lightblue",
  color="red", main=" ", pch=21, box=T) # for
  Figure 1-9
```

1.5.3 PIE CHARTS AND BAR GRAPHS

Pie charts and bar graphs are used with count data that describe the prevalence of each of a number of categories in the sample. Alternatively, they can display the percentage or proportion (see Section 1.6.1 for the precise definition and notation) of each category in the sample. Examples include the counts (or percentages or proportions) of different ethnic or education or income categories, the market share of different car companies at a given point in time, the popularity of different car colors, and so on. When the heights of the bars are arranged in a decreasing order, the bar graph is also called a **Pareto chart**. The Pareto chart is a key tool in improvement programs, where it is often used to represent the most common sources of defects in a manufacturing process, or the most frequent reasons for customer complaints, and so on.

The pie chart is perhaps the most widely used statistical chart in the business world and is particularly popular in the mass media. It is a circular chart,

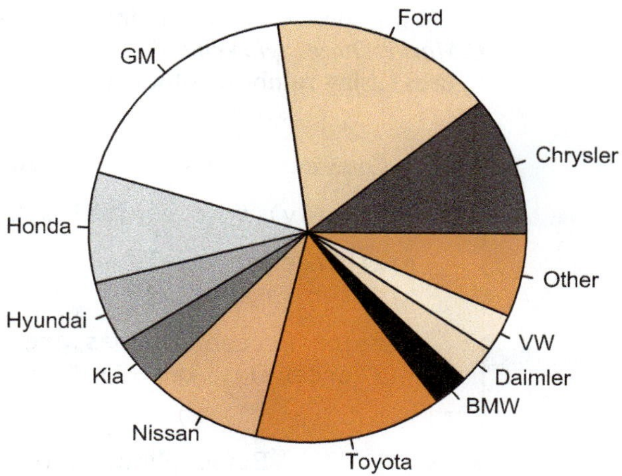

Figure 1-10 Pie chart of light vehicle market share data.

where the sample or the population is represented by a circle (pie) that is divided into sectors (slices) whose sizes represent proportions. The pie chart in Figure 1-10 displays information on the November 2011 light vehicle market share of car companies.[6]

It has been pointed out, however, that it is difficult to compare different sections of a given pie chart, or to compare data across different pie charts such as the light vehicle market share of car companies at two different time points. According to *Stevens' power law*,[7] length is a better scale to use than area. The bar graph achieves improved visual perception by using bars of height proportional to the proportion it represents. In that sense, a bar graph is similar to a histogram with area adjusted to one. The bar graph for the aforementioned light vehicle market share data is shown in Figure 1-11.

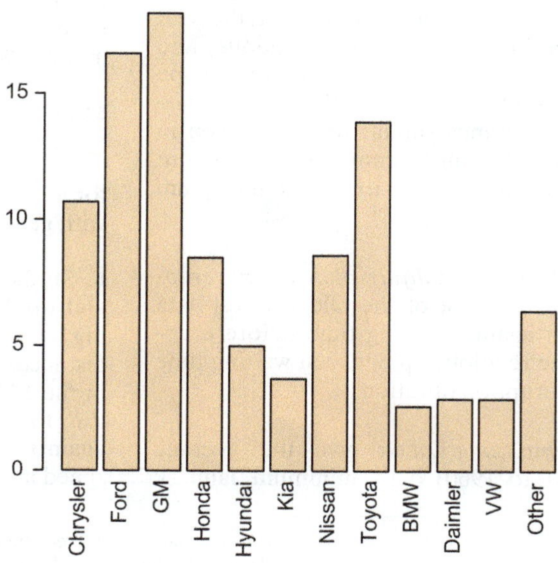

Figure 1-11 Bar graph of light vehicle market share data.

---

[6] http://wardsauto.com/datasheet/us-light-vehicle-sales-and-market-share-company-2004–2013

[7] S. S. Stevens (1957). On psychophysical law. *Psychological Review*. 64(3): 153–181.

With the data read into the data frame *lv*, [e.g., by *lv* = *read.table*("*MarketShareLightVeh.txt*", *header* = *T*)], the R commands used to generate these figures (using rainbow colors) are:

R Commands for Figures 1-10 and 1-11

```
attach(lv) # so variables can be referred to by name

pie(Percent, labels=Company, col=rainbow(length(Percent)))
  # for Figure 1-10

barplot(Percent, names.arg=Company, col=rainbow(length
  (Percent)), las=2) # for Figure 1-11
```

REMARK 1.5-2 Bar graphs can also be displayed horizontally. Try *barplot(Percent, names.arg* = *Company, col* = *rainbow(length(Percent)), horiz* = *T, las* = *1)* for a horizontal version of Figure 1-11. See also Exercise 17 below for a variation of the Pareto chart. ◁

Exercises

1. Use *cs* = *read.table("Concr.Strength.1s.Data.txt", header* = *T)* to read into the R object *cs* data on 28-day compressive-strength measurements of concrete cylinders using water/cement ratio 0.4.[8] Then use the commands *attach(cs); hist(Str, freq* = *FALSE); lines(density(Str)); stem(Str)* to produce a histogram with the smooth histogram superimposed, and a stem and leaf plot.

2. Use the commands *attach(faithful); hist(waiting); stem(waiting)* to produce a basic histogram of the Old Faithful data on waiting times before eruptions, and the corresponding stem and leaf plot. Is the shape of the stem and leaf plot similar to that of the histogram? Next, use commands similar to those given in Remark 1.5-1 to color the histogram, to superimpose a colored smooth histogram, and to add a histogram title.

3. Use the commands *attach(faithful); plot(waiting, eruptions)* to produce a scatterplot of the Old Faithful data on eruption duration against waiting time before eruption. Comment on the relationship between waiting time before eruption and eruption duration.

4. The data in *Temp.Long.Lat.txt* give the average (over the years 1931 to 1960) daily minimum January temperature in degrees Fahrenheit with the latitude and longitude of 56 US cities.[9]

(a) Construct a scatterplot matrix of the data. Does longitude or latitude appear to be the better predictor of a city's temperature? Explain in terms of this plot.

(b) Construct a 3D scatterplot of the data. Does longitude or latitude appear to be the better predictor of a city's temperature? Explain in terms of this plot.

5. Import the bear measurements data into the R data frame *br* as described in Section 1.5.2, and use the command

```
scatterplot3d(br[6:8], pch=21,
    bg=c("red","green")[unclass(br$Sex)])
```

for a 3D scatterplot of neck girth, chest girth, and weight with gender identification.[10]

6. Studies of the relationship between a vehicle's speed and the braking distance have led to changes in speeding laws and car design. The R data set *cars* has braking distances for cars traveling at different speeds recorded in the 1920s. Use the commands *attach(cars); plot(speed, dist)* to produce a basic scatterplot of the braking distance against speed. Comment on the relationship between speed and breaking distance.

[8] V. K. Alilou and M. Teshnehlab (2010). Prediction of 28-day compressive strength of concrete on the third day using artificial neural networks, *International Journal of Engineering (IJE)*, 3(6).

[9] J. L. Peixoto (1990). A property of well-formulated polynomial regression models. *American Statistician*, 44: 26–30.

[10] Braking distance is the distance required for a vehicle to come to a complete stop after its brakes have been activated.

7. Read the data on the average stopping times (on a level, dry stretch of highway, free from loose material) of cars and trucks at various speeds into the data frame *bd* by *bd* = *read.table("SpeedStopCarTruck.txt", header* = *T)*. Then, use commands similar to those for Figure 1-6, given in Section 1.5.2, to plot the data using colors to differentiate between cars and trucks. Add a legend to the plot.

8. Determining the age of trees is of interest to both individual land owners and to the Forest Service of the US Department of Agriculture. The simplest (though not the most reliable) way of determining the age of a tree is to use the relationship between a tree's diameter at breast height (4–5 ft) and age. Read the data on the average age of three types of trees at different diameter values into the data frame *ad* by *ad* = *read.table("TreeDiamAAge3Stk.txt", header* = *T)*. Then, use the commands *attach(ad); plot(diam, age, pch*=*21, bg*=*c("red", "green", "blue")[unclass(tree)]); legend(x*=*35, y*=*250, pch = c(21, 21, 21), col = c("red", "green", "blue"), legend = c("SMaple", "ShHickory", "WOak"))* to plot the data and add a legend. Comment on any difference in the growth patterns of the three different trees.

9. Read the data on Robot reaction times to simulated malfunctions into the data frame *t* by *t*=*read.table("RobotReactTime.txt", header*= *T)*. Copy the reaction times of Robot 1 into the vector *t1* by *attach(t); t1* = *Time[Robot==1]*. Using the commands given in (1.5.1) do:

(a) A basic histogram with a smooth histogram superimposed for the reaction times of Robot 1.

(b) A stem and leaf plot for the reaction times of Robot 1.

10. Data from an article reporting conductivity (μS/cm) measurements of surface water (X) and water in the sediment at the bank of a river (Y), taken at 10 points during winter, can be found in *ToxAssesData.txt*.[11] Read the data into the data frame *Cond* using a command similar to that given in Exercise 7, and construct a basic scatterplot of Y on X using the commands given in Exercise 6. Does it appear that the surface conductivity can be used for predicting sediment conductivity?

11. Is rainfall volume a good predictor of runoff volume? Data from an article considering this question can be found in *SoilRunOffData.txt*.[12] Read the data into the data frame *Rv* using a command similar to that given in the Exercise 7, and construct a basic scatterplot of the data using the commands given in Exercise 6. Does it appear that the rainfall volume is useful for predicting the runoff volume?

12. Read the projected data on the electricity consumed in an industrial plant in 30 consecutive 30-day periods, together with the average temperature and amount of production (in tons) into the data frame *el* using the commands given for Figure 1-9, Section 1.5.2, and use the commands given in the same section to construct a scatterplot matrix for the data. Which of the variables *temperature* and *production* is a better single predictor for the amount of electricity consumed?

13. The R data set *airquality* contains daily ozone measurements, temperature, wind, solar radiation, month, and day. Use the command *pairs(airquality[1:5])* to produce a scatterplot matrix for the first five variables of this data set, and answer the following questions:

(a) Is it more likely to have higher ozone levels on hot days?

(b) Is it more likely to have higher ozone levels on windy days?

(c) What seems to happen to ozone levels when there is increased solar radiation?

(d) Which month seems to have the highest ozone levels?

14. Data from an article investigating the effect of auxin-cytokinin interaction on the organogenesis of haploid geranium callus can be found in *AuxinKinetinWeight.txt*.[13] Read the data into the data frame *Ac* using a command similar to that given in Exercise 7, and use the commands given in Section 1.5.2 to

(a) construct a scatterplot matrix for the data, and

(b) construct a 3D scatterplot for the data.

Comment on the usefulness of the variables *auxin* and *kinetin* as predictors of the callus weight.

15. The R data set *mtcars* contains data on the weight, displacement, and mileage of cars. Use the commands *library(scatterplot3d); attach(mtcars); scatterplot3d(wt, disp, mpg, pch*=*21, highlight.3d*=*T, type*= *"h", box*=*T, main* = *" ")* for a variation of the 3D scatterplot. Repeat the command replacing *box*=*T* by *box*=*F*.

16. Read the projected numbers and types of accidents of US residents age 16–24 years, found in *AccidentTypes.txt*, into the frame *At* using a command similar to that given in Exercise 7. Construct a bar graph and a pie chart for this data using the commands given in Section 1.5.3.

---

[11] M. Latif and E. Licek (2004). Toxicity assessment of wastewaters, river waters, and sediments in Austria using cost-effective microbiotests, *Environmental Toxicology*, 19(4): 302–308.

[12] M. E. Barrett et al. (1995). *Characterization of Highway Runoff in Austin, Texas, Area.* Center for Research in Water Resourses, University of Texas at Austin, Tech. Rep.# CRWR 263.

[13] M. M. El-Nil, A. C. Hildebrandt, and R. F. Evert (1976). Effect of auxin-cytokinin interaction on organogenesis in haploid callus of *Pelargonium hortorum*, *In Vitro* 12(8): 602–604.

17. Read the projected percents and reasons why people in the Boston area are late for work into the data frame *lw* using the command *lw = read.table("ReasonsLateForWork.txt", sep = ",", header = T)*.

(a) Construct a bar graph and a pie chart for this data using the commands given in Section 1.5.3.

(b) The bar graph constructed in part (a) above is actually a Pareto chart, since the bar heights are arranged in decreasing order. Use the commands *attach(lw); plot(c(0, 6), c(0, 100), pch=" ", xlab=" ", ylab=" ", xaxt="n", yaxt="n"); barplot(Percent, width=0.8, names.arg=Reason, col=rainbow(length(Percent)), las=2, add=T); lines(seq(0.5, 5.5, 1), cumsum(Percent), col="red")* to construct a variation of the Pareto chart, which also displays the cumulative percentages.

1.6 Proportions, Averages, and Variances

Typically scientists want to learn about certain quantifiable aspects, called *parameters*, of the variable, or statistical population, of interest. The most common parameters are the proportion, the average, and the variance. In this section we discuss these parameters for finite populations. Sample versions of these parameters will also be discussed.

1.6.1 POPULATION PROPORTION AND SAMPLE PROPORTION

When the variable of interest is categorical, such as Defective or Non-Defective, Strength of Opinion, Type of Car, and so on, then interest lies in the proportion of population (or sample) units in each of the categories. Graphical methods for visualizing proportions were presented in Section 1.5.3. Here we introduce the formal definition and notation for the population proportion and the sample proportion, and illustrate the sampling variability of the sample proportion.

If the population has N units, and N_i units are in category i, then the population proportion of category i is

Definition of Population Proportion

$$p_i = \frac{\#\{\text{population units in category } i\}}{\#\{\text{population units}\}} = \frac{N_i}{N}. \tag{1.6.1}$$

If a sample of size n is taken from this population, and n_i sample units are in category i, then the **sample proportion** of category i is

Definition of Sample Proportion

$$\widehat{p}_i = \frac{\#\{\text{sample units in category } i\}}{\#\{\text{sample units}\}} = \frac{n_i}{n}. \tag{1.6.2}$$

Example 1.6-1

A car manufacturer receives a shipment of $N = 10{,}000$ navigation systems that are to be installed as a standard feature in the next line of luxury cars. Of concern is a type of satellite reception malfunction. $N_1 = 100$ systems have this malfunction (category 1) and $N_2 = 9900$ do not (category 2). For quality control purposes, a sample of $n = 1000$ systems is taken. After examination, it is found that $n_1 = 8$ systems in the sample have this malfunction and $n_2 = 992$ do not. Give the population and sample proportions for the two categories.

Solution

According to the formula (1.6.1), the population proportions for the two categories are

$$p_1 = \frac{100}{10{,}000} = 0.01, \quad p_2 = \frac{9900}{10{,}000} = 0.99.$$

According to the formula (1.6.2), the sample proportions for the two categories are

$$\widehat{p}_1 = \frac{8}{1000} = 0.008, \quad \widehat{p}_2 = \frac{992}{1000} = 0.992.$$

■

As already suggested in Example 1.2-4, sample properties of the variable of interest approximate (though, in general, will not be identical to) corresponding population properties. In particular,

> The sample proportion $\widehat{p}$ approximates but is, in general, different from the population proportion p.

The next example further illustrates the quality of the approximation of p by $\widehat{p}$, while also illustrating the sampling variability of $\widehat{p}$.

Example 1.6-2

Use R to obtain five samples of size 1,000 from the population of 10,000 navigation systems of Example 1.6-1, and to compute the sample proportions for the two categories in each sample.

Solution

We begin by forming the statistical population, that is, by assigning the value 1 to each of the 100 systems with reception malfunction and the number 2 to each of the 9900 systems with no reception malfunction. The R commands for defining an object (vector) x in R representing the statistical population of 100 1's and 9900 2's, for obtaining a simple random sample of 1000 from this population, and for calculating the two sample proportions are:

```
x = c(rep(1, 100), rep(2, 9900)) # set the statistical
  population in x                                    (1.6.3)
y = sample(x, size=1000) # set the sample of size
  1000 in y                                          (1.6.4)
table(y)/length(y) # compute the sample proportions  (1.6.5)
```

Repeating the set of commands (1.6.4), (1.6.5) five times gives the following pairs of sample proportions, all of which approximate the population proportions of (0.01, 0.99): (0.013, 0.987), (0.012, 0.988), (0.008, 0.992), (0.014, 0.986), (0.01, 0.99). ■

1.6.2 POPULATION AVERAGE AND SAMPLE AVERAGE

Consider a population consisting of N units, and let $v_1, v_2, \ldots, v_N$ denote the values in the statistical population corresponding to some variable of interest. Then the **population average** or **population mean**, denoted by μ, is simply the arithmetic average of all numerical values in the statistical population. That is,

Definition of Population Mean

$$\mu = \frac{1}{N}\sum_{i=1}^{N} v_i. \qquad (1.6.6)$$

If the random variable X denotes the value of the variable of a randomly selected population unit, then a synonymous term for the population mean is **expected value** of X, or **mean value** of X, and is denoted by μ_X or $E(X)$.

If a sample of size n is randomly selected from the population, and if $x_1, x_2, \ldots, x_n$ denote the variable values corresponding to the sample units (note that a different symbol is used to denote the sample values), then the **sample average** or **sample mean** is simply

Definition of Sample Mean

$$\overline{x} = \frac{1}{n}\sum_{i=1}^{n} x_i. \tag{1.6.7}$$

Example 1.6-3

A company of $N = 10{,}000$ employees initiates an employee productivity study in which the productivity of each employee is rated on a scale from 1 to 5. Suppose that 300 of the employees are rated 1, 700 are rated 2, 4000 are rated 3, 4000 are rated 4, and 1000 are rated 5. A pilot study into the degree of employee satisfaction at work interviews 10 randomly selected employees of this company. The productivity ratings of the 10 selected employees are

$$x_1 = 2,\ \ x_2 = x_3 = x_4 = 3,\ \ x_5 = x_6 = x_7 = x_8 = 4,\ \ x_9 = x_{10} = 5.$$

(a) Describe the statistical population for the variable productivity rating.

(b) Letting the random variable X denote the productivity rating of a randomly selected employee, compute the mean (or expected) value $E(X)$ of X.

(c) Compute the sample mean of the productivity ratings of the 10 selected employees.

Solution

(a) The statistical population consists of 10,000 productivity ratings, $v_1, v_2, \ldots, v_{10{,}000}$, which are

$$\begin{aligned} v_i &= 1,\ \ i = 1, \ldots, 300,\\ v_i &= 2,\ \ i = 301, \ldots, 1000,\\ v_i &= 3,\ \ i = 1001, \ldots, 5000,\\ v_i &= 4,\ \ i = 5001, \ldots, 9000,\\ v_i &= 5,\ \ i = 9001, \ldots, 10{,}000. \end{aligned}$$

(b) According to the expression (1.6.6), the expected value of X (which is also the population average rating) is

$$E(X) = \frac{1}{10{,}000}\sum_{i=1}^{10{,}000} v_i = \frac{1}{10{,}000}(1 \times 300 + 2 \times 700$$
$$+ 3 \times 4000 + 4 \times 4000 + 5 \times 1000) = 3.47.$$

(c) Finally, according to the expression (1.6.7), the sample of 10 productivity ratings yields a sample mean of

$$\overline{x} = \frac{1}{10}\sum_{i=1}^{10} x_i = 3.7.$$

Example 1.2-4 already highlights the fact that sample properties of the variable of interest approximate (though, in general, will not be identical to) corresponding population properties. In particular,

> The sample mean $\overline{x}$ approximates but is, in general, different from the population mean μ.

To further illustrate the quality of the approximation of μ by $\overline{x}$ and to illustrate the sampling variability of $\overline{x}$, we will obtain five samples of size 10 from the population of 10,000 employees of Example 1.6-3 and for each sample we will compute the sample mean.

Example 1.6-4

Use R to obtain five samples of size 10 from the population of 10,000 employees of Example 1.6-3 and to compute the sample mean for each sample.

Solution

Setting

```
x=c(rep(1, 300), rep(2, 700), rep(3, 4000), rep(4, 4000),
  rep(5, 1000))
```

for the statistical population given in the solution of Example 1.6-3, and repeating the commands

```
y=sample(x, size=10); mean(y)
```

five times gives, for example the following sample means: 3.7, 3.6, 2.8, 3.4, 3.2. ■

Example 1.6-5 demonstrates the simple, but very important, fact that *proportions can be expressed as means* or, in other words,

> A proportion is a special case of mean.

Example 1.6-5

A certain county has 60,000 US citizens of voting age, 36,000 of whom are in favor of expanding the use of solar energy. Of the 50 such citizens who are randomly selected in a statewide public opinion poll, 28 are in favor of expanding the use of solar energy.

(a) Express the proportion of citizens who are in favor of expanding solar energy, $p = 36{,}000/60{,}000 = 0.6$, as the expected value of a random variable X.

(b) Express the sample proportion of citizens in the sample who are in favor of expanding solar energy, $\widehat{p} = 28/50 = 0.56$, as a sample mean.

Solution

(a) The characteristic of interest here is qualitative (in favor or not in favor), but we can convert it to a variable by setting 0 for "not in favor" and 1 for "in favor." With this variable, the statistical population consists of 24,000 0's and 36,000 1's:

$$v_i = 0, \quad i = 1, \ldots, 24{,}000; \quad v_i = 1, \quad i = 24{,}001, \ldots, 60{,}000.$$

Letting the random variable X denote a randomly selected value from this statistical population, we have (according to the expression (1.6.6))

$$\mu_X = \frac{1}{60{,}000} \sum_{i=1}^{60{,}000} v_i = \frac{36{,}000}{60{,}000} = 0.6.$$

(b) Next, the sample of 50 citizens corresponds to a sample from the statistical population with 22 0's and 28 1's:

$$x_i = 0, \quad i = 1, \ldots, 22; \quad x_i = 1, \quad i = 23, \ldots, 50.$$

According to the expression (1.6.7), the sample mean is

$$\overline{x} = \frac{1}{50} \sum_{i=1}^{50} x_i = \frac{28}{50} = 0.56.$$

The above exposition pertains to univariate variables. The population mean and the sample mean for bivariate or multivariate variables is given by averaging each coordinate separately. Moreover, the above definition of a population mean assumes a finite population. The definition of population mean for an infinite or conceptual population, such as that of the cement mixtures in Example 1.1-1, case study 3, will be given in Chapter 3. The definition of sample mean remains the same regardless of whether or not the sample has been drawn from a finite or an infinite population.

1.6.3 POPULATION VARIANCE AND SAMPLE VARIANCE

The population variance and *standard deviation* offer a quantification of the intrinsic variability of the population. Quantification of the intrinsic variability is of interest as a quality measure in manufacturing. Indeed, the main characteristic(s) of a high-quality product should vary as little as possible from one unit of the product to another (i.e., the corresponding statistical population should have as low an intrinsic variability as possible). For example, while high average gas mileage is a desirable characteristic of a certain car, it is also desirable that different cars of the same make and model achieve similar gas mileage.

Consider a population consisting of N units, and let $v_1, v_2, \ldots, v_N$ denote the values in the statistical population corresponding to some variable. Then the **population variance**, denoted by σ^2, is defined as

Definition of Population Variance

$$\sigma^2 = \frac{1}{N} \sum_{i=1}^{N} (v_i - \mu)^2 \qquad \textbf{(1.6.8)}$$

where μ is the population mean. If the random variable X denotes the value of the variable of a randomly selected population unit, then the population variance is also called the variance of the random variable X, and we write σ_X^2, or $\mathrm{Var}(X)$.

The variance of a random variable X, or of its underlying population, quantifies the extent to which the values in the statistical population differ from the population mean. As its expression in (1.6.8) indicates, the population variance is the average squared distance of members of the statistical population from the population mean. As it is an average squared distance, it goes without saying that the variance of a random variable can never be negative. Some simple algebra reveals the following alternative expression for the population variance.

Computational Formula for Population Variance

$$\sigma^2 = \frac{1}{N}\sum_{i=1}^{N} v_i^2 - \mu^2 \qquad (1.6.9)$$

Expression (1.6.9) is more convenient for calculating the variance.

The positive square root of the population variance is called the population **standard deviation** and is denoted by σ:

Definition of Population Standard Deviation

$$\sigma = \sqrt{\sigma^2} \qquad (1.6.10)$$

The standard deviation of a random variable X, or of its underlying statistical population, is expressed in the same units as the variable itself, whereas the variance is expressed in squared units. For example, a variable measured in inches will have a standard deviation measured in inches, but variance measured in square inches. For this reason, the standard deviation is often a preferable measure of the intrinsic variability.

If a sample of size n is randomly selected from the population, and if $x_1, x_2, \ldots, x_n$ denote the variable values corresponding to the sample units, then the **sample variance** is

Definition of Sample Variance

$$S^2 = \frac{1}{n-1}\sum_{i=1}^{n}(x_i - \overline{x})^2 \qquad (1.6.11)$$

REMARK 1.6-1 Dividing by $n-1$ (instead of n) in (1.6.11) is a source of intrigue to anyone who sees this formula for the first time. The typical explanation, offered in textbooks and classrooms alike, is given in terms of the statistical parlance of **degrees of freedom**: Because the definition of S^2 involves the **deviations** of each observation from the sample mean, that is, $x_1 - \overline{x}, x_2 - \overline{x}, \ldots, x_n - \overline{x}$, and because these deviations sum to zero, that is,

$$\sum_{i=1}^{n}(x_i - \overline{x}) = 0, \qquad (1.6.12)$$

there are $n-1$ degrees of freedom, or independent quantities (deviations) that determine S^2. This is not completely satisfactory because a relation similar to (1.6.12) holds also at the population level (simply replace n, x_i, and $\overline{x}$ by N, v_i, and μ). A more complete explanation is that if a large number of investigators each select a random sample of size n with replacement and the sample variances that they obtain are averaged, then this average sample variance will be almost identical to the population variance. (See Exercise 10.) This property of S^2, called **unbiasedness**, requires that the sum of squared deviations be divided by $n-1$, not n. Unbiasedness is a desirable property of estimators and it will be discussed in more detail in Chapter 6. ◁

The positive square root of the sample variance is called the **sample standard deviation** and is denoted by S:

Definition of Sample Standard Deviation

$$S = \sqrt{S^2} \qquad (1.6.13)$$

A computational formula for S^2 is

Computational Formula for Sample Variance

$$S^2 = \frac{1}{n-1}\left[\sum_{i=1}^{n} x_i^2 - \frac{1}{n}\left(\sum_{i=1}^{n} x_i\right)^2\right] \tag{1.6.14}$$

As we did with the sample mean and sample proportion, we emphasize here that

S^2 and S approximate but are, in general, different from σ^2 and σ.

Example 1.6-6

(a) Find the variance and standard deviation of the statistical population

$$v_i = 0, \quad i = 1, \ldots, 24{,}000; \quad v_i = 1, \quad i = 24{,}001, \ldots, 60{,}000.$$

corresponding to the 60,000 US citizens of voting age of Example 1.6-5.
(b) Find the sample variance and standard deviation of the sample

$$x_i = 0, \quad i = 1, \ldots, 22; \quad x_i = 1, \quad i = 23, \ldots, 50$$

from the above statistical population.

Solution

(a) Using the computational formula (1.6.9), the population variance is

$$\sigma^2 = \frac{1}{N}\sum_{i=1}^{N} v_i^2 - \mu^2 = \frac{36{,}000}{60{,}000} - (0.6)^2 = 0.6(1 - 0.6) = 0.24.$$

The population standard deviation is $\sigma = \sqrt{0.24} = 0.49$.

(b) Next, using the computational formula (1.6.14) for the sample variance, we have

$$S^2 = \frac{1}{n-1}\left[\sum_{i=1}^{n} x_i^2 - \frac{1}{n}\left(\sum_{i=1}^{n} x_i\right)^2\right] = \frac{1}{49}\left[28 - \frac{1}{50}28^2\right] = 0.25,$$

and thus, the sample standard deviation is $S = \sqrt{0.25} = 0.5$. ■

The sample variance and sample standard deviation in the above example provide good approximations to the population variance and standard deviation, respectively. The next example provides further insight into the quality of these approximations and also the sampling variability of the sample variance and sample standard deviation.

Example 1.6-7

Use R to obtain five samples of size 50 from the statistical population given in Example 1.6-6, and to compute the sample variance for each sample.

Solution

Setting $x =$ c(rep(0, 24,000), rep(1, 36,000)) for the statistical population, and repeating the commands

```
y=sample(x, size=50); var(y); sd(y)
```

five times gives, for example, the following five pairs of sample variances and standard deviations: (0.2143, 0.4629), (0.2404, 0.4903), (0.2535, 0.5035), (0.2551, 0.5051), (0.2514, 0.5014). ■

The definitions of the population variance and standard deviation given in this section assume a finite population. The more general definition, applicable to any population, will be given in Chapter 3. The definitions of the sample variance and standard deviation remain the same regardless of whether or not the sample has been drawn from a finite or an infinite population.

Exercises

1. A polling organization samples 1000 adults nationwide and finds that the average duration of daily exercise is 37 minutes with a standard deviation of 18 minutes.

(a) The correct notation is for the number 37 is (choose one): (i) $\overline{x}$, (ii) μ.

(b) The correct notation is for the number 18 is (choose one): (i) S, (ii) σ.

(c) Of the 1000 adults in the sample 72% favor tougher penalties for persons convicted of drunk driving. The correct notation for the number 0.72 is (choose one): (i) $\widehat{p}$, (ii) p.

2. In its year 2000 census, the United States Census Bureau found that the average number of children of all married couples is 2.3 with a standard deviation of 1.6.

(a) The correct notation is for the number 2.3 is (choose one): (i) $\overline{x}$, (ii) μ.

(b) The correct notation is for the number 1.6 is (choose one): (i) S, (ii) σ.

(c) According to the same census, 17% of all adults chose not to marry. The correct notation for the number 0.17 is (choose one): (i) $\widehat{p} = 0.17$, (ii) $p = 0.17$.

3. A data set of 14 ozone measurements (Dobson units) taken at different times from the lower stratosphere, between 9 and 12 miles (15 and 20 km) altitude, can be found in *OzoneData.txt*. What proportion of these measurements falls below 250? What does this sample proportion estimate?

4. Use *cs = read.table("Concr.Strength.1s.Data.txt", header = T)* to read into the R object *cs* data on 28-day compressive-strength measurements of concrete cylinders using water/cement ratio 0.4 (see footnote 5 in Exercise 1 in Section 1.5). Then use the commands *attach(cs); sum(Str <= 44)/length(Str); sum(Str >= 47)/length(Str)* to obtain the proportion of measurements that are less than or equal to 44, and greater than or equal to 47. What do these sample proportions estimate?

5. Refer to Example 1.6-3.

(a) Use the information on the statistical population of productivity ratings given in the example to calculate the population variance and standard deviation.

(b) Use the sample of 10 productivity ratings given in the example to calculate the sample variance and standard deviation.

6. Use R commands to obtain a simple random sample of size 50 from the statistical population of productivity ratings given in Example 1.6-3, and calculate the sample mean and sample variance. Repeat this for a total of five times, and report the five pairs of $(\overline{x}, S^2)$.

7. Refer to Exercise 1 in Section 1.4.

(a) For the statistical population corresponding to the number of scratches of the 500 tin plates described in the exercise ($N_0 = 190$ 0's, $N_1 = 160$ 1's and $N_2 = 150$ 2's), find the population mean, the population variance, and the population standard deviation.

(b) A simple random sample of $n = 100$ from the above statistical population consists of $n_0 = 38$ 0's, $n_1 = 33$ 1's, and $n_2 = 29$ 2's. Find the sample mean, the sample variance, and the sample standard deviation.

8. Set the statistical population of Exercise 7 in the R object (vector) *x* by *x = c(rep(0, 190), rep(1, 160), rep(2, 150))*.

(a) The R commands *y = sample(x, 100); table(y)/100* select a sample of size $n = 100$ and compute the proportions for the three categories. Repeat these commands a total of five times, report the results, and give the population proportions that the sample proportions estimate.

(b) The R commands *y = sample(x, 100); mean(y); var(y); sd(y)* select a sample of size $n = 100$ and compute the sample mean, sample variance, and sample standard deviation. Repeat these commands a total of five times, report the results, and give the values of the population parameters that are being estimated.

9. The outcome of a roll of a die is a random variable X that can be thought of as resulting from a simple random selection of one number from $1,\ldots,6$.

(a) Compute μ_X and σ_X^2, either by hand or using R.

(b) Select a sample of size 100 with replacement from the finite population $1,\ldots,6$, and compute the sample mean and sample variance. The R commands for doing so are:

```
x=sample(1:6, 100, replace=T) # T for
  TRUE -- capitalization is necessary
mean(x); var(x)
```

Comment on how well the sample mean and variance approximates the true population parameters.

(c) Use the R command *table(x)/100*, where x is the sample of size $n = 100$ obtained in part (b), to obtain the sample proportions of $1,\ldots,6$. Are they all reasonably close to 1/6?

10. Setting 1 for Heads and 0 for Tails, the outcome X of a flip of a coin can be thought of as resulting from a simple random selection of one number from $\{0, 1\}$.

(a) Compute the variance σ_X^2 of X.

(b) The possible samples of size two, taken with replacement from the population $\{0, 1\}$, are $\{0, 0\}, \{0, 1\}, \{1, 0\}, \{1, 1\}$. Compute the sample variance for each of the possible four samples.

(c) Consider the statistical population consisting of the four sample variances obtained in part (b), and let Y denote the random variable resulting from a simple random selection of one number from this statistical population. Compute $E(Y)$.

(d) Compare σ_X^2 and $E(Y)$. If the sample variance in part (b) was computed according to a formula that divides by n instead of $n - 1$, how would σ_X^2 and $E(Y)$ compare?

11. A random sample of 5 cars of type A that were test driven yielded the following gas mileage on the highway: 29.1, 29.6, 30, 30.5, 30.8. A random sample of 5 cars of type B yielded the following gas mileage when test driven under similar conditions: 21, 26, 30, 35, 38.

(a) For each type of car, estimate the population mean gas mileage.

(b) For each type of car, estimate the population variance of gas mileage.

(c) On the basis of the above analysis, rank the two types of car in terms of quality. Justify your answer.

12. Consider a statistical population consisting of N values $v_1,\ldots,v_N$, and let μ_v, σ_v^2, σ_v denote the population mean value, variance, and standard deviation.

(a) Suppose that the v_i values are coded to $w_1,\ldots,w_N$, where $w_i = c_1 + v_i$, where c_1 is a known constant. Show that the mean value, variance, and standard deviation of the statistical population $w_1,\ldots,w_N$ are

$$\mu_w = c_1 + \mu_v, \quad \sigma_w^2 = \sigma_v^2, \quad \sigma_w = \sigma_v.$$

(*Hint.* The computational formula for the variance is not very convenient for this derivation. Use instead (1.6.8). The same is true for parts (b) and (c).)

(b) Suppose that the v_i values are coded to $w_1,\ldots,w_N$, where $w_i = c_2 v_i$, where c_2 is a known constant. Show that the mean value, variance, and standard deviation of the statistical population $w_1,\ldots,w_N$ are

$$\mu_w = c_2\mu_v, \quad \sigma_w^2 = c_2^2\sigma_v^2, \quad \sigma_w = |c_2|\sigma_v.$$

(c) Suppose that the v_i values are coded to $w_1,\ldots,w_N$, where $w_i = c_1 + c_2 v_i$, where c_1, c_2 are known constants. Show that the mean value, variance, and standard deviation of the statistical population $w_1,\ldots,w_N$ are

$$\mu_w = c_1 + c_2\mu_v, \quad \sigma_w^2 = c_2^2\sigma_v^2, \quad \sigma_w = |c_2|\sigma_v.$$

13. Consider a sample $x_1,\ldots,x_n$ from some statistical population, and let $\bar{x}$, S_x^2, and S_x denote the sample mean, sample variance, and sample standard deviation.

(a) Suppose that the x_i values are coded to $y_1,\ldots,y_n$, where $y_i = c_1+x_i$, where c_1 is a known constant. Show that the sample mean, sample variance, and sample standard deviation of $y_1,\ldots,y_n$ are

$$\bar{y} = c_1 + \bar{x}, \quad S_y^2 = S_x^2, \quad S_y = S_x.$$

(b) Suppose that the x_i values are coded to $y_1,\ldots,y_n$, where $y_i = c_2 x_i$, where c_2 is a known constant. Show that the sample mean, sample variance, and sample standard deviation of $y_1,\ldots,y_n$ are

$$\bar{y} = c_2\bar{x}, \quad S_y^2 = c_2^2 S_x^2, \quad S_y = |c_2| S_x.$$

(c) Suppose that the x_i values are coded to $y_1,\ldots,y_n$, where $y_i = c_1+c_2 x_i$, where c_1, c_2 are known constants. Show that the sample mean, sample variance, and sample standard deviation of the statistical population $y_1,\ldots,y_n$ are

$$\bar{y} = c_1 + c_2\bar{x}, \quad S_y^2 = c_2^2 S_x^2, \quad S_y = |c_2| S_x.$$

14. The noon-time temperature of seven randomly selected days of August in a coastal site of Spain gives $\bar{x} = 31C^o$ and $S = 1.5C^o$. The formula for converting a measurement in the Celsius scale to the Fahrenheit scale is $F^o = 1.8C^o + 32$. Find the sample mean and variance of the seven temperature measurements when expressed in the Fahrenheit scale.

15. Consider the sample $X_1 = 81.3001$, $X_2 = 81.3015$, $X_3 = 81.3006$, $X_4 = 81.3011$, $X_5 = 81.2997$, $X_6 = 81.3005$, $X_7 = 81.3021$. Code the data by subtracting 81.2997 and multiplying by 10,000. Thus the coded data are 4, 18, 9, 14, 0, 8, 24. It is given that the sample variance of the coded data is $S_Y^2 = 68.33$. Find the sample variance of the original data.

16. The following data show the starting salaries, in \$1000 per year, for a sample of 15 senior engineers:

```
152 169 178 179 185 188 195 196 198 203 204 209 210 212 214
```

(a) Assuming that the 15 senior engineers represent a simple random sample from the population of senior engineers, estimate the population mean and variance.

(b) Give the sample mean and variance for the data on second-year salaries for the same group of engineers if

(i) if each engineer gets a \$5000 raise, and

(ii) if each engineer gets a 5% raise.

1.7 Medians, Percentiles, and Boxplots

Percentiles are used mainly for continuous variables, or discrete-valued variables if the divisions between values are fine enough, as, for example, SAT scores. The definition of percentiles for finite populations (the only type of populations considered in this chapter) is the same as that for sample percentiles. For this reason, only the sample percentiles will be discussed in this section. Population percentiles for infinite populations will be defined in Chapter 3.

Let $x_1, \ldots, x_n$ be a simple random sample from a continuous population distribution. Roughly speaking, the $(1-\alpha)100$th **sample percentile** divides the sample in two parts, the part having the $(1-\alpha)100\%$ smaller values, and the part having the $\alpha 100\%$ larger values. For example, the 90th sample percentile [note that $90 = (1 - 0.1)100$] separates the upper (largest) 10% from the lower 90% of values in the data set. The 50th sample percentile is also called the **sample median** and is denoted by $\tilde{x}$; it is the value that separates the upper or largest 50% from the lower or smallest 50% of the data. The 25th, the 50th, and the 75th sample percentiles are also called **sample quartiles**, as they divide the sample into roughly four equal parts. We also refer to the 25th and the 75th sample percentiles as the **lower sample quartile (q_1)** and **upper sample quartile (q_3)**, respectively. The precise (computational) definition of sample quartiles is given in Definition 1.7-2, but for now we give the following definition of a different measure of variability.

Definition 1.7-1

The **sample interquartile range**, or **sample IQR**, defined as

$$IQR = q_3 - q_1$$

is an estimator of the population IQR, which is a measure of variability.

Sample percentiles serve as estimators of corresponding population percentiles. For a precise definition of sample percentiles we need to introduce notation for the **ordered** sample values, or **order statistics**: The sample values $x_1, \ldots, x_n$ arranged in increasing order are denoted

Notation for the Order Statistics

$$x_{(1)}, x_{(2)}, \ldots, x_{(n)} \tag{1.7.1}$$

Because they have been ordered in increasing order, it follows that $x_{(1)}$ is the smallest observation and $x_{(n)}$ is the largest. In particular, $x_{(1)} \leq x_{(2)} \leq \cdots \leq x_{(n)}$.

We begin by identifying each $x_{(i)}$ as an estimator of a population percentile. Following that, we give precise (computational) definitions of the sample median and the upper and lower quartiles.

Definition 1.7-2

Let $x_{(1)}, x_{(2)}, \ldots, x_{(n)}$ denote the ordered sample values in a sample of size n. Then $x_{(i)}$, the ith smallest sample value, is taken to be the $100\left(\frac{i-0.5}{n}\right)$-th **sample percentile**. Sample percentiles estimate the corresponding population percentiles.

Example 1.7-1

A simple random sample of size 10, drawn from the statistical population of the 50 black bears' weight measurements used in Example 1.2-5, is:

154 158 356 446 40 154 90 94 150 142

Give the order statistics, and state the population percentiles they estimate.

Solution

The R command

```
sort(c(154, 158, 356, 446, 40, 154, 90, 94, 150, 142))
```

returns the order statistics: 40, 90, 94, 142, 150, 154, 154, 158, 356, 446. These order statistics estimate the 5th, 15th, 25th, 35th, 45th, 55th, 65th, 75th, 85th, and 95th population percentiles, respectively. For example, $x_{(3)} = 94$ is the $100(3-0.5)/10 =$ 25th percentile and estimates the corresponding population percentile. ■

As the above example demonstrates, it is possible that none of the order statistics corresponds to a sample percentile of interest. For example, none of the order statistics in Example 1.7-1 corresponds to the median or the 90th percentile. In general, if the sample size is even, none of the order statistics will be the sample median, and if the sample size is not of the form 6 + (a multiple of 4), none of the order statistics will equal the quartiles. R uses an interpolation algorithm for evaluating any sample percentile from a given data set. With data in the object x, the commands

R Commands for Percentiles

```
median(x)
quantile(x, 0.25)
quantile(x, c(0.3, 0.7, 0.9))
summary(x)
```

(1.7.2)

give, respectively, the median, the 25th percentile, the 30th, 70th, and 90th percentiles, and a **five number summary** of the data consisting of $x_{(1)}$, q_1, $\tilde{x}$, q_3, and $x_{(n)}$.

Example 1.7-2

Using the sample of 10 black bear weights given in Example 1.7-1, estimate the population median, 70th, 80th, and 90th percentiles.

Solution

Putting the sample values in the object w as described in Example 1.7-1, the R command *quantile(w, c(0.5, 0.7, 0.8, 0.9))* returns 152.0, 155.2, 197.6, 365.0 for the sample median, 70th, 80th, and 90th percentiles, respectively. ■

The five number summary of the data given by the *summary(x)* command in R is the basis for the **boxplot**, which is a simple but effective visual description of the main features of a data set $x_1, \ldots, x_n$. A boxplot displays the central 50% of the data with a box, the lower (or left) edge of which is at q_1 and the upper (or right) edge at q_3. A line inside the box represents the median. The lower 25% and upper 25% of the data are represented by lines (or *whiskers*) that extend from each edge of the box. The lower whisker extends from q_1 until the smallest observation within 1.5 interquartile ranges from q_1. The upper whisker extends from q_3 until the largest observation within 1.5 interquartile ranges from q_3. Observations farther from the box than the whiskers' ends (i.e., smaller than $q_1 - 1.5 \times \text{IQR}$ or larger than $q_3 + 1.5 \times \text{IQR}$) are called **outliers**, and are plotted individually. The construction of a boxplot is demonstrated in the following example.

Example 1.7-3

Scientists have been monitoring the ozone hole since 1980. A data set of 14 ozone measurements (Dobson units) taken from the lower stratosphere, between 9 and 12 miles (15 and 20 km) altitude, can be found in *OzoneData.txt*. Give the five number summary of this data and construct the boxplot.

Solution

Reading this data into the R object *oz*, the command *summary(oz)* gives the five number summary of this data as: $x_{(1)} = 211.0$, $q_1 = 247.8$, $\tilde{x} = 272.5$, $q_3 = 292.2$, $x_{(14)} = 446.0$. The interquartile range is $\text{IQR} = 292.2 - 247.8 = 44.4$, and $q_3 + 1.5 \times \text{IQR} = 358.8$. Thus, the two largest observations, which are the 395 and 446.0, are outliers. The boxplot of this data, shown in Figure 1-12, was generated by the R command

```
boxplot(oz, col= "grey").
```

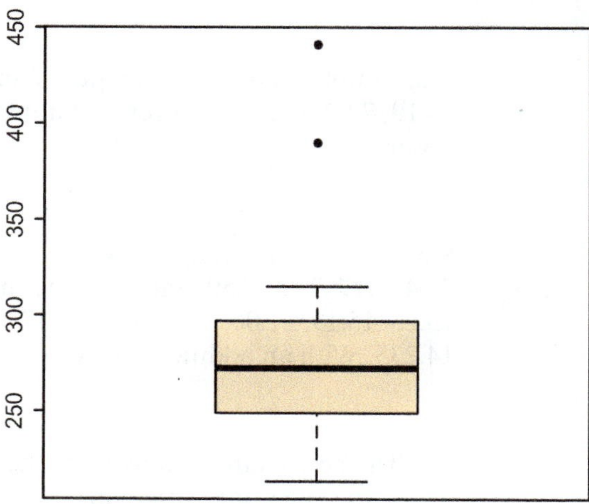

Figure 1-12 Boxplot of ozone data.

■

In what follows we give computational definitions for the sample median and the lower and upper quartiles.

Definition 1.7-3

Let $x_{(1)}, x_{(2)}, \ldots, x_{(n)}$ denote the order statistics. Then

1. The **sample median** is defined as

$$\tilde{x} = \begin{cases} x_{\left(\frac{n+1}{2}\right)}, & \text{if } n \text{ is odd} \\ \dfrac{x_{\left(\frac{n}{2}\right)} + x_{\left(\frac{n}{2}+1\right)}}{2}, & \text{if } n \text{ is even} \end{cases}$$

2. The **sample lower quartile** is defined as

$$q_1 = \text{Median of smaller half of the data values}$$

where, if n is even the smaller half of the values consists of the smallest $n/2$ values, and if n is odd the smaller half consists of the smallest $(n+1)/2$ values. Similarly, the **sample upper quartile** is defined as

$$q_3 = \text{Median of larger half of the data values}$$

where, if n is even the larger half of the values consists of the largest $n/2$ values, and if n is odd the larger half consists of the largest $(n+1)/2$ values.

Thus, when the sample size is even, the sample median is defined by interpolating between the nearest sample percentiles. Similarly, when the sample size is not of the form $6 +$ (a multiple of 4), the above definition uses interpolation to define the sample lower and upper quartiles. This interpolation is convenient for hand calculations but is different from the interpolation used by R. For example, the R command *summary(1:10)* yields the first and third quartiles for the numbers $1, \ldots, 10$ as $q_1 = 3.25$ and $q_3 = 7.75$, while the rule of Definition 1.7-3 gives $q_1 = 3$, $q_3 = 8$. However the R command *summary(1:11)* yields the first and third quartiles of the numbers $1, \ldots, 11$ as $q_1 = 3.5$ and $q_3 = 8.5$, respectively, which is exactly what the rule of Definition 1.7-3 gives.

Example 1.7-4

The sample values of a sample of size $n = 8$ are 9.39, 7.04, 7.17, 13.28, 7.46, 21.06, 15.19, 7.50. Find the lower and upper quartiles. Repeat the same with an additional observation of 8.20.

Solution

Since n is even and $n/2 = 4$, q_1 is the median of the smallest four values, which are 7.04, 7.17, 7.46, 7.50, and q_3 is the median of the largest four values which are 9.39, 13.28, 15.19, 21.06. Thus $q_1 = (7.17 + 7.46)/2 = 7.315$, and $q_3 = (13.28 + 15.19)/2 = 14.235$. With an additional observation of 8.20, so $n = 9$, $q_1 = 7.46$, $q_3 = 13.28$. ■

The next example illustrates the similarities and differences between the sample mean and the sample median.

Example 1.7-5 Let the sample values of a sample of size $n = 5$, be $x_1 = 2.3$, $x_2 = 3.2$, $x_3 = 1.8$, $x_4 = 2.5$, $x_5 = 2.7$. Find the sample mean and the sample median. Repeat the same after changing the x_2 value from 3.2 to 4.2.

Solution

The sample mean is

$$\overline{x} = \frac{2.3 + 3.2 + 1.8 + 2.5 + 2.7}{5} = 2.5.$$

For the median, we first order the values from smallest to largest: 1.8, 2.3, 2.5, 2.7, 3.2. Since the sample size here is odd, and $(n+1)/2 = 3$, the median is

$$\widetilde{x} = x_{(3)} = 2.5,$$

which is the same as the mean. Changing the x_2 value from 3.2 to 4.2 we get

$$\overline{x} = 2.7, \quad \widetilde{x} = 2.5.$$

This example illustrates the point that the value of $\overline{x}$ is affected by extreme observations (*outliers*), where as the median is not.

Exercises

1. The following is a stem and leaf display of $n = 40$ solar intensity measurements (integers in watts/m^2) on different days at a location in southern Australia. The (optional) first column of the stem and leaf plot contains a leaf count in a cumulative fashion from the top down to the stem that contains the median and also from the bottom up to the stem that contains the median. The stem containing the median has its own leaf count, shown in parentheses. Thus, $18 + 4 + 18$ equals the sample size.

4	67	3 3 6 7
8	68	0 2 2 8
11	69	0 1 9
18	70	0 1 4 7 7 9 9
(4)	71	5 7 7 9
18	72	0 0 2 3
14	73	0 1 2 4 4 5
8	74	0 1 3 6 6 6
2	75	0 8

(a) Obtain the sample median and the 25th and the 75th percentiles.

(b) Obtain the sample interquartile range.

(c) What sample percentile is the 19th ordered value?

2. Read the data on robot reaction times to simulated malfunctions into the data frame *t* by *t = read.table("RobotReactTime.txt", header = T)*. Read the reaction times of Robot 1 into the vector *t1* by *attach(t); t1 = Time[Robot==1]*, and sort the data (i.e., arrange it from smallest to largest) by *sort(t1)*. Using the sorted data and hand calculations

(a) estimate the population median and the 25th and the 75th percentiles,

(b) estimate the population interquartile range, and

(c) find the percentile of the 19th ordered value.

3. The site given in Exercise 2 also gives the reaction times of Robot 2. Use commands similar to those given in Exercise 2 to read the reaction times of Robot 2 into the vector *t2*.

(a) Use the R command *summary* given in (1.7.2) to obtain the five number summary of this data set.

(b) Use the R command *quantile* given in (1.7.2) get the sample 90th percentile.

(c) Use the R command "boxplot" given in Example 1.7-3 to construct a boxplot for the data. Are there any outliers?

4. Enter the solar intensity measurements of Exercise 1 into the R object *si* with *si = read.table("SolarIntensAuData.txt", header = T)*. Use R commands to

(a) construct a boxplot of the solar intensity measurements, and

(b) obtain the 30th, 60th, and 90th sample percentiles.

1.8 Comparative Studies

Comparative studies aim at discerning and explaining differences between two or more populations. In this section we introduce the basic concepts and jargon associated with such studies.

1.8.1 BASIC CONCEPTS AND COMPARATIVE GRAPHICS

The comparison of two methods of cloud seeding for hail and fog suppression at international airports, the comparison of two or more cement mixtures in terms of compressive strength, and the comparison of the effectiveness of three cleaning products in removing four different types of stains (which are mentioned as case studies 2, 3, and 4 in Example 1.1-1) are examples of comparative studies.

Comparative studies have their own jargon. Thus, the comparison of three cement mixtures in terms of their compressive strength is a **one-factor** study, the factor being *cement mixture*; this factor enters the study at three **levels**, and the **response variable** is *cement strength*. In one-factor studies the levels of the factor are also called **treatments**. The study comparing the effectiveness of three cleaning products in removing four different types of stains is a **two-factor** study where the factor *cleaning product* has three levels and the factor *stain* has four levels; the response variable is the degree of stain removal. In two-factor studies, treatments correspond to the different factor-level combinations; see Figure 1-13. Thus, a two-factor study where factor A enters with a levels and factor B enters with b levels involves $a \times b$ treatments.

Example 1.8-1

A study will compare the level of radiation emitted by five kinds of cellphones at each of three volume settings. State the factors involved in this study, the number of levels for each factor, the total number of populations or treatments, and the response variable.

Solution

The two factors involved in this study are *type of cellphone* (factor 1) and *volume setting* (factor 2). Factor 1 has five levels, and factor 2 has three levels. The total number of populations is $5 \times 3 = 15$, and the response variable is *level of radiation*. ■

Different treatments (factor levels or factor-level combinations) correspond to different populations. The complete description of these populations, however, also involves the **experimental units**, which are the units on which measurements are made.

Example 1.8-2

(a) In the study that compares the cleaning effectiveness of cleaning products on different types of stains, experimental units are pieces of fabric.

(b) In the study that compares the effect of temperature and humidity on the yield of a chemical reaction, experimental units are aliquots of materials used in the reaction.

(c) In studying the effectiveness of a new diet in reducing weight, experimental units are the subjects participating in the study. ■

Figure 1-13 Treatments, or factor-level combinations, in a two-factor study.

	Factor B			
Factor A	1	2	3	4
1	Tr_{11}	Tr_{12}	Tr_{13}	Tr_{14}
2	Tr_{21}	Tr_{22}	Tr_{23}	Tr_{24}

Comparisons of the different populations typically focus either on comparisons of means, proportions, medians, or variances. Comparisons of means (also proportions and medians) is typically based on differences, while comparisons of variances are typically based on ratios. For example, the comparison of two different cloud seeding methods may be based on

$$\overline{x}_1 - \overline{x}_2, \tag{1.8.1}$$

where $\overline{x}_1$ and $\overline{x}_2$ are the sample mean rainfalls produced by methods 1 and 2, respectively.

The difference in (1.8.1) is the simplest type of a **contrast**. In general, contrasts may involve differences not only of individual means but also of certain linear combinations of means. The following example illustrates different contrasts that may be of interest in one-factor studies.

Example 1.8-3

A study is aimed at comparing the mean tread life of four types of high-performance tires designed for use at higher speeds. Specify three different types of contrasts that may be of interest.

Solution

Let $\overline{x}_1, \ldots, \overline{x}_4$ denote sample mean tread lives obtained from samples of size $n_1, \ldots, n_4$ from each of the four types of tires.

(a) If tire type 1 is currently manufactured, and tire types 2, 3, and 4 are experimental, interest may lie in the contrasts

$$\overline{x}_1 - \overline{x}_2, \quad \overline{x}_1 - \overline{x}_3, \quad \overline{x}_1 - \overline{x}_4.$$

These types of contrasts are common in the so-called *control versus treatment* studies.

(b) If tire types 1 and 2 are made by manufacturer A, while tire types 3 and 4 are made by manufacturer B, interest may lie in the contrast

$$\frac{\overline{x}_1 + \overline{x}_2}{2} - \frac{\overline{x}_3 + \overline{x}_4}{2},$$

which compares the two brands made by manufacturer A to the two brands made by manufacturer B.

(c) An overall comparison of the four types of tires is typically based on the contrasts

$$\overline{x}_1 - \overline{x}, \quad \overline{x}_2 - \overline{x}, \quad \overline{x}_3 - \overline{x}, \quad \overline{x}_4 - \overline{x},$$

where $\overline{x} = (\overline{x}_1 + \overline{x}_2 + \overline{x}_3 + \overline{x}_4)/4$. The contrast $\overline{x}_i - \overline{x}$ of the ith sample mean to the average of all four sample means is called the **effect** of level i of the factor *tire type*. ■

The *sample effects*, defined in Example 1.8-3 part (c), are typically denoted by $\widehat{\alpha}_i$:

Sample Effect of Level i in a One-Factor Design

$$\widehat{\alpha}_i = \overline{x}_i - \overline{x} \tag{1.8.2}$$

The sample contrasts and sample effects estimate their population counterparts. For example, the sample effects in (1.8.2) estimate the k population effects

$$\alpha_i = \mu_i - \mu, \quad \text{where } \mu = \frac{1}{k}\sum_{i=1}^{k} \mu_i. \tag{1.8.3}$$

For example, if the four types of high-performance tires have mean tread lives $\mu_1 = 16$, $\mu_2 = 13$, $\mu_3 = 14$, and $\mu_4 = 17$, the overall mean tread life is $\mu = (16+13+14+17)/4 = 15$, and the effects of the tire types are $\alpha_1 = 16-15 = 1$, $\alpha_2 = 13-15 = -2$, $\alpha_3 = 14-15 = -1$, and $\alpha_4 = 17-15 = 2$. Note that the tire effects sum to zero.

Additional contrasts, relevant in two-factor studies, will be given in Section 1.8.4.

The **comparative boxplot** and the **comparative bar graph** are commonly used for visualizing population differences in one-factor studies. The comparative boxplot consists of side-by-side individual boxplots for the data sets from each population; it is useful for providing a visual impression of differences in the median and percentiles. Example 1.8-4 provides the context for Figure 1-14 and the R commands for constructing it.

Example 1.8-4

Comparative boxplots in R. Iron concentration measurements from four ore formations are given in *FeData.txt*. Construct a comparative boxplot and comment on possible concentration differences.

Solution

Use *fe = read.table("FeData.txt", header = T)* to import the data into the R data frame *fe*, and also the following commands:

```
w=stack(fe)  # stacks data and assigns indices
boxplot(w$values~w$ind, col=rainbow(4))
  # constructs the boxplot
```

The comparative boxplot suggests that the fourth iron ore formation has higher, on average, iron concentration than the other three. (It should always be kept in mind that the differences at the data level, which the comparative boxplot suggest, are only approximations to the population level differences.) ■

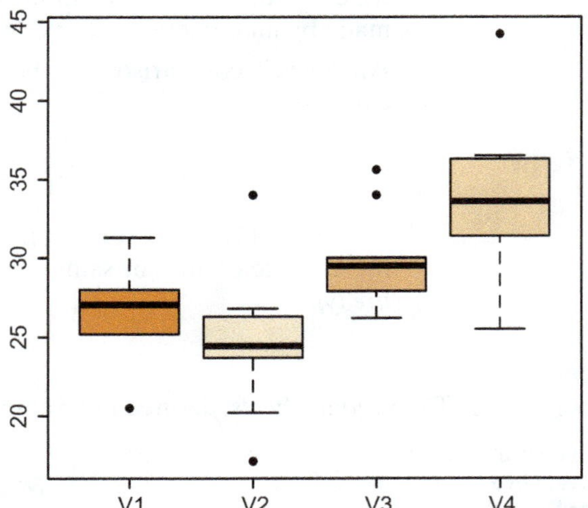

Figure 1-14 Comparative boxplot for iron concentration data.

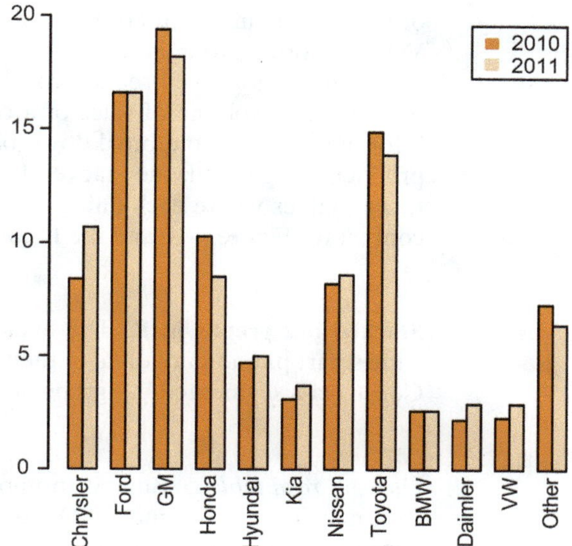

Figure 1-15 Comparative bar graph for light vehicle market share data.

The comparative bar graph generalizes the bar graph in that it plots several bars for each category. Each bar represents the category's proportion in one of the populations being compared; different colors are used to distinguish bars that correspond to different populations. Example 1.8-5 provides the context for Figure 1-15 and the R commands for constructing it.

Example 1.8-5

Comparative bar graphs in R. The light vehicle market share of car companies for November 2010 and 2011 is given in *MarketShareLightVehComp.txt*.[14] Construct a comparative bar graph and comment on possible changes in the companies' light vehicle market share.

Solution

Use the *read.table* command to import the data into the R data frame *lv2* (see Example 1.8-4), and also the following commands:

```
m=rbind(lv2$Percent_2010, lv2$Percent_2011)
  # creates a data matrix
barplot(m, names.arg=lv2$Company, ylim=c(0, 20),
  col=c("darkblue", "red"), legend.text= c("2010","2011"),
  beside=T, las=2) # constructs the bar graph
```

Figure 1-15 makes it easy to discern the changes in the companies' market shares. In particular, Chrysler had the biggest market share gain over this one-year period. ■

Bar graphs are also used to represent how a quantity other than a proportion varies across certain categories. Most often the quantity represented is a count and the category is a time period, such as how the number of visitors at Napa Valley, or the volume of sales of a certain product, varies across the months or seasons of the year. A **stacked bar graph** (also called **segmented bar graph**) is a visualization technique that can also incorporate information about an additional classification

[14] Data from http://wardsauto.com/datasheet/us-light-vehicle-sales-and-market-share-company-2004–2013.

of the units being counted. For example, in addition to classifying visitors at Napa Valley according to month of visit, a stacked bar graph can also display information about the nationality breakdown of the visitors; similarly, in addition to showing the quarterly volume of sales of a company, a stacked bar graph can also display information about the breakdown of the quarterly volume into sales of particular products. In general, the stacked bar graph is useful for studying two-way tables, that is, tables where each unit is classified in two ways. Example 1.8-6 provides the context for Figure 1-16 and the R commands for constructing it.

Example 1.8-6

Stacked bar graphs in R. The data file *QsalesSphone.txt* shows simulated world-wide smart phone sales data, in thousands of units, categorized by year and quarter. Construct a segmented bar graph and comment on its features.

Solution

Use the *read.table* command to import the data into the R object *qs*, and form a data matrix by m = rbind(qs$Q1, qs$Q2, qs$Q3, qs$Q4). The stacked bar graph is constructed with the command

```
barplot(m, names.arg=qs$Year, ylim=c(0, 40000),
  col=c("green", "blue", "yellow", "red"))
```

and the legend is added with the command

```
legend("topleft", pch=c(22, 22, 22, 22), col=c("green",
  "blue", "yellow", "red"), legend=c("Quarter 1",
  "Quarter 2", "Quarter 3", "Quarter 4"))
```

Figure 1-16 makes it apparent that more units were sold in the fourth quarter of each year than in any other quarter. This may be due to a strategy of increased product promotion during that quarter. ■

In comparative studies with two or more factors, it is of interest to also examine how the different factors **interact** in influencing the response. This is discussed in Section 1.8.4.

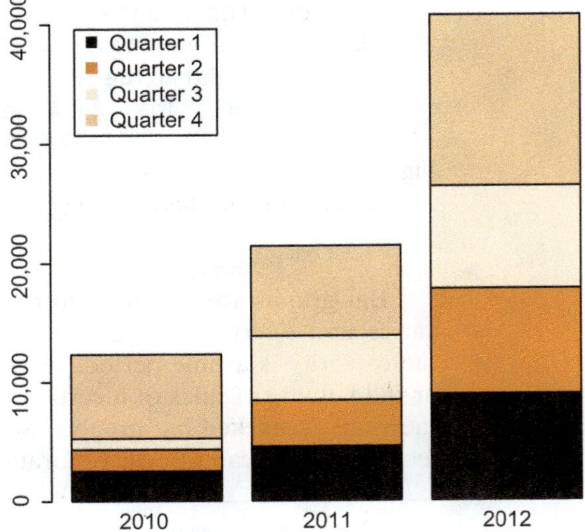

Figure 1-16 Stacked bar graph of annual smart phone sales.

1.8.2 LURKING VARIABLES AND SIMPSON'S PARADOX

In order to avoid comparing apples with oranges, the experimental units assigned to different treatments must be as similar (or homogenous) as possible. For example, if the age of the fabric is a factor that affects the response in Example 1.8-2 part (a), then, unless the ages of the fabrics that are assigned to different treatments are homogenous, the comparison of treatments will be distorted.

To guard against such distorting effects of other possible factors, also called *lurking variables*, it is recommended that the allocation of units to treatments be **randomized**. Randomization helps mitigate the distorting effects, called **confounding** in technical parlance, by equalizing the distribution of lurking variables across treatments.

Example 1.8-7

(a) Randomizing the allocation of fabric pieces to the different treatments (i.e., combinations of cleaning product and type of stain) avoids confounding the factors of interest (*cleaning product* and *stain*) with the potentially influential factor *age of fabric*.

(b) In the study of Example 1.8-2 part (b), the acidity of the materials used in the reaction might be another factor affecting the yield. Randomizing the allocation of the experimental units, that is, the materials, to the different treatments avoids confounding the factors of interest (*temperature* and *humidity*) with the potentially influential factor *acidity*. ■

The distortion caused by lurking variables in the comparison of proportions is called **Simpson's paradox**. Some examples of Simpson's paradox follow.

Example 1.8-8

1. *Batting averages:* The overall batting average of baseball players Derek Jeter (New York Yankees) and David Justice (Atlanta Braves) during the years 1995 and 1996 were 0.310 and 0.270, respectively. This seems to show that Jeter is more effective at bat than Justice. However, if we take into consideration each year's performance for the two players, the conclusion is not so straightforward:

	1995	1996	Combined
Derek Jeter	12/48 or 0.250	183/582 or 0.314	195/630 or 0.310
David Justice	104/411 or 0.253	45/140 or 0.321	149/551 or 0.270

In both 1995 and 1996, Justice had a higher batting average than Jeter, even though his overall batting average is lower. This appears paradoxical because the combined or overall average is not computed by a simple average of each year's average.[15]

2. *Kidney stone treatment:* This is a real-life example from a medical study comparing the success rates of two treatments for kidney stones.[16] The first table shows the overall success rates for Treatment *A* (all open procedures) and Treatment *B* (percutaneous nephrolithotomy):

Treatment *A*	Treatment *B*
78% (273/350)	83% (289/350)

[15] The batting averages used in this example are from Ken Ross (2004). *A Mathematician at the Ballpark: Odds and Probabilities for Baseball Fans* (paperback). Pi Press.

[16] C. R. Charig, D. R. Webb, S. R. Payne, and O. E. Wickham (1986). Comparison of treatment of renal calculi by operative surgery, percutaneous nephrolithotomy, and extracorporeal shock wave lithotripsy. *Br Med J (Clin Res Ed)* 292(6524): 879–882.

The table seems to show Treatment B is more effective. However, if we include data about kidney stone size, a different picture emerges:

	Treatment A	Treatment B
Small Stones	93% (81/87)	87% (234/270)
Large Stones	73% (192/263)	69% (55/80)
Both	78% (273/350)	83% (289/350)

The information about stone size has reversed our conclusion about the effectiveness of each treatment. Now Treatment A is seen to be more effective in both cases. In this example the lurking variable (or confounding variable) of stone size was not previously known to be important until its effects were included. ■

1.8.3 CAUSATION: EXPERIMENTS AND OBSERVATIONAL STUDIES

A study is called a **statistical experiment** if the investigator controls the allocation of units to treatments or factor-level combination, and this allocation is done in a randomized fashion. Thus, the studies mentioned in Example 1.8-7 are statistical experiments. Similarly, the study on the effectiveness of a new diet in reducing weight becomes a statistical experiment if the allocation of the participating subjects to the *control group* (which is the group using the standard diet) and the treatment group (which is the group using the new diet) is done in a randomized fashion.

Though desirable, randomization is not always possible.

Example 1.8-9

1. It is not possible to assign subjects to different levels of smoking in order to study the effects of smoking.
2. It does not make sense to assign random salary increases in order to study the effects of salary on productivity.
3. It may not be possible to randomly assign parents to different types of disciplinary actions in order to study the actions' effects on teenage delinquency. ■

When the allocation of units to treatments is not controlled by the investigator, and thus the allocation is not randomized, the study is called **observational**. Observational studies cannot be used for establishing **causation** because the lack of randomization allows potentially influential lurking variables to be confounded with the factors being studied.

For example, even a strong relation between salary increases and employee productivity does not imply that salary increases cause increased productivity (it could be vice versa). Similarly, a strong relation between spanking and anti-social behavior in children does not imply that spanking causes anti-social behavior (it could be vice versa). Causality can be established only through experimentation. This is why experimentation plays a key role in industrial production and especially in the area of quality improvement of products. In particular, *factorial experimentation* (which is discussed in the next section) was vigorously advocated by W. Edwards Deming (1900–1993). Having said that, one should keep in mind that observational studies have yielded several important insights and facts. For example, studies involving the effect of smoking on health are observational, but the link they have established between the two is one of the most important issues of public health.

1.8.4 FACTORIAL EXPERIMENTS: MAIN EFFECTS AND INTERACTIONS

A statistical experiment involving several factors is called a *factorial experiment* if all factor-level combinations are considered. Thus, in a two-factor factorial experiment, where factor A enters with a levels and factor B enters with b levels, $a \times b$ samples are collected, one from each factor-level combination. For example, the two-factor study portrayed in Figure 1-13 is a factorial experiment if all eight treatments are included in the study.

In factorial experiments with two or more factors, it is not enough to consider possible differences between the levels of each factor separately. Comparative boxplots for the levels of each individual factor fail to capture possible *synergistic effects* among the levels of different factors. Such synergistic effects, called *interactions* in statistical parlance, may result in some factor-level combinations yielding improved or diminished response levels far beyond what can be explained by any differences between the levels of the individual factors.

Example 1.8-10

An experiment considers two types of corn, used for bio-fuel, and two types of fertilizer. The table in Figure 1-17 gives the population mean yields for the four combinations of seed type and fertilizer type. It is seen that the fertilizer factor has different effects on the mean yields of the two seeds. For example, fertilizer II improves the mean yield of seed A by $111 - 107 = 4$, while the mean yield of seed B is improved by $110 - 109 = 1$. Moreover, the best yield is obtained by using fertilizer II on seed A (synergistic effect) even though the average yield of seed A over both fertilizers, which is $\overline{\mu}_{1\cdot} = (107 + 111)/2 = 109$, is lower than the average yield of seed B which is $\overline{\mu}_{2\cdot} = (109 + 110)/2 = 109.5$. ■

Definition 1.8-1
When a change in the level of factor A has different effects on the levels of factor B we say that there is **interaction** between the two factors. The absence of interaction is called **additivity**.

Under additivity there is an indisputably best level for each factor and the best factor-level combination is that of the best level of factor A with the best level of factor B. To see this, suppose that the mean values in the fertilizer and seed experiment of Example 1.8-10 are as in Figure 1-18. In this case, changing to fertilizer II has the same effect on both seeds (an increase of 4 in the mean yield). Similarly, it can be said that seed B is better than seed A because it results in higher yield (by two units) regardless of the fertilizer used. Thus, there is a clear-cut better level for factor A (seed B) and a clear-cut better level for factor B (fertilizer II), and the best results (highest yield) is achieved by the factor-level combination that corresponds to the two best levels (seed B and fertilizer II in this case).

Figure 1-17 A 2×2 design with interaction (non-additive design).

	Fertilizer		Row Averages	Main Row Effects
	I	II		
Seed A	$\mu_{11} = 107$	$\mu_{12} = 111$	$\overline{\mu}_{1\cdot} = 109$	$\alpha_1 = -0.25$
Seed B	$\mu_{21} = 109$	$\mu_{22} = 110$	$\overline{\mu}_{2\cdot} = 109.5$	$\alpha_2 = 0.25$
Column Averages	$\overline{\mu}_{\cdot 1} = 108$	$\overline{\mu}_{\cdot 2} = 110.5$	$\overline{\mu}_{\cdot\cdot} = 109.25$	
Main Column Effects	$\beta_1 = -1.25$	$\beta_2 = 1.25$		

Figure 1-18 A 2 × 2 design with no interaction (additive design).

	Fertilizer		Row Averages	Main Row Effects
	I	II		
Seed *A*	$\mu_{11} = 107$	$\mu_{12} = 111$	$\overline{\mu}_{1\cdot} = 109$	$\alpha_1 = -1$
Seed *B*	$\mu_{21} = 109$	$\mu_{22} = 113$	$\overline{\mu}_{2\cdot} = 111$	$\alpha_2 = 1$
Column Averages	$\overline{\mu}_{\cdot 1} = 108$	$\overline{\mu}_{\cdot 2} = 112$	$\overline{\mu}_{\cdot\cdot} = 110$	
Main Column Effects	$\beta_1 = -2$	$\beta_2 = 2$		

Under additivity, the comparison of the levels of each factor is typically based on the so-called **main effects**. The *main row effects*, denoted by α_i, and *main column effects*, denoted by β_j, are defined as

Main Row and Column Effects

$$\alpha_i = \overline{\mu}_{i\cdot} - \overline{\mu}_{\cdot\cdot}, \quad \beta_j = \overline{\mu}_{\cdot j} - \overline{\mu}_{\cdot\cdot} \tag{1.8.4}$$

Figures 1-17 and 1-18 show the main row and the main column effects in these two fertilizer-seed designs.

Under additivity, the cell means μ_{ij} are given in terms of their overall average, $\overline{\mu}_{\cdot\cdot}$ and the main row and column effects in an additive manner:

Cell Means under Additivity

$$\mu_{ij} = \overline{\mu}_{\cdot\cdot} + \alpha_i + \beta_j \tag{1.8.5}$$

For example, in the additive design of Figure 1-18, $\mu_{11} = 107$ equals the sum of the main effect of row 1, $\alpha_1 = -1$, plus the main effect of column 1, $\beta_1 = -2$, plus the overall mean, $\overline{\mu}_{\cdot\cdot} = 110$; similarly, $\mu_{12} = 111$ equals the sum of the main effect of row 1, $\alpha_1 = -1$, plus the main effect of column 2, $\beta_2 = 2$, plus the overall mean, $\overline{\mu}_{\cdot\cdot} = 110$, and so on.

When there is interaction between the two factors, the cell means are not given by the additive relation (1.8.5). The discrepancy/difference between the left and right-hand sides of this relation quantifies the interaction effects:

Interaction Effects

$$\gamma_{ij} = \mu_{ij} - \left(\overline{\mu}_{\cdot\cdot} + \alpha_i + \beta_j\right) \tag{1.8.6}$$

Example 1.8-11

Compute the interaction effects in the design of Example 1.8-10.

Solution

Using the information shown in Figure 1-17, we have

$$\gamma_{11} = \mu_{11} - \overline{\mu}_{\cdot\cdot} - \alpha_1 - \beta_1 = 107 - 109.25 + 0.25 + 1.25 = -0.75$$

$$\gamma_{12} = \mu_{12} - \overline{\mu}_{\cdot\cdot} - \alpha_1 - \beta_2 = 111 - 109.25 + 0.25 - 1.25 = 0.75$$

$$\gamma_{21} = \mu_{21} - \overline{\mu}_{\cdot\cdot} - \alpha_2 - \beta_1 = 109 - 109.25 - 0.25 + 1.25 = 0.75$$

$$\gamma_{22} = \mu_{22} - \overline{\mu}_{\cdot\cdot} - \alpha_2 - \beta_2 = 110 - 109.25 - 0.25 - 1.25 = -0.75.$$

■

Figure 1-19 Data notation in a 2 × 4 factorial experiment.

	Factor B			
Factor A	1	2	3	4
1	x_{11k}, $k = 1, \ldots, n_{11}$	x_{12k}, $k = 1, \ldots, n_{12}$	x_{13k}, $k = 1, \ldots, n_{13}$	x_{14k}, $k = 1, \ldots, n_{14}$
2	x_{21k}, $k = 1, \ldots, n_{21}$	x_{22k}, $k = 1, \ldots, n_{22}$	x_{23k}, $k = 1, \ldots, n_{23}$	x_{24k}, $k = 1, \ldots, n_{24}$

Data from a two-factor factorial experiment are typically denoted using three subscripts as shown in Figure 1-19. Thus, the first two subscripts correspond to the factor-level combination and the third subscript enumerates the observations, the number of which may differ among the different treatments. Sample versions of the main effects and interactions are defined similarly using the cell means

Sample Mean of Observations in Cell (i, j)

$$\overline{x}_{ij} = \frac{1}{n_{ij}} \sum_{k=1}^{n_{ij}} x_{ijk}, \tag{1.8.7}$$

instead of the population means μ_{ij}. The formulas for the *sample main row effects*, $\widehat{\alpha}_i$, and the *sample main column effects*, $\widehat{\beta}_j$, are

Sample Main Row and Column Effects

$$\widehat{\alpha}_i = \overline{x}_{i\cdot} - \overline{x}_{\cdot\cdot}, \quad \widehat{\beta}_j = \overline{x}_{\cdot j} - \overline{x}_{\cdot\cdot} \tag{1.8.8}$$

Also, in analogy to (1.8.6), estimates of the interaction effects, the *sample interaction effects*, are obtained by

Sample Interaction Effects

$$\widehat{\gamma}_{ij} = \overline{x}_{ij} - \left(\overline{x}_{\cdot\cdot} + \widehat{\alpha}_i + \widehat{\beta}_j\right) \tag{1.8.9}$$

These calculations, and the relevant notation, are illustrated in Figure 1-20.

Note that $\overline{x}_{\cdot\cdot}$ can be obtained either as a column average (i.e., the average of $\overline{x}_{1\cdot}$, $\overline{x}_{2\cdot}$ and $\overline{x}_{3\cdot}$), or as a row average (i.e., the average of $\overline{x}_{\cdot 1}, \overline{x}_{\cdot 2}$ and $\overline{x}_{\cdot 3}$), or as the average of the nine cell sample means $\overline{x}_{ij}$, (i.e., $\overline{x}_{\cdot\cdot} = (1/9)\sum_{i=1}^{3}\sum_{j=1}^{3}\overline{x}_{ij}$).

The following example demonstrates the calculation of the sample main and interaction effects with R.

Figure 1-20 Calculation of the sample main row effects in a 3 × 3 design.

	Column Factor				
Row Factor	1	2	3	Row Averages	Main Row Effects
1	$\overline{x}_{11}$	$\overline{x}_{12}$	$\overline{x}_{13}$	$\overline{x}_{1\cdot}$	$\widehat{\alpha}_1 = \overline{x}_{1\cdot} - \overline{x}_{\cdot\cdot}$
2	$\overline{x}_{21}$	$\overline{x}_{22}$	$\overline{x}_{23}$	$\overline{x}_{2\cdot}$	$\widehat{\alpha}_2 = \overline{x}_{2\cdot} - \overline{x}_{\cdot\cdot}$
3	$\overline{x}_{31}$	$\overline{x}_{32}$	$\overline{x}_{33}$	$\overline{x}_{3\cdot}$	$\widehat{\alpha}_3 = \overline{x}_{3\cdot} - \overline{x}_{\cdot\cdot}$
Column Averages	$\overline{x}_{\cdot 1}$	$\overline{x}_{\cdot 2}$	$\overline{x}_{\cdot 3}$	$\overline{x}_{\cdot\cdot}$	
Main Column Effects	$\widehat{\beta}_1 = \overline{x}_{\cdot 1} - \overline{x}_{\cdot\cdot}$	$\widehat{\beta}_2 = \overline{x}_{\cdot 2} - \overline{x}_{\cdot\cdot}$	$\widehat{\beta}_3 = \overline{x}_{\cdot 3} - \overline{x}_{\cdot\cdot}$		

Example 1.8-12

Cell means, main effects, and interaction effects in R. Figure 1-21 contains data on the amount of rainfall, in inches, in select target areas of Tasmania with and without cloud seeding during the different seasons.[17,18] Use R to compute the cell means, and the main and interaction effects for the factors seed and season.

Solution

Import the data into R using *cs=read.table("CloudSeed2w.txt", header=T)* and use the following commands:

```
mcm=tapply(cs$rain, cs[,c(2, 3)], mean) # the matrix
  of cell means
alphas=rowMeans(mcm)-mean(mcm) # the vector of main row effects
betas=colMeans(mcm)-mean(mcm) # the vector main column effects
gammas=t(t(mcm-mean(mcm)-alphas) -betas) # the matrix of
  interaction effects
```

The computed interaction effects are

	Season			
Seeded	Autumn	Spring	Summer	Winter
no	0.0298	-0.0883	-0.1345	0.1930
yes	-0.0298	0.0883	0.1345	-0.1930

The computed main effects for seeded and unseeded are -0.0352 and 0.0352, respectively, while the computed main effects for Autumn, Spring, Summer, and Winter are 0.4802, -0.0017, -0.9355, and 0.4570, respectively. ■

As we did in other sections of this chapter, we stress again that the sample versions of the main effects and interactions approximate but, in general, they are not equal to their population counterparts. In particular, the sample interaction effects will not be equal to zero even if the design is additive. The **interaction plot** is a useful graphical technique for assessing whether the sample interaction effects are

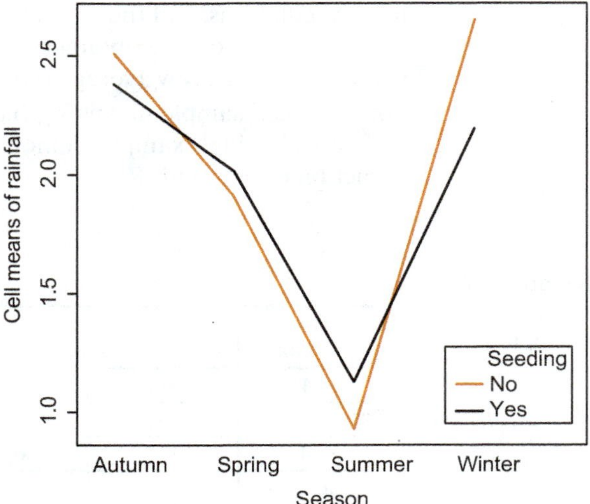

Figure 1-21 Interaction plot for the cloud seeding data of Example 1.8-12.

[17] A. J. Miller et al. (1979). Analyzing the results of a cloud-seeding experiment in Tasmania, *Communications in Statistics—Theory & Methods*, A8(10): 1017–1047.

[18] See also the related article by A. E. Morrison et al. (2009). On the analysis of a cloud-seeding dataset over Tasmania, *American Meteorological Society*, 48: 1267–1280.

sufficiently different from zero to imply a non-additive design. For each level of one factor, say factor B, the interaction plot traces the cell means along the levels of the other factor. If the design is additive, these traces (also called *profiles*) should be approximately parallel. The interaction plot for the cloud-seeding data of Example 1.8-12, shown in Figure 1-21, was generated with the commands:

R Commands for the Interaction Plot of Figure 1-21

```
attach(cs) # so variables can be referred to by name
interaction.plot(season, seeded, rain, col=c(2,3), lty = 1,
  xlab="Season", ylab="Cell Means of Rainfall", trace.label
  ="Seeding")
```

The crossing of the traces (or profiles) seen in Figure 1-21 is typically indicative of interaction.

Factor interaction is prevalent in everyday life as it is in sciences. For example, different spices may interact with different types of food to enhance taste, and different wines interact with different appetizers. In agriculture, different types of fertilization may interact with different types of soil as well as different levels of watering. The June 2008 issue of *Development* features research suggesting interaction between two transcription factors that regulate the development and survival of retinal ganglion cells. A quality of service (QoS) IEEE article[19] considers the impact of several factors on total throughput and average delay as measures of service delivery. Due to interactions, the article concludes, the factors cannot be studied in isolation. Finally, in product and industrial design it is typical to consider the potential impact of a large number of factors and their interactions on a number of quality characteristics of a product, or aspects of a product. In car manufacturing, for example, quality aspects range from the car's handling to the door's holding ability for remaining open when the car is parked uphill. Optimization of such quality characteristics is only possible through factorial experimentation.

Exercises

1. An experiment is conducted to determine the optimal time and temperature combination for baking a cake. The response variable of interest is taste. Four batches of cake will be baked separately at each combination of baking times (25 and 30 minutes) and temperature settings ($275^o F$, $300^o F$, and $325^o F$).

(a) What are the experimental units?

(b) What are the factors in this experiment?

(c) State the levels of each factor.

(d) List all the treatments in this experiment.

(e) Is the response variable qualitative or quantitative?

2. An experiment to assess the effect of watering on the life span of a certain type of root system incorporates three watering regimens.

(a) How many populations involved in the study?

(b) The population(s) involved is (are) hypothetical: True or false?

(c) The variable of interest is qualitative: True or false?

(d) What is considered a *treatment* in this study?

(e) Suppose the experiment will be carried out in three different locations. It is known that specific location characteristics (e.g., temperature and soil conditions) also affect the life span of the root systems.

(i) Does this change the number of populations involved in the study?

(ii) List the factors involved in this experiment and their levels.

3. A quantification of coastal water quality converts measurements on several pollutants (arsenic in oyster shells, mercury, etc.) to a water quality index with values from 1 to 10. An investigation into the after-clean-up water

[19] K. K. Vadde and Syrotiuk (2004). Factor interaction on service delivery in mobile ad hoc networks, *Selected Areas in Communications*, 22: 1335–1346.

quality of a lake analyzes water samples collected from five areas encompassing the two beaches on the eastern shore and the three beaches on the western shore. Let μ_1 and μ_2 denote the mean water quality index for the beaches on the eastern shore, and μ_3, μ_4, and μ_5 be the mean water quality index for the beaches on the western shore.

(a) Write the contrasts that represent the effects of each of the five areas.

(b) Write the contrast for comparing the water quality around the two beaches on the eastern shore with that around the three beaches on the western shore.

4. An article reports on the results of a cloud-seeding experiment.[20] The question of interest is whether cloud seeding with silver nitrate increases rainfall. Out of 52 clouds, 26 were randomly selected for seeding, with the remaining 26 serving as controls. The rainfall measurements, in acre-feet, are given in *CloudSeedingData.txt*. Use the R commands given in Example 1.8-4 to construct a comparative boxplot and comment on possible differences in rainfall between seeded and unseeded clouds.

5. For its new generation of airplanes, a commercial airline is considering three new designs of the control panel as possible alternatives to the current design with the aim of improving the pilot's response time to emergency displays. Letting μ_1 denote the mean pilot response time to simulated emergency conditions with the current design, and μ_2, μ_3, and μ_4 denote the mean response times with the three new designs, write the control versus treatment contrasts.

6. Rural roads with little or no evening lighting use reflective paint to mark the lanes on highways. It is suspected that the currently used paint does not maintain its reflectivity over long periods of time. Three new types of reflective paint are now available and a study is initiated to compare all four types of paint.

(a) How many populations are involved in the study?

(b) What is considered a *treatment* in this study?

(c) Letting μ_1 denote the mean time the currently used paint maintains its reflectivity, and μ_2, μ_3, and μ_4 denote the corresponding means for the three new types of paint, write the control versus treatment contrasts.

7. The researcher in charge of the study described in Exercise 6 identifies four locations of the highway, and for each location she designates four sections of length six feet to serve as the experimental units on which the paints will be applied. It is known that specific aspects of each location (e.g., traffic volume and road conditions) also affect the duration of the reflectivity of the paints.

(a) Does this change the number of populations involved in the study?

(b) List the factors involved in this experiment, the levels of each factor, and the treatments.

8. The ignition times of two types of material used for children's clothing are measured to the nearest hundredth of a second. The 25 measurements from material type *A* and 28 measurements from material type *B* are given in *IgnitionTimesData.txt*. Read the data into the data file *ig* and use the *boxplot* command given in Example 1.8-4, with ig$Time~ig$Type instead of w$values~w$ind, to construct a comparative boxplot and comment on possible differences in the ignition times of the two material types.

9. Wildlife conservation officials collected data on black bear weight during the period September to November. After sedation, the weight and gender (among other measurements) were obtained for a sample of 50 black bears. The data can be found in *bearWeightgender.txt*.[21] Construct a comparative boxplot and comment on the differences between female and male black bear sample weights.

10. Read the projected data on reasons why people in the Boston, MA, and Buffalo, NY, areas are late for work, found in *ReasonsLateForWork2.txt*, into the data frame *lw* using the *read.table* command given in Exercise 17 in Section 1.5. Then, use commands similar to those in Example 1.8-5 to construct a comparative bar graph. What are the biggest differences you notice?

11. Import the data on monthly online and catalog sales of a company into the R object *oc* using *oc = read.table("MonthlySalesOC.txt", header = T)*.

(a) Use R commands similar to those in Example 1.8-5 to construct a bar graph comparing the online and catalog volumes of sale.

(b) Use R commands similar to those in Example 1.8-6 to construct a stacked bar graph showing the breakdown of the total volume of sales into online and catalog.

(c) Comment on the relative advantages of each of the two types of plots.

12. In the context of Exercise 2 part (e), the researcher proceeds to assign a different watering regimen to each of the three locations. Comment on whether or not the above allocation of treatments to units (root systems) will avoid confounding the effect of the watering levels with the location factor. Explain your answer and describe a possibly better allocation of treatments to units.

13. The researcher mentioned in Exercise 7 proceeds to randomly assign a type of paint to each of the four locations. It is known that specific aspects of each location (e.g., traffic volume and road conditons) also affect the duration of the reflectivity of the paints. Comment on whether or not the above allocation of paints (treatments) to the road segments (experimental units) will avoid confounding with the treatment effect with the location

[20] J. Simpson, A. Olsen, and J. C. Eden (1975). *Technometrics*, 17: 161–166.

[21] The data is a subset of a data set contributed to Minitab by Dr. Gary Alt.

factor. Explain your answer and describe a possibly better allocation of treatments to units.

14. A study is initiated to compare the effect of two levels of fertilization and two levels of watering on the yield per bushel for a variety of corn. One hundred bushels are to be grown under each of the four combinations of fertilization and watering.

(a) How many populations are involved in this study?

(b) The population(s) involved is (are) hypothetical: True or false?

(c) The variable of interest is qualitative: True or false?

(d) List the factors and their levels involved in this study.

(e) Suppose the experiment will be carried out on two farms, one using traditional pest control practices and one that uses organic practices. To avoid confounding the factors in this study with the potentially influential factor of pest control practices, all fertilization and watering levels must be applied to both farms. True or false?

15. The 1973 admission rates of men and women applying to graduate school in different departments of the University of California at Berkeley are as follows:

Major	Men		Women	
	Applicants	% Admitted	Applicants	% Admitted
A	825	62%	108	82%
B	560	63%	25	68%
C	325	37%	593	34%
D	417	33%	375	35%
E	191	28%	393	24%
F	272	6%	341	7%

(a) What are the overall admission rates of men and women applying to graduate programs at Berkeley?

(b) UC Berkeley was actually sued for bias against women applying to graduate school on the basis of the overall admission rates. Do you agree that the above overall admission rates suggest gender bias in Berkeley's graduate admissions?

(c) Are the overall averages appropriate indicators for gender bias in this case? Why or why not?

16. Pygmalion was a mythical king of Cyprus who sculpted a figure of the ideal woman and then fell in love with his own creation. The Pygmalion effect in psychology refers to a situation where high expectations of a supervisor translates into improved performance by subordinates. A study was conducted in an army training camp using a company of male recruits and one of female recruits. Each company had two platoons. One platoon in each company was randomly selected to be the Pygmalion platoon. At the conclusion of basic training, soldiers took a battery of tests. The following table gives the population mean scores for female (F) and male (M) soldiers. PP denotes the Pygmalion platoon and CP denotes the control platoon.

	CP	PP
F	$\mu_{11} = 8$	$\mu_{12} = 13$
M	$\mu_{21} = 10$	$\mu_{22} = 12$

(a) Is this an additive design? Justify your answer.

(b) Compute the main gender effects and the main Pygmalion effect.

(c) Compute the interaction effects.

17. A soil scientist is considering the effect of soil pH level on the breakdown of a pesticide residue. Two pH levels are considered in the study. Because pesticide residue breakdown is affected by soil temperature, four different temperatures are included in the study.

	Temp A	Temp B	Temp C	Temp D
pH I	$\mu_{11} = 108$	$\mu_{12} = 103$	$\mu_{13} = 101$	$\mu_{14} = 100$
pH II	$\mu_{21} = 111$	$\mu_{22} = 104$	$\mu_{23} = 100$	$\mu_{24} = 98$

(a) Draw the interaction plot by hand with pH being the trace factor.

(b) Is there interaction between factors pH and temperature? Use the interaction plot to justify your answer.

(c) Compute the main pH effects and the main temperature effects.

(d) Compute the interaction effects.

18. The file *SpruceMothTrap.txt* contains data on the number of moths caught in moth traps using different lures and placed at different locations on spruce trees.[22] Use R to:

(a) Compute the cell means and the main and interaction effects for the factors location and lure.

(b) Construct an interaction plot with the levels of the factor location being traced. Comment on the main effects of the factors location and lure, and on their interaction effects.

19. The data file *AdLocNews.txt* contains the number of inquiries regarding ads placed in a local newspaper. The ads are categorized according to the day of the week and in which section of the newspaper they appeared. Use R to:

(a) Compute the cell means and the main and interaction effects for the factors day and newspaper section. Is there an overall best day to place a newspaper ad? Is there an overall best newspaper section to place the ad in?

(b) Construct an interaction plot with the levels of the factor day being traced. Construct an interaction plot with section being the trace factor. What have you learned from these interaction plots?

[22] Data based on "Two-way ANOVA?" *Talk Stats*, April 22, 2012, http://www.talkstats.com/showthread.php/25167-Two-way-ANOVA

1.9 The Role of Probability

The most common probability questions deal with flipping a fair coin or picking cards from a deck.[23] Flipping a fair coin once can be thought of as taking a simple random sample of size one from the population consisting of {*Heads, Tails*}. By the definition of simple random sampling, there is a 50% chance of heads. The coin-flipping paradigm leads to more complicated probability questions by simply increasing the number of flips.

Example 1.9-1

(a) What are the chances of one heads in two flips of a fair coin? This can be thought of as taking a sample of size two with replacement from the population consisting of {*Heads, Tails*} and asking for the chances of the sample containing the item heads only once.

(b) What are the chances of 4 heads, or of 10 heads, or of 18 heads in 20 flips of a fair coin? Again, this can be rephrased in terms of a sample of size 20 taken with replacement from the population {*Heads, Tails*}. ■

Other examples of probability questions, not related to games of chance, are given in the following example.

Example 1.9-2

(a) If 75% of citizens of voting age are in favor of introducing incentives for the use of solar energy, what are the chances that in a sample of 1000, at least 650 will be in favor of such incentives?

(b) If 5% of electrical components have a certain defect, what are the chances that a batch of 500 such components will contain less than 20 defective ones?

(c) If 60% of all batteries last more than 1500 hours in operation, what are the chances that a sample of 100 batteries will contain at least 80 that last more than 1500 hours?

(d) If the highway mileage achieved by the 2011 Toyota Prius cars has population mean and standard deviation of 51 and 1.5 miles per gallon, respectively, what are the chances that in a sample of size 10 cars, the average highway mileage is less than 50 miles per gallon? ■

The probability questions in Examples 1.9-1 and 1.9-2 highlight what is true of all probability questions. Namely, in probability theory one assumes that all relevant information about the population is known and seeks to assess the chances that a sample will possess certain properties of interest. This, of course, is opposite from statistics where one uses sample-level information to infer properties of the population. For example, a statistical counterpart of the battery-life question in Example 1.9-2 would be: "If 80 batteries out of a sample of 100 last more than 1500 hours, can we conclude that the corresponding population proportion is more than 60%?" The reverse actions of probability and statistics are illustrated in Figure 1-22.

In spite of this difference, statistical inference itself would not be possible without probability. That probability is such an indispensable tool for statistics is seen by considering the meaning of the expression *statistical proof*. A statistical proof is proof beyond reasonable doubt. This is the only kind of proof that statistics can provide because the sample (on which statistical proofs are based) is only a small

[23] The field of probability known today as *classical probability* arose from the study of games of chance.

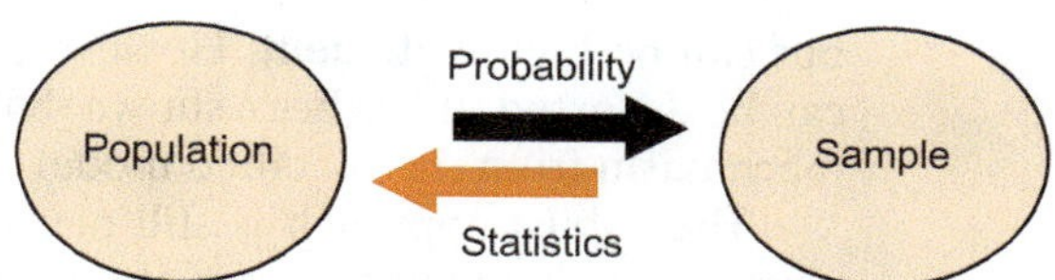

Figure 1-22 The reverse actions of probability and statistics.

part of the population. Probability is the tool for establishing statistical proof that a population has a certain property; for example that there is interaction between the two factors in a factorial experiment. This is done by assuming that the population does not have the property in question (for example that the design is additive), and calculating the chances of obtaining the kind of sample we observed (for example the kind of interaction plot the data produced). If the chances are small enough, we conclude that we have statistical proof for the existence of the property in question. This process of establishing a statistical proof is demonstrated in the following simple example.

Example 1.9-3

In 20 flips of a coin, 18 heads result. Should the fairness of the coin be dismissed?

Solution

Since it is not impossible to have 18 heads in 20 flips of a fair coin, one can never be sure whether the coin is fair or not. The decision in such cases is based on the knowledge, which is provided by probability, of the likelihood of an outcome *at least as extreme* as the one observed. In the present example, outcomes at least as extreme as the one observed are 18, 19, or 20 heads in 20 flips. Since 18 or more heads in 20 flips of a fair coin is quite unlikely (the chances are 2 in 10,000), one can claim there is statistical proof, or proof beyond reasonable doubt, that the coin is not fair. ■

1.10 Approaches to Statistical Inference

The main approaches to statistical inference can be classified into *parametric*, *robust*, *nonparametric*, and *Bayesian*.

The parametric approach relies on modeling aspects of the mechanism underlying the generation of the data.

Example 1.10-1

Predicting the failure time on the basis of stress applied hinges on the *regression model* and the *distribution* of the *intrinsic error* (words in italics are technical terms to be clarified in Chapter 4). A parametric approach might specify a *linear regression model* and the *normal distribution* for the intrinsic error. ■

In the parametric approach, models are described in terms of unknown *model parameters*. Hence the name *parametric*. In the above example, the slope and intercept of the linear function that models the relation between failure time and stress are model parameters; specification of the (intrinsic) error distribution typically introduces further model parameters. In the parametric approach, model parameters are assumed to coincide with population parameters, and thus they become the focus of the statistical inference. If the assumed parametric model is a good approximation to the data-generation mechanism, then the parametric inference is not only valid but can be highly efficient. However, if the approximation is not good, the results can be distorted. It has been shown that even small deviations of the data-generation mechanism from the specified model can lead to large biases.

The robust approach is still parametric in flavor, but its main concern is with procedures that guard against aberrant observations such as outliers.

The nonparametric approach is concerned with procedures that are valid under minimal modeling assumptions. Some procedures are both nonparametric and robust, so there is overlap between these two approaches. In spite of their generality, the efficiency of nonparametric procedures is typically very competitive compared to parametric ones that employ correct model assumptions.

The Bayesian approach is quite different from the first three as it relies on modeling prior beliefs/information about aspects of the population. The increase of computational power and efficiency of algorithms have made this approach attractive for dealing with some complex problems in different areas of application.

In this book we will develop, in a systematic way, parametric and nonparametric procedures, with passing reference to robustness issues, for the most common ("bread-and-butter") applications of statistics in the sciences and engineering.

Chapter 2

INTRODUCTION TO PROBABILITY

2.1 Overview

The field of probability known as *classical probability* arose from the need to quantify the likelihood of occurrence of certain events associated with games of chance. Most games of chance are related to sampling experiments. For example, rolling a die five times is equivalent to taking a sample of size five, with replacement, from the population $\{1, 2, 3, 4, 5, 6\}$, while dealing a hand of five cards is equivalent to taking a simple random sample of size five from the population of 52 cards. This chapter covers the basic ideas and tools used in classical probability. This includes an introduction to *combinatorial methods* and to the concepts of *conditional probability* and *independence*.

More modern branches of probability deal with *modeling* the randomness of phenomena such as the number of earthquakes, the amount of rainfall, the lifetime of a given electrical component, or the relation between education level and income. Such models, and their use for calculating probabilities, will be discussed in Chapters 3 and 4.

2.2 Sample Spaces, Events, and Set Operations

Any action whose outcome is random, such as counting the number of heads in ten flips of a coin or recording the number of disabled vehicles on a motorway during a snowstorm, is a (*random* or *probabilistic*) experiment.

Definition 2.2-1
The set of all possible outcomes of an experiment is called the **sample space** of the experiment and will be denoted by $\mathcal{S}$.

Example 2.2-1

(a) Give the sample space of the experiment that selects two fuses and classifies each as non-defective or defective.

(b) Give the sample space of the experiment that selects two fuses and records how many are defective.

(c) Give the sample space of the experiment that records the number of fuses inspected until the second defective is found.

Solution

(a) The sample space of the first experiment can be represented as

$$\mathcal{S}_1 = \{NN, ND, DN, DD\},$$

where N denotes a non-defective fuse and D denotes a defective fuse.

(b) When only the number of defective fuses is recorded, the sample space is

$$\mathcal{S}_2 = \{0, 1, 2\}.$$

The outcome 0 means that none of the two examined fuses are defective, the outcome 1 means that either the first or the second of the selected fusses is defective (but not both), and the outcome 2 means that both fuses are defective.

(c) For the experiment that records the number of fuses examined until the second defective is found, the sample space is

$$\mathcal{S}_3 = \{2, 3, \ldots\}.$$

Note that 0 and 1 are not possible outcomes because one needs to examine at least two fuses in order to find two defective fuses. ■

Example 2.2-2

An undergraduate student from a particular university is selected and his/her opinion about a proposal to expand the use of solar energy is recorded on a scale from 1 to 10.

(a) Give the sample space of this experiment.

(b) How does the sample space differ from the statistical population?

Solution

(a) When the opinion of only one student is recorded, the sample space is $\mathcal{S} = \{1, 2, \ldots, 10\}$.

(b) The statistical population for this sampling experiment is the collection of opinion ratings from the entire student body of that university. The sample space is smaller in size since each of the possible outcomes is listed only once. ■

Example 2.2-3

Three undergraduate students from a particular university are selected and their opinions about a proposal to expand the use of solar energy are recorded on a scale from 1 to 10.

(a) Describe the sample space of this experiment. What is the size of this sample space?

(b) Describe the sample space if only the average of the three responses is recorded. What is the size of this sample space?

Solution

(a) When the opinions of three students are recorded the set of all possible outcomes consists of the triplets (x_1, x_2, x_3), where $x_1 = 1, 2, \ldots, 10$ denotes the response of the first student, $x_2 = 1, 2, \ldots, 10$ denotes the response of the second student, and $x_3 = 1, 2, \ldots, 10$ denotes the response of the third student. Thus, the sample space is described as

$$\mathcal{S}_1 = \{(x_1, x_2, x_3) : x_1 = 1, 2, \ldots, 10,\ x_2 = 1, 2, \ldots, 10,\ x_3 = 1, 2, \ldots, 10\}.$$

There are $10 \times 10 \times 10 = 1000$ possible outcomes.

(b) The easiest way to describe the sample space, $\mathcal{S}_2$, when the three responses are averaged is to say that it is the collection of *distinct* averages $(x_1 + x_2 + x_3)/3$ formed from the 1000 triplets of $\mathcal{S}_1$. The word "distinct" is emphasized because the sample space lists each individual outcome only once, whereas several triplets might result in the same average. For example, the triplets $(5, 6, 7)$ and $(4, 6, 8)$ both yield an average of 6. Determining the size of $\mathcal{S}_2$ can be done, most easily, with the following R commands:

```
S1=expand.grid(x1=1:10, x2=1:10, x3=1:10) # lists all
  triplets in S1

length(table(rowSums(S1))) # gives the number of different
  sums
```

The last command[1] gives the desired answer, which is 28.

In experiments with many possible outcomes, investigators often classify individual outcomes into distinct categories. This is done for convenience in summarizing and interpreting the results. For example, in the context of the experiment of Example 2.2-2(a), the investigator may wish to classify the opinion ratings into low ($L = \{0, 1, 2, 3\}$), medium ($M = \{4, 5, 6\}$) and high ($H = \{7, 8, 9, 10\}$). Such subsets of the sample space (i.e., collections of individual outcomes) are called **events**. An event consisting of only one outcome is called a **simple event**. Events consisting of more than one outcome are called **compound**.

Events can be described either by listing the individual outcomes comprising them or in a descriptive manner. For example, in selecting one card from a deck of cards, the event $A = \{$the card is a spade$\}$ can also be described by listing the 13 spade cards. Also, when tossing a coin five times and recording the number of heads, the event $E = \{$at most 3 heads$\}$ can also be described by listing the outcomes that comprise it, which are $\{0, 1, 2, 3\}$.

We say that a particular event A has **occurred** if the outcome of the experiment is a member of A. In this parlance, the sample space of an experiment is an event which always occurs when the experiment is performed.

Because events are sets, the usual set operations are relevant for probability theory. *Venn diagram* illustrations of the basic set operations are given in Figure 2-1.

The operations of union and intersection can also be defined for any number of events. Verbally, the **union** $A_1 \cup \cdots \cup A_k$ is the event consisting of all outcomes that make up the events $A_1, \ldots, A_k$. The union is also referred to as the event that happens when A_1 or A_2 or ... or A_k happens (the "or" is used in its nonexclusive sense here) or as the event that happens when *at least one* of $A_1, \ldots, A_k$ happens. The **intersection** $A_1 \cap \cdots \cap A_k$ is the event consisting of all outcomes that are common to all events $A_1, \ldots, A_k$. The intersection is also referred to as the event that

[1] The R command *rowMeans(S1)* can be used instead of *rowSums(S1)*.

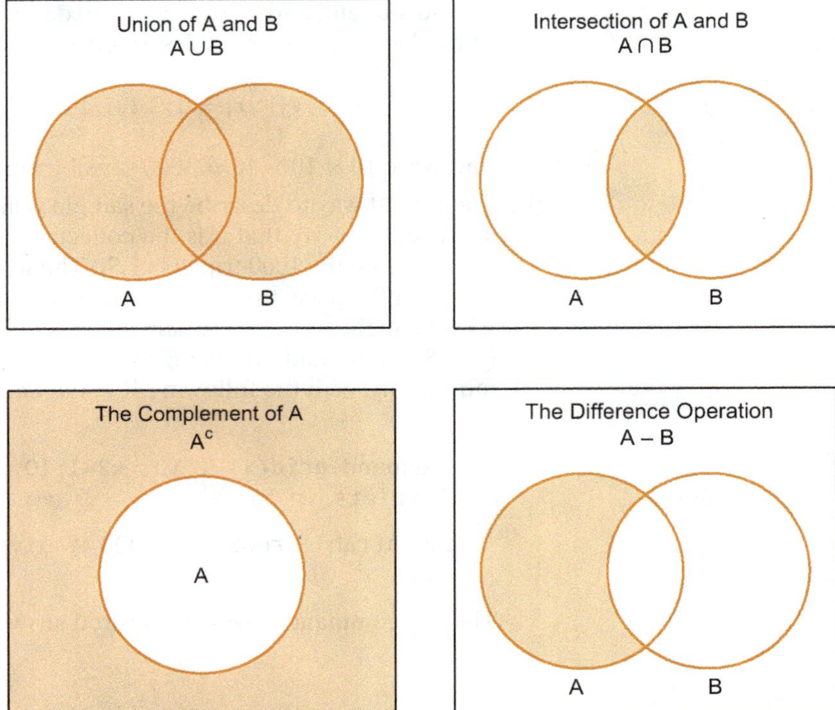

Figure 2-1 Venn diagram illustrations of basic set operations.

happens when A_1 and A_2 and ... and A_k happen or as the event that happens when *all* of $A_1, \ldots, A_k$ happen. The **complement** A^c of A is the event that consists of all outcomes that are not in A. Alternatively, A^c is the event that happens when A does not happen. The **difference** $A - B$ is the event consisting of those outcomes in A that are not in B. Alternatively, $A - B$ is the event that happens when A happens and B does not happen, that is, $A - B = A \cap B^c$.

Two events, A, B, are called **disjoint** or **mutually exclusive** if they have no outcomes in common and therefore they cannot occur together. In mathematical notation, A, B are disjoint if $A \cap B = \emptyset$, where $\emptyset$ denotes the empty set. The empty event can be thought of as the complement of the sample space, $\emptyset = \mathcal{S}^c$. Finally, we say that an event A is a **subset** of an event B if all outcomes of A are also outcomes of B. Alternatively, A is a subset of B if the occurrence of A implies the occurrence of B. The mathematical notation indicating that A is a subset of B is $A \subseteq B$. Figure 2-2 illustrates disjoint A, B, and $A \subseteq B$.

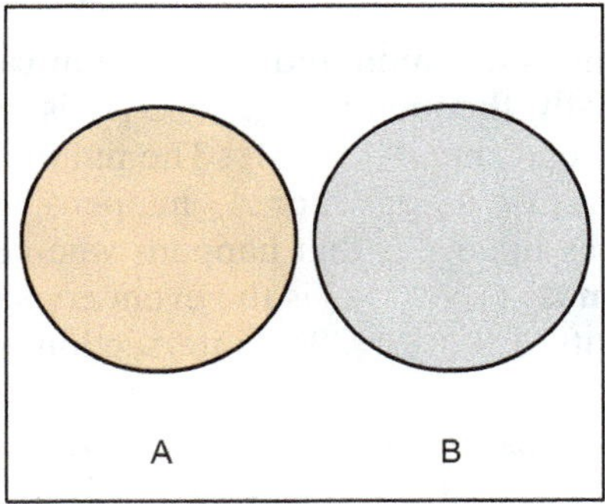

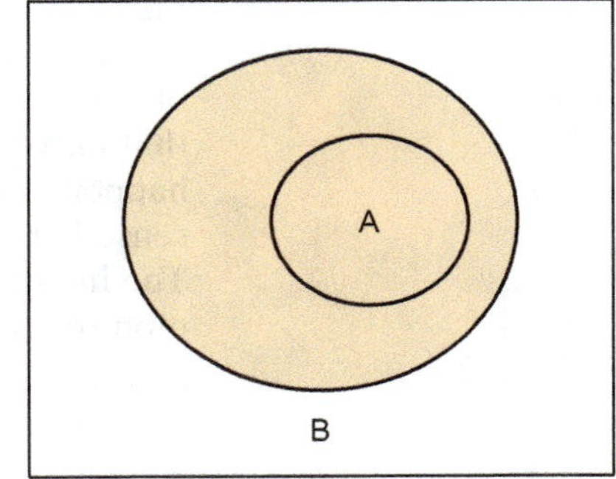

Figure 2-2 Venn diagram illustrations of A, B disjoint (left), and $A \subseteq B$ (right).

The mathematical notation expressing that x is an element of an event E is $x \in E$. The usual way of establishing that an event A is a subset of an event B is to show that if $x \in A$ then $x \in B$ is also true. The usual way of establishing that events A and B are equal is to show that $A \subseteq B$ and $B \subseteq A$.

Example 2.2-4

Resistors manufactured by a machine producing 1kΩ resistors are unacceptable if they are not within 50Ω of the nominal value. Four such resistors are tested.

(a) Describe the sample space of this experiment.

(b) Let E_i denote the event that the ith resistor tests acceptable. Are the E_is mutually exclusive?

(c) Let A_1 denote the event that all resistors test acceptable and A_2 denote the event that exactly one resistor tests unacceptable. Give verbal descriptions of the events $B_1 = A_1 \cup A_2$ and $B_2 = A_1 \cap A_2$.

(d) Express A_1 and A_2 in terms of the E_is.

Solution

(a) Setting 1 when a resistor tests acceptable and 0 when it tests unacceptable, the sample space is $\mathcal{S} = \{(x_1, x_2, x_3, x_4) : x_i = 0 \text{ or } 1, i = 1, 2, 3, 4\}$.

(b) The event E_i consists of all outcomes (x_1, x_2, x_3, x_4) in $\mathcal{S}$ with $x_i = 1$. For example, $E_1 = \{(1, x_2, x_3, x_4) : x_i = 0 \text{ or } 1, i = 2, 3, 4\}$. It follows that the events E_i are not disjoint. For example, the outcome $(1, 1, 1, 1)$ is contained in all of them.

(c) The event $B_1 = A_1 \cup A_2$ happens when at most one resistor tests unacceptable. The event $B_2 = A_1 \cap A_2$ is the empty event, since A_1 and A_2 are disjoint.

(d) That all resistors test acceptable means that all E_i happen. Thus, $A_1 = E_1 \cap E_2 \cap E_3 \cap E_4 = \{(1, 1, 1, 1)\}$. Exactly one resistor tests unacceptable means that either the first resistor tests unacceptable and the rest test acceptable, the second tests unacceptable and the rest test acceptable, the third tests unacceptable and the rest test acceptable, or the fourth tests unacceptable and the rest test acceptable. Translating the above into mathematical notation we have

$$A_2 = F_1 \cup F_2 \cup F_3 \cup F_4,$$

where F_i is the event that the ith resistor tests unacceptable and the others test acceptable. For example, $F_1 = E_1^c \cap E_2 \cap E_3 \cap E_4$, $F_2 = E_2^c \cap E_1 \cap E_3 \cap E_4$, and so forth. ■

Example 2.2-5

In measuring the diameter of a cylinder the sample space (in cm) is $\mathcal{S} = \{x : 5.3 \le x \le 5.7\}$. Let $E_1 = \{x : x > 5.4\}$ and $E_2 = \{x : x < 5.6\}$. Describe the events $E_1 \cup E_2$, $E_1 \cap E_2$, and $E_1 - E_2$.

Solution

$E_1 \cup E_2 = \mathcal{S}$, $E_1 \cap E_2 = \{x : 5.4 < x < 5.6\}$, and $E_1 - E_2 = \{x : 5.6 \le x \le 5.7\}$. ■

The event operations conform to the following laws:

Commutative Laws:

$$A \cup B = B \cup A, \quad A \cap B = B \cap A.$$

Associative Laws:

$$(A \cup B) \cup C = A \cup (B \cup C), \quad (A \cap B) \cap C = A \cap (B \cap C).$$

Distributive Laws:

$$(A \cup B) \cap C = (A \cap C) \cup (B \cap C), \quad (A \cap B) \cup C = (A \cup C) \cap (B \cup C).$$

De Morgan's Laws:

$$(A \cup B)^c = A^c \cap B^c, \quad (A \cap B)^c = A^c \cup B^c.$$

These laws can be demonstrated with Venn diagrams (see Exercises 6 and 7) but can also be shown formally by showing that the event on the left side of each equation is a subset of the event on the right side and vice versa. As an illustration of this type of argument, we will show the first of the distributive laws. To do so note that $x \in (A \cup B) \cap C$ is the same as $x \in C$ and $x \in A \cup B$, which is the same as $x \in C$ and $x \in A$ or $x \in B$, which is the same as $x \in C$ and $x \in A$ or $x \in C$ and $x \in B$, which is the same as $x \in (A \cap C) \cup (B \cap C)$. This shows that $(A \cup B) \cap C \subseteq (A \cap C) \cup (B \cap C)$. Since the sequence of arguments is reversible, it also follows that $(A \cap C) \cup (B \cap C) \subseteq (A \cup B) \cap C$ and thus $(A \cap C) \cup (B \cap C) = (A \cup B) \cap C$.

Example 2.2-6

In telecommunications, a *handoff* or *handover* is when a cellphone call in progress is redirected from its current cell (called source) to a new cell (called target). For example, this may happen when the phone is moving away from the area covered by the source cell and entering the area covered by the target cell. A random sample of 100 cellphone users is selected and their next phone call is categorized according to its duration and the number of handovers it undergoes. The results are shown in the table below.

	Number of Handovers		
Duration	0	1	> 1
> 3	10	20	10
< 3	40	15	5

Let A and B denote the events that a phone call undergoes one handover and a phone call lasts less than three minutes, respectively.

(a) How many of the 100 phone calls belong in $A \cup B$, and how many in $A \cap B$?

(b) Describe in words the sets $(A \cup B)^c$ and $A^c \cap B^c$. Use these descriptions to confirm the first of De Morgan's laws.

Solution

(a) The union $A \cup B$ consists of the 80 phone calls that either undergo one handover or last less than three minutes (or both) and so are categorized either in the column with heading 1 or in the second row of the table. The intersection $A \cap B$ consists of the 15 phone calls that undergo one handover and last less than three minutes and so are categorized both in the column with heading 1 and in the second row of the table.

(b) In words, the complement $(A \cup B)^c$ consists of the 20 phone calls that are not in $A \cup B$, that is, the phone calls that undergo either zero or more than one handovers and last more than three minutes. The intersection $A^c \cap B^c$ consists of the phone calls that do not undergo one handover (so they undergo either zero or more than one handovers) and do not last less than three minutes (so they last more than three minutes). Thus $(A \cup B)^c$ and $A^c \cap B^c$ are the same, in accordance to the first of De Morgan's laws. ■

Exercises

1. Give the sample space for each of the following experiments.

(a) A die is rolled twice and the outcomes are recorded.

(b) A die is rolled twice and the sum of the outcomes is recorded.

(c) From a shipment of 500 iPods, 6 of which have a click wheel problem, a simple random sample of 30 ipods is taken and the number found to have the click wheel problem is recorded.

(d) Fuses are inspected until the first defective fuse is found. The number of fuses inspected is recorded.

2. In a certain community, 40% of the households subscribe to a local newspaper, 30% subscribe to a newspaper of national circulation, and 60% subscribe to at least one of the two types of newspapers. Let E_1 denote the event that a randomly chosen household subscribes to a local newspaper, and let E_2 denote the corresponding event for the national newspaper.

(a) Make a Venn Diagram showing events E_1 and E_2 and shade the region representing the 60% of the households that subscribe to at least one of the two types of newspapers.

(b) Make a Venn Diagram showing events E_1 and E_2 and shade the event that a randomly selected household subscribes to both types of newspapers.

(c) Make a Venn Diagram showing events E_1 and E_2 and shade the event that a randomly selected household subscribes only to a local newspaper.

3. An engineering firm is considering the possibility of establishing a branch office in Toronto and one in Mexico City. Let T be the event that the firm will establish a branch office in Toronto and M be the event that the firm will establish a branch office in Mexico City.

(a) Express each of the events described below in terms of set operations on T and M.

(i) The firm establishes a branch office in both cities.

(ii) The firm establishes a branch office in neither of the cities.

(iii) The firm establishes a branch office in exactly one of the cities.

(b) For each of the three subparts of part (a), draw a Venn diagram that shows the events T and M and shade the event described.

4. Sketch two Venn diagrams like the one in Figure 2-3. On the first shade the set $(A - B) \cup (B - A)$, and on the second shade the event $(A \cup B) - (A \cap B)$. Are they the same?

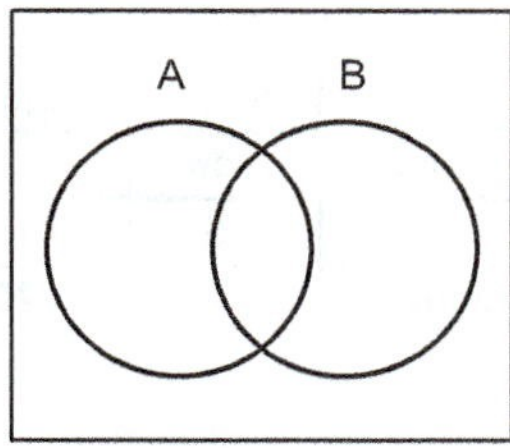

Figure 2-3 Generic Venn diagram with two events.

5. In testing the lifetime of components, the sample space is the set $\mathcal{S} = \{x{:}x > 0\}$ of positive real numbers. Let A be the event that the next component tested will last less than 75 time units and B the event that it will last more than 53 time units. In mathematical notation, $A = \{x{:}x < 75\}$, and $B = \{x{:}x > 53\}$. Describe each of the events (a) A^c, (b) $A \cap B$, (c) $A \cup B$, and (d) $(A - B) \cup (B - A)$, both in words and in mathematical notation.

6. Prove the second of De Morgan's Laws by sketching two Venn diagrams like the one in Figure 2-3. On the first shade the event $(A \cap B)^c$, on the second shade the event $A^c \cup B^c$, and then confirm that they are the same.

7. Sketch two Venn diagrams like the one in Figure 2-4. On the first shade the event $(A \cap B) \cup C$, and on the second shade the event $(A \cup C) \cap (B \cup C)$. Are they the same?

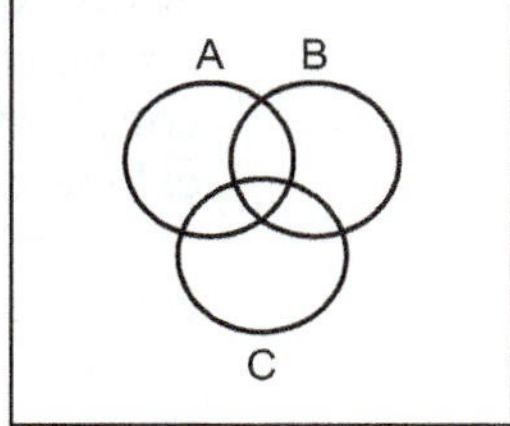

Figure 2-4 Generic Venn diagram with three events.

8. Prove that the pairs of event given in Exercises 4, 6, and 7 are equal by means of a sequence of formal (logical) arguments showing that, in each case, each event is a subset of the other.

9. In measuring the diameter of a cylinder to the nearest millimeter the sample space (in cm) is $\mathcal{S} = \{5.3, 5.4, 5.5, 5.6, 5.7\}$. Five cylinders are randomly selected and their diameters are measured to the nearest millimeter.

(a) Describe the sample space of this experiment. What is the size of this sample space?

(b) Describe the sample space if only the average of the five measurements is recorded, and use R commands similar to those used in Example 2.2-3 to determine the size of this sample space.

10. A random sample of 100 polycarbonate plastic disks are categorized according to their hardness and shock absorption. The results are shown in the table below.

	Shock Absorption	
Hardness	low	high
low	5	16
high	9	70

A disk is selected at random. Define the events E_1 = {the disk has low hardness}, E_2 = {the disk has low shock absorption}, E_3 = {the disk has low shock absorption or low hardness}.

(a) How many of the 100 disks belong in each of the three events?

(b) Make two drawings of a Venn diagram showing the events E_1 and E_2. On the first drawing shade the event $(E_1 \cap E_2)^c$ and on the second shade the event $E_1^c \cup E_2^c$. Confirm the second of De Morgan's Laws for these events.

(c) Describe in words the events $E_1 \cap E_2$, $E_1 \cup E_2$, $E_1 - E_2$, and $(E_1 - E_2) \cup (E_2 - E_1)$.

(d) How many of the 100 disks belong in each of the events in part (c)?

2.3 Experiments with Equally Likely Outcomes

2.3.1 DEFINITION AND INTERPRETATION OF PROBABILITY

In any given experiment we might be interested in assessing the likelihood, or chance, of occurrence of an outcome or, more generally, of an event. The **probability** of an event E, denoted by $P(E)$, is used to quantify the likelihood of occurrence of E by assigning a number from the interval $[0, 1]$. Higher numbers indicate that the event is more likely to occur. A probability of 1 indicates that the event will occur with certainty, while a probability of 0 indicates that the event will not occur.

The likelihood of occurrence of an event is also quantified as a *percent*, or in terms of the *odds*. The expression "the odds of winning are two to one" means that a win is twice as likely as a non-win, which means that the probability of a win is about 0.67. The percent quantification has a more direct correspondence to probability. For example, the expression "there is a 70% chance of rain this afternoon" means that the probability of rain this afternoon is 0.7. The use of percents is suggestive of the **limiting relative frequency** interpretation of probability, which is based on the conceptual model of repeated replications of the experiment under the same conditions. For example, if we conceptualize the collection of all days with identical meteorological conditions as today, the statement that there is a 70% chance of rain can be interpreted to mean that it would rain in 70% of the days with identical conditions as today.

In general, if $N_n(E)$ denotes the number of occurrences of the event E in n repetitions of the experiment, the limiting relative frequency approach interprets $P(E)$ as the limiting value, as n gets larger and larger (i.e., $n \to \infty$) of the ratio

$$\frac{N_n(E)}{n}.$$

Even though the limiting relative frequency interpretation is intuitively appealing, it cannot serve as a formal definition of probability because there is no guarantee that the limit of $N_n(E)/n$ exists. For example, why is it true that in a long sequence of coin tosses, the proportion of heads tends to 1/2? For this reason, the modern approach to probability sets forth a number of *axioms* that any assignment of probabilities to events must satisfy, and from these axioms derive all properties of probability. These axioms are given in Section 2.4. That the limit of $N_n(E)/n$ exists is a consequence of the *Law of Large Numbers* given in Chapter 5.

Even if the existence of the limit of $N_n(E)/n$ were to be accepted as an axiom, the limiting relative frequency interpretation of probability is not a practical method for assigning probabilities to events. In modern probability theory this assignment is based on *probability models* deemed suitable for each experiment. The simplest such model pertains to experiments with a finite number of *equally likely outcomes*, such as those used in games of chance. The definition and assignment of probabilities for such models is discussed next. Other probability models are discussed in Chapters 3 and 4.

For experiments that have a finite number of equally likely outcomes, including those that take the form of simple random sampling from a finite population, the assignment of probabilities to individual outcomes is straightforward and intuitive: If we denote by N the finite number of outcomes of such an experiment, then the probability of each outcome is $1/N$. This is important enough to be highlighted:

Probability of Each of N Equally Likely Outcomes

If the sample space of an experiment consists of N outcomes that are equally likely to occur, then the probability of each outcome is $1/N$.

For example, a toss of a fair coin has two equally likely outcomes so the probability of each outcome (heads or tails) is 1/2, the roll of a die has six equally likely outcomes so the probability of each outcome is 1/6, and in drawing a card at random from a deck of 52 cards each card has probability 1/52 of being drawn.

Having the probability of each individual outcome, it is straightforward to assign probabilities to events consisting of several outcomes. If $N(E)$ denotes the number of outcomes that constitute the event E, then the probability of E is

Assignment of Probabilities in the Case of N Equally Likely Outcomes

$$P(E) = \frac{N(E)}{N} \tag{2.3.1}$$

For example, in rolling a die, the probability of an even outcome is 3/6, and in drawing a card at random from a deck of 52 cards, the probability of drawing an ace is 4/52. Two additional examples follow.

Example 2.3-1

The efficiency of laser diodes (measured at 25°C in mW per mA) varies from 2 to 4. In a shipment of 100, the numbers having efficiency 2, 2.5, 3, 3.5, and 4 are 10, 15, 50, 15, and 10, respectively. One laser diode is randomly selected. Find the probabilities of the events $E_1 = \{$the selected laser diode has efficiency 3$\}$ and $E_2 = \{$the selected laser diode has efficiency at least 3$\}$.

Solution

Here there are $N = 100$ equally likely outcomes. Moreover, $N(E_1) = 50$ and $N(E_2) = 75$. Thus, $P(E_1) = 0.5$ and $P(E_2) = 0.75$. ■

Example 2.3-2

Roll two dice separately (or one die twice). Find the probability of the event that the sum of the two sides is seven.

Solution

When two dice are rolled, there are $N = 36$ equally likely outcomes. The event $A = \{\text{sum of two sides} = 7\}$ consists of the outcomes (1,6), (2,5), (3,4),

(4,3), (5,2), and (6,1). Thus $N(A) = 6$ and, according to the formula (2.3.1), $P(A) = 6/36 = 1/6$. ■

2.3.2 COUNTING TECHNIQUES

While the method of assigning probabilities to events of experiments with equally likely outcomes is straightforward, the implementation of this method is not straightforward if N is large and/or the event A is complicated. For example, to find the probability that five cards, randomly selected from a deck of 52 cards, will form a full house (three of a kind and two of a kind) we need to be able to determine how many 5-card hands are possible and how many of those constitute a full house. Such determination requires specialized counting techniques, which are presented in this section.

We begin with the most basic counting technique, from which all results follow.

Fundamental Principle of Counting

> If a task can be completed in two stages, if stage 1 has n_1 outcomes, and if stage 2 has n_2 outcomes, regardless of the outcome in stage 1, then the task has $n_1 n_2$ outcomes.

The task in the fundamental principle of counting can be an experiment and the stages can be subexperiments, or the stages can be experiments in which case the task is that of performing the two experiments in succession. For example, the task of rolling two dice has $6 \times 6 = 36$ outcomes. The rolling of two dice can be the experiment with subexperiments the rolling of each die, or each of the two die rolls can be an experiment in which case the task is to roll the two dice in succession.

Example 2.3-3

For each of the following tasks specify the stages, the number of outcomes of each stage, and the number of outcomes of the task.

(a) Select a plumber and an electrician from three plumbers and two electricians available in the yellow pages.

(b) Select two items from an assembly line, and classify each item as defective (0) or non-defective (1).

(c) Select a first and a second place winner from a group of four finalists.

Solution

(a) Stage 1 can be the selection of a plumber and stage 2 the selection of an electrician. Then $n_1 = 3, n_2 = 2$, and thus the task has $n_1 n_2 = 6$ possible outcomes.

(b) The outcome of stage 1 can be either 0 or 1. Similarly for stage 2. Thus, this task has $2 \times 2 = 4$ outcomes.

(c) Stage 1 can be the selection of the second place winner, and stage 2 the selection of the first place winner. Then $n_1 = 4, n_2 = 3$, and thus the task has $n_1 n_2 = 12$ possible outcomes. ■

The stages of all three tasks of Example 2.3-3 involve sampling. In part (a) different sets of subjects are sampled in the two stages, in part (b) the same set (which is $\{0, 1\}$) is sampled with replacement, and in part (c) the same set of subjects is sampled without replacement.

The fundamental principle of counting generalizes in a straightforward manner.

Generalized Fundamental Principle of Counting

> If a task can be completed in k stages and stage i has n_i outcomes, regardless of the outcomes of the previous stages, then the task has $n_1 n_2 \cdots n_k$ outcomes.

Example 2.3-4

For each of the following tasks specify the stages, the number of outcomes of each stage and the number of outcomes of the task.

(a) Select a plumber, an electrician, and a remodeler from three plumbers, two electricians, and four remodelers available in the yellow pages.

(b) Form a binary sequence of length 10 (i.e., a 10-long sequence of 0's and 1's).

(c) Form a string of seven characters such that the first three are letters and the last four are numbers.

(d) Select a first, second, and third place winner from a group of four finalists.

Solution

(a) This task consists of three stages with number of outcomes $n_1 = 3, n_2 = 2$, and $n_3 = 4$. Thus the task has $3 \times 2 \times 4 = 24$ outcomes.

(b) This task consists of 10 stages with each stage having two possible outcomes, either 0 or 1. Thus, the task has $2^{10} = 1024$ outcomes.

(c) This task consists of seven stages. Each of the first three stages involves selecting one of the 26 letters, so $n_1 = n_2 = n_3 = 26$. Each of last four stages involves selecting one of the 10 numbers, so $n_4 = \cdots = n_7 = 10$. Thus the task has $26^3 \times 10^4 = 175{,}760{,}000$ outcomes.

(d) This task consists of three stages with number of outcomes $n_1 = 4, n_2 = 3$, and $n_3 = 2$. Thus the task has $4 \times 3 \times 2 = 24$ outcomes. ■

When the stages of a task involve sampling without replacement from the same set of units (objects or subjects), as in part (c) of Example 2.3-3 or part (d) of Example 2.3-4, we may or may not want to distinguish between the outcomes of the different stages. This is exemplified in the following.

Example 2.3-5

In selecting a first and second place winner from the group of four finalists consisting of Niki, George, Sophia, and Martha, the outcomes (George, Sophia) and (Sophia, George) are two of the 12 distinct outcomes mentioned in Example 2.3-3(c). However, if both winners will receive the same prize, so there is no distinction between the first and second place winner, the two outcomes will be counted as one (since, in this case, George and Sophia will each receive the same prize). Similarly, if all three winners mentioned in Example 2.3-4(d) receive the same prize there is no need to distinguish between the outcomes of the different stages because any arrangement of, for example, (Niki, George, Sophia) identifies the three equal winners. ■

Definition 2.3-1

If the k stages of a task involve sampling one unit each, without replacement, from the same group of n units, then:

1. If a distinction is made between the outcomes of the stages, we say the outcomes are **ordered**. Otherwise we say the outcomes are **unordered**.
2. The ordered outcomes are called **permutations** of k units. The number of permutations of k units selected from a group of n units is denoted by $P_{k,n}$.
3. The unordered outcomes are called **combinations** of k units. The number of combinations of k units selected from a group of n units is denoted by $\binom{n}{k}$.

The formula for $P_{k,n}$ follows readily from the generalized fundamental principle of counting. To see how, reason as follows: The task making an ordered selection of k units out of a set of n units consists of k stages. Stage 1, which corresponds to selecting the first unit, has $n_1 = n$ possible outcomes. Stage 2, which corresponds to selecting the second unit from the remaining $n-1$ units (remember that the sampling is without replacement), has $n_2 = n - 1$ possible outcomes, and so forth until stage k which has $n_k = n - k + 1$ outcomes. Hence, according to the generalized principle of counting,

Number of Permutations of k Units Selected from n

$$P_{k,n} = n(n-1)\cdots(n-k+1) = \frac{n!}{(n-k)!} \tag{2.3.2}$$

where, for a nonnegative integer m, the notation $m!$ is read m *factorial* and is defined as

$$m! = m(m-1)\cdots(2)(1).$$

For $k = n$, formula (2.3.2) with the convention that $0! = 1$, yields the number of different permutations (or arrangements or orderings) of n units among themselves.

Number of Permutations of n Units Among Themselves

$$P_{n,n} = n! \tag{2.3.3}$$

Example 2.3-6

(a) The lineup or batting order is a list of the nine baseball players for a team in the order they will bat during the game. How many lineups are possible?

(b) The Department of Tranpostation (DOT) plans to assign six civil engineers to oversee six interstate safety design projects. How many different assignments of civil engineers to projects are possible?

Solution

(a) There are $P_{9,9} = 9! = 362{,}880$ possible lineups.

(b) There are $P_{6,6} = 6! = 720$ different assignments. ■

The number of combinations (unordered selections, or groups) of k units that can be drawn from a set of n units does not follow directly from the generalized fundamental principle of counting, but it does follow from the permutation formulas (2.3.2) and (2.3.3). To see how, consider the specific task of selecting a group of three out of a set of four units (e.g., a group of 3 equal winners from a group of 4 finalists). Note now that, by (2.3.3), each group of three yields $P_{3,3} = 3! = 6$ permutations. This means that the number of permutations will be six times the number of combinations. By formula (2.3.2), the number of permutations of 3 that can be obtained from a set of 4 is $P_{3,4} = 4!/(4-3)! = 4! = 24$. Hence, the number of combinations of 3 that can be obtained from a set of 4 is $\binom{4}{3} = P_{3,4}/P_{3,3} = 24/6 = 4$.

In general we have the formula

Number of Combinations of k Units Selected from n

$$\binom{n}{k} = \frac{P_{k,n}}{P_{k,k}} = \frac{n!}{k!(n-k)!} \tag{2.3.4}$$

Because the numbers $\binom{n}{k}$, $k = 1, \ldots, n$, are used in the *binomial theorem* (see Exercise 18), they are referred to as the **binomial coefficients**.

Example 2.3-7

Two cards will be selected from a deck of 52 cards.

(a) How many outcomes are there if the first card will be given to player 1 and the second card will be given to player 2?

(b) How many outcomes are there if both cards will be given to player 1?

Solution

(a) In this case it makes sense to distinguish the outcome (Ace, King), meaning that player 1 gets the ace and player 2 gets the king, from the outcome (King, Ace). Thus, we are interested in the number of permutations of two cards selected from 52 cards. According to (2.3.2) the number of (ordered) outcomes is $P_{2,52} = 52 \times 51 = 2652$.

(b) In this case it can be argued that the order in which the two cards are received is not relevant, since both result in the same two-card hand. Thus, we are interested in the number of combinations of two cards selected from 52 cards. According to (2.3.4) the number of (unordered) outcomes is $\binom{n}{2} = 2652/2 = 1326$.

Example 2.3-8

(a) How many binary sequences (i.e., sequences of 0's and 1's) of length 10 with exactly four 1's can be formed?

(b) If a binary sequence of length 10 is selected at random, what is the probability that it has four 1's?

Solution

(a) A 10-long binary sequence of four 1's (and hence six 0's) is determined from the location of the four 1's in the 10-long sequence (all other locations in the sequence have 0's). Thus the problem is that of selecting four out of the 10 locations in the sequence. The answer is $\binom{10}{4} = 210$.

(b) Using also the result of Example 2.3-4(b), the probability is $\binom{10}{4}/2^{10} = 210/1024 = 0.2051$.

Example 2.3-9

(a) How many binary sequences of length 10 with exactly four 1's that are not consecutive can be formed?

(b) If a binary sequence of length 10 having four 1's is selected at random, what is the probability that the four 1's are nonconsecutive?

Solution

(a) Any binary sequence of length 10 with four nonconsecutive 1's is formed by selecting four of the seven spaces created by the six 0's, which are shown below as wedges.

$$\wedge\, 0 \wedge 0 \wedge 0 \wedge 0 \wedge 0 \wedge 0 \wedge$$

Thus, the answer is $\binom{7}{4} = 35$.

(b) Using also the result of Example 2.3-8(a), the probability is $\binom{7}{4}/\binom{10}{4} = 35/210 = 0.1667$. ■

Example 2.3-10

In the game of poker each player receives five cards dealt from a deck of 52 cards. Full house refers to a five-card hand consisting of three of a kind and two of a kind. An example of a full house is a hand consisting of three 10's and two 5's. Find the probability that a randomly dealt five-card hand is a full house.

Solution

Since the five-card hand is randomly dealt, the set of equally likely outcomes is that of all five-card hands. To determine the probability we need to find the number of possible outcomes and the number of outcomes that constitute a full house. First, the number of all five-card hands is $\binom{52}{5} = 2{,}598{,}960$. To find the number of hands that constitute a full house, think of the task of forming a full house as consisting of two stages. Stage 1 consists of choosing two cards of the same kind, and stage 2 consists of choosing three cards of the same kind. Stage 1 can be completed in $\binom{13}{1}\binom{4}{2} = (13)(6) = 78$ ways. (This is because stage 1 can be thought of as consisting of two substages: first selecting a kind from the available 13 kinds and then selecting two from the four cards of the selected kind.) For each outcome of stage 1, the task of stage 2 becomes that of selecting three of a kind from one of the remaining 12 kinds. This can be completed in $\binom{12}{1}\binom{4}{3} = 48$ ways. Thus there are $(78)(48) = 3744$ possible full houses. It follows that the probability of dealing a full house is $3744/2{,}598{,}960 = 1.4406 \times 10^{-3}$. ■

We are often interested in dividing n units into more than two groups. The number of such divisions can be obtained through the generalized fundamental principle of counting, and use of the formula (2.3.4). To fix ideas, suppose that eight mechanical engineers will be divided into three groups of three, two, and three to work on design projects A, B, and C, respectively. The task of assigning the eight engineers to the three projects consists of two stages. Stage 1 selects three of the engineers to work on project A, and stage 2 selects two of the remaining five engineers to work on project B. (The remaining three engineers are then assigned to work on project C.) Stage 1 has $n_1 = \binom{8}{3}$ outcomes, and stage 2 has $n_2 = \binom{5}{2}$ outcomes. Hence, the task of assigning three of the eight engineers to project A, two to project B, and three to project C has

$$\binom{8}{3}\binom{5}{2} = \frac{8!}{3!5!}\frac{5!}{2!3!} = \frac{8!}{3!2!3!} = 560 \qquad \textbf{(2.3.5)}$$

possible outcomes. We will use the notation $\binom{n}{n_1,n_2,\ldots,n_r}$ for the number of ways n units can be divided into r groups of sizes $n_1, n_2, \ldots, n_r$. Thus, the outcome of the calculation in (2.3.5) can be written as $\binom{8}{3,2,3} = 560$.

A generalization of the line of reasoning leading to (2.3.5) yields the following result:

Number of Arrangements of n Units into r Groups of Sizes $n_1, n_2, \ldots, n_r$

$$\binom{n}{n_1,n_2,\ldots,n_r} = \frac{n!}{n_1!n_2!\cdots n_r!} \tag{2.3.6}$$

Because the numbers $\binom{n}{n_1,n_2,\ldots,n_r}$, with $n_1 + n_2 + \cdots + n_r = n$, are used in the *multinomial theorem* (see Exercise 19), they are referred to as the **multinomial coefficients.**

Example 2.3-11

The clock rate of a CPU (central processing unit) chip refers to the frequency, measured in megahertz (MHz), at which it functions reliably. CPU manufacturers typically categorize (*bin*) CPUs according to their clock rate and charge more for CPUs that operate at higher clock rates. A chip manufacturing facility will test and bin each of the next 10 CPUs in four clock rate categories denoted by G_1, G_2, G_3, and G_4.

(a) How many possible outcomes of this binning process are there?

(b) How many of the outcomes have three CPUs classified as G_1, two classified as G_2, two classified as G_3, and three classified as G_4?

(c) If the outcomes of the binning process are equally likely, what is the probability of the event described in part (b)?

Solution

(a) The binning process consists of 10 stages, each of which has four possible outcomes. Hence, by the generalized fundamental principle of counting, there are $4^{10} = 1{,}048{,}576$ possible outcomes.

(b) The number of possible outcomes is

$$\binom{10}{3,2,2,3} = \frac{10!}{3!2!2!3!} = 25{,}200.$$

(c) The probability is $25,200/1,048,576 = 0.024$. ■

2.3.3 PROBABILITY MASS FUNCTIONS AND SIMULATIONS

In many sampling experiments, even though the units are selected with equal probability, the sample space of the random variable recorded consists of outcomes that are not equally likely. For example, the outcomes of the experiment that records the sum of two die rolls, that is, $\{2, 3, \ldots, 12\}$, are not equally likely, since the probability of a 2 (and also of a 12) is 1/36, while the probability of a seven is six times that, as derived in Example 2.3-2. As another example, if an undergraduate student is randomly selected and his/her opinion regarding a proposed expansion of the use of solar energy is rated on a scale from 1 to 10, each student has an equal chance of being selected, but the individual outcomes of the sample space (which are $\{1, 2, \ldots, 10\}$) will not be equally likely.

Definition 2.3-2
The **probability mass function** (**PMF** for short) of an experiment that records the value of a discrete random variable X, or simply the PMF of X, is a list of the probabilities $p(x)$ for each value x of the sample space $\mathcal{S}_X$ of X.

Example 2.3-12

A simple random sample of size $n = 3$ is drawn from a batch of 10 product items. If three of the 10 items are defective, find the PMF of the random variable $X = \{\text{number of defective items in the sample}\}$.

Solution
By the definition of simple random sampling, each of the $\binom{10}{3}$ samples are equally likely. Thus, the probabilities for each outcome $\mathcal{S}_X = \{0, 1, 2, 3\}$ can be calculated as:

$$P(X = 0) = \frac{\binom{7}{3}}{\binom{10}{3}}, \quad P(X = 1) = \frac{\binom{3}{1}\binom{7}{2}}{\binom{10}{3}},$$

$$P(X = 2) = \frac{\binom{3}{2}\binom{7}{1}}{\binom{10}{3}}, \quad P(X = 3) = \frac{\binom{3}{3}}{\binom{10}{3}}.$$

Thus, the PMF of X is:

x	0	1	2	3
p(x)	0.292	0.525	0.175	0.008

Figure 2-5 shows the PMF of Example 2.3-12 as a bar graph.

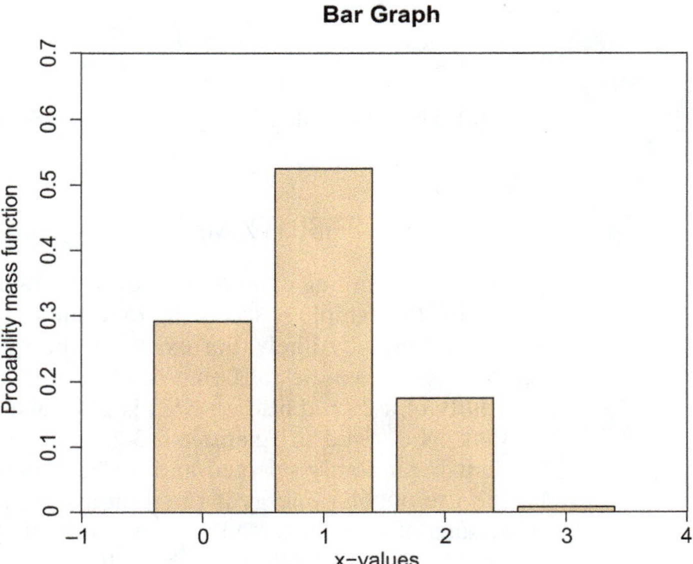

Figure 2-5 Bar graph for the PMF of Example 2.3-12.

Example 2.3-13

Roll two dice separately (or one die twice). Find the probability mass function of the experiment that records the sum of the two die rolls.

Solution

This experiment records the value of the variable $X = \{\text{sum of two die rolls}\}$, whose possible values are $2, 3, \ldots, 12$. Counting which of the 36 equally likely outcomes of the two die rolls result in each of the possible values of X (see Example 2.3-2), we obtain the following PMF of the experiment (or of X):

x	2	3	4	5	6	7	8	9	10	11	12
$p(x)$	1/36	2/36	3/36	4/36	5/36	6/36	5/36	4/36	3/36	2/36	1/36

This can also be obtained with the following R commands:

R Commands for the PMF of the Sum of Two Die Rolls

```
S=expand.grid(X1=1:6, X2=1:6)
table(S$X1+S$X2)/36
```

Once the PMF of an experiment is obtained, it is easier to compute probabilities of more complicated events using it than by counting the number of equally likely outcomes that comprise these events. For example, it is easier to compute the probability of the event $E = \{\text{the sum of two die rolls is at least 10}\}$ by the sum $p(10) + p(11) + p(12)$, where $p(x)$ is the PMF found in Example 2.3-13, than by the formula $N(E)/36$. That the two are the same follows from the properties of probability presented in Section 2.4, but can also be verified here by noting that $N(E) = N(E_{10}) + N(E_{11}) + N(E_{12})$, where E_{10}, E_{11}, and E_{12} denote the events that the sum of the two die rolls is 10, 11, and 12, respectively. Hence,

$$P(E) = \frac{N(E)}{36} = \frac{N(E_{10})}{36} + \frac{N(E_{11})}{36} + \frac{N(E_{12})}{36} = p(10) + p(11) + p(12) = \frac{6}{36}.$$

Moreover, having the PMF of an experiment and access to a software package, one can **simulate** the experiment. This means that one can obtain outcomes from the sample space of the experiment without actually performing the experiment. The following example illustrates the use of R for simulating the experiment of Example 2.3-13 ten times.

Example 2.3-14

Simulating an experiment with R. Use the PMF obtained in Example 2.3-13 to simulate 10 repetitions of the experiment that records the sum of two die rolls.

Solution

Note first that the R command *c(1:6, 5:1)/36* produces the PMF given in Example 2.3-13. For sampling from the sample space $\{2, 3, \ldots, 12\}$ ten times use the R command

```
sample(2:12, size=10, replace=T, prob=c(1:6, 5:1)/36)    (2.3.7)
```

If one sets the seed to 111 by *set.seed(111)* (setting the seed to the same value ensures reproducibility of the results), the above command yields the 10 numbers

9 4 6 5 6 6 7 5 6 7

These numbers represent the outcome of rolling a pair of dice ten times and each time summing the two die rolls (without rolling any dice!). Repeating the R command in (2.3.7) without setting the seed will give different sets of 10 numbers. ■

The simulation carried out in Example 2.3-14 involves random sampling, with replacement, from the sample space of the experiment using its probability mass function. This kind of sampling, which is random but not simple random, is called **probability sampling**, or **sampling from a probability mass function**. When the sample space is considered as the population from which we sample with replacement, it is called a **sample space population**.

Simulations can be used to gain understanding of different properties of the system, as well as for empirical verification of certain results. For example, setting the seed to 111 with *set.seed(111)*, the R command

```
table(sample(2:12, size=10000, replace=T,
  prob=c(1:6, 5:1)/36))/10000
```
(2.3.8)

yields the following relative frequencies for each number in the sample space

2	3	4	5	6	7	8	9	10	11	12
0.0275	0.0561	0.0833	0.1083	0.1366	0.1686	0.1367	0.1137	0.0865	0.0567	0.0260

Figure 2-6 shows the probability mass function of Example 2.3-13 (line graph in color) and the above relative frequencies (bar graph/histogram). Since all relative frequencies are good approximations to corresponding probabilities, we have empirical confirmation of the limiting relative frequency interpretation of probability.

Because of the aforementioned advantages of working with a probability mass function, Chapter 3 presents *probability models*, which are classes of probability

Figure 2-6 Histogram of relative frequencies and line graph of the PMF.

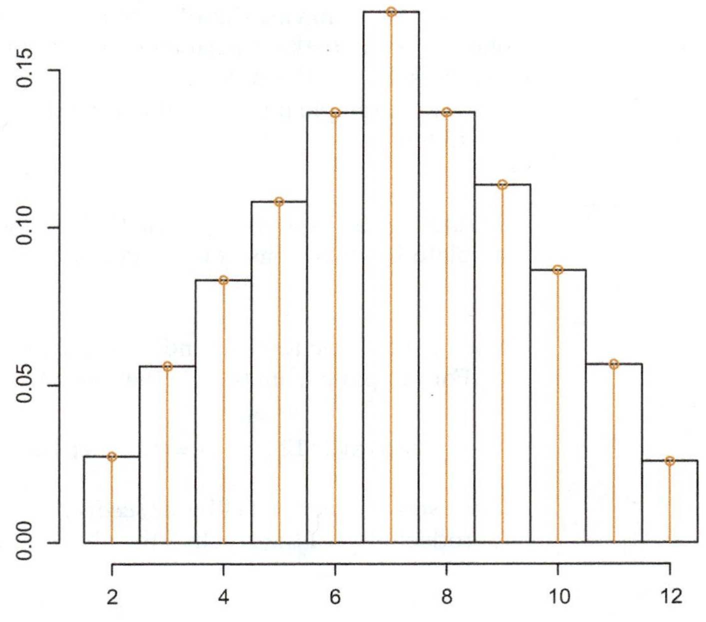

mass functions, that are relevant to the sample space of the most prevalent types of experiments used in sciences and engineering.

Exercises

1. In electronics, a *wafer* is a thin slice of semiconductor material used in the fabrication of integrated circuits and other micro-devices. They are formed of highly pure crystalline material, which is *doped* (i.e., impurity atoms are added) to modulate its electrical properties. The doping is either n-type or p-type. Moreover, the doping is either light or heavy (one dopant atom per 100 million atoms, or per ten thousand atoms, respectively). The following table shows a batch of 10 wafers broken down into the four categories.

	Degree of Doping	
Type of Doping	light	heavy
n-type	2	3
p-type	3	2

One wafer is selected at random. Let E_1 denote the event that the selected wafer is n-type, and E_2 the event that the wafer is heavily doped. Find the probabilities $P(E_1)$, $P(E_2)$, $P(E_1 \cap E_2)$, $P(E_1 \cup E_2)$, $P(E_1 - E_2)$, and $P((E_1 - E_2) \cup (E_2 - E_1))$.

2. Refer to Exercise 1.

(a) Select two wafers, at random and with replacement, from the batch of 10 wafers given in the exercise.

(i) Give the sample space for the experiment that records the doping type of the two wafers and the probability for each outcome.

(ii) Give the sample space of the experiment that records the number of n-type wafers among the two selected and the corresponding probability mass function.

(b) Select four wafers, at random and with replacement, from the batch of 10 wafers given in the exercise.

(i) Give a verbal description of the sample space for the experiment that records the doping type of the four wafers, find the size of the sample space using the R commands *G=expand.grid(W1=0:1,W2=0:1,W3=0:1,W4=0:1); length(G$W1)*, and give the probability of each outcome.

(ii) Give the sample space of the experiment that records the number of n-type wafers among the four selected and the corresponding PMF using the additional R commands *attach(G); table((W1+W2+W3+W4)/4)/length(W1)*.

(iii) Use the probability mass function to find the probability of at most one n-type wafer among the four selected.

3. Soil or water pH is measured on a scale of 0–14. A pH reading below 7 is considered acidic, while a pH reading above 7 is basic. The pH level of water provided by a type of spray tank irrigator is equally likely to be either 6.8, 6.9, 7.0, 7.1 or 7.2. Let E_1 denote the event that at the next irrigation the water pH level measurement is at most 7.1, and E_2 the event that the water pH level is at least 6.9. Find the probabilities $P(E_1)$, $P(E_2)$, $P(E_1 \cap E_2)$, $P(E_1 \cup E_2)$, $P(E_1 - E_2)$, and $P((E_1 - E_2) \cup (E_2 - E_1))$.

4. The following two questions pertain to the spray tank irrigator of Exercise 3.

(a) The water pH level is measured over the next two irrigations.

(i) Give the sample space of this experiment, and its size, using R commands *t=seq(6.8, 7.2, 0.1); G=expand.grid(X1=t,X2=t); G; length(G$X1)*.

(ii) Give the sample space of the experiment that records the average of the two pH measurements, and the corresponding probability mass function, using the additional R commands *attach(G); table((X1+X2)/2)/length(X1)*.

(b) Using R commands similar to the above, give the probability mass function of the experiment that records the average of the pH measurements taken over the next five irrigations.

5. The R command *S=expand.grid(X1=1:6, X2=1:6)* generates the sample space of two die rolls. The additional R command *attach(S); Y=(X1==6)+(X2==6)* generates the number of times a six occurs for each of the 36 outcomes in the sample space. Finally, the additional R commands *pr=table(Y)/36; pr* generates the probability mass function for the experiment that records the number of times a six occurs in two rolls of a die.

(a) Use the PMF obtained and R commands similar to those given in (2.3.7) to simulate 10 replications of the experiment that records the number of times a six occurs in two die rolls.

(b) Use the PMF obtained and R commands similar to those given in (2.3.8), namely *x= sample(0:2, size=10000, replace=T, prob=pr); table(x)/10000*, to obtain the relative frequencies for each outcome in the sample space of the experiment that records the number of times a six occurs in two die rolls, based on 10,000 replications.

(c) Use the R command *hist(x,seq(-0.5, 2.5, 1), freq=F); lines(0:2, pr, type="p", col="red"); lines(0:2, pr, type="h", col="red")* to construct a histogram of the relative frequencies and line graph of the probability mass function. This figure provides empirical verification of which property?

6. A test consists of five true-false questions.

(a) In how many ways can it be completed? (*Hint.* The task of answering five true-false questions consists of five stages.)

(b) A correctly answered question receives 1 point, while an incorrectly answered question gets 0. Give the sample space for the experiment that records the test score.

(c) A reasonable model for the answers given by a student who has not studied assumes that each question is marked T or F by flipping a coin. Thus, any 5-long binary sequence of 0's and 1's, that is, points received in each of the five questions, is equally likely. Let X denote the test score of such a student. Find the PMF of X. (*Hint.* The probability that $X = k$ is the number of binary sequences that sum to k divided by the total number of binary sequences, which is your answer in part (a). This can be found, simultaneously for all k, with the following R commands: *S=expand.grid(X1=0:1, X2=0:1, X3=0:1, X4=0:1, X5=0:1); attach(S); table(X1+X2+X3+X4+X5)/length(X1)*)

7. An information technology company will assign four electrical engineers to four different JAVA programming projects (one to each project). How many different assignments are there?

8. In many countries the license plates consist of a string of seven characters such that the first three are letters and the last four are numbers. If each such string of seven characters is equally likely, what is the probability that the string of three letters begins with a W and the string of four numbers begins with a 4? (*Hint.* Assume an alphabet of 26 letters. The number of possible such license plates is found in Example 2.3-4(c).)

9. Twelve individuals want to form a committee of four.

(a) How many committees are possible?

(b) The 12 individuals consist of 5 biologists, 4 chemists, and 3 physicists. How many committees consisting of 2 biologists, 1 chemist, and 1 physicist are possible?

(c) In the setting of part (b), if all committees are equally likely, what is the probability the committee formed will consist of 2 biologists, 1 chemist, and 1 physicist?

10. Answer the following questions.

(a) A team of 5 starters will be selected from 10 basketball players. How many selections are there?

(b) Ten basketball players will be divided into two teams for a practice game. How many divisions of the 10 players into two teams of 5 are there?

(c) If each of 12 individuals shakes hands with everybody else, how many handshakes take place?

11. A path going from the lower left corner of the grid in Figure 2-7 to the upper right corner can be represented as a binary sequence having four 0's and four 1's, with each 0 representing a move to the right and each 1 representing a move upwards.

(a) How many paths going from the lower left corner to the upper right corner of this grid are there?

(b) How many paths going from the lower left corner to the upper right corner of this grid and passing through the circled point are there?

(c) If a path is selected at random, what is the probability it will pass through the circled point?

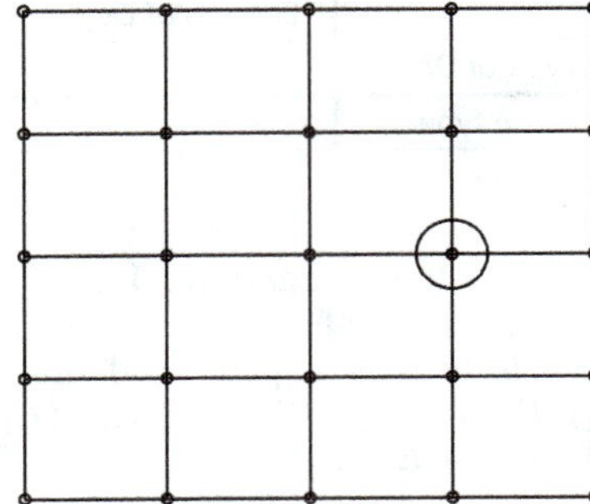

Figure 2-7 A 4×4 grid for Exercise 11.

12. A communication system consists of 13 antennas arranged in a line. The system functions as long as no two nonfunctioning antennas are next to each other. Suppose five antennas stop functioning.

(a) How many different arrangements of the five nonfunctioning antennas result in the system being functional? (*Hint.* This question and the next are related to Example 2.3-9.)

(b) If the arrangement of the five nonfunctioning antennas is equally likely, what is the probability the system is functioning?

13. Five of the 15 school buses of a particular school district will be selected for thorough inspection. Suppose four of the buses have developed a slight defect since their last inspection (the steering wheel shakes when braking).

(a) How many possible selections are there?

(b) How many selections contain exactly three buses with the defect?

(c) If the five buses are selected by simple random sampling, what is the probability the sample includes exactly three of the buses with the defect?

(d) If the buses are selected by simple random sampling, what is the probability all five buses are free of the defect?

14. A forest contains 30 moose of which six are captured, tagged, and released. A certain time later, five of the 30 moose are captured.

(a) How many samples of size five are possible?

(b) How many samples of size five, which include two of the six tagged moose, are possible?

(c) If the five captured moose represent a simple random sample drawn from the 30 moose (six of which are tagged), find the probability that (i) two of the five captured moose are tagged and (ii) none of the five captured moose is tagged.

15. A simple random sample of size five is selected from 52 cards. Find the probability of each of the following events.

(a) The five-card hand contains all four 2's.

(b) The five-card hand contains two aces and two 7's.

(c) The five-card hand contains three kings and the other two cards are of different denomination.

16. The information technology company mentioned in Exercise 7 has 10 EE majors working as interns. Two interns will be assigned to work on each of the four programing projects, and the remaining two interns will be assigned to another project. How many possible assignments of the 10 interns are there?

17. After testing, asphalt shingles are classified as high, medium, or low grade, and different grade shingles are sold under different warranties.

(a) In how many ways can the next 15 shingles be classified into high, medium, or low grade? (*Hint.* Think of a task consisting of 15 stages, with each stage having 3 outcomes.

(b) How many classifications have three, five, and seven shingles in the high, medium, and low grade categories, respectively?

(c) If the classifications are all equally likely, what is the probability of the event in part (b)?

18. The **binomial theorem** states that

$$(a+b)^n = \sum_{k=0}^{n} \binom{n}{k} a^k b^{n-k}.$$

(a) Use the binomial theorem to show that $\sum_{k=0}^{n} \binom{n}{k} = 2^n$. (*Hint.* $2^n = (1+1)^n$.)

(b) Expand $(a^2+b)^4$.

19. The **multinomial theorem** states that

$$(a_1 + \cdots + a_r)^n = \sum_{n_1+\cdots+n_r=n} \binom{n}{n_1, n_2, \ldots, n_r} a_1^{n_1} a_2^{n_2} \cdots a_r^{n_r},$$

where $\sum_{n_1+\cdots+n_r=n}$ denotes the sum over all nonnegative integers $n_1, n_2, \ldots, n_r$, which sum to n. Using the multinomial theorem, expand $(a_1^2 + 2a_2 + a_3)^3$.

2.4 Axioms and Properties of Probabilities

The previous section introduced probability in the intuitive context of experiments having a finite number of equally likely outcomes. For more general contexts we have the following definition.

Definition 2.4-1

For an experiment with sample space S, probability is a function that assigns a number, denoted by $P(E)$, to any event E so that the following **axioms** hold:

Axiom 1: $0 \le P(E) \le 1$

Axiom 2: $P(S) = 1$

Axiom 3: For any sequence of disjoint events $E_1, E_2, \ldots$ the probability of their union equals the sum of their probabilities, that is,

$$P\left(\bigcup_{i=1}^{\infty} E_i\right) = \sum_{i=1}^{\infty} P(E_i).$$

Axiom 1 states that the probability that an event will occur is some number between 0 and 1, something that is already known from the discussion in Section 2.3. Axiom 2 states that the sample space will occur with probability one, which is intuitive since the sample space is an event that contains all possible outcomes. Axiom 3 states that for any countable collection of disjoint events, the probability that at least one of them will occur is the sum of their individual probabilities.

Proposition 2.4-1 The three axioms imply the following properties of probability:

1. The empty set, $\emptyset$, satisfies $P(\emptyset) = 0$.
2. For any finite collection, $E_1, \ldots, E_m$, of disjoint events

$$P(E_1 \cup E_2 \cup \cdots \cup E_m) = P(E_1) + P(E_2) + \cdots + P(E_m).$$

3. If $A \subseteq B$ then $P(A) \leq P(B)$.
4. $P(A^c) = 1 - P(A)$, where A^c is the complement of A. ■

Part (1) of Proposition 2.4-1 is quite intuitive: Since the empty set does not contain any outcomes, it will never occur and thus its probability should be zero. Part (2) of Proposition 2.4-1 states that Axiom 2.4.3 applies also to a finite collection of disjoint events. In addition to being intuitive, parts (1) and (2) of Proposition 2.4-1 can be derived from the axioms; an outline of these derivations is given in Exercise 13.

Part (3) of Proposition 2.4-1 follows by noting that if $A \subseteq B$, then $B = A \cup (B - A)$ and the events A, $B - A$ are disjoint; see Figure 2-8. Hence, by part (2),

$$P(B) = P(A \cup (B - A)) = P(A) + P(B - A)$$

and since $P(B - A) \geq 0$ it follows that $P(A) \leq P(B)$.

Part (4) of Proposition 2.4-1 states that if we know that the probability event A will occur is 0.6, then the probability event A will not occur must be $1 - 0.6 = 0.4$. For example, if the probability that a Hershey's Kiss will land on its base when tossed is 0.6, then the probability that it will not land on its base must be $1 - 0.6 = 0.4$. As an additional example, if we know that the probability a die roll will result in 3 is 1/6, then the probability a die roll will not result in 3 must be $1 - 1/6 = 5/6$. To derive this property in general, note that $\mathcal{S} = A \cup A^c$ and that A, A^c are disjoint. Hence, $1 = P(\mathcal{S}) = P(A) + P(A^c)$, or $P(A^c) = 1 - P(A)$.

A particularly useful consequence of Axiom 2.4.3 (and/or part (2) of Proposition 2.4-1) is that the probability of an event E equals the sum of the probabilities of each outcome included in E. For the special case where the event in question is the

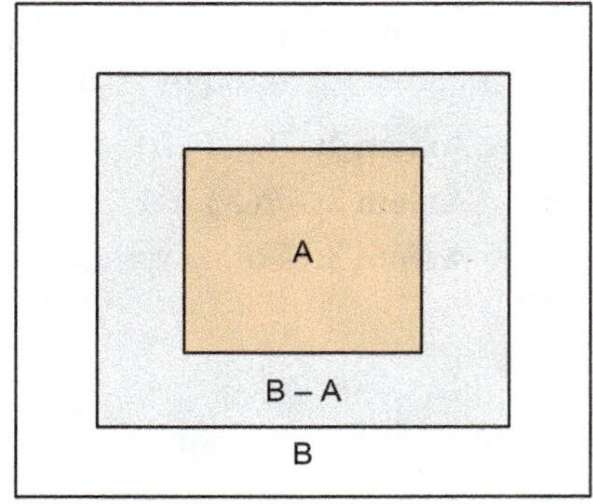

Figure 2-8 Venn diagram showing B as the disjoint union of A and $B - A$.

entire sample space, it implies that the sum of the individual probabilities of each outcome is 1. In mathematical notation, if $s_1, s_2, \ldots$ denote the possible outcomes of an experiment, we have

$$P(E) = \sum_{\text{all } s_i \text{ in } E} P(\{s_i\}) \tag{2.4.1}$$

$$1 = \sum_{\text{all } s_i \text{ in } \mathcal{S}} P(\{s_i\}) \tag{2.4.2}$$

The use of these properties is illustrated in the next example.

Example 2.4-1

The PMF of the experiment that records the number of heads in four flips of a coin, which can be obtained with the R commands *attach(expand.grid(X1=0:1, X2=0:1, X3=0:1, X4=0:1)); table(X1+X2+X3+X4)/length(X1)*, is

x	0	1	2	3	4
p(x)	0.0625	0.25	0.375	0.25	0.0625

Thus, if the random variable X denotes the number of heads in four flips of a coin then the probability of, for example, two heads is $P(X = 2) = p(2) = 0.375$.

(a) What does relation (2.4.2) imply for the sum of the probabilities given by the probability mass function?

(b) What is $P(X \geq 2)$, that is, the probability that the number of heads will be at least 2?

Solution

(a) The probabilities in the PMF sum to one.

(b) The event $[X \geq 2] = \{\text{the number of heads is at least 2}\}$ consists of the outcomes 2, 3, and 4. By relation (2.4.1),

$$P(X \geq 2) = 0.375 + 0.25 + 0.0625 = 0.6875.$$

Use of the property in part (3) of Proposition 2.4-1 is key for the solution in the next example.

Example 2.4-2

The **reliability** of a system is defined as the probability that a system will function correctly under stated conditions. A system's reliability depends on the reliability of its components as well as the way the components are arranged. A type of communications system works if at least half of its components work. Suppose it is possible to add a sixth component to such a system having five components. Show that the resulting six-component system has improved reliability.

Solution

Let E_5 denote the event that the five-component system works and E_6 denote the event that the system with the additional component works. Since these types of systems work if at least half of their components work, we have

$$E_5 = \{\text{at least three of the five components work}\},$$

$$E_6 = \{\text{at least three of the six components work}\}.$$

However, E_6 can be written as $E_6 = E_5 \cup B$, where

$$B = \{\text{two of the original five components work and the additional component works}\}.$$

Because the events E_5 and B are disjoint, part (2) of Proposition 2.4-2 implies that $P(E_6) \geq P(E_5)$. This shows that the six-component system is at least as reliable as the five-component system. ■

The point of Example 2.4-2 is not that the addition of another component will always improve reliability; see Exercise 9.

Some additional implications of the three axioms of probability are given in the following proposition.

Proposition 2.4-2 The three axioms imply the following properties of probability:

1. $P(A \cup B) = P(A) + P(B) - P(A \cap B)$.
2. $P(A \cup B \cup C) = P(A) + P(B) + P(C) - P(A \cap B) - P(A \cap C) - P(B \cap C) + P(A \cap B \cap C)$. ■

Example 2.4-3 In a certain community, 60% of the families own a dog, 70% own a cat, and 50% own both a dog and a cat. If a family is selected at random, calculate the probabilities of the following events.

(a) The family owns a dog but not a cat.

(b) The family owns a cat but not a dog.

(c) The family owns at least one of the two kinds of pets.

Solution

Define the events A = {family owns a dog} and B = {family owns a cat}. The intersection and differences of these events are shown in the Venn diagram of Figure 2-9.

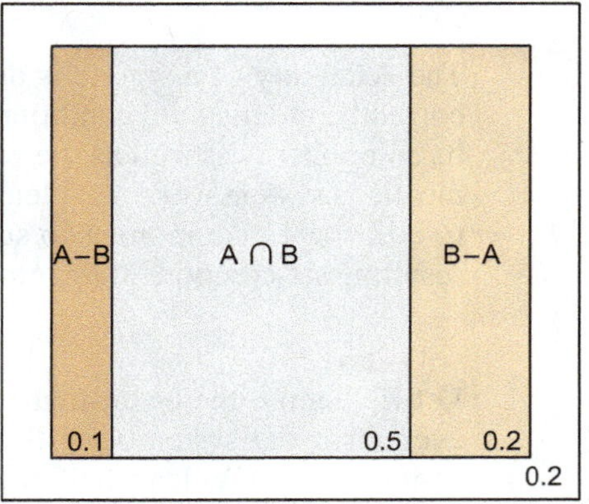

Figure 2-9 Venn diagram of events and probabilities for Example 2.4-3.

(a) This event is represented as $A - B$. Since a family that owns a dog will either not own a cat or will also own a cat, the event A is the union of $A - B$ and $A \cap B$ and these two events are disjoint. Hence,

$$P(A) = P(A - B) + P(A \cap B). \tag{2.4.3}$$

From this it follows that $P(A - B) = P(A) - P(A \cap B) = 0.6 - 0.5 = 0.1$.

(b) This event is represented as $B - A$. A line of reasoning similar to the above leads to

$$P(B) = P(B - A) + P(A \cap B) \tag{2.4.4}$$

from which it follows that $P(B - A) = P(B) - P(A \cap B) = 0.7 - 0.5 = 0.2$.

(c) This event is represented as $A \cup B$. Since a family that owns at least one of the two kinds of pets will own only a dog, only a cat, or both a dog and a cat, $A \cup B$ is the union of $A - B$, $B - A$, and $A \cap B$ and these three events are disjoint. Hence,

$$\begin{aligned} P(A \cup B) &= P(A - B) + P(A \cap B) + P(B - A) \\ &= 0.1 + 0.5 + 0.2 = 0.8. \end{aligned} \tag{2.4.5}$$

Another way of finding this probability is to apply directly part (1) of Proposition 2.4-2:

$$P(A \cup B) = P(A) + P(B) - P(A \cap B) = 0.6 + 0.7 - 0.5 = 0.8.$$

■

The solution to the above example provides insight into the formula in part (1) of Proposition 2.4-2. From relations (2.4.3) and (2.4.4), which hold in general, it follows that when the probabilities of A and B are added, the probability of the intersection is counted twice, that is,

$$P(A) + P(B) = P(A - B) + 2P(A \cap B) + P(B - A).$$

Therefore, in order to obtain the expression for the probability of $A \cup B$ given in (2.4.5), which also holds in general, the probability of the intersection must be subtracted.

The formula in part (2) of Proposition 2.4-2 can be justified similarly: By adding the probabilities of A, B, and C, the probabilities of the pairwise intersections have been added twice. Therefore, the probabilities of $A \cap B$, $A \cap C$, and $B \cap C$ must be subtracted. Now notice that the intersection $A \cap B \cap C$ is included in all three events and their pairwise intersections. Thus, its probability has been added three times and subtracted three times and, therefore, it needs to be added back. This type of *inclusion-exclusion* argument can also be used for the probability of the union of more than three events.

Exercises

1. A person is selected at random from the population of Verizon wireless subscribers. Let A be the event that the chosen subscriber has friends or family added to his/her plan, and B denote the event that the subscriber has unlimited text messaging. Extensive records suggest that $P(A) = 0.37$, $P(B) = 0.23$, and $P(A \cup B) = 0.47$. Find $P(A \cap B)$.

2. The events $A_1, A_2, \ldots, A_m$ are said to form a **partition** of the sample space of an experiment if they are disjoint and their union equals the sample space.

(a) If all m events of such a partition are equally likely, what is their common probability?

(b) If $m = 8$ find the probability of $A_1 \cup A_2 \cup A_3 \cup A_4$.

3. A new generation of hybrid vehicles achieves highway gas mileage in the range of 50 to 53 MPG. The gas mileage of three such cars, during a pre-specified 100-mile drive, will be rounded to the nearest integer, resulting in the sample space $\mathcal{S} = \{(x_1, x_2, x_3) : x_i = 50, 51, 52, 53\}$, $i = 1, 2, 3$.

(a) Assume that the outcomes of the sample space $\mathcal{S}$ are equally likely, and use R commands similar to those used in Example 2.4-1 to find the probability mass function of the experiment that records the average mileage the three cars achieve.

(b) Use the PMF obtained in part (a) to compute the probability that the average gas mileage is at least 52 MPG.

4. The PMF of the sum of two die rolls, found in Example 2.3-13, is

x	2	3	4	5	6	7	8	9	10	11	12
p(x)	1/36	2/36	3/36	4/36	5/36	6/36	5/36	4/36	3/36	2/36	1/36

(a) For each of the following events specify the outcomes that belong to them, and use relation (2.4.1) to find their probabilities.
  (i) $E_1 = \{$the sum of the two die rolls is at least 5$\}$.
  (ii) $E_2 = \{$the sum of the two die rolls is no more than 8$\}$.
  (iii) $E_3 = E_1 \cup E_2$, $E_4 = E_1 - E_2$, and $E_5 = E_1^c \cap E_2^c$.

(b) Recalculate the probability of E_3 using part (1) of Proposition 2.4-2.

(c) Recalculate the probability of E_5 using De Morgan's first law, the probability of E_3, and part (4) of Proposition 2.4-1.

5. A telecommunications company classifies transmission calls by their duration as brief (< 3 minutes) or long (> 3 minutes) and by their type as voice (V), data (D), or fax (F). From an extensive data base, the company has come up with the following probabilities for the category of a random (e.g. the next) transmission call.

	Type of Transmission Call		
Duration	V	D	F
> 3	0.25	0.10	0.07
< 3	0.30	0.15	0.13

(a) For each of the following events specify the category outcomes that belong to them, and find their probabilities using relation (2.4.1).
  (i) $E_1 =$ the next call is a voice call.
  (ii) $E_2 =$ the next call is brief.
  (iii) $E_3 =$ the next call is a data call.
  (iv) $E_4 = E_1 \cup E_2$, and $E_5 = E_1 \cup E_2 \cup E_3$.

(b) Recalculate the probability of E_4 using part (1) of Proposition 2.4-2.

(c) Recalculate the probability of E_5 using part (2) of Proposition 2.4-2.

6. Each of the machines A and B in an electronics fabrication plant produces a single batch of 50 electrical components per hour. Let E_1 denote the event that, in any given hour, machine A produces a batch with no defective components, and E_2 denote the corresponding event for machine B. The probabilities of E_1, E_2, and $E_1 \cap E_2$ are 0.95, 0.92, and 0.88, respectively. Express each of the following events as set operations on E_1 and E_2, and find their probabilities.

(a) In any given hour, only machine A produces a batch with no defects.

(b) In any given hour, only machine B produces a batch with no defects.

(c) In any given hour, exactly one machine produces a batch with no defects.

(d) In any given hour, at least one machine produces a batch with no defects.

7. The electronics fabrication plant in Exercise 6 has a third machine, machine C, which is used in periods of peak demand and is also capable of producing a batch of 50 electrical components per hour. Let E_1, E_2 be as in Exercise 6, and E_3 be the corresponding event for machine C. The probabilities of E_3, $E_1 \cap E_3$, $E_2 \cap E_3$, and $E_1 \cap E_2 \cap E_3$ are 0.9, 0.87, 0.85, and 0.82, respectively. Find the probability that at least one of the machines will produce a batch with no defectives.

8. The monthly volume of book sales from the online site of a bookstore is categorized as below expectations (0), in line with expectations (1), or above expectations (2). The monthly volume of book sales from the brick and mortar counterpart of the bookstore is categorized similarly. The following table gives the probabilities of the nine possible outcomes of an experiment that records the monthly volume of sales categories.

	Brick and Mortar Sales		
Online Sales	0	1	2
0	0.10	0.04	0.02
1	0.08	0.30	0.06
2	0.06	0.14	0.20

(a) Find the probabilities of each of the following events.
  (i) $E_1 =$ the online sales volume category is ≤ 1.
  (ii) $E_2 =$ the brick and mortar sales volume category is ≤ 1.
  (iii) $E_3 = E_1 \cap E_2$.

(b) Find the probability mass function for the experiment that records only the online monthly volume of sales category.

9. A type of communications system works if at least half of its components work. Suppose it is possible to add a fifth component to such a system having four components. Show that the resulting five-component system is not necessarily more reliable. (*Hint.* In the notation of Example 2.4-2, it suffices to show that $E_4 \not\subset E_5$. Expressing E_5 as the union of {at least three of the original four components work} and {two of the original four components work and the additional component works}, it can be seen that the event {two of the original four components work and the additional component does not work} is contained in E_4 but not in E_5.)

10. Consider the game where two dice, die A and die B, are rolled. We say that die A wins, and write $A > B$, if the outcome of rolling A is larger than that of rolling B. If both rolls result in the same number it is a tie.

(a) Find the probability of a tie.

(b) Find the probability that die A wins.

11. Efron's dice. Using arbitrary numbers on the sides of dice can have surprising consequences for the game described in the Exercise 10. Efron's dice are sets of dice with the property that for each die there is another that beats it with larger probability when the game of Exercise 10 is played. An example of a set of four Efron's dice is as follows:

- Die A: four 4's and two 0's
- Die B: six 3's
- Die C: four 2's and two 6's
- Die D: three 5's and three 1's

(a) Specify the events $A > B$, $B > C$, $C > D$, $D > A$.

(b) Find the probabilities that $A > B$, $B > C$, $C > D$, $D > A$.

12. Let's make a deal. In the game Let's Make a Deal, the host asks a participant to choose one of three doors. Behind one of the doors is a big prize (e.g., a car), while behind the other two doors are minor prizes (e.g., a blender). After the participant selects a door, the host opens one of the other two doors (knowing it is not the one having the big prize). The host does not show the participant what is behind the door the participant chose. The host asks the participant to either

(a) stick with his/her original choice, or

(b) select the other of the remaining two closed doors.

Find the probability that the participant will win the big prize for each of the strategies (a) and (b).

13. Using only the three axioms of probability, prove parts (1) and (2) of Proposition 2.4-1. (*Hint.* For part (1) apply Axiom 2.4.3 to the sequence of events $E_1 = \mathcal{S}$ and $E_i = \emptyset$ for $i = 2, 3, \ldots$, the union of which is $\mathcal{S}$. This results in the equation $P(\mathcal{S}) = \sum_{i=1}^{\infty} P(E_i) = P(\mathcal{S}) + \sum_{i=2}^{\infty} P(\emptyset)$. Now complete the argument. For part (2) apply Axiom 2.4.3 to the sequence of events $E_1, \ldots, E_n$, and $E_i = \emptyset$ for $i = n+1, n+2, \ldots$, the union of which is $\cup_{i=1}^{n} E_i$. This results in the equation $P(\cup_{i=1}^{n} E_i) = \sum_{i=1}^{\infty} P(E_i) = P(\cup_{i=1}^{n} E_i) + \sum_{i=n+1}^{\infty} P(\emptyset)$. Now complete the argument.)

2.5 Conditional Probability

Conditional probability refers to the probability computed when some partial information concerning the outcome of the experiment is available.

Example 2.5-1

A card drawn at random from a deck of 52 cards is observed to be a face card. Given this partial information, what is the probability the card is a king?

Solution

Since four of the 12 face cards are kings, it is intuitive that the desired probability is 4/12 or 1/3. ■

If we let A denote the event that the card drawn is a face card and B denote the event that the card drawn is a king, the probability obtained in Example 2.5-1 is called the *conditional probability that B occurs given that A occurred* and is denoted by $P(B|A)$.

The intuitive derivation of the conditional probability in Example 2.5-1 rests on the following basic principle for computing conditional probabilities in experiments with equally likely outcomes.

Basic Principle for Conditional Probabilities

> Given the information that an experiment with sample space $\mathcal{S}$ resulted in the event A, conditional probabilities are calculated by
>
> - replacing the sample space $\mathcal{S}$ by A, and
> - treating the outcomes of the new sample space A as equally likely.
>
> **(2.5.1)**

Example 2.5-2

Two dice are rolled and their sum is observed to be 7. Given this information, what is the conditional probability that one of the two die rolls was a 3?

Solution

To follow the basic principle for computing conditional probabilities, let the sample space $\mathcal{S}$ be the 36 equally likely outcomes of rolling two dice. The event $A =$ {the sum of the two die rolls is 7} consists of the outcomes (1,6), (2,5), (3,4), (4,3), (5,2), and (6,1), which now constitute the new sample space. Because the outcomes in this new sample space are equally likely, the conditional probability of each outcome is 1/6. Since a die roll of 3 occurs in two of the six equally likely outcomes, it follows that the desired conditional probability is 2/6 or 1/3. ■

The basic principle (2.5.1) can be used even when the number of equally likely outcomes is not known.

Example 2.5-3

Example 2.4-3 gave the percentages of families in a certain community that own a dog, a cat, or both as 60%, 70%, and 50%, respectively. If a randomly selected family owns a dog, what is the probability it also owns a cat?

Solution

Let A denote the event that the family owns a dog and B the event that the family owns a cat. In Example 2.4-3 it was found that $P(A - B) = 0.1$. Since $A = (A - B) \cup (A \cap B)$ (see also Figure 2-9), it follows that among families who own a dog, the ratio of families who also own a cat to those who do not is 5 to 1. Therefore, the conditional probability that the family owns a cat given that it owns a dog is $P(B|A) = 5/6$. ■

The reasoning used in Example 2.5-3 generalizes to any two events A and B: Given that the event A occurred, B will also occur if the outcome belongs in $A \cap B$. According to the basic principle (2.5.1), however, A is the new sample space and the ratio of the probability of $A \cap B$ to $A - B$ remains the same. Since $A \cap B$ and $A - B$ are complementary events relative to the new sample space, that is, given that A occurred, then either $A \cap B$ or $A - B$ must also have occurred, we arrive at the following definition.

Definition 2.5-1

The **definition of conditional probability** states that for any two events A, B with $P(A) > 0$,

$$P(B|A) = \frac{P(A \cap B)}{P(A)}.$$

Example 2.5-4

The probability that the life of a product does not exceed t time units is $1 - \exp(-0.1t)$. Given that a product has lasted 10 time units, what is the probability it will fail in the next 5 time units?

Solution

Let A denote the event that a product's life exceeds 10 time units and B denote the event that the life will not exceed 15 time units. The events A, A^c, and B are shown in the diagram of Figure 2-10. According to Definition 2.5-1, in order to find $P(B|A)$ it suffices find $P(A)$ and $P(A \cap B)$. By the formula given, the probabilities of A^c, that is, that the life of a product does not exceed 10 time units, and of B are

$$P(A^c) = 1 - \exp(-0.1 \times 10) = 0.632 \quad \text{and} \quad P(B) = 1 - \exp(-0.1 \times 15) = 0.777.$$

Noting that $A^c \subset B$ (see Figure 2-10), it follows that

$$B = (B \cap A^c) \cup (B \cap A) = A^c \cup (B \cap A).$$

Hence, since A^c and $(B \cap A)$ are disjoint, $P(B) = P(A^c) + P(B \cap A)$ from which it follows that

$$P(B \cap A) = P(B) - P(A^c) = 0.777 - 0.632 = 0.145.$$

Hence, since $P(A) = 1 - P(A^c) = 1 - 0.632 = 0.368$, we have

$$P(B|A) = \frac{P(B \cap A)}{P(A)} = \frac{0.145}{0.368} = 0.394.$$

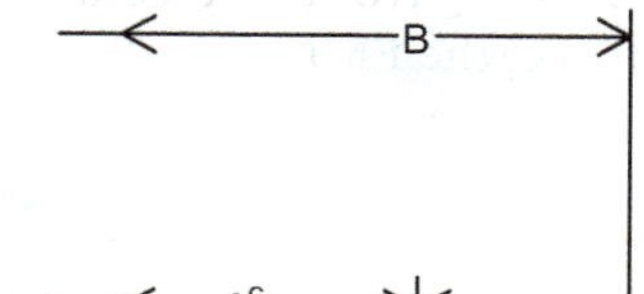

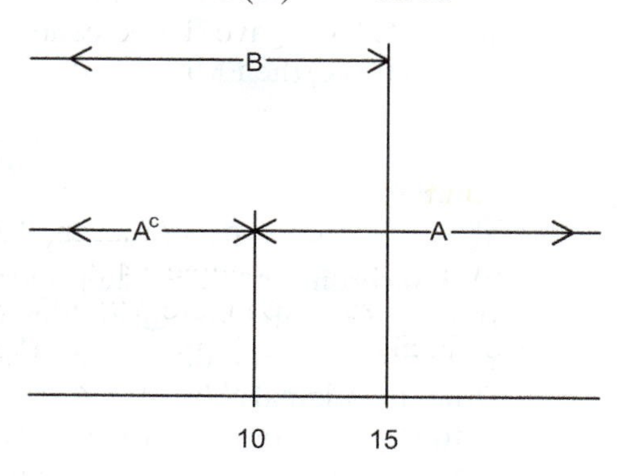

Figure 2-10 Diagram showing the events used in Example 2.5-4.

Example 2.5-5

A supermarket has regular (manned) checkout lines and self checkout lines. Let X take the value 0, 1, or 2 depending on whether there are no customers, between 1 and 10 customers, or more than 10 customers, respectively, in the self checkout lines in any given five-minute period. Let Y be the corresponding variable for the regular (manned) checkout lines. The probabilities for each of the nine possible outcomes of the experiment that records X and Y for a five-minute period are given in the following table.

	Regular Checkout		
Self Checkout	0	1	2
0	0.17	0.02	0.01
1	0.15	0.225	0.125
2	0.06	0.105	0.135

Let A_0, B_0 be the events that $X = 0$ and $Y = 0$, respectively.

(a) Find the probability of A_0.

(b) Find the conditional probability of B_0 given A_0.

(c) Find the probability mass function of the random variable X.

Solution

(a) The event $A_0 = [X = 0]$ can only occur together with one of the events $B_0 = [Y = 0]$, $B_1 = [Y = 1]$, or $B_2 = [Y = 2]$. In mathematical notation, we have

$$A_0 = (A_0 \cap B_0) \cup (A_0 \cap B_1) \cup (A_0 \cap B_2),$$

and the events $(A_0 \cap B_0)$, $(A_0 \cap B_1)$, and $(A_0 \cap B_2)$ are disjoint. Hence, by part (2) of Proposition 2.4-1,

$$P(A_0) = P(A_0 \cap B_0) + P(A_0 \cap B_1) + P(A_0 \cap B_2) = 0.17 + 0.02 + 0.01 = 0.2.$$

(b)

$$P(B_0|A_0) = \frac{P(B_0 \cap A_0)}{P(A_0)} = \frac{0.17}{0.2} = 0.85.$$

(c) In part (a) we found that $P(X = 0) = 0.2$. Working similarly, we find that

$$P(X = 1) = 0.15 + 0.225 + 0.125 = 0.5 \quad \text{and}$$

$$P(X = 2) = 0.060 + 0.105 + 0.135 = 0.3.$$

Hence, the PMF of X is

x	0	1	2
p(x)	0.2	0.5	0.3

According to the basic principle for conditional probabilities (2.5.1), the conditional probabilities given the information that an event A has occurred are probabilities from a new experiment in which the sample space has been reduced from the original $\mathcal{S}$ to A. One way of simulating outcomes from this new experiment is to generate outcomes from the original experiment and ignore those outcomes that are not in A. For example, given the information that the roll of a die is an even number, the sample space shrinks from $\mathcal{S} = \{1, \ldots, 6\}$ to $A = \{2, 4, 6\}$. In the absence of a random number generator, outcomes from this reduced sample space can be obtained by repeated die rolls and ignoring the outcomes that are not in A.

Example 2.5-6

Fair game with an unfair coin. Suppose that when a biased coin is flipped it results in heads with probability p. A fair game with such an unfair coin can be played as follows: Flip the coin twice. If the outcome is (H, H) or (T, T) ignore the outcome and flip the coin two more times. Repeat until the outcome of the two flips is either (H, T) or (T, H). Code the first of these outcomes as 1 and the second as 0. Prove that the probability of getting a 1 equals 0.5.

Solution

Ignoring the outcomes (H, H) and (T, T) is equivalent to conditioning on the event $A = \{(H, T), (T, H)\}$. Thus we will be done if we show that the conditional probability of (H, T) given A is 0.5. Using the definition of conditional probability we have

$$P((H,T)|A) = \frac{P((H,T)\cap A)}{P(A)} = \frac{P((H,T))}{P(A)} = \frac{p(1-p)}{p(1-p)+(1-p)p} = 0.5.$$

2.5.1 THE MULTIPLICATION RULE AND TREE DIAGRAMS

Though the axioms and properties of probability do not contain an explicit formula for calculating the intersection of events, part (1) of Proposition 2.4-2 contains implicitly the following formula

$$P(A \cap B) = P(A) + P(B) - P(A \cup B). \quad \textbf{(2.5.2)}$$

Note that (2.5.2) requires three pieces of information, which are $P(A)$, $P(B)$, and $P(A \cup B)$, for the calculation of $P(A \cap B)$. Definition 2.5-1 of conditional probability yields the following alternative (multiplicative) formula that uses only two pieces of information.

Multiplication Rule for Two Events

$$P(A \cap B) = P(A)P(B|A) \quad \textbf{(2.5.3)}$$

Example 2.5-7

Two consecutive traffic lights have been synchronized to make a run of green lights more likely. In particular, if a driver finds the first light to be red, the second light will be green with probability 0.9, and if the first light is green the second will be green with probability 0.7. If the probability of finding the first light green is 0.6, find the probability that a driver will find both lights green.

Solution

Let A denote the event that the first light is green and B the corresponding event for the second light. The question concerns the probability of the intersection of A and B. From the multiplication rule we obtain

$$P(A \cap B) = P(A)P(B|A) = 0.6 \times 0.7 = 0.42.$$

This multiplicative formula generalizes to more than two events.

Multiplication Rule for Three Events

$$P(A \cap B \cap C) = P(A)P(B|A)P(C|A \cap B) \quad \textbf{(2.5.4)}$$

To prove the multiplication rule for three events apply the definition of conditional probability to its right-hand side to get

$$P(A)\frac{P(A \cap B)}{P(A)}\frac{P(A \cap B \cap C)}{P(A \cap B)} = P(A \cap B \cap C).$$

See Exercise 13 for the extension of the multiplication rule to several events.

Example 2.5-8

Pick three cards from a deck. Find the probability that the first draw is an ace, the second draw is a king, and the third draw is a queen.

Solution

Let A = {first draw results in ace}, B = {second draw results in king}, and C = {third draw results in queen}. Thus we want to calculate $P(A \cap B \cap C)$. From the multiplication rule for three events we have

$$P(A \cap B \cap C) = P(A)P(B|A)P(C|A \cap B) = \frac{4}{52}\frac{4}{51}\frac{4}{50} = 0.000483.$$

Example 2.5-9

Of the customers entering a department store 30% are men and 70% are women. The probability a male shopper will spend more than \$50 is 0.4, and the corresponding probability for a female shopper is 0.6. The probability that at least one of the items purchased is returned is 0.1 for male shoppers and 0.15 for female shoppers. Find the probability that the next customer to enter the department store is a woman who will spend more than \$50 on items that will not be returned.

Solution

Let W = {customer is a woman}, B = {the customer spends >\$50}, and R = {at least one of the purchased items is returned}. We want the probability of the intersection of W, B, and R^c. By the formula in (2.5.4), this probability is given by

$$P(W \cap B \cap R^c) = P(W)P(B|W)P(R^c|W \cap B) = 0.7 \times 0.6 \times 0.85 = 0.357.$$

The multiplication rule typically applies in situations where the events whose intersection we wish to compute are associated with different stages of an experiment. For example, the experiment in Example 2.5-9 consists of three stages, which are (a) record customer's gender, (b) record amount spent by customer, and (c) record whether any of the items purchased are subsequently returned. Therefore, by the generalized fundamental principle of counting, this experiment has $2 \times 2 \times 2 = 8$ different outcomes. Each outcome is represented by a path going from left to right in the tree diagram of Figure 2-11. The numbers appearing along each path are the (conditional) probabilities of going from each outcome of a stage to the outcomes of the next stage. The probability of each outcome of the experiment is the product of the numbers appearing along the path that represents it. For example, the probability found in Example 2.5-9 is that of the outcome represented by the path that defines the bottom boundary of the tree diagram. Tree diagrams provide additional insight into the probability structure of the experiment and facilitate the

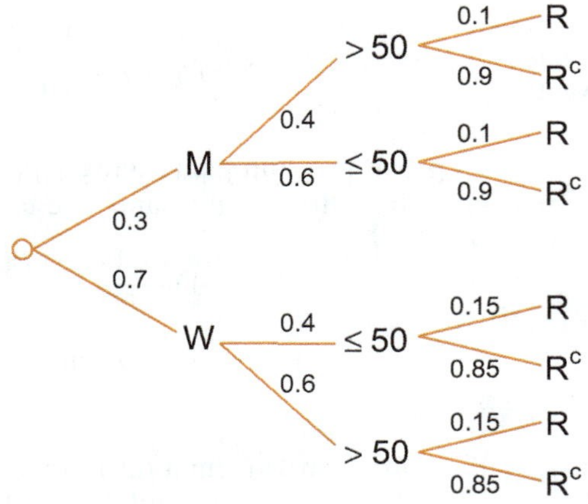

Figure 2-11 Tree diagram for Example 2.5-9.

computation of probabilities. For example, the probability that a purchase will result in at least one item being returned is found by summing the probabilities of the four events represented by the four paths leading to an R in Figure 2-11:

$$0.3 \times 0.4 \times 0.1 + 0.3 \times 0.6 \times 0.1 + 0.7 \times 0.4 \times 0.15 + 0.7 \times 0.6 \times 0.15 = 0.135. \tag{2.5.5}$$

The rule for constructing tree diagrams is as follows: Start with the stage of the experiment for which unconditional probabilities for its possible outcomes are given. Call this stage 1. From the origin (i.e., the little circle at the very left of Figure 2-11) draw branches to each of the possible outcomes of stage 1. Next, there are conditional probabilities, given the outcome of stage 1, for the outcomes of the next stage of the experiment. Call this stage 2. From each outcome of stage 1 draw branches to each of the possible outcomes of stage 2, and so forth.

2.5.2 LAW OF TOTAL PROBABILITY AND BAYES' THEOREM

The probability computed in (2.5.5) is an example of a *total probability*. It is called total because it is obtained by summing the probabilities of the individual outcomes of the experiment, that is, paths in the tree diagram, whose end result is $R = \{\text{at least one of the purchased items is returned}\}$. Grouping the paths into those that pass through M and those that pass through W (see Figure 2-11), the terms in the left-hand side of (2.5.5) can be written as

$$0.3(0.4 \times 0.1 + 0.6 \times 0.1) + 0.7(0.4 \times 0.15 + 0.6 \times 0.15) = P(M)P(R|M) + P(W)P(R|W) = P(R). \tag{2.5.6}$$

This is a simple form of the Law of Total Probability.

In general, the Law of Total Probability is a formula for calculating the probability of an event B, when B arises in connection with events $A_1, \ldots, A_k$, which constitute a *partition* of the sample space (i.e., they are disjoint and make up the entire sample space); see Figure 2-12. If the probability of each A_i and the conditional probability of B given each A_i are all known, the Law of Total Probability expresses the probability of B as

Law of Total Probability

$$P(B) = P(A_1)P(B|A_1) + \cdots + P(A_k)P(B|A_k) \tag{2.5.7}$$

The events $A_1, \ldots, A_k$ can also be thought of as a stratification of the population. In the simple example of the Law of Total Probability given in (2.5.6), the events $M = \{\text{customer is a man}\}$ and $W = \{\text{customer is a woman}\}$, which play the role of A_1 and A_2 (so $k = 2$), form a stratification of the population of customers. That M and W form a partition of the sample space also follows from the fact that every path in the tree diagram of Figure 2-11 passes either through M or through W.

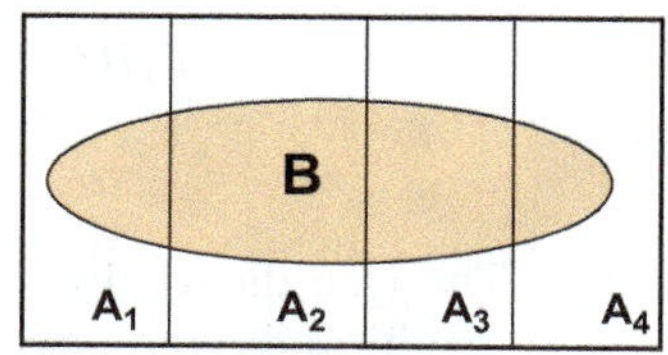

Figure 2-12 An event B arising in connection with events $A_1, \ldots, A_4$.

To prove the Law of Total Probability use the multiplication rule, and the fact that $A_1, \ldots, A_k$ form a partition of the sample space, to write the right-hand side of (2.5.7) as

$$P(B \cap A_1) + \cdots + P(B \cap A_k) = P[(B \cap A_1) \cup \cdots \cup (B \cap A_k)] = P(B),$$

where the first equality follows because the events $B \cap A_i$ are disjoint, and the second follows because $(B \cap A_1) \cup \cdots \cup (B \cap A_k) = B$; see also Figure 2-12.

REMARK 2.5-1 The number of the events A_i in the partition may also be (countably) infinite. Thus, if $A_i \cap A_j = \emptyset$, for all $i \neq j$, and $\cup_{i=1}^{\infty} A_i = \mathcal{S}$, the Law of Total Probability states that

$$P(B) = \sum_{i=1}^{\infty} P(A_i)P(B|A_i)$$

holds for any event B. ◁

Example 2.5-10

Two friends will be dealt a card each. The two cards will be drawn from a standard deck of 52 cards at random and without replacement. If neither gets an ace, the full deck is reshuffled and two cards are again drawn without replacement. The game ends when at least one of the two friends gets an ace. The ones with ace win a prize. Is this a fair game?

Solution

The game will be fair if the probability of an ace in the second draw is 4/52, which is the probability of an ace in the first draw. Let B denote the event that the second draw results in an ace, let A_1 denote the event that the first draw results in an ace, and let A_2 be the complement of A_1. Then, according to the Law of Total Probability,

$$P(B) = P(A_1)P(B|A_1) + P(A_2)P(B|A_2) = \frac{4}{52}\frac{3}{51} + \frac{48}{52}\frac{4}{51} = \frac{4}{52}.$$

Thus the game is fair. ■

Example 2.5-11

Use the information given in Example 2.5-7 regarding the two consecutive synchronized traffic lights to complete the following.

(a) Find the probability that a driver will find the second traffic light green.

(b) Recalculate the probability of part (a) through a tree diagram for the experiment that records whether or not a car stops at each of the two traffic lights.

Solution

(a) Let A and B denote the events that a driver will find the first and the second, respectively, traffic lights green. Because the events A and A^c constitute a partition of the sample space, according to the Law of Total Probability

$$\begin{aligned} P(B) &= P(A)P(B|A) + P(A^c)P(B|A^c) \\ &= 0.6 \times 0.7 + 0.4 \times 0.9 = 0.42 + 0.36 = 0.78. \end{aligned}$$

(b) The tree diagram is given in Figure 2-13. The experiment has two outcomes resulting in the second light being green which are represented by the

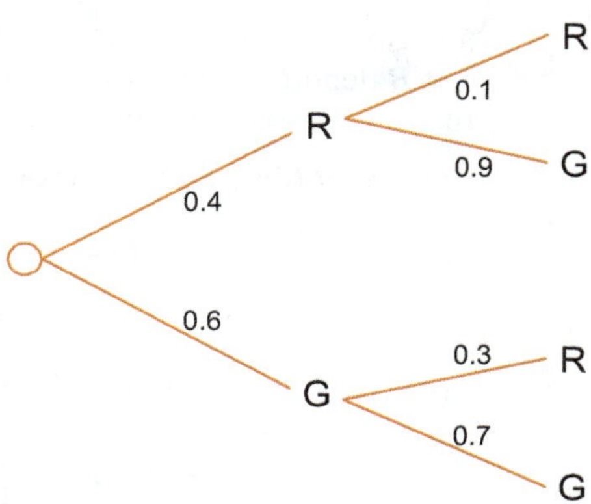

Figure 2-13 Tree diagram for Example 2.5-11.

paths with the pairs of probabilities (0.6, 0.7) and (0.4, 0.9). The sum of the probabilities of these two outcomes is $0.6 \times 0.7 + 0.4 \times 0.9 = 0.42 + 0.36 = 0.78$. ■

Bayes' Theorem applies to the same context, and with the same information, used in the Law of Total Probability. Thus there is a partition $A_1, \ldots, A_k$ of the sample space and an event B, as shown in Figure 2-12. The probabilities of the events A_i are given, and so are the conditional probabilities of B given that an A_i has occurred. Bayes' Theorem answers the question: Given that B has occurred, what is the probability that a particular A_j has occurred? The answer is provided by the following formula.

Bayes' Theorem

$$P(A_j|B) = \frac{P(A_j)P(B|A_j)}{\sum_{i=1}^{k} P(A_i)P(B|A_i)} \tag{2.5.8}$$

A proof of the formula follows by first writing $P(A_j|B) = P(A_j \cap B)/P(B)$ and then applying the multiplication rule in the numerator and the Law of Total Probability in the denominator.

Example 2.5-12

In the setting of Example 2.5-11, find the probability that a passing car encountered a green first light given that it encountered a green second light.

Solution

With the events A and B as defined in Example 2.5-11, we want to find $P(A|B)$. Using Bayes' theorem we obtain

$$\begin{aligned} P(A|B) &= \frac{P(A)P(B|A)}{P(A)P(B|A) + P(A^c)P(B|A^c)} \\ &= \frac{0.6 \times 0.7}{0.6 \times 0.7 + 0.4 \times 0.9} = \frac{0.42}{0.78} = 0.538. \end{aligned}$$

■

Example 2.5-13

Suppose that 5% of all men and 0.25% of all women are color-blind. A person is chosen at random from a community having 55% women and 45% men.

(a) What is the probability that the person is color-blind?

(b) If the chosen person is color-blind, what is the probability that the person is male?

Solution

Let B denote the event that the selected person is color-blind, A_1 denote the event that the person is a man, and A_2 denote the event that the person is a woman.

(a) According to the Law of Total Probability,

$$P(B) = P(A_1)P(B|A_1) + P(A_2)P(B|A_2)$$
$$= 0.45 \times 0.05 + 0.55 \times 0.0025 = 0.0239.$$

(b) According to Bayes' Theorem,

$$P(A_1|B) = \frac{P(A_1)P(B|A_1)}{\sum_{i=1}^{2} P(A_i)P(B|A_i)} = \frac{0.45 \times 0.05}{0.0239} = 0.9424.$$

Exercises

1. The probability that a phone call will last more than t minutes is $(1+t)^{-2}$. Given that a particular phone call has not ended in two minutes, what is the probability it will last more than three minutes?

2. The probability that two or more of a system's 10 components fail between consecutive inspections is 0.005, while the probability that only one component fails is 0.1. When two or more components fail, a re-evaluation of the system is initiated during which all failed components, and those deemed unreliable, are replaced. Otherwise, components are replaced upon their failure. Find the probability that a system re-evaluation occurs before any component is individually replaced. (*Hint.* Let B = {system re-evaluation occurs}, C = {a component is individually replaced}, and consider a new experiment with reduced sample space $A = B \cup C$. The desired probability is the probability of B in this new experiment. See also Example 2.5-6.)

3. The moisture content of batches of a chemical substance is measured on a scale from 1 to 3, while the impurity level is recorded as either low (1) or high (2). Let X and Y denote the moisture content and the impurity level, respectively, of a randomly selected batch. The probabilities for each of the six possible outcomes of the experiment that records X and Y for a randomly selected batch are given in the following table.

	Impurity Level	
Moisture	1	2
1	0.132	0.068
2	0.24	0.06
3	0.33	0.17

Let A and B be the events that $X = 1$ and $Y = 1$, respectively.

(a) Find the probability of A.

(b) Find the conditional probability of B given A.

(c) Find the probability mass function of the random variable X.

4. Two major brands of flat screen TVs control 50% and 30% of the market, respectively. Other brands have the remaining 20% of the market. It is known that 10% of brand 1 TVs require warranty repair work, as do 20% of brand 2 and 25% of different brand TVs.

(a) Find the probability that the next flat screen TV purchased is a brand 1 TV which will need warranty repair work.

(b) Make a tree diagram for the experiment that records the brand of the next flat screen TV to be sold, and whether or not it will require warranty repair work, and mark the given probabilities on the different paths of the tree diagram.

(c) Use the tree diagram to find the probability that the next flat screen TV to be sold will need warranty repair work.

5. An article on vehicle choice behavior[2] gives the following information about the US car and truck market. The ratio of cars to trucks is 36/64. Among the cars sold 42% are made in the US, while 58% are imports. The corresponding percentages for trucks sold are 70% and 30%.

(a) Find the probability that the next auto consumer will buy an imported car. (*Hint.* 36% of the sales are cars.)

[2] K. E. Train and C. Winston (2007). Vehicle choice behavior and the declining market share of US automakers, *International Economic Review*, 48(4): 1469–1796.

(b) It is also known that 35% of the consumers who get an import choose to lease it, while 20% of those getting a US-made vehicle lease. Make a tree diagram for the experiment that records the decisions made in the various vehicle acquisition stages, which are (1) car or truck, (2) US made or imported, and (3) buy or lease, and mark the given probabilities on the different paths of the tree diagram.

(c) Use the tree diagram to compute the probability that the next auto consumer will choose to lease his/her chosen vehicle.

6. A particular consumer product is being assembled on production lines A, B, and C, and packaged in batches of 10. Each day, the quality control team selects a production line by probability sampling with probabilities $P(A) = P(B) = 0.3$ and $P(C) = 0.4$ and inspects a randomly drawn batch from the selected production line. The probability that no defects are found in a batch selected from production line A is 0.99, and the corresponding probabilities for production lines B and C are 0.97 and 0.92, respectively. A tree diagram may be used for answering the following questions.

(a) What is the probability that a batch from production line A is inspected and no defects are found?

(b) Answer the above question for production lines B and C.

(c) What is the probability that no defects are found in any given day?

(d) Given that no defects were found in a given day, what is the probability the inspected batch came from production line C?

7. Fifteen percent of all births involve Cesarean (C) section. Ninety-eight percent of all babies survive delivery, whereas when a C section is performed the baby survives with probability 0.96.

(a) Make a tree diagram and mark the given information on the appropriate paths of the diagram. (Note that the probabilities for certain paths are not given.)

(b) What is the probability that a baby will survive delivery if a C section is not performed? (*Hint.* Use the tree diagram and the remaining information given to set up an equation.)

8. Thirty percent of credit card holders carry no monthly balance, while 70% do. Of those card holders carrying a balance, 30% have annual income \$20,000 or less, 40% between \$20,001 and \$50,000, and 30% over \$50,000. Of those card holders carrying no balance, 20%, 30%, and 50% have annual incomes in these three respective categories.

(a) What is the probability that a randomly chosen card holder has annual income \$20,000 or less?

(b) If this card holder has an annual income that is \$20,000 or less, what is the probability that (s)he carries a balance?

9. You ask your roommate to water a sickly plant while you are on vacation. Without water the plant will die with probability 0.8 and with water it will die with probability 0.1. With probability 0.85, your roommate will remember to water the plant.

(a) What is the probability that your plant is alive when you return? (You may use a tree diagram.)

(b) If the plant is alive when you return, what is the probability that your roommate remembered to water it?

10. A batch of 10 fuses contains three defective ones. A sample of size two is taken at random and without replacement.

(a) Find the probability that the sample contains no defective fuses.

(b) Let X be the random variable denoting the number of defective fuses in the sample. Find the probability mass function of X.

(c) Given that $X = 1$, what is the probability that the defective fuse was the first one selected?

11. A city's police department plans to enforce speed limits by using radar traps at four different locations. During morning rush hour, the radar traps at locations L_1, L_2, L_3, L_4 are operated with probabilities 0.4, 0.3, 0.2, and 0.3, respectively. A person speeding to work has probabilities of 0.2, 0.1, 0.5, and 0.2, respectively, of passing through these locations.

(a) What is the probability the speeding person will receive a speeding ticket?

(b) If the person received a speeding ticket while speeding to work, what is the probability that he/she passed through the radar trap at location L_2?

12. Seventy percent of the light aircraft that disappear while in flight in a certain country are subsequently discovered. Of the aircraft that are discovered, 60% have an emergency locator, whereas 10% of the aircraft not discovered have such a locator. Suppose a light aircraft disappears while in flight.

(a) What is the probability that it has an emergency locator and it will not be discovered?

(b) What is the probability that it has an emergency locator?

(c) If it has an emergency locator, what is the probability that it will not be discovered?

13. Prove the following generalization of the multiplication rule.

$$P(E_1 \cap E_2 \cap \cdots \cap E_n) = P(E_1)P(E_2|E_1)P(E_3|E_1 \cap E_2) \cdots P(E_n|E_1 \cap E_2 \cap \cdots \cap E_{n-1}).$$

2.6 Independent Events

If a coin is flipped twice, knowing that the first flip is heads does not change the probability that the second flip will be heads. This captures the notion of independent events. Namely, events A and B are *independent* if the knowledge that A occurred does not change the probability of B occurring. In mathematical notation, this is expressed as

$$P(B|A) = P(B). \tag{2.6.1}$$

Primarily for reasons of symmetry, the definition of independence of two events is given in terms of their intersection.

Definition 2.6-1

Events A and B are called **independent** if

$$P(A \cap B) = P(A)P(B).$$

If events A and B are not independent, then they are **dependent**.

By the multiplication rule (2.5.3), $P(A \cap B) = P(A)P(B|A)$, provided $P(A) > 0$. Thus, $P(A \cap B) = P(A)P(B)$ holds if and only if relation (2.6.1) holds. A similar argument implies that, if $P(B) > 0$, Definition 2.6-1 is equivalent to

$$P(A|B) = P(A). \tag{2.6.2}$$

Typically, independent events arise in connection with experiments that are performed *independently*, or in connection with *independent* repetitions of the same experiment. By **independent experiments** or **independent repetitions** of the same experiment we mean that there is no mechanism through which the outcome of one experiment will influence the outcome of the other. The independent repetitions of an experiment are typically sub-experiments of an experiment, such as the individual flips of a coin in an experiment consisting of n coin flips, or the selection of each individual unit in simple random sampling of n units from a very large/conceptual population.

Example 2.6-1

(a) A die is rolled twice. Let A = {outcome of first roll is even} and B = {outcome of second roll is either a 1 or a 3}. Are the events A and B independent?

(b) Two electronic components are selected from the production line for thorough inspection. It is known that 90% of the components have no defects. Find the probability that the two inspected components have no defects.

Solution

(a) The two rolls of a die can realistically be assumed to be independent repetitions of the same experiment. Therefore, since event A pertains to the first roll, while event B pertains to the second roll, we may conclude that A and B are independent. Alternatively, by Definition 2.6-1, A and B are independent if the probability of their intersection is the product of their probabilities. Assume that the 36 possible outcomes of the two die rolls are equally likely. Since $A \cap B$ has $3 \times 2 = 6$ outcomes, it follows that $P(A \cap B) = 6/36 = 1/6$. Also, $P(A)P(B) = (1/2)(1/3) = 1/6$. Hence, A and B are independent.

(b) The two selected components are a simple random sample of size two from the conceptual population of such components. Let A = {the first inspected component has no defects} and B = {the second inspected component has no defects}. Since the sub-experiments that select each of the two components are independent, so are the events A and B. Therefore, $P(A \cap B) = P(A)P(B) = 0.9^2 = 0.81$.

When A and B do not arise in connection with independent experiments, their independence can only be verified through Definition 2.6-1.

Example 2.6-2

A card is drawn at random from an ordinary deck of 52 cards. Let A and B denote the events that the card is a five and the card is a spade, respectively. Are the events A and B independent?

Solution

The events A and B are independent if the probability of their intersection is the product of their probabilities. Since $P(A \cap B) = 1/52$, and $P(A)P(B) = (4/52)(13/52) = 1/52$, it follows that A and B are independent.

Whenever independence seems a reasonably realistic assumption, assuming independence can facilitate the computation of probabilities.

Example 2.6-3

A laundromat's aging washing machine and clothes dryer are being replaced. The probability a new washing machine will require warranty service is 0.22. The corresponding probability for a new dryer is 0.15. What is the probability that both machines will require warranty service?

Solution

Let experiments 1 and 2 record whether or not the washing machine and the dryer, respectively, require warranty service. The problem statement does not provide sufficient information to compute the desired probability without the assumption of independence. Assuming that the two experiments are independent, it follows that the events A = {the washing machine requires warranty service} and B = {the dryer requires warranty service} are independent. Hence,

$$P(A \cap B) = P(A)P(B) = (0.22)(0.15) = 0.033.$$

Some basic properties of independent events are given next.

Proposition 2.6-1

1. If A and B are independent, then so are A^c and B.
2. The empty set, $\emptyset$, and the sample space, S, are independent from any other set.
3. Disjoint events are not independent unless the probability of one of them is zero.

To see why part (1) of Proposition 2.6-1 is true, use the fact that B is the union of the disjoint events $B \cap A$ and $B \cap A^c$, and the independence of A and B, to write $P(B) = P(B)P(A) + P(B \cap A^c)$. Now bring $P(B)P(A)$ on the left side of this equation to get $P(B)[1 - P(A)] = P(B \cap A^c)$. Hence, $P(B)P(A^c) = P(B \cap A^c)$, which implies their independence. The first statement of part (2) of the proposition is true because, for any event A, $A \cap \emptyset = \emptyset$. Hence, $0 = P(A \cap \emptyset) = P(\emptyset)P(A) = 0$. The second statement of part (2) of the proposition follows from part (1), since $S = \emptyset^c$. Part (3) of the proposition follows by noting that if A and B are disjoint, then $P(A \cap B) = P(\emptyset) = 0$. Thus, A and B cannot be independent unless $P(A) = 0$ or $P(B) = 0$.

Example 2.6-4

The proportion of female voters who strongly support the exploration of all alternative forms of energy production is the same as the proportion of all voters who strongly support the exploration of all alternative forms of energy production. For a person selected at random from the population of voters, let F and E denote the events that the selected voter is female and the selected voter strongly supports the exploration of all alternative forms of energy production, respectively.

(a) Are the events E and F independent?

(b) Is the proportion of male voters who strongly support the exploration of all alternative forms of energy production the same as the corresponding proportion of female voters?

Solution

(a) Translated into mathematical notation, the first sentence of the problem statement is written as $P(E|F) = P(E)$. According to the discussion following Definition 2.6-1 (see relations (2.6.1) and (2.6.2)), this implies that E and F are independent.

(b) Let M be the event that a randomly selected voter is male. Since $M = F^c$, the independence of E and F shown in part (a) and part (1) of Proposition 2.6-1 imply that M and E are independent. According to relations (2.6.1) and/or (2.6.2), this is equivalent to $P(E|M) = P(E)$. Using the result of part (a), this implies $P(E|M) = P(E|F)$. Translated into words, this relationship is stated as the proportion of male voters who strongly support the exploration of all alternative forms of energy production is the same as the corresponding proportion among all female voters. ■

It would appear that events E_1, E_2, E_3 are independent if E_1 is independent from E_2 and E_3, and E_2 is independent from E_3 or, in mathematical notation, if

$$\left.\begin{aligned} P(E_1 \cap E_2) &= P(E_1)P(E_2), \\ P(E_1 \cap E_3) &= P(E_1)P(E_3), \\ P(E_2 \cap E_3) &= P(E_2)P(E_3). \end{aligned}\right\} \tag{2.6.3}$$

However, this *pairwise independence* does not imply

$$P(E_1 \cap E_2 \cap E_3) = P(E_1)P(E_2)P(E_3); \tag{2.6.4}$$

see Exercise 8. It is also possible that (2.6.4) holds but one of the relations in (2.6.3) does not hold. This is demonstrated next.

Example 2.6-5

Roll a die once and record the outcome. Define the events $E_1 = \{1, 2, 3\}$, $E_2 = \{3, 4, 5\}$, $E_3 = \{1, 2, 3, 4\}$. Verify that

$$P(E_1 \cap E_2 \cap E_3) = P(E_1)P(E_2)P(E_3)$$

and also that E_1 and E_2 are not independent.

Solution

Since $P(E_1 \cap E_2 \cap E_3) = P(\{3\}) = 1/6$ and $P(E_1)P(E_2)P(E_3) = (1/2)(1/2)(4/6) = 1/6$, it follows that $P(E_1 \cap E_2 \cap E_3) = P(E_1)P(E_2)P(E_3)$. Next, $P(E_1 \cap E_2) = P(\{3\}) = 1/6$, which is not equal to $P(E_1)P(E_2) = (1/2)(1/2) = 1/4$. Thus, E_1 and E_2 are not independent. ■

The above discussion leads to the following definition for the independence of three events.

Definition 2.6-2

Independence of three events. The events E_1, E_2, E_3 are **(mutually) independent** if all three relations in (2.6.3) hold and (2.6.4) holds.

The specification *mutually* serves to distinguish the concept of Definition 2.6-1 from that of pairwise independence. In this book we will use *independence* to mean *mutual independence.*

Of course, the concept of independence extends to more than three events. The events $E_1, \ldots, E_n$ are said to be **independent** if for every subset $E_{i_1}, \ldots, E_{i_k}$, $k \leq n$,

$$P(E_{i_1} \cap E_{i_2} \cap \cdots \cap E_{i_k}) = P(E_{i_1})P(E_{i_2}) \cdots P(E_{i_k}).$$

If $E_1, E_2, \ldots, E_n$ are independent, then so are their complements. This is similar to the corresponding property for two events (part (1) of Proposition 2.6-1). Moreover, any one of the n independent events will be independent from events formed from all the others. For instance, in the case of three independent events, E_1, E_2, and E_3, the event E_1 is independent of events such as $E_2 \cup E_3$, $E_2^c \cup E_3$, etc. See Exercise 9.

Example 2.6-6

At 25^oC, 20% of a certain type of laser diodes have efficiency below 0.3 mW/mA. For five diodes, selected by simple random sampling from a large population of such diodes, find the probability of the following events.

(a) All five have efficiency above 0.3 at 25^oC.

(b) Only the second diode selected has efficiency below 0.3 at 25^oC.

(c) Exactly one of the five diodes has efficiency below 0.3 at 25^oC.

(d) Exactly two of the five diodes have efficiency below 0.3 at 25^oC.

Solution

Define the events $A_i = $ {the ith diode has efficiency below 0.3}, $i = 1, \ldots, 5$. Because the five sub-experiments, each consisting of selecting one diode and measuring its efficiency, are independent, so are the events $A_1, \ldots, A_5$. Hence we have:

(a)

$$P(A_1^c \cap \cdots \cap A_5^c) = P(A_1^c) \cdots P(A_5^c) = 0.8^5 = 0.328.$$

(b)

$$P(A_1^c \cap A_2 \cap A_3^c \cap A_4^c \cap A_5^c) = P(A_1^c)P(A_2)P(A_3^c)P(A_4^c)P(A_5^c)$$
$$= (0.2)(0.8^4) = 0.082.$$

(c) This event is the union of the disjoint events E_i = {only the ith diode has efficiency below 0.3 at 25^oC}. A calculation similar to that of part (b) yields that $P(E_i) = 0.082$ for all $i = 1, \ldots, 5$. Thus, the requested probability is

$$P(E_1 \cup \cdots \cup E_5) = P(E_1) + P(E_2) + P(E_3) + P(E_4) + P(E_5)$$
$$= 5 \times 0.082 = 0.41.$$

(d) This event is the union of $\binom{5}{2} = 10$ disjoint events, each of which has the same probability as the event A = {only the first and the second have efficiency below 0.3 at 25^oC}. Thus, the requested probability is

$$10 \times P(A) = 10 \times 0.2^2 \times 0.8^3 = 0.205.$$

2.6.1 APPLICATIONS TO SYSTEM RELIABILITY

In Example 2.4-2 it was mentioned that the reliability of a system, which is defined as the probability that a system will function correctly under stated conditions, depends on the reliability of its components as well as the way the components are arranged. The two basic types of component arrangements are in *series* and in *parallel*. These are depicted in Figure 2-14. A system (or part of a system) whose components are arranged in series works if all its components work. For instance, the four wheels of an automobile represent an arrangement in series since the automobile cannot be driven with a flat tire. A system (or part of a system) whose components are arranged in parallel works if at least one of its components works. For instance, three photocopying machines in an office represent an arrangement in parallel since a photocopying request can be carried out if at least one of the three machines is working. Thus, arranging components in parallel is a way of building redundancy into the system in order to improve its reliability.

The assumption that components fail independently is often used for calculating the reliability of a system from the probability of failure of its components.

Example 2.6-7

The three components of the series system shown in the left panel of Figure 2-14 fail with probabilities $p_1 = 0.1$, $p_2 = 0.15$, and $p_3 = 0.2$, respectively, independently of each other. What is the probability the system will fail?

Solution

Let A denote the event that the system fails. The probability of A is computed most easily by computing first the probability of A^c. Since components fail independently, and thus the events that they do not fail are also independent, we have

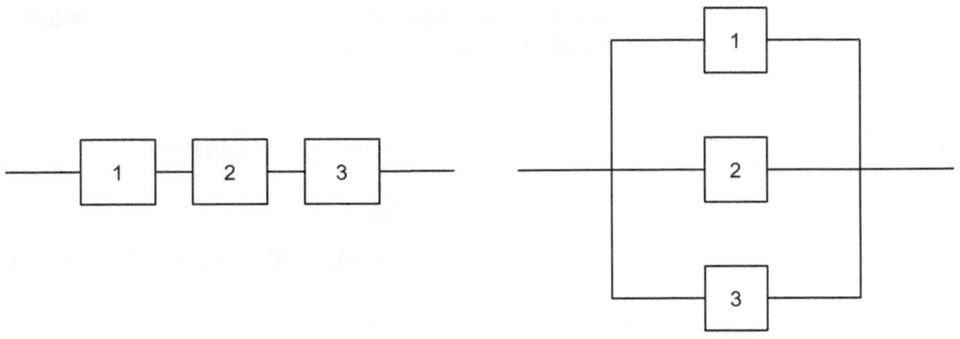

Figure 2-14 Components connected in series (left) and in parallel (right).

$$P(A^c) = P(\text{no component fails}) = (1 - 0.1)(1 - 0.15)(1 - 0.2) = 0.612.$$

Using part (4) of Proposition 2.4-1, it follows that $P(A) = 1 - P(A^c) = 0.388$. ■

Example 2.6-8

The three components of the parallel system shown in the right panel of Figure 2-14 function with probabilities $p_1 = 0.9$, $p_2 = 0.85$, and $p_3 = 0.8$, respectively, independently of each other. What is the probability the system functions?

Solution

Let A denote the event that the system functions and A_i denote the event that component i functions, $i = 1, 2, 3$. Because the components are connected in parallel, $A = A_1 \cup A_2 \cup A_3$. Using part (2) of Proposition 2.4-2, and the independence of the events A_i, we have

$$\begin{aligned} P(A) &= P(A_1) + P(A_2) + P(A_3) - P(A_1 \cap A_2) - P(A_1 \cap A_3) \\ &\quad - P(A_2 \cap A_3) + P(A_1 \cap A_2 \cap A_3) \\ &= 0.9 + 0.85 + 0.8 - 0.9 \times 0.85 - 0.9 \times 0.8 - 0.85 \times 0.8 + 0.9 \times 0.85 \times 0.8 \\ &= 0.997. \end{aligned}$$

An alternative way of computing the probability that the system functions is to compute first the probability that the system does not function. Because it is a parallel system, it does not function only if all three components do not function. Thus, by the independence of the events A_i and hence of their complements,

$$P(A^c) = P(A_1^c)P(A_2^c)P(A_3^c) = 0.1 \times 0.15 \times 0.2 = 0.003,$$

which yields $P(A) = 1 - 0.003 = 0.997$, as before. This alternative method is much more expedient for computing the reliability of parallel systems with more than three components. ■

A series system has no redundancy in the sense that it functions only if all its components function. A parallel system has the maximum possible redundancy since it functions if at least one of its components functions. A **k-out-of-n system** functions if at least k of its n components functions. For example, the engine of a V8 car may be designed so that the car can be driven if at least four of its eight cylinders are firing, in which case it is a 4-out-of-8 system. With this terminology, a series system is an n-out-of-n system, while a parallel system is a 1-out-of-n system.

Example 2.6-9

Find the reliability of a 2-out-of-3 system whose three components function with probabilities $p_1 = 0.9$, $p_2 = 0.85$, and $p_3 = 0.8$, respectively, independently of each other.

Solution

Let A denote the event that the system functions, and A_i denote the event that component i functions, $i = 1, 2, 3$. Because it is a 2-out-of-3 system, it functions if only components 1 and 2 function, or only components 1 and 3 function, or only components 2 and 3 function, or all components function. In mathematical notation, $A = (A_1 \cap A_2) \cup (A_1 \cap A_3) \cup (A_2 \cap A_3) \cup (A_1 \cap A_2 \cap A_3)$. Because these events are disjoint,

$$\begin{aligned} P(A) &= P(A_1 \cap A_2 \cap A_3^c) + P(A_1 \cap A_2^c \cap A_3) + P(A_1^c \cap A_2 \cap A_3) + P(A_1 \cap A_2 \cap A_3) \\ &= 0.9 \times 0.85 \times 0.2 + 0.9 \times 0.15 \times 0.8 + 0.1 \times 0.85 \times 0.8 + 0.9 \times 0.85 \times 0.8 \\ &= 0.941, \end{aligned}$$

where the second equality follows by the independence of the events A_i. ■

Exercises

1. In a batch of 10 laser diodes, two have efficiency below 0.28, six have efficiency between 0.28 and 0.35, and two have efficiency above 0.35. Two diodes are selected at random and without replacement. Are the events E_1 = {the first diode selected has efficiency below 0.28} and E_2 = {the second diode selected has efficiency above 0.35} independent? Justify your answer.

2. In the context of Exercise 2.5-3, are the events $[X = 1]$ and $[Y = 1]$ independent? Justify your answer.

3. A simple random sample of 10 software widgets are chosen for installation. If 10% of this type of software widgets have connectivity problems, find the probability of each of the following events.

(a) None of the 10 have connectivity problems.

(b) The first widget installed has connectivity problems but the rest do not.

(c) Exactly one of the 10 has connectivity problems.

4. An experiment consists of inspecting fuses as they come off a production line until the first defective fuse is found. Assume that each fuse is defective with a probability of 0.01, independently of other fuses. Find the probability that a total of eight fuses are inspected.

5. Quality control engineers monitor the number of nonconformances per car in an automobile production facility. Each day, a simple random sample of four cars from the first assembly line and a simple random sample of three cars from the second assembly line are inspected. The probability that an automobile produced in the first shift has zero nonconformances is 0.8. The corresponding probability for the second shift is 0.9. Find the probability of the events (a) zero nonconformances are found in the cars from the first assembly line in any given day, (b) the corresponding event for the second assembly line, and (c) zero nonconformances are found in any given day. State any assumptions you use.

6. An athlete is selected at random from the population of student athletes in a small private high school, and the athlete's gender and sports preference is recorded. Define the events M = {the student athlete is male}, F = {the student athlete is female}, and T= {the student athlete prefers track}. We are told that the proportion of male athletes who prefer track is the same as the proportion of student athletes who prefer track or, in mathematical notation, $P(T|M) = P(T)$. Can we conclude that the proportion of female athletes who prefer track is the same as the proportion of student athletes who prefer track, or $P(T|F) = P(T)$? Justify your answer.

7. Some information regarding the composition of the student athlete population in the high school mentioned in Exercise 6 is given in the table below. For example, 65% of the student athletes are male, 50% of the student athletes play basketball, and female athletes do not play football. For a student athlete selected at random, the events F = {the student athlete is female} and T= {the student athlete prefers track} are independent.

	Football	Basketball	Track	Total
Male	0.3			0.65
Female	0			
Total	0.3	0.5	0.2	

(a) Fill in the remaining entries of the above table.

(b) If a randomly selected student athlete prefers basketball, what is the probability that the student athlete is female?

(c) Are the events F and B= {the student athlete prefers basketball} independent?

8. Roll a die twice and record the two outcomes. Let E_1 = {the sum of the two outcomes is 7}, E_2={the outcome of the first roll is 3}, E_3={the outcome of the second roll is 4}. Show that E_1, E_2, E_3 are pairwise independent but (2.6.4) does not hold.

9. Show that if E_1, E_2, E_3 are independent, then E_1 is independent from $E_2 \cup E_3$. (*Hint.* By the Distributive Law, $P(E_1 \cap (E_2 \cup E_3)) = P((E_1 \cap E_2) \cup (E_1 \cap E_3))$. Using the formula for the probability of the union of two events (part (1) of Proposition 2.4-2) and the independence of E_1, E_2, E_3, write this as $P(E_1)P(E_2) + P(E_1)P(E_3) - P(E_1)P(E_2 \cap E_3)$ and finish the proof.)

10. The system of components shown in Figure 2-15 below functions as long as components 1 and 2 both function or components 3 and 4 both function. Each of the four components functions with probability 0.9 independently of the others. Find the probability that the system functions.

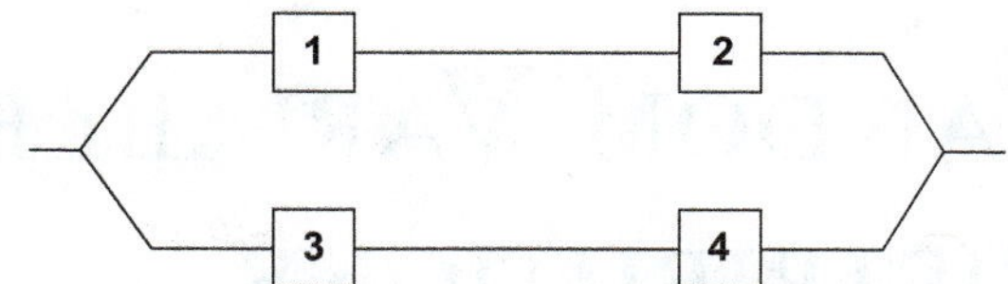

Figure 2-15 System of four components.

11. Find the reliability of a 3-out-of-4 system if each of its four components functions with probability 0.9 independently of the others.

Chapter 3

RANDOM VARIABLES AND THEIR DISTRIBUTIONS

3.1 Introduction

The **probability distribution** of a random variable specifies how the total probability of its sample space, which is 1, is distributed within the range of values of the sample space. We say that we know the probability distribution of a random variable if we know the probability with which its value will fall in any given interval. The probability mass function, or PMF, which was introduced in Section 2.3.3, is a way of describing the probability distribution of a discrete random variable. This chapter introduces the *probability density function*, or *PDF*, as the continuous variable version of the probability mass function, and the *cumulative distribution function*, or *CDF*, which is another way of describing the probability distribution of a (discrete or continuous) random variable. The PMF and PDF are used to extend the notions of expected (or mean) value and variance to more general random variables, such as variables with an infinite sample space, while the CDF is used to define *percentiles* of continuous random variables. Finally, this chapter introduces the most common *probability models* for both discrete and continuous random variables. Similar concepts for bivariate and multivariate random variables will be discussed in the next chapter.

3.2 Describing a Probability Distribution

3.2.1 RANDOM VARIABLES, REVISITED

The concept of a *random variable* was introduced in Section 1.4 as the numerical description of a unit's characteristic(s) when the unit has been selected at random from a population of interest, and was generalized to the outcome of any action or process that generates a random numerical outcome. A more formal definition of a random variable can be given using concepts introduced in Chapter 2.

Definition 3.2-1
A **random variable** is a function (or rule) that associates a number with each outcome of the sample space of a random experiment.

For example, in a sampling experiment where observations $X_1, \ldots, X_n$ are collected from a population, the sample mean, $\overline{X}$, the sample variance, S^2, and a sample proportion $\widehat{p}$ (such as the proportion of observations that are greater than 25) are random variables.

The notions of *discrete* and *continuous* random variables were also introduced in Section 1.4. More formally, we have

Definition 3.2-2
A **discrete random variable** is a random variable whose sample space has a finite or at most a countably infinite number of values.

Example 3.2-1

The following three examples of discrete random variables arise in sample inspection experiments used for product quality control.

(a) Ten laser diodes are randomly selected from the production line and the number of those with efficiency above 3 mW per mA at 25^oC is recorded. The resulting random variable is discrete with finite sample space $\mathcal{S} = \{0, 1, \ldots, 10\}$.

(b) Ten laser diodes are randomly selected from a shipment of 100 and the number of those with efficiency above 3 mW per mA at 25^oC is recorded. Assuming the shipment contains at least 10 laser diodes with efficiency above 3, the resulting random variable is discrete with finite sample space $\mathcal{S} = \{0, 1, \ldots, 10\}$, same as the sample space in part (a).

(c) The efficiency of laser diodes is measured, as they come off the production line, until 10 diodes with efficiency above 3 are found. Let X denote the total number of diodes inspected until the tenth diode with efficiency above 3 is found. Then X is a discrete random variable with infinite sample space $\mathcal{S}_X = \{10, 11, 12, \ldots\}$. ■

The following is an example of a random variable that is not discrete.

Example 3.2-2

In accelerated life testing, products are operated under harsher conditions than those encountered in real life. Consider the experiment where one such product is tested until failure, and let X denote the time to failure. The sample space of this experiment, or of X, is $\mathcal{S}_X = [0, \infty)$. ■

The reason why X of the above example is not discrete is because its sample space is not *countably* infinite (i.e., it cannot be enumerated). As an indication that the numbers in $[0, \infty)$ cannot be enumerated, note that it is impossible to identify which number comes after 0. Even finite intervals, such as $[0, 1]$ contain uncountably infinite many numbers.

Definition 3.2-3
A random variable X is called **continuous** if it can take any value within a finite or infinite interval of the real number line $(-\infty, \infty)$.

Examples of experiments resulting in continuous variables include measurements of length, weight, strength, hardness, life time, pH, or concentration of contaminants in soil or water samples.

REMARK 3.2-1

(a) Although a continuous variable can take any possible value in an interval, its measured value cannot. This is because no measuring device has infinite resolution. Thus, continuous variables do not exist in real life; they are only ideal versions of the discretized variables that are measured. Nevertheless, the study of continuous random variables is meaningful as it provides useful, and quite accurate, approximations to probabilities pertaining to their discretized versions.

(b) If the underlying population of units is finite, there is a finite number of values that a variable can possibly take, regardless of whether it is thought of in its ideal continuous state or in its discretized state. For example, an experiment investigating the relation between height, weight, and cholesterol level of men 55–65 years old records these three continuous variables for a sample of the aforementioned finite population. Even if we think of these variables as continuous (i.e., non-discretized), the number of different values that each of them can take cannot exceed the number of existing 55–65 year old men. Even in such cases the model of a continuous random variable offers both convenience and accurate approximation of probability calculations. ◁

3.2.2 THE CUMULATIVE DISTRIBUTION FUNCTION

A concise description of the probability distribution of a random variable X, either discrete or continuous, can be achieved through its *cumulative distribution function.*

Definition 3.2-4
The **cumulative distribution function**, or **CDF**, of a random variable X gives the probability of events of the form $[X \leq x]$, for all numbers x.

The CDF of a random variable X is typically denoted by a capital letter, most often F in this book. Thus, in mathematical notation, the CDF of X is defnined as

$$F_X(x) = P(X \leq x), \tag{3.2.1}$$

for all numbers x of the real number line $(-\infty, \infty)$. When no confusion is possible, the CDF of X will simply be denoted as $F(x)$, that is, the subscript X will be omitted.

Proposition 3.2-1 The cumulative distribution function, F, of any (i.e., discrete or continuous) random variable X satisfies the following basic properties:

1. It is non-decreasing: If $a \leq b$ then $F(a) \leq F(b)$.
2. $F(-\infty) = 0$, $F(\infty) = 1$.
3. If $a < b$ then $P(a < X \leq b) = F(b) - F(a)$. ■

To show the first property, note that if it is known that the event $[X \leq a]$ occurred then the event $[X \leq b]$ has also occurred. Thus, $[X \leq a] \subseteq [X \leq b]$ and hence, $P(X \leq a) \leq P(X \leq b)$, which is equivalent to $F(a) \leq F(b)$. Property 2 follows by noting that the event $[X \leq -\infty]$ never happens, whereas the event $[X \leq \infty]$ happens always. It follows that $[X \leq -\infty] = \emptyset$ whereas $[X \leq \infty] = \mathcal{S}_X$, and hence $F(-\infty) = P(\emptyset) = 0$ whereas $F(\infty) = P(\mathcal{S}_X) = 1$. Finally, property 3 follows by noting that the event $[X \leq b]$ is the union of the disjoint events $[X \leq a]$ and $[a < X \leq b]$.

Therefore, $P(X \le b) = P(X \le a) + P(a < X \le b)$, or $F(b) = F(a) + P(a < X \le b)$, or $P(a < X \le b) = P(X \le b) - P(X \le a)$.

As stated in the introduction of this chapter, the probability distribution of a random variable X is known if the probability of events of the form $[a < X \le b]$ is known for all $a < b$. Thus, property 3 of Proposition 3.2-1 implies that the CDF describes completely the probability distribution of a random variable.

Example 3.2-3

The PMF of a random variable X is

x	1	2	3	4
$p(x)$	0.4	0.3	0.2	0.1

Find the CDF F of X.

Solution

The given PMF implies that the sample space of X is $\mathcal{S}_X = \{1, 2, 3, 4\}$, that $P(X = x) = p(x)$ for $x = 1, \ldots, 4$, and that $P(X = x) = 0$ for all other x values. The key to finding the CDF is to re-express the "cumulative" probabilities $P(X \le x)$ in terms of the probabilities $P(X = x)$ for x in the sample space. Note first that $P(X \le 1) = P(X = 1)$; this is because X does not take values < 1. Also, $P(X \le 2) = P([X = 1] \cup [X = 2])$, which is because if $X \le 2$ then either $X = 1$ or $X = 2$. For similar reasons, $P(X \le 3) = P([X \le 2] \cup [X = 3])$ and $P(X \le 4) = P([X \le 3] \cup [X = 4])$. Using now the additivity property of probability (i.e., the probability of the union of disjoint events equal the sum of their probabilities) and the PMF of X we can compute $F(x)$ for all x in $\mathcal{S}_X$:

$$F(1) = P(X \le 1) = P(X = 1) = 0.4, \tag{3.2.2}$$

$$F(2) = P(X \le 2) = F(1) + P(X = 2) = 0.4 + 0.3 = 0.7, \tag{3.2.3}$$

$$F(3) = P(X \le 3) = F(2) + P(X = 3) = 0.7 + 0.2 = 0.9, \tag{3.2.4}$$

$$F(4) = P(X \le 4) = F(3) + P(X = 4) = 0.9 + 0.1 = 1. \tag{3.2.5}$$

It remains to determine $F(x)$ for x values that are not in $\mathcal{S}_X$. Again, the key is to re-express the cumulative probabilities $P(X \le x)$ for x not in $\mathcal{S}_X$ in terms of cumulative probabilities $P(X \le x)$ for x in $\mathcal{S}_X$. The end result and brief explanations are:

$$\left.\begin{array}{ll} F(x) = 0 & \text{for all } x < 1 \text{ (because } [X \le x] = \emptyset), \\ F(x) = F(1) = 0.4 & \text{for all } 1 \le x < 2 \text{ (because } [X \le x] = [X \le 1]), \\ F(x) = F(2) = 0.7 & \text{for all } 2 \le x < 3 \text{ (because } [X \le x] = [X \le 2]), \\ F(x) = F(3) = 0.9 & \text{for all } 3 \le x < 4 \text{ (because } [X \le x] = [X \le 3]), \\ F(x) = F(4) = 1 & \text{for all } 4 \le x \text{ (because } [X \le x] = [X \le 4]). \end{array}\right\} \tag{3.2.6}$$

The function $F(x)$ is plotted in Figure 3-1. Functions with plots such as that of Figure 3-1 are called **step** or **jump** functions. ■

The derivation of the CDF from the PMF in Example 3.2-3 also suggests that the PMF can be obtained from the CDF by reversing the process. This reverse process is summarized in the table below.

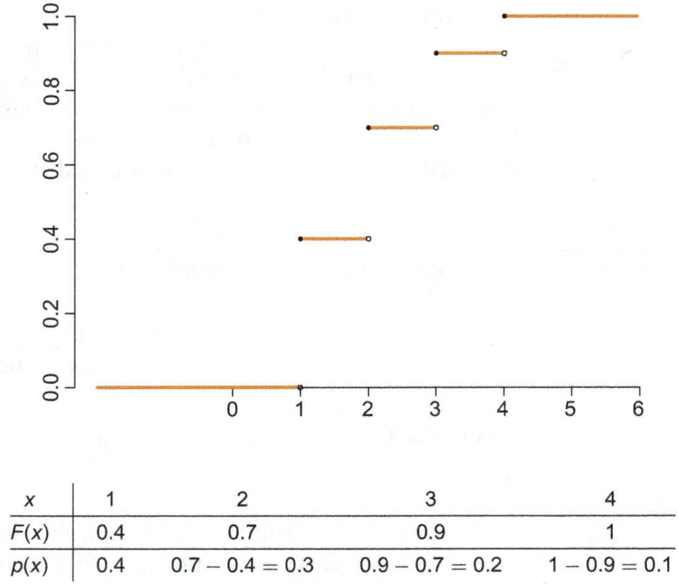

Figure 3-1 The CDF of the random variable of Example 3.2-3.

x	1	2	3	4
$F(x)$	0.4	0.7	0.9	1
$p(x)$	0.4	$0.7 - 0.4 = 0.3$	$0.9 - 0.7 = 0.2$	$1 - 0.9 = 0.1$

Some of the key features of the CDF of Example 3.2-3 are true for the CDF of any random variable X whose sample space $\mathcal{S}_X$ is a subset of the integers (or, more generally, $\mathcal{S}_X = \{x_1, x_2, \ldots\}$ with $x_1 < x_2 < \cdots$). In particular, the CDF F of any such random variable is a step function with jumps occurring only at the values x of $\mathcal{S}_X$, while the flat regions of F correspond to regions where X takes no values. Moreover, the size of the jump at each x of $\mathcal{S}_X$ equals $p(x) = P(X = x)$. Thus, there is a connection between the PMF and the CDF, and one can be obtained from the other. These facts are stated formally in the following proposition.

Proposition 3.2-2 Let $x_1 < x_2 < \cdots$ denote the possible values of the discrete random variable X arranged in an increasing order. Then

1. F is a step function with jumps occurring only at the values x of $\mathcal{S}_X$, while the flat regions of F correspond to regions where X takes no values. The size of the jump at each x of $\mathcal{S}_X$ equals $p(x) = P(X = x)$.
2. The CDF can be obtained from the PMF through the formula

$$F(x) = \sum_{x_i \le x} p(x_i).$$

3. The PMF can be obtained from the CDF as

$$p(x_1) = F(x_1), \quad \text{and } p(x_i) = F(x_i) - F(x_{i-1}) \quad \text{for } i = 2, 3, \ldots.$$

4. The probability of events of the form $[a < X \le b]$ is given in terms of the PMF as

$$P(a < X \le b) = \sum_{a < x_i \le b} p(x_i),$$

and in terms of the CDF as

$$P(a < X \le b) = F(b) - F(a).$$

■

In view of part (2) of Proposition 3.2-2, the CDF property $F(\infty) = 1$ (see property 2 of Proposition 3.2-1) can be restated as

$$\sum_{x_i \in \mathcal{S}_X} p(x_i) = 1, \tag{3.2.7}$$

that is, the values of the PMF sum to 1. Of course, (3.2.7) can be independently justified in terms of Axiom 2.4.2 of probabilities.

3.2.3 THE DENSITY FUNCTION OF A CONTINUOUS DISTRIBUTION

A continuous random variable X cannot have a PMF. The reason for this is

$$P(X = x) = 0, \text{ for any value } x. \tag{3.2.8}$$

This rather counterintuitive fact can be demonstrated in terms of the continuous random variable X that records the outcome of selecting a number at random from the interval $[0, 1]$. The selection is random in the sense that any two subintervals of $[0, 1]$ of equal length, such as $[0, 0.1]$ and $[0.9, 1]$, are equally likely to contain the selected number. This implies that

$$P(X \text{ in an interval of length } l) = l. \tag{3.2.9}$$

For example, $P(0 < X < 0.5) = 0.5$ follows because $P(0 < X < 0.5) = P(0.5 < X < 1)$, since the two intervals are of equal length, and $P(0 < X < 0.5) + P(0.5 < X < 1) = 1$, since they are disjoint and their union is the entire sample space. Relation (3.2.9) implies (3.2.8) because a single number is an interval of zero length.

The random variable used to demonstrate (3.2.8) is the simplest named continuous random variable.

Definition 3.2-5

Uniform in $[0, 1]$ random variable. Select a number from $[0, 1]$ so that any two subintervals of $[0, 1]$ of equal length are equally likely to contain the selected number, and let X denote the selected number. Then we say that X has the **uniform** in $[0, 1]$ distribution and denote this by writing $X \sim U(0, 1)$.

Relation (3.2.9) implies that the probability distribution of the uniform in $[0, 1]$ random variable is known. In fact, if $X \sim U(0, 1)$ its CDF is

$$F_X(x) = P(X \le x) = \begin{cases} 0 & \text{if } x < 0 \\ x & \text{if } 0 \le x \le 1 \\ 1 & \text{if } x > 1. \end{cases} \tag{3.2.10}$$

Note how the plot of this CDF, shown in Figure 3-2, differs from the CDF plot of a discrete random variable.

In addition to the CDF, the probability distribution of a continuous random variable can be described in terms of its *probability density function*.

Definition 3.2-6

The **probability density function**, or **PDF**, of a continuous random variable X is a nonnegative function f_X (thus, $f_X(x) \ge 0$, for all x), with the property that $P(a < X < b)$ equals the area under it and above the interval $[a, b]$. Thus,

$$P(a < X < b) = \text{area under } f_X \text{ between } a \text{ and } b. \tag{3.2.11}$$

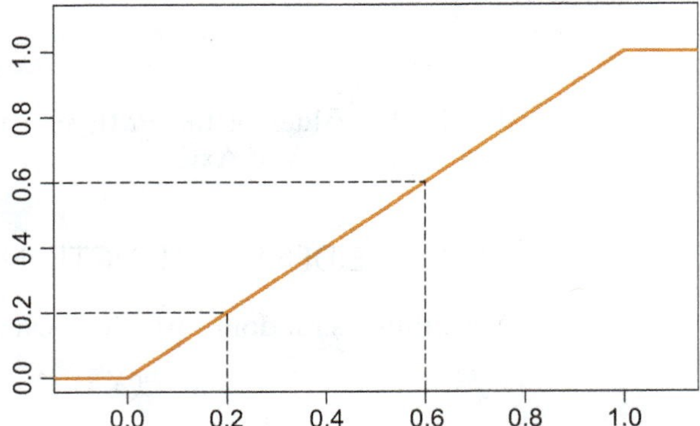

Figure 3-2 CDF of the uniform in [0, 1] random variable.

Some typical shapes of probability density functions are presented in Figure 3-3. A **positively skewed** distribution is also called **skewed to the right**, and a **negatively skewed** distribution is also called **skewed to the left**.

The area under a curve and above an interval is illustrated in Figure 3-4. Since the area under a curve is found by integration, we have

Probability of an Interval in Terms of the PDF

$$P(a < X < b) = \int_a^b f_X(x)\, dx \tag{3.2.12}$$

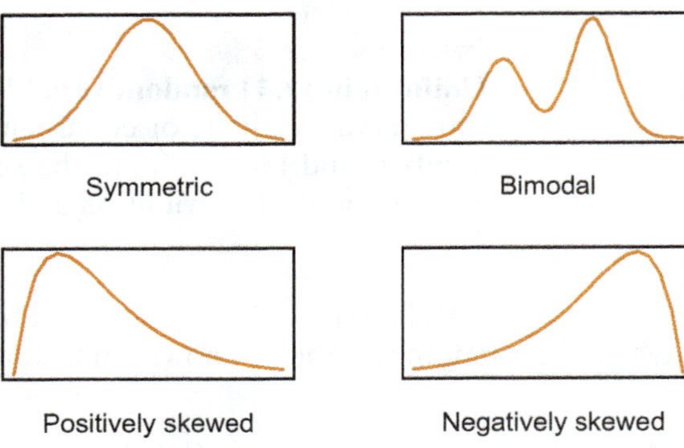

Figure 3-3 Typical shapes of PDFs.

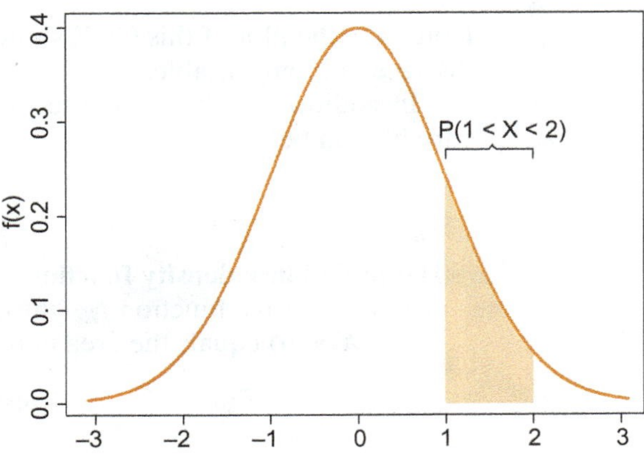

Figure 3-4 $P(a < X < b)$ as the area under the PDF above the interval $[a, b]$.

This is basically why integration is needed in probability theory. For a nonnegative function f to be a probability density function, the total area under the curve it defines must equal 1.

Total Area Under the Curve of a PDF Must Equal 1

$$\int_{-\infty}^{\infty} f(x)\,dx = 1 \tag{3.2.13}$$

Example 3.2-4

If $X \sim U(0,1)$, show that the PDF of X is

$$f_X(x) = \begin{cases} 0 & \text{if } x < 0 \\ 1 & \text{if } 0 \le x \le 1 \\ 0 & \text{if } x > 1. \end{cases}$$

Solution

We need to show that (3.2.11) holds. Since all the area under this function corresponds to the interval $[0,1]$ (see also Figure 3-5), the area above any interval $(a,b]$ equals the area above the intersection of $[a,b]$ with $[0,1]$. Thus, it suffices to show that (3.2.11) holds for intervals $[a,b]$ with $0 \le a < b \le 1$. For such intervals,

$$\int_a^b f_X(x)\,dx = \int_a^b 1\,dx = b - a.$$

By (3.2.9) it is also true that $P(a < X < b) = b - a$. Thus, (3.2.11) holds. ■

Because the area above an interval of length zero is zero for any curve, it follows that (3.2.8) is true for any continuous random variable X. Thus, we have the following result.

Proposition 3.2-3

If X is a continuous random variable,

$$P(a < X < b) = P(a \le X \le b) = F(b) - F(a).$$ ■

For example, if $X \sim U(0,1)$, then

$$P(0.2 < X < 0.6) = P(0.2 \le X \le 0.6) = 0.6 - 0.2 = 0.4$$

and, of course, $P(0.2 < X \le 0.6) = P(0.2 \le X < 0.6) = 0.4$.

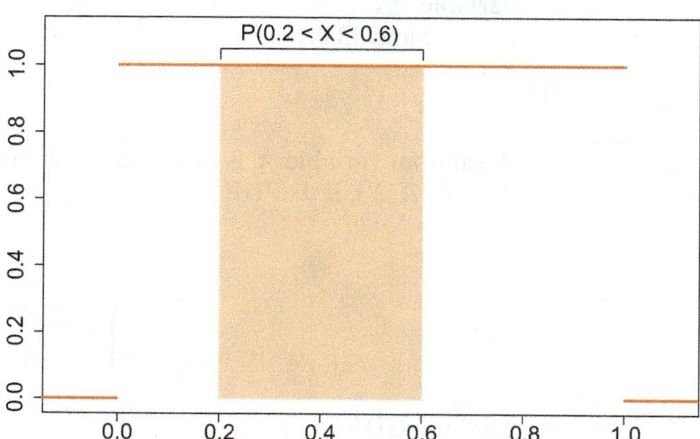

Figure 3-5 PDF of the uniform in $[0,1]$ random variable.

REMARK 3.2-2 Proposition 3.2-3 is true only for the idealized version of a continuous random variable. As we have pointed out, in real life, all continuous variables are measured on a discrete scale. The PMF of the discretized random variable that is actually measured is readily approximated from the formula

$$P(x - \Delta x \le X \le x + \Delta x) \approx 2f_X(x)\Delta x,$$

where Δx denotes a small number. Thus, if Y denotes the discrete measurement of the continuous random variable X, and if Y is measured to three decimal places with the usual rounding, then

$$P(Y = 0.123) = P(0.1225 < X < 0.1235) \approx f_X(0.123)(0.001).$$

Moreover, breaking up the interval $[a, b]$ into n small subintervals $[x_k - \Delta x_k, x_k + \Delta x_k]$, $k = 1, \ldots, n$, we have

$$P(a \le Y \le b) \approx \sum_{k=1}^{n} 2f_X(x_k)\Delta x_k.$$

Since the summation on the right approximates the integral $\int_a^b f_X(x)dx$, the above confirms the approximation

$$P(a \le X \le b) \approx P(a \le Y \le b),$$

namely, that the distribution of the discrete random variable Y is approximated by that of its idealized continuous version. ◁

Proposition 3.2-4 If X is a continuous random variable with PDF f and CDF F, then

(a) The CDF can be obtained from the PDF through the formula

$$F(x) = \int_{-\infty}^{x} f(y)\,dy. \tag{3.2.14}$$

(b) The PDF can be obtained from the CDF through the formula

$$f(x) = F'(x) = \frac{d}{dx}F(x). \tag{3.2.15}$$

■

Part (a) of Proposition 3.2-4 follows from relation (3.2.12) by setting $-\infty$ for a and x for b. Part (b) of the proposition is a consequence of the Fundamental Theorem of Calculus.

The reader easily can verify that the CDF and PDF of a uniform in $[0, 1]$ random variable, given in relation (3.2.10) and in Example 3.2-4, respectively, satisfy relation (3.2.14) and (3.2.15) except for $x = 0$ or 1, where the derivative of the CDF does not exist.

Example 3.2-5 A random variable X is said to have the **uniform in $[A, B]$** distribution, denoted by $X \sim U(A, B)$, if its PDF is

$$f(x) = \begin{cases} 0 & \text{if } x < A \\ \dfrac{1}{B - A} & \text{if } A \le x \le B \\ 0 & \text{if } x > B. \end{cases}$$

Find the CDF $F(x)$.

Solution

Note first that since $f(x) = 0$ for $x < A$, we also have $F(x) = 0$ for $x < A$. This and relation (3.2.14) imply that for $A \leq x \leq B$,

$$F(x) = \int_A^x \frac{1}{B-A} dy = \frac{x-A}{B-A}.$$

Finally, since $f(x) = 0$ for $x > B$, it follows that $F(x) = F(B) = 1$ for $x > B$. ■

Example 3.2-6

If the life time T, measured in hours, of a randomly selected electrical component has PDF $f_T(t) = 0$ for $t < 0$, and $f_T(t) = 0.001 \exp(-0.001t)$ for $t \geq 0$, find the probability the component will last between 900 and 1200 hours of operation.

Solution

Using (3.2.13),

$$\begin{aligned} P(900 < T < 1200) &= \int_{900}^{1200} 0.001 e^{-0.001x} dx \\ &= e^{-0.001(900)} - e^{-0.001(1200)} = e^{-0.9} - e^{-1.2} = 0.1054. \end{aligned}$$

Alternatively, one can first find the CDF and use Proposition 3.2-3. By (3.2.14),

$$F_T(t) = \int_{-\infty}^t f_T(s) ds = \int_0^t 0.001 e^{-0.001s} ds = 1 - e^{-0.001t}, \ t > 0.$$

Thus, by Proposition 3.2-3,

$$\begin{aligned} P(900 < T < 1200) &= F_T(1200) - F_T(900) \\ &= \left[1 - e^{-0.001(1200)}\right] - \left[1 - e^{-0.001(900)}\right] = 0.1054. \end{aligned}$$ ■

It is often more convenient to work with the CDF. In the above example, one can use the CDF to find any probability of the form $P(a < T < b)$ without further integration. An additional advantage of working with the CDF is demonstrated in the following examples.

Example 3.2-7

In the context of Example 3.2-6, let $\widetilde{T}$ be the life time, measured in minutes, of the randomly selected electrical component. Find the PDF of $\widetilde{T}$.

Solution

The easiest way to solve this type of problem is to first find the CDF of $\widetilde{T}$ and then use relation (3.2.15) to find the PDF from the CDF. Noting that $\widetilde{T} = 60T$, where T is the life time measured in hours, we have

$$F_{\widetilde{T}}(x) = P(\widetilde{T} \leq x) = P(60T \leq x) = P\left(T \leq \frac{x}{60}\right) = F_T\left(\frac{x}{60}\right).$$

Hence, since $F'_T(t) = f_T(t) = 0.001 \exp(-0.001t)$, for $t > 0$, it follows that for $x > 0$,

$$f_{\widetilde{T}}(x) = \frac{d}{dx} F_{\widetilde{T}}(x) = \frac{d}{dx} F_T\left(\frac{x}{60}\right) = \frac{1}{60} f_T\left(\frac{x}{60}\right) = \frac{0.001}{60} \exp\left(-\frac{0.001}{60} x\right).$$ ■

Example 3.2-8

Let X denote the amount of time a statistics reference book on a two-hour reserve at the engineering library is checked out by a randomly selected student. Suppose that X has density function

$$f(x) = \begin{cases} \frac{1}{\log(4)} \frac{1}{1+x} & 0 \le x \le 3 \\ 0 & \text{otherwise.} \end{cases}$$

For books returned after two hours, students are charged a fine of \$2.00 plus \$1.00 for each additional 15-minute delay.

(a) Find the probability that a student checking out the book will be charged a fine.

(b) Given that a student has been charged a fine, what is the probability the fine is at least \$3.00?

Solution

(a) The formula for the CDF F of X is

$$F(x) = \int_0^x \frac{1}{\log(4)} \frac{1}{1+t}\,dt = \frac{1}{\log(4)} \int_1^{1+x} \frac{1}{y}\,dy = \frac{\log(1+x)}{\log(4)} \quad \text{for } 0 \le x \le 3,$$

$F(x) = 0$ for $x \le 0$, and $F(x) = 1$ for $x \ge 3$. Hence, the desired probability is $P(X > 2) = 1 - F(2) = 1 - \log(3)/\log(4) = 0.2075$.

(b) The fine is at least \$3.00 if $X > 2.25$. The desired conditional probability is

$$P(X > 2.25|X > 2) = \frac{P([X > 2.25] \cap [X > 2])}{P(X > 2)} = \frac{P(X > 2.25)}{P(X > 2)}$$

$$= \frac{1 - F(2.25)}{1 - F(2)} = \frac{0.1498}{0.2075} = 0.7218.$$

Example 3.2-9

Suppose that a point is selected at random from a circle centered at the origin and having radius 6. Thus, the probability of the point lying in a region A of this circle is proportional to the area of A. Find the PDF of the distance D of this point from the origin.

Solution

The range of values of the random variable D is clearly $[0, 6]$. We will first find the CDF $F_D(d) = P(D \le d)$ of D. Note that since the probability that the point lies in the given circle of radius 6 is 1 and since the area of this circle is $\pi 6^2$, the probability of the point lying in a region A of this circle equals the area of A divided by $\pi 6^2$. Thus, since $D \le d$ happens if and only if the selected point lies in a circle of radius d centered at the origin, we have

$$F_D(d) = \frac{\pi d^2}{\pi 6^2} = \frac{d^2}{6^2}$$

for $0 \le d \le 6$. It follows that the PDF of D is

$$f_D(d) = F'_D(d) = d/18, \quad 0 \le d \le 6,$$

and zero otherwise.

Probability Sampling from a PDF In Section 2.3.3 we saw that it is possible to simulate the experiment of a discrete random variable by probability sampling from its probability mass function. Simulating the experiment of a continuous random variable is also possible. R commands for probability sampling from a probability density function will be given separately for each class of PDFs that will be discussed in Section 3.5. For the class of uniform PDFs, the R command is as follows:

R Command for Simulating the Uniform PDF

```
runif(n, A, B) # returns a random sample of size n drawn from
  the uniform(A, B) distribution
```

For example, the R command *set.seed(111); runif(4, 10, 15)* returns the four numbers 12.96491 13.63241 11.85211 12.57462, which represent a random sample of size 4 from the uniform(10, 15) distribution. (Repeated applications of the *runif(4, 10, 15)* part of the command will give different samples of size 4; using both parts of the command will always result in the same four numbers.) The default values of A and B are 0 and 1, respectively. Thus, *set.seed(200); runif(5)* and *set.seed(200); runif(5, 0, 1)* return the same sample of size 5 drawn from the uniform(0, 1) distribution.

Simulations are used extensively in statistics as they offer insight on properties of samples drawn from different PDFs. In Exercise 11, simulations are used to provide numerical evidence for the fact that a sample's histogram approximates the PDF from which the sample was drawn.

Exercises

1. Answer the following questions.

(a) Check whether or not each of $p_1(x)$, $p_2(x)$ is a legitimate probability mass function.

x	0	1	2	3
$p_1(x)$	0.3	0.3	0.5	−0.1

x	0	1	2	3
$p_2(x)$	0.1	0.4	0.4	0.1

(b) Find the value of the multiplicative constant k so $p(x)$ given in the following table is a legitimate probability mass function.

x	0	1	2	3
$p(x)$	$0.2k$	$0.3k$	$0.4k$	$0.2k$

2. A metal fabricating plant currently has five major pieces under contract each with a deadline for completion. Let X be the number of pieces completed by their deadlines, and suppose its PMF $p(x)$ is given by

x	0	1	2	3	4	5
$p(x)$	0.05	0.10	0.15	0.25	0.35	0.10

(a) Find and plot the CDF of X.

(b) Use the CDF to find the probability that between one and four pieces, inclusive, are completed by their deadline.

3. Let Y denote the cost, in hundreds of dollars, incurred to the metal fabricating plant of Exercise 2 due to missing deadlines. Suppose the CDF of Y is

$$F_Y(y) = \begin{cases} 0 & y < 0 \\ 0.2 & 0 \le y < 1 \\ 0.7 & 1 \le y < 2 \\ 0.9 & 2 \le y < 3 \\ 1 & 3 \le y. \end{cases}$$

(a) Plot the CDF and find the probability that the cost from delays will be at least \$200.00.

(b) Find the probability mass function of Y.

4. A simple random sample of size $n = 3$ is drawn from a batch of ten product items. If three of the 10 items are defective, find the PMF and the CDF of the random variable $X = \{\text{number of defective items in the sample}\}$.

5. Answer the following questions.

(a) Check whether or not each of $f_1(x)$, $f_2(x)$ is a legitimate probability density function

$$f_1(x) = \begin{cases} 0.5(3x - x^3) & 0 < x < 2 \\ 0 & \text{otherwise.} \end{cases}$$

$$f_2(x) = \begin{cases} 0.3(3x - x^2) & 0 < x < 2 \\ 0 & \text{otherwise.} \end{cases}$$

(b) Let X denote the resistance of a randomly chosen resistor, and suppose that its PDF is given by

$$f(x) = \begin{cases} kx & \text{if } 8 \le x \le 10 \\ 0 & \text{otherwise.} \end{cases}$$

(i) Find k and the CDF of X, and use the CDF to calculate $P(8.6 \le X \le 9.8)$.

(ii) Find the conditional probability that $X \le 9.8$ given that $X \ge 8.6$.

6. Let $X \sim U(0,1)$. Show that $Y = 3+6X \sim U(3,9)$, that is, that it has the uniform in [3, 9] distribution defined in Example 3.2-5. (*Hint.* Find the CDF of Y and show it has the form of the CDF found in the solution of Example 3.2-5.)

7. Let $X \sim U(0,1)$, and set $Y = -\log(X)$. Give the sample space of Y, and find the CDF and PDF of Y. (*Hint.* $F_Y(y) = P(Y \le y) = P(X \ge \exp(-y))$.)

8. The cumulative distribution function of checkout duration X, measured in minutes, in a certain supermarket is

$$F(x) = \frac{x^2}{4} \quad \text{for } x \text{ between 0 and 2,}$$

$F(x) = 0$ for $x \le 0$, and $F(x) = 1$ for $x > 2$.

(a) Find the probability that the duration is between 0.5 and 1 minute.

(b) Find the probability density function $f(x)$.

(c) Let Y denote the checkout duration measured in seconds. Find the CDF and PDF of Y.

9. In a game of darts, a player throws the dart and wins $X = 30/D$ dollars, where D is the distance in inches of the dart from the center of the dartboard. Suppose a player throws the dart in such a way that it lands in a randomly selected point on the 18-inch diameter dartboard. Thus, the probability that it lands in any region of the dartboard is proportional to the region's area, and the probability that it lands in the dartboard is 1.

(a) Find the probability that the player will win more than \$10.00.

(b) Find the PDF of X.

10. The time X in hours for a certain plumbing manufacturer to deliver a custom made fixture is a random variable with PDF

$$f(x) = \begin{cases} 0.02e^{-0.02(x-48)} & \text{if } x \ge 48 \\ 0 & \text{otherwise.} \end{cases}$$

An architect overseeing a renovation orders a custom made fixture to replace the old one, which unexpectedly broke. If the ordered fixture arrives within three days no additional cost is incurred, but for every day beyond that an additional cost of \$200.00 is incurred.

(a) Find the probability that no additional cost is incurred.

(b) Find the probability that the additional cost incurred is between \$400 and \$800.

11. Use the R commands *set.seed(111); hist(runif(100), freq=F)* to generate a sample of size 100 from the uniform(0, 1) distribution and to plot its histogram, and the additional R command *curve(dunif, 0, 1, add=T)* to superimpose the uniform(0, 1) PDF on the graph. Does the histogram provide a reasonable approximation to the uniform(0, 1) PDF? Repeat the set of commands using samples of size 1000, 10,000 and 100,000. For what sample size(s) would you say the histogram provides a reasonable approximation to the PDF?

3.3 Parameters of Probability Distributions

This section introduces certain *summary parameters* that are useful for describing prominent features of the distribution of a random variable. The parameters we will consider are the *mean value*, also referred to as the *average value* or *expected value*, the *variance*, and *standard deviation*. These generalize the corresponding quantities defined in Chapter 1. For continuous random variables, we will also consider *percentiles*, such as the *median*, which are commonly used as additional parameters to describe the location, variability, and shape of a continuous distribution.

3.3.1 EXPECTED VALUE

Discrete Random Variables Let X be a discrete random variable with sample space $\mathcal{S}_X$, which can possibly be infinite, and let $p(x) = P(X = x)$ denote its probability mass function. Then, the **expected value**, $E(X)$ or μ_X, of X is defined as

General Definition of Expected Value

$$E(X) = \mu_X = \sum_{x \text{ in } \mathcal{S}_X} xp(x) \tag{3.3.1}$$

The **mean value** of an arbitrary discrete population is the same as the expected value of the random variable it underlies.

This definition generalizes definition (1.6.6) of Section 1.6.2 because it also applies to random variables that are not necessarily obtained through simple random sampling from a finite population. For example, if X denotes the number of heads in 10 flips of a coin, X is not obtained by simple random sampling from the numbers $0, \ldots, 10$, and thus its expected value cannot be computed from (1.6.6). As another example, X can be obtained by simple random sampling from an infinite population; see Example 3.3-2 below. Finally, (3.3.1) applies also to random variables with infinite sample space; see Example 3.3-3 below.

Example 3.3-1

Suppose the population of interest is a batch of $N = 100$ units, 10 of which have some type of defect, received by a distributor. An item is selected at random from the batch and is inspected. Let X take the value 1 if the selected unit has the defect and 0 otherwise. Use formula (3.3.1) to compute the expected value of X, and show that result coincides with the expected value computed according to the definition (1.6.6).

Solution

The sample space of X is $\mathcal{S}_X = \{0, 1\}$ and its PFM is $p(0) = P(X = 0) = 0.9$, $p(1) = P(X = 1) = 0.1$. Thus, according to (3.3.1),

$$\mu_X = 0 \times 0.9 + 1 \times 0.1 = 0.1.$$

Let v_i, $i = 1, 2, \ldots, 100$, where 90 v_i are 0 and 10 are 1, be the statistical population. Then, according to (1.6.6),

$$\mu_X = \frac{1}{100} \sum_{i=1}^{100} v_i = \frac{(90)(0) + (10)(1)}{100} = 0.1.$$

Thus, both definitions give the same mean value for X. ■

The result of Example 3.3-1 is true whenever X is obtained by simple random sampling from any finite population. To see this let $v_1, v_2, \ldots, v_N$ denote the N values in the underlying statistical population, and let $\mathcal{S}_X = \{x_1, \ldots, x_m\}$ be the sample space of X. (Thus, $x_1, \ldots, x_m$ are the distinct values among $v_1, \ldots, v_N$.) Also, let n_j denote the number of times that the distinct value x_j is repeated in the statistical population, so that the PMF of X is given by $p(x_j) = P(X = x_j) = n_j/N$. In this case, the expressions for the expected value of X according to definitions (1.6.6) and (3.3.1) are, respectively,

$$\mu_X = \frac{1}{N} \sum_{i=1}^{N} v_i \quad \text{and} \quad \mu_X = \sum_{j=1}^{m} x_j p(x_j). \tag{3.3.2}$$

That the two expressions in (3.3.2) are equivalent follows by noting that $\sum_{i=1}^{N} v_i = \sum_{j=1}^{m} n_j x_j$.

Example 3.3-2 Select a product item from the production line and let X take the value 1 or 0 as the product item is defective or not. Let p be the proportion of defective items in the conceptual population of this experiment. Find $E(X)$ in terms of p.

Solution

The random variable in this experiment is similar to that of Example 3.3-1 except for the fact that the population of all product items is infinite and conceptual. Thus, definition (1.6.6) cannot be used. The sample space of X is $\mathcal{S}_X = \{0, 1\}$ and its PMF is $p(0) = P(X = 0) = 1 - p$, $p(1) = P(X = 1) = p$. Thus, according to (3.3.1),

$$E(X) = \sum_{x \text{ in } \mathcal{S}_X} xp(x) = 0(1-p) + 1p = p.$$

Thus, for $p = 0.1$ the answer is similar to that of Example 3.3-1. ■

Example 3.3-3 Consider the experiment where product items are being inspected for the presence of a particular defect until the first defective product item is found. Let X denote the total number of items inspected. Suppose a product item is defective with probability p, $p > 0$, independently of other product items. Find $E(X)$ in terms of p.

Solution

The sample space of X is $\mathcal{S}_X = \{1, 2, 3, \ldots\}$. Since items are defective or not independently of each other, the PMF $p(x) = P(X = x)$ is

$$\begin{aligned} p(x) &= P(\text{the first } x-1 \text{ items are not defective and the } x\text{th is defective}) \\ &= (1-p)^{x-1}p. \end{aligned}$$

Note that the geometric series $\sum_{x=1}^{\infty}(1-p)^{x-1} = \sum_{s=0}^{\infty}(1-p)^s$ equals $1/p$, so the PMF sums to one as indeed it should. According to (3.3.1),

$$\begin{aligned} E(X) &= \sum_{x \text{ in } \mathcal{S}_X} xp(x) = \sum_{x=1}^{\infty} x(1-p)^{x-1}p \\ &= \sum_{x=1}^{\infty}(x-1+1)(1-p)^{x-1}p \quad \text{(add and subtract 1)} \\ &= \sum_{x=1}^{\infty}(x-1)(1-p)^{x-1}p + \sum_{x=1}^{\infty}(1-p)^{x-1}p \\ &= \sum_{x=1}^{\infty}(x-1)(1-p)^{x-1}p + 1 \quad \text{(since the PMF sums to 1)} \\ &= \sum_{x=0}^{\infty} x(1-p)^{x}p + 1 \quad \text{(change of summation index)} \end{aligned}$$

Since for $x = 0$ the term is zero, the last infinite series can start from $x = 1$. Moreover, since $(1-p)$ is a common factor to all terms, we obtain

$$E(X) = (1-p)\sum_{x=1}^{\infty} x(1-p)^{x-1}p + 1 = (1-p)E(X) + 1.$$

Solving $E(X) = (1-p)E(X) + 1$ for $E(X)$ yields $E(X) = p^{-1}$. ■

Even for finite populations, definition (3.3.1) is preferable to (1.6.6) for two reasons. First, taking a *weighted* average of the values in the sample space is simpler/easier than averaging the values of the underlying statistical population because the sample space is often much smaller (has fewer values) than the statistical population. Second, definition (3.3.1) affords an abstraction of the random variable in the sense that it disassociates it from its underlying population and refers X to an equivalent experiment involving probability sampling from the sample space population. For example, when referred to their sample space populations, the random variables X in Examples 3.3-1 and 3.3-2 correspond to identical sampling experiments. This abstraction will be very useful in Sections 3.4 and 3.5 where we will introduce models for probability distributions.

Continuous Random Variables The **expected value** or **mean value** of a continuous random variable X with probability density function $f(x)$ is defined by

Definition of Expected Value for Continuous X

$$E(X) = \mu_X = \int_{-\infty}^{\infty} x f(x) dx \tag{3.3.3}$$

provided the integral exists. As in the discrete case, the **mean value** of the population underlying X is used synonymously with the mean or expected value of X. The approximation of integrals by sums, as we saw in Remark 3.2-2, helps connect the definitions of expected value for discrete and continuous random variables.

Example 3.3-4

If the PDF of X is $f(x) = 2x$ for $0 \le x \le 1$ and 0 otherwise, find $E(X)$.

Solution

According to definition (3.3.3),

$$E(X) = \int_{-\infty}^{\infty} x f(x) dx = \int_0^1 2x^2 dx = \frac{2}{3}.$$

Example 3.3-5

Let $X \sim U(0, 1)$, that is, X has the uniform in $[0, 1]$ distribution (see Example 3.2-4). Show that $E(X) = 0.5$.

Solution

Using definition (3.3.3) and the PDF of a uniform in $[0, 1]$ random variable, given in Example 3.2-4, it follows that

$$E(X) = \int_{-\infty}^{\infty} x f(x) dx = \int_0^1 x dx = 0.5.$$

Example 3.3-6

The time T, in days, required for the completion of a contracted project is a random variable with PDF $f_T(t) = 0.1 \exp(-0.1t)$ for $t > 0$ and 0 otherwise. Find the expected value of T.

Solution

Using definition (3.3.3),

$$E(T) = \int_{-\infty}^{\infty} t f_T(t) dt = \int_0^{\infty} t\, 0.1 e^{-0.1t}\, dt$$

$$= -te^{-0.1t}\Big|_0^{\infty} + \int_0^{\infty} e^{-0.1t}\, dt = -\frac{1}{0.1} e^{-0.1t}\Big|_0^{\infty} = 10.$$

R can also be used for numerical integration. The commands for evaluating the integrals

$$\int_0^5 \frac{1}{(x+1)\sqrt{x}}dx, \quad \int_1^\infty \frac{1}{x^2}dx, \quad \text{and} \quad \int_{-\infty}^{\infty} e^{-|x|}dx$$

are given below:

R Commands for Function Definition and Integration

```
f=function(x){1/((x+1)*sqrt(x))}; integrate(f, lower=0,
  upper=5)
f=function(x){1/x**2}; integrate(f, lower=1, upper=Inf)
f=function(x){exp(-abs(x))}; integrate(f, lower=-Inf,
  upper=Inf)
```

In particular the answer to Example 3.3-6 can also be found with the R command

```
g=function(x){x*0.1*exp(-0.1*x)}; integrate(g, lower=0,
  upper=Inf)
```

Mean Value of a Function of a Random Variable If the random variable of interest, Y, is a function of another random variable, X, whose distribution is known, the expected value of Y can be computed using the PMF or PDF of X without finding first the PMF or PDF of Y. The formulas for doing so are given in the next proposition.

Proposition 3.3-1

1. If X is discrete with sample space $\mathcal{S}_X$ and $h(x)$ is a function on $\mathcal{S}_X$, the mean value of $Y = h(X)$ can be computed using the PMF $p_X(x)$ of X as

Mean Value of a Function of a Discrete Random Variable X

$$E(h(X)) = \sum_{x \text{ in } \mathcal{S}_X} h(x)p_X(x).$$

2. If X is continuous and $h(x)$ is a function, the expected value of $Y = h(X)$ can be computed using the PDF $f_X(x)$ of X as

Mean Value of a Function of a Continuous Random Variable X

$$E(h(X)) = \int_{-\infty}^{\infty} h(x)f(x)dx.$$

3. If the function $h(x)$ is linear, that is, $h(x) = ax + b$, so $Y = aX + b$, then

Mean Value of a Linear Function of a General Random Variable X

$$E(h(X)) = aE(X) + b.$$

■

Example 3.3-7

A bookstore purchases three copies of a book at \$6.00 each and sells them for \$12.00 each. Unsold copies are returned for \$2.00 each. Let $X = \{\text{number of copies sold}\}$ and $Y = \{\text{net revenue}\}$. If the PMF of X is

x	0	1	2	3
$p_X(x)$	0.1	0.2	0.2	0.5

find the expected value of Y.

Solution

The net revenue can be expressed as a function of the number of copies sold, that is, as $Y = h(X) = 12X + 2(3 - X) - 18 = 10X - 12$. For instructive purposes, $E(Y)$ will be computed in three ways. First, note that the PMF of Y is

y	−12	−2	8	18
$p_Y(y)$	0.1	0.2	0.2	0.5

Thus, using definition (3.3.1),

$$E(Y) = \sum_{\text{all } y \text{ values}} y p_Y(y) = (-12)(0.1) + (-2)(0.2) + (8)(0.2) + (18)(0.5) = 9.$$

Alternatively, $E(Y)$ can be computed, without first finding the PMF of Y, through the formula in part (1) of Proposition 3.3-1:

$$E(Y) = \sum_{\text{all } x \text{ values}} h(x) p_X(x) = (-12)(0.1) + (-2)(0.2) + (8)(0.2) + (18)(0.5) = 9.$$

Finally, since $Y = 10X - 12$ is a linear function of X, part (3) of Proposition 3.3-1 implies that $E(Y)$ can be computed using only the value of $E(X)$. Since $E(X) = \sum_x x p_X(x) = 2.1$, we have $E(Y) = 10(2.1) - 12 = 9$. ■

Example 3.3-8

Let $Y \sim U(A, B)$, that is, Y has the uniform in $[A, B]$ distribution (see Example 3.2-5). Show that $E(Y) = (B + A)/2$.

Solution

This computation can be done using definition (3.3.3) and the PDF of a uniform in $[A, B]$ random variable, which is given in Example 3.2-5 (the interested reader is encouraged to do this computation). Alternatively, $E(Y)$ can be found through the formula in part (3) of Proposition 3.3-1 using the fact that if $X \sim U(0, 1)$, then

$$Y = A + (B - A)X \sim U(A, B). \tag{3.3.4}$$

Relation (3.3.4) can be verified by finding the CDF of Y and showing it has the form of the CDF found in the solution of Example 3.2-5 (see also Exercise 6 in Section 3.2). Thus,

$$E(Y) = A + (B - A)E(X) = A + \frac{B - A}{2} = \frac{B + A}{2}.$$ ■

Example 3.3-9

The time T, in days, required for the completion of a contracted project is a random variable with PDF $f_T(t) = 0.1 \exp(-0.1t)$ for $t > 0$ and 0 otherwise. Suppose the contracted project must be completed in 15 days. If $T < 15$ there is a cost of \$5$(15 - T)$ and if $T > 15$ there is a cost of \$10$(T - 15)$. Find the expected value of the cost.

Solution

Define the functions $h(t) = 5(15 - t)$ if $t < 15$, and $h(t) = 10(t - 15)$ if $t > 15$, and let $Y = h(T)$ denote the cost. According to part (3) of Proposition 3.3-1,

$$E(Y) = \int_{-\infty}^{\infty} h(t) f_T(t) dx = \int_0^{15} 5(15 - t) 0.1 e^{-0.1t} dt + \int_{15}^{\infty} 10(t - 15) 0.1 e^{-0.1t} dt$$
$$= 36.1565 + 22.313 = 58.4695.$$

This answer can also be found by summing the outputs of the following two R commands:

```
g=function(x){5*(15-x)*0.1*exp(-0.1*x)}; integrate(g, lower=0,
  upper=15)
g=function(x){10*(x-15)*0.1*exp(-0.1*x)}; integrate(g, lower=15,
  upper=Inf)
```

3.3.2 VARIANCE AND STANDARD DEVIATION

The **variance** σ_X^2, or $\text{Var}(X)$, of a random variable X is defined as

General Definition of Variance of a Random Variable *X*

$$\sigma_X^2 = E\left[(X - \mu_X)^2\right] \tag{3.3.5}$$

where $\mu_X = E(X)$ is the expected value of X. The variance of an arbitrary discrete population is the same as the variance of the random variable it underlies. A computationally simpler formula (also called the *short-cut formula*) for σ_X^2 is

Short-cut Formula for Variance of a Random Variable *X*

$$\sigma_X^2 = E(X^2) - [E(X)]^2 \tag{3.3.6}$$

In terms of the PMF $p(x)$ of X, if X is discrete with sample space $\mathcal{S}_X$ or the PDF $f(x)$ of X, if X is continuous, (3.3.5) can be written, respectively, as

$$\sigma_X^2 = \sum_{x \text{ in } \mathcal{S}_X} (x - \mu_X)^2 p(x) \quad \text{and} \quad \sigma_X^2 = \int_{-\infty}^{\infty} (x - \mu_X)^2 f_X(x)\, dx. \tag{3.3.7}$$

These alternative expressions for σ_X^2 follow from parts (1) and (2) of Proposition 3.3-1, respectively, with $h(x) = (x - \mu_X)^2$. Similarly, expressing $E(X^2)$ according to parts (1) and (2) of Proposition 3.3-1 with $h(x) = x^2$ yields the following alternative expression for the short-cut formula (3.3.6) for discrete and continuous random variables, respectively:

$$\sigma_X^2 = \sum_{x \text{ in } \mathcal{S}_X} x^2 p(x) - \mu_X^2 \quad \text{and} \quad \sigma_X^2 = \int_{-\infty}^{\infty} x^2 f_X(x) dx - \mu_X^2. \tag{3.3.8}$$

The **standard deviation** of X is defined to be the positive square root, σ, of σ^2:

Definition of Standard Deviation

$$\sigma_X = \sqrt{\sigma_X^2} \tag{3.3.9}$$

Example 3.3-10

Select a product from the production line and let X take the value 1 or 0 as the product is defective or not. If p is the probability that the selected item is defective, find $\mathrm{Var}(X)$ in terms of p.

Solution

In Example 3.3-2 we saw that $E(X) = p$, where p denotes the proportion of defective items. Next, because X take only the values 0 or 1 it follows that $X^2 = X$. Hence, $E(X^2) = E(X) = p$. Using the short-cut formula (3.3.6) we obtain

$$\sigma_X^2 = E(X^2) - [E(X)]^2 = p - p^2 = p(1-p).$$

■

Example 3.3-11

Roll a die and let X denote the outcome. Find $\mathrm{Var}(X)$.

Solution

The expected value of X is $\mu_X = (1 + \cdots + 6)/6 = 3.5$. Using the short-cut formula for the variance we have

$$\sigma_X^2 = E(X^2) - \mu_X^2 = \sum_{j=1}^{6} x_j^2 p_j - \mu_X^2 = \frac{91}{6} - 3.5^2 = 2.917.$$

■

Example 3.3-12

Consider the experiment where product items are being inspected for the presence of a particular defect until the first defective product item is found. Let X denote the total number of items inspected. Suppose a product item is defective with probability p, $p > 0$, independently of other product items. Find σ_X^2 in terms of p.

Solution

The PMF and the mean value of X were found in Example 3.3-3 to be $p(k) = P(X = k) = (1-p)^{k-1}p$, for $k = 1, 2, \ldots$, and $\mu_X = 1/p$. Next, setting $q = 1 - p$,

$$\begin{aligned}
E(X^2) &= \sum_{k=1}^{\infty} k^2 q^{k-1} p = \sum_{k=1}^{\infty} (k-1+1)^2 q^{k-1} p \quad \text{(add and subtract 1)} \\
&= \sum_{k=1}^{\infty} (k-1)^2 q^{k-1} p + \sum_{k=1}^{\infty} 2(k-1) q^{k-1} p + \sum_{k=1}^{\infty} q^{k-1} p \quad \text{(expand the square)} \\
&= \sum_{k=1}^{\infty} k^2 q^k p + 2\sum_{k=1}^{\infty} k q^k p + 1 \quad \text{(change summation index; PDF sums to 1)} \\
&= qE(X^2) + 2qE(X) + 1.
\end{aligned}$$

Using $E(X) = 1/p$ and solving the equation $E(X^2) = qE(X^2) + 2qE(X) + 1$ for $E(X^2)$ yields $E(X^2) = (q+1)/p^2 = (2-p)/p^2$. Hence, by (3.3.6)

$$\sigma_X^2 = E(X^2) - [E(X)]^2 = \frac{2-p}{p^2} - \frac{1}{p^2} = \frac{1-p}{p^2}.$$

■

Example 3.3-13

Let $X \sim U(0,1)$, that is, X has the uniform in $[0,1]$ distribution (see Example 3.2-4). Show $\mathrm{Var}(X) = 1/12$.

Solution

We have $E(X) = \int_0^1 x\,dx = 0.5$, as was also found in Example 3.3-5. Moreover,

$$E(X^2) = \int_0^1 x^2\,dx = 1/3,$$

so that, by the short-cut formula (3.3.6), $\sigma_X^2 = 1/3 - 0.5^2 = 1/12$. ■

Example 3.3-14

Let X have PDF $f_X(x) = 0.1\exp(-0.1x)$ for $x > 0$ and 0 otherwise. Find the variance and standard deviation of X.

Solution

From Example 3.3-6 we have $E(X) = \int_0^\infty 0.1\,x\,e^{-0.1x}\,dx = 1/0.1$. Next,

$$E(X^2) = \int_{-\infty}^{\infty} x^2 f_X(x)\,dx = \int_0^\infty x^2 0.1e^{-0.1x}\,dx$$

$$= -x^2e^{-0.1x}\Big|_0^\infty + \int_0^\infty 2xe^{-0.1x}\,dx = \frac{2}{0.1^2},$$

since the last integral equals $(2/0.1)E(X)$. Thus, by (3.3.6), we have

$$\sigma_X^2 = E(X^2) - [E(X)]^2 = \frac{2}{0.1^2} - \frac{1}{0.1^2} = 100, \quad \text{and} \quad \sigma_X = 10.$$

Note that the standard deviation of this random variable equals its mean value. ■

Variance and Standard Deviation of a Linear Transformation

Proposition 3.3-2

Variance and Standard Deviation of a Linear Transformation

If the variance of X is σ_X^2 and $Y = a + bX$, then

$$\sigma_Y^2 = b^2\sigma_X^2, \quad \sigma_Y = |b|\sigma_X$$

■

Example 3.3-15

A bookstore purchases three copies of a book at \$6.00 each and sells each at \$12.00 each. Unsold copies are returned for \$2.00. The PMF of X = {number of copies sold} is given in Example 3.3-7. Find the variance of X and of the net revenue $Y = 10X - 12$.

Solution

The mean value of X was found in Example 3.3-7 to be $E(X) = 2.1$. Next,

$$E(X^2) = 0^2 \times 0.1 + 1^2 \times 0.2 + 2^2 \times 0.2 + 3^2 \times 0.5 = 5.5.$$

Thus, $\sigma_X^2 = 5.5 - 2.1^2 = 1.09$. Using Proposition 3.3-2, the variance of Y is

$$\sigma_Y^2 = 10^2\sigma_X^2 = 109.$$

■

Example 3.3-16 Let $Y \sim U(A, B)$, that is, Y has the uniform in $[A, B]$ distribution (see Example 3.2-5). Show that $\text{Var}(Y) = (B - A)^2/12$.

Solution

In Example 3.3-13 we found that if $X \sim U(0, 1)$, then $\text{Var}(X) = 1/12$. Using the additional fact that $Y = A + (B - A)X \sim U(A, B)$, as was done in Example 3.3-8, Proposition 3.3-2 yields

$$\text{Var}(Y) = (B - A)^2\text{Var}(X) = \frac{(B - A)^2}{12}.$$

3.3.3 POPULATION PERCENTILES

The precise definition of a *population percentile* or *percentile of a random variable* involves the cumulative distribution function and will be given only for continuous random variables. While the definition of population percentiles appears quite different from that of sample percentiles given in Section 1.7, it should be kept in mind that sample percentiles estimate corresponding population percentiles.

Definition 3.3-1

Let X be a continuous random variable with CDF F and α a number between 0 and 1. The **100(1 − α)-th percentile** (or **quantile**) of X is the number, denoted by x_α, with the property

$$F(x_\alpha) = P(X \leq x_\alpha) = 1 - \alpha.$$

In particular:

1. The 50th percentile, which corresponds to $\alpha = 0.5$ and is denoted by $x_{0.5}$, is called the **median** and is also denoted by $\tilde{\mu}_X$. The defining property of $\tilde{\mu}_X$ is

$$F(\tilde{\mu}_X) = 0.5. \tag{3.3.10}$$

2. The 25th percentile, which corresponds to $\alpha = 0.75$ and is denoted by $x_{0.75}$, is called the **lower quartile** and is also denoted by Q_1. The defining property of Q_1 is

$$F(Q_1) = 0.25. \tag{3.3.11}$$

3. The 75th percentile, which corresponds to $\alpha = 0.25$ and is denoted by $x_{0.25}$, is called the **upper quartile** and is also denoted by Q_3. The defining property of Q_3 is

$$F(Q_3) = 0.75. \tag{3.3.12}$$

The defining property of each percentile also serves as the equation whose solution determines the value of the percentile. For example, the defining property of the median means that $\tilde{\mu}$ is the point where the graph of F crosses the horizontal line at 0.5. This is illustrated in the left panel of Figure 3-6 using the CDF $F(x) = 1 - e^{-x}$ for $x \geq 0$, and $F(x) = 0$ for $x < 0$. For the cumulative distribution functions we will consider there is only one point of contact between the horizontal line at 0.5 and the CDF, and thus $\tilde{\mu}_X$ is the unique solution of $F(\tilde{\mu}_X) = 0.5$. The median could also be

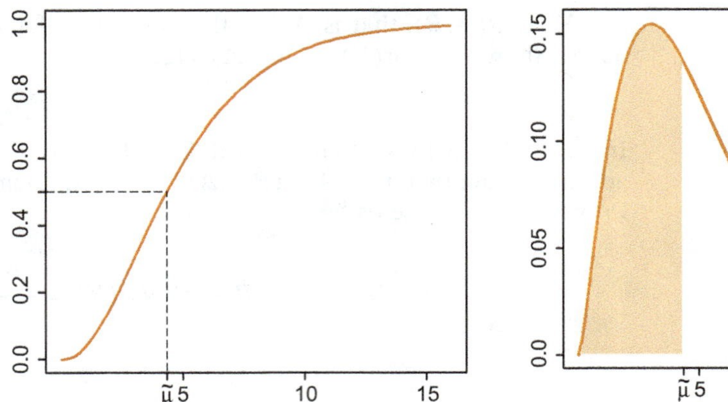

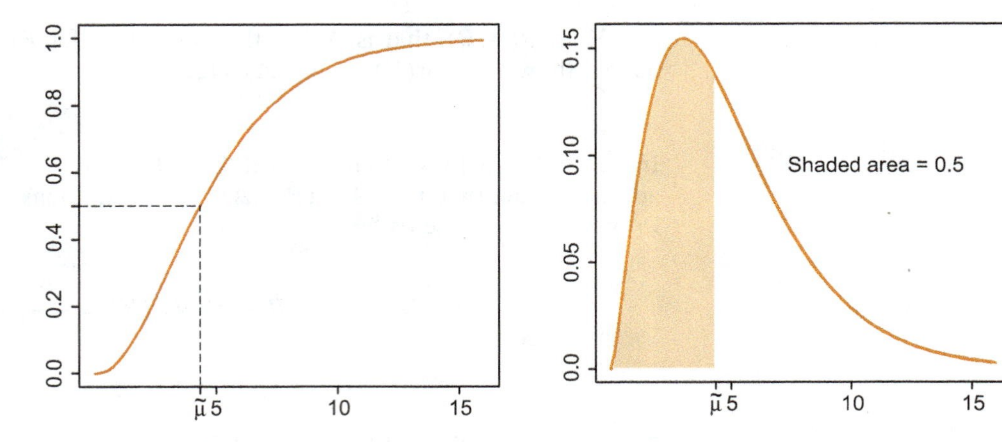

Figure 3-6 Left: The horizontal line $y = 0.5$ crosses $F(x)$ at $x = \tilde{\mu}$. Right: The area under the PDF is split into two equal parts at $\tilde{\mu}$.

defined as the point $\tilde{\mu}$ which splits the area under the PDF of X into two equal parts. This is illustrated in the right panel of Figure 3-6 for the PDF $f(x) = e^{-x}$ for $x \geq 0$, and $f(x) = 0$ for $x < 0$, which corresponds to the CDF used in the left panel. Similar comments apply for the other percentiles. For example, the 95th percentile is found as the unique solution of its defining equation, and it has the property that it splits the area under the PDF into two parts, with the left part having area 0.95 and the right part having area 0.05.

Example 3.3-17

Suppose X has PDF $f(x) = e^{-x}$ for $x \geq 0$, and $f(x) = 0$ for $x < 0$. Find the median and the 95th percentile of X.

Solution

The median of X is the unique solution of the equation

$$F(\tilde{\mu}) = 0.5,$$

where $F(x) = 0$ for $x < 0$, and $F(x) = \int_0^x e^{-s} ds = 1 - e^{-x}$ for $x > 0$. Thus, the above equation becomes

$$1 - e^{-\tilde{\mu}} = 0.5,$$

or $e^{-\tilde{\mu}} = 0.5$, or $-\tilde{\mu} = \log(0.5)$, or

$$\tilde{\mu} = -\log(0.5) = 0.693.$$

Similarly, the 95th percentile is found by solving $F(x_{0.05}) = 0.95$, or $1 - e^{-x_{0.05}} = 0.95$, or $e^{-x_{0.05}} = 0.05$, or

$$x_{0.05} = -\log(0.05) = 2.996.$$

Example 3.3-18

If $X \sim U(A, B)$, find the median and the 90th percentile of X.

Solution

The CDF of a uniform in $[A, B]$ distribution was found in Example 3.2-5 to be $F(x) = 0$ for $x < A$,

$$F(x) = \frac{x - A}{B - A} \quad \text{for } A \leq x \leq B,$$

and $F(x) = F(B) = 1$ for $x > B$. Thus, the median is the unique solution to the equation

$$\frac{\tilde{\mu}_X - A}{B - A} = 0.5,$$

which yields $\tilde{\mu}_X = A + 0.5(B - A)$. Similarly, the 90th percentile is the unique solution to the equation $(x_{0.1} - A)/(B - A) = 0.9$, which yields $x_{0.1} = A + 0.9(B - A)$. ■

Rewriting the median of the uniform in $[A, B]$ random variable X as $\tilde{\mu}_X = (A + B)/2$ reveals that, for this random variable, the median equals the mean; see Example 3.3-8. In general, it is true that for random variables having a symmetric distribution the median equals the mean. For random variables having a positively skewed distribution the mean is greater than the median and for random variables having a negatively skewed distribution, the mean is smaller than the median.

Like the mean, percentiles are measures of *location* in the sense of identifying points of interest of a continuous distribution. In addition, percentiles help define measures of spread (or variability), which serve as alternatives to the standard deviation. The most common such measure of spread is the (population) interquartile range, whose definition is the direct analogue of the sample interquartile range defined in Section 1.7.

Definition 3.3-2

The **interquartile range**, abbreviated by **IQR**, is the distance between the 25th and 75th percentile:

$$IQR = Q_3 - Q_1.$$

Example 3.3-19

Let X have PDF $f(x) = 0.001e^{-0.001x}$ for $x \geq 0$, and $f(x) = 0$ for $x < 0$. Find a general expression for the $100(1-\alpha)$-th percentile of X in terms of α, for α between 0 and 1, and use it to find the interquartile range of X.

Solution

The CDF of X is $F(x) = 1 - e^{-0.001x}$ for $x > 0$ and zero otherwise; see Example 3.2-6. Hence, according to Definition 3.3-1, the $100(1-\alpha)$-th percentile is the unique solution to $1 - e^{-0.001x_\alpha} = 1 - \alpha$, which is

$$x_\alpha = -\frac{\log(\alpha)}{0.001}.$$

Using this formula, $Q_3 = x_{0.25} = -\log(0.25)/0.001 = 1386.29$, and $Q_1 = x_{0.75} = -\log(0.75)/0.001 = 287.68$. Thus, the IQR $= 1098.61$. ■

Exercises

1. A simple random sample of three items is selected from a shipment of 20 items of which four are defective. Let X be the number of defective items in the sample.

(a) Find the PMF of X.

(b) Find the mean value and variance of X.

2. Let X have PMF

x	1	2	3	4
p(x)	0.4	0.3	0.1	0.2

(a) Calculate $E(X)$ and $E(1/X)$.

(b) In a win-win game, the player will win a monetary prize, but has to decide between the fixed price of \$1000/$E(X)$ and the random price of \$1000/X, where the random variable X has the PMF given above. Which choice would you recommend the player make?

3. A customer entering an electronics store will buy a flat screen TV with probability 0.3. Sixty percent of the customers buying a flat screen TV will spend \$750.00 and 40% will spend \$400.00. Let X denote the amount spent on flat screen TVs by two random customers entering the store.

(a) Find the PMF of X. (*Hint.* $S_X = \{0, 400, 750, 800, 1150, 1500\}$ and $P(X = 400) = 2 \times 0.7 \times 0.3 \times 0.4$.)

(b) Find the mean value and variance of X.

4. A metal fabricating plant currently has five major pieces under contract, each with a deadline for completion. Let X be the number of pieces completed by their deadlines. Suppose that X is a random variable with PMF $p(x)$ given by

x	0	1	2	3	4	5
p(x)	0.05	0.10	0.15	0.25	0.35	0.10

(a) Compute the expected value and variance of X.

(b) For each piece completed by the deadline, the plant receives a bonus of \$15,000. Find the expected value and variance of the total bonus amount.

5. The life time X, in months, of certain equipment is believed to have PDF

$$f(x) = (1/100)xe^{-x/10}, \quad x > 0 \quad \text{and} \quad f(x) = 0, \quad x \le 0.$$

Using R commands for the needed integrations, find $E(X)$ and σ_X^2.

6. Consider the context of Example 3.3-9 where there is a cost associated with either early (i.e., before 15 days) or late (i.e., after 15 days) completion of the project. In an effort to reduce the cost, the company plans to start working on the project five days after the project is commissioned. Thus, the cost due to early or late completion of the project is given by $\widetilde{Y} = h(\widetilde{T})$, where $\widetilde{T} = T + 5$, and the function h is $h(\tilde{t}) = 5(15 - \tilde{t})$ if $\tilde{t} < 15$, and $h(\tilde{t}) = 10(\tilde{t} - 15)$ if $\tilde{t} > 15$. The PDF of T is $f_T(t) = 0.1\exp(-0.1t)$ for $t > 0$, and 0 otherwise.

(a) Find the PDF of $f_{\widetilde{T}}(\tilde{t})$ of $\widetilde{T}$. (*Hint.* First find the CDF $F_{\widetilde{T}}(\tilde{t})$ of $\widetilde{T}$.)

(b) Use R commands similar to those given in Example 3.3-9 to find the expected cost, $E(\widetilde{T})$. Does the company's plan to delay the work on the project reduce the expected cost?

7. The CDF function of the checkout duration, X, in a certain supermarket, measured in minutes, is $F(x) = 0$ for $x \le 0$, $F(x) = 1$ for $x > 2$, and

$$F(x) = \frac{x^2}{4} \quad \text{for } x \text{ between 0 and 2}.$$

(a) Find the median and the interquartile range of the checkout duration.

(b) Find $E(X)$ and σ_X. You may use R commands for the needed integrations.

8. The length of time X, in hours, that a statistics reference book on a two-hour reserve at the engineering library is checked out by a randomly selected student has PDF

$$f(x) = \begin{cases} \dfrac{1}{\log(4)} \dfrac{1}{1+x} & 0 \le x \le 3 \\ 0 & \text{otherwise.} \end{cases}$$

For books returned after two hours, students are charged a fine of \$2.00 plus 6 cents times the number of minutes past the two hours.

(a) Let $Y = 60X$ be the amount of time, in minutes, the book is checked out. Find the PDF of Y. (*Hint.* First find the CDF of Y using the CDF of X found in Example 3.2-8.)

(b) Let V be the fine amount, in cents, that a random student checking out the book will pay. Find $E(V)$ and σ_V^2. You may use R commands for the needed integrations. (*Hint.* $V = h(Y)$, where $h(y) = 0$ for $0 \le y \le 120$, and $h(y) = 200 + 6(y - 120), y > 120$.)

(c) Give the mean value and variance of the fine amount expressed in dollars.

9. Plumbing suppliers typically ship packages of plumbing supplies containing many different combinations of items such as pipes, sealants, and drains. Almost invariably there are one or more parts in the shipment that are not correct: the part may be defective, missing, not the one that was ordered, etc. In this question the random variable of interest is the proportion P of parts in a shipment, selected at random, that are not correct. A family of distributions for modeling a random variable P, where P is a proportion, has the probability density function

$$f_P(p) = \theta p^{\theta - 1}, \quad 0 < p < 1, \quad \theta > 0.$$

(a) Find $E(P)$ and σ_P^2 in terms of the parameter θ.

(b) Find the CDF of P in terms of the parameter θ.

(c) Find the interquartile range of P in terms of the parameter θ.

3.4 Models for Discrete Random Variables

Considering each random variable as being obtained by probability sampling from its own sample space leads to a classification of sampling experiments, and the corresponding variables, into classes. Random variables within each class share a common probability mass function up to unknown *parameters*. These classes of probability distributions are also called **probability models.** In this section we describe the four main types of probability models for discrete random variables and the practical contexts to which they apply.

3.4.1 THE BERNOULLI AND BINOMIAL DISTRIBUTIONS

The Bernoulli Distribution A **Bernoulli trial** or **experiment** is one whose outcome can be classified as either a *success* or a *failure*. The **Bernoulli random variable** X takes the value 1 if the outcome is a success, and the value 0 if it is a failure.

Example 3.4-1

Examples of Bernoulli random variables.

1. The prototypical Bernoulli experiment is a flip of a coin, with heads and tails being success and failure, respectively.
2. In an experiment where a product is selected from the production line, the Bernoulli random variable X takes the value 1 or 0 as the product is defective (success) or not (failure).
3. In an experiment where a product undergoes accelerated life testing (see Example 3.2-2), the Bernoulli random variable X can take the value 1 if the product lasts more than 1500 hours in operation (success) and 0 otherwise.
4. In an experiment where two fuses are examined for the presence of a defect, the Bernoulli random variable X can take the value 1 if none of the two fuses have the defect and 0 otherwise. ■

If the probability of success is p and that of failure is $1 - p$, the PMF and CDF of X are

x	0	1
$p(x)$	$1-p$	p
$F(x)$	$1-p$	1

The expected value and variance of a Bernoulli random variable X have already been derived in Examples 3.3-2 and 3.3-10, before the random variable in these examples was identified as Bernoulli. The results from these examples are summarized below for convenience.

$$\mu_X = p, \quad \sigma_X^2 = p(1-p). \tag{3.4.1}$$

Example 3.4-2

The probability that an electronic product will last more than 5500 time units is 0.1. Let X take the value 1 if a randomly selected product lasts more than 5500 time units and the value 0 otherwise. Find the mean value and variance of X.

Solution

Here X is Bernoulli with probability of success $p = 0.1$. Thus, according to (3.4.1), $\mu_X = 0.1$ and $\sigma_X^2 = 0.1 \times 0.9 = 0.09$. ■

The Binomial Distribution Suppose n Bernoulli experiments, each having probability of success equal to p, are performed independently. Taken together, the n independent Bernoulli experiments constitute a **binomial experiment**. The **binomial random variable** Y is the total number of successes in the n Bernoulli trials.

The prototypical binomial experiment consists of n flips of a coin, with the binomial random variable Y being the total number of heads, but independent repetitions of the other Bernoulli trials mentioned in Example 3.4-1 also lead to binomial experiments and corresponding binomial random variables. For example, if n products are randomly selected from the production line, the inspection of each of them for the presence of a defect constitutes a Bernoulli trial and, assuming that products are defective or not independently of each other, the total number of defective products among the n examined is a binomial random variable.

If X_i denotes the Bernoulli random variable associated with the ith Bernoulli trial, that is,

$$X_i = \begin{cases} 1 & \text{if } i\text{th experiment results in success} \\ 0 & \text{otherwise} \end{cases} \qquad \text{for } i = 1, \ldots, n,$$

then the binomial random variable Y equals

$$Y = \sum_{i=1}^{n} X_i \,. \tag{3.4.2}$$

The binomial random variable Y takes the value 0 if all n Bernoulli trials result in failure, and takes the value n if all Bernoulli trials result in success. The sample space of Y is $\mathcal{S}_Y = \{0, 1, \ldots, n\}$. The probability distribution of a binomial random variable is controlled by two parameters, the number of trials n and the common probability of success in each of the n Bernoulli trials. If $Y \sim \text{Bin}(n,p)$, which means that Y is a binomial random variable with parameters n and p, its PMF $p(y) = P(Y = y)$ is given by the formula

PMF of the Binomial Distribution

$$P(Y = y) = \binom{n}{y} p^y (1-p)^{n-y}, \quad y = 0, 1, \ldots, n \tag{3.4.3}$$

To justify this formula note first that, by the assumption of independence, the probability that the n Bernoulli trials result in a sequence with exactly y 1's is $p^y(1-p)^{n-y}$, and then argue that the number of such sequences is $\binom{n}{y}$; see also Example 2.3-8. Figure 3-7 shows three binomial PMFs for $n = 20$. Note that for $p = 0.5$, the PMF is symmetric about 10, while those for $p = 0.3$ and $p = 0.7$ are mirror images of each other.

There is no closed form expression for the binomial CDF $P(Y \le y)$, but Table A.1 in the appendix gives the CDF for $n = 5, 10, 15$, and 20 and selected values of p. Both the PMF and the CDF of the binomial(n,p) distribution can be obtained, for any n and p, with the following R commands:

R Commands for the Binomial PMF and CDF

```
dbinom(y, n, p) # gives the PMF P(Y = y) for y = 0,1,...,n
pbinom(y, n, p) # gives the CDF P(Y ≤ y) for y = 0,1,...,n
```

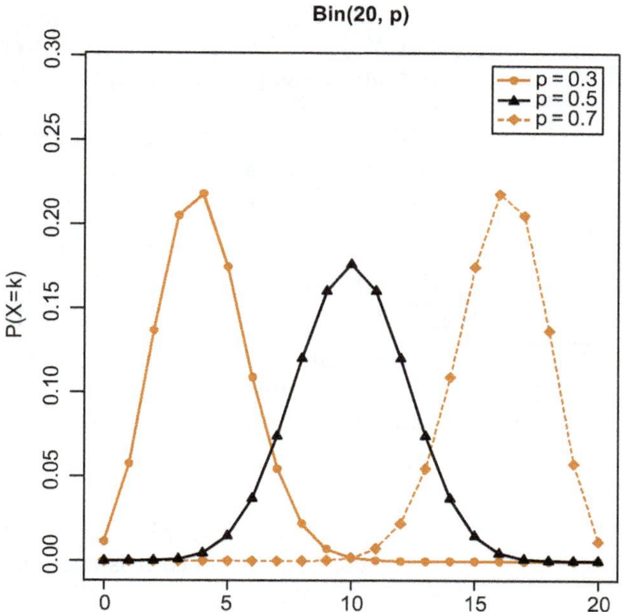

Figure 3-7 Some binomial PMFs.

In the above R commands, y can be a vector of integers from 0 to n. For example,

- The command *dbinom(4, 10, 0.5)* returns 0.2050781, which is the probability of four heads in 10 flips of a fair coin, or $P(Y = 4)$ for $Y \sim \text{Bin}(10, 0.5)$.
- The command *dbinom(0:10, 10, 0.5)* returns the entire PMF of $Y \sim \text{Bin}(10, 0.5)$. Thus, probability sampling from the Bin(10, 0.5) *PMF* can be done either with the command *sample* that was used in Example 2.3-14, or with the new command *rbinom*. For example, *sample(0:10, size=5, replace=T, prob=dbinom(0:10, 10, 0.5))* and *rbinom(5, 10, 0.5)* both give five numbers that represent the number of heads in five sets of 10 coin flips.
- The commands *sum(dbinom(4:7, 10, 0.5))* and *pbinom(7, 10, 0.5)-pbinom(3, 10, 0.5)* both give 0.7734375, which is the probability $P(3 < Y \leq 7) = F(7) - F(3)$.

The mean value and variance of a binomial X with parameters n, p are

Mean and Variance of the Binomial Distribution

$$E(X) = np, \quad \sigma_X^2 = np(1 - p) \tag{3.4.4}$$

For $n = 1$ the binomial random variable is just a Bernoulli random variable, and the above formulas reduce to the mean and variance given in (3.4.1).

Example 3.4-3

Physical traits such as eye color are determined from a pair of genes, with one gene inherited from the mother and one from the father. Each gene can be either dominant (D) or recessive (R). People with gene pairs (DD), (DR), and (RD) are alike in that physical trait. Assume that a child is equally likely to inherit either of the two genes from each parent. If both parents are hybrid with respect to a particular trait (i.e., both have pairs of genes (DR) or (RD)), find the probability that three of their four children will be like their parents with respect to this trait.

Solution

Let X denote the number of children among the four offspring that are like their parents in that physical trait. Each child represents a Bernoulli trial with probability of success (meaning that the child shares the physical trait) $p = P(DD) + P(RD) + P(DR) = 0.25 + 0.25 + 0.25 = 0.75$. Since it is reasonable to assume that the four Bernoulli trials are independent, $X \sim \text{Bin}(4, 0.75)$. Thus, by formula (3.4.3),

$$P(X = 3) = \binom{4}{3} 0.75^3 0.25^1 \cong 0.422.$$

Example 3.4-4

Suppose 70% of all purchases in a certain store are made with credit card. Let X denote the number of credit card uses in the next 10 purchases. Find (a) the expected value and variance of X, and (b) the probability that $P(5 \le X \le 8)$.

Solution

(a) Each purchase represents a Bernoulli trial where success means use of credit card. Since it is reasonable to assume that the 10 Bernoulli trials are independent, $X \sim \text{Bin}(10, 0.7)$. Thus, by formula (3.4.4),

$$E(X) = np = 10(0.7) = 7, \quad \sigma_X^2 = 10(0.7)(0.3) = 2.1.$$

(b) Next, using property 3 of Proposition 3.2-1 and Table A.1 we have

$$P(5 \le X \le 8) = P(4 < X \le 8) = F(8) - F(4)$$
$$= 0.851 - 0.047 = 0.804.$$

The R command *pbinom(8, 10, 0.7)-pbinom(4, 10, 0.7)* returns 0.8033427. Alternatively, this probability can be calculated as

$$P(5 \le X \le 8) = P(X = 5) + P(X = 6) + P(X = 7) + P(X = 8)$$
$$= \binom{10}{5} 0.7^5 0.3^5 + \binom{10}{6} 0.7^6 0.3^4 + \binom{10}{7} 0.7^7 0.3^3 + \binom{10}{8} 0.7^8 0.3^2$$
$$= 0.103 + 0.200 + 0.267 + 0.233 = 0.803. \tag{3.4.5}$$

The R command *sum(dbinom(5:8, 10, 0.7))* returns the same answer as the previous R command. In absence of a software package like R, however, the alternative calculation (3.4.5) is more labor intensive.

Example 3.4-5

Suppose that in order for the defendant to be convicted in a jury trial, at least eight of the 12 jurors must enter a guilty vote. Assume each juror makes the correct decision with probability 0.7 independently of other jurors. If 40% of the defendants in such jury trials are innocent, what is the proportion of correct verdicts?

Solution

The proportion of correct verdicts is $P(B)$, where $B = \{\text{jury renders the correct verdict}\}$. If $A = \{\text{defendant is innocent}\}$ then, according to the Law of Total Probability,

$$P(B) = P(B|A)P(A) + P(B|A^c)P(A^c) = P(B|A)0.4 + P(B|A^c)0.6.$$

Next, let X denote the number of jurors who reach the correct verdict in a particular trial. Here, each juror represents a Bernoulli trial where success means that the juror reached the correct verdict. Since the Bernoulli trials are independent, $X \sim \text{Bin}(12, 0.7)$. Note further that the correct verdict is "not guilty" if the defendant is innocent and "guilty" otherwise. Thus,

$$P(B|A) = P(X \geq 5) = 1 - \sum_{k=0}^{4} \binom{12}{k} 0.7^k 0.3^{12-k} = 0.9905, \quad \text{and}$$

$$P(B|A^c) = P(X \geq 8) = \sum_{k=8}^{12} \binom{12}{k} 0.7^k 0.3^{12-k} = 0.724.$$

It follows that

$$P(B) = P(B|A)0.4 + P(B|A^c)0.6 = 0.8306.$$

3.4.2 THE HYPERGEOMETRIC DISTRIBUTION

The hypergeometric model applies to situations where a simple random sample of size n is taken from a finite population of N units of which M_1 are labeled 1 and the rest, which are $M_2 = N - M_1$, are labeled 0. The number X of units labeled 1 in the sample is a **hypergeometric** random variable with parameters M_1, M_2, and n.

Sampling from finite populations is relevant in several contexts including ecology; see Example 3.4-7. The prototypical engineering application of the hypergeometric distribution is that of quality control at the distributor level: A batch of N product items arrives at a distributor. The distributor draws a simple random sample of size n and inspects each for the presence of a particular defect. The hypergeometric random variable X is the number of defective items in the sample.

In this prototypical hypergeometric experiment, each product item represents a Bernoulli trial where success corresponds to the product item being defective. The probability of success is the same in all Bernoulli trials and equals $p = M_1/N$. This follows by a generalization of the argument in Example 2.5-10 where it is shown that the probability of success in the first and second draw are the same. If X_i is the Bernoulli random variable corresponding to the ith product item, the hypergeometric random variable X equals

$$X = \sum_{i=1}^{n} X_i, \tag{3.4.6}$$

which is similar to the expression (3.4.2) for the binomial random variable. A hypergeometric experiment, however, differs from a binomial experiment in that the successive Bernoulli trials are not independent. This is because the conditional probability of success in the second Bernoulli trial given success in the first is different from their common (unconditional) probability of success.

The number of defective items in the sample cannot exceed the total number M_1 of defective items, and of course it cannot exceed the sample size n. For example, if a batch of size $N = 10$ product items has $M_1 = 5$ defective items and a sample of size $n = 6$ is drawn, the number of defective items in the sample cannot exceed five. In the same example, the number of defective items in the sample of size $n = 6$

cannot be zero. This is because there are only five non-defective items; hence, a sample of size six will have at least one defective item. Thus, in general, the sample space of hypergeometric(M_1, M_2, n) variable X may be a subset of $\{0, 1, \ldots, n\}$. The precise subset is typically clear from the context, but in mathematical notation it is expressed as

$$\mathcal{S}_X = \{\max(0, n - M_2), \ldots, \min(n, M_1)\}, \tag{3.4.7}$$

where $\max(a_1, a_2)$ and $\min(a_1, a_2)$ denote the larger and the smaller, respectively, of the two numbers a_1 and a_2.

By the definition of simple random sampling, all $\binom{M_1+M_2}{n}$ samples of size n are equally likely to be selected. Since there are $\binom{M_1}{x}\binom{M_2}{n-x}$ samples having exactly x defective items, it follows that the PMF of a hypergeometric(M_1, M_2, n) random variable X is (see also Example 2.3-12)

Probability Mass Function of the Hypergeometric Distribution

$$P(X = x) = \frac{\binom{M_1}{x}\binom{M_2}{n-x}}{\binom{M_1+M_2}{n}} \tag{3.4.8}$$

As in the binomial case we have $P(X = x) = 0$ if $x > n$. In addition we now have that $P(X = x) = 0$ if $x > M_1$, since the sample cannot contain more 1's than the population, or if $n - x > M_2$, since the sample cannot contain more 0's than the population. This can be restated equivalently as $P(X = x) = 0$ if x does not belong in the sample space given in (3.4.7).

Figure 3-8 shows the hypergeometric PMF for $n = 10$, $N = 60$, and different values of M_1. Note that for $M_1 = 30$, so $p = M_1/N = 0.5$, the PMF is symmetric about 5, while those for $M_1 = 15$ and $M_1 = 45$ are mirror images of each other.

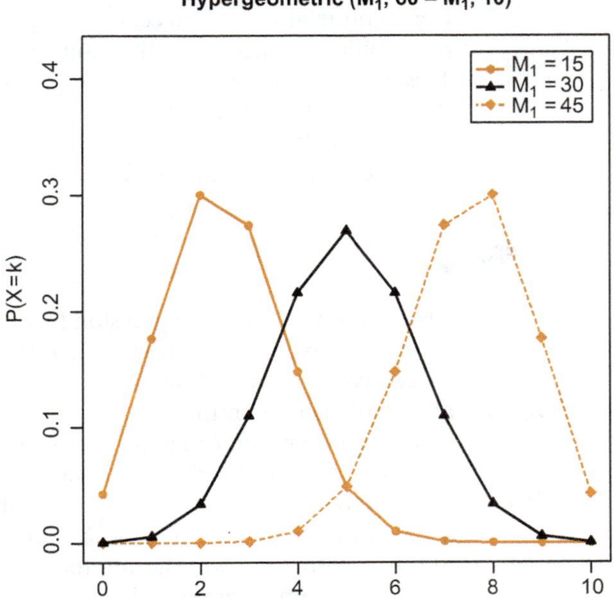

Figure 3-8 Some hypergeometric PMFs.

There is no closed form expression for the CDF of the hypergeometric random variable. Both the PMF and the CDF of the hypergeometric(M_1, M_2, n) random variable X can be obtained, for any values of the parameters M_1, M_2, and n, with the following R commands:

R Commands for the Hypergeometric PMF and CDF

```
dhyper(x,M1,M2,n) # gives the PMF P(X = x) for x in SX given
  in (3.4.7).

phyper(x,M1,M2,n) # gives the CDF P(X ≤ x) for x in SX given
  in (3.4.7).
```

In the above R commands, x can be a vector of integers from $\mathcal{S}_X$. For example, if Y is the hypergeometric(30, 30, 10) random variable,

- The command *dhyper(4, 30, 30, 10)* returns 0.2158315 for the value of $P(Y = 4)$.
- The command *dhyper(0:10, 30, 30, 10)* returns the entire PMF of Y. Thus, probability sampling from the hypergeometric(30, 30, 10) PMF can be done either with the command *sample* that was used in Example 2.3-14, or with the new command *rhyper*. For example, *sample(0:10, size=5, replace=T, prob=dhyper(0:10, 30, 30, 10))* and *rhyper(5, 30, 30, 10)* both give five numbers that represent a random sample from the hypergeometric(30, 30, 10) PMF.
- The commands *sum(dhyper(4:7, 30, 30, 10))* and *phyper(7, 30, 30, 10)-phyper(3, 30, 30, 10)* both give 0.8106493, which is the probability $P(3 < Y \le 7) = F(7) - F(3)$.

The mean value and variance of a hypergeometric($M_1, N - M_1, n$) random variable X are

Mean and Variance of the Hypergeometric Distribution

$$\mu_X = n\frac{M_1}{N}, \quad \sigma_X^2 = n\frac{M_1}{N}\left(1 - \frac{M_1}{N}\right)\frac{N-n}{N-1} \tag{3.4.9}$$

Example 3.4-6

Twelve refrigerators have been returned to the distributor because of a high-pitched oscillating noise. Suppose that four of the 12 have a defective compressor and the rest less-serious problems. Six refrigerators are selected at random for problem identification. Let X be the number of those found with a defective compressor. Give the sample space of X, and find $P(X = 3)$ as well as the expected value and variance of X.

Solution

Here $N = 12$, $n = 6$, and $M_1 = 4$. Thus, the possible values of X are $\mathcal{S}_X = \{0, 1, 2, 3, 4\}$. Using formula (3.4.8),

$$P(X = 3) = \frac{\binom{4}{3}\binom{8}{3}}{\binom{12}{6}} = 0.2424.$$

Next, using formula (3.4.9),

$$E(X) = 6\frac{4}{12} = 2, \quad \text{Var}(X) = 6\left(\frac{1}{3}\right)\left(\frac{2}{3}\right)\frac{12-6}{12-1} = \frac{8}{11}.$$

The most common method for estimating the size of wildlife populations is the so-called *capture-recapture* method. It consists of taking a sample of animals (i.e., capturing them), then tagging and releasing them. On a later occasion, after the tagged animals have had a chance to reintegrate into their community, a second sample is taken. The number of tagged animals in the second sample is used to estimate the size of the wildlife population.

Example 3.4-7

The capture-recapture method. A forest contains 30 elk of which 10 are captured, tagged, and released. A certain time later, five of the 30 elk are captured. Find the probability that two of the five captured elk are tagged. What assumptions are you making?

Solution

Assume that the five captured elk constitute a simple random sample from the population of 30 elk. In particular, it is assumed that each elk, whether tagged or not, has the same probability of being captured. Under this assumption, the number X of tagged elk among the five captured elk is a hypergeometric random variable with $M_1 = 10$, $M_2 = 20$, and $n = 5$. Thus, according to formula (3.4.8),

$$P(X = 2) = \frac{\binom{10}{2}\binom{20}{3}}{\binom{30}{5}} = 0.360.$$

Binomial Approximation to Hypergeometric Probabilities As mentioned in connection to relation (3.4.6), a hypergeometric random variable differs from a binomial only in that the Bernoulli trials that comprise it are not independent. However, if the population size N is large and the sample size n is small, the dependence of the Bernoulli trials will be weak. For example, if $N = 1000$ and $M_1 = 100$, the conditional probability of success in the second trial given success in the first trial, which is $99/999 = 0.099$, is not very different from the unconditional probability of success, which is $100/1000 = 0.1$. In such cases, the hypergeometric PMF can be well approximated by a binomial PMF with $p = M_1/N$ and the same n. Note also that the formula for the hypergeometric mean is the same as that for the binomial mean with $p = M_1/N$, and the formula for the hypergeometric variance differs from that of the binomial by the multiplicative factor $\frac{N-n}{N-1}$, which is close to 1 if N is large and n is small. The factor $\frac{N-n}{N-1}$ is called the **finite population correction factor.**

The practical usefulness of this approximation does not rest on the fact that binomial probabilities are simpler to compute. With software packages like R, hypergeometric probabilities can be computed just as readily as binomial ones. Instead, by treating a hypergeometric random variable as binomial, hypergeometric probabilities can be computed, to a good approximation, without knowledge of the population size.

To gain some insight into how the quality of the binomial approximation to hypergeometric probabilities improves as the population size N increases relative to the sample size n, let X be hypergeometric$(M_1, N-M_1, n)$, and Y be binomial$(n, p = M_1/N)$. Then,

- If $M_1 = 5$, $N = 20$, and $n = 10$, $P(X = 2) = \frac{\binom{5}{2}\binom{15}{8}}{\binom{20}{10}} = 0.3483$.
- If $M_1 = 25$, $N = 100$, and $n = 10$, $P(X = 2) = \frac{\binom{25}{2}\binom{75}{8}}{\binom{100}{10}} = 0.2924$.
- If $M_1 = 250$, $N = 1000$, and $n = 10$, $P(X = 2) = \frac{\binom{250}{2}\binom{750}{8}}{\binom{1000}{10}} = 0.2826$.

In all cases $p = M_1/N = 0.25$, so $P(Y = 2) = 0.2816$, which provides a reasonably good approximation to the third hypergeometric probability.

If X is a hypergeometric($M_1, N - M_1, n$) random variable and Y is a binomial($n, p = M_1/N$) random variable, the rule of thumb we will use in this book for applying the binomial approximation to hypergeometric probabilities is

Rule of Thumb for Using the Binomial Approximation to Hypergeometric Probabilities

$$\text{If } \frac{n}{N} \le 0.05, \quad \text{then} \quad P(X = x) \simeq P(Y = x) \tag{3.4.10}$$

3.4.3 THE GEOMETRIC AND NEGATIVE BINOMIAL DISTRIBUTIONS

The Geometric Distribution A **geometric** experiment is one where independent Bernoulli trials, each with the same probability p of success, are performed until the occurrence of the first success. The geometric(p) random variable X is the total number of trials up to and including the first success in such a geometric experiment.

The prototypical engineering application of the geometric distribution is that of quality control at the production level: Product items are being inspected as they come off the production line until the first one with a certain defect is found. The geometric random variable X is the total number of items inspected.

The sample space of a geometric(p) random variable X is $\mathcal{S}_X = \{1, 2, \ldots\}$. Note that 0 is not in the sample space since at least one item must be inspected in order to find the first defective item. On the other hand, this sample space is open-ended on the right because the probability $P(X = x)$ is positive for any value x; see the PMF below. The event $X = x$ means that the first $x - 1$ Bernoulli trials resulted in failure, while the xth Bernoulli trial resulted in success. Hence, by the independence of the Bernoulli trials, we arrive at the following formula for the PMF of the geometric(p) distribution; see also Example 3.3-3:

PMF of the Geometric Distribution

$$P(X = x) = (1 - p)^{x-1}p, \quad x = 1, 2, 3, \ldots \tag{3.4.11}$$

Figure 3-9 shows the geometric PMF for different values of p. Using the formula for the partial sums of a geometric series, which is $\sum_{y=0}^{x} a^y = (1 - a^{x+1})/(1 - a)$, it follows that

$$F(x) = \sum_{y \le x} P(Y = y) = p \sum_{y=1}^{x} (1 - p)^{y-1} = p \sum_{y=0}^{x-1} (1 - p)^y,$$

where the last equality follows by a change of the summation variable. Hence the CDF of the geometric(p) distribution is given by the following formula.

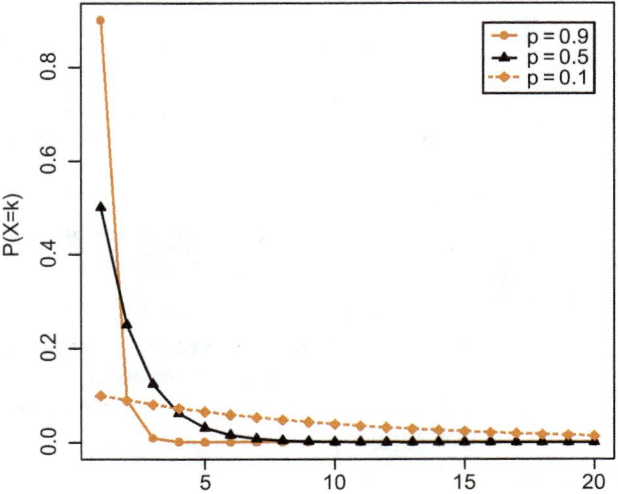

Figure 3-9 Some geometric PMFs.

CDF of the Geometric Distribution

$$F(x) = 1 - (1 - p)^x, \quad x = 1, 2, 3, \ldots \tag{3.4.12}$$

The mean value and variance of a geometric random variable X are derived in Examples 3.3-3 and 3.3-12, respectively, and summarized below.

Mean and Variance of the Geometric Distribution

$$E(X) = \frac{1}{p}, \quad \sigma^2 = \frac{1-p}{p^2} \tag{3.4.13}$$

The Negative Binomial Distribution A **negative binomial** experiment is one where independent Bernoulli trials, each with the same probability p of success, are performed until the occurrence of the rth success. The negative binomial(r, p) random variable Y is the total number of trials up to and including the rth success in such a negative binomial experiment.

The sample space of the negative binomial(r, p) random variable Y is $\mathcal{S}_Y = \{r, r+1, \ldots\}$. For $r = 1$, the negative binomial(r, p) experiment reduces to the geometric(p) experiment. In fact, if X_1 is the geometric(p) random variable that counts the number of trials until the first success, X_2 is the geometric(p) random variable that counts the additional number of trials until the second success, and so forth, the negative binomial(r, p) random variable Y can be expressed in terms of these geometric(p) random variables as

$$Y = \sum_{i=1}^{r} X_i. \tag{3.4.14}$$

The PMF $P(Y = y)$, $y = r, r+1, \ldots$, of the negative binomial(r, p) random variable Y is

PMF of the Negative Binomial Distribution

$$P(Y = y) = \binom{y-1}{r-1} p^r (1-p)^{y-r} \tag{3.4.15}$$

To see how this formula is derived argue as follows: Any particular outcome sequence has r successes and $y - r$ failures, and thus its probability is $p^r(1-p)^{y-r}$.

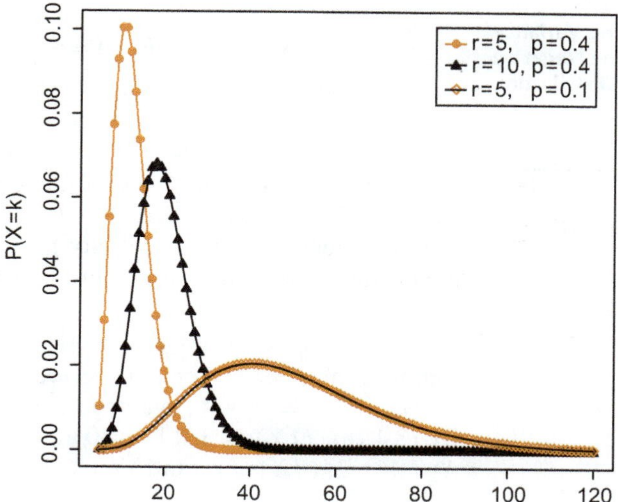

Figure 3-10 Some negative binomial PMFs.

Hence the formula follows by noting that there are $\binom{y-1}{r-1}$ binary sequences of length $y-1$ with exactly $r-1$ successes. Figure 3-10 shows negative binomial(r,p) PMFs for different values of r and p, all shifted to the origin for easier comparison.

The PMF and the CDF of the negative binomial(r,p) random variable Y can be obtained with the following R commands:

R Commands for the Negative Binomial PMF and CDF

```
dnbinom(x, r, p) # gives the PMF P(Y=r+x) for x=0,1,2,...
pnbinom(x, r, p) # gives the CDF P(Y≤r+x) for x=0,1,2,...
```

In the above R commands x, which represents the number of failures until the rth success, can be a vector of integers from $\{0,1,\ldots\}$. For example, if Y is a negative binomial$(5,0.4)$,

- The command *dnbinom(6, 5, 0.4)* returns 0.1003291 for the value of $P(Y=11)$.
- The command *dnbinom(0:15, 5, 0.4)* returns the values of $P(Y=5),\ldots,P(Y=20)$.
- The commands *sum(dnbinom(0:15, 5, 0.4))* and *pnbinom(15, 5, 0.4)* both return 0.949048, which is the value of $P(Y\le 20)$.

Because of its infinite sample space, only the new R command *rnbinom* is available for probability sampling.

R Command for Simulating Negative Binomial Experiment

```
r+rnbinom(k, r, p) # gives a sample of k negative
  binomial(r, p) random variables
```

It can be shown that the mean and variance of a negative binomial(r,p) random variable Y are

Mean and Variance of the Negative Binomial Distribution

$$E(Y) = \frac{r}{p}, \quad \sigma_Y^2 = r\frac{1-p}{p^2} \tag{3.4.16}$$

Example 3.4-8

Items are being inspected as they come off the production line until the third defective item is found. Let X denote the number of non-defective items found. If an item is defective with probability $p = 0.1$ independently of other items, find the mean value and variance of X and $P(X = 15)$.

Solution

The total number of items inspected until the third defective item is found, which is given by $Y = 3 + X$, is a negative binomial with parameters $r = 3$ and $p = 0.1$. By (3.4.16), $E(Y) = 3/0.1 = 30$ and $\text{Var}(Y) = 3 \times 0.9/(0.1^2) = 270$. Hence, since $X = Y - 3$, $E(X) = 27$ and $\text{Var}(X) = 270$. Next, using formula (3.4.15),

$$P(X = 15) = P(Y = 18) = \binom{18-1}{3-1} \times 0.1^3 \times 0.9^{18-3} = 0.028.$$

The R command *dnbinom(15, 3, 0.1)* returns 0.02800119. ■

REMARK 3.4-1 As in Example 3.4-8, the outcome recorded in a negative binomial(r, p) experiment is often the total number X of failures until the rth success. X and $Y = r + X$ are both referred to as negative binomial(r, p) random variables. In particular the R command for the negative binomial PMF gives the PMF of X, and PMFs plotted in Figure 3-10 correspond to X. ◁

Example 3.4-9

Three electrical engineers toss coins to see who pays for coffee. If all three match, they toss another round. Otherwise the "odd person" pays for coffee.

(a) Find the probability that a round of tossing will result in a match (that is, either three heads or three tails).

(b) Let X be the number of times they toss coins until the odd person is determined. Name the probability distribution of X, and compute the probability $P(X \geq 3)$.

(c) Find the expected value and variance of X.

Solution

(a) The probability that all three match is $0.5^3 + 0.5^3 = 0.25$.

(b) X has the geometric distribution with $p = 0.75$. Using formula (3.4.12) we have

$$P(X \geq 3) = 1 - P(X \leq 2) = 1 - \left[1 - (1 - 0.75)^2\right] = 1 - 0.9375 = 0.0625.$$

The R command *1-pnbinom(1, 1, 0.75)* also returns 0.0625 for $P(X \geq 3)$.

(c) Using formula (3.4.13),

$$E(X) = \frac{1}{0.75} = 1.333 \quad \text{and} \quad \sigma_X^2 = \frac{1 - 0.75}{0.75^2} = 0.444.$$ ■

Example 3.4-10

Two athletic teams, A and B, play a best-of-three series of games (i.e., the first team to win two games is the overall winner). Suppose team A is the stronger team and will win any game with probability 0.6, independently from other games. Find the probability that the stronger team will be the overall winner.

Solution

Let X be the number of games needed for team A to win twice. Then X has the negative binomial distribution with $r = 2$ and $p = 0.6$. Team A will win the series if $X = 2$ or $X = 3$. Since these two events are disjoint, formula (3.4.15) with $r = 2$ gives

$$\begin{aligned} P(\text{Team } A \text{ wins the series}) &= P(X = 2) + P(X = 3) \\ &= \binom{1}{1} 0.6^2(1 - 0.6)^{2-2} + \binom{2}{1} 0.6^2(1 - 0.6)^{3-2} \\ &= 0.36 + 0.288 = 0.648. \end{aligned}$$

3.4.4 THE POISSON DISTRIBUTION

The Model and Its Applications A random variable X that takes values $0, 1, 2, \ldots$ is said to be a Poisson random variable with parameter λ, denoted by $X \sim \text{Poisson}(\lambda)$, if its PMF is given by

PMF of the Poisson Distribution

$$P(X = x) = e^{-\lambda} \frac{\lambda^x}{x!}, \quad x = 0, 1, 2, \ldots \tag{3.4.17}$$

for some $\lambda > 0$, where $e = 2.71828\ldots$ is the base of the natural logarithm. That $p(x) = P(X = x)$ given above is a proper PMF (i.e., the probabilities sum to 1) is easily seen from the fact that $e^{\lambda} = \sum_{k=0}^{\infty}(\lambda^k/k!)$. Figure 3-11 shows the Poisson PMF for three different values of λ.

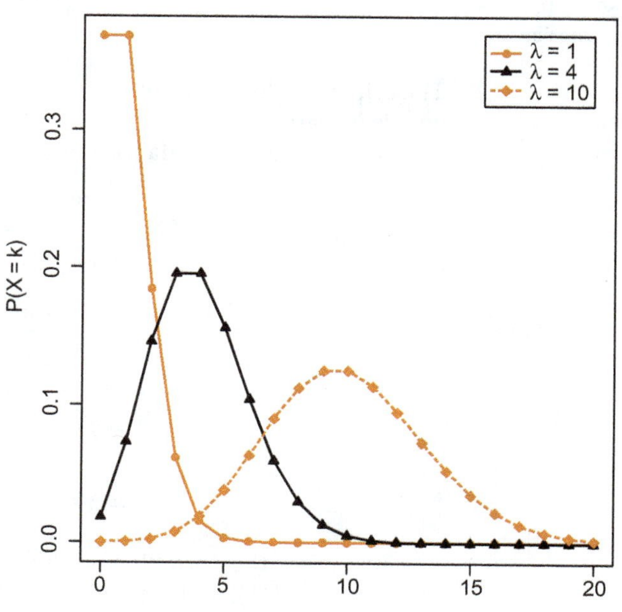

Figure 3-11 Poisson PMFs for different values of λ.

The Poisson cumulative distribution function does not have a closed form expression. Its value, for selected values of λ and x, is given in Table A.2. The R commands for the Poisson(λ) PMF and CDF and for simulating a Poisson experiment are:

R Commands for the Poisson PMF, CDF, and Simulation

```
dpois(x, λ) # gives the PMF P(X=x) for x integer
ppois(x, λ) # gives the CDF P(X≤x) for all x
rpois(n, λ) # gives a sample of n Poisson(λ) random variables
```

In the above R commands x can be a vector of integers from $\{0, 1, \ldots\}$. For example, if X is Poisson(5),

- The command *dpois(6, 5)* returns 0.1462228 for the value of $P(X = 6)$.
- The command *dpois(0:10, 5)* returns the values of $P(X = 0), \ldots, P(X = 10)$.
- The commands *sum(dpois(6:10, 5))* and *ppois(10, 5)-ppois(5, 5)* both return the value of $P(6 \le X \le 10)$, which is 0.3703441.

The Poisson distribution is used to model the probability that a number of certain events occur in a specified period of time. The type of events whose occurrences are thus modeled must occur at random and at a rate that does not change with time. The Poisson distribution can also be used for the number occurrences of events occurring in other specified intervals such as distance, area, or volume.

The parameter λ in (3.4.17) specifies the "average" number of occurrences in the given interval (of time, area, or space). In particular, if $X \sim$ Poisson(λ) then

Mean and Variance of the Poisson Distribution

$$\mu_X = \lambda, \quad \sigma_X^2 = \lambda \tag{3.4.18}$$

Thus, if a random variable has the Poisson distribution, then its expected value equals its variance.

To derive the formula for the expected value write

$$\begin{aligned} E(X) &= \sum_{x=0}^{\infty} \frac{x e^{-\lambda} \lambda^x}{x!} \\ &= \lambda \sum_{x=1}^{\infty} \frac{e^{-\lambda} \lambda^{x-1}}{(x-1)!} \\ &= \lambda e^{-\lambda} \sum_{y=0}^{\infty} \frac{\lambda^y}{y!} \quad \text{by letting } y = x - 1 \\ &= \lambda. \end{aligned}$$

To derive the formula for the variance, first use a similar technique to get $E(X^2) = \lambda(\lambda + 1)$, and then apply the formula $\text{Var}(X) = E(X^2) - [E(X)]^2$.

Example 3.4-11

Let $X \sim \text{Poisson}(4)$. Use Table A.2 to find $P(X \le 5)$, $P(3 \le X \le 6)$, and $P(X \ge 8)$.

Solution

All probabilities are obtained using the cumulative probabilities listed under $\lambda = 4.0$ in Table A.2. The first probability is given directly from the table as $P(X \le 5) = 0.785$. The second and third probabilities must first be expressed in terms of cumulative probabilities before we can use Table A.2. Thus, the second probability is given by

$$P(3 \le X \le 6) = P(2 < X \le 6) = P(X \le 6) - P(X \le 2) = 0.889 - 0.238 = 0.651,$$

and the third one is given by

$$P(X \ge 8) = 1 - P(X \le 7) = 1 - F(7) = 1 - 0.949 = 0.051.$$

■

Example 3.4-12

Suppose that a person taking Vitamin C supplements contracts an average of three colds per year and that this average increases to five colds per year for persons not taking Vitamin C supplements. Suppose further that the number of colds a person contracts in a year is a Poisson random variable.

(a) Find the probability of no more than two colds for a person taking supplements and a person not taking supplements.

(b) Suppose 70% of the population takes Vitamin C supplements. Find the probability that a randomly selected person will have no more than two colds in a given year.

(c) Suppose that a randomly selected person contracts no more than two colds in a given year. What is the probability that person takes Vitamin C supplements?

Solution

(a) Let X_1 denote the number of colds contracted by a person taking Vitamin C supplements and X_2 denote the number of colds contracted by a person not taking supplements. We are given that X_1, X_2 are Poisson random variables with mean values 3, 5, respectively. Therefore, by (3.4.18), $X_1 \sim \text{Poisson}(3)$ and $X_2 \sim \text{Poisson}(5)$. Hence, from Table A.2,

$$P(X_1 \le 2) = 0.423, \quad P(X_2 \le 2) = 0.125.$$

The R commands *ppois(2, 3)* and *ppois(2, 5)* return 0.4231901 and 0.1246520 for $P(X_1 \le 2)$ and $P(X_2 \le 2)$, respectively.

(b) Let X denote the number of colds contracted by a person, and let A denote the event that this person takes Vitamin C supplements. By the Law of Total Probability,

$$P(X \le 2) = (0.423)(0.7) + (0.125)(0.3) = 0.334.$$

(c) Using Bayes' Theorem, the desired probability is calculated as

$$P(A|X \le 2) = \frac{(0.423)(0.7)}{0.334} = 0.887.$$

■

One of the earliest uses of the Poisson distribution was in modeling the number of alpha particles emitted from a radioactive source during a given period of time. Today it has a tremendous range of applications in such diverse areas as insurance, tourism traffic engineering, demography, forestry, and astronomy. For example, the Poisson random variable X can be

1. the number of fish caught by an angler in an afternoon,
2. the number of new potholes in a stretch of I80 during the winter months,
3. the number of disabled vehicles abandoned on I95 in a year,
4. the number of earthquakes (or other natural disasters) in a region of the United States in a month,
5. the number of wrongly dialed telephone numbers in a given city in an hour,
6. the number of freak accidents, such as falls in the shower, in a given time period,
7. the number of vehicles that pass a marker on a roadway in a given time period,
8. the number of marriages, or the number of people who reach the age of 100,
9. the distribution of trees in a forest, and
10. the distribution of galaxies in a given region of the sky.

As seen from these applications of the Poisson model, the random phenomena that the Poisson distribution models differ from those of the previously discussed distributions in that they are not outcomes of sampling experiments from a well-understood population. Consequently, the Poisson PMF is derived by arguments that are different from the ones used for deriving the PMFs of the previously discussed distributions (which use the counting techniques of Chapter 2 and the concept of independence). Instead, the Poisson PMF is derived as the limit of the binomial PMF (see Proposition 3.4-1 below) and can also be obtained as a consequence of certain postulates governing the random occurrence of events (see the following discussion about the Poisson process).

Poisson Approximation to Binomial Probabilities The enormous range of applications of the Poisson random variable is, to a large extent, due to the following proposition stating that it can be used as an approximation to binomial random variables.

Proposition 3.4-1 A binomial experiment where the number of trials n is large ($n \geq 100$), the probability p of success in each trial is small ($p \leq 0.01$), and the product np is not large ($np \leq 20$), can be modeled (to a good approximation) by a Poisson distribution with $\lambda = np$. In particular, if $Y \sim \text{Bin}(n, p)$, with $n \geq 100$, $p \leq 0.01$, and $np \leq 20$, then the approximation

Poisson Approximation to Binomial Probabilities

$$P(Y \geq k) \simeq P(X \geq k)$$

holds for all $k = 0, 1, 2, \ldots, n$, where $X \sim \text{Poisson}(\lambda = np)$.

Proof of Proposition We will show that as $n \to \infty$ and $p \to 0$ in such a way that $np \to \lambda$, some $\lambda > 0$, then

$$P(Y = k) = \binom{n}{k} p^k (1-p)^{n-k} \to e^{-\lambda} \frac{\lambda^k}{k!}, \quad \text{as } n \to \infty, \tag{3.4.19}$$

holds for all $k = 0, 1, 2, \ldots$. The proof makes use of Stirling's formula for approximating $n!$ for large n: $n! \simeq \sqrt{2\pi n}(\frac{n}{e})^n$, or more precisely

$$n! = \sqrt{2\pi n}\left(\frac{n}{e}\right)^n e^{\lambda_n} \quad \text{where} \quad \frac{1}{12n+1} < \lambda_n < \frac{1}{12n}.$$

Using this, the left-hand side of (3.4.19) can be approximated by

$$\begin{aligned} P(Y = k) &\approx \frac{\sqrt{2\pi n}n^n e^{-n}}{k!\sqrt{2\pi(n-k)}(n-k)^{n-k}e^{-(n-k)}}p^k(1-p)^{n-k} \\ &= \frac{\sqrt{n}n^{n-k}e^{-k}}{k!\sqrt{n-k}(n-k)^{n-k}}(np)^k(1-p)^{n-k} \end{aligned} \tag{3.4.20}$$

by canceling out the $\sqrt{2\pi}$ in the numerator and denominator, simplifying the exponent of e, and multiplying and dividing by n^k. Note now that the ratio $\sqrt{n}/\sqrt{n-k}$ tends to 1 as $n \to \infty$ with k remaining fixed, that $(np)^k \to \lambda^k$ as $n \to \infty$, $np \to \lambda$, and k remains fixed, that

$$\begin{aligned} (1-p)^{n-k} &= \left(1 - \frac{np}{n}\right)^{n-k} \\ &= \left(1 - \frac{np}{n}\right)^n \left(1 - \frac{np}{n}\right)^{-k} \to e^{-\lambda} \cdot 1 = e^{-\lambda} \end{aligned}$$

as $n \to \infty$, $np \to \lambda$, and k remains fixed, and that

$$\left(\frac{n}{n-k}\right)^{n-k} = \left(1 + \frac{k}{n-k}\right)^{n-k} \to e^k$$

as $n \to \infty$ and k remains fixed. Substituting these into (3.4.20) establishes (3.4.19). ■

This proposition justifies the use of the Poisson random variable for modeling occurrences of random events such as car accidents: Each person getting in his or her car to drive to work each morning has a very small chance of getting into an accident. Assuming each driver acts independently we have a large number of Bernoulli trials with a small probability of success (i.e., accident). As a consequence of Proposition 3.4-1, the number of accidents in a given day is modeled as a Poisson random variable.

The same rationale can be used for modeling the number of earthquakes in a month by dividing the month into small time intervals and thinking of each interval as a Bernoulli trial, where success is the occurrence of an earthquake in that interval. Since the probability of success in each interval is small and the number of intervals is large, the number of earthquakes in a given month is modeled as a Poisson random variable. The following discussion of the Poisson process provides the conditions needed for giving rigorous support to this type of argument.

To illustrate the convergence of the binomial(n, p) probabilities to those of the Poisson$(\lambda = np)$ distribution as n increases and p decreases in such a way that np remains constant, consider the binomial random variables

$$\begin{aligned} &Y_1 \sim \text{Bin}(9, 1/3), && Y_2 \sim \text{Bin}(18, 1/6), \\ &Y_3 \sim \text{Bin}(30, 0.1), \quad \text{and} && Y_4 \sim \text{Bin}(60, 0.05). \end{aligned}$$

Note that in all cases $\lambda = np = 3$. Figure 3-12 shows the PMFs of these four binomial random variables and the PMF of the approximating Poisson(3) random variable.

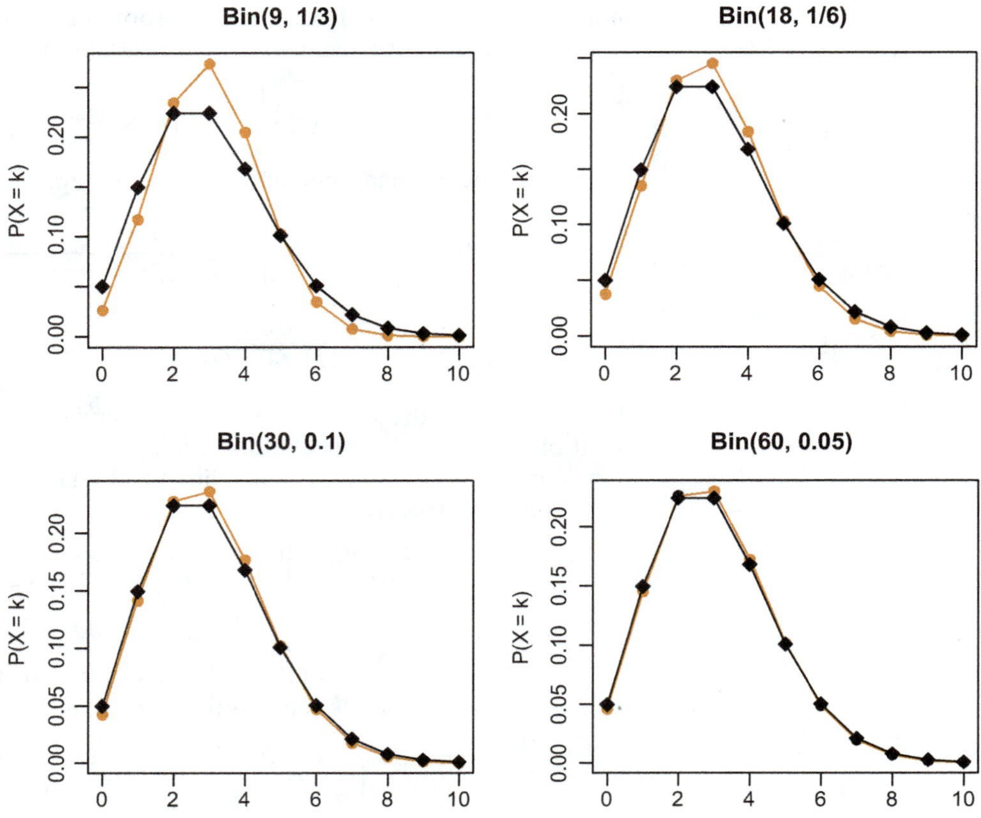

Figure 3-12 Binomial PMFs (circles) superimposed on the Poisson(3) PMF (diamonds).

For a numerical illustration of the quality of approximation, the binomial CDF at $x = 2$ in each case is compared with the corresponding CDF of the Poisson(3) distribution:

$$P(Y_1 \le 2) = 0.3772, \qquad P(Y_2 \le 2) = 0.4027,$$
$$P(Y_3 \le 2) = 0.4114, \quad \text{and} \quad P(Y_4 \le 2) = 0.4174.$$

The approximating Poisson(3) probability is

$$P(X \le 2) = e^{-3}\left(1 + 3 + \frac{3^2}{2}\right) = 0.4232,$$

which is reasonably close to the value of $P(Y_4 \le 2)$. Note, however, that the first two conditions on n and p mentioned in Proposition 3.4-1 are not satisfied for any of the four binomial random variables.

Example 3.4-13

Due to a serious defect, a car manufacturer issues a recall of $n = 10{,}000$ cars. Let $p = 0.0005$ be the probability that a car has the defect, and let Y be the number of defective cars. Find (a) $P(Y \ge 10)$ and (b) $P(Y = 0)$.

Solution

Here each car represents a Bernoulli trial with success if the car has the defect and failure if it does not. Thus, Y is a binomial random variable with $n = 10{,}000$, and $p = 0.0005$. Note that the three conditions on n and p mentioned in Proposition 3.4-1 for the approximation of the binomial probabilities by corresponding Poisson probabilities are satisfied. Let $X \sim \text{Poisson}(\lambda = np = 5)$. For part (a) write

$$P(Y \geq 10) \simeq P(X \geq 10) = 1 - P(X \leq 9) = 1 - 0.968,$$

where the value of the CDF in the last equality was obtained from Table A.2. Similarly, for part (b) write

$$P(Y = 0) \simeq P(X = 0) = e^{-5} = 0.007.$$

■

Example 3.4-14

Suppose the monthly suicide rate in a certain county is 1 per 100,000 people. Give an approximation to the probability that in a city of 500,000 in this county there will be no more than six suicides in the next month.

Solution

Let Y denote the number of suicides in that city during the next month. We want to approximate the probability $P(Y \leq 6)$. Here $Y \sim \text{Bin}(n = 500{,}000,\ p = 10^{-5})$ so the three conditions mentioned in Proposition 3.4-1 are satisfied. Hence, letting $X \sim \text{Poisson}(\lambda = np = 5)$, and using Table A.2, it follows that

$$P(Y \leq 6) \simeq P(X \leq 6) = 0.762.$$

■

The Poisson Process All examples of Poisson random variable pertain to the number of events occurring in a fixed time period (fish caught in an afternoon, potholes during the winter months, etc). Often, however, events are recorded at the time they occur as time unfolds. This requires that time itself become an integral part of the notation describing the data records. Letting time 0 denote the start of observations we set

$$X(t) = \text{number of events occurring in the time interval } (0, t]. \quad \textbf{(3.4.21)}$$

Definition 3.4-1

Viewed as a function of time, the number of occurrences $X(t)$, $t \geq 0$, is called a Poisson process if the following postulates are satisfied.

1. The probability of exactly one occurrence in a time period of length h is equal to $\alpha h + o(h)$, for some $\alpha > 0$, where the quantity $o(h)$ satisfies $o(h)/h \to 0$, as $h \to 0$.
2. The probability of more than one occurrence in a short time period of length h is equal to $o(h)$.
3. For any set of nonoverlapping time intervals A_i, $i = 1, \ldots, n$, the events $E_i = [k_i \text{ events occur in } A_i]$, $i = 1, \ldots, n$, where the k_i are any integers, are mutually independent.

The parameter α in the first postulate specifies the *rate* of the occurrences or, synonymously, the average number of occurrences per unit of time. Thus, the first postulate states that the rate at which events occur is constant in time. The second postulate means that the events are rare, in the sense that it is highly unlikely that two will occur simultaneously. Finally, the third postulate specifies that in disjoint time intervals events occur independently. When these postulates hold we have the following proposition.

Proposition 3.4-2

If $X(t)$, $t \geq 0$, is a Poisson(α) process, then

1. For each fixed t_0, the random variable $X(t_0)$, which counts the number of occurrences in $(0, t_0]$, has the Poisson distribution with parameter $\lambda = \alpha \times t_0$. Thus,

$$P(X(t_0) = k) = e^{-\alpha t_0} \frac{(\alpha t_0)^k}{k!}, \quad k = 0, 1, 2, \ldots. \tag{3.4.22}$$

2. For any two positive numbers $t_1 < t_2$ the random variable $X(t_2) - X(t_1)$, which counts the number of occurrences in $(t_1, t_2]$, has the Poisson distribution with parameter $\lambda = \alpha(t_2 - t_1)$. Thus, the PMF of $X(t_2) - X(t_1)$ is given by (3.4.22) with t_0 replaced by $(t_2 - t_1)$.
3. For any two positive numbers $t_1 < t_2$ the random variable $X(t_2) - X(t_1)$ is independent from $X(s)$ for all $s \leq t_1$. ■

A noteworthy implication of part (2) of this proposition is that the zero time point of a Poisson process can be any arbitrary time point. In other words, one may start recording the events which happen after time t_1, completely ignoring anything that happened up to that point, and still get a Poisson process with the same rate.

The proof of Proposition 3.4-2 is based on the fact that Poisson probabilities are obtained as limits of binomial probabilities (Proposition 3.4-1). Indeed, if the postulates of Definition 3.4-1 are satisfied, then by dividing an interval into a large number of small subintervals of equal length, the total number of occurrences in that interval can be thought of as a binomial random variable made up of the sum of Bernoulli random variables, each of which corresponds to one of the small subintervals. Since by Proposition 3.4-1 the Poisson probability mass function is obtained as the limit of binomial probability mass functions, it can be argued that the total number of occurrences is a Poisson random variable. While making this argument rigorous is not beyond the scope of this book, such a proof adds little to the understanding of the Poisson process and thus will not be presented.

Example 3.4-15

Continuous inspection of electrolytic tin plate yields on average 0.2 imperfections per minute. Find each of the following:

(a) The probability of one imperfection in three minutes.

(b) The probability of at least two imperfections in five minutes.

(c) The probability of at most one imperfection in 0.25 hours.

Solution

Let $X(t)$ denote the number of imperfections found in $(0, t]$, where the time t is expressed in minutes.

(a) Here $\alpha = 0.2$, $t = 3$, so that $X(3) \sim \text{Poisson}(\lambda = \alpha t = 0.6)$. Thus, using Table A.2,

$$P(X(3) = 1) = P(X(3) \leq 1) - P(X(3) \leq 0) = 0.878 - 0.549 = 0.329.$$

(b) Here $\alpha = 0.2$, $t = 5$, so that $X(5) \sim \text{Poisson}(\lambda = \alpha t = 1.0)$. Thus, using Table A.2,

$$P(X(5) \geq 2) = 1 - P(X(5) \leq 1) = 1 - 0.736 = 0.264.$$

(c) Here $\alpha = 0.2$, $t = 15$, so that $X(15) \sim \text{Poisson}(\lambda = \alpha t = 3.0)$. Thus, using Table A.2,

$$P(X(15) \leq 1) = 0.199.$$

■

Example 3.4-16

People enter a department store according to a Poisson process with rate α per hour. It is known that 30% of those entering the store will make a purchase of \$50.00 or more. Find the probability mass function of the number of customers who will make purchases of \$50.00 or more during the next hour.

Solution

Let X denote the number of people entering the store during the next hour and Y the number of those who make a purchase of \$50.00 or more. The information given implies that $X \sim \text{Poisson}(\alpha)$, and that the conditional distribution of Y given $X = n$ is binomial$(n, 0.3)$. Thus,

$$P(Y = k|X = n) = \binom{n}{k} (0.3)^k (0.7)^{n-k} \quad \text{for } n \geq k,$$

and $P(Y = k|X = n) = 0$ for $n < k$, since the number of customers spending \$50.00 or more cannot exceed the number of customers entering the department store. Thus, by the Law of Total Probability,

$$\begin{aligned}
P(Y = k) &= \sum_{m=0}^{\infty} P(Y = k|X = k + m)P(X = k + m) \\
&= \sum_{m=0}^{\infty} \binom{k+m}{k} (0.3)^k (0.7)^m e^{-\alpha} \frac{\alpha^{k+m}}{(k+m)!} \\
&= \sum_{m=0}^{\infty} e^{-0.3\alpha} \frac{(0.3\alpha)^k}{k!} e^{-0.7\alpha} \frac{(0.7\alpha)^m}{m!} \\
&= e^{-0.3\alpha} \frac{(0.3\alpha)^k}{k!} \sum_{m=0}^{\infty} e^{-0.7\alpha} \frac{(0.7\alpha)^m}{m!} \\
&= e^{-0.3\alpha} \frac{(0.3\alpha)^k}{k!}, \quad k = 0, 1, 2, \ldots,
\end{aligned}$$

where the last equality follows by the fact that a PMF sums to 1. Thus, $Y \sim \text{Poisson}(0.3\alpha)$.

■

Example 3.4-17

Let $X(t)$ be a Poisson process with rate α. It is given that $X(1) = n$. Show that the conditional distribution of $X(0.4)$ is binomial$(n, 0.4)$. In words, if we know that n events occurred in the interval $(0, 1]$, then the number of events that occurred in the interval $(0, 0.4]$ is a binomial$(n, 0.4)$ random variable.

Solution

For $k = 0, 1, \ldots, n$, the events

$$[X(0.4) = k] \cap [X(1) = n] \quad \text{and} \quad [X(0.4) = k] \cap [X(1) - X(0.4) = n - k]$$

are identical as they both express the fact that k events occurred in $(0, 0.4]$ and $n-k$ events occurred in $(0.4, 1]$. Thus,

$$
\begin{aligned}
P(X(0.4) = k|X(1) = n) \\
&= \frac{P([X(0.4) = k] \cap [X(1) = n])}{P(X(1) = 1)} \\
&= \frac{P([X(0.4) = k] \cap [X(1) - X(0.4) = n-k])}{P(X(1) = 1)} \\
&= \frac{P(X(0.4) = k)P(X(1) - X(0.4) = n-k)}{P(X(1) = 1)} \qquad \text{(by part (3) of Proposition 3.4-2)} \\
&= \frac{[e^{-0.4\alpha}(0.4\alpha)^k/k!]e^{-(1-0.4)\alpha}[(1-0.4)\alpha]^{n-k}/(n-k)!}{e^{-\alpha}\alpha^n/n!} \\
&\qquad \text{(by part (2) of Proposition 3.4-2)} \\
&= \frac{n!}{k!(n-k)!} 0.4^k (1-0.4)^{n-k},
\end{aligned}
$$

which is the PMF of the binomial$(n, 0.4)$ distribution. ■

Exercises

1. Grafting, the uniting of the stem of one plant with the stem or root of another, is widely used commercially to grow the stem of one variety that produces fine fruit on the root system of another variety with a hardy root system. For example, most sweet oranges grow on trees grafted to the root of a sour orange variety. Suppose each graft fails independently with probability 0.3. Five grafts are scheduled to be performed next week. Let X denote the number of grafts that will fail next week.

(a) The random variable X is (choose one)
 (i) binomial (ii) hypergeometric (iii) negative binomial (iv) Poisson.

(b) Give the sample space and PMF of X.

(c) Give the expected value and variance of X.

(d) Suppose that the cost of each failed graft is \$9.00. Find:
 (i) The probability that the cost from failed grafts will exceed \$20.00.
 (ii) The expected value and the variance of the cost from failed grafts.

2. Suppose that 30% of all drivers stop at an intersection having flashing red lights when no other cars are visible. Of 15 randomly selected drivers coming to an intersection under these conditions, let X denote the number of those who stop.

(a) The random variable X is (choose one)
 (i) binomial (ii) hypergeometric (iii) negative binomial (iv) Poisson.

(b) Give the expected value and variance of X.

(c) Find the probabilities $P(X = 6)$ and $P(X \geq 6)$. You may use R commands.

3. A company sells small, colored binder clips in packages of 20 and offers a money-back guarantee if two or more of the clips are defective. Suppose a clip is defective with probability 0.01, independently of other clips. Let X denote the number of defective clips in a package of 20.

(a) The distribution of the random variable X is (choose one)
 (i) binomial (ii) hypergeometric (iii) negative binomial (iv) Poisson.

(b) Specify the value of the parameter(s) of the chosen distribution and use R commands to find the probability that a package sold will be refunded.

4. A test consists of 10 true-false questions. Suppose a student answers the questions by flipping a coin. Let X denote the number of correctly answered questions.

(a) Give the expected value and variance of X.

(b) Find the probability the student will answer correctly exactly 5 of the questions.

(c) Find the probability the student will answer correctly at most 5 of the questions. Use the CDF to answer this question.

(d) Let $Y = 10 - X$. In words, what does Y represent?

(e) Use the CDF to find $P(2 \leq Y \leq 5)$.

5. The probability that a letter will be delivered within three working days is 0.9. You send out 10 letters on Tuesday to invite friends for dinner. Only those who receive the invitation by Friday (i.e., within 3 working days) will come. Let X denote the number of friends who come to dinner.

(a) The random variable X is (choose one)
(i) binomial (ii) hypergeometric (iii) negative binomial (iv) Poisson.

(b) Give the expected value and variance of X.

(c) Determine the probability that at least 7 friends will come.

(d) A catering service charges a base fee of \$100 plus \$10 for each guest coming to the party. What is the expected value and variance of the catering cost?

6. Suppose that in order for the defendant to be convicted in a military court the majority of the nine appointed judges must enter a guilty vote. Assume that a judge enters a guilty vote with probability 0.1 or 0.9 if the defendant is innocent or guilty, respectively, independently of other judges. Assume also that 40% of the defendants in such trials are innocent.

(a) What proportion of all defendants is convicted?

(b) What is the proportion of correct verdicts?

7. In the grafting context of Exercise 1, suppose that grafts are done one at a time and the process continues until the first failed graft. Let X denote the number of grafts up to and including the first failed graft.

(a) The random variable X is (choose one)
(i) binomial (ii) hypergeometric (iii) negative binomial (iv) Poisson.

(b) Give the sample space and PMF of X.

(c) Give the expected value and variance of X.

8. In the context of quality control, a company manufacturing bike helmets decides that helmets be inspected until the fifth helmet having a particular type of flaw is found. The total number X of helmets inspected will be used to decide whether or not the production process is under control. Assume that each helmet has the flaw with probability 0.05 independently of other helmets.

(a) The random variable X is (choose one)
(i) binomial (ii) hypergeometric (iii) negative binomial (iv) Poisson.

(b) Give the sample space and PMF of X.

(c) Use R commands to find the probability that $X > 35$.

9. Two athletic teams, A and B, play a best-of-five series of games (i.e., the first team to win three games is the overall winner). Suppose team A is the better team and will win any game with probability 0.6, independently from other games.

(a) Find the probability that the better team will be the overall winner.

(b) A similar question was answered in Example 3.4-10 for a best-of-three series. Compare the two probabilities and provide an intuitive explanation for why one of the two probabilities is larger.

10. Average run length. To control the quality of a manufactured product, samples of the product are taken at specified inspection time periods and a quality characteristic is measured for each product. If the average measurement falls below a certain predetermined threshold, the process is declared out of control and is interrupted. The number of inspections between successive interruptions of the process is called a *run length*. The expected value of the random variable X = run length is called the *average run length*. Suppose the probability that an inspection will result in the process being interrupted is 0.01.

(a) The random variable X is (choose one)
(i) binomial (ii) hypergeometric (iii) negative binomial (iv) Poisson.

(b) Give the sample space and PMF of X.

(c) What is the average run length?

11. In the context of Exercise 10, suppose that after five interruptions the process undergoes a major evaluation. Suppose also that inspections happen once every week. Let Y denote the number of weeks between successive major evaluations.

(a) The random variable Y is (choose one)
(i) binomial (ii) hypergeometric (iii) negative binomial (iv) Poisson.

(b) Find the expected value and variance of Y.

12. Suppose that six of the 15 school buses in a particular school district have developed a slight defect since their last inspection (the steering wheel shakes when braking). Five buses are to be selected for thorough inspection. Let X denote the number of buses among the five that are inspected that have the defect.

(a) The random variable X is (choose one)
(i) binomial (ii) hypergeometric (iii) negative binomial (iv) Poisson.

(b) Give the sample space and formula for the PMF of X.

(c) Use R commands to find $P(2 \leq X \leq 4)$.

(d) Give the expected value and variance of X.

13. A distributor receives a new shipment of 20 iPods. He draws a random sample of five iPods and thoroughly inspects the click wheel of each of them. Suppose that the shipment contains three iPods with a malfunctioning click wheel. Let X denote the number of iPods with a defective click wheel in the sample of five.

(a) The random variable X is (choose one)
(i) binomial (ii) hypergeometric (iii) negative binomial (iv) Poisson.

(b) Give the sample space and the formula for the PMF of X.

(c) Compute $P(X = 1)$.

(d) Find the expected value and variance of X.

14. In a study of a lake's fish population, scientists capture fish from the lake, then tag and release them. Suppose that over a period of five days, 200 fish of a certain type are tagged and released. As part of the same study, 20 such fish are captured three days later. Let X denote the number of tagged fish among the 20 captured. Suppose it is known that the lake has 1000 fish of this particular type.

(a) The distribution of the random variable X is (choose one)
(i) binomial (ii) hypergeometric (iii) negative binomial (iv) Poisson.

(b) Use R commands to find $P(X \leq 4)$.

(c) Which distribution from those listed in part (a) can be used as an approximation to the distribution of X?

(d) Using the approximate distribution, give an approximation to the probability $P(X \leq 4)$, and compare it with the exact probability found in part (b).

15. In a shipment of 10,000 of a certain type of electronic component, 300 are defective. Suppose that 50 components are selected at random for inspection, and let X denote the number of defective components found.

(a) The distribution of the random variable X is (choose one)
(i) binomial (ii) hypergeometric (iii) negative binomial (iv) Poisson.

(b) Use R commands to find $P(X \leq 3)$.

(c) Which distribution from those listed in part (a) can be used as an approximation to the distribution of X?

(d) Using the approximate distribution, give an approximation to the probability $P(X \leq 3)$, and compare it with the exact probability found in part (b).

16. A particular website generates income when people visiting the site click on ads. The number of people visiting the website is modeled as a Poisson process with rate $\alpha = 30$ per second. Of those visiting the site, 60% click on an ad. Let Y denote the number of those who will click on an ad over the next minute.

(a) The distribution of the random variable Y is (choose one)
(i) binomial (ii) hypergeometric (iii) negative binomial (iv) Poisson.
(*Hint.* See Example 3.4-16.)

(b) Give the mean and variance of Y.

(c) Use R commands to find the probability that $Y > 1100$.

17. Structural loads are forces applied to a structure or its components. Loads cause stresses that can lead to structural failure. It has been suggested that the occurrence of live (or probabilistic) structural loads over time in aging concrete structures can be modeled by a Poisson process with a rate of two occurrences per year. Find the probability that more than two loads will occur during the next quarter of a year.

18. During a typical Pennsylvania winter, I80 averages 1.6 potholes per 10 miles. A certain county is responsible for repairing potholes in a 30-mile stretch of the interstate. Let X denote the number of potholes the county will have to repair at the end of next winter.

(a) The distribution of the random variable X is (choose one)
(i) binomial (ii) hypergeometric (iii) negative binomial (iv) Poisson.

(b) Give the expected value and variance of X.

(c) Find $P(4 < X \leq 9)$.

(d) The cost of repairing a pothole is \$5000. If Y denotes the county's pothole repair expense for next winter, find the mean value and variance of Y.

19. A typesetting agency used by a scientific journal employs two typesetters. Let X_1 and X_2 denote the number of errors committed by typesetter 1 and 2, respectively, when asked to typeset an article. Suppose that X_1 and X_2 are Poisson random variables with expected values 2.6 and 3.8, respectively.

(a) What is the variance of X_1 and of X_2?

(b) Suppose that typesetter 1 handles 60% of the articles. Find the probability that the next article will have no errors.

(c) If an article has no typesetting errors, what is the probability it was typeset by the second typesetter?

20. An engineer at a construction firm has a subcontract for the electrical work in the construction of a new office building. From past experience with this electrical subcontractor, the engineer knows that each light switch that is installed will be faulty with probability $p = 0.002$ independent of the other switches installed. The building will have $n = 1500$ light switches in it. Let X be the number of faulty light switches in the building.

(a) The distribution of the random variable X is (choose one)
(i) binomial (ii) hypergeometric (iii) negative binomial (iv) Poisson.

(b) Use R commands to find $P(4 \leq X \leq 8)$.

(c) Which distribution from those listed in part (a) can be used as an approximation to the distribution of X?

(d) Using the approximate distribution, give an approximation to the probability $P(4 \leq X \leq 8)$, and compare it with the exact probability found in part (b).

(e) Compute the exact and approximate probability of no faulty switches.

21. Suppose that a simple random sample of 200 is taken from the shipment of 10,000 electronic components of Exercise 15, which contais 300 defective components, and let Y denote the number of defective components in the sample.

(a) The random variable Y has a hypergeometric(M_1, M_2, n) distribution, which can be approximated by a binomial(n, p) distribution, which can be approximated by a Poisson(λ) distribution. Specify the parameters of each distribution mentioned in the last sentence.

(b) Use R commands to compute the exact probability $P(Y \leq 10)$, as well as the two approximations to this probability mentioned in part (a).

22. Let X be the random variable that counts the number of events in each of the following cases.

(a) The number of fish caught by an angler in an afternoon.

(b) The number of disabled vehicles abandoned on I95 in a year.

(c) The number of wrongly dialed telephone numbers in a given city in an hour.

(d) The number of people who reach the age of 100 in a given city.

For each case explain how the Poisson approximation to the binomial distribution can be used to justify the use of the Poisson model for X, and discuss the assumptions that are needed for this justification.

23. Let $X(t)$ be a Poisson process with rate α.

(a) Use words to justify that the events

$$[X(t) = 1] \cap [X(1) = 1] \text{ and}$$
$$[X(t) = 1] \cap [X(1) - X(t) = 0]$$

are the same

(b) Use Proposition 3.4-2 to find the probability of the event in (a) when $\alpha = 2$ and $t = 0.6$.

(c) It is given that $X(1) = 1$, that is, only one event in the time interval $[0, 1]$. Let T denote the time the event occurred, and let t be between 0 and 1.

(i) Use words to justify that the events $T \leq t$ and $X(t) = 1$ are the same.

(ii) Show that the conditional distribution of T, given that $X(1) = 1$, is uniform in $[0, 1]$ by showing that $P(T \leq t | X(1) = 1) = t$.

3.5 Models for Continuous Random Variables

The simplest continuous distribution, which is the uniform, was introduced in Definition 3.2-5, extended in Example 3.2-5, and further studied in Examples 3.2-4, 3.3-8, and 3.3-16. This section presents in some detail two other useful classes of continuous distributions, the *exponential* and the *normal*. Three additional families of distributions, commonly used in reliability theory, are briefly introduced in the exercises.

Unlike the discrete random variables discussed in Section 3.4, where the nature of the experiment determines the type of probability model under fairly transparent assumptions, we often have no indication as to which probability model will best describe the true distribution of a particular continuous random variable. For example, there may be no a priori knowledge that the probability density function of the life time of a randomly chosen electrical component has the form assumed in Example 3.2-6. For this reason, this section also presents a diagnostic procedure that helps assess the *goodness-of-fit* of a particular probability model to a data set.

3.5.1 THE EXPONENTIAL DISTRIBUTION

A random variable X is said to be an **exponential**, or to have the exponential distribution with parameter λ, denoted by $X \sim \text{Exp}(\lambda)$, if its PDF is

$$f(x) = \begin{cases} \lambda e^{-\lambda x} & \text{if } x \geq 0 \\ 0 & \text{if } x < 0. \end{cases}$$

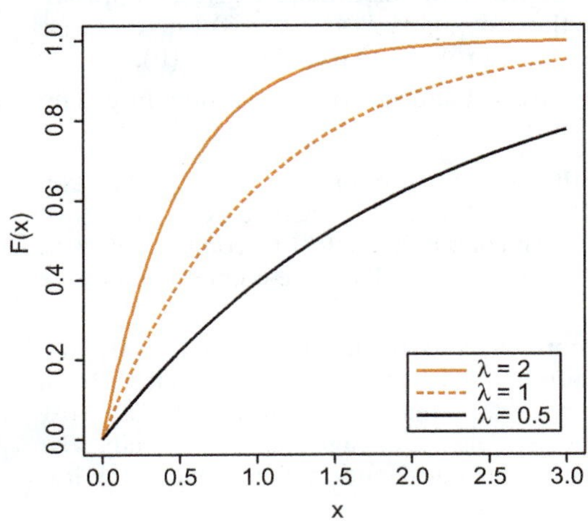

Figure 3-13 PDFs (left panel) and CDFs (right panel) of three exponential distributions.

The exponential distribution is used in reliability theory as the simplest model for the life time of equipment. (See Exercises 13 and 14 for generalizations of the exponential distribution.) Moreover, as discussed below, the time until the next event of a Poisson process follows the exponential distribution. Thus, the exponential distribution models a wide range of *waiting times*, such as the time for the next customer to arrive at a service station, the time until the next bank or investment firm failure, the time until the next outbreak of hostilities, the time until the next earthquake, or the time until the next component of a multi-component system fails.

The PDF used in Example 3.2-6 is exponential with $\lambda = 0.001$. With an integration similar to the one used in that example, it follows that the CDF of the Exp(λ) distribution is

$$F(x) = \begin{cases} 1 - e^{-\lambda x} & \text{if } x \geq 0 \\ 0 & \text{if } x < 0. \end{cases} \tag{3.5.1}$$

Figure 3-13 presents plots of the PDF and CDF of the exponential distribution for different values of the parameter λ.

Some R commands related to the exponential distribution are given in Exercise 4.

Examples 3.3-6 and 3.3-14 find the mean value and variance, respectively, of an exponential distribution with $\lambda = 0.1$, while Example 3.3-19 finds the percentiles of an exponential distribution with $\lambda = 0.001$. The same type of calculations yield the following formulas for a general λ:

Mean, Variance, and Percentiles of the Exponential Distribution

$$\mu = \frac{1}{\lambda}, \quad \sigma^2 = \frac{1}{\lambda^2}, \quad x_\alpha = -\frac{\log(\alpha)}{\lambda} \tag{3.5.2}$$

Example 3.5-1

Suppose the useful life time, in years, of a personal computer (PC) is exponentially distributed with parameter $\lambda = 0.25$. A student entering a four-year undergraduate program inherits a two-year-old PC from his sister who just graduated. Find the probability the useful life time of the PC the student inherited will last at least until the student graduates.

Solution

Let X denote the useful life time of the PC. The PC has already operated for two years and we want the probability it will last at least four more years. In mathematical notation this is expressed as $P(X > 2+4|X > 2)$. Using the definition of conditional probability and the form of the CDF of an exponential random variable we have

$$\begin{aligned} P(X > 2+4|X > 2) &= \frac{P([X > 2+4] \cap [X > 2])}{P(X > 2)} \\ &= \frac{P([X > 2+4])}{P(X > 2)} = \frac{e^{-0.25\times(2+4)}}{e^{-0.25\times 2}} \\ &= e^{-0.25\times 4}. \end{aligned}$$

Since $P(X > 4)$ also equals $e^{-0.25\times 4}$, it follows that the two-year-old PC has the same probability of lasting until the student graduates as a brand new PC would. ■

A nonnegative random variable X is said to have the **memoryless property**, also called the **no-aging property**, if for all $s, t > 0$,

Memoryless Property of a Random Variable

$$P(X > s+t|X > s) = P(X > t) \qquad \textbf{(3.5.3)}$$

By a calculation similar to the one done in Example 3.5-1 it follows that the exponential random variable has the memoryless property. In fact, it can be shown that the exponential is the only distribution with the memoryless property.

The Poisson-Exponential Connection For a Poisson process, let T_1 be the time the first event occurs, and for $i = 2, 3, \ldots$, let T_i denote the time elapsed between the occurrence of the $(i-1)$-st and the ith event. For example, $T_1 = 3$ and $T_2 = 5$ means that the first occurrence of the Poisson process happened at time 3 and the second at time 8. The times $T_1, T_2, \ldots$ are called *interarrival times.*

Proposition 3.5-1 If $X(s)$, $s \geq 0$, is a Poisson process with rate α, the interarrival times have the exponential distribution with PDF $f(t) = \alpha e^{-\alpha t}$, $t > 0$.

Proof Let T_1 be the first arrival time. To find the PDF of T_1 we will first find $1 - F_{T_1}(t) = P(T_1 > t)$. This is done by noting that event $T_1 > t$ is equivalent to the event $X(t) = 0$ since both are equivalent to the statement that no event occurred in the interval $(0, t]$. Thus,

$$P(T_1 > t) = P(X(t) = 0) = e^{-\alpha t}.$$

Hence, $F_{T_1}(t) = P(T_1 \leq t) = 1 - e^{-\alpha t}$, and upon differentiation we find that the PDF of T_1 is as specified in the proposition. To show that T_2, the second interarrival time, has the same distribution note that, by Proposition 3.4-2, if we start recording the events that occur after time T_1 we obtain a new Poisson process for which time zero is set to T_1 and has the same rate α. Since in this new Poisson process T_2 is the first interarrival time, it follows that T_2 has the same distribution as T_1. That all interarrival times have the same distribution follows by a similar argument. ■

Example 3.5-2

User log-ons to a college's computer network can be modeled as a Poisson process with a rate of 10 per minute. If the system's administrator begins tracking the number of log-ons at 10:00 a.m., find the probability that the first log-on recorded occurs between 10 and 20 seconds after that.

Solution

With time zero set at 10:00 a.m., let T_1 denote the time, in minutes, of the first arrival. Since by Proposition 3.5-1 $T_1 \sim \text{Exp}(10)$, the CDF formula given in (3.5.1) yields

$$P\left(\frac{10}{60} < T < \frac{20}{60}\right) = e^{-10\times(10/60)} - e^{-10\times(20/60)} = 0.1532.$$

3.5.2 THE NORMAL DISTRIBUTION

A random variable is said to have the **standard normal** distribution if its PDF and CDF, which are denoted (universally) by ϕ and Φ, respectively, are

$$\phi(z) = \frac{1}{\sqrt{2\pi}}e^{-z^2/2} \quad \text{and} \quad \Phi(z) = \int_{-\infty}^{z} \phi(x)\,dx$$

for $-\infty < z < \infty$. A standard normal random variable is denoted by Z. Note that the PDF ϕ is symmetric about zero; see Figure 3-14.

A random variable X is said to have the **normal** distribution, with parameters μ and σ, denoted by $X \sim N(\mu, \sigma^2)$, if its PDF and CDF are

$$f(x) = \frac{1}{\sigma}\phi\left(\frac{x-\mu}{\sigma}\right) \quad \text{and} \quad F(x) = \Phi\left(\frac{x-\mu}{\sigma}\right)$$

for $-\infty < x < \infty$. Thus,

$$f(x) = \frac{1}{\sqrt{2\pi\sigma^2}}\exp(-[x-\mu]^2/[2\sigma^2]),$$

which is symmetric about μ. Thus, μ is both the mean, the median, and the mode of X. The parameter σ is the standard deviation of X. For $\mu = 0$ and $\sigma = 1$, X is standard normal and is denoted by Z.

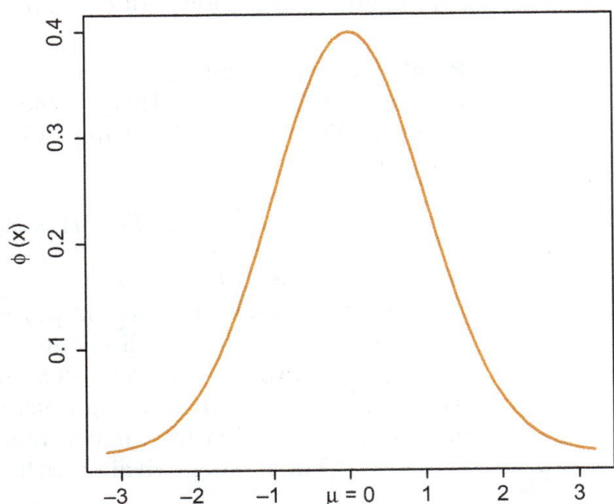

Figure 3-14 PDF of the $N(0, 1)$ distribution.

The normal PDF is difficult to integrate and will not be used for calculating probabilities by integration. Moreover, the CDF does not have a closed form expression. The R commands for the normal(μ, σ^2) PDF, CDF, and percentiles and for simulating normal samples are as follows:

R Commands for the Normal(μ, σ^2) Distribution

```
dnorm(x, μ, σ) # gives the PDF for x in (−∞, ∞)

pnorm(x, μ, σ) # gives the CDF for x in (−∞, ∞)

qnorm(s, μ, σ) # gives the s100th percentile for s in (0,1)

rnorm(n, μ, σ) # gives a sample of n normal(μ, σ²) random
  variables
```

In the above R commands both x and s can be vectors. For example, if $X \sim N(5, 16)$,

- *dnorm(6, 5, 4)* returns 0.09666703 for the value of the PDF of X at $x = 6$.
- *pnorm(c(3, 6), 5, 4)* returns the values of $P(X \leq 3)$ and $P(X \leq 6)$.
- *qnorm(c(0.9, 0.99), 5, 4)* returns 10.12621 and 14.30539 for the 90th and 99th percentile of X, respectively.

The standard normal PDF $\Phi(z)$ is tabulated in Table A.3 for values of z from 0 to 3.09 in increments of 0.01. For the rest of this section we will learn how to use Table A.3 not only for finding probabilities and percentiles of the standard normal random variable, but for any other normal random variable. The ability to use only one table, that for the standard normal, for finding probabilities and percentiles of any normal random variable is due to an interesting property of the normal distribution, which is given in the following proposition.

Proposition 3.5-2 If $X \sim N(\mu, \sigma^2)$ and a, b are any real numbers, then

$$a + bX \sim N(a + b\mu,\ b^2\sigma^2). \tag{3.5.4}$$

■

The new element of this proposition is that a linear transformation of a normal random variable is also a normal random variable. That the mean value and variance of the transformed variable, $Y = a+bX$, are $a+b\mu$ and $b^2\sigma^2$ follows from Propositions 3.3-1 and 3.3-2, respectively, so there is nothing new in these formulas.

Finding Probabilities We first illustrate the use of Table A.3 for finding probabilities associated with the standard normal random variable.

Example 3.5-3 Let $Z \sim N(0, 1)$. Find (a) $P(-1 < Z < 1)$, (b) $P(-2 < Z < 2)$, and (c) $P(-3 < Z < 3)$.

Solution

In Table A.3, z-values are listed in two decimal places, with the second decimal place identified in the top row of the table. Thus, the z-value 1 is identified by 1.0 in the left column of the table and 0.00 in the top row of the table. The probability $\Phi(1) =$

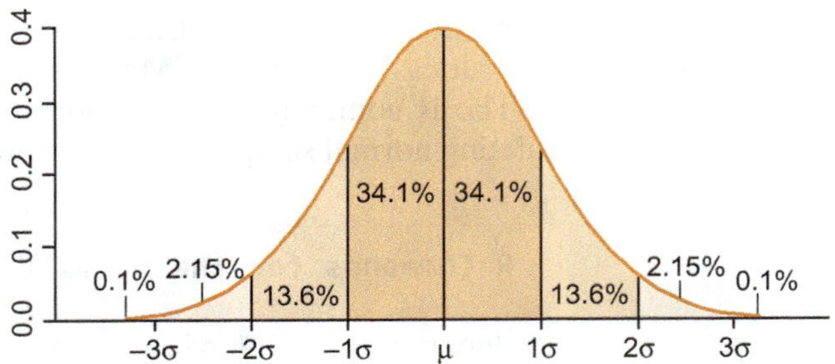

Figure 3-15 Illustration of the 68-95-99.7% property of normal distributions.

$P(Z \leq 1)$ is the number that corresponds to the row and column identified by 1.0 and 0.00, which is 0.8413. Since negative values are not listed in Table A.3, $\Phi(-1) = P(Z \leq -1)$ is found by exploiting the fact that the standard normal distribution is symmetric about zero. This means that the area under the $N(0, 1)$ PDF to the left of -1 is equal to the area under it to the right of 1; see Figure 3-14. Hence,

$$\Phi(-1) = 1 - \Phi(1),$$

and the same relation holds with any positive number substituting 1. Thus, the answer to part (a) is

$$\begin{aligned} P(-1 < Z < 1) &= \Phi(1) - \Phi(-1) = 0.8413 - (1 - 0.8413) \\ &= 0.8413 - 0.1587 = 0.6826. \end{aligned}$$

Working similarly, we find the following answers for parts (b) and (c):

$$\begin{aligned} P(-2 < Z < 2) &= \Phi(2) - \Phi(-2) = 0.9772 - 0.0228 = 0.9544, \quad \text{and} \\ P(-3 < Z < 3) &= \Phi(3) - \Phi(-3) = 0.9987 - 0.0013 = 0.9974. \end{aligned}$$

Thus, approximately 68% of the values of a standard normal random variable fall within one standard deviation from its mean, approximately 95% fall within two standard deviations of its mean, and approximately 99.7% of its values fall within three standard deviations of its mean. This is known as the 68-95-99.7% rule. (See also Figure 3-15.) ■

The use of Table A.3 for finding probabilities associated with any normal random variable is made possible through the following corollary to Proposition 3.5-2.

Corollary 3.5-1

If $X \sim N(\mu, \sigma^2)$, then

1. $\dfrac{X - \mu}{\sigma} \sim N(0, 1)$, and
2. $P(a \leq X \leq b) = \Phi\left(\frac{b-\mu}{\sigma}\right) - \Phi\left(\frac{a-\mu}{\sigma}\right)$.

To show how the corollary follows from Proposition 3.5-2, first apply formula (3.5.4) with $a = -\mu$ and $b = 1$ to see that if $X \sim N(\mu, \sigma^2)$, then

$$X - \mu \sim N(0, \sigma^2).$$

A second application of the formula (3.5.4), now on the normal random variable $X - \mu$ with $a = 0$ and $b = 1/\sigma$, yields

$$\frac{X - \mu}{\sigma} \sim N(0, 1).$$

In words, part (1) of Corollary 3.5-1 means that any normal random variable, X, can be **standardized** (i.e., transformed to a standard normal random variable, Z), by subtracting from it its mean and dividing by its standard deviation. This implies that any event of the form $a \le X \le b$ can be expressed in terms of the standardized variable:

$$[a \le X \le b] = \left[\frac{a - \mu}{\sigma} \le \frac{X - \mu}{\sigma} \le \frac{b - \mu}{\sigma}\right].$$

Thus, part (2) of Corollary 3.5-1 follows from

$$P(a \le X \le b) = P\left(\frac{a - \mu}{\sigma} \le \frac{X - \mu}{\sigma} \le \frac{b - \mu}{\sigma}\right) = \Phi\left(\frac{b - \mu}{\sigma}\right) - \Phi\left(\frac{a - \mu}{\sigma}\right),$$

where the last equality follows from the fact that $(X - \mu)/\sigma$ has the standard normal distribution.

Example 3.5-4

Let $X \sim N(1.25, 0.46^2)$. Find (a) $P(1 \le X \le 1.75)$ and (b) $P(X > 2)$.

Solution

A direct application of part (2) of Corollary 3.5-1, yields

$$\begin{aligned} P(1 \le X \le 1.75) &= \Phi\left(\frac{1.75 - 1.25}{0.46}\right) - \Phi\left(\frac{1 - 1.25}{0.46}\right) \\ &= \Phi(1.09) - \Phi(-0.54) = 0.8621 - 0.2946 = 0.5675. \end{aligned}$$

Working similarly for the event in part (b), we have

$$P(X > 2) = P\left(Z > \frac{2 - 1.25}{0.46}\right) = 1 - \Phi(1.63) = 0.0516.$$

■

Another consequence of Corollary 3.5-1 is that the 68-95-99.7% rule of the standard normal seen in Example 3.5-3 applies for any normal random variable $X \sim N(\mu, \sigma^2)$:

$$\begin{aligned} P(\mu - 1\sigma < X < \mu + 1\sigma) &= P(-1 < Z < 1) = 0.6826, \\ P(\mu - 2\sigma < X < \mu + 2\sigma) &= P(-2 < Z < 2) = 0.9544, \quad \text{and} \\ P(\mu - 3\sigma < X < \mu + 3\sigma) &= P(-3 < Z < 3) = 0.9974. \end{aligned}$$

Finding Percentiles According to the notation introduced in Definition 3.3-1, the $(1-\alpha)$-100th percentile of Z will be denoted by z_α. Thus, the area under the standard

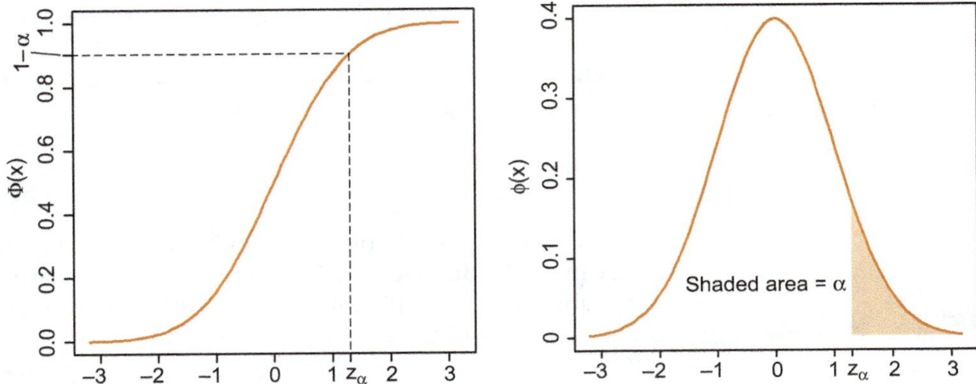

Figure 3-16 The percentile z_α in relation to the standard normal CDF (left panel) and PDF (right panel).

normal PDF to the right of z_α is α, as shown in the right panel of Figure 3-16. The left panel of this figure illustrates the defining property of z_α, that is,

$$\Phi(z_\alpha) = 1 - \alpha,$$

which is used to find z_α. Since the function Φ does not have a closed form expression, we use Table A.3 to solve this equation by first locating $1-\alpha$ in the body of the table and then reading z_α from the margins. If the exact value of $1-\alpha$ does not exist in the main body of the table, then an approximation is used. This process is demonstrated in the following example.

Example 3.5-5

Find the 95th percentile of Z.

Solution

Here $\alpha = 0.05$, so $1-\alpha = 0.95$. However, the exact number 0.95 does not exist in the body of Table A.3. So we use the entry that is closest to but larger than 0.95 (which is 0.9505), as well as the entry that is closest to but smaller than 0.95 (which is 0.9495), and approximate $z_{0.05}$ by averaging the z-values that correspond to these two closest entries: $z_{0.05} \simeq (1.64 + 1.65)/2 = 1.645$. ■

The use of Table A.3 for finding percentiles of any normal random variable is made possible through the following corollary to Proposition 3.5-2.

Corollary 3.5-2

Let $X \sim N(\mu, \sigma^2)$, and let x_α denote the $(1-\alpha)$-100th percentile of X. Then,

$$x_\alpha = \mu + \sigma z_\alpha. \tag{3.5.5}$$

For the proof of this corollary it must be shown that $P(X \le \mu + \sigma z_\alpha) = 1 - \alpha$. But this follows by an application of part (2) of Corollary 3.5-1 with $a = -\infty$ and $b = \mu + \sigma z_\alpha$:

$$P(X \le \mu + \sigma z_\alpha) = \Phi(z_\alpha) - \Phi(-\infty) = 1 - \alpha - 0 = 1 - \alpha.$$

Example 3.5-6

Let $X \sim N(1.25, 0.46^2)$. Find the 95th percentile, $x_{0.05}$, of X.

Solution

From (3.5.5) we have

$$x_{0.05} = 1.25 + 0.46 z_{0.05} = 1.25 + (0.46)(1.645) = 2.01.$$

The Q-Q Plot As already mentioned, most experiments resulting in the measurement of a continuous random variable provide little insight as to which probability model best describes the distribution of the measurements. Thus, several procedures have been devised to test the *goodness-of-fit* of a particular model to a random sample obtained from some population. Here we discuss a very simple graphical procedure, called the **Q-Q plot**, as it applies for checking the goodness-of-fit of the normal distribution.

The basic idea of the Q-Q plot is to plot the sample percentiles, which are the ordered sample values, against with the corresponding percentiles of the assumed model distribution. Since sample percentiles estimate corresponding population percentiles, if the assumed model distribution is a good approximation to the true population distribution, the plotted points should fall approximately on a straight line of angle 45° that passes through the origin.

For example, in a sample of size 10 the order statistics are the 5th, 15th, ..., 95th sample percentiles; see Definition 1.7-2. To check if this sample could have come from the standard normal distribution, the sample percentiles would be plotted against the standard normal percentiles, which can be obtained from the R command *qnorm(seq(0.05, 0.95, 0.1))*. In fact, the sample percentiles would be plotted against the standard normal percentiles even for checking if the sample could have come from a normal(μ, σ^2), for unspecified μ and σ. This is because the normal(μ, σ^2) percentiles, x_α, are related to the normal$(0, 1)$ percentiles, z_α, through $x_\alpha = \mu + \sigma z_\alpha$, which is a linear relationship. Thus, if the normal model is correct, the plotted points would fall on a straight line, though not necessarily the 45° line through the origin.

In R there is a customized command for the normal Q-Q plot. With the data in the R object x, two versions of the command are as follows:

R Commands for the Normal Q-Q Plot

```
qqnorm(x); qqline(x, col=2)

qqnorm(x, datax=T); qqline(x, datax=T, col=2)
```

The first version has the sample percentiles on the y-axis, and the second puts them on the x-axis.

The two plots in Figure 3-17, which are based on simulated samples of size 50, illustrate the extent to which the plotted points conform to a straight line when the data have indeed come from a normal distribution, or do not conform to a straight line when the data have come from an exponential distribution. The R commands that generated the left panel of Figure 3-17 are *set.seed(111); x=rnorm(50); qqnorm(x, datax=T); qqline(x, datax=T, col=2)*, and those that generated the right panel are *x=rexp(50); qqnorm(x, datax=T); qqline(x, datax=T, col=2)*.

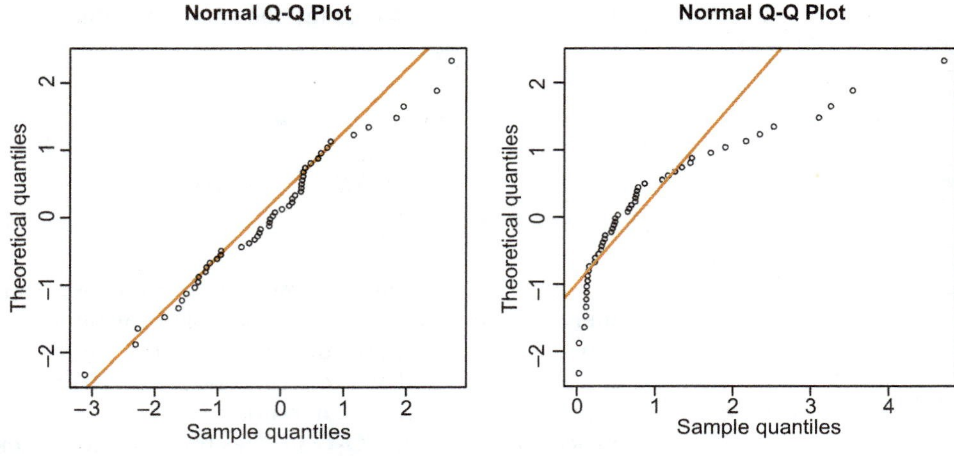

Figure 3-17 Normal Q-Q plots with normal data (left) and exponential data (right).

Exercises

1. The lifespan of a car battery averages six years. Suppose the battery lifespan follows an exponential distribution.

(a) Find the probability that a randomly selected car battery will last more than four years.

(b) Find the variance and the 95th percentile of the battery lifespan.

(c) Suppose a three-year-old battery is still going strong.
 (i) Find the probability the battery will last an additional five years.
 (ii) How much longer is this battery expected to last?

2. The number of wrongly dialed phone calls you receive can be modeled as a Poisson process with the rate of one per month.

(a) Find the probability that it will take between two and three weeks to get the first wrongly dialed phone call.

(b) Suppose that you have not received a wrongly dialed phone call for two weeks. Find the expected value and variance of the additional time until the next wrongly dialed phone call.

3. Justify that the no-aging or memoryless property of the exponential random variable X, stated in (3.5-3), can be equivalently restated as

$$P(X \leq s+t|X \geq s) = 1 - \exp\{-\lambda t\}.$$

4. Use the R command *set.seed(111); hist(rexp(10000), breaks=35, freq=F)* to generate a sample of size 10,000 from the exponential(1) distribution and to plot its histogram, and the additional R command *curve(dexp, 0, 8, add=T)* to superimpose the exponential(1) PDF on the graph. Does the histogram provide a reasonable approximation to the PDF? Repeat the above set of commands with a sample size of 1000, using *breaks=27*. Comment on how well this histogram approximates the PDF.

5. The yield strength (ksi) for A36 steel is normally distributed with $\mu = 43$ and $\sigma = 4.5$.

(a) What is the 25th percentile of the distribution of A36 steel strength?

(b) What strength value separates the strongest 10% from the others?

(c) What is the value of c such that the interval $(43 - c, 43 + c)$ includes 99% of all strength values?

(d) What is the probability that at most three of 15 independently selected A36 steels have strength less than 43?

6. The mean weight of frozen yogurt cups in an ice cream parlor is 8 oz. Suppose the weight of each cup served is normally distributed with standard deviation 0.5 oz, independently of others.

(a) What is the probability of getting a cup weighing more than 8.64 oz?

(b) What is the probability of getting a cup weighing more than 8.64 oz three days in a row?

7. The resistance for resistors of a certain type is a random variable X having the normal distribution with mean 9 ohms and standard deviation 0.4 ohms. A resistor is acceptable if its resistance is between 8.6 and 9.8 ohms.

(a) What is the probability that a randomly chosen resistor is acceptable?

(b) What is the probability that out of four randomly and independently selected resistors, two are acceptable?

8. Admission officers in Colleges A and B use SAT scores as their admission criteria. SAT scores are normally distributed with mean 500 and standard deviation 80. College A accepts people whose scores are above 600, and College B accepts the top 1% of people in terms of their SAT scores.

(a) What percentage of high school seniors can get into College A?

(b) What is the minimum score needed to get accepted by College B ?

9. The finished inside diameter of a piston ring is normally distributed with a mean of 10 cm and a standard deviation of 0.03 cm.

(a) Above what value of inside diameter will 85.08% of the piston rings fall?

(b) What is the probability that the diameter of a randomly selected piston will be less than 10.06?

10. A machine manufactures tires with a tread thickness that is normally distributed with mean 10 millimeters (mm) and standard deviation 2 mm. The tire has a 50,000-mile warranty. In order to last for 50,000 miles the tread thickness must be at least 7.9 mm. If the thickness of tread is measured to be less than 7.9 mm, then the tire is sold as an alternative brand with a warranty of less than 50,000 miles.

(a) Find the expected proportion of tires sold under the alternative brand.

(b) The demand for the alternative brand of tires is such that 30% of the total output should be sold under the alternative brand name. What should the critical thickness, originally 7.9 mm, be set at in order to meet the demand?

11. Answer the following questions.

(a) Use the R command *x=runif(50)* to generate a simulated sample of size 50 from the uniform$(0, 1)$ distribution and use commands like those given in Section 3.5.2 to construct a normal Q-Q plot. Could the simulated sample of 50 have come from a normal distribution? Explain.

(b) Use the R command *x=rgamma(50, 1, 1)* to generate a simulated sample of size 50 from the gamma$(1, 1)$ distribution (see Exercise 13) and use commands like those given in Section 3.5.2 to construct a normal Q-Q plot. Could the simulated sample of 50 have come from a normal distribution? Explain.

Probability Models Used in Reliability Theory

12. A random variable T is said to have the **log-normal**$(\mu_{\ln}, \sigma_{\ln})$ distribution if $\log T \sim N(\mu_{\ln}, \sigma^2_{\ln})$, where log is the natural logarithm. The mean value and variance of T are

$$\mu_T = e^{\mu_{\ln}+\sigma^2_{\ln}/2}, \quad \sigma^2_T = e^{2\mu_{\ln}+\sigma^2_{\ln}}\left(e^{\sigma^2_{\ln}} - 1\right).$$

The log-normal$(0, 1)$ distribution is called the **standard log-normal** distribution.

(a) Show that if T has the log-normal$(\mu_{\ln}, \sigma_{\ln})$ distribution, its CDF is given by

$$F_T(t) = \Phi\left(\frac{\log t - \mu_{\ln}}{\sigma_{\ln}}\right) \quad \text{for } t > 0,$$

and $F_T(t) = 0$ for $t < 0$. (*Hint.* $F_T(t) = P(T \le t) = P(\log T \le \log t)$, and $\log T \sim N(\mu_{\ln}, \sigma^2_{\ln})$.)

(b) Use the R commands *curve(dlnorm(x, 0, 1), 0, 10, col=1, ylab="Log-Normal PDFs")*, *curve(dlnorm(x, 1, 1), 0, 10, add=T, col=2)*, and *curve(dlnorm(x, 1.5, 1), 0, 10, add=T, col=3)* to superimpose the plots of three log-normal PDFs corresponding to different parameter values. Superimpose the plots of the corresponding three log-normal CDFs by making appropriate changes to these commands (*dlnorm* changes to *plnorm*, and PDFs changes to CDFs).

(c) Using the formulas given above, compute the mean and variance of the log-normal$(0, 1)$, log-normal$(5, 1)$ and log-normal$(5, 2)$ distributions.

(d) The R command *qlnorm(0.95)*, which is equivalent to *qlnorm(0.95, 0, 1)*, gives the 95th percentile of the standard log-normal distribution. Verify that the R commands *log(qlnorm(0.95))* and *qnorm(0.95)* return the same value, which is the 95th percentile of the standard normal distribution, and provide an explanation for this.

13. A random variable T has a **gamma** distribution with *shape* parameter $\alpha > 0$ and *scale* parameter $\beta > 0$ if its PDF is zero for negative values and

$$f_T(t) = \frac{1}{\beta^\alpha \Gamma(\alpha)} t^{\alpha-1} e^{-t/\beta} \quad \text{for } t \ge 0,$$

where Γ is the *gamma function* defined by $\Gamma(\alpha) = \int_0^\infty t^{\alpha-1}e^{-t}dt$. The most useful properties of the gamma function are: $\Gamma(1/2) = \pi^{1/2}$, $\Gamma(\alpha) = (\alpha - 1)\Gamma(\alpha - 1)$, for $\alpha > 1$, and $\Gamma(r) = (r - 1)!$ for an integer $r \ge 1$. The mean and variance of a gamma(α, β) distribution are given by

$$\mu_T = \alpha\beta, \quad \sigma^2_T = \alpha\beta^2.$$

When $\alpha = 1$ we get the family of exponential distributions with $\lambda = 1/\beta$. Additionally, for $\alpha = r$, with r integer ≥ 1, we get the family of **Erlang** distributions, which models the time until the rth occurrence in a Poisson process. Finally, the **chi-square** distribution with ν degrees of freedom, where $\nu \ge 1$ is an integer, denoted by χ^2_ν, corresponds to $\alpha = \nu/2$ and $\beta = 2$.

(a) Use the R commands *curve(dgamma(x, 1, 1), 0, 7, ylab="Gamma PDFs")*, *curve(dgamma(x, 2, 1), 0, 7, add=T, col=2)*, and *curve(dgamma(x, 4, 1), 0, 7, add=T, col=3)* to superimpose the plots of three gamma PDFs corresponding to different parameter values. Superimpose the plots of the corresponding three gamma CDFs by making appropriate changes to these commands (*dgamma* changes to *pgamma*, and PDFs changes to CDFs).

(b) Using the formulas given above, compute the mean and variance of the gamma(2, 1), gamma(2, 2), gamma(3, 1), and gamma(3, 2) distributions.

(c) Use the R command *qgamma(0.95, 2, 1)* to find the 95th percentile of the gamma(2, 1) distribution. Making appropriate changes to this command, find the 95th percentile of the gamma(2, 2), gamma(3, 1), and gamma(3, 2) distributions.

14. A random variable T is said to have a **Weibull** distribution with *shape* parameter $\alpha > 0$ and *scale* parameter $\beta > 0$ if its PDF is zero for $t < 0$ and

$$f_T(t) = \frac{\alpha}{\beta^\alpha} t^{\alpha-1} e^{-(t/\beta)^\alpha} \quad \text{for } t \geq 0.$$

The CDF of a Weibull(α, β) distribution has the following closed form expression:

$$F_T(t) = 1 - e^{-(t/\beta)^\alpha}.$$

When $\alpha = 1$ the Weibull PDF reduces to the exponential PDF with $\lambda = 1/\beta$. The mean and variance of a Weibull(α, β) distribution are given by

$$\mu_T = \beta\Gamma\left(1 + \frac{1}{\alpha}\right),$$

$$\sigma_T^2 = \beta^2 \left\{ \Gamma\left(1 + \frac{2}{\alpha}\right) - \left[\Gamma\left(1 + \frac{1}{\alpha}\right)\right]^2 \right\},$$

where Γ is the gamma function defined in Exercise 13.

(a) Use the R commands *curve(dweibull(x, 0.5, 1), 0, 4)*, *curve(dweibull(x, 1, 1), 0, 4, add=T, col="red")*, *curve(dweibull(x, 1.5, 1), 0, 4, add=T, col="blue")*, and *curve(dweibull(x, 2, 1), 0, 4, add=T, col="green")* to superimpose four Weibull PDFs, noting that the second corresponds to the exponential(1) distribution.

(b) One of the imbedded functions in R is the gamma function. Use the R commands *10*gamma(1+1/0.2)* and *10**2*(gamma(1+2/0.2)-gamma(1+1/0.2)**2)* to find the mean and variance, respectively, of the Weibull(0.2, 10) distribution.

(c) Use the formula for the Weibull CDF given above to find $P(20 \leq T < 30)$, where $T \sim$ Weibull (0.2, 10). Confirm your answer with the R command *pweibull(30, 0.2, 10)-pweibull(20, 0.2, 10)*.

(d) Find the 95th percentile of T having the Weibull(0.2, 10) distribution by solving the equation $F_T(t_{0.05}) = 0.95$, where F_T is the Weibull CDF given above with parameters $\alpha = 0.2$ and $\beta = 10$. Confirm your answer with the R command *qweibull(0.95, 0.2, 10)*.

Chapter 4

Jointly Distributed Random Variables

4.1 Introduction

When experiments record multivariate observations (see Section 1.4), the behavior of each individual variable is, typically, not the primary focus of the investigation. For example, studies of atmospheric turbulence may focus on understanding and quantifying the degree of relationship between the components X, Y, and Z of wind velocity; studies of automobile safety may focus on the relationship between the velocity X and stopping distance Y under different road and weather conditions; and understanding the relationship between the diameter at breast height X and age of a tree can lead to an equation for predicting age from the (easier to measure) diameter.

In this chapter we will introduce, among other things, the notion of *correlation*, which serves as a quantification of the relationship between two variables, and the notion of a *regression function*, which forms the basis for predicting one variable from another. These concepts follow from the *joint distribution* of the random variables. Moreover, the joint distribution of the observations in a simple random sample leads to the distribution of statistics, such as the sample average, which forms the basis of statistical inference. Formulas for the mean and variance of sums will be derived, while a more complete discussion of the distribution of sums will be given in the next chapter. Finally, some of the most common probability models for joint distributions will be presented.

4.2 Describing Joint Probability Distributions

4.2.1 The Joint and Marginal PMF

Definition 4.2-1
The **joint**, or **bivariate**, probability mass function (PMF) of the jointly discrete random variables X and Y is defined as

$$p(x, y) = P(X = x, Y = y).$$

If $\mathcal{S} = \{(x_1, y_1), (x_2, y_2), \ldots\}$ is the sample space of (X, Y), Axioms 2.4.1 and 2.4.2 of probability imply

$$p(x_i, y_i) \geq 0 \text{ for all } i, \quad \text{and} \quad \sum_{\text{all } (x_i, y_i) \in \mathcal{S}} p(x_i, y_i) = 1. \tag{4.2.1}$$

Moreover, by part (2) of Proposition 2.4-1,

$$P(a < X \leq b, c < Y \leq d) = \sum_{i: a < x_i \leq b, c < y_i \leq d} p(x_i, y_i). \tag{4.2.2}$$

In the context of joint distributions, the distributions of individual variables are called **marginal distributions**. (Recall the *marginal histograms* used in Section 1.5.2.) The marginal PMFs of X and Y are obtained as

Obtaining the Marginal PMFs from the Joint PMF

$$p_X(x) = \sum_{y \in \mathcal{S}_Y} p(x, y), \quad p_Y(y) = \sum_{x \in \mathcal{S}_X} p(x, y) \tag{4.2.3}$$

Example 4.2-1

Let X, Y have the joint PMF as shown in the following table.

	$p(x, y)$	y = 1	y = 2
x	1	0.034	0.134
	2	0.066	0.266
	3	0.100	0.400

This PMF is illustrated in Figure 4-1.

(a) Find $P(0.5 < X \leq 2.5, 1.5 < Y \leq 2.5)$ and $P(0.5 < X \leq 2.5)$.

(b) Find the marginal PMF of Y.

Solution

(a) By relation (4.2.2), $P(0.5 < X \leq 2.5, 1.5 < Y \leq 2.5)$ is the sum of $p(x_i, y_i)$ for all (x_i, y_i) such that $0.5 < x_i \leq 2.5$ and $1.5 < y_i \leq 2.5$. These two conditions are satisfied for the (x, y) pairs (1, 2) and (2, 2). Thus,

$$P(0.5 < X \leq 2.5, 1.5 < Y \leq 2.5) = p(1, 2) + p(2, 2) = 0.134 + 0.266 = 0.4.$$

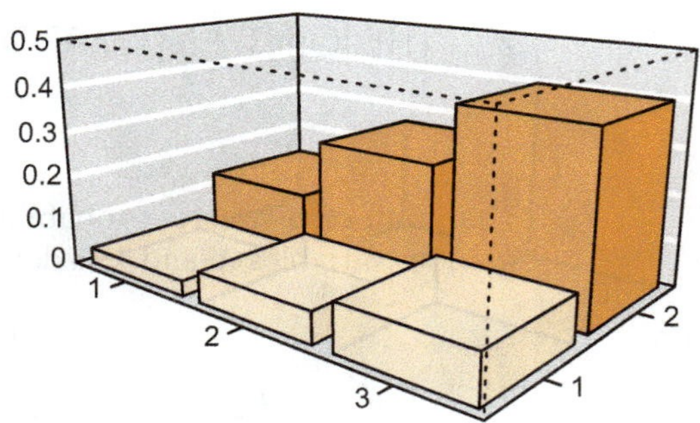

Figure 4-1 3D barplot of the bivariate PMF of Example 4.2-1.

Next, by (4.2.2) again, $P(0.5 < X \le 2.5) = P(0.5 < X \le 2.5, -\infty < Y < \infty)$ is the sum of $p(x_i, y_i)$ for all (x_i, y_i) such that $0.5 < x_i \le 2.5$. This condition is satisfied for the (x, y) pairs (1, 1), (1, 2), (2, 1) and (2, 2). Thus,

$$P(0.5 < X \le 2.5) = p(1,1) + p(1,2) + p(2,1) + p(2,2)$$
$$= 0.034 + 0.134 + 0.066 + 0.266 = 0.5.$$

(b) $P(Y = 1) = p_Y(1)$ is the sum of $p(x_i, y_i)$ for all (x_i, y_i) such that $y_i = 1$. This condition is satisfied for the (x, y) pairs (1, 1), (2, 1) and (3, 1). Thus,

$$p_Y(1) = 0.034 + 0.066 + 0.100 = 0.2,$$

which also follows by a direct application of (4.2.3). In terms of Figure 4-1, $p_Y(1)$ is the height of the block obtained by stacking the three light-colored blocks. Similarly,

$$p_Y(2) = 0.134 + 0.266 + 0.400 = 0.8,$$

which is the height of the block formed by stacking the three dark-colored blocks in Figure 4-1.

By the formula in (4.2.3), the marginal PMF of X in the above example is found by summing the rows in the joint PMF table:

$$p_X(1) = 0.034 + 0.134 = 0.168, \quad p_X(2) = 0.066 + 0.266 = 0.332,$$
$$p_X(3) = 0.1 + 0.4 = 0.5.$$

The table below reproduces the joint PMF of Example 4.2-1 and has an additional column on the right to display the PMF of X as well as an additional row at the bottom to display the PMF of Y.

	$p(x, y)$	y = 1	2	$p_X(x)$
	1	0.034	0.134	0.168
x	2	0.066	0.266	0.332
	3	0.100	0.400	0.500
	$p_Y(y)$	0.200	0.800	1.000

This method of displaying the marginal PMFs, that is, in the margins of the joint PMF table, justifies their name.

If $X_1, X_2, \ldots, X_m$ are jointly discrete, their **joint** or **multivariate** PMF is defined as

$$p(x_1, x_2, \ldots, x_n) = P(X_1 = x_1, X_2 = x_2, \ldots, X_n = x_n).$$

Example 4.2-2

In a batch of 12 laser diodes, three have efficiency below 0.28, four have efficiency between 0.28 and 0.35, and five have efficiency above 0.35. Three diodes are selected at random and without replacement. Let X_1, X_2, and X_3 denote, respectively, the number of diodes with efficiency below 0.28, between 0.28 and 0.35, and above 0.35 in the sample. Find the joint PMF of X_1, X_2, and X_3, and the marginal PMF of X_1.

Solution
The sample space of this experiment consists of triplets (x_1, x_2, x_3) of nonnegative integers satisfying $x_1 + x_2 + x_3 = 3$. Because all samples of size three are equally likely, application of the Generalized Fundamental Principle of Counting yields the following probabilities

$$p(0,0,3) = \frac{\binom{3}{0}\binom{4}{0}\binom{5}{3}}{\binom{12}{3}} = \frac{10}{220}, \quad p(0,1,2) = \frac{\binom{3}{0}\binom{4}{1}\binom{5}{2}}{\binom{12}{3}} = \frac{40}{220}$$

$$p(0,2,1) = \frac{\binom{3}{0}\binom{4}{2}\binom{5}{1}}{\binom{12}{3}} = \frac{30}{220}, \quad p(0,3,0) = \frac{\binom{3}{0}\binom{4}{3}\binom{5}{0}}{\binom{12}{3}} = \frac{4}{220}$$

$$p(1,0,2) = \frac{\binom{3}{1}\binom{4}{0}\binom{5}{2}}{\binom{12}{3}} = \frac{30}{220}, \quad p(1,1,1) = \frac{\binom{3}{1}\binom{4}{1}\binom{5}{1}}{\binom{12}{3}} = \frac{60}{220}$$

$$p(1,2,0) = \frac{\binom{3}{1}\binom{4}{2}\binom{5}{0}}{\binom{12}{3}} = \frac{18}{220}, \quad p(2,0,1) = \frac{\binom{3}{2}\binom{4}{0}\binom{5}{1}}{\binom{12}{3}} = \frac{15}{220}$$

$$p(2,1,0) = \frac{\binom{3}{2}\binom{4}{1}\binom{5}{0}}{\binom{12}{3}} = \frac{12}{220}, \quad p(3,0,0) = \frac{\binom{3}{3}\binom{4}{0}\binom{5}{0}}{\binom{12}{3}} = \frac{1}{220}.$$

By analogy to the formula (4.2.3), $P(X_1 = 0) = p_{X_1}(0)$ is the sum of $p(x_1, x_2, x_3)$ for all (x_1, x_2, x_3) such that $x_1 = 0$. This condition is satisfied for the (x_1, x_2, x_3) triplets (0, 0, 3), (0, 1, 2), (0, 2, 1), (0, 3, 0). Thus,

$$p_{X_1}(0) = \frac{10}{220} + \frac{40}{220} + \frac{30}{220} + \frac{4}{220} = \frac{84}{220}.$$

Similarly, we obtain

$$p_{X_1}(1) = \frac{30}{220} + \frac{60}{220} + \frac{18}{220} = \frac{108}{220},$$

$$p_{X_1}(2) = \frac{15}{220} + \frac{12}{220} = \frac{27}{220},$$

$$p_{X_1}(3) = \frac{1}{220}.$$

In the above example, the random variable X_1 = {the number of diodes with efficiency below 0.28} in a simple random sample of size three, taken from 12 diodes only three of which have efficiency below 0.28, is hypergeometric(3, 9, 3). The R command *dhyper(0:3, 3, 9, 3)* returns 0.381818182, 0.490909091, 0.122727273, and 0.004545455 for the PMF of X_1, which confirms the PMF found in Example 4.2-2.

4.2.2 THE JOINT AND MARGINAL PDF

Definition 4.2-2
The **joint** or **bivariate** density function of the jointly continuous random variables X and Y is a nonnegative function $f(x, y)$ with the property that the probability that (X, Y) will take a value in a region A of the x-y plane equals the volume under the surface defined by $f(x, y)$ and above the region A.

Since the volume under a surface is found by integration, we have

Volume Under the Entire Surface Defined by $f(x, y)$ Is 1

$$\int_{-\infty}^{\infty}\int_{-\infty}^{\infty} f(x,y)\,dx\,dy = 1 \tag{4.2.4}$$

and the probability that (X, Y) will take a value in a region A of the plane is

Probability that (X, Y) Lies in the Region A

$$P((X,Y)\in A) = \int\int_A f(x,y)\,dx\,dy \tag{4.2.5}$$

Taking A to be the rectangle $A = (a,b]\times(c,d] = \{(x,y): a < x \le b,\ c < y \le d\}$ we obtain

Probability of (X, Y) Lying in a Rectangle

$$P(a \le X \le b, c \le Y \le d) = \int_a^b \int_c^d f(x,y)\,dy\,dx \tag{4.2.6}$$

Finally, the marginal PDFs of X and Y are obtained from their joint PDF as

Obtaining the Marginal PDFs from the Joint PDF

$$f_X(x) = \int_{-\infty}^{\infty} f(x,y)\,dy, \qquad f_Y(y) = \int_{-\infty}^{\infty} f(x,y)\,dx \tag{4.2.7}$$

Example 4.2-3

Consider the bivariate density function

$$f(x,y) = \begin{cases} \frac{12}{7}(x^2+xy) & 0 \le x,y \le 1 \\ 0 & \text{otherwise.} \end{cases}$$

(a) Find the probability that $X > Y$.
(b) Find the probability that $X \le 0.6$ and $Y \le 0.4$.
(c) Find the marginal PDF of X and Y.

Solution

(a) The desired probability can be found by integrating f over the region $A = \{(x,y): 0 \le y \le x \le 1\}$. Note that A is not a rectangle, so we use (4.2.5):

$$P(X > Y) = \frac{12}{7}\int_0^1\int_0^x (x^2+xy)\,dy\,dx = \frac{9}{14}.$$

(b) Using (4.2.6) we have

$$P(X \le 0.6,\ Y \le 0.4) = \frac{12}{7}\int_0^{0.6}\int_0^{0.4}(x^2+xy)\,dy\,dx = 0.0741.$$

(c) Using (4.2.7), we have that for $0 \le x \le 1$,

$$f_X(x) = \int_0^1 \frac{12}{7}(x^2+xy)dy = \frac{12}{7}x^2 + \frac{6}{7}x,$$

and $f_X(x) = 0$ for x not in $[0,1]$. Similarly, the marginal PDF of Y is given by

$$f_Y(y) = \int_0^1 \frac{12}{7}(x^2 + xy)dx = \frac{4}{7} + \frac{6}{7}y$$

for $0 \le y \le 1$, and $f_Y(y) = 0$ for y not in $[0,1]$. ■

The **joint** or **multivariate** probability density function of the continuous $(X_1, X_2, \ldots, X_n)$ is a nonnegative function $f(x_1, x_2, \ldots, x_n)$ such that

$$\int_{-\infty}^{\infty} \cdots \int_{-\infty}^{\infty} f(x_1, x_2, \ldots, x_n)\, dx_1 \cdots dx_n = 1 \quad \text{and}$$

$$P((X_1, X_2, \ldots, X_n) \in B) = \int \cdots \int_B f(x_1, x_2, \ldots, x_n)\, dx_1 \cdots dx_n, \qquad \textbf{(4.2.8)}$$

where B is a region in n-dimensional space. A formula analogous to (4.2.7) exists in the multivariate case, and its use is demonstrated in the following example.

Example 4.2-4

Let X_1, X_2, X_3 have the joint PDF given by

$$f(x_1, x_2, x_3) = e^{-x_1}e^{-x_2}e^{-x_3} \quad \text{for } x_1 > 0, x_2 > 0, x_3 > 0,$$

and $f(x_1, x_2, x_3) = 0$ if one or more of the x_i is negative.

(a) Find an expression for $P(X_1 \le t_1, X_2 \le t_2)$.
(b) Find $F_{X_1}(t_1)$, the marginal CDF of X_1.
(c) Find $f_{X_1}(t_1)$, the marginal PDF of X_1.

Solution

(a) Because the event $X_1 \le t_1$ and $X_2 \le t_2$ is equivalent to the event $0 \le X_1 \le t_1$, $0 \le X_2 \le t_2$, and $0 \le X_3 \le \infty$, it follows that

$$P(X_1 \le t_1, X_2 \le t_2) = \int_0^{t_1}\int_0^{t_2}\int_0^{\infty} e^{-x_1}e^{-x_2}e^{-x_3}\, dx_3\, dx_2\, dx_1$$

$$= \int_0^{t_1} e^{-x_1}\, dx_1 \int_0^{t_2} e^{-x_2}\, dx_2 \int_0^{\infty} e^{-x_3}\, dx_3 = (1 - e^{-t_1})(1 - e^{-t_2}).$$

(b) Because the event $X_1 \le t_1$ is equivalent to the event $0 \le X_1 \le t_1, 0 \le X_2 \le \infty$, and $0 \le X_3 \le \infty$, and $F_{X_1}(t_1) = P(X_1 \le t_1)$, it follows that

$$F_{X_1}(t_1) = \int_0^{t_1}\int_0^{\infty}\int_0^{\infty} e^{-x_1}e^{-x_2}e^{-x_3}\, dx_3\, dx_2\, dx_1 = 1 - e^{-t_1}.$$

(c) The marginal PDF of X_1 can be obtained by differentiating its marginal CDF. This gives $f_{X_1}(t_1) = e^{-t_1}$. Alternatively, by the formula analogous to (4.2.7),

$$f_{X_1}(x_1) = \int_0^{\infty}\int_0^{\infty} e^{-x_1}e^{-x_2}e^{-x_3}\, dx_3\, dx_2 = e^{-x_1}.$$

■

Exercises

1. Let X be the number of daily purchases of a luxury item from a factory outlet location and Y be the daily number of purchases made online. Let the values 1, 2, and 3 denote the number of purchases less than five, at least five but less than 15, and 15 or more, respectively. Suppose the joint PMF of X and Y is

	$p(x,y)$	y: 1	2	3
x	1	0.09	0.12	0.13
	2	0.12	0.11	0.11
	3	0.13	0.10	0.09

(a) Find the probability of each of the events $(X > 1, Y > 2)$, $(X > 1 \text{ or } Y > 2)$, and $(X > 2, Y > 2)$. (*Hint.* List the outcomes, i.e., the (x, y)-values, that comprise each event and sum the corresponding probabilities.)

(b) Find the marginal PMF of X and that of Y.

2. The joint PMF of X, the amount of drug administered to a randomly selected laboratory rat, and Y, the number of tumors the rat develops, is

	$p(x,y)$	y: 0	1	2
x	0.0	0.388	0.009	0.003
	1.0	0.485	0.010	0.005
	2.0	0.090	0.006	0.004

(a) Find the marginal PMF of X and that of Y.

(b) What is the probability that a randomly selected rat has (i) one tumor, (ii) at least one tumor?

(c) Given that a randomly selected rat has received the 1.0 mg/kg drug dosage, what is the probability that it has (i) no tumor, (ii) at least one tumor?

3. A local diner offers entrees in three prices, \$8.00, \$10.00, and \$12.00. Diner customers are known to tip either \$1.50, \$2.00, or \$2.50 per meal. Let X denote the price of the meal ordered, and Y denote the tip left, by a random customer. The joint PMF of X and Y is

	$p(x,y)$	y: \$1.50	\$2.00	\$2.50
x	\$8.00	0.3	0.12	0
	\$10.00	0.15	0.135	0.025
	\$12.00	0.03	0.15	0.09

(a) Find $P(X \le 10, Y \le 2)$ and $P(X \le 10, Y = 2)$.

(b) Compute the marginal PMFs of X and Y.

(c) Given that a customer has left a tip of \$2.00, find the probability that the customer ordered a meal of \$10.00 or less.

4. The *joint cumulative distribution function*, or *joint CDF*, of the random variables X and Y is defined as $F(x, y) = P(X \le x, Y \le y)$. Let X and Y be the random variables of Exercise 1.

(a) Make a table for the $F(x, y)$ at the possible (x, y) values that (X, Y) takes.

(b) The marginal CDFs of X and Y can be obtained from their joint CDF as $F_X(x) = F(x, \infty)$, and $F_Y(y) = F(\infty, y)$. Use these formulas to find the marginal CDFs of X and Y.

(c) It can be shown that the joint PMF can be obtained from the joint CDF as

$$P(X = x, Y = y) = F(x, y) - F(x, y - 1) - F(x - 1, y) + F(x - 1, y - 1).$$

(This is more complicated than the formula $P(X = x) = F_X(x) - F_X(x - 1)$ for the univariate case!) Use this formula to compute $P(X = 2, Y = 2)$, and confirm your answer from the PMF given in Exercise 1.

5. Let X_1, X_2, and X_3 denote the number of customers in line for self checkout, for regular checkout, and for express (15 items of less) checkout, respectively. Let the values 0, 1, and 2 denote zero customers, one customer, and two or more customers, respectively. Suppose the joint PMF, $p(x_1, x_2, x_3)$, of (X_1, X_2, X_3) is given in the table below. Find the marginal PMFs of X_1, X_2, and X_3.

		$p(0, x_2, x_3)$			$p(1, x_2, x_3)$			$p(2, x_2, x_3)$		
		x_3			x_3			x_3		
		1	2	3	1	2	3	1	2	3
x_2	0	0.030	0.027	0.024	0.030	0.027	0.024	0.040	0.036	0.032
	1	0.033	0.042	0.039	0.033	0.042	0.039	0.044	0.056	0.052
	2	0.024	0.033	0.048	0.024	0.033	0.048	0.032	0.044	0.064

6. When being tested, an integrated circuit (IC) is considered as a black box that performs certain designed functions. Four ICs will be randomly selected from a shipment of 15 and will be tested for static voltages, external components associated with the IC, and dynamic operation. Let X_1, X_2, and X_3 be the number of ICs in the sample that fail the first, second, and third test, respectively, and X_4 be the number of ICs in the sample that do not fail any of the tests. Suppose that, if tested, three of the 15 ICs would fail only the first test, two would fail only the second test, one would fail only the third test, and nine would not fail any of the three tests.

(a) Specify the sample space of $(X_1, \ldots, X_4)$.

(b) Find the joint PMF of X_1, X_2, and X_3.

7. Let the random variables X and Y have the joint PDF given below:

$$f(x, y) = kxy^2 \quad \text{for } 0 \leq x \leq 2,\ x \leq y \leq 3.$$

(a) Find the constant k. (*Hint.* Use the property that the volume under the entire surface defined by $f(x, y)$ is 1.)

(b) Find the joint CDF of X and Y.

8. Let the random variables X and Y have the joint PDF given below:

$$f(x, y) = \begin{cases} 2e^{-x-y} & 0 \leq x \leq y < \infty \\ 0 & \text{otherwise.} \end{cases}$$

(a) Find $P(X + Y \leq 3)$.

(b) Find the marginal PDFs of Y and X.

4.3 Conditional Distributions

4.3.1 CONDITIONAL PROBABILITY MASS FUNCTIONS

For jointly discrete (X, Y), the concept of a *conditional PMF* is an extension of the concept of conditional probability of an event. If x is one of the possible values that X can take, then the conditional probability that Y takes the value y given that $X = x$ is

$$P(Y = y|X = x) = \frac{P(X = x, Y = y)}{P(X = x)} = \frac{p(x, y)}{p_X(x)}.$$

The above relation follows simply from the definition of conditional probability, but when we think of it as a function of y, with y ranging in the sample space $\mathcal{S}_Y$ of Y, while keeping x fixed, we call it the *conditional PMF* of Y given the information that $X = x$:

Definition of Conditional PMF of Y given $X = x$

$$p_{Y|X=x}(y) = \frac{p(x, y)}{p_X(x)}, \quad y \in \mathcal{S}_Y \tag{4.3.1}$$

for $p_X(x) > 0$. Similarly, the conditional PMF of X given $Y = y$ is defined as $p_{X|Y=y}(x) = p(x, y)/p_Y(y)$, $x \in \mathcal{S}_X$, for $p_Y(y) > 0$.

If the joint PMF of (X, Y) is given in a table form, $p_{Y|X=x}(y)$ is found by dividing the joint probabilities in the row that corresponds to x by the marginal probability that $X = x$.

Example 4.3-1

A robot performs two tasks, welding joints and tightening bolts. Let X be the number of defective welds and Y be the number of improperly tightened bolts per car. The joint and marginal PMFs of X and Y are given in the table below:

		y				
	$p(x, y)$	0	1	2	3	$p_X(x)$
	0	0.84	0.03	0.02	0.01	0.9
x	1	0.06	0.01	0.008	0.002	0.08
	2	0.01	0.005	0.004	0.001	0.02
	$p_Y(y)$	0.91	0.045	0.032	0.013	1.0

Find the conditional PMF of Y given $X = 0$.

Solution

The conditional PMF of Y given $X = 0$ is obtained by dividing each joint probability in the row that corresponds to $x = 0$ by the marginal probability that $X = 0$:

y	0	1	2	3
$p_{Y\mid X=0}(y)$	0.9333	0.0333	0.0222	0.0111

The next example illustrates the computation of conditional PMFs without the use of a table of joint probabilities.

Example 4.3-2

Let $X(t)$ be a Poisson process with rate α. Find the conditional PMF of $X(0.6)$ given $X(1) = n$ (i.e., given that there are n occurrences in the time period $[0, 1]$).

Solution

Because $0 \le X(0.6) \le X(1)$, and we are given that $X(1) = n$, it follows that the possible values of $X(0.6)$ are $0, 1, \ldots, n$. For $m = 0, 1, \ldots, n$, we have

$$P(X(0.6) = m|X(1) = n) = \frac{P(X(0.6) = m, X(1) - X(0.6) = n - m)}{P(X(1) = n)}. \tag{4.3.2}$$

By the properties of Poisson processes, the events $[X(0.6) = m]$ and $[X(1)-X(0.6) = n - m]$ are independent. Moreover, $X(0.6) \sim \text{Poisson}(\alpha \times 0.6)$ and, according to Proposition 3.4-2, part (b), $X(1)-X(0.6) \sim \text{Poisson}(\alpha(1-0.6))$. Thus, the numerator of (4.3.2) becomes

$$e^{-\alpha\times 0.6}\frac{(\alpha \times 0.6)^m}{m!}e^{-\alpha\times(1-0.6)}\frac{(\alpha \times (1-0.6))^{n-m}}{(n-m)!} = \frac{e^{-\alpha}\alpha^n}{m!(n-m)!}0.6^m(1-0.6)^{n-m}.$$

Finally, the denominator of (4.3.2) is $e^{-\alpha}\alpha^n/n!$. Hence,

$$P(X(0.6) = m|X(1) = n) = \binom{n}{m}0.6^m(1-0.6)^{n-m},$$

which is the Bin$(n, 0.6)$ PMF.

A conditional PMF is a proper PMF and, as such, it has the same basic properties:

Basic Properties of Conditional PMFs

$$p_{Y|X=x}(y) \ge 0, \quad y \in \mathcal{S}_Y, \quad \text{and} \sum_y p_{Y|X=x}(y) = 1. \tag{4.3.3}$$

Because a conditional PMF is a proper PMF, it makes sense to consider the *conditional expected value* and the *conditional variance* of, say, Y when the value of X is given.

Example 4.3-3

Let X and Y be as in Example 4.3-1. The conditional PMF of Y given $X = 0$ was found there to be

y	0	1	2	3
$p_{Y\mid X=0}(y)$	0.9333	0.0333	0.0222	0.0111

Calculate the conditional expected value and variance of Y given that $X = 0$.

Solution

The conditional expected value of Y given $X = 0$ is

$$E(Y|X=0) = 0 \times (0.9333) + 1 \times (0.0333) + 2 \times (0.0222) + 3 \times (0.0222) = 0.111.$$

To compute the conditional variance, $\text{Var}(Y|X = 0)$, we first compute

$$E(Y^2|X=0) = 0 \times (0.9333) + 1 \times (0.0333) + 4 \times (0.0222) + 9 \times (0.0222) = 0.222,$$

Thus, using $\text{Var}(Y|X = 0) = E(Y^2|X = 0) - [E(Y|X = 0)]^2$, we obtain

$$\text{Var}(Y|X=0) = 0.222 - (0.111)^2 = 0.2097.$$

The definition of conditional PMF is equivalent to the relation

Multiplication Rule for Joint Probabilities

$$p(x, y) = p_{Y|X=x}(y)p_X(x) \tag{4.3.4}$$

which is a direct analogue of the multiplication rule (2.5.3). Using (4.3.4), the formula (4.2.3) for the marginal PMF of Y can be written as

Law of Total Probability for Marginal PMFs

$$p_Y(y) = \sum_{x \in \mathcal{S}_X} p_{Y|X=x}(y)p_X(x) \tag{4.3.5}$$

which is a version of the Law of Total Probability (2.5.7).

Example 4.3-4

Let X take the value 0, 1, or 2 depending on whether there are no customers, between 1 and 10 customers, and more than 10 customers in the regular (manned) checkout lines of a supermarket. Let Y be the corresponding variable for the self checkout lines. An extensive study undertaken by the management team of the supermarket resulted in the following conditional distributions of Y given $X = x$, and the marginal distribution of X:

y	0	1	2
$p_{Y\|X=0}(y)$	0.85	0.10	0.05
$p_{Y\|X=1}(y)$	0.30	0.45	0.25
$p_{Y\|X=2}(y)$	0.20	0.35	0.45

x	0	1	2
$p_X(x)$	0.20	0.50	0.30

(a) Use the Law of Total Probability to find the marginal PMF of Y.

(b) Use the multiplication rule for joint probabilities to tabulate the joint distribution of X and Y.

Solution

(a) According to formula (4.3.5), $p_Y(y)$ is found by multiplying the entries in the y column of the table by the corresponding entry in the marginal PMF of X and summing the products. Thus, $p_Y(0)$ is found by multiplying 0.85, 0.3, and 0.2 by 0.2, 0.5, and 0.3, respectively, and summing the products:

$$p_Y(0) = 0.85 \times 0.2 + 0.3 \times 0.5 + 0.2 \times 0.3 = 0.38.$$

Similarly, $p_Y(1) = 0.1 \times 0.2 + 0.45 \times 0.5 + 0.35 \times 0.3 = 0.35$ and $p_Y(2) = 0.05 \times 0.2 + 0.25 \times 0.5 + 0.45 \times 0.3 = 0.27$.

(b) According to formula (4.3.4), the x-row in the table of joint probabilities, that is, $p(x, y)$ for $y = 0, 1, 2$, is found by multiplying the $p_{Y|X=x}(y)$-row in the table of conditional probabilities by $p_X(x)$. Thus, the $p_{Y|X=0}(y)$-row is multiplied by 0.2, the $p_{Y|X=1}(y)$-row is multiplied by 0.5 and the $p_{Y|X=2}(y)$-row is multiplied by 0.3:

		y		
	$p(x, y)$	0	1	2
	0	0.170	0.020	0.010
x	1	0.150	0.225	0.125
	2	0.060	0.105	0.135

4.3.2 CONDITIONAL PROBABILITY DENSITY FUNCTIONS

In analogy with the definition in the discrete case, if (X, Y) are continuous with the joint PDF f, and marginal PDFs f_X, f_Y, the **conditional PDF** of Y given $X = x$ is defined to be

Definition of the Conditional PDF of Y Given $X = x$

$$f_{Y|X=x}(y) = \frac{f(x, y)}{f_X(x)} \tag{4.3.6}$$

if $f_X(x) > 0$. Similarly, the conditional PDF of X given $Y = y$ is defined as $f_{X|Y=y}(x) = f(x, y)/f_Y(y)$, $x \in \mathcal{S}_X$, for $f_Y(y) > 0$.

Example 4.3-5

The joint PDF of X and Y is $f(x, y) = 0$ if either x or y is < 0, and

$$f(x, y) = \frac{e^{-x/y}e^{-y}}{y} \quad \text{for } x > 0,\ y > 0.$$

Find $f_{X|Y=y}(x)$.

Solution

The marginal PDF of Y is

$$f_Y(y) = \int_0^\infty \frac{1}{y} e^{-x/y} e^{-y} dx = e^{-y} \int_0^\infty \frac{1}{y} e^{-x/y} dx = e^{-y}$$

for $y > 0$, and $f_Y(y) = 0$ otherwise. Thus, for $y > 0$,

$$f_{X|Y=y}(x) = \frac{f(x, y)}{f_Y(y)} = \frac{1}{y} e^{-x/y} \quad \text{for } x > 0,$$

and $f_{X|Y=y}(x) = 0$ otherwise.

The conditional PDF is a proper PDF and, as such, it has the same basic properties: $f_{Y|X=x}(y) \geq 0$, and

Conditional Probabilities in Terms of the Conditional PDF

$$P(a < Y < b|X = x) = \int_a^b f_{Y|X=x}(y)\,dy. \quad \textbf{(4.3.7)}$$

Thus, as in the discrete case, it makes sense to consider the conditional expected value and the conditional variance of Y given that $X = x$.

A remarkable aspect of relation (4.3.7) should not go unnoticed: The definition of conditional probabilities given in Chapter 2, namely $P(B|A) = P(B \cap A)/P(A)$, requires $P(A) > 0$. If $P(A) = 0$, the definition does not apply. As we have seen, when X is continuous, $P(X = x) = 0$ for any value x. Thus, the conditional probability $P(a < Y < b|X = x)$ cannot be evaluated according to the definition given in Chapter 2.

Example 4.3-6

Let X, Y have the joint PDF given in Example 4.3-5.

(a) Find $P(X > 1|Y = 3)$.

(b) Find the conditional mean and variance of X given that $Y = 3$.

Solution

(a) According to Example 4.3-5, $f_{X|Y=3}(x) = 3^{-1}e^{-x/3}$ for $x > 0$. Thus,

$$P(X > 1|Y = 3) = \int_1^\infty \frac{1}{3}e^{-x/3}\,dx = e^{-1/3}.$$

Alternatively, the same answer can be obtained by recognizing $f_{X|Y=3}(x)$ as the PDF of the exponential distribution with parameter $\lambda = 1/3$, and using the formula for the exponential CDF given in (3.5.1).

(b) The conditional expected value of X given $Y = 3$ is

$$E(X|Y = 3) = \int_{-\infty}^\infty x f_{X|Y=y}(x)\,dx = \int_0^\infty x\frac{1}{3}e^{-x/3}\,dx = 3.$$

Alternatively, the same result can be obtained by applying directly the formula for the mean value of the exponential distribution given in (3.5.2). The formula for the variance of the exponential distribution given in the same relation yields $\text{Var}(X|Y = 3) = 9$. ■

The definition of the conditional PDF is equivalent to the relation

Multiplication Rule for Joint PDFs

$$f(x, y) = f_{Y|X=x}(y)f_X(x) \quad \textbf{(4.3.8)}$$

which is the continuous variable version of the multiplication rule. Using (4.3.8), the formula (4.2.7) for the marginal PDF of Y can be written as

Law of Total Probability for Marginal PDFs

$$f_Y(y) = \int_{-\infty}^\infty f_{Y|X=x}(y)f_X(x)dx \quad \textbf{(4.3.9)}$$

Example 4.3-7

Let X be the force (in hundreds of pounds) applied to a randomly selected beam and Y the time to failure of the beam. Suppose that the PDF of X is

$$f_X(x) = \frac{1}{\log(6) - \log(5)} \frac{1}{x} \quad \text{for } 5 \le x \le 6$$

and zero otherwise, and that the conditional distribution of Y, given that a force $X = x$ is applied, is exponential($\lambda = x$). Thus,

$$f_{Y|X=x}(y) = xe^{-xy} \quad \text{for } y \ge 0,$$

and $f_{Y|X=x}(y) = 0$ for $y < 0$. Find the joint PDF of (X, Y), and the marginal PDF of Y.

Solution

Using the multiplication rule for joint probabilities given in (4.3.8),

$$f(x, y) = f_{Y|X=x}(y) f_X(x) = \frac{1}{\log(6) - \log(5)} e^{-xy}.$$

Next, using the Law of Total Probability for marginal PDFs given in (4.3.9),

$$\begin{aligned} f_Y(y) &= \int_{-\infty}^{\infty} f(x, y)\, dx = \int_5^6 \frac{1}{\log(6) - \log(5)} e^{-xy}\, dx \\ &= \frac{1}{\log(6) - \log(5)} \frac{1}{y} \left(e^{-5y} - e^{-6y}\right) \end{aligned}$$

for $y \ge 0$, and $f_Y(y) = 0$ otherwise. ■

4.3.3 THE REGRESSION FUNCTION

The conditional expected value of Y given that $X = x$,

$$\mu_{Y|X}(x) = E(Y|X = x), \tag{4.3.10}$$

when considered as a function of x, is called the **regression function** of Y on X. Thus, "regression function" is synonymous to *conditional mean value function*. Formulas for calculating the regression function for discrete and continuous random variables are

Regression Function for Jointly Discrete (X, Y)

$$\mu_{Y|X}(x) = \sum_{y \in \mathcal{S}_Y} y\, p_{Y|X=x}(y), \quad x \in \mathcal{S}_X \tag{4.3.11}$$

Regression Function for Jointly Continuous (X, Y)

$$\mu_{Y|X}(x) = \int_{-\infty}^{\infty} y f_{Y|X=x}(y)\, dy, \quad x \in \mathcal{S}_X \tag{4.3.12}$$

Example 4.3-8

Let X and Y be as in Example 4.3-4. Find the regression function of Y on X.

Solution

Using the conditional PMFs of Y given $X = x$, given in Example 4.3-4, we have

$$E(Y|X = 0) = \sum_{y=0}^{2} y p_{Y|X=0}(y) = 0 \times 0.85 + 1 \times 0.10 + 2 \times 0.05 = 0.2,$$

$$E(Y|X = 1) = \sum_{y=0}^{2} y p_{Y|X=1}(y) = 0 \times 0.30 + 1 \times 0.45 + 2 \times 0.25 = 0.95,$$

$$E(Y|X = 2) = \sum_{y=0}^{2} y p_{Y|X=2}(y) = 0 \times 0.20 + 1 \times 0.35 + 2 \times 0.45 = 1.25.$$

Thus, in a table form, the regression function of Y on X is:

x	0	1	2
$\mu_{Y\|X}(x)$	0.2	0.95	1.25

The information that this regression function makes visually apparent, and that was not easily discernable from the joint probability mass function, is that if the regular checkout lines are long, you can expect long self-checkout lines as well. ■

Example 4.3-9

Suppose (X, Y) have joint PDF

$$f(x, y) = \begin{cases} 24xy & 0 \le x \le 1, 0 \le y \le 1, x + y \le 1 \\ 0 & \text{otherwise.} \end{cases}$$

Find the regression function of Y on X.

Solution

The marginal PDF of X is

$$f_X(x) = \int_0^{1-x} 24xy\, dy = 12x(1 - x)^2$$

for $0 \le x \le 1$ and zero otherwise. This gives

$$f_{Y|X=x}(y) = \frac{f(x, y)}{f_X(x)} = 2\frac{y}{(1 - x)^2}.$$

Thus, $E(Y|X = x) = \int_0^{1-x} y f_{Y|X=x}(y)\, dy = \frac{2}{3}(1 - x)$. ■

As a consequence of the Law of Total Probability for marginal PMFs and PDFs, given in (4.3.5) and (4.3.9), respectively, the expected value of Y can be obtained as the expected value of the regression function. This is called the **Law of Total Expectation**.

Law of Total Expectation

$$E(Y) = E[E(Y|X)] \quad \textbf{(4.3.13)}$$

Explicit formulas for discrete and continuous random variables are as follows:

Law of Total Expectation for Discrete Random Variables

$$E(Y) = \sum_{x \in \mathcal{S}_X} E(Y|X = x)p_X(x) \quad \textbf{(4.3.14)}$$

Law of Total Expectation for Continuous Random Variables

$$E(Y) = \int_{-\infty}^{\infty} E(Y|X = x)f_X(x)\,dx \quad \textbf{(4.3.15)}$$

Example 4.3-10

Use the regression function of Y on X and the marginal PMF of X,

x	0	1	2
$\mu_{Y\|X}(x)$	0.2	0.95	1.25

and

x	0	1	2
$p_X(x)$	0.2	0.5	0.3

which were given in Examples 4.3-8 and 4.3-4, respectively, in order to find $E(Y)$.

Solution

Using the formula in (4.3.14), we have

$$\begin{aligned} E(Y) &= E(Y|X = 0)p_X(0) + E(Y|X = 1)p_X(1) + E(Y|X = 2)p_X(2) \\ &= 0.2 \times 0.2 + 0.95 \times 0.5 + 1.25 \times 0.3 = 0.89. \end{aligned}$$

Of course, we obtain the same result using the marginal distribution of Y, which is found in Example 4.3-4: $E(Y) = 0 \times 0.38 + 1 \times 0.35 + 2 \times 0.27 = 0.89$. ■

Example 4.3-11

Use the regression function of Y on X, and the marginal PDF of X,

$$E(Y|X = x) = \frac{2}{3}(1 - x) \quad \text{and} \quad f_X(x) = 12x(1 - x)^2, \quad 0 \le x \le 1,$$

which were found in Example 4.3-9, in order to find $E(Y)$.

Solution

Using the formula in (4.3.15), we have

$$E(Y) = \int_0^1 \frac{2}{3}(1 - x)12x(1 - x)^2\,dx = \frac{24}{3}\int_0^1 x(1 - x)^3\,dx.$$

The R commands *f=function(x){x*(1-x)**3}; integrate(f, 0, 1)* give 0.05 for the value of the above integral. Thus, $E(Y) = 0.4$. ■

The following example shows that the Law of Total Expectation, (4.3.13), can be applied without knowledge of the marginal PMF or PDF of X, that is, without use of (4.3.14) or (4.3.15).

Example 4.3-12

Let Y denote the age of a tree, and let X denote the tree's diameter at breast height. Suppose that, for a particular type of tree, the regression function of Y on X is $\mu_{Y|X}(x) = 5 + 0.33x$ and that the average diameter of such trees in a given forested area is 45 cm. Find the mean age of this type of tree in the given forested area.

Solution

According to the Law of Total Expectation, given in (4.3.13), and the property of expected values given in part (3) of Proposition 3.3-1, we have

$$E(Y) = E[E(Y|X)] = E(5 + 0.33X) = 5 + 0.33E(X) = 5 + 0.33 \times 45 = 19.85.$$

■

An interesting variation of the Law of Total Expectation occurs when Y is a Bernoulli random variable, that is, it takes the value 1 whenever an event B happens and zero otherwise. In this case we have $E(Y) = P(B)$ and, similarly, $E(Y|X = x) = P(B|X = x)$. Hence, in this case, (4.3.14) and (4.3.15) can be written as

$$P(B) = \sum_{x \in \mathcal{S}_X} P(B|X = x)p_X(x) \quad \text{and} \quad P(B) = \int_{-\infty}^{\infty} P(B|X = x)f_X(x)\,dx \qquad \textbf{(4.3.16)}$$

The first expression in (4.3.16) is just the Law of Total Probability; see (2.5.7).

4.3.4 INDEPENDENCE

The notion of *independence* of random variables is an extension of the notion of independence of events. The random variables X and Y are **independent** if any event defined in terms of X is independent of any event defined in terms of Y. In particular, X and Y are independent if the events $[X \le x]$ and $[Y \le y]$ are independent for all x and y, that is, if

Definition of Independence of Two Random Variables

$$P(X \in A,\ Y \in B) = P(X \in A)P(Y \in B) \qquad \textbf{(4.3.17)}$$

holds for any two sets (subsets of the real line) A and B.

Proposition 4.3-1

1. The jointly discrete random variables X and Y are independent if and only if

Condition for Independence of Two Discrete Random Variables

$$p_{X,Y}(x, y) = p_X(x)p_Y(y) \qquad \textbf{(4.3.18)}$$

holds for all x, y, where $p_{X,Y}$ is the joint PMF of (X, Y) and p_X, p_Y are the marginal PMFs of X, Y, respectively.

2. The jointly continuous random variables X and Y are independent if and only if

Condition for Independence of Two Continuous Random Variables

$$f_{X,Y}(x, y) = f_X(x)f_Y(y) \qquad \textbf{(4.3.19)}$$

holds for all x, y, where $f_{X,Y}$ is the joint PDF of (X, Y) and f_X, f_Y are the marginal PDFs of X, Y, respectively. ■

Example 4.3-13

Consider the joint distribution of the two types of errors, X and Y, a robot makes, as given in Example 4.3-1. Are X, Y independent?

Solution

Using the table that displays the joint and marginal PMFs given in Example 4.3-1 we have

$$p(0,0) = 0.84 \neq p_X(0)p_Y(0) = (0.9)(0.91) = 0.819.$$

This suffices to conclude that X and Y are not independent. ■

If the jointly discrete X and Y are independent then, by part (1) of Proposition 4.3-1,

$$p(x,y) = p_Y(y)p_X(x).$$

On the other hand, the multiplication rule for joint PMFs states that

$$p(x,y) = p_{Y|X=x}(y)p_X(x)$$

is always true. Thus, when X and Y are independent it must be that $p_Y(y) = p_{Y|X=x}(y)$ for all x in the sample space of X. Similarly, if the jointly continuous X and Y are independent then, by part (2) of Proposition 4.3-1 and the multiplication rule for joint PDFs (4.3.8), $f_Y(y) = f_{Y|X=x}(y)$. This argument is the basis for the following result.

Proposition 4.3-2

If X and Y are jointly discrete, each of the following statements implies, and is implied by, their independence.

1. $p_{Y|X=x}(y) = p_Y(y)$.
2. $p_{Y|X=x}(y)$ does not depend on x, that is, is the same for all possible values of X.
3. $p_{X|Y=y}(x) = p_X(x)$.
4. $p_{X|Y=y}(x)$ does not depend on y, that is, is the same for all possible values of Y.

Each of the above statements with PDFs replacing PMFs implies, and is implied by, the independence of the jointly continuous X and Y. ■

Example 4.3-14

A system is made up of two components, A and B, connected in parallel. Let X take the value 1 or 0 if component A works or not, and Y take the value 1 or 0 if component B works or not. From the repair history of the system it is known that the conditional PMFs of Y given $X = 0$ and $X = 1$ are

	y	
	0	1
$p_{Y\|X=0}(y)$	0.01	0.99
$p_{Y\|X=1}(y)$	0.01	0.99

Are X and Y independent?

Solution

From the table of conditional probabilities, it is seen that the conditional PMF of Y given $X = 0$ is the same as its conditional PMF given $X = 1$. By part (2) of Proposition 4.3-2, we conclude that X and Y are independent. ■

Example 4.3-15

For a cylinder selected at random from the production line, let X be the cylinder's height and Y the cylinder's radius. Suppose X, Y have a joint PDF

$$f(x,y) = \begin{cases} \frac{3}{8}\frac{x}{y^2} & \text{if } 1 \le x \le 3, \quad \frac{1}{2} \le y \le \frac{3}{4} \\ 0 & \text{otherwise.} \end{cases}$$

Are X and Y independent?

Solution

The marginal PDF of X is

$$f_X(x) = \int_{-\infty}^{\infty} f(x,y)\,dy = \int_{.5}^{.75} \left(\frac{3}{8}\frac{x}{y^2}\right) dy = \frac{x}{4}$$

for $1 \le x \le 3$ and zero otherwise. The marginal PDF of Y is

$$f_Y(y) = \int_{-\infty}^{\infty} f(x,y)\,dx = \int_{1}^{3} \left(\frac{3}{8}\frac{x}{y^2}\right) dx = \frac{3}{2}\frac{1}{y^2}$$

for $0.5 \le y \le 0.75$ and zero otherwise. Since

$$f(x,y) = f_X(x) f_Y(y),$$

we conclude that X and Y are independent.

It is instructive to also consider the conditional PDF of Y given $X = x$, which is

$$f_{Y|X=x}(y) = \frac{f(x,y)}{f_X(x)} = \frac{3}{2}\frac{1}{y^2}$$

for $0.5 \le y \le 0.75$ and zero otherwise. It is seen that this expression does not depend on the value x; in fact it is seen that $f_{Y|X=x}(y) = f_Y(y)$. Thus, by the PDF version of either part (1) or part (2) of Proposition 4.3-2, we again conclude that X, Y are independent. ■

In Example 4.3-15 the joint PDF can be written as $f(x,y) = g(x)h(y)$, where $g(x) = (3/8)x$ for $1 \le x \le 3$ and zero otherwise and $h(y) = 1/y^2$ for $0.5 \le y \le 0.75$ and zero otherwise. In such cases, one may conclude that X and Y are independent without finding their marginal PDFs; see Exercise 12.

The following proposition summarizes some important properties of independent random variables.

Proposition 4.3-3

Let X and Y be independent. Then,

1. The regression function $E(Y|X = x)$ of Y on X is constant, that is, does not depend on the value of X, and equals $E(Y)$.
2. $g(X)$ and $h(Y)$ are independent for any functions g, h.
3. $E(g(X)h(Y)) = E(g(X))E(h(Y))$ holds for any functions g, h. ■

Part (1) of Proposition 4.3-3 follows from the computational formulas (4.3.11) and (4.3.12) of the regression function and Proposition 4.3-2, which asserts that if X

and Y are independent the conditional distribution of Y given the value of X is the same as the marginal distribution of Y. Part (2) is self-evident. Part (3) of Proposition 4.3-3 will also be shown in Example 4.4-4 of the next section, but it is instructive to give here a proof based on the Law of Total Expectation. Using the form of the Law given in (4.3.13), that is, $E(Y) = E[E(Y|X)]$, with $g(X)h(Y)$ in place of Y we have

$$E(g(X)h(Y)) = E[E(g(X)h(Y)|X)] = E[g(X)E(h(Y)|X)],$$

where the second equality holds by the fact that given the value of X, the value of $g(X)$ is also known and thus $E(g(X)h(Y)|X) = g(X)E(h(Y)|X)$ follows from part (3) of Proposition 3.3-1. Next, since X and $h(Y)$ are also independent (as it follows from part (2) of Proposition 4.3-3), the regression function $E(h(Y)|X)$ of $h(Y)$ on X equals $E(h(Y))$, and one more application of part 3.3-1 of Proposition 3.3-1 yields

$$E[g(X)E(h(Y)|X)] = E[g(X)E(h(Y))] = E(g(X))E(h(Y)),$$

showing that $E(g(X)h(Y)) = E(g(X))E(h(Y))$.

Example 4.3-16

Consider the two-component system described in Example 4.3-14, and suppose that the failure of component A incurs a cost of \$500.00, while the failure of component B incurs a cost of \$750.00. Let C_A and C_B be the costs incurred by the failures of components A and B, respectively. Are C_A and C_B independent?

Solution

The random variable C_A takes values 500 and 0 depending on whether component A fails or not. Thus, $C_A = 500(1 - X)$, where X takes the value 1 if component A works and the value 0 of it does not. Similarly $C_B = 750(1 - Y)$, where Y takes the value 1 or 0 if component B works or not. In Example 4.3-14 it was seen that X and Y are independent. Thus, by part (2) of Proposition 4.3-3, C_A and C_B are also independent. ■

Example 4.3-17

Let the height, X, and radius, Y, both measured in centimeters, of a cylinder randomly selected from the production line have the joint PDF given in Example 4.3-15.

(a) Find the expected volume of a randomly selected cylinder.

(b) Let X_1, Y_1 be the height and radius of the cylinder expressed in inches. Are X_1 and Y_1 independent?

Solution

(a) In Example 4.3-15 we saw that X and Y are independent with marginal PMFs $f_X(x) = x/4$ for $1 \leq x \leq 3$ and zero otherwise, and $f_Y(y) = 3/(2y^2)$ for $0.5 \leq y \leq 0.75$ and zero otherwise. Since the volume is given by πXY^2, an application of part (3) of Proposition 4.3-3 gives

$$\begin{aligned} E\left[\pi XY^2\right] &= \pi E(X)E(Y^2) \\ &= \pi \int_1^3 x f_X(x)\,dx \int_{0.5}^{0.75} y^2 f_Y(y)\,dy \\ &= \pi \frac{13}{6}\frac{3}{8} = \pi \frac{13}{16}. \end{aligned}$$

(b) Since X_1 and Y_1 are (linear) functions of X and Y, respectively, the independence of X and Y implies that X_1 and Y_1 are also independent. ■

The concept of independence extends to several random variables in a straightforward manner. In particular, conditions (4.3.18) and (4.3.19) extend as follows: The jointly discrete random variables $X_1, X_2, \ldots, X_n$ are independent if and only if

Condition for Independence of Several Discrete Random Variables

$$p(x_1, x_2, \ldots, x_n) = p_{X_1}(x_1) \cdots p_{X_n}(x_n)$$

and the jointly continuous $X_1, X_2, \ldots, X_n$ are independent if and only if

Condition for Independence of Several Continuous Random Variables

$$f(x_1, x_2, \ldots, x_n) = f_{X_1}(x_1) \cdots f_{X_n}(x_n)$$

hold for all $x_1, \ldots, x_n$. If $X_1, X_2, \ldots, X_n$ are independent and also have the same distribution (which is the case of a simple random sample from an infinite/hypothetical population) they are called **independent and identically distributed**, or **iid** for short.

Exercises

1. Let X denote the monthly volume of book sales from the online site of a bookstore, and let Y denote the monthly volume of book sales from its brick and mortar counterpart. The possible values of X and Y are 0, 1, or 2, in which 0 represents a volume that is below expectations, 1 represents a volume that meets expectations, and 2 represents a volume above expectations. The joint PMF $p(x, y)$ of (X, Y) appears in the table.

		y		
		0	1	2
	0	0.06	0.04	0.20
x	1	0.08	0.30	0.06
	2	0.10	0.14	0.02

(a) Find the marginal PMFs of X and Y, and use them to determine if X and Y are independent. Justify your answer.

(b) Compute the conditional PMFs, $p_{Y|X=x}(y)$, for $x = 0, 1, 2$, and use them to determine if X and Y are independent. Justify your answer. (*Hint.* Proposition 4.3-2.)

(c) Compute the conditional variance, $\mathrm{Var}(Y|X = 1)$, of Y given $X = 1$.

2. Let X, Y have the joint PMF given in Exercise 1.

(a) Find the regression function Y on X.

(b) Use the Law of Total Expectation to find $E(Y)$.

3. Let X, Y be as in Exercise 3 in Section 4.2.

(a) Find the regression function Y on X.

(b) Use the Law of Total Expectation to find $E(Y)$.

(c) Is the amount of tip left independent of the price of the meal? Justify your answer in terms of the regression function. (*Hint.* Use part (1) of Proposition 4.3-3.)

4. Consider the information given in Exercise 2 in Section 4.2.

(a) What is the conditional PMF of the number of tumors for a randomly selected rat in the 1.0 mg/kg drug dosage group?

(b) Find the regression function of Y, the number of tumors present on a randomly selected laboratory rat, on X, the amount of drug administered to the rat.

(c) Use the Law of Total Expectation to find $E(Y)$.

5. Let X take the value 0 if a child between 4 and 5 years of age uses no seat belt, 1 if he or she uses a seat belt, and 2 if it uses a child seat for short-distance car commutes. Also, let Y take the value 0 if a child survived a motor vehicle accident and 1 if he or she did not. Accident records from a certain state suggest the following conditional PMFs of Y given $X = x$ and marginal distribution of X:[1]

[1] The effectiveness of seat belts in preventing fatalities is considered by the National Highway Traffic Safety Administration; see http://www.nhtsa.gov/search?q=SEAT+BELT&x=25&y=4.

y	0	1
$p_{Y\|X=0}(y)$	0.69	0.31
$p_{Y\|X=1}(y)$	0.85	0.15
$p_{Y\|X=2}(y)$	0.84	0.16

x	0	1	2
$p_X(x)$	0.54	0.17	0.29

(a) Use the table of conditional PMFs of Y given $X = x$ to conclude whether or not X and Y are independent. Justify your answer.

(b) Make a table for the joint PMF of (X, Y), showing also the marginal PMFs, and use it to conclude whether or not X and Y are independent. Justify your answer.

6. Consider the information given in Exercise 5.

(a) Find the regression function, $\mu_{Y|X}(x)$, of Y on X.

(b) Use the Law of Total Expectation to find $E(Y)$.

7. The moisture content of batches of a chemical substance is measured on a scale from 1 to 3, while the impurity level is recorded as either low (1) or high (2). Let X and Y denote the moisture content and the impurity level of a randomly selected batch, respectively. Use the information given in the table to answer parts (a)–(d).

	y	
	1	2
$p_{Y\|X=1}(y)$	0.66	0.34
$p_{Y\|X=2}(y)$	0.80	0.20
$p_{Y\|X=3}(y)$	0.66	0.34

x	1	2	3
$P(X = x)$	0.2	0.3	0.5

(a) Find $E(Y|X = 1)$ and $\text{Var}(Y|X = 1)$.

(b) Tabulate the joint PMF of X and Y.

(c) What is the probability that the next batch received will have a low impurity level?

(d) Suppose the next batch has a low impurity level. What is the probability that the level of its moisture content is 1?

8. Consider the information given in Exercise 7.

(a) Find the regression function, $\mu_{Y|X}(x)$, of Y on X.

(b) Use the Law of Total Expectation to find $E(Y)$.

9. Let X be the force applied to a randomly selected beam for 150 hours, and let Y take the value 1 or 0 depending on whether the beam fails or not. The random variable X takes the values 4, 5, and 6 (in 100-lb units) with probability 0.3, 0.5, and 0.2, respectively. Suppose that the probability of failure when a force of $X = x$ is applied is

$$P(Y = 1|X = x) = \frac{(-0.8 + 0.04x)^4}{1 + (-0.8 + 0.04x)^4}$$

(a) Tabulate the joint PMF of X and Y. Are X and Y independent?

(b) Find the average force applied to beams that fail ($E(X|Y = 1)$), and the average force applied to beams that do not fail ($E(X|Y = 0)$).

10. It is known that, with probability 0.6, a new laptop owner will install a wireless Internet connection at home within a month. Let X denote the number (in hundreds) of new laptop owners in a week from a certain region, and let Y denote the number among them who install a wireless connection at home within a month. Suppose that the PMF of X is

x	0	1	2	3	4
$p_X(x)$	0.1	0.2	0.3	0.25	0.15

(a) Argue that given $X = x$, $Y \sim \text{Bin}(n = x, p = 0.6)$, and find the joint PMF of (X, Y).

(b) Find the regression function of Y on X.

(c) Use the Law of Total Expectation to find $E(Y)$.

11. The joint PDF of X and Y is $f(x, y) = x + y$ for $0 < x < 1$, $0 < y < 1$, and $f(x, y)$ otherwise.

(a) Find $f_{Y|X=x}(y)$ and use it to compute $P(0.3 < Y < 0.5|X = x)$.

(b) Use (4.3.16) to compute $P(0.3 < Y < 0.5)$.

12. Criterion for independence. X and Y are independent if and only if

$$f_{X,Y}(x, y) = g(x)h(y) \tag{4.3.20}$$

for some functions g and h (which need not be PDFs). [An important point to keep in mind when applying this criterion is that condition (4.3.20) implies that the region of (x, y) values where $f(x, y)$ is positive has to be a rectangle, i.e., it has to be of the form $a \le x \le b$, $c \le y \le d$, where a, c may also be $-\infty$ and b, d may also be ∞.] Use this criterion to determine if X and Y are independent in each of the following cases.

(a) The joint PDF of X and Y is $f(x, y) = 6e^{-2x}e^{-3y}$ for $0 < x < \infty$, $0 < y < \infty$ and zero otherwise.

(b) The joint PDF of X and Y is $f(x, y) = 24xy$ for $0 < x + y < 1$ and zero otherwise.

(c) The joint PDF of X and Y is $f(x, y) = \frac{e^{-x/y}e^{-y}}{y}$ for $0 < x < \infty$, $0 < y < \infty$ and zero otherwise.

13. Let T_i, $i = 1, 2$, denote the first two interarrival times of a Poisson process $X(s)$, $s \ge 0$, with rate α. (So, according to Proposition 3.5-1, both T_1 and T_2 have an exponential distribution with PDF $f(t) = \alpha e^{-\alpha t}$, $t > 0$.) Show that T_1 and T_2 are independent. (*Hint.* Argue that $P(T_2 > t|T_1 = s) = P(\text{No events in } (s, s + t]|T_1 = s)$, and use the third postulate in definition 3.4-1 of a Poisson process to justify that it equals $P(\text{No events in } (s, s + t])$. Express this as $P(X(s + t) - X(s) = 0)$ and use part (2) of Proposition 3.4-2 to obtain that it equals $e^{-\alpha t}$. This shows that $P(T_2 > t|T_1 = s)$, and hence the conditional density of T_2 given $T_1 = s$ does not depend on s.)

14. During a typical Pennsylvania winter, potholes along I80 occur according to a Poisson process averaging 1.6 per 10 miles. A certain county is responsible for repairing potholes in a 30-mile stretch of I80. At the end of winter the repair crew starts inspecting for potholes from one end of the 30-mile stretch. Let T_1 be the distance (in miles) to the first pothole, and T_2 the distance from the first pothole the second one.

(a) If the first pothole found is 8 miles from the start, find the probability that the second pothole will be found between 14 and 19 miles from the start. (*Hint.* Argue the desired probability is that of T_2 taking value between $14 - 8 = 6$ and $19 - 8 = 11$ miles. According to Proposition 3.5-1, T_2 is exponential(0.16).)

(b) Let $X = T_1$ and $Y = T_1 + T_2$. Find the regression function of Y on X. (*Hint.* $E(T_1 + T_2|T_1 = x) = x + E(T_2|T_1 = x)$. You may use the result from Exercise 13 stating that T_1 and T_2 are independent.)

15. Let X and Y have the joint PDF of Example 4.3-5. Use the form of the conditional PDF of X given $Y = y$ for $y > 0$, derived there, to conclude whether or not X and Y are independent. (*Hint.* Use part (4) of Proposition 4.3-2.)

16. Let X be the force (in hundreds of pounds) applied to a randomly selected beam and Y the time to failure of the beam. Suppose that the PDF of X is

$$f_X(x) = \frac{1}{\log(6) - \log(5)} \frac{1}{x} \quad \text{for } 5 \le x \le 6$$

and zero otherwise, and that the conditional distribution of Y given that a force $X = x$ is applied is exponential ($\lambda = x$). Thus,

$$f_{Y|X=x}(y) = xe^{-xy} \quad \text{for } y \ge 0,$$

and $f_{Y|X=x}(y) = 0$ for $y < 0$.

(a) Find the regression function of Y on X, and give the numerical value of $E(Y|X = 5.1)$. (*Hint.* Use the formula for the mean value of an exponential random variable.)

(b) Use the Law of Total Expectation to find $E(Y)$.

17. A type of steel has microscopic defects that are classified on a continuous scale from 0 to 1, with 0 the least severe and 1 the most severe. This is called the defect index. Let X and Y be the static force at failure and the defect index, respectively, for a particular type of structural member made from this steel. For a member selected at random, X and Y are jointly distributed random variables with joint PDF

$$f(x,y) = \begin{cases} 24x & \text{if } 0 \le y \le 1 - 2x \quad \text{and} \quad 0 \le x \le .5 \\ 0 & \text{otherwise.} \end{cases}$$

(a) Sketch the support of this PDF, that is, the region of (x, y) values where $f(x, y) > 0$.

(b) Are X and Y independent? Justify your answer in terms the support of the PDF sketched above.

(c) Find each of the following: $f_X(x)$, $f_Y(y)$, $E(X)$, and $E(Y)$.

18. Consider the context of Exercise 17.

(a) It is given that the marginal density of X is $f_X(x) = \int_0^{1-2x} 24x\, dy = 24x(1 - 2x)$, $0 \le x \le 0.5$. Find $f_{Y|X=x}(y)$ and the regression function $E(Y|X = x)$. Plot the regression function and give the numerical value of $E(Y|X = 0.3)$.

(b) Use the Law of Total Expectation to find $E(Y)$.

4.4 Mean Value of Functions of Random Variables

4.4.1 THE BASIC RESULT

As in the univariate case the expected value and, consequently, the variance of a statistic, that is, a function of random variables, can be obtained without having to first obtain its distribution. The basic result follows.

Proposition 4.4-1

1. Let (X, Y) be discrete with joint PMF $p(x, y)$. The expected value of a function, $h(X, Y)$, of (X, Y) is computed by

Mean Value of a Function of Discrete Random Variables

$$E[h(X,Y)] = \sum_{x \in \mathcal{S}_X} \sum_{y \in \mathcal{S}_Y} h(x,y)p(x,y)$$

2. Let (X, Y) be continuous with joint PDF $f(x, y)$. The expected value of a function, $h(X, Y)$, of (X, Y) is computed by

Mean Value of a Function of Continuous Random Variables

$$E[h(X,Y)] = \int_{-\infty}^{\infty}\int_{-\infty}^{\infty} h(x,y)f(x,y)dx\,dy$$

■

The variance of $h(X, Y)$ is computed by

Variance of a Function of Two Random Variables

$$\sigma^2_{h(X,Y)} = E[h^2(X,Y)] - [E[h(X,Y)]]^2 \qquad \textbf{(4.4.1)}$$

where, according to parts (1) and (2) of Proposition 4.4-1,

$$E[h^2(X,Y)] = \sum_x \sum_y h^2(x,y)p_{X,Y}(x,y)$$

$$E[h^2(X,Y)] = \int_{-\infty}^{\infty}\int_{-\infty}^{\infty} h^2(x,y)f_{X,Y}(x,y)\,dx\,dy$$

in the discrete and continuous case, respectively.

The formulas in Proposition 4.4-1 extend directly to functions of more than two random variables. For example, in the discrete case, the expected value of the statistic $h(X_1, \ldots, X_n)$ is computed by

$$E[h(X_1,\ldots,X_n)] = \sum_{x_1}\cdots\sum_{x_n} h(x_1,\ldots,x_n)p(x_1,\ldots,x_n),$$

where p denotes the joint PMF of $X_1, \ldots, X_n$, while in the continuous case, the expected value of $h(X_1, \ldots, X_n)$ is computed by

$$E[h(X_1,\ldots,X_n)] = \int_{-\infty}^{\infty}\cdots\int_{-\infty}^{\infty} h(x_1,\ldots,x_n)f(x_1,\ldots,x_n)dx_1\cdots dx_n.$$

Example 4.4-1

A photo processing website receives compressed files of images with $X \times Y$ pixels where X and Y are random variables. At compression factor 10:1, 24 bits-per-pixel images result in compressed images of $Z = 2.4XY$ bits. Find the expected value and variance of Z when the joint PMF of X and Y is

	$p(x, y)$	y = 480	600	900
	640	0.15	0.1	0.15
x	800	0.05	0.2	0.1
	1280	0	0.1	0.15

Solution

The formula in part (1) of Proposition 4.4-1, with $h(x, y) = xy$ yields

$$\begin{aligned} E(XY) &= 640 \times 480 \times 0.15 + 640 \times 600 \times 0.1 + 640 \times 900 \times 0.15 \\ &\quad + 800 \times 480 \times 0.05 + 800 \times 600 \times 0.2 + 800 \times 900 \times 0.1 \\ &\quad + 1280 \times 480 \times 0 + 1280 \times 600 \times 0.1 + 1280 \times 900 \times 0.15 \\ &= 607{,}680. \end{aligned}$$

The same formula yields

$$\begin{aligned}E[(XY)^2] &= 640^2 \times 480^2 \times 0.15 + 640^2 \times 600^2 \times 0.1 + 640^2 \times 900^2 \times 0.15 \\ &\quad +800^2 \times 480^2 \times 0.05 + 800^2 \times 600^2 \times 0.2 + 800^2 \times 900^2 \times 0.1 \\ &\quad +1280^2 \times 480^2 \times 0 + 1280^2 \times 600^2 \times 0.1 + 1280^2 \times 900^2 \times 0.15 \\ &= 442{,}008{,}576{,}000.\end{aligned}$$

It follows that the variance of XY is

$$\text{Var}(XY) = 442{,}008{,}576{,}000 - 607{,}680^2 = 72{,}733{,}593{,}600.$$

Finally, the expected value and variance of $Z = 2.4XY$ are $E(Z) = 2.4E(XY) = 1{,}458{,}432$ and $\text{Var}(Z) = 2.4^2\text{Var}(XY) = 418{,}945{,}499{,}136$. ■

Example 4.4-2 A system consists of components A and B connected in series. If the two components fail independently, and the time to failure for each component is a uniform(0, 1) random variable, find the expected value and variance of the time to failure of the system.

Solution

Because the two components are connected in series, if X and Y denote the times to failure of components A and B, respectively, the time to failure of the system is the smaller of X and Y. Thus, we want the expected value and variance of the function $T = \min\{X, Y\}$. These are most easily found by first finding the CDF of the random variable T, and differentiating the CDF to get the PDF. Note first that for any number t between 0 and 1, the event $[T > t]$ means that both $[X > t]$ and $[Y > t]$ are true. Thus,

$$P(T > t) = P(X > t, Y > t) = P(X > t)P(Y > t) = (1 - t)(1 - t) = 1 - 2t + t^2,$$

where the second equality holds by the fact that the events $[X > t]$ and $[Y > t]$ are independent, and the third equality uses the fact that X and Y have the uniform distribution. Thus, if $F_T(t)$ and $f_T(t)$ denote the CDF and PDF of T then, for $0 < t < 1$, we have

$$F_T(t) = P(T \le t) = 1 - P(T > t) = 2t - t^2 \quad \text{and} \quad f_T(t) = \frac{d}{dt}F_T(t) = 2 - 2t.$$

Hence,

$$E(T) = \int_0^1 tf_T(t)dt = 1 - \frac{2}{3} = \frac{1}{3}, \quad E(T^2) = \int_0^1 t^2 f_T(t)dt = \frac{2}{3} - \frac{2}{4} = \frac{1}{6},$$

which yields $\text{Var}(T) = 1/6 - (1/3)^2 = 0.05556$.

Alternatively, the mean and variance of T can be found by considering T as a function $h(X, Y) = \min\{X, Y\}$ of X and Y, and using part (2) of Proposition 4.4-1:

$$
\begin{aligned}
E[\min\{X,Y\}] &= \int_0^1 \int_0^1 \min\{x,y\}\,dx\,dy \\
&= \int_0^1 \left[\int_0^y \min\{x,y\}\,dx + \int_y^1 \min\{x,y\}\,dx\right] dy \\
&= \int_0^1 \left[\int_0^y x\,dx + \int_y^1 y\,dx\right] dy \\
&= \int_0^1 \left[\frac{1}{2}y^2 + y(1-y)\right] dy \;=\; \frac{1}{2}\frac{1}{3} + \frac{1}{2} - \frac{1}{3} = \frac{1}{3}.
\end{aligned}
$$

Next, with similar steps as above we obtain

$$
\begin{aligned}
E[\min\{X,Y\}^2] &= \int_0^1 \left[\int_0^y x^2\,dx + \int_y^1 y^2\,dx\right] dy \\
&= \int_0^1 \left[\frac{1}{3}y^3 + y^2(1-y)\right] dy \;=\; \frac{1}{3}\frac{1}{4} + \frac{1}{3} - \frac{1}{4} = \frac{1}{6}.
\end{aligned}
$$

Thus, $\text{Var}(\min\{X,Y\}) = 1/6 - (1/3)^2 = 0.05556$. ■

The next two examples deal with the expected value of the sum of two variables and the expected value of the product of two independent random variables, respectively.

Example 4.4-3

Show that for any two random variables

$$E(X+Y) = E(X) + E(Y).$$

Solution

Assume that X and Y are jointly discrete; the proof in the continuous case is similar. Then, according to part (1) of Proposition 4.4-1,

$$
\begin{aligned}
E(X+Y) &= \sum_{x\in\mathcal{S}_X}\sum_{y\in\mathcal{S}_Y}(x+y)p(x,y) \\
&= \sum_{x\in\mathcal{S}_X}\sum_{y\in\mathcal{S}_Y}xp(x,y) + \sum_{x\in\mathcal{S}_X}\sum_{y\in\mathcal{S}_Y}yp(x,y) \quad \text{(separate terms)} \\
&= \sum_{x\in\mathcal{S}_X}\sum_{y\in\mathcal{S}_Y}xp(x,y) + \sum_{y\in\mathcal{S}_Y}\sum_{x\in\mathcal{S}_X}yp(x,y) \\
&\quad \text{(interchange summations in second term)} \\
&= \sum_{x\in\mathcal{S}_X}x\sum_{y\in\mathcal{S}_Y}p(x,y) + \sum_{y\in\mathcal{S}_Y}y\sum_{x\in\mathcal{S}_X}p(x,y) \\
&= \sum_{x\in\mathcal{S}_X}xp_X(x) + \sum_{y\in\mathcal{S}_Y}yp_Y(y) \qquad \text{(definition of marginal PMFs)} \\
&= E(X) + E(Y).
\end{aligned}
$$

■

Example 4.4-4

If X and Y are independent, show that, for any functions g and h,

$$E[g(X)h(Y)] = E[g(X)]E[h(Y)].$$

Solution

Assume that X and Y are jointly continuous; the proof in the discrete case is similar. Then, according to part (2) of Proposition 4.4-1,

$$\begin{aligned} E[g(X)h(Y)] &= \int_{-\infty}^{\infty}\int_{-\infty}^{\infty} g(x)h(y)f(x,y)\,dx\,dy \\ &= \int_{-\infty}^{\infty}\int_{-\infty}^{\infty} g(x)h(y)f_X(x)f_Y(y)\,dx\,dy \quad \text{(by independence)} \\ &= \int_{-\infty}^{\infty} g(x)f_X(x)\,dx \int_{-\infty}^{\infty} h(y)f_Y(y)\,dy = E[g(X)]E[g(Y)]. \end{aligned}$$

The same result was obtained in Proposition 4.3-3 with a different method. ■

4.4.2 EXPECTED VALUE OF SUMS

Proposition 4.4-2

Let $X_1, \ldots, X_n$ be any n random variables (i.e., they may be discrete or continuous, independent or dependent), with marginal means $E(X_i) = \mu_i$. Then

Expected Value of a Linear Combination of Random Variables

$$E(a_1X_1 + \cdots + a_nX_n) = a_1\mu_1 + \cdots + a_n\mu_n$$

holds for any constants $a_1, \ldots, a_n$. ■

The proof of this proposition is similar to the proof of Example 4.4-3, the result of which it generalizes. In particular, applications of Proposition 4.4-2 with $n = 2$, $a_1 = 1$, and $a_2 = -1$, and with $n = 2$, $a_1 = 1$, and $a_2 = 1$ yield, respectively,

$$E(X_1 - X_2) = \mu_1 - \mu_2 \quad \text{and} \quad E(X_1 + X_2) = \mu_1 + \mu_2. \tag{4.4.2}$$

Corollary 4.4-1

If the random variables $X_1, \ldots, X_n$ have common mean μ, that is, if $E(X_1) = \cdots = E(X_n) = \mu$, then

Expected Value of the Average and the Total

$$E(\overline{X}) = \mu \quad \text{and} \quad E(T) = n\mu \tag{4.4.3}$$

where $\overline{X} = (1/n)\sum_i X_i$ and $T = n\overline{X} = \sum_i X_i$.

The proof of this corollary follows by an application of Proposition 4.4-2 with $a_1 = \cdots = a_n = 1/n$ and $a_1 = \cdots = a_n = 1$, for the mean and total sum, respectively. If the X_i in Corollary 4.4-1 are Bernoulli with probability of success p, then $\mu = p$ and $\overline{X} = \widehat{p}$, the sample proportion of successes. Thus, we obtain

Expected Value of the Sample Proportion

$$E(\widehat{p}) = p \qquad (4.4.4)$$

Moreover, since $T = X_1 + \cdots + X_n \sim \text{Bin}(n,p)$, Corollary 4.4-1 provides an alternative (easier!) proof that the expected value of a $\text{Bin}(n,p)$ random variable is $E(T) = np$.

Example 4.4-5

In a typical evening, a waiter serves four tables that order alcoholic beverages and three that do not.

(a) The tip left at a table that orders alcoholic beverages is a random variable with mean $\mu_1 = 20$ dollars. Find the expected value of the total amount of tips the waiter will receive from the four tables that order alcoholic beverages.

(b) The tip left at a table where no alcoholic beverages are ordered is a random variable with mean $\mu_2 = 10$ dollars. Find the expected value of the total amount of tips the waiter will receive from the three tables where no alcoholic beverages are ordered.

(c) Find the expected value of the total amount of tips the waiter will receive in a typical evening.

Solution

(a) Let $X_1, \ldots, X_4$ denote the tips left at the four tables that order alcoholic beverages. The X_i's have a common mean value of $\mu_1 = 20$. Thus, according to Corollary 4.4-1, the expected value of the total amount, $T_1 = \sum_{i=1}^{4} X_i$, of tips is $E(T_1) = 4 \times 20 = 80$.

(b) Let Y_1, Y_2, Y_3 denote the tips left at the three tables where no alcoholic beverages are ordered. The Y_i's have a common mean value of $\mu_2 = 10$. Thus, according to Corollary 4.4-1, the expected value of the total amount, $T_2 = \sum_{i=1}^{3} Y_i$, of tips is $E(T_2) = 3 \times 10 = 30$.

(c) The total amount of tips the waiter will receive in a typical evening is $T = T_1 + T_2$, where T_1 and T_2 are the total tips received from tables with and without alcoholic beverages. Thus, according to (4.4.2), $E(T) = E(T_1) + E(T_2) = 80 + 30 = 110$. ■

The following proposition, which gives the expected value of the sum of a random number of random variables, has interesting applications.

Proposition 4.4-3

Suppose that N is an integer-valued random variable, and the random variables X_i are independent from N and have common mean value μ. Then,

Expected Value of a Sum of a Random Number of Random Variables

$$E\left(\sum_{i=1}^{N} X_i\right) = E(N)\mu$$

■

The proof of this proposition follows by a combination of the Law of Total Expectation and the formula for the expected value of sums, but the details will not be presented.

Example 4.4-6

Let N denote the number of people entering a department store in a typical day, and let X_i denote the amount of money spent by the ith person. Suppose the X_i have a common mean of \$22.00, independently from the total number of customers N. If N is a Poisson random variable with parameter $\lambda = 140$, find the expected amount of money spent in the store in a typical day.

Solution

The total amount, T, of money spent in the store in a typical day is the sum of the amounts X_i, $i = 1, \ldots, N$, spent by each of the N people that enter the store, that is, $T = \sum_{i=1}^{N} X_i$. The information that $N \sim \text{Poisson}(\lambda = 140)$ implies that $E(N) = 140$. Since the conditions stated in Proposition 4.4-3 are satisfied, it follows that

$$E(T) = E(N)E(X_1) = 140 \times 22 = 3080.$$

4.4.3 THE COVARIANCE AND THE VARIANCE OF SUMS

In the previous section we saw that the same simple formula for the expected value of a linear combination of random variables holds regardless of whether or not the random variables are independent. Dependence, however, does affect the formula for the variance of sums. To see why, let's consider the variance of $X + Y$:

$$\begin{aligned}\text{Var}(X+Y) &= E\left\{\left[X+Y-E(X+Y)\right]^2\right\} \\ &= E\left\{\left[(X-E(X))+(Y-E(Y))\right]^2\right\} \\ &= E\left[(X-E(X))^2+(Y-E(Y))^2+2(X-E(X))(Y-E(Y))\right] \\ &= \text{Var}(X)+\text{Var}(Y)+2E\left[(X-E(X))(Y-E(Y))\right]. \end{aligned} \qquad \textbf{(4.4.5)}$$

If X and Y are independent then, part (3) of Proposition 4.3-3 (or Example 4.4-4) with $g(X) = X - E(X)$ and $h(Y) = Y - E(Y)$ implies

$$\begin{aligned} E\left[(X-E(X))(Y-E(Y))\right] &= E\left[X-E(X)\right]E\left[Y-E(Y)\right] \\ &= \left[E(X)-E(X)\right]\left[E(Y)-E(Y)\right] = 0. \end{aligned} \qquad \textbf{(4.4.6)}$$

Thus, if X and Y are independent the formula for the variance of $X + Y$ simplifies to $\text{Var}(X + Y) = \text{Var}(X) + \text{Var}(Y)$.

The quantity $E\left[(X - E(X))(Y - E(Y))\right]$ that appears in formula (4.4.5) is called the **covariance** of X and Y, and is denoted by $\text{Cov}(X, Y)$ or $\sigma_{X,Y}$:

Definition of and Short-cut Formula for the Covariance

$$\begin{aligned}\sigma_{X,Y} &= E\left[(X-\mu_X)(Y-\mu_Y)\right] \\ &= E(XY) - \mu_X\mu_Y \end{aligned} \qquad \textbf{(4.4.7)}$$

where μ_X and μ_Y are the marginal expected values of X and Y, respectively. The second equality in (4.4.7) is a computational formula for the covariance, similar to the computational (short-cut) formula, $\sigma_X^2 = E[(X - \mu_X)^2] = E(X^2) - \mu_X^2$, for the variance.

The formula for the variance of the sum of two random variables derived in (4.4.5), and a corresponding formula for the difference of two random variables, will be used often in the chapters to follow. For this reason, these formulas and their extensions to the sums of several random variables are highlighted in the following proposition.

Proposition 4.4-4

1. Let σ_1^2, σ_2^2 denote the variances of X_1, X_2, respectively. Then
   (a) If X_1, X_2 are independent (or just $\text{Cov}(X_1, X_2) = 0$),

$$\text{Var}(X_1 + X_2) = \sigma_1^2 + \sigma_2^2 \quad \text{and} \quad \text{Var}(X_1 - X_2) = \sigma_1^2 + \sigma_2^2.$$

   (b) If X_1, X_2 are dependent,

$$\text{Var}(X_1 - X_2) = \sigma_1^2 + \sigma_2^2 - 2\text{Cov}(X_1, X_2)$$
$$\text{Var}(X_1 + X_2) = \sigma_1^2 + \sigma_2^2 + 2\text{Cov}(X_1, X_2).$$

2. Let $\sigma_1^2, \ldots, \sigma_m^2$ denote the variances of $X_1, \ldots, X_m$, respectively, and $a_1, \ldots, a_m$ be any constants. Then
   (a) If $X_1, \ldots, X_m$ are independent (or just $\text{Cov}(X_i, X_j) = 0$, for all $i \neq j$),

$$\text{Var}(a_1X_1 + \cdots + a_mX_m) = a_1^2\sigma_1^2 + \cdots + a_m^2\sigma_m^2.$$

   (b) If $X_1, \ldots, X_m$ are dependent,

$$\text{Var}(a_1X_1 + \cdots + a_mX_m) = a_1^2\sigma_1^2 + \cdots + a_m^2\sigma_m^2 + \sum_i \sum_{j \neq i} a_i a_j \sigma_{ij}$$

■

According to part (1) of Proposition 4.4-4, the variances of $X + Y$ and $X - Y$ are the same if X and Y are independent, but differ if their covariance is different from zero. Because part (1a) appears counterintuitive at first sight, the following example offers numerical verification of it based on the fact that when the sample size is large enough the sample variance is a good approximation to the population variance.

Example 4.4-7

Simulation-based verification of part (1a) of Proposition 4.4-4. Let X, Y be independent uniform(0, 1) random variables. Generate a random sample of size 10,000 of $X+Y$ values, and a random sample of size 10,000 of $X-Y$ values and compute the sample variances of the two samples. Argue that this provides numerical evidence in support of part (1a) of Proposition 4.4-4. (See also Exercise 13 for a numerical verification of part (1b) of Proposition 4.4-4).

Solution

The R commands

```
set.seed=111; x=runif(10000); y=runif(10000)
```

generate a random sample of size 10,000 X values and a random sample of size 10,000 Y values. (*set.seed=111* was used in order to have reproducibility of the results.) The additional R commands

```
var(x + y); var(x - y)
```

yield 0.167 and 0.164 (rounded to 3 decimal places) for the sample variances of a sample of 10,000 $X + Y$ values and a sample of 10,000 $X - Y$ values, respectively.

Repeating the above commands with the seed set to 222 and to 333 yields pairs of sample variances of (0.166, 0.167) and (0.168, 0.165). This suggests that the sample variances of $X + Y$ values approximate the same quantity as the sample variances of $X - Y$ values, supporting the statement of part (1a) of Proposition 4.4-4 that $\text{Var}(X + Y) = \text{Var}(X - Y) = 2/12 = 0.1667$ (where we also used the fact that $\text{Var}(X) = \text{Var}(Y) = 1/12$; see Example 3.3-13). ■

Example 4.4-8

Let X take the value 0, 1, or 2 depending on whether there are no customers, between one and 10 customers, and more than 10 customers in the regular (manned) checkout lines of a supermarket. Let Y be the corresponding variable for the self checkout lines. Find $\text{Var}(X + Y)$, when the joint PMF of X and Y is given by

		y			
	p(x, y)	0	1	2	$p_X(x)$
	0	0.17	0.02	0.01	0.20
x	1	0.150	0.225	0.125	0.50
	2	0.060	0.105	0.135	0.30
	$p_Y(y)$	0.38	0.35	0.27	

Solution

In order to use formula (4.4.5) we will need to compute σ_X^2, σ_Y^2, and σ_{XY}. As a first step we compute $E(X)$, $E(Y)$, $E(X^2)$, $E(Y^2)$ and $E(XY)$:

$$E(X) = \sum_x xp_X(x) = 1.1, \quad E(Y) = \sum_y yp_Y(y) = 0.89,$$

$$E(X^2) = \sum_x x^2 p_X(x) = 1.7, \quad E(Y^2) = \sum_y y^2 p_Y(y) = 1.43,$$

and, according to part (1) of Proposition 4.4-1,

$$E(XY) = \sum_x \sum_y xyp(x, y) = 0.225 + 2 \times 0.125 + 2 \times 0.105 + 4 \times 0.135 = 1.225.$$

Thus, $\sigma_X^2 = 1.7 - 1.1^2 = 0.49$, $\sigma_Y^2 = 1.43 - 0.89^2 = 0.6379$ and, according to the computational formula (4.4.7), $\text{Cov}(X, Y) = 1.225 - 1.1 \times 0.89 = 0.246$. Finally, by (4.4.5),

$$\text{Var}(X + Y) = 0.49 + 0.6379 + 2 \times 0.246 = 1.6199.$$ ■

Example 4.4-9

Using a geolocation system, a dispatcher sends messages to two trucks sequentially. Suppose the joint PDF of the response times X_1 and X_2, measured in seconds, is $f(x_1, x_2) = \exp(-x_2)$ for $0 \le x_1 \le x_2$, and $f(x_1, x_2) = 0$ otherwise. Find $\text{Cov}(X_1, X_2)$.

Solution

Will use the computational formula $\text{Cov}(X_1, X_2) = E(X_1X_2) - E(X_1)E(X_2)$. First, the marginal PDFs of X_1 and X_2 are,

$$f_{X_1}(x_1) = \int_{x_1}^{\infty} e^{-x_2}\, dx_2 = e^{-x_1} \quad \text{and} \quad f_{X_2}(x_2) = \int_0^{x_2} e^{-x_2}\, dx_1 = x_2 e^{-x_2},$$

respectively. Thus,

$$E(X_1) = \int_0^\infty x f_{X_1}(x)\,dx = \int_0^\infty x e^{-x} dx = 1. \quad \text{and}$$

$$E(X_2) = \int_0^\infty y f_{X_2}(y)\,dy = \int_0^\infty y^2 e^{-y} dy = 2$$

follow by integration techniques similar to those in Examples 3.3-6 and 3.3-14, or by the R command *f=function(x){x**2*exp(-x)}; integrate(f, 0, Inf)* for the second integral and a similar one for the first. Next, according to part (4.4-1) of Proposition 4.4-1,

$$E(X_1X_2) = \int_0^\infty \int_0^\infty x_1 x_2 f(x_1, x_2)\,dx_1\,dx_2 = \int_0^\infty \int_0^{x_2} x_1 x_2 e^{-x_2}\,dx_1\,dx_2$$

$$= \int_0^\infty x_2 e^{-x_2} \int_0^{x_2} x_1\,dx_1\,dx_2 = \int_0^\infty 0.5 x_2^2 e^{-x_2}\,dx_2 = 1.$$

Thus, we obtain $\text{Cov}(X_1, X_2) = 1 - 1 \cdot 2 = -1$. ■

An important special case of part (2a) of Proposition 4.4-4 has to do with the variance of the sample mean and the sample sum. This is given next.

Corollary 4.4-2 Let $X_1, \ldots, X_n$ be iid (i.e., a simple random sample from an infinite population) with common variance σ^2. Then,

Variance of the Average and the Sum

$$\text{Var}(\overline{X}) = \frac{\sigma^2}{n} \quad \text{and} \quad \text{Var}(T) = n\sigma^2 \tag{4.4.8}$$

where $\overline{X} = n^{-1} \sum_{i=1}^n X_i$ and $T = \sum_{i=1}^n X_i$.

If the X_i in Corollary 4.4-2 are Bernoulli with probability of success p, then $\sigma^2 = p(1-p)$ and $\overline{X} = \widehat{p}$, the sample proportion of successes. Thus, we obtain

Variance of the Sample Proportion

$$\text{Var}(\widehat{p}) = \frac{p(1-p)}{n} \tag{4.4.9}$$

Moreover, since $T = X_1 + \cdots + X_n \sim \text{Bin}(n, p)$, Corollary 4.4-2 provides an alternative (easier!) proof that the variance of a $\text{Bin}(n, p)$ random variable is $\text{Var}(T) = np(1-p)$.

Proposition 4.4-5 **Properties of covariance.**

1. $\text{Cov}(X, Y) = \text{Cov}(Y, X)$.
2. $\text{Cov}(X, X) = \text{Var}(X)$.

3. If X, Y are independent, then $\text{Cov}(X, Y) = 0$.
4. $\text{Cov}(aX + b, cY + d) = ac\text{Cov}(X, Y)$ for any real numbers a, b, c, and d.

Proof of Proposition 4.4-5. Parts (1) and (2) follow immediately from the definition of covariance, while part (3) is already proved in relation (4.4.6). For part (4) note that $E(aX + b) = aE(X) + b$ and $E(cY + d) = cE(Y) + d$. Hence,

$$\begin{aligned}\text{Cov}(aX + b, cY + d) &= E\left\{[aX + b - E(aX + b)][cY + d - E(cY + d)]\right\}\\ &= E\left\{[aX - aE(X)][cY - cE(Y)]\right\}\\ &= E\left\{a[X - E(X)]c[Y - E(Y)]\right\}\\ &= ac\text{Cov}(X, Y).\end{aligned}$$

■

Example 4.4-10

Consider the information given in Example 4.4-9, but assume that the response times are given in milliseconds. If $(\widetilde{X}_1, \widetilde{X}_2)$ denote the response times in milliseconds, find $\text{Cov}(\widetilde{X}_1, \widetilde{X}_2)$.

Solution

The new response times are related to those of Example 4.4-9 by $(\widetilde{X}_1, \widetilde{X}_2) = (1000X_1, 1000X_2)$. Hence, according to part (4) of Proposition 4.4-5,

$$\text{Cov}(\widetilde{X}_1, \widetilde{X}_2) = \text{Cov}(1000X_1, 1000X_2) = -1{,}000{,}000.$$

■

The next example shows that $\text{Cov}(X, Y)$ can be zero even when X and Y are not independent. An additional example of this is given in Exercise 8 in Section 4.5.

Example 4.4-11

If X, Y have the joint PMF given by

	$p(x, y)$	y = 0	1	
	−1	1/3	0	1/3
x	0	0	1/3	1/3
	1	1/3	0	1/3
		2/3	1/3	1.0

find $\text{Cov}(X, Y)$. Are X and Y independent?

Solution

Since $E(X) = 0$, use of the computational formula $\text{Cov}(X, Y) = E(XY) - E(X)E(Y)$ gives $\text{Cov}(X, Y) = E(XY)$. However, the product XY takes the value zero with probability 1. Thus, $\text{Cov}(X, Y) = E(XY) = 0$. Finally, X and Y are not independent because $p(0, 0) = 0 \neq p_X(0)p_Y(0) = 2/9$.

■

Exercises

1. Due to promotional sales, an item is sold at 10% or 20% below its regular price of \$150. Let X and Y denote the selling prices of the item at two online sites, and let their joint PMF be

	$p(x, y)$	y = 150	135	120
	150	0.25	0.05	0.05
x	135	0.05	0.2	0.1
	120	0.05	0.1	0.15

If a person checks both sites and buys from the one listing the lower price, find the expected value and variance of the price the person pays. (*Hint.* The price the person pays is expressed as $\min\{X, Y\}$.)

2. A system consists of components A and B connected in parallel. Suppose the two components fail independently, and the time to failure for each component is a uniform$(0, 1)$ random variable.

(a) Find the PDF of the time to failure of the system. (*Hint.* If X, Y are the times components A, B fail, respectively, the system fails at time $T = \max\{X, Y\}$. Find the CDF of T from $P(T \le t) = P(X \le t)P(Y \le t)$ and the CDF of a uniform(0, 1) random variable. Then find the PDF by differentiation.)

(b) Find the expected value and variance of the time to failure of the system.

3. The joint distribution of X = height and Y = radius of a cylinder is $f(x, y) = 3x/(8y^2)$ for $1 \le x \le 3, 0.5 \le y \le 0.75$ and zero otherwise. Find the variance of the volume of a randomly selected cylinder. (*Hint.* The volume of the cylinder is given by $h(X, Y) = \pi Y^2 X$. In Example 4.3-17 it was found that $E[h(X, Y)] = (13/16)\pi$.)

4. In a typical week a person takes the bus five times in the morning and three times in the evening. Suppose the waiting time for the bus in the morning has mean 3 minutes and variance 2 minutes^2, while the waiting time in the evening has mean 6 minutes and variance 4 minutes^2.

(a) Let X_i denote the waiting time in the ith morning of the week, $i = 1, \ldots, 5$, and let Y_j denote the waiting time in the jth evening of the week. Express the total waiting time as a linear combination of these X's and Y's.

(b) Find the expected value and the variance of the total waiting time in a typical week. State any assumptions needed for the validity of your calculations.

5. Two towers are constructed, each by stacking 30 segments of concrete vertically. The height (in inches) of a randomly selected segment is uniformly distributed in the interval (35.5, 36.5).

(a) Find the mean value and the variance of the height of a randomly selected segment. (*Hint.* See Examples 3.3-8 and 3.3-16 for the mean and variance of a uniform random variable.)

(b) Let $X_1, \ldots, X_{30}$ denote the heights of the segments used in tower 1. Find the mean value and the variance of the height of tower 1. (*Hint.* Express the height of tower 1 as the sum of the X's.)

(c) Find the mean value and the variance of the difference of the heights of the two towers. (*Hint.* Set $Y_1, \ldots, Y_{30}$ for the heights of the segments used in tower 2, and express the height of tower 2 as the sum of the Y's.)

6. Let N denote the number of accidents per month in all locations of an industrial complex, and let X_i denote the number of injuries reported for the ith accident. Suppose that the X_i are independent random variables having common expected value of 1.5 and are independent from N. If $E(N) = 7$, find the expected number of injuries in a month.

7. In a typical evening, a waiter serves N_1 tables that order alcoholic beverages and N_2 tables that do not. Suppose N_1, N_2 are Poisson random variables with parameters $\lambda_1 = 4$, $\lambda_2 = 6$, respectively. Suppose the tips, X_i, left at tables that order alcoholic beverages have common mean value of \$20.00, while the tips, Y_j, left at tables that do not order alcoholic beverages have a common mean value of \$10.00. Assuming that the tips left are independent from the total number of tables being served, find the expected value of the total amount in tips received by the waiter in a typical evening. (*Hint.* Use Proposition 4.4-3)

8. Suppose (X, Y) have the joint PDF

$$f(x, y) = \begin{cases} 24xy & 0 \le x \le 1, 0 \le y \le 1, x + y \le 1 \\ 0 & \text{otherwise.} \end{cases}$$

Find Cov(X, Y). (*Hint.* Use the marginal PDF of X, which was derived in Example 4.3-9, and note that by the symmetry of the joint PDF in x, y, it follows that the marginal PDF of Y is the same as that of X.)

9. Suppose the random variables Y, X, and ϵ are related through the model

$$Y = 9.3 + 1.5X + \varepsilon,$$

where ε has zero mean and variance $\sigma_\varepsilon^2 = 16$, $\sigma_X^2 = 9$, and X, ε are independent. Find the covariance of Y and X and that of Y and ε. (*Hint.* Write $\text{Cov}(X, Y) = \text{Cov}(X, 9.3 + 1.5X + \varepsilon)$ and use part (4) of Proposition 4.4-5. Use a similar process for $\text{Cov}(\varepsilon, Y)$.)

10. Using the information on the joint distribution of meal price and tip given in Exercise 3 in Section 4.3, find the expected value and the variance of the total cost of the meal (entree plus tip) for a randomly selected customer.

11. Consider the information given in Exercise 1 in Section 4.3 on the joint distribution of the volume, X, of online monthly book sales, and the volume, Y, of monthly book sales from the brick and mortar counterpart of a bookstore. An approximate formula for the monthly profit, in thousands of dollars, of the bookstore is $8X + 10Y$. Find the expected value and variance of the monthly profit of the bookstore.

12. Consider the information given in Exercise 7 in Section 4.3 regarding the level of moisture content and impurity of chemical batches. Such batches are used to prepare a particular substance. The cost of preparing the substance is $C = 2\sqrt{X}+3Y^2$. Find the expected value and variance of the cost of preparing the substance.

13. Let X, Y, and Z be independent uniform(0, 1) random variables, and set $X_1 = X + Z$, $Y_1 = Y + 2Z$.

(a) Find $\text{Var}(X_1 + Y_1)$ and $\text{Var}(X_1 - Y_1)$. (*Hint.* Find $\text{Var}(X_1)$, $\text{Var}(Y_1)$, $\text{Cov}(X_1, Y_1)$, and use part (1) of Proposition 4.4-4.)

(b) Using R commands similar to those used in Example 4.4-7, generate a sample of size 10,000 of $X_1 + Y_1$ values and a sample of size 10,000 of $X_1 - Y_1$ values and compute the two sample variances. Argue that this provides numerical evidence in support of part (a) above and also for part (1b) of Proposition 4.4-4.

14. On the first day of a wine-tasting event three randomly selected judges are to taste and rate a particular wine before tasting any other wine. On the second day the same three judges are to taste and rate the wine after tasting other wines. Let X_1, X_2, X_3 be the ratings, on a 100-point scale, on the first day, and Y_1, Y_2, Y_3 be the ratings on the second day. We are given that the variance of each X_i is $\sigma_X^2 = 9$, the variance of each Y_i is $\sigma_Y^2 = 4$, the covariance $\text{Cov}(X_i, Y_i) = 5$ for all $i = 1, 2, 3$, and $\text{Cov}(X_i, Y_j) = 0$ for all $i \neq j$. Find the variance of the combined rating $\overline{X} + \overline{Y}$. (*Hint.* The formula in part (4) of Proposition 4.4-5 generalizes to the following formula for the covariance of two sums of random variables:

$$\text{Cov}\left(\sum_{i=1}^{m} a_i X_i, \sum_{j=1}^{n} b_j Y_j\right) = \sum_{i=1}^{m}\sum_{j=1}^{n} a_i b_j \text{Cov}\left(X_i, Y_j\right).$$

Use this formula to find the $\text{Cov}(\overline{X}, \overline{Y})$, keeping in mind that $\text{Cov}(X_i, Y_j) = 0$ if $i \neq j$. Then use the formula in part (1) of Propositon 4.4-4 and Corollary 4.4-2 for the variance of the sample average.)

15. Let X be a hypergeometric random variable with parameters n, M_1, and M_2. Use Corollary 4.4-1 to give an alternative (easier) derivation of the formula for $E(X)$. (*Hint.* See the derivation of the expected value of a $\text{Bin}(n, p)$ random variable following Corollary 4.4-1.)

16. Let X have the negative binomial distribution with parameters r and p. Thus, X counts the total number of Bernoulli trials until the rth success. Next, let X_1 denote the number of trials up to and including the first success, let X_2 denote the number from the first success up to and including the second success, and so on, so that X_r denotes the number of trials from the $(r - 1)$-st success up to and including the rth success. Note that the X_i's are iid having the geometric distribution, $X = X_1 + \cdots + X_r$. Use Corollary 4.4-1 and Proposition 4.4-4 to derive the expected value and variance of the negative binomial random variable X. (*Hint.* The expected value and variance of a geometric random variable are derived in Examples 3.3-3 and 3.3-12.)

4.5 Quantifying Dependence

When two random variables X and Y are not independent, they are dependent. Of course, there are various degrees of dependence, ranging from very strong to very weak. In this section we introduce *correlation* as a means for quantifying dependence. First we introduce the concept of monotone dependence, make the distinction between positive and negative dependence, and illustrate the role of the covariance in characterizing the distinction. Then we present Pearson's correlation coefficient and discuss its interpretation.

4.5.1 POSITIVE AND NEGATIVE DEPENDENCE

We say that X and Y are **positively dependent**, or **positively correlated**, if "large" values of X are associated with "large" values of Y and "small" values of X are associated with "small" values of Y. (Here, "large" means "larger than average" and "small" means "smaller than average.") For example, the variables X = height and Y = weight of a randomly selected adult male are positively dependent. In the opposite case, that is, when "large" values of X are associated with "small" values of Y and "small" values of X are associated with "large" values of Y, we say that X and Y are **negatively dependent** or **negatively correlated.** An example of negatively dependent variables is X = stress applied and Y = time to failure. If the variables are either positively or negatively dependent, their dependence is called **monotone.**

It should be clear that if the dependence is positive then the regression function, $\mu_{Y|X}(x) = E(Y|X = x)$, of Y on X is an increasing function of x. For example, if we consider X = height and Y = weight then, due to the positive dependence of these variables, we have $\mu_{Y|X}(1.82) < \mu_{Y|X}(1.90)$, that is, the average weight of men 1.82 meters tall is smaller than the average weight of men 1.90 meters tall (1.82 meters is about 6 feet). Similarly, if the dependence is negative then $\mu_{Y|X}(x)$ is decreasing, and if the dependence is monotone then $\mu_{Y|X}(x)$ is monotone.

The fact that covariance can be used for identifying a monotone dependence as being positive or negative is less obvious, but here is the rule: The monotone dependence is positive or negative if the covariance takes a positive or negative value, respectively.

In order to develop an intuitive understanding as to why the sign of covariance identifies the nature of a monotone dependence, consider a finite population of N units, let $(x_1, y_1), (x_2, y_2), \ldots, (x_N, y_N)$ denote the values of a bivariate characteristic of interest for each of the N units, and let (X, Y) denote the bivariate characteristic of a randomly selected unit. Then (X, Y) has a discrete distribution taking each of the possible values $(x_1, y_1), \ldots, (x_N, y_N)$ with probability $1/N$. In this case the covariance formula in definition (4.4.7) can be written as

$$\sigma_{X,Y} = \frac{1}{N}\sum_{i=1}^{N}(x_i - \mu_X)(y_i - \mu_Y), \tag{4.5.1}$$

where $\mu_X = \frac{1}{N}\sum_{i=1}^{N} x_i$ and $\mu_Y = \frac{1}{N}\sum_{i=1}^{N} y_i$ are the marginal expected values of X and Y, respectively. Suppose now X and Y are positively correlated. Then, X-values larger than μ_X are associated with Y-values that are larger than μ_Y, and X-values smaller than μ_X are associated with Y-values smaller than μ_Y.Thus, the products

$$(x_i - \mu_X)(y_i - \mu_Y), \tag{4.5.2}$$

which appear in the summation of relation (4.5.1), will tend to be positive, resulting in a positive value for $\sigma_{X,Y}$. Similarly, if X and Y are negatively correlated, the products in (4.5.2) will tend to be negative, resulting in a negative value for $\sigma_{X,Y}$.

However, the usefulness of covariance in quantifying dependence does not extend beyond its ability to characterize the nature of monotone dependence. This is because a successful measure of dependence should be scale-free. For example, the strength of dependence of the variables (Height, Weight) should not depend on whether the variables are measured in meters and kilograms or feet and pounds. According to part (4) of Proposition 4.4-5, however, the value of the covariance is scale-dependent and thus cannot serve as a quantification of dependence.

4.5.2 PEARSON'S (OR LINEAR) CORRELATION COEFFICIENT

It turns out that a simple adjustment of the covariance makes its value scale-free, and leads to the most commonly used quantification of dependence, the *(linear) correlation coefficient*, also known as *Pearson's correlation coefficient* in honor of its inventor.

Definition 4.5-1

The **Pearson's (or linear) correlation coefficient** of X and Y, denoted by $\text{Corr}(X, Y)$ or $\rho_{X,Y}$, is defined as

$$\rho_{X,Y} = \text{Corr}(X, Y) = \frac{\text{Cov}(X, Y)}{\sigma_X \sigma_Y},$$

where σ_X, σ_Y are the marginal standard deviations of X, Y, respectively.

The following proposition summarizes some properties of the correlation coefficient.

Proposition 4.5-1

1. If a and c are either both positive or both negative, then

$$\text{Corr}(aX + b, cY + d) = \rho_{X,Y}.$$

If a and c are of opposite signs, then

$$\text{Corr}(aX + b, cY + d) = -\rho_{X,Y}.$$

2. $-1 \leq \rho_{X,Y} \leq 1$, and
   (a) if X, Y are independent then $\rho_{X,Y} = 0$.
   (b) $\rho_{X,Y} = 1$ or -1 if and only if $Y = aX + b$ for some numbers a, b with $a \neq 0$.

■

The properties listed in Proposition 4.5-1 imply that correlation is indeed a successful measure of *linear* dependence. The properties in part (1) mean that it has the desirable property of being scale-free. The properties in part (2) make it possible to develop a feeling for the strength of linear dependence implied by a $\rho_{X,Y}$-value. Thus, if the variables are independent, $\rho_{X,Y} = 0$, while $\rho_{X,Y} = \pm 1$ happens if and only if X and Y have the strongest possible *linear* dependence (that is, knowing one amounts to knowing the other). The scatterplots in Figure 4-2 correspond to

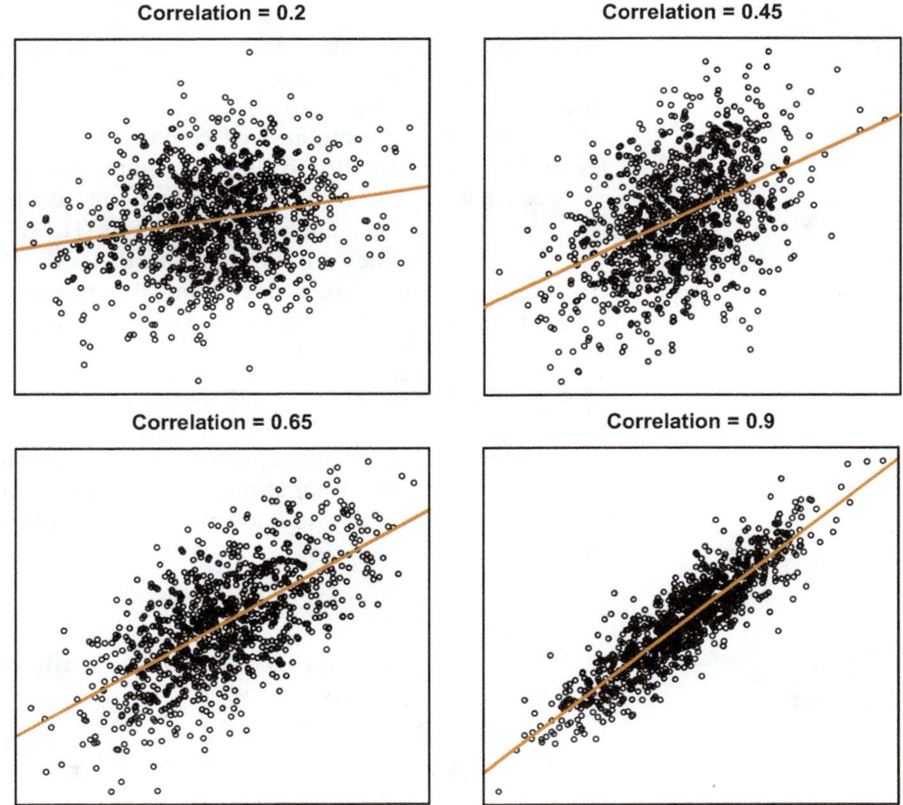

Figure 4-2 Scatterplots corresponding to different correlation coefficients.

artificially constructed populations of size 1000, having correlation ranging from 0.2 to 0.9; the line through each scatterplot is the corresponding regression function of Y on X.

Example 4.5-1

In a reliability context a randomly selected electronic component will undergo an accelerated failure time test. Let X_1 take the value 1 if the component lasts less than 50 hours and zero otherwise, and X_2 take the value 1 if the component lasts between 50 and 90 hours and zero otherwise. The probabilities that a randomly selected component will last less than 50 hours, between 50 and 90 hours, and more than 90 hours are 0.2, 0.5, and 0.3, respectively. Find ρ_{X_1,X_2}.

Solution

We will first find the covariance of X_1 and X_2 using the short-cut formula $\sigma_{X_1X_2} = E(X_1X_2) - E(X_1)E(X_2)$. Next, because the sample space of (X_1, X_2) is $\{(1,0),(0,1),(0,0)\}$, it follows that the product X_1X_2 is always equal to zero and hence $E(X_1X_2) = 0$. Because the marginal distribution of each X_i is Bernoulli, we have that

$$E(X_1) = 0.2, \quad E(X_2) = 0.5, \quad \sigma^2_{X_1} = 0.16, \quad \text{and } \sigma^2_{X_2} = 0.25.$$

Combining these calculations we find $\sigma_{X_1X_2} = 0 - 0.2 \times 0.5 = -0.1$. Finally, using the definition of correlation,

$$\rho_{X_1,X_2} = \frac{\sigma_{X_1X_2}}{\sigma_{X_1}\sigma_{X_2}} = \frac{-0.1}{0.4 \times 0.5} = -0.5.$$

■

Example 4.5-2

Using a geolocation system, a dispatcher sends messages to two trucks sequentially. Suppose the joint PDF of the response times X_1 and X_2, measured in seconds, is $f(x_1, x_2) = \exp(-x_2)$ for $0 \le x_1 \le x_2$, and $f(x_1, x_2) = 0$ otherwise.

(a) Find $\text{Corr}(X_1, X_2)$.

(b) If $(\widetilde{X}_1, \widetilde{X}_2)$ denote the response times in milliseconds, find $\text{Corr}(\widetilde{X}_1, \widetilde{X}_2)$.

Solution

(a) In Example 4.4-9 we saw that $\text{Cov}(X_1, X_2) = -1$. In the same example, we found that the marginal PDFs and means of X_1 and X_2 are

$$f_{X_1}(x_1) = e^{-x_1}, \quad f_{X_2}(x_2) = x_2e^{-x_2}, \quad E(X_1) = 1, \quad \text{and} \quad E(X_2) = 2.$$

Using these marginal PDFs, we find

$$E(X_1^2) = \int_0^\infty x^2 f_{X_1}(x)\,dx = 2 \quad \text{and} \quad E(X_2^2) = \int_0^\infty x^2 f_{X_2}(x)\,dx = 6.$$

The above integrations can also be performed in R; for example the value of $E(X_2^2)$ can be found with the R command *f=function(x){x**3*exp(-x)}; integrate(f, 0, Inf)*. Combining the above results, the standard deviations of X_1 and X_2 are obtained as

$$\sigma_{X_1} = \sqrt{2 - 1^2} = 1 \quad \text{and} \quad \sigma_{X_2} = \sqrt{6 - 2^2} = 1.414.$$

Thus, from the definition of the correlation coefficient,

$$\rho_{X_1, X_2} = \frac{-1}{1 \cdot 1.414} = -0.707.$$

(b) Using properties of the covariance, in Example 4.4-10 we saw that $\text{Cov}(\widetilde{X}_1, \widetilde{X}_2) = -1{,}000{,}000$. Moreover, from the properties of standard deviation (see Proposition 3.3-2) we have $\sigma_{\widetilde{X}_1} = 1000$ and $\sigma_{\widetilde{X}_2} = 1414$. It follows that $\rho_{\widetilde{X}_1, \widetilde{X}_2} = -1{,}000{,}000/(1000 \cdot 1414) = -0.707$. Thus, $\rho_{\widetilde{X}_1, \widetilde{X}_2} = \rho_{X_1, X_2}$ as stipulated by Proposition 4.5-1, part (1). ■

Pearson's Correlation as a Measure of Linear Dependence It should be emphasized that correlation measures only *linear dependence*. In particular, it is possible to have the strongest possible dependence, that is, knowing one amounts to knowing the other, but if the relation between X and Y is not linear, then the correlation will not equal 1, as the following example shows.

Example 4.5-3

Let X have the uniform in $(0, 1)$ distribution, and $Y = X^2$. Find $\rho_{X,Y}$.

Solution

First we will find the covariance through the short-cut formula $\text{Cov}(X, Y) = E(XY) - E(X)E(Y)$. Note that since $Y = X^2$, we have $XY = X^3$. Thus, $E(XY) = E(X^3) = \int_0^1 x^3\,dx = 1/4$. Also, since $E(Y) = E(X^2) = 1/3$ and $E(X) = 1/2$, we obtain

$$\text{Cov}(X, Y) = \frac{1}{4} - \frac{1}{2}\frac{1}{3} = \frac{1}{12}.$$

Next, $\sigma_X = 1/\sqrt{12}$ (see Example 3.3-13), and $\sigma_Y = \sqrt{E(X^4) - [E(X^2)]^2} = \sqrt{1/5 - 1/9} = 2/3\sqrt{5}$. Combining the above results we obtain

$$\rho_{X,Y} = \frac{\text{Cov}(X, Y)}{\sigma_X \sigma_Y} = \frac{3\sqrt{5}}{2\sqrt{12}} = 0.968.$$

A similar set of calculations reveals that with X as before and $Y = X^4$, $\rho_{X,Y} = 0.866$. ■

Note that in the above example, knowing X amounts to knowing Y and, conversely, X is given as the positive square root of Y. However, although monotone, the relationship between X and Y is not linear.

Definition 4.5-2

Two variables having zero correlation are called **uncorrelated**.

Independent variables are uncorrelated, but uncorrelated variables are not necessarily independent; see Example 4.4-11. In general, if the dependence is not monotone, that is, neither positive nor negative, it is possible for two variables to have zero correlation even though they are very strongly related; see Exercise 8.

Sample Versions of the Covariance and Correlation Coefficient If $(X_1, Y_1), \ldots, (X_n, Y_n)$ is a sample from the bivariate distribution of (X, Y), the **sample covariance**, denoted by $\widehat{\text{Cov}}(X, Y)$ or $S_{X,Y}$, and **sample correlation coefficient**, denoted by $\widehat{\text{Corr}}(X, Y)$ or $r_{X,Y}$, are defined as

Sample Versions of Covariance and Correlation Coefficient

$$S_{X,Y} = \frac{1}{n-1}\sum_{i=1}^{n}(X_i - \overline{X})(Y_i - \overline{Y})$$
$$r_{X,Y} = \frac{S_{X,Y}}{S_X S_Y} \tag{4.5.3}$$

where $\overline{X}$, S_X and $\overline{Y}$, S_Y are the (marginal) sample mean and sample standard deviation of the X-sample and Y-sample, respectively.

In Chapter 1 it was repeatedly stressed that sample versions of the population parameters estimate but, in general, they are not equal to the corresponding population parameters. In particular, $\widehat{\text{Cov}}(X, Y)$ and $\widehat{\text{Corr}}(X, Y)$ estimate $\text{Cov}(X, Y)$ and $\text{Corr}(X, Y)$, respectively, but, in general, they are not equal to them.

A computational formula for the sample covariance is

$$S_{X,Y} = \frac{1}{n-1}\left[\sum_{i=1}^{n} X_i Y_i - \frac{1}{n}\left(\sum_{i=1}^{n} X_i\right)\left(\sum_{i=1}^{n} Y_i\right)\right].$$

If the X_i values are in the R object x and the Y_i values are in y, the R commands for computing $S_{X,Y}$ and $r_{X,Y}$ are

R Commands for Covariance and Correlation

```
cov(x, y) # gives S_X,Y
cor(x, y) # gives r_X,Y
```

Example 4.5-4

To calibrate a method for measuring lead concentration in water, the method was applied to 12 water samples with known lead content. The concentration measurements, y, and the known concentration levels, x, are given below.

x	5.95	2.06	1.02	4.05	3.07	8.45	2.93	9.33	7.24	6.91	9.92	2.86
y	6.33	2.83	1.65	4.37	3.64	8.99	3.16	9.54	7.11	7.10	8.84	3.56

Compute the sample covariance and correlation coefficient.

Solution

With this data, $\sum_{i=1}^{12} X_i = 63.79$, $\sum_{i=1}^{12} Y_i = 67.12$, and $\sum_{i=1}^{12} X_i Y_i = 446.6939$. Thus,

$$S_{X,Y} = \frac{1}{11}\left[446.6939 - \frac{1}{12}63.79 \times 67.12\right] = 8.172.$$

Moreover, $\sum_{i=1}^{12} X_i^2 = 440.302$ and $\sum_{i=1}^{12} Y_i^2 = 456.745$, so that, using the computational formula for the sample variance given in (1.6.14), $S_X^2 = 9.2$ and $S_Y^2 = 7.393$. Thus,

$$r_{X,Y} = \frac{8.172}{\sqrt{9.2}\sqrt{7.393}} = 0.99.$$

Entering the data into the R objects x and y by *x=c(5.95, 2.06, 1.02, 4.05, 3.07, 8.45, 2.93, 9.33, 7.24, 6.91, 9.92, 2.86)* and *y=c(6.33, 2.83, 1.65, 4.37, 3.64, 8.99, 3.16, 9.54, 7.11, 7.10, 8.84, 3.56)*, the above values for the covariance and correlation coefficient can be found with the R commands *cov(x, y)* and *cor(x, y)*, respectively. ■

REMARK 4.5-1 As Example 4.5-3 demonstrated, linear correlation is not a good measure of the non-linear dependence of two variables. Two different types of correlation coefficients, *Kendall's* τ and *Spearman's* ρ, are designed to capture correctly the strength of non-linear (but monotone) dependence. Detailed descriptions of these correlation coefficients is beyond the scope of this book. ◁

Exercises

1. Using the joint distribution, given in Exercise 1 in Section 4.3, of the volume of monthly book sales from the online site, X, and the volume of monthly book sales from the brick and mortar counterpart, Y, of a bookstore, compute the linear correlation coefficient of X and Y.

2. Consider the information given in Exercise 2 in Section 4.2 regarding the amount, X, of drug administered to a randomly selected laboratory rat, and number, Y, of tumors the rat develops.

(a) Would you expect X and Y to be positively or negatively correlated? Explain your answer, and confirm it by computing the covariance.

(b) Compute the linear correlation coefficient of X and Y.

3. An article[2] reports data on X = distance between a cyclist and the roadway center line and Y = the separation distance between the cyclist and a passing car, from ten streets with bike lanes. The paired distances (X_i, Y_i) are determined by photography and are given below in feet.

x	12.8	12.9	12.9	13.6	14.5	14.6	15.1	17.5	19.5	20.8
y	5.5	6.2	6.3	7.0	7.8	8.3	7.1	10.0	10.8	11.0

(a) Compute $S_{X,Y}$, S_X^2, S_Y^2, and $r_{X,Y}$.

(b) Indicate how the quantities in part (a) would change if the distances had been given in inches.

4. Use *ta=read.table("TreeAgeDiamSugarMaple.txt", header=T)* to import the data set of diameter–age measurements for 27 sugar maple trees into the R data frame *ta*, and *x=ta$Diamet; y=ta$Age* to copy the diameter and age values into the R objects x and y, respectively.

(a) Would you expect the diameter and age to be positively or negatively correlated? Explain your answer, and confirm it by doing a scatterplot of the data (*plot(x,y)*).

(b) Compute the sample covariance and linear correlation of diameter and age using R commands. On the basis of the scatterplot are you satisfied that linear correlation correctly captures the strength of the diameter–age dependence?

5. Import the bear data into R with the R command *br=read.table("BearsData.txt", header=T)*, and form a data frame consisting only of the measurements with the R commands *attach(br); bd=data.frame(Head.L, Head.W, Neck.G, Chest.G, Weight)*.[3] The R command *cor(bd)* returns a matrix of the pairwise correlations of all variables. (The matrix is symmetric because $r_{X,Y} = r_{Y,X}$, and its diagonal elements are 1 because the correlation of a variable with itself is 1.) Using this correlation matrix, which would you say are the two best single predictors of the variable Weight?

6. Select two products from a batch of 10 containing three defective and seven non-defective products. Let $X = 1$ or 0 as the first selection from the 10 products is defective or not, and $Y = 1$ or 0 as the second selection (from the nine remaining products) is defective or not.

(a) Find the marginal distribution of X.

(b) Find the conditional distributions of Y given each of the possible values of X.

(c) Use the results in parts (a) and (b), and the multiplication rule for joint probability mass functions given in (4.3.4), to find the joint distribution of X and Y.

(d) Find the marginal distribution of Y. Is it the same as that of X?

(e) Find the covariance and the linear correlation coefficient of X and Y.

[2] B. J. Kroll and M. R. Ramey (1977). Effects of bike lanes on driver and bicyclist behavior, *Transportation Eng. J.*, 243–256.

[3] This data set is a subset of a data set contributed to Minitab by Gary Alt.

7. Consider the context of Exercise 17 in Section 4.3, so that the variables X = static force at failure, Y = defect index, have joint PDF

$$f(x,y) = \begin{cases} 24x & \text{if } 0 \le y \le 1-2x \quad \text{and} \quad 0 \le x \le .5 \\ 0 & \text{otherwise.} \end{cases}$$

(a) It is given that the marginal density of X is $f_X(x) = \int_0^{1-2x} 24x\,dy = 24x(1-2x)$, $0 \le x \le 0.5$, and the marginal density of Y is $f_Y(y) = \int_0^{(1-y)/2} 24x\,dx = 3(1-y)^2$, $0 \le y \le 1$. Find σ_X^2, σ_Y^2, and $\sigma_{X,Y}$.

(b) Find the linear correlation coefficient, $\rho_{X,Y}$, of X and Y.

(c) Find the regression function of Y on X. Taking this into consideration, comment on the appropriateness of $\rho_{X,Y}$ as a measure of the dependence between X and Y.

8. Let X have the uniform in $(-1,1)$ distribution and let $Y = X^2$. Using calculations similar to those in Example 4.5-3, show that $\rho_{X,Y} = 0$.

9. Let X be defined by the probability density function

$$f(x) = \begin{cases} -x & -1 < x \le 0 \\ x & 0 < x \le 1 \\ 0 & \text{otherwise.} \end{cases}$$

(a) Define $Y = X^2$ and find $\text{Cov}(X,Y)$.

(b) Without doing any calculations, find the regression function $E(Y|X=x)$ (*Hint.* When the value of X is given, the value of Y is known).

(c) On the basis of the regression function found above, comment on the appropriateness of the linear correlation coefficient as a measure of dependence between X and Y.

4.6 Models for Joint Distributions

4.6.1 HIERARCHICAL MODELS

The multiplication rule for joint probability mass functions given in (4.3.4) expresses the joint PMF of X and Y as the product of the conditional PMF of Y given $X = x$ and the marginal PMF of X, that is, $p(x,y) = p_{Y|X=x}(y)p_X(x)$. Similarly, the multiplication rule for joint PDFs given in (4.3.8) states that $f(x,y) = f_{Y|X=x}(y)f_X(x)$.

The *principle of hierarchical modeling* uses the multiplication rules in order to specify the joint distribution of X and Y by first specifying the conditional distribution of Y given $X = x$, and then specifying the marginal distribution of X. Thus, a hierarchical model consists of

$$Y|X = x \sim F_{Y|X=x}(y), \quad X \sim F_X(x), \tag{4.6.1}$$

where the conditional distribution of Y given $X = x$, $F_{Y|X=x}(y)$, and the marginal distribution of X, $F_X(x)$, can depend on additional parameters. (The description of the hierarchical model in (4.6.1) uses CDFs in order to include both discrete and continuous random variables.) Examples of hierarchically specified joint distributions have already been seen in Examples 4.3-4 and 4.3-7, and in Exercises 5, 7, 9, and 10 in Section 4.3. An additional example follows.

Example 4.6-1

Let X be the number of eggs an insect lays and Y the number of eggs that survive. Suppose each egg survives with probability p, independently of other eggs. Use the principle of hierarchical modeling to describe a reasonable model for the joint distribution of X and Y.

Solution

The principle of hierarchical modeling can be applied in this context as follows. First we can model the number of eggs X an insect lays as a Poisson random variable. Second, since each egg survives with probability p, independently of other eggs, if we are given the number of eggs $X = x$ the insect lays it is reasonable to model the number of eggs that survive as a binomial random variable with x trials and probability of success p. Thus, we arrive at the hierarchical model

$$Y|X = x \sim \text{Bin}(x,p), \quad X \sim \text{Poisson}(\lambda),$$

which leads to a joint PMF of (X, Y), which, for $y \leq x$, is:

$$p(x,y) = p_{Y|X=x}(y)p_X(x) = \binom{x}{y}p^y(1-p)^{x-y}\frac{e^{-\lambda}\lambda^x}{x!}.$$

The hierarchical approach to modeling offers a way to specify the joint distribution of a discrete and a continuous variable; see Exercise 2. Finally, the class of hierarchical models includes the bivariate normal distribution which, because of its importance, is revisited in Section 4.6.3.

Example 4.6-2

The bivariate normal distribution. X and Y are said to have a bivariate normal distribution if their joint distribution is specified according to the hierarchical model

$$Y|X = x \sim N\left(\beta_0 + \beta_1(x - \mu_X), \sigma_\varepsilon^2\right) \quad \text{and} \quad X \sim N(\mu_X, \sigma_X^2). \tag{4.6.2}$$

Give an expression of the joint PDF of X and Y.

Solution

The hierarchical model (4.6.2) implies that the conditional distribution of Y given that $X = x$ is normal with mean $\beta_0 + \beta_1(x - \mu_X)$ and variance σ_ε^2. Plugging this mean and variance into the form of the normal PDF we obtain

$$f_{Y|X=x}(y) = \frac{1}{\sqrt{2\pi\sigma_\varepsilon^2}} \exp\left\{-\frac{(y - \beta_0 - \beta_1(x - \mu_X))^2}{2\sigma_\varepsilon^2}\right\}.$$

In addition, the hierarchical model (4.6.2) specifies that the marginal distribution of X is normal with mean μ_X and variance σ_X^2. Thus,

$$f_X(x) = \frac{1}{\sqrt{2\pi\sigma_X^2}} \exp\left\{-\frac{(x - \mu_X)^2}{2\sigma_X^2}\right\}.$$

It follows that the joint PDF of (X, Y), which is given by the product $f_{Y|X=x}(y)f_X(x)$, takes the form

$$f_{X,Y}(x,y) = \frac{1}{2\pi\sigma_\varepsilon\sigma_X} \exp\left\{-\frac{(y - \beta_0 - \beta_1(x - \mu_X))^2}{2\sigma_\varepsilon^2} - \frac{(x - \mu_X)^2}{2\sigma_X^2}\right\}. \tag{4.6.3}$$

4.6.2 REGRESSION MODELS

Regression models are used whenever the primary objective of the study is to understand the nature of the regression function of a variable Y on another variable X. A study of the speed, X, of an automobile and the stopping distance, Y; or a study of the diameter at breast height, X, and age of a tree, Y; or a study of the stress applied, X, and time to failure, Y, are examples of such studies. In regression studies Y is called the **response variable**, and X is interchangeably referred to as the **covariate**, or the **independent variable**, or the **predictor**, or the **explanatory variable**. Because interest lies in the conditional mean of Y given $X = x$, regression models specify the conditional distribution of Y given $X = x$ while the marginal distribution of X, which is of little interest in such studies, is left unspecified.

Regression models are similar to, but more general than, hierarchical models since they specify only the conditional distribution of Y given $X = x$ and leave the marginal distribution of X unspecified. In fact, regression models allow the covariate X to be nonrandom because in some studies the investigator selects the values of the covariate in a deterministic fashion. An even more general type of a regression model is one that specifies only the form of the regression function, without specifying the conditional distribution of Y given $X = x$.

In this section we will introduce the simple linear regression model and the normal simple linear regression model, setting the stage for revisiting the bivariate normal distribution in Section 4.6.3.

The Simple Linear Regression Model The **simple linear regression model** specifies that the regression function of Y on X is linear, that is,

$$\mu_{Y|X}(x) = \alpha_1 + \beta_1 x, \qquad \textbf{(4.6.4)}$$

and the conditional variance of Y given $X = x$, denoted by σ_ε^2, is the same for all values x. The latter is known as the **homoscedasticity** assumption. In this model, α_1, β_1, and σ_ε^2 are unknown parameters. The regression function (4.6.4) is often written as

$$\mu_{Y|X}(x) = \beta_0 + \beta_1(x - \mu_X), \qquad \textbf{(4.6.5)}$$

where μ_X is the marginal mean value of X, and β_0 is related to α_1 through $\beta_0 = \alpha_1 + \beta_1\mu_X$. The straight line defined by the equation (4.6.4) (or (4.6.5)) is called the **regression line**. Figure 4-3 illustrates the meaning of the slope of the regression line. Basically, the slope expresses the change in the average or mean value of Y when the value of X changes by one unit. Thus, if $\beta_1 > 0$ then X and Y are positively correlated, and if $\beta_1 < 0$ then X and Y are negatively correlated. If $\beta_1 = 0$ then X and Y are uncorrelated, in which case X is not relevant for predicting Y. The above discussion hints of a close connection between the slope in a simple linear regression model and the covariance/correlation of X and Y. This connection is made precise in Proposition 4.6-3.

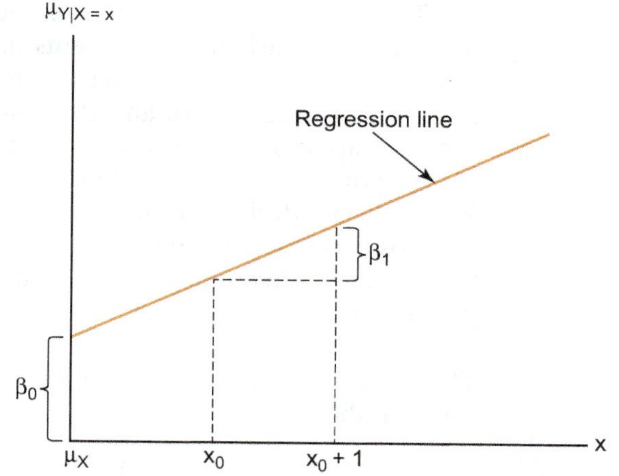

Figure 4-3 Illustration of regression parameters.

The reason the seemingly more complicated expression (4.6.5) is sometimes preferred over (4.6.4) is that the parameter β_0 equals the marginal mean of Y.

Proposition 4.6-1 The marginal expected value of Y is given by

$$E(Y) = \alpha_1 + \beta_1 \mu_X \quad \text{and} \quad E(Y) = \beta_0$$

when the simple linear regression model is parametrized as in (4.6.4) and (4.6.5), respectively.

Proof Consider the parametrization in (4.6.5). Using the Law of Total Expectation (4.3.13), we obtain

$$\begin{aligned} E(Y) &= E[E(Y|X)] \\ &= E[\beta_0 + \beta_1(X - \mu_X)] \\ &= \beta_0 + \beta_1 E(X - \mu_X) = \beta_0. \end{aligned}$$

The expression $E(Y) = \alpha_1 + \beta_1 \mu_X$ is obtained similarly using the parametrization in (4.6.4). ■

The simple linear regression model is commonly (and equivalently) given in the so-called **mean plus error** form. If X has mean $\mu_X = E(X)$, its mean plus error form is

$$X = \mu_X + \varepsilon,$$

where $\varepsilon = X - \mu_X$ is called the **(intrinsic) error variable**. In statistics, the term *error variable* is generally used to denote a random variable with zero mean. The mean plus error expression of the response variable Y in a general regression setting is of the form

$$Y = E(Y|X) + \varepsilon,$$

with the error variable given by $\varepsilon = Y - E(Y|X)$. For the simple linear regression model, where $E(Y|X)$ is given by either (4.6.4) or (4.6.5), the mean plus error form is

Mean Plus Error Form of the Simple Linear Regression Model

$$Y = \alpha_1 + \beta_1 X + \varepsilon, \quad \text{or} \quad Y = \beta_0 + \beta_1(X - \mu_X) + \varepsilon \qquad \textbf{(4.6.6)}$$

The mean plus error representation of the response variable suggests that the intrinsic error variable ε represents the *uncertainty* regarding the value of Y given the value of X. (See the statement regarding the conditional variance of Y given $X = x$ following (4.6.4), and Proposition 4.6-2.) In addition, the mean plus error representation of Y is useful for deriving properties of the simple linear regression model. For example, the result of Proposition 4.6-1 for the marginal mean value of Y can also be derived from (4.6.6) along with the result of the mean value of sums (Proposition 4.4-2); see Exercise 6. The mean plus error representation of Y will also be used in the derivation of Proposition 4.6-3. But first we state without proof the properties of the intrinsic error variable.

Proposition 4.6-2 The intrinsic error variable, ε, has zero mean and is uncorrelated from the explanatory variable X:

$$E(\varepsilon) = 0 \quad \text{and} \quad \text{Cov}(\varepsilon, X) = 0.$$

Moreover, the variance of ε, σ_ε^2, is the conditional variance of Y given the value of X. ■

Proposition 4.6-3 If the regression function of Y on X is linear (so (4.6.4) or, equivalently, (4.6.5) holds), then we have the following:

1. The marginal variance of Y is

$$\sigma_Y^2 = \sigma_\varepsilon^2 + \beta_1^2 \sigma_X^2. \tag{4.6.7}$$

2. The slope β_1 is related to the covariance, $\sigma_{X,Y}$, and the correlation, $\rho_{X,Y}$, by

$$\beta_1 = \frac{\sigma_{X,Y}}{\sigma_X^2} = \rho_{X,Y}\frac{\sigma_Y}{\sigma_X}. \tag{4.6.8}$$

Proof Using the mean plus error representation (4.6.6), the fact that adding (or subtracting) a constant, which in this case is α_1, does not change the variance, and the formula for the variance of a sum, we have

$$\begin{aligned}\text{Var}(Y) &= \text{Var}(\alpha_1 + \beta_1 X + \varepsilon) = \text{Var}(\beta_1 X + \varepsilon)\\ &= \text{Var}(\beta_1 X) + \text{Var}(\varepsilon) + 2\text{Cov}(\beta_1 X, \varepsilon)\\ &= \beta_1^2 \text{Var}(X) + \sigma_\varepsilon^2,\end{aligned}$$

since $\text{Cov}(\beta_1 X, \varepsilon) = \beta_1 \text{Cov}(X, \varepsilon) = 0$ by the fact that ε and X are uncorrelated. For the second part, it suffices to show the first equality, that is, that $\text{Cov}(X, Y) = \beta_1 \sigma_X^2$, since the second is equivalent. Using again the mean plus error representation of Y and the linearity property of covariance (part (4) of Proposition 4.4-5),

$$\begin{aligned}\text{Cov}(X, Y) &= \text{Cov}(X, \alpha_1 + \beta_1 X + \varepsilon)\\ &= \text{Cov}(X, \beta_1 X) + \text{Cov}(X, \varepsilon)\\ &= \beta_1 \text{Cov}(X, X) = \beta_1 \text{Var}(X),\end{aligned}$$

since $\text{Cov}(X, \varepsilon) = 0$ and $\text{Cov}(X, X) = \text{Var}(X)$. ■

REMARK 4.6-1 **Sample version of the regression line.** Proposition 4.6-3 suggests an estimator of β_1. Indeed, if $(X_1, Y_1), \ldots, (X_n, Y_n)$ is a sample from the bivariate distribution of (X, Y), $\sigma_{X,Y}$ can be estimated by $S_{X,Y}$ and σ_X^2 can be estimated by S_X^2. Hence, if (X, Y) satisfy the simple linear regression model, then, according to the first equation in (4.6.8), β_1 can be estimated by

$$\widehat{\beta}_1 = \frac{S_{X,Y}}{S_X^2}. \tag{4.6.9}$$

Moreover, from Proposition 4.6-1 we have $\alpha_1 = E(Y) - \beta_1 \mu_X$, which suggests that α_1 can be estimated by

$$\widehat{\alpha}_1 = \overline{Y} - \widehat{\beta}_1 \overline{X}. \tag{4.6.10}$$

These empirically derived estimators of the slope and intercept will be rederived in Chapter 6 using the principle of *least squares*. ◁

The Normal Simple Linear Regression Model The **normal regression model** specifies that the conditional distribution of Y given $X = x$ is normal,

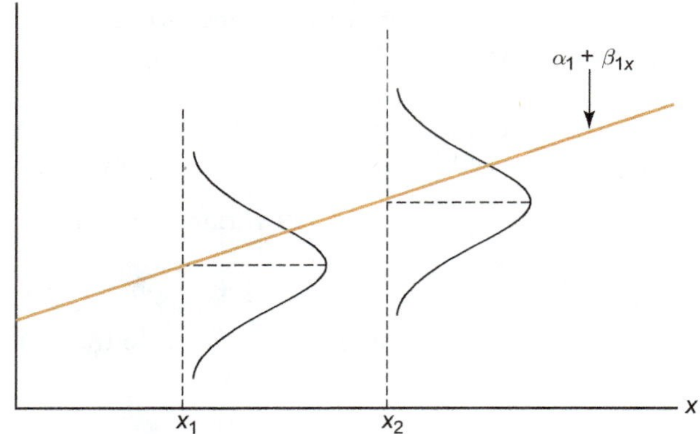

Figure 4-4 Illustration of intrinsic scatter in regression.

$$Y|X = x \sim N\left(\mu_{Y|X}(x), \sigma_\varepsilon^2\right), \tag{4.6.11}$$

where $\mu_{Y|X}(x)$ is a given function of x, typically depending on unknown parameters.

The **normal simple linear regression model** specifies, in addition, that the regression function $\mu_{Y|X}(x)$ in (4.6.11) is linear, that is, that (4.6.5) or, equivalently, (4.6.4) holds. The normal simple linear regression model is also written as

$$Y = \alpha_1 + \beta_1 x + \varepsilon, \quad \text{with} \quad \varepsilon \sim N(0, \sigma_\varepsilon^2), \tag{4.6.12}$$

where the first part of the model can also be written as $Y = \beta_0 + \beta_1(X - \mu_X) + \varepsilon$. The intrinsic error variable ε expresses the conditional variability of Y around its conditional mean given $X = x$, as Figure 4-4 illustrates.

Quadratic and more complicated normal regression models are also commonly used. The advantages of such models are (a) it is typically easy to *fit* such a model to data (i.e., estimate the model parameters from the data), and (b) such models offer easy interpretation of the effect of X on the expected value of Y.

Example 4.6-3

Suppose that $Y|X = x \sim N(5 - 2x, 16)$, that is, given $X = x$, Y has the normal distribution with mean $\mu_{Y|X}(x) = 5 - 2x$ and variance $\sigma_\varepsilon^2 = 16$, and let $\sigma_X = 3$.

(a) Find σ_Y^2 and $\rho_{X,Y}$.

(b) If Y_1 is an observation to be taken when X has been observed to take the value 1, find the 95th percentile of Y_1.

Solution

(a) Using (4.6.7) we have $\sigma_Y^2 = 16 + (-2)^2 3^2 = 52$. Next, using Proposition 4.6-3, $\rho_{X,Y} = \beta_1(\sigma_X/\sigma_Y) = -2(3/\sqrt{52}) = -0.832$.

(b) Because X has been observed to take the value 1, $Y_1 \sim N(3, 4^2)$, where the mean is computed from the given formula $\mu_{Y|X}(x) = 5 - 2x$ with $x = 1$. Thus, the 95th percentile of Y_1 is $3 + 4z_{0.05} = 3 + 4 \times 1.645 = 9.58$. ■

4.6.3 THE BIVARIATE NORMAL DISTRIBUTION

The bivariate normal distribution was already introduced in Example 4.6-2, where the joint PDF of (X, Y) was derived as a product of the conditional PDF of Y given $X = x$ times the PDF of the marginal distribution of X, which is specified to be normal with mean μ_X and σ_X^2. It is worth pointing out that in the hierarchical modeling of Example 4.6-2 the conditional distribution of Y given $X = x$ is specified as

$$Y|X = x \sim N\left(\beta_0 + \beta_1(x - \mu_X), \sigma_\varepsilon^2\right),$$

which is precisely the normal simple linear regression model. Because of its connection to the normal simple linear regression model, as well as some additional properties, the bivariate normal distribution is considered to be the most important bivariate distribution.

A more common and useful form of the joint PDF of (X, Y) is

$$f(x,y) = \frac{1}{2\pi\sigma_X\sigma_Y\sqrt{1-\rho^2}} \exp\left\{\frac{-1}{1-\rho^2}\left[\frac{\widetilde{x}^2}{2\sigma_X^2} - \frac{\rho\widetilde{x}\widetilde{y}}{\sigma_X\sigma_Y} + \frac{\widetilde{y}^2}{2\sigma_Y^2}\right]\right\}, \qquad \textbf{(4.6.13)}$$

where $\widetilde{x} = x - \mu_X$, $\widetilde{y} = y - \mu_Y$, and ρ is the correlation coefficient between X and Y. This form of the PDF can be derived from the expression given in Example 4.6-2 and some careful algebra using Propositions 4.6-1 and 4.6-3; see Exercise 10. Figure 4-5 shows the joint PDFs of two marginally $N(0, 1)$ random variables with $\rho = 0$ (left panel) and $\rho = 0.5$ (right panel).

The expression of the PDF given in (4.6.13) makes it apparent that a bivariate normal distribution is completely specified by the mean values and variances of X and Y and the covariance of X and Y, that is, by μ_X, μ_Y, σ_X^2, σ_Y^2, and $\sigma_{X,Y}$. The two variances and the covariance are typically arranged in a symmetric matrix, called the **covariance matrix**:

$$\mathbf{\Sigma} = \begin{pmatrix} \sigma_X^2 & \sigma_{X,Y} \\ \sigma_{X,Y} & \sigma_Y^2 \end{pmatrix} \qquad \textbf{(4.6.14)}$$

An alternative form of the bivariate normal PDF, one that is expressed in terms of matrix operations, is given in Exercise 10.

The bivariate normal PDF and CDF are available in the R package *mnormt*, which needs to be installed with the R command *install.packages("mnormt")*. To use it, the package must be evoked with the R command *library(mnormt)* in each new R session. Once the package has been evoked, the PDF and CDF of the bivariate normal distribution with parameters μ_X, μ_Y, σ_X^2, σ_Y^2, and $\sigma_{X,Y}$ evaluated at (x, y), are obtained with the following R commands:

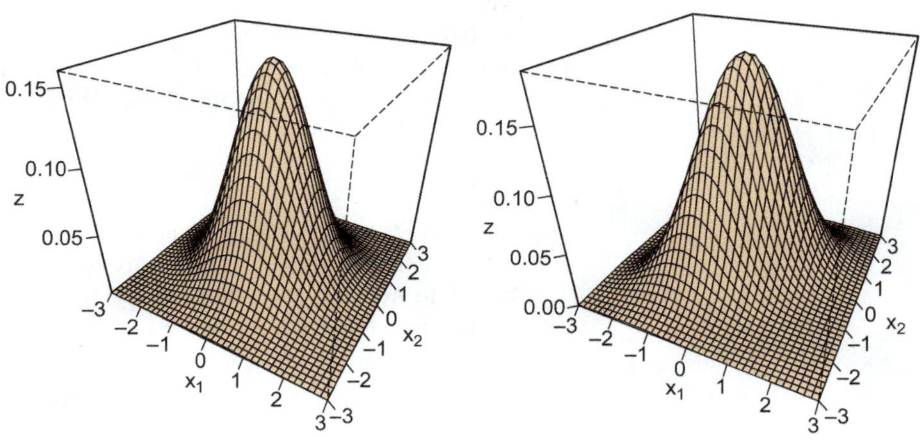

Figure 4-5 Joint PMFs of marginally $N(0, 1)$ random variables: $\rho = 0$ (left panel) and $\rho = 0.5$ (right panel).

R Commands for the Bivariate Normal PMF and CDF

```
dmnorm(c(x, y), c(μX, μY), matrix(c(σX², σX,Y, σX,Y, σY²),2)) # for
  the PDF

pmnorm(c(x, y), c(μX, μY), matrix(c(σX², σX,Y, σX,Y, σY²),2)) # for
  the CDF
```

Proposition 4.6-4 Let (X, Y) have a bivariate normal distribution with parameters $\mu_X, \mu_Y, \sigma_X^2, \sigma_Y^2$ and $\sigma_{X,Y}$. Then we have the following:

1. The marginal distribution of Y is also normal.
2. If X and Y are uncorrelated then they are independent.
3. If X and Y are independent normal random variables, their joint distribution is bivariate normal with parameters $\mu_X, \mu_Y, \sigma_X^2, \sigma_Y^2$, and $\sigma_{X,Y} = 0$.
4. Any linear combination of X and Y has a normal distribution. In particular

$$aX + bY \sim N(a\mu_X + b\mu_Y, a^2\sigma_X^2 + b^2\sigma_Y^2 + 2ab\text{Cov}(X, Y)).$$

Part (1) of Proposition 4.6-4 follows by the fact that the joint PDF $f(x, y)$ of X and Y is symmetric in x and y (this is most easily seen from the form of the PDF given in (4.6.13)) together with the fact that, according to the hierarchical definition of the bivariate normal distribution, the marginal distribution of X is normal. Hence, the marginal PDF of Y, $f_Y(y) = \int_{-\infty}^{\infty} f(x, y)\, dx$, is the PDF of a normal distribution because this holds for the marginal PDF $f_X(x) = \int_{-\infty}^{\infty} f(x, y)\, dy$ of X. Part (2) of Proposition 4.6-4 follows by noting that if $\rho = 0$ the joint PDF given in (4.6.13) becomes a product of a function of x times a function of y. Part (3) follows upon writing the product of the two normal PDFs, which is the joint PDF of the independent X and Y, and checking that it has the form given in (4.6.13) with $\rho = 0$. The proof of part (4) of the proposition will not be given here as it requires techniques not covered in this book.

Example 4.6-4 Suppose that $Y|X = x \sim N(5 - 2x, 16)$, that is, given $X = x$, Y has the normal distribution with mean $\mu_{Y|X}(x) = 5 - 2x$ and variance $\sigma_\varepsilon^2 = 16$, and let $\sigma_X = 3$.

(a) Let Y_1 and Y_2 be observations to be taken, independently from each other, when X has been observed to take the value 1 and 2, respectively. Find the probability that $Y_1 > Y_2$.

(b) Assume in addition that X has the normal distribution with mean 2 (and variance 9, as mentioned above). Use R commands to find the probability $P(X \leq 0, Y \leq 2)$.

Solution

(a) In Example 4.6-3 we saw that $Y_1 \sim N(3, 4^2)$. Similarly, we find that $Y_2 \sim N(1, 4^2)$. Because Y_1 and Y_2 are independent, their joint distribution is bivariate normal, according to part (3) of Proposition 4.6-4. By part (4) of Proposition 4.6-4, $Y_1 - Y_2 \sim N(3 - 1, 4^2 + 4^2) = N(2, 32)$. Thus,

$$P(Y_1 > Y_2) = P(Y_1 - Y_2 > 0) = 1 - \Phi\left(\frac{-2}{\sqrt{32}}\right) = 0.638.$$

(b) X and Y have a bivariate normal distribution with parameters $\mu_X = 2$, $\mu_Y = 5 - 2\mu_X = 1$ (by Proposition 4.6-1), $\sigma_X^2 = 9$, $\sigma_Y^2 = 16 + 4 \times 9 = 52$ (see (4.6.7)), and $\sigma_{X,Y} = -2 \times 9 = -18$ (by Proposition 4.6-3). Thus, assuming that the package *mnormt* has been installed (and evoked in the current R session with the command *library(mnormt)*), the R command *pmnorm(c(0, 2), c(2, 1), matrix(c(9, -18, -18, 52), 2))* gives 0.021 (rounded to 3 decimal places) for the value of $P(X \leq 0, Y \leq 2)$. ■

4.6.4 THE MULTINOMIAL DISTRIBUTION

The multinomial distribution arises in cases where a basic experiment that has r possible outcomes is repeated independently n times. For example, the basic experiment can be life testing of an electronic component, with $r = 3$ possible outcomes: 1 if the life time is short (less than 50 time units), 2 if the life time is medium (between 50 and 90 time units), or 3 if the life time is long (exceeds 90 time units). When this basic experiment is repeated n times, one typically records

$$N_1, \ldots, N_r, \tag{4.6.15}$$

where N_j = the number of times outcome j occurred. If the r possible outcomes of the basic experiment have probabilities $p_1, \ldots, p_r$, the joint distribution of the random variables $N_1, \ldots, N_r$ is said to be **multinomial** with n trials and probabilities $p_1, \ldots, p_r$.

Note that, by their definition, $N_1, \ldots, N_r$ and $p_1, \ldots, p_r$ satisfy

$$N_1 + \cdots + N_r = n \quad \text{and} \quad p_1 + \cdots + p_r = 1. \tag{4.6.16}$$

For this reason, N_r is often omitted as superfluous (since $N_r = n - N_1 - \cdots - N_{r-1}$), and the multinomial distribution is defined to be the distribution of $(N_1, \ldots, N_{r-1})$. With this convention, if $r = 2$ (i.e., there are only two possible outcomes, which can be labeled "success" and "failure") the multinomial distribution reduces to the binomial distribution.

If $N_1, \ldots, N_r$ have the multinomial$(n, p_1, \ldots, p_r)$ distribution, then their joint PMF is

$$P(N_1 = x_1, \ldots, N_r = x_r) = \frac{n!}{x_1! \cdots x_r!} p_1^{x_1} \cdots p_r^{x_r} \tag{4.6.17}$$

if $x_1 + \cdots + x_r = n$ and zero otherwise. The multinomial PMF can be obtained for any set of r nonnegative integers $x_1, \ldots, x_r$, and $p_1, \ldots, p_r$ with the following R command:

R Command for the Multinomial PMF

```
dmultinom(c(x1, ..., xr), prob=c(p1, ..., pr)) # gives the PMF
  P(N1 = x1, ..., Nr = xr)
```

Example 4.6-5

The probabilities that a certain electronic component will last less than 50 hours in continuous use, between 50 and 90 hours, or more than 90 hours, are $p_1 = 0.2$, $p_2 = 0.5$, and $p_3 = 0.3$, respectively. The time to failure of eight such electronic components will be recorded. Find the probability that one of the eight will last less than 50 hours, five will last between 50 and 90 hours, and two will last more than 90 hours.

Solution

Set N_1 for the number of these components that last less than 50 hours, N_2 for the number that last between 50 and 90 hours, and N_3 for the number that last more than 90 hours. Then (N_1, N_2, N_3) have the multinomial$(n, 0.2, 0.5, 0.3)$ distribution and, according to (4.6.17),

$$P(N_1 = 1, N_2 = 5, N_3 = 2) = \frac{8!}{1!5!2!} 0.2^1 0.5^5 0.3^2 = 0.0945.$$

The R command *dmultinom(c(1, 5, 2), prob=c(0.2, 0.5, 0.3))* gives the same value. ■

If $N_1, \ldots, N_r$ have the multinomial$(n, p_1, \ldots, p_r)$ distribution then the marginal distribution of each N_j is binomial(n, p_j). The easiest way to see this is to call the outcome of the basic experiment a "success" if outcome j occurs, and a "failure" if outcome j does not occur. Then N_j counts the number of successes in n independent trials when the probability of success in each trial is p_j. Moreover, the covariance between an N_i and an N_j can be shown to be $-np_ip_j$. These results are summarized in the following proposition.

Proposition 4.6-5 If $N_1, \ldots, N_r$ have the multinomial distribution with parameters n, r, and $p_1, \ldots, p_r$, the marginal distribution of each N_i is binomial with probability of success equal to p_i, that is, $N_i \sim \text{Bin}(n, p_i)$. Thus,

$$E(N_i) = np_i \quad \text{and} \quad Var(N_i) = np_i(1 - p_i).$$

Moreover, for $i \neq j$ the covariance of N_i and N_j is

$$Cov(N_i, N_j) = -np_ip_j.$$ ■

Example 4.6-6 In the context of Example 4.6-5, set N_1 for the number of these components that last less than 50 hours, N_2 for the number that last between 50 and 90 hours, and N_3 for the number that last more than 90 hours.

(a) Find the probability that exactly one of the eight electronic components will last less than 50 hours.

(b) Find the covariance of N_2 and N_3 and explain, at an intuitive level, why this covariance is negative.

(c) Find $\text{Var}(N_2 + N_3)$ and $\text{Cov}(N_1, N_2 + N_3)$.

Solution

(a) According to Proposition 4.6-5, $N_1 \sim \text{Bin}(8, 0.2)$. Hence,

$$P(N_1 = 1) = \binom{8}{1} 0.2^1 \times 0.8^7 = 0.3355.$$

(b) According to Proposition 4.6-5, $\text{Cov}(N_2, N_3) = -8 \times 0.5 \times 0.3 = -1.2$. At an intuitive level, the negative covariance can be explained as follows. Since $N_1 + N_2 + N_3 = 8$, it follows that if N_2 takes a small value then the probability that N_1 and N_3 take a larger value increases; similarly, if N_2 takes a large value then it is more likely that N_1 and N_3 will take a small value. This means that the dependence between any two of N_1, N_2, and N_3 is negative, and hence the covariance is negative.

(c) It is instructive to compute the variance of $N_2 + N_3$ in two ways. First,

$$\begin{aligned}\text{Var}(N_2 + N_3) &= \text{Var}(N_2) + \text{Var}(N_3) + 2\text{Cov}(N_2, N_3)\\ &= 8 \times 0.5 \times 0.5 + 8 \times 0.3 \times 0.7 - 2 \times 1.2 = 1.28,\end{aligned}$$

where the second equality in the above relation used the formula for the variance of the binomial distribution and the covariance found in part (b). An alternative way of finding the variance of $N_2 + N_3$ is to use the fact that $N_2 + N_3$ is the number of components that last more than 50 hours. Hence, $N_2 + N_3 \sim \text{Bin}(8, 0.8)$ and, using the formula for the variance of a binomial random variable, $\text{Var}(N_1 + N_2) = 8 \times 0.8 \times 0.2 = 1.28$. Finally, using the properties of covariance,

$$\begin{aligned}\text{Cov}(N_1, N_2 + N_3) &= \text{Cov}(N_1, N_2) + \text{Cov}(N_1, N_3)\\ &= -8 \times 0.2 \times 0.5 - 8 \times 0.2 \times 0.3 = -1.28.\end{aligned}$$

Exercises

1. In an accelerated life testing experiment, different batches of n equipment are operated under different stress conditions. Because the stress level is randomly set for each batch, the probability, P, with which an equipment will last more than T time units is a random variable. In this problem it is assumed that P is discrete, taking the values 0.6, 0.8, and 0.9 with corresponding probabilities 0.2, 0.5, and 0.3. Let Y denote the number of equipment from a randomly selected batch that last more than T time units.

(a) Use the principle of hierarchical modeling to specify the joint distribution of (P, Y). (*Hint.* Given that $P = p$, $Y \sim \text{Bin}(n, p)$.)

(b) Find the marginal PMF of Y when $n = 3$.

2. Consider the same setting as in Exercise 1, except now P is assumed here to have the uniform(0, 1) distribution.

(a) Use the principle of hierarchical modeling to specify the joint density of $f_{P,Y}(p, y)$ of (P, Y) as the product of the conditional PMF of Y given $P = p$ times the marginal PDF of P.

(b) Find the marginal PMF of Y. (*Hint.* The marginal PDF of Y is still given by integrating $f_{P,Y}(p, y)$ over p. You may use $\int_0^1 \binom{n}{k} p^k (1-p)^{n-k}\, dp = \frac{1}{n+1}$ for $k = 0, \ldots, n$.)

3. In the context of the normal simple linear regression model

$$Y|X = x \sim N(9.3 + 1.5x, 16),$$

let Y_1, Y_2 be independent observations corresponding to $X = 20$ and $X = 25$, respectively.

(a) Find the 95th percentile of Y_1.

(b) Find the probability that $Y_2 > Y_1$. (*Hint.* See Example 4.6-4.)

4. Consider the information given in Exercise 3. Suppose further that the marginal mean and variance of X are $E(X) = 24$ and $\sigma_X^2 = 9$, and the marginal variance of Y is $\sigma_Y^2 = 36.25$.

(a) Find the marginal mean of Y. (*Hint.* Use Proposition 4.6-1, or the Law of Total Expectation given in 4.3.15.)

(b) Find the covariance and the linear correlation coefficient of X and Y. (*Hint.* Use Proposition 4.6-3.)

5. Consider the information given in Exercise 3. Suppose further that the marginal distribution of X is normal with $\mu_X = 24$ and $\sigma_X^2 = 9$.

(a) Give the joint PDF of X, Y.

(b) Use R commands to find $P(X \le 25, Y \le 45)$.

6. Use the second mean plus error expression of the response variable given in (4.6.6) and Proposition 4.4-2 to derive the formula $E(Y) = \beta_0$. (*Hint.* Recall that any (intrinsic) error variable has mean value zero.)

7. The exponential regression model. The exponential regression model is common in reliability studies investigating how the expected life time of a product changes with some operational stress variable X. This model assumes that the life time, Y, has an exponential distribution whose parameter λ depends on the value x of X. We write $\lambda(x)$ to indicate the dependence of the parameter λ on the value of the stress variable X. An example of such a regression model is

$$\log \lambda(x) = \alpha + \beta x.$$

Suppose that in a reliability study, the stress variable X is uniformly distributed in the interval $(2, 6)$, and the above exponential regression model holds with $\alpha = 4.2$ and $\beta = 3.1$.

(a) Find the expected life time of a randomly selected product. (*Hint.* Given $X = x$, the expected life time is $1/\lambda(x) = 1/\exp(\alpha + \beta x)$. Use the Law of Total Expectation. Optionally, R may be used for the integration.)

(b) Give the joint PDF of (X, Y). (*Hint.* Use the principle of hierarchical modeling.)

8. Suppose that 60% of the supply of raw material kits used in a chemical reaction can be classified as recent, 30% as moderately aged, 8% as aged, and 2% unusable. Sixteen kits are randomly chosen to be used for 16 chemical reactions. Let N_1, N_2, N_3, N_4 denote the number of chemical reactions performed with recent, moderately aged, aged, and unusable materials.

(a) Find the probability that exactly one of the 16 planned chemical reactions will not be performed due to unusable raw materials.

(b) Find the probability that 10 chemical reactions will be performed with recent materials, 4 with moderately aged materials, and 2 with aged materials.

(c) Use an R command to recalculate the probabilities in part (b).

(d) Find $\text{Cov}(N_1 + N_2, N_3)$ and explain, at an intuitive level, why it is reasonable for the covariance to be negative.

(e) Find the variance of $N_1 + N_2 + N_3$. (*Hint.* Think of $N_1 + N_2 + N_3$ as binomial.)

9. An extensive study undertaken by the National Highway Traffic Safety Administration reported that 17% of children between the ages of five and six use no seat belt, 29% use a seat belt, and 54% use a child seat. In a sample of 15 children between five and six let N_1, N_2, N_3 be the number of children using no seat belt, a seat belt, and a child seat, respectively.

(a) Find the probability that exactly 10 children use a child seat.

(b) Find the probability that exactly 10 children use a child seat and five use a seat belt.

(c) Find $\text{Var}(N_2 + N_3)$ and $\text{Cov}(N_1, N_2 + N_3)$.

10. This exercise connects the form of bivariate normal PDF obtained in (4.6.3) through the principle of hierarchical modeling with its more common form given in (4.6.13). It also gives an alternative form of the PDF using matrix operations. For simplicity, ρ denotes $\rho_{X,Y}$.

(a) Show that $1 - \rho^2 = \sigma_\varepsilon^2/\sigma_Y^2$. (*Hint.* From Proposition 4.6-3 we have $\rho^2 = \beta_1^2\sigma_X^2/\sigma_Y^2$. Now use (4.6.7) to show that $\sigma_\varepsilon^2 = \sigma_Y^2 - \rho^2\sigma_Y^2$ and finish the proof.)

(b) Using the result of part (a), which implies $\sigma_X\sigma_Y\sqrt{1-\rho^2} = \sigma_\varepsilon\sigma_X$, and making additional use of the relationships given in Propositions 4.6-1 and 4.6-3, show that the form of the joint PDF given in (4.6.13) is equivalent to the form given in (4.6.3).

(c) Let $\mathbf{\Sigma}$ be the variance-covariance matrix given in (4.6.14). Use matrix operations to show that an equivalent form of the joint PDF of (X, Y) is

$$\frac{1}{2\pi\sqrt{|\mathbf{\Sigma}|}}\exp\left\{-\frac{1}{2}(x-\mu_X, y-\mu_Y)\mathbf{\Sigma}^{-1}\begin{pmatrix} x-\mu_X \\ y-\mu_Y \end{pmatrix}\right\},$$

where $|\mathbf{\Sigma}|$ denotes the determinant of $\mathbf{\Sigma}$.

Chapter 5

Some Approximation Results

5.1 Introduction

Chapter 1 introduced the sample mean or average, the sample variance, the sample proportion, and sample percentiles. In each case it was stressed that these statistics approximate but are, in general, different from the true population parameters they estimate. Moreover, we have accepted as true, based on intuition, that the bigger the sample size the better the approximation; for example, the numerical verification of Corollary 4.4-4, offered in Example 4.4-7, is based on this intuition. Evidence supporting this intuition is provided by the formulas for the variances of $\overline{X}$ and $\widehat{p}$, which decrease as the sample size increases. The *Law of Large Numbers*, or *LLN* for short, stated in the second section of this chapter, is an explicit assertion that this intuition is in fact true. The LLN is stated for the sample mean, but similar results hold for all statistics we consider.

Though the LLN justifies the approximation of the population mean by the sample mean, it does not offer guidelines for determining how large the sample size should be for a desired quality of the approximation. This requires knowledge of the distribution of the sample mean. Except for a few cases, such as when sampling from a normal population (see Section 5.3.2), the exact distribution of the sample mean is very difficult to obtain. This is discussed further in Section 5.3. In Section 5.4 we present the *Central Limit Theorem*, or *CLT* for short, which provides an approximation to the distribution of sums or averages. Moreover, the CLT provides the foundation for approximating the distribution of other statistics, such as the regression coefficients, which will be used in the chapters to follow.

5.2 The LLN and the Consistency of Averages

The limiting relative frequency definition of probability suggests that $\widehat{p}$ becomes a more accurate estimator of p as the sample size increases. In other words, the

$$\text{error of estimation } |\widehat{p} - p|$$

converges to zero as the sample size increases. A more precise term is to say that $\widehat{p}$ *converges in probability* to p, which means that the probability that the error of estimation exceeds ϵ tends to zero for any $\epsilon > 0$. This is written as

$$P(|\widehat{p} - p| > \epsilon) \to 0 \quad \text{as } n \to \infty. \tag{5.2.1}$$

Whenever an estimator converges in probability to the quantity it is supposed to estimate, we say that the estimator is **consistent**. The LLN, stated below, asserts that averages possess the consistency property.

Theorem 5.2-1 **The Law of Large Numbers.** Let $X_1, \ldots, X_n$ be independent and identically distributed and let g be a function such that $-\infty < E[g(X_1)] < \infty$. Then,

$$\frac{1}{n}\sum_{i=1}^{n} g(X_i) \text{ converges in probability to } E[g(X_1)],$$

that is, for any $\epsilon > 0$,

$$P\left(\left|\frac{1}{n}\sum_{i=1}^{n} g(X_i) - E[g(X_1)]\right| > \epsilon\right) \to 0 \quad \text{as } n \to \infty. \tag{5.2.2}$$

If g is the identity function, that is, $g(x) = x$, this theorem asserts that for any $\epsilon > 0$

$$P\left(|\overline{X} - \mu| > \epsilon\right) \to 0 \quad \text{as } n \to \infty, \tag{5.2.3}$$

that is, $\overline{X} = n^{-1}\sum_{i=1}^{n} X_i$ is a consistent estimator of the population mean $\mu = E(X_1)$, provided μ is finite. Since $\widehat{p}$ is the average of independent Bernoulli random variables, whose mean value is p, we see that relation (5.2.1) is a special case of (5.2.3).

The consistency property, which the Law of Large Numbers (and its various ramifications) guarantees, is so basic and indispensable that all estimators used in this book, and indeed all estimators used in statistics, have this property. For example, the numerical verification of Corollary 4.4-4 offered in Example 4.4-7 is possible because of the consistency of the sample variance.

If we also assume that the common variance of the $g(X_i)$ is finite, then the proof of the Law of Large Numbers is a simple consequence of the following inequality, which is useful in its own right.

Lemma 5.2-1 **Chebyshev's inequality.** Let the random variable Y have mean value μ_Y and variance $\sigma_Y^2 < \infty$. Then, for any $\epsilon > 0$,

$$P(|Y - \mu_Y| > \epsilon) \le \frac{\sigma_Y^2}{\epsilon^2}.$$

In words, Chebyshev's inequality makes an explicit connection between the variance of a random variable and (an upper bound on) the likelihood that the random variable will differ "much" from its mean: The smaller the variance, the less likely it is for the variable to differ "much" from its mean, and this likelihood tends to zero if the variance tends to zero. Recall now that the mean of the sample mean is the population mean, that is, $E(\overline{X}) = \mu$, regardless of the sample size (see (4.4.3)), but its variance is

$$\text{Var}(\overline{X}) = \frac{\sigma^2}{n}$$

where σ^2 is the population variance; see (4.4.8). As long as σ^2 is finite, the variance of $\overline{X}$ tends to zero as the sample size increases. Hence, Chebyshev's inequality implies the probability of the error of estimation exceeding ϵ, that is, $P(|\overline{X} - \mu| > \epsilon)$, tends to zero for any $\epsilon > 0$. This is the gist of the proof of the consistency of the sample mean in the case of a finite population variance. The technical proof, also in the more general context of the average of the $g(X_i)$, follows.

Proof of the Law of Large Numbers (assuming also a finite variance). We will use Chebyshev's inequality with $Y = n^{-1}\sum_{i=1}^{n} g(X_i)$. Thus,

$$\mu_Y = \frac{1}{n}\sum_{i=1}^{n} E[g(X_i)] = E[g(X_1)] \quad \text{and} \quad \sigma_Y^2 = \text{Var}\left(\frac{1}{n}\sum_{i=1}^{n} g(X_i)\right) = \frac{\sigma_g^2}{n},$$

where $\sigma_g^2 = \text{Var}[g(X_i)]$. Hence, by Chebyshev's inequality we have that for any $\epsilon > 0$,

$$P\left(\left|\frac{1}{n}\sum_{i=1}^{n} g(X_i) - E[g(X_1)]\right| > \epsilon\right) \le \frac{\sigma_g^2}{n\epsilon^2} \to 0 \quad \text{as} \quad n \to \infty.$$

Though it is a fundamental result, the usefulness of the LLN has its limitations: While it asserts that as the sample size increases, sample averages approximate the population mean more accurately, it provides no guidance regarding the quality of the approximation. In addition to helping prove the LLN (in the case of a finite variance), Chebyshev's inequality provides some information about the quality of the approximation but only in the sense of probability bounds. The following example illustrates these points.

Example 5.2-1

Cylinders are produced in such a way that their height is fixed at 5 centimeters (cm), but the radius of their base is uniformly distributed in the interval (9.5 cm, 10.5 cm). The volume of each of the next 100 cylinders to be produced will be measured, and the 100 volume measurements will be averaged.

(a) What will the approximate value of this average be?

(b) What can be said about the probability that the average of the 100 volume measurements will be within 20 cm^3 from its population mean?

Solution

Let X_i, $i = 1, \ldots, 100$, and $\overline{X}$ denote the volume measurements and their average, respectively.

(a) By the LLN, $\overline{X}$ should be approximately equal to the expected volume of a randomly selected cylinder. Since the volume is given by $X = \pi R^2 h$ cm^3 with $h = 5$ cm, the expected volume of a randomly selected cylinder is

$$E(X) = 5\pi E\left(R^2\right) = 5\pi \int_{9.5}^{10.5} r^2\, dr$$

$$= 5\pi \frac{1}{3} r^3 \Big|_{9.5}^{10.5} = 5\pi \frac{1}{3}\left(10.5^3 - 9.5^3\right) = 1572.105.$$

Thus, the value of $\overline{X}$ should be "close" to 1572.105.

(b) We are interested in an assessment of the probability

$$P(1572.105 - 20 \le \overline{X} \le 1572.105 + 20) = P(1552.105 \le \overline{X} \le 1592.105).$$

Since the LLN does not provide any additional information about the quality of the approximation of μ by $\overline{X}$, we turn to Chebyshev's inequality. Note that the event $1552.105 \le \overline{X} \le 1592.105$ is the complement of the event $\left|\overline{X} - 1572.105\right| > 20$. Since Chebyshev's inequality provides an upper bound for the probability of the later event, which is

$$P\left(\left|\overline{X} - 1572.105\right| > 20\right) \le \frac{\mathrm{Var}(\overline{X})}{20^2},$$

it follows that it also provides a lower bound for the probability of the former event

$$P\left(1552.105 \le \overline{X} \le 1592.105\right) = 1 - P\left(\left|\overline{X} - 1572.105\right| > 20\right)$$
$$\ge 1 - \frac{\mathrm{Var}(\overline{X})}{20^2}. \qquad \textbf{(5.2.4)}$$

It remains to compute the variance of $\overline{X}$. Since

$$E\left(X^2\right) = 5^2\pi^2 E\left(R^4\right) = 5^2\pi^2 \int_{9.5}^{10.5} r^4\, dr = 5^2\pi^2 \frac{1}{5} r^5 \Big|_{9.5}^{10.5} = 2{,}479{,}741,$$

the variance of X is $\sigma^2 = 2{,}479{,}741 - 1572.105^2 = 8227.06$. Hence, $\mathrm{Var}(\overline{X}) = \sigma^2/n = 82.27$. Substituting into (5.2.4), we obtain

$$P\left(1552.105 \le \overline{X} \le 1592.105\right) \ge 1 - \frac{82.27}{400} = 0.79. \qquad \textbf{(5.2.5)}$$

Thus, it can be said that the probability of the average of the 100 volume measurements being within 20 cm^3 from its population mean is at least 0.79. ■

In general, Chebyshev's inequality provides a lower bound to probabilities of the form

$$P(\mu - C \le \overline{X} \le \mu + C),$$

for any constant C. These lower bounds are valid for any sample size n and for samples drawn from any population, provided the population variance is finite. Because these lower bounds apply so generally, they can be quite conservative for some distributions. For example, if the volume measurements in Example 5.2-1 are normally distributed, thus $X_1, \ldots, X_{100}$ are iid $N(1572.105, 8227.06)$, then $\overline{X} \sim N(1572.105, 82.27)$ (this is a consequence of Proposition 4.6-4; see also Corollary 5.3-1). Using this fact, the exact value of the probability in (5.2.5), which can be found with the R command *pnorm(1592.105, 1572.105, sqrt(82.27)) – pnorm(1552.105, 1572.105, sqrt(82.27))*, is 0.97. Because the lower bounds obtained from Chebyshev's inequality can underestimate the true probability (considerably for some distributions), they are not useful for answering practical questions involving the sample size required for a specified level of accuracy in the estimation of μ by $\overline{X}$. To properly address such questions, knowledge (or approximate knowledge) of the distribution of averages is required; see Example 5.3-6. The rest of this chapter deals with this issue, but first we discuss the assumptions of finite mean and finite variance that underly all developments.

The Assumptions of a Finite Mean and a Finite Variance The LLN requires the existence of a finite mean, while the simple proof of Theorem 5.2-1, based in Chebyshev's inequality, requires the stronger assumption of a finite variance. In a first course in probability and statistics many students wonder how it is possible for a random variable not to have a finite mean or to have a finite mean but infinite variance, how one becomes aware of such abnormalities when confronted with a real data set, and what the consequences are of ignoring evidence of such abnormalities. The following paragraphs give brief answers to these questions.

First, it is easy to construct examples of distributions having infinite mean, or having finite mean but infinite variance. Consider the functions

$$f_1(x) = x^{-2},\ 1 \le x < \infty, \quad \text{and} \quad f_2(x) = 2x^{-3},\ 1 \le x < \infty,$$

and both are zero for $x < 1$. It is easy to see that both are probability density functions (both are nonnegative and integrate to 1). Let X_1 have PDF f_1 and X_2 have PDF f_2. Then the mean value (and hence the variance) of X_1 is infinite, while the mean value of X_2 is 2 but its variance is infinite:

$$E(X_1) = \int_1^{\infty} x f_1(x)\,dx = \infty, \quad E(X_1^2) = \int_1^{\infty} x^2 f_1(x)\,dx = \infty$$

$$E(X_2) = \int_1^{\infty} x f_2(x)\,dx = 2, \quad E(X_2^2) = \int_1^{\infty} x^2 f_2(x)\,dx = \infty$$

The most famous abnormal distribution is the (standard) **Cauchy** distribution, whose PDF is

$$f(x) = \frac{1}{\pi}\frac{1}{1+x^2}, \quad -\infty < x < \infty. \tag{5.2.6}$$

Note that this PDF is symmetric about zero, and thus its median is zero. However, its mean does not exist in the sense that the integral $\int_{-\infty}^{\infty} x f(x)\,dx$, with $f(x)$ given in (5.2.6), is undefined. Hence, its variance cannot be defined; however, $\int_{-\infty}^{\infty} x^2 f(x)\,dx = \infty$.

If the sample comes from a distribution without a finite mean, then the LLN does not hold. In particular, if the mean is $\pm\infty$, $\overline{X}$ diverges to $\pm\infty$ as the sample size tends to ∞. If the mean of the distribution does not exist, $\overline{X}$ need not converge to any constant and it need not diverge; see Exercise 1 in Section 5.4 for a numerical demonstration of this fact using samples from the Cauchy distribution. If the mean exists and is finite then, by the Law of Large Numbers, $\overline{X}$ converges to the mean. However, assessment of the accuracy in the estimation of μ by $\overline{X}$ requires that the variance be finite too.

Distributions with infinite mean, or infinite variance, are described as *heavy tailed*, a term justified by the fact that there is much more area under the tails (i.e., the extreme ends) of the PDF than for distributions with a finite variance. As a consequence, samples obtained from heavy tailed distributions are much more likely to contain outliers. If large outliers exist in a data set, it might be a good idea to focus on estimating another quantity, such as the median, which is well defined also for heavy tailed distributions.

Exercises

1. Using Chebyshev's inequality:

(a) Show that any random variable X with mean μ and variance σ^2 satisfies

$$P(|X - \mu| > a\sigma) \leq \frac{1}{a^2},$$

that is, that the probability X differs from its mean by more than a standard deviations cannot exceed $1/a^2$.

(b) Supposing $X \sim N(\mu, \sigma^2)$, compute the exact probability that X differs from its mean by more than a standard deviations for $a = 1, 2$, and 3, and compare the exact probabilities with the upper bounds provided by Chebyshev's inequality.

2. The life span of an electrical component has the exponential distribution with parameter $\lambda = 0.013$. Let $X_1, \ldots, X_{100}$ be a simple random sample of 100 life times of such components.

(a) What will be the approximate value of their average $\overline{X}$?

(b) What can be said about the probability that $\overline{X}$ will be within 15.38 units from the population mean? (*Hint.* The mean and variance of a random variable having the exponential distribution with parameter λ are $1/\lambda$ and $1/\lambda^2$, respectively.)

3. Let $X_1, \ldots, X_{10}$ be independent Poisson random variables having mean 1.

(a) Use Chebyshev's inequality to find a lower bound on the probability that $\overline{X}$ is within 0.5 from its mean, that is, $P\left(0.5 \leq \overline{X} \leq 1.5\right)$. (*Hint.* The probability in question can be written as $1 - P(|\overline{X} - 1| > 0.5)$; see Example 5.2-1.)

(b) Use the fact that $\sum_{i=1}^{10} X_i$ is a Poisson random variable with mean 10 (see Example 5.3-1) to find the exact value of the probability given in part (a). Compare the exact probability to the lower bound obtained in part (a). (*Hint.* The R command *ppois(x, λ)* gives the value of the Poisson(λ) CDF at x.)

5.3 Convolutions

5.3.1 WHAT THEY ARE AND HOW THEY ARE USED

In probability and statistics, the *convolution* of two independent random variables refers to the distribution of their sum. Alternatively, the convolution refers to formulas for the PDF/PMF and the CDF of their sum; see (5.3.3). The next two examples find the convolution of two independent Poisson random variables and the convolution of two independent binomial random variables having the same probability of success.

Example 5.3-1

Sum of independent Poisson random variables. If $X \sim \text{Poisson}(\lambda_1)$ and $Y \sim \text{Poisson}(\lambda_2)$ are independent random variables, show that

$$X + Y \sim \text{Poisson}(\lambda_1 + \lambda_2).$$

Solution

We will find the distribution of $Z = X + Y$ by first finding its conditional distribution given $X = k$ and subsequent application of the Law of Total Probability for marginal PMFs given in relation (4.3.5). Note that given the information that $X = k$, the possible values of Z are $k, k+1, k+2, \ldots$. For $n \geq k$

$$\begin{aligned} P(Z = n|X = k) &= \frac{P(Z = n, X = k)}{P(X = k)} = \frac{P(Y = n - k, X = k)}{P(X = k)} \\ &= \frac{P(Y = n - k)P(X = k)}{P(X = k)} = P(Y = n - k) \\ &= e^{-\lambda_2} \frac{\lambda_2^{n-k}}{(n-k)!}, \end{aligned}$$

where the third equality above follows from the independence of X and Y. Next,

$$P(Z = n) = \sum_{k=0}^{n} P(Z = n|X = k)p_X(k) = \sum_{k=0}^{n} e^{-\lambda_2}\frac{\lambda_2^{n-k}}{(n-k)!}e^{-\lambda_1}\frac{\lambda_1^k}{k!}$$

$$= e^{-(\lambda_1+\lambda_2)}\sum_{k=0}^{n}\frac{\lambda_1^k\lambda_2^{n-k}}{k!(n-k)!}$$

$$= \frac{e^{-(\lambda_1+\lambda_2)}}{n!}\sum_{k=0}^{n}\frac{n!}{k!(n-k)!}\lambda_1^k\lambda_2^{n-k}$$

$$= \frac{e^{-(\lambda_1+\lambda_2)}}{n!}(\lambda_1+\lambda_2)^n,$$

which shows that $X + Y \sim \text{Poisson}(\lambda_1 + \lambda_2)$. ■

Example 5.3-2

Sum of independent binomial random variables. If $X \sim \text{Bin}(n_1,p)$ and $Y \sim \text{Bin}(n_2,p)$ are independent binomial random variables with common probability of success, show that

$$X + Y \sim \text{Bin}(n_1 + n_2, p).$$

Solution

This problem can be done with steps similar to those used in Example 5.3-1; see Exercise 1 in Section 4.3. Alternatively, recalling that a binomial random variable arises as the number of successes in a number of independent Bernoulli trials each of which has the same probability of success, it can be argued that $Z = X_1 + X_2 \sim \text{Bin}(n_1+n_2,p)$ because Z is the number of successes in n_1+n_2 independent Bernoulli trials each of which has the same probability of success. ■

By an inductive argument, Example 5.3-2 also implies that if $X_i \sim \text{Bin}(n_i,p)$, $i = 1,\ldots,k$, are independent then $X_1+\cdots+X_k \sim \text{Bin}(n,p)$, where $n = n_1+\cdots+n_k$. Similarly, Example 5.3-1 and an inductive argument yields that the sum of several independent Poisson random variables is also a Poisson random variable with mean equal to the sum of their means. Moreover, in Proposition 4.6-4 it was seen that the sum of multivariate normal random variables has a normal distribution. Unfortunately, such nice examples are exceptions to the rule. In general, the distribution of the sum of two independent random variables need not resemble the distribution of the variables being summed. For example, the sum of two independent binomial random variables is binomial only if they share a common probability of success; if the probabilities of success are different, the distribution of their sum is none of the common types of discrete distributions we considered in Chapter 3. Also, as the next example shows, the sum of two independent uniform random variables is not a uniform random variable.

Example 5.3-3

The sum of two uniforms. If X_1 and X_2 are independent random variables having the uniform in $(0, 1)$ distribution, find the distribution of $X_1 + X_2$.

Solution

We will first find the cumulative distribution function, $F_{X_1+X_2}(y) = P(X_1+X_2 \le y)$, of $X_1 + X_2$, for $0 < y < 2$. The probability density function will follow by differentiation. A general method for finding the cumulative distribution of the sum of two independent random variables is to condition on one of them, say X_1, and then use a version of the Law of Total Expectation given in (4.3.16). Using this formula with X_1 in place of X, and B being the event that the sum $X_1 + X_2$ is less than or equal to y, that is, $[X_1 + X_2 \le y]$, we obtain

$$\begin{aligned} F_{X_1+X_2}(y) &= \int_{-\infty}^{\infty} P(X_1 + X_2 \le y | X_1 = x_1) f_{X_1}(x_1)\, dx_1 \\ &= \int_{-\infty}^{\infty} P(X_2 \le y - x_1 | X_1 = x_1) f_{X_1}(x_1)\, dx_1 \\ &= \int_{-\infty}^{\infty} P(X_2 \le y - x_1) f_{X_1}(x_1)\, dx_1 \end{aligned}$$

by the independence of X_1 and X_2. Replacing $P(X_2 \le y - x_1)$ by $F_{X_2}(y - x_1)$, we obtain

$$F_{X_1+X_2}(y) = \int_{-\infty}^{\infty} F_{X_2}(y - x_1) f_{X_1}(x_1)\, dx_1 \tag{5.3.1}$$

From the fact that X_1 and X_2 are nonnegative it follows that, in the integral in (5.3.1), x_1 has to be less than y (so the upper limit of the integral can be replaced by y). Moreover, since X_1 and X_2 are uniform in $(0, 1)$, $f_{X_1}(x_1) = 1$ if $0 < x_1 < 1$ and zero otherwise, and when $y - x_1 < 1$, $F_{X_2}(y - x_1) = y - x_1$, while when $y - x_1 > 1$, $F_{X_2}(y - x_1) = 1$. Hence, if $y < 1$ the upper limit in the integral in (5.3.1) can be replaced by y, and we have

$$F_{X_1+X_2}(y) = \int_0^y (y - x_1)\, dx_1 = \frac{1}{2}y^2.$$

If $1 < y < 2$, the upper limit in the integral in (5.3.1) can be replaced by 1, and we have

$$\begin{aligned} F_{X_1+X_2}(y) &= \int_0^1 F_{X_2}(y - x_1)\, dx_1 = \int_0^{y-1} F_{X_2}(y - x_1)\, dx_1 + \int_{y-1}^1 F_{X_2}(y - x_1)\, dx_1 \\ &= \int_0^{y-1} dx_1 + \int_{y-1}^1 (y - x_1)\, dx_1 = y - 1 + y[1 - (y - 1)] - \frac{1}{2}x_1^2\Big|_{y-1}^1 \\ &= 2y - \frac{1}{2}y^2 - 1. \end{aligned}$$

Differentiating the cumulative distribution function gives the following PDF of $X_1 + X_2$:

$$f_{X_1+X_2}(y) = \begin{cases} y & \text{if } 0 \le y \le 1 \\ 2 - y & \text{if } 1 \le y \le 2. \end{cases} \tag{5.3.2}$$

■

Formula (5.3.1) gives the cumulative distribution function of the sum of two independent random variables X_1 and X_2 and is called the **convolution** of the distributions F_{X_1} and F_{X_2}. The convolution also refers to the expression giving the PDF of the sum of the independent random variables X_1 and X_2, which is obtained by differentiating (5.3.1):

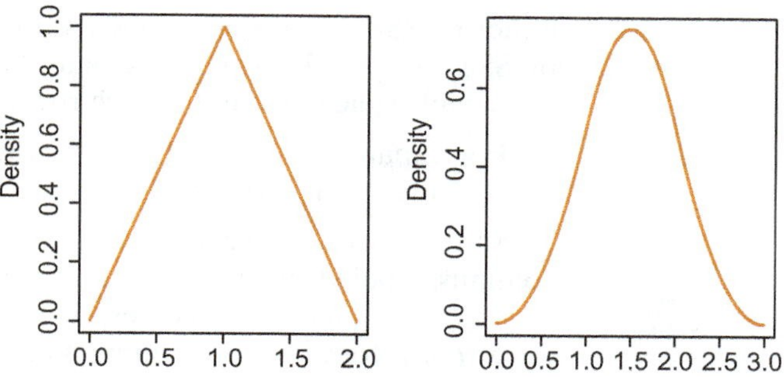

Figure 5-1 The convolution PDF of two (left panel) and of three (right panel) uniform PDFs.

Convolution of the PDFs f_1 and f_2

$$f_{X_1+X_2}(y) = \int_{-\infty}^{\infty} f_{X_2}(y - x_1) f_{X_1}(x_1)\, dx_1 \tag{5.3.3}$$

The left panel of Figure 5-1 shows the PDF (5.3.2). It is quite clear that the convolution of two uniform(0, 1) PDFs is very different from the PDF of a uniform.

Applying either of the convolution formulas shows that the distribution of the sum of two independent exponential random variables is not an exponential random variable (see Exercise 2 in Section 5.4). In general the distribution of the sum of two random variables need not resemble the distribution of either variable.

The convolution formula (5.3.3) can be used recursively to find the distribution of the sum of several independent random variables. The right panel of Figure 5-1 shows the convolution of three uniform PDFs. It is quite clear that the distribution of the sum of three independent uniform random variables is different from that of the sum of two, as well as from that of a uniform random variable.

A version of the convolution formulas applies to discrete random variables as well. In fact, one such version was used in Example 5.3-1 to find the distribution of the sum of two independent Poisson random variables. Again, convolution formulas for two discrete random variables can be applied recursively to find the distribution of the sum of several independent discrete random variables. Such formulas, however, may be difficult or impractical to use for calculating probabilities.

As an alternative to deriving formulas, computer evaluation of the convolution of two random variables is possible. The following example demonstrates the use of R for computing the convolution of two binomial random variables with different probabilities of success (a case for which we have not derived a formula for the convolution).

Example 5.3-4

The PMF and CDF of $X + Y$ with R. If X and Y are independent with $X \sim \text{Bin}(3, 0.3)$ and $Y \sim \text{Bin}(4, 0.6)$, find the convolution distribution (the PMF and CDF) of $X + Y$ using R.

Solution

First we create the sample space of (X, Y), which consists of the $4 \times 5 = 20$ pairs of (x, y) values with $x = 0, 1, 2, 3$ and $y = 0, 1, 2, 3, 4$. This is done with the R command

```
S = expand.grid(X=0:3,Y=0:4)
```

The first column of S contains the x-values of the 20 (x, y) pairs, and the second column of S contains the y-values of the pairs. The first column of S can be accessed either by S$X or by S[,1]. Similarly, the second column of S can be accessed either by S$Y or by S[,2]. Next, we create the joint probabilities $p(x, y)$ for each (x, y) in the sample space, that is, for each row of S. This is done with the R commands

```
P = expand.grid(px=dbinom(0:3, 3, .3),  py=dbinom(0:4, 4, .6));
  P$pxy  =  P$px*P$py
```

The first of the above commands creates a matrix P having 20 rows and two columns, labeled *px* and *py*, that contain the marginal probabilities $(p_X(x), p_Y(y))$ for the (x, y) pair in the corresponding row of S. The second command, *P$pxy* = *P$px*P$py*, forms a new column in the P matrix that contains the joint probabilities $p(x, y) = p_X(x)p_Y(y)$ for each (x, y) pair in S. The additional commands

```
attach(P); attach(S)
```

allow the columns of P and S to be accessed simply by their name. With the sample space in S and the joint probabilities in the column pxy of P, all joint probabilities pertaining to the variables X and Y can be calculated. For example,

```
sum(pxy[which(X + Y==4)])
```

gives 0.266328, which is the probability $P(X + Y = 4)$;

```
pz=rep(0,8); for(i in 1:8)pz[i]=sum(pxy[which(X + Y==i - 1)]); pz
```

returns 0.009 0.064 0.191 0.301 0.266 0.132 0.034 0.003, which are the values of the PMF $p_Z(z)$ of $Z = X + Y$ for $z = 0, \ldots, 7$ (the probabilities are rounded to three decimal places);

```
sum(pxy[which(X + Y<=4)])
```

gives 0.830872, which is the cumulative probability $P(X + Y \leq 4)$; and

```
Fz=rep(0,8); for(i in 1:8)Fz[i]=sum(pxy[which(X + Y<=i-1)]); Fz
```

returns 0.009 0.073 0.264 0.565 0.831 0.963 0.997 1.000, which are the values (again rounded to three decimals) of the CDF $F_Z(z) = P(X + Y \leq x)$, for $z = 0, \ldots, 7$. Finally,

```
sum(pxy[which(3<X + Y & X + Y<=5)])
```

gives 0.398, which is the probability $P(3 < X + Y \leq 5) = F_Z(5) - F_Z(3)$. ■

The main points of this section are (a) the distribution of the sum of two independent random variables need not resemble the distribution of the individual variables, and (b) as the number of random variables that are summed increases so does the difficulty in using both the convolution formulas and the R code for finding the exact distribution of the sums.

5.3.2 THE DISTRIBUTION OF $\overline{X}$ IN THE NORMAL CASE

The following proposition follows by a recursive application of part (4) of Proposition 4.6-4 for the case of independent normal random variables, but is highlighted here because of the importance of the normal distribution for statistical inference.

Proposition 5.3-1 Let $X_1, X_2, \ldots, X_n$ be independent and normally distributed random variables, $X_i \sim N(\mu_i, \sigma_i^2)$, and let $Y = a_1X_1 + \cdots + a_nX_n$ be a linear combination of the X_i. Then

$$Y \sim N(\mu_Y, \sigma_Y^2), \quad \text{where } \mu_Y = a_1\mu_1 + \cdots + a_n\mu_n, \quad \sigma_Y^2 = a_1^2\sigma_1^2 + \cdots + a_n^2\sigma_n^2.$$ ■

Example 5.3-5 Two airplanes, A and B, are traveling parallel to each other in the same direction at independent speeds of X_1 km/hr and X_2 km/hr, respectively, such that $X_1 \sim N(495, 8^2)$ and $X_2 \sim N(510, 10^2)$. At noon, plane A is 10 km ahead of plane B. Let D denote the distance by which plane A is ahead of plane B at 3:00 p.m. (Thus D is negative if plane B is ahead of plane A.)

(a) What is the distribution of D?

(b) Find the probability that at 3:00 p.m. plane A is still ahead of plane B.

Solution

The distance by which plane A is ahead at 3:00 p.m. is given by $D = 3X_1 - 3X_2 + 10$. According to Proposition 5.3-1, the answer to part (a) is

$$D \sim N(3 \times 495 - 3 \times 510 + 10,\ 9 \times 64 + 9 \times 100) = N(-35, 1476).$$

Hence, the answer to part (b) is

$$P(D > 0) = 1 - \Phi\left(\frac{35}{\sqrt{1476}}\right) = 0.181.$$

■

Corollary 5.3-1 Let $X_1, \ldots, X_n$ be iid $N(\mu, \sigma^2)$, and let $\overline{X}$ be the sample mean. Then

$$\overline{X} \sim N(\mu_{\overline{X}}, \sigma_{\overline{X}}^2), \quad \text{where } \mu_{\overline{X}} = \mu, \quad \sigma_{\overline{X}}^2 = \frac{\sigma^2}{n}.$$

The next example demonstrates the use of Corollary 5.3-1 for determining, in the case of a normal distribution with known variance, the sample size needed to ensure that the sample mean achieves a satisfactory approximation to the population mean.

Example 5.3-6 It is desired to estimate the mean of a normal population whose variance is known to be $\sigma^2 = 9$. What sample size should be used to ensure that $\overline{X}$ lies within 0.3 units of the population mean with probability 0.95?

Solution

In probabilistic notation, we want to determine the sample size n so that $P(|\overline{X} - \mu| < 0.3) = 0.95$. According to Corollary 5.3-1,

$$\frac{\overline{X} - \mu}{\sigma/\sqrt{n}} \sim N(0, 1).$$

Using this, and rewriting $P(|\overline{X} - \mu| < 0.3) = 0.95$ as

$$P\left(\left|\frac{\overline{X} - \mu}{\sigma/\sqrt{n}}\right| < \frac{0.3}{\sigma/\sqrt{n}}\right) = P\left(|Z| < \frac{0.3}{\sigma/\sqrt{n}}\right) = 0.95,$$

where $Z \sim N(0,1)$, it follows that $0.3/(\sigma/\sqrt{n}) = z_{0.025}$. This is because $z_{0.025}$ is the only number that satisfies $P(|Z| < z_{0.025}) = 0.95$. Solving for n gives

$$n = \left(\frac{1.96\sigma}{0.3}\right)^2 = 384.16.$$

Thus, using $n = 385$ will satisfy the desired precision objective. ■

REMARK 5.3-1 The solution to Example 5.3-6 is not completely satisfactory because, typically, σ is unknown. Of course, σ can be estimated by the sample standard deviation S. More details on this will be given in Chapter 7, where the determination of the required sample size for satisfactory approximation of the mean will be discussed in more detail. ◁

Exercises

1. Let $X \sim \text{Bin}(n_1,p)$, $Y \sim \text{Bin}(n_2,p)$ be independent and let $Z = X + Y$.

(a) Find the conditional PMF of Z given that $X = k$.

(b) Use the result in part (a), and the Law of Total Probability for marginal PMFs as was done in Example 5.3-1, to provide an analytical proof of Example 5.3-2, namely that $Z \sim \text{Bin}(n_1+n_2,p)$. (*Hint.* You will need the combinatorial identity $\binom{n_1+n_2}{k} = \sum_{i=0}^{n_1}\binom{n_1}{i}\binom{n_2}{k-i}$.)

2. Let X_1 and X_2 be two independent exponential random variables with mean $\mu = 1/\lambda$. (Thus, their common density is $f(x) = \lambda\exp(-\lambda x), x > 0$.). Use the convolution formula (5.3.3) to find the PDF of the sum of two independent exponential random variables.

3. Let X_1, X_2, X_3 be independent normal random variables with common mean $\mu_1 = 60$ and common variance $\sigma_1^2 = 12$, and Y_1, Y_2, Y_3 be independent normal random variables with common mean $\mu_2 = 65$ and common variance $\sigma_2^2 = 15$. Also, X_i and Y_j are independent for all i and j.

(a) Specify the distribution of $X_1 + X_2 + X_3$, and find $P(X_1 + X_2 + X_3 > 185)$.

(b) Specify the distribution of $\overline{Y} - \overline{X}$, and find $P(\overline{Y} - \overline{X} > 8)$.

4. Each of 3 friends bring one flashlight containing a fresh battery for their camping trip, and they decide to use one flashlight at a time. Let X_1, X_2, and X_3 denote the lives of the batteries in each of the 3 flashlights, respectively. Suppose that they are independent normal random variables with expected values $\mu_1 = 6$, $\mu_2 = 7$, and $\mu_3 = 8$ hours, and variances $\sigma_1^2 = 2$, $\sigma_2^2 = 3$, and $\sigma_3^2 = 4$, respectively.

(a) Find the 95th percentile of the total duration of the flashlights.

(b) Calculate the probability that the flashlights will last a total of less than 25 hours.

(c) Suppose that the 3 friends have five camping trips that year and each time they start with the same types of fresh batteries as above. Find the probability that the batteries last more than 25 hours exactly 3 of the 5 times.

5. It is desired to estimate the mean diameter of steel rods so that, with probability 0.95, the error of estimation will not exceed 0.005 cm. It is known that the distribution of the diameter of a randomly selected steel rod is normal with standard deviation 0.03 cm. What sample size should be used?

5.4 The Central Limit Theorem

Because, in general, finding the exact distribution of the sum or average of a large number of random variables is impractical, it would be very desirable to have a simple way to approximate it. Such an approximation is made possible by the *Central*

Limit Theorem, or *CLT* for short. In all that follows, $\stackrel{\cdot}{\sim}$ is read as "is approximately distributed as."

Theorem 5.4-1 **The Central Limit Theorem.** Let $X_1, \ldots, X_n$ be iid with mean μ and a finite variance σ^2. Then for large enough n ($n \geq 30$ for our purposes),

1. $\overline{X}$ has approximately a normal distribution with mean μ and variance σ^2/n, that is,

$$\overline{X} \stackrel{\cdot}{\sim} N\left(\mu, \frac{\sigma^2}{n}\right).$$

2. $T = X_1 + \ldots + X_n$ has approximately a normal distribution with mean $n\mu$ and variance $n\sigma^2$, that is,

$$T = X_1 + \ldots + X_n \stackrel{\cdot}{\sim} N\left(n\mu, n\sigma^2\right).$$

REMARK 5.4-1 The quality of the approximation increases with n, and also depends on the population distribution. For example, data from skewed populations require a larger sample size than data from, say, the uniform distribution. Moreover, the presence of really extreme outliers might indicate non-finite population variance, in which case the CLT does not hold; see the discussion on the assumptions of a finite mean and a finite variance at the end of Section 5.2. For the rest of this book we will always assume that data sets have been drawn from a population with a finite variance, and, as a rule of thumb, will apply the CLT whenever $n \geq 30$. ◁

The CLT is a really amazing result that explains the central role of the normal distribution in probability and statistics. Indeed, the importance to statistics of being able to approximate the distribution of averages (or sums) cannot be overstated. For this reason, the Central Limit Theorem is considered the most important theorem in probability and statistics. The convergence of the distribution of the average of exponential random variables to the normal distribution is demonstrated in Figure 5-2.

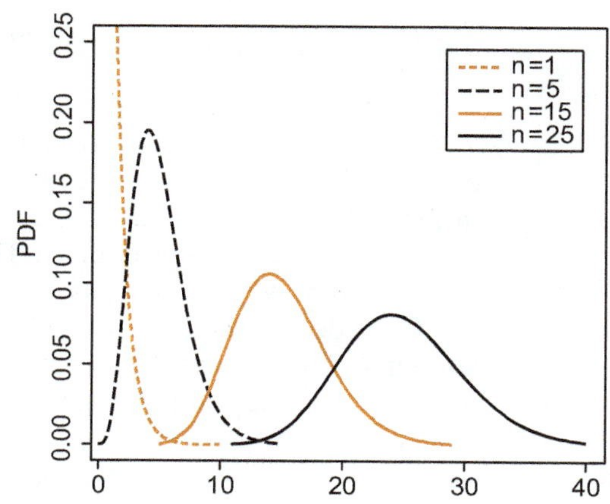

Figure 5-2 The distribution of the sum of n exponential($\lambda = 1$) random variables.

Example 5.4-1

The number of units serviced in a week at a certain service facility is a random variable having mean 50 and variance 16. Find an approximation to the probability that the total number of units to be serviced at the facility over the next 36 weeks is between 1728 and 1872.

Solution

Let $X_1, \ldots, X_{36}$ denote the number of units that are serviced in each of the next 36 weeks, and assume they are iid. Set $T = \sum_{i=1}^{36} X_i$ for the total number of units serviced. Then $E(T) = 36 \times 50 = 1800$ and $\text{Var}(T) = 36 \times 16 = 576$. Since the sample size is ≥ 30, according to the CLT the distribution of T is approximately normal with mean 1800 and variance 576. Thus,

$$P(1728 < T < 1872) = P\left(\frac{-72}{\sqrt{576}} < \frac{T - 1800}{\sqrt{576}} < \frac{72}{\sqrt{576}}\right)$$

$$\simeq \Phi(3) - \Phi(-3) = 0.997.$$

■

Example 5.4-2

The level of impurity in a randomly selected batch of chemicals is a random variable with μ =4.0% and σ =1.5%. For a random sample of 50 batches, find

(a) an approximation to the probability that the average level of impurity is between 3.5% and 3.8%, and

(b) an approximation to the 95th percentile of the average impurity level.

Solution

Let $X_1, \ldots, X_{50}$ denote the levels of impurity in each of the 50 batches, and let $\overline{X}$ denote their average. Since the sample size is ≥ 30, according to the CLT, $\overline{X} \stackrel{\cdot}{\sim} N(4.0,\ 1.5^2/50) = N(4.0,\ 0.045)$. The probability for part (a) and percentile for part (b) will be approximated according to this distribution. Thus, the answer to part (a) is

$$P(3.5 < \overline{X} < 3.8) \simeq P\left(\frac{3.5 - 4.0}{\sqrt{0.045}} < Z < \frac{3.8 - 4.0}{\sqrt{0.045}}\right)$$

$$\simeq \Phi(-0.94) - \Phi(-2.36) = 0.1645,$$

and the answer to part (b) is

$$\overline{x}_{0.05} \simeq 4.0 + z_{0.05}\sqrt{1.5^2/50} = 4.35.$$

■

5.4.1 THE DEMOIVRE-LAPLACE THEOREM

The DeMoivre-Laplace Theorem, which is the earliest form of the Central Limit Theorem, pertains to the normal approximation of binomial probabilities. It was proved first by DeMoivre in 1733 for $p = 0.5$ and extended to general p by Laplace in 1812. Because of the prevalence of the Bernoulli distribution in the experimental sciences, the DeMoivre-Laplace Theorem continues to be stated separately even though it is now recognized as a special case of the CLT.

Consider n replications of a Bernoulli experiment with probability of success p, and let T denote the total number of successes. Thus, $T \sim \text{Bin}(n, p)$. The relevance of the CLT for approximating binomial probabilities becomes clear if T is expressed as a sum,

$$T = X_1 + \cdots + X_n,$$

of the individual Bernoulli variables. Since the $X_1, \ldots, X_n$ are iid with $E(X_i) = p$ and $\text{Var}(X_i) = p(1-p)$ we have the following consequence of the Central Limit Theorem.

Theorem 5.4-2 **DeMoivre-Laplace.** If $T \sim \text{Bin}(n, p)$ then, for large enough n,

$$T \overset{\cdot}{\sim} N\left(np, np(1-p)\right).$$

The general condition $n \geq 30$ for achieving acceptable quality in the approximation can be specialized for the binomial distribution as follows:

Sample Size Requirement for Approximating Binomial Probabilities by Normal Probabilities

$$np \geq 5 \quad \text{and} \quad n(1-p) \geq 5$$

Figure 5-3 demonstrates that the probability mass function of the binomial tends to become symmetric as the sample size increases.

The Continuity Correction Whenever the Central Limit Theorem is used to approximate probabilities of a discrete distribution, the approximation is improved by the so-called *continuity correction*. To explain how this correction works let $X \sim \text{Bin}(10, 0.5)$ and suppose we are interested in using the DeMoivre-Laplace Theorem to approximate $P(X \leq 5)$. Since the sample size requirements are satisfied, we would approximate $P(X \leq 5)$ by $P(Y \leq 5)$, where $Y \sim N(5, 2.5)$. Figure 5-4 shows the bar graph of the Bin(10, 0.5) PMF with the $N(5, 2.5)$ PDF superimposed. In bar graphs of PMFs the area of each bar corresponds to individual probabilities. For example, $P(X = 5)$ equals the area of the bar centered at $x = 5$, half of which is shown colored. It follows that the probability $P(X \leq 5)$ equals the sum of the areas of the bars centered at $0, 1, \ldots, 5$. Approximating this by $P(Y \leq 5)$, which is the area under the normal PDF to the left of $x = 5$, leaves the area in color (i.e., half of $P(X = 5)$) unaccounted for.

The continuity correction consists of using the area under the normal PDF to the left of 5.5 as an approximation to $P(X \leq 5)$. The improvement in approximation is remarkable. Indeed, $P(X \leq 5) = 0.623$ would be approximated by $P(Y \leq 5) = 0.5$ without the continuity correction, and by $P(Y \leq 5.5) = 0.624$ with the continuity correction.

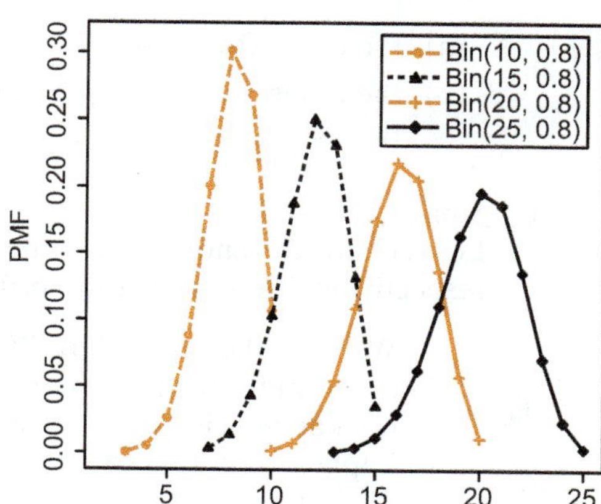

Figure 5-3 Binomial PMFs with $p = 0.8$ and increasing n.

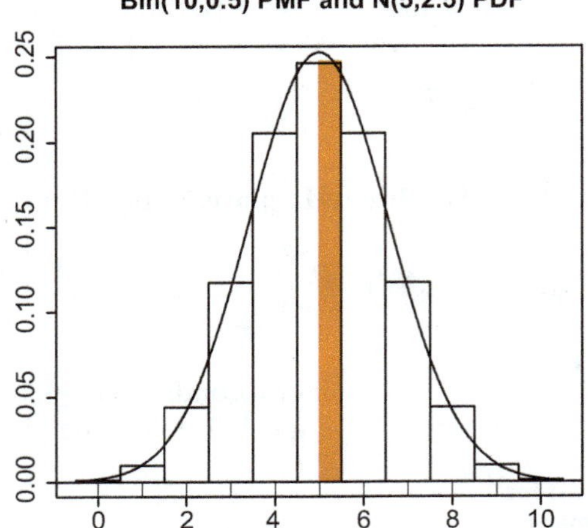

Figure 5-4 Approximation of $P(X \leq 5)$ with and without continuity correction.

In general, if X is a discrete random variable taking integer values and Y is the approximating normal random variable, probabilities and cumulative probabilities of X are approximated by

$$P(X = k) \simeq P(k - 0.5 < Y < k + 0.5) \text{ and } P(X \leq k) \simeq P(Y \leq k + 0.5). \quad \textbf{(5.4.1)}$$

Application of the continuity correction to the DeMoivre-Laplace Theorem yields the following approximation to the cumulative probabilities of $X \sim \text{Bin}(n,p)$: If the conditions $np \geq 5$ and $n(1-p) \geq 5$ hold, then

$$P(X \leq k) \simeq P(Y \leq k + 0.5) = \Phi\left(\frac{k + 0.5 - np}{\sqrt{np(1-p)}}\right), \quad \textbf{(5.4.2)}$$

where Y is a random variable having the normal distribution with mean and variance equal to the mean and variance of the Binomial random variable X, that is, $Y \sim N(np, np(1-p))$.

Example 5.4-3

A college basketball team plays 30 regular season games, 16 of which are against class A teams and 14 are against class B teams. The probability that the team will win a game is 0.4 if the team plays against a class A team and 0.6 if the team plays against a class B team. Assuming that the results of different games are independent, approximate the probability that

(a) the team will win at least 18 games, and

(b) the number of wins against class B teams is smaller than that against class A teams.

Solution

Let X_1 and X_2 denote the number of wins against class A teams and class B teams, respectively. Then, $X_1 \sim \text{Bin}(16, 0.4)$, and $X_2 \sim \text{Bin}(14, 0.6)$.

(a) We want the probability $P(X_1 + X_2 \geq 18)$. Since the probability of success is different for the two binomial distributions, the exact distribution of $X_1 + X_2$ is not known. However, since 16×0.4 and 14×0.4 are both ≥ 5, the DeMoivre-Laplace Theorem can be used to approximate the individual distributions of X_1 and X_2,

$$X_1 \stackrel{\cdot}{\sim} N(6.4, 3.84), \quad X_2 \stackrel{\cdot}{\sim} N(8.4, 3.36), \tag{5.4.3}$$

where $6.4 = 16 \times 0.4$, $3.84 = 16 \times 0.4 \times 0.6$, $8.4 = 14 \times 0.6$, and $3.36 = 14 \times 0.6 \times 0.4$. Consequently, since X_1 and X_2 are independent, by Proposition 5.3-1,

$$X_1 + X_2 \stackrel{\cdot}{\sim} N(6.4 + 8.4, 3.84 + 3.36) = N(14.8, 7.20).$$

Hence, using also continuity correction, the needed approximation for part (a) is

$$P(X_1 + X_2 \geq 18) = 1 - P(X_1 + X_2 \leq 17) \simeq 1 - \Phi\left(\frac{17.5 - 14.8}{\sqrt{7.2}}\right)$$
$$= 1 - \Phi(1.006) = 1 - 0.843 = 0.157.$$

(b) Using again the approximation to the individual distributions of X_1 and X_2 given in (5.4.3), the fact that X_1 and X_2 are independent, and Proposition 5.3-1, we have

$$X_2 - X_1 \stackrel{\cdot}{\sim} N(8.4 - 6.4, 3.84 + 3.36) = N(2, 7.20).$$

Hence, using also continuity correction, the needed approximation for part (b) is

$$P(X_2 - X_1 < 0) \simeq \Phi\left(\frac{-0.5 - 2}{\sqrt{7.2}}\right) = \Phi(-0.932) = 0.176.$$

Example 5.4-4

Suppose that 10% of a certain type of component last more than 600 hours in operation. For $n = 200$ components, let X denote the number of those that last more than 600 hours. Approximate the probabilities (a) $P(X \leq 30)$, (b) $P(15 \leq X \leq 25)$, and (c) $P(X = 25)$.

Solution

Here X has a binomial distribution with $n = 200$ and $p = 0.1$. Since $200 \times 0.1 = 20$, the sample size conditions for the application of the DeMoivre-Laplace Theorem are met. Using also the continuity correction, we have:

(a) $P(X \leq 30) \simeq \Phi\left(\frac{30.5 - 20}{\sqrt{18}}\right) = \Phi(2.47) = 0.9932.$

(b) To apply the DeMoivre-Laplace Theorem for approximating this probability, it is necessary to first express it as

$$P(15 \leq X \leq 25) = P(X \leq 25) - P(X \leq 14)$$

and then apply the DeMoivre-Laplace approximation to each probability on the right hand side. Thus,

$$P(15 \leq X \leq 25) \simeq \Phi\left(\frac{25.5 - 20}{\sqrt{18}}\right) - \Phi\left(\frac{14.5 - 20}{\sqrt{18}}\right)$$
$$= 0.9032 - 0.0968 = 0.8064.$$

(c) To apply the DeMoivre-Laplace Theorem for approximating this probability, it is necessary to first express it as

$$P(X = 25) = P(X \le 25) - P(X \le 24)$$

and then apply the DeMoivre-Laplace approximation to each probability on the right hand side. Thus,

$$P(X = 25) \simeq \Phi\left(\frac{25.5 - 20}{\sqrt{18}}\right) - \left(\frac{24.5 - 20}{\sqrt{18}}\right) = 0.9032 - 0.8554 = 0.0478.$$

Exercises

1. A random variable is said to have the (standard) Cauchy distribution if its PDF is given by (5.2.6). This exercise uses computer simulations to demonstrate that a) samples from this distribution often have extreme outliers (a consequence of the heavy tails of the distribution), and (b) the sample mean is prone to the same type of outliers. (In fact, for any sample size, the sample mean has the standard Cauchy distribution, implying that the LLN and CLT do not apply for samples from a Cauchy distribution.)

(a) The R commands *x=rcauchy(500); summary(x)* generate a random sample of size 500 from the Cauchy distribution and display the sample's five number summary; see Section 1.7. Report the five number summary and the interquartile range, and comment on whether or not the smallest and largest order statistics are outliers. Repeat this 10 times.

(b) The R commands *m=matrix(rcauchy(50000), nrow=500); xb=apply(m, 1, mean); summary(xb)* generate the matrix m that has 500 rows, each of which is a sample of size $n = 100$ from the Cauchy distribution, compute the 500 sample means and store them in xb, and display the five number summary of xb. Repeat these commands 10 times, and report the 10 sets of five number summaries. Compare with the 10 sets of five number summaries from part (a), and comment on whether or not the distribution of the averages seems to be as prone to extreme outliers as that of the individual observations.

2. Let $X_1, \ldots, X_{30}$ be independent Poisson random variables having mean 1.

(a) Use the CLT, with and without continuity correction, to approximate the probability $P(X_1 + \cdots + X_{30} \le 35)$. (*Hint.* The R command *pnorm(z)* gives $\Phi(z)$, the value of the standard normal CDF at z.)

(b) Use the fact that $X_1 + \cdots + X_{30}$ is a Poisson random variable (see Example 5.3-1) to find the exact value of the probability given in part (a). Compare the two approximations obtained in part (a) to the exact probability. (*Hint.* The R command *ppois(x, λ)* gives the value of the Poisson(λ) CDF at x).

3. Suppose that the waiting time for a bus, in minutes, has the uniform in $(0, 10)$ distribution. In five months a person catches the bus 120 times. Find an approximation to the 95th percentile of the person's total waiting time. (*Hint.* The mean and variance of a uniform(0, 10) distribution are 5 and 100/12, respectively; see Examples 3.3-8 and 3.3-13.)

4. Suppose the stress strengths of two types of materials follow the gamma distribution (see Exercise 13 in Section 3.5) with parameters $\alpha_1 = 2$, $\beta_1 = 2$ for type 1 and $\alpha_2 = 1$, $\beta_2 = 3$ for type two. Let $\overline{X}_1$ and $\overline{X}_2$ be average stress strength measurements corresponding to samples of sizes $n_1 = 36$ specimens of type 1 material and $n_2 = 42$ specimens of type 2 material, respectively.

(a) Specify the (approximate) distributions of $\overline{X}_1$, $\overline{X}_2$, and $\overline{X}_1 - \overline{X}_2$. Justify your answers.

(b) Find the (approximate) probability that $\overline{X}_1$ will be larger than $\overline{X}_2$.

5. Two towers are constructed, each by stacking 30 segments of concrete vertically. The height (in inches) of a randomly selected segment is uniformly distributed in the interval (35.5, 36.5). A roadway can be laid across the 2 towers provided the heights of the 2 towers are within 4 inches of each other. Find the probability that the roadway can be laid. Be careful to justify the steps in your argument, and state whether the probability is exact or approximate.

6. Using the information on the joint distribution of meal price and tip given in Exercise 3 in Section 4.3, answer the following question: If a waitress serves 70 customers in an evening, find an approximation to the probability that her tips for the night exceed \$120. (*Hint.* The mean and variance of the tip from a random customer are 1.8175 and 0.1154.)

7. When a randomly selected number A is rounded off to its nearest integer R_A, it is reasonable to assume that the round-off error $A - R_A$ is uniformly distributed in $(-0.5, 0.5)$. If 50 numbers are rounded off to the nearest integer and then averaged, approximate the probability that the resulting average differs from the exact average of the 50 numbers by more than 0.1.

8. Components that are critical for the operation of electrical systems are replaced immediately upon failure. Suppose that the life time of a certain such component has mean and standard deviation of 100 and 30 time units, respectively. How many of these components must be in stock to ensure a probability of at least 0.95 for the system to be in continuous operation for at least the next 3000 time units? (*Hint.* If $T = X_1 + \cdots + X_n$ is the combined duration of n components, we want $P(T > 3000) = 0.95$. This means that 3000 is the 5th percentile of T. Using the CLT to approximate the 5th percentile of T leads to a quadratic equation for the square root of n, that is, an equation of the form $\alpha x^2 + \beta x + \gamma = 0$, with x being the square root of n. The roots of such an equation can be found with the R command *polyroot(c(*γ, β, α*))*.)

9. An optical company uses a vacuum deposition method to apply a protective coating to certain lenses. The coating is built up one layer at a time. The thickness of a given layer is a random variable with mean $\mu = 0.5$ microns and standard deviation $\sigma = 0.2$ microns. The thickness of each layer is independent of the others and all layers have the same thickness distribution. In all, 36 layers are applied.

(a) What is the approximate distribution of the coating thickness? Cite the appropriate theorem to justify your answer.

(b) The company has determined that a minimum thickness of 16 microns for the entire coating is necessary to meet all warranties. Consequently, each lens is tested and additional layers are applied if the lens does not have at least a 16-micron-thick coat. What proportion of lenses must have additional layers applied?

10. A batch of 100 steel rods passes inspection if the average of their diameters falls between 0.495 cm and 0.505 cm. Let μ and σ denote the mean and standard deviation, respectively, of the diameter of a randomly selected rod. Answer the following questions assuming that $\mu = 0.503$ cm and $\sigma = 0.03$ cm.

(a) What is the (approximate) probability the inspector will accept (pass) the batch?

(b) Over the next 6 months 40 batches of 100 will be delivered. Let X denote the number of batches that will pass inspection.

(i) State the exact distribution of X, and use R to find the probability $P(X \leq 30)$.

(ii) Use the DeMoivre-Laplace Theorem, with and without continuity correction, to approximate $P(X \leq 30)$. Comment on the quality of the approximation provided by the two methods.

11. Suppose that only 60% of all drivers wear seat belts at all times. In a random sample of 500 drivers let X denote the number of drivers who wear seat belt at all times.

(a) State the exact distribution of X and use R to find $P(270 \leq X \leq 320)$.

(b) Use the DeMoivre-Laplace Theorem, with and without continuity correction, to approximate $P(270 \leq X \leq 320)$. Comment on the quality of the approximation provided by the two methods.

12. A machine manufactures tires with a tread thickness that is normally distributed with mean 10 millimeters (mm) and standard deviation 2 mm. The tire has a 50,000-mile warranty. For the tire to last 50,000 miles, the manufacturer's guidelines specify that the tread thickness must be at least 7.9 mm. If the thickness of tread is measured to be less than 7.9 mm, then the tire is sold as an alternative brand with a warranty of less than 50,000 miles. Give an approximation to the probability that in a batch of 100 tires there are no more than 10 rejects.

13. Items produced in assembly line A are defect free with probability 0.9, and those produced in assembly line B are defect free with probability 0.99. A sample of 200 items from line A and a sample of 1000 from line B are inspected.

(a) Give an approximation to the probability that the total number of defective items found is at most 35. (*Hint.* See Example 5.4-3.)

(b) Use R commands similar to those used in Example 5.3-4 to find the exact probability of part (a).

Chapter 6

Fitting Models to Data

6.1 Introduction

In Chapter 1 we saw that estimation of population parameters, such as proportion, mean, variance, and percentiles, is achieved by using the corresponding sample quantities. Similarly, in Chapter 4 we saw sample versions of the covariance and Pearson's correlation coefficients that estimate the corresponding population quantities. This approach to estimation, which is called interchangeably **empirical**, **model-free**, or **nonparametric**, is universal in the sense that it applies to all types of population distributions.

When a model for the distribution of the data is assumed, it is typically of interest to estimate the parameters of the assumed model. For example,

(a) if it can be reasonably assumed that the data came from a uniform distribution it would be of interest to estimate the two endpoints,

(b) if it can be reasonably assumed that the data came from a gamma or a Weibull distribution, both of which are governed by parameters denoted by α and β (see Exercises 13 and 14 in Section 3.5), it would be of interest to estimate these two parameters, and

(c) if it can be reasonably assumed that the data came from the normal simple linear regression model, it would be of interest to estimate the regression line (i.e., the slope and the intercept) and the intrinsic error variance.

In statistical jargon, estimating the parameters of a particular model from a data set is called *fitting* the model to the data. Three methods of fitting models to data are (a) the *method of moments*, (b) the *method of maximum likelihood*, and (c) the *method of least squares*. The last is most commonly used for fitting regression models.

Estimation of the model parameters leads to an alternative way for estimating population parameters, called *model-based estimation*; this and the notion of *unbiased estimation* are discussed in Section 6.2. The aforementioned three methods for fitting models to data are presented in Section 6.3. Model-based estimation of population parameters can differ from the empirical, or model-free, estimation discussed in Chapter 1; moreover, the three methods for fitting models will occasionally produce different estimators of model parameters. Thus, another learning objective of this chapter is to develop criteria for selecting the best among different estimators of the same (model or population) parameter; this is the subject of Section 6.4.

6.2 Some Estimation Concepts

6.2.1 UNBIASED ESTIMATION

The Greek letter θ will be used as a generic notation for any model or population parameter(s) that we are interested in estimating. Thus, if we are interested in the population mean value, then $\theta = \mu$, and if we are interested in the population mean value and variance, then $\theta = (\mu, \sigma^2)$. The expression **true value of** θ refers to the (unknown to us) population value of θ.

When a sample is denoted in capital letters, such as $X_1, \ldots, X_n$, the X_i's are considered random variables, that is, before their values are observed. The observed sample values, or data, are denoted in lowercase letters, that is, $x_1, \ldots, x_n$.

A quantity used to estimate the true value of a parameter θ is denoted by $\widehat{\theta}$. Because $\widehat{\theta}$ is computed from the sample, it is a function of it. This is emphasized by writing

$$\widehat{\theta} = \widehat{\theta}(X_1, \ldots, X_n) \quad \text{or} \quad \widehat{\theta} = \widehat{\theta}(x_1, \ldots, x_n).$$

In the former case, $\widehat{\theta}$ is called an **estimator**, and in the latter, an **estimate**. Thus, an estimator is a random variable, while an estimate is an observed value.

The distribution of an estimator $\widehat{\theta}$ depends on the true value of θ (and perhaps the true value of additional parameters). For example, suppose that $X_1, \ldots, X_n$ is a sample from a $N(\mu, \sigma^2)$ population and the true values of the parameters are $\mu = 8.5$, $\sigma^2 = 18$. Then, the estimator of $\theta = \mu$ is $\widehat{\theta} = \overline{X}$ and, according to Corollary 5.3-1,

$$\overline{X} \sim N\left(8.5, \frac{18}{n}\right).$$

Thus, in this case, the distribution of $\widehat{\theta}$ depends on the true value of θ and the true value of the additional parameter σ^2. We also write

$$E_{\mu=8.5}\left(\overline{X}\right) = 8.5 \quad \text{and} \quad \text{Var}_{\sigma^2=18}(\overline{X}) = \frac{18}{n}$$

to emphasize the dependence of the mean and variance of $\overline{X}$ on the true values of the parameters. Similar notation will be used to emphasize the dependence of the mean and variance of any estimator $\widehat{\theta}$ on the true value(s) of the relevant parameter(s).

An estimator $\widehat{\theta}$ of θ is called **unbiased** for θ if $E(\widehat{\theta}) = \theta$ or, according to the notation just introduced, if

Definition of an Unbiased Estimator

$$E_\theta\left(\widehat{\theta}\right) = \theta \qquad (6.2.1)$$

The difference $E_\theta(\widehat{\theta}) - \theta$ is called the **bias** of $\widehat{\theta}$ and is denoted by bias$(\widehat{\theta})$:

Definition of the Bias of an Estimator

$$\text{bias}(\widehat{\theta}) = E_\theta\left(\widehat{\theta}\right) - \theta \qquad (6.2.2)$$

Actually, the correct notation is $\text{bias}_\theta(\widehat{\theta})$ but we will use bias$(\widehat{\theta})$ for simplicity.

As established in Corollary 4.4-1, and relation (4.4.4), the estimators $\overline{X}$ and $\widehat{p}$ are unbiased for μ and p, respectively. That is,

$$E_p(\widehat{p}) = p \quad \text{and} \quad E_\mu(\overline{X}) = \mu.$$

Moreover, the least squares estimators $\widehat{\beta}_1$ and $\widehat{\alpha}_1$ in the simple linear regression model, which will be given in Section 6.3.3 (see also Remark 4.6-1), are also unbiased. The next proposition shows that the sample variance, S^2, is also an unbiased estimator for σ^2.

Proposition 6.2-1 Let $X_1, \ldots, X_n$ be iid with (common) variance σ^2. Then the expected value of the sample variance $S^2 = (n-1)^{-1} \sum_{i=1}^{n} \left(X_i - \overline{X}\right)^2$ equals σ^2. That is,

Expected Value of the Sample Variance

$$E\left(S^2\right) = \sigma^2$$

Proof of Proposition 6.2-1: Assume without loss of generality that the population mean is zero, that is, $E(X_i) = 0$ for all $i = 1, \ldots, n$. By straightforward algebra we obtain $\sum_i \left(X_i - \overline{X}\right)^2 = \sum_i X_i^2 - n\overline{X}^2$. Now using the facts

$$E(X_i^2) = \text{Var}(X_i) = \sigma^2 \quad \text{and} \quad E(\overline{X}^2) = \text{Var}(\overline{X}) = \frac{\sigma^2}{n},$$

we obtain

$$E\left(\sum_{i=1}^{n} \left(X_i - \overline{X}\right)^2\right) = E\left(\sum_{i=1}^{n} X_i^2 - n\overline{X}^2\right) = n\sigma^2 - n\frac{\sigma^2}{n} = (n-1)\sigma^2.$$

It follows that $E(S^2) = (n-1)^{-1} E(\sum_i \left(X_i - \overline{X}\right)^2) = (n-1)^{-1}(n-1)\sigma^2 = \sigma^2$. ■

The **estimation error** of an estimator $\widehat{\theta}$ for θ is defined as

Definition of Estimation Error

$$\widehat{\theta} - \theta \tag{6.2.3}$$

Unbiased estimators have zero bias, which means that there is no tendency to overestimate or underestimate the true value of θ. Thus, though with any given sample $\widehat{\theta}$ may underestimate or overestimate the true value of θ, the estimation errors average to zero. In particular, the unbiasedness of S^2, implied by Proposition 6.2-1, means that if a large number of samples of size n, any $n \geq 2$, are taken from any population (e.g., Poisson, normal, exponential, etc.) and the sample variance is computed for each sample, the average of these sample variances will be very close to the population variance; equivalently, the average of the estimation errors will be very close to zero. This is illustrated in the computer activity of Exercise 8.

While unbiasedness is a desirable property, it is not indispensable. What justifies the use of biased estimators is the fact that their bias is often small and tends to zero as the sample size increases. (An estimator whose bias does not tend to zero does not possess the indispensable property of consistency, and would not be used!) An example of a commonly used biased estimator is the sample standard deviation. The bias of the sample standard deviation, and the fact that its bias decreases as the sample size increases, are also demonstrated in the computer activity of Exercise 8.

The **standard error** of an estimator $\widehat{\theta}$ is an alternative, but widely used, term for the estimator's standard deviation:

Standard Error of an Estimator $\widehat{\theta}$

$$\sigma_{\widehat{\theta}} = \sqrt{\text{Var}_\theta\left(\widehat{\theta}\right)} \tag{6.2.4}$$

In accordance with the notation explained above, the subscript θ on the right hand side of (6.2.4) indicates the dependence of the variance of $\widehat{\theta}$ on the true value of θ. An estimator/estimate of the standard error, is called the **estimated standard error** and is denoted by $S_{\widehat{\theta}}$.

Example 6.2-1

(a) Give the standard error and the estimated standard error of the estimator $\widehat{p} = X/n$, where $X \sim \text{Bin}(n, p)$.

(b) Given the information that there are 12 successes in 20 trials, compute the estimate of p and the estimated standard error.

Solution

(a) The standard error and the estimated standard error of $\widehat{p}$ are, respectively,

$$\sigma_{\widehat{p}} = \sqrt{\frac{p(1-p)}{n}} \quad \text{and} \quad S_{\widehat{p}} = \sqrt{\frac{\widehat{p}(1-\widehat{p})}{n}}.$$

(b) With the given the information, $\widehat{p} = 12/20 = 0.6$ and

$$S_{\widehat{p}} = \sqrt{\frac{\widehat{p}(1-\widehat{p})}{n}} = \sqrt{\frac{0.6 \times 0.4}{20}} = 0.11.$$

Example 6.2-2

(a) Let $\overline{X}$, S^2 be the sample mean and variance of a simple random sample of size n from a population with mean μ and variance σ^2, respectively. Give the standard error and the estimated standard error of $\overline{X}$.

(b) Given the information that $n = 36$ and $S = 1.3$, compute the estimated standard error of $\overline{X}$.

Solution

(a) The standard error and the estimated standard error of $\overline{X}$ are, respectively,

$$\sigma_{\overline{X}} = \frac{\sigma}{\sqrt{n}} \quad \text{and} \quad S_{\overline{X}} = \frac{S}{\sqrt{n}}.$$

(b) With the given the information, the estimated standard error of $\overline{X}$ is

$$S_{\overline{X}} = \frac{1.3}{\sqrt{36}} = 0.22.$$

Example 6.2-3

Let $\overline{X}_1$, S_1^2 be the sample mean and variance of a simple random sample of size m from a population with mean μ_1 and variance σ_1^2, respectively, and $\overline{X}_2$, S_2^2 be the sample mean and variance of a simple random sample of size n from a population with mean μ_2 and variance σ_2^2, respectively.

(a) Show that $\overline{X}_1 - \overline{X}_2$ is an unbiased estimator of $\mu_1 - \mu_2$.

(b) Assume the two samples are independent, and give the standard error and the estimated standard error of $\overline{X}_1 - \overline{X}_2$.

Solution

(a) From the properties of expectation we have

$$E(\overline{X}_1 - \overline{X}_2) = E(\overline{X}_1) - E(\overline{X}_2) = \mu_1 - \mu_2,$$

which shows that $\overline{X}_1 - \overline{X}_2$ is an unbiased estimator of $\mu_1 - \mu_2$.

(b) Recall that if two variables are independent, the variance of their difference is the sum of their variances; see Proposition 4.4-4. Thus, the standard error and estimated standard error of $\overline{X}_1 - \overline{X}_2$ are

$$\sigma_{\overline{X}_1 - \overline{X}_2} = \sqrt{\frac{\sigma_1^2}{m} + \frac{\sigma_2^2}{n}} \quad \text{and} \quad S_{\overline{X}_1 - \overline{X}_2} = \sqrt{\frac{S_1^2}{m} + \frac{S_2^2}{n}}.$$

■

6.2.2 MODEL-FREE VS MODEL-BASED ESTIMATION

As mentioned in Section 6.1, if a model for the population distribution has been assumed, the focus of estimation shifts to the model parameters. This is because estimation of the model parameters entails estimation of the entire distribution, and hence estimation of any other population quantity of interest. For example, if $X_1, \ldots, X_n$ can be assumed to have come from a $N(\mu, \sigma^2)$ distribution, the method of moments and the method of maximum likelihood estimate $\theta = (\mu, \sigma^2)$ by $\widehat{\theta} = (\overline{X}, S^2)$.[1] Hence, the population distribution is estimated by $N(\overline{X}, S^2)$. This has the following consequences:

(a) There is no need to use a histogram of the data because the density is estimated by the $N(\overline{X}, S^2)$ density. (Of course, histograms and Q-Q plots are indispensable for checking the appropriateness of an assumed model.)

(b) Because the $(1 - \alpha)$-100th percentile of a normal population is expressed as $\mu + \sigma z_\alpha$ (see Corollary 3.5-2), it may be estimated by $\overline{X} + S z_\alpha$; in particular, the median is also estimated by $\overline{X}$.

(c) Because $P(X \le x) = \Phi((x - \mu)/\sigma)$ such probabilities may be estimated by $\Phi((x - \overline{X})/S)$.

The estimators of the density, percentiles, and probabilities in parts (a), (b), and (c), respectively, which are appropriate only if the normality assumption is correct, are examples of **model-based** estimators. They can be used instead of the model-free estimators of Chapter 1, which are, respectively, the histogram, sample percentiles, and sample proportions (i.e., $\#\{X_i \le x; i = 1, \ldots, n\}/n$ in this case).

Such model-based estimators of the density, percentiles, and probabilities can similarly be constructed if $X_1, \ldots, X_n$ is assumed to have come from any other distribution, such as exponential, gamma, Weibull, and so forth.

Example 6.2-4

(a) Let $X_1, \ldots, X_n$ represent n weekly counts of earthquakes in North America, and assume they have the Poisson(λ) distribution. Find a model-based estimator of the population variance.

(b) Let $X_1, \ldots, X_n$ represent waiting times of a random sample of n passengers of a New York commuter train, and assume they have the uniform$(0, \theta)$ distribution. Find a model-based estimator of the population mean waiting time.

Solution

(a) From Section 3.4.4 we have that the variance of the Poisson(λ) distribution equals its mean (and both equal λ). Both the method of moments and the method of maximum likelihood estimate λ by $\widehat{\lambda} = \overline{X}$ (see Example 6.3-2

[1] This is not quite true, as both methods of estimation yield $[(n-1)/n]S^2$ as the estimator of σ^2. For simplicity, however, we will ignore this difference.

and Exercise 6 in Section 6.3). Thus, the Poisson model-based estimator of the variance is the sample mean $\overline{X}$.

(b) From Example 3.3-8 we have that the mean of the uniform$(0, \theta)$ distribution is $\mu = \theta/2$. Hence, if $\widehat{\theta}$ is an estimator of θ, a model-based estimator for the mean of the uniform$(0, \theta)$ distribution is $\widehat{\mu} = \widehat{\theta}/2$. The maximum likelihood estimator for θ (derived in Example 6.3-6) is $\widehat{\theta} = X_{(n)} = \max\{X_1, \ldots, X_n\}$. Thus, the model-based estimator of the population mean in this case is $\widehat{\mu} = X_{(n)}/2$. ■

REMARK 6.2-1 The method of moments estimator of θ in the uniform$(0, \theta)$ distribution is $2\overline{X}$; see Example 6.3-1. Using this estimator of θ, the model-based estimator of the population mean in part (b) of Example 6.2-4 is $\overline{X}$, that is, the same as the model-free estimator. ◁

If the model assumption is correct then, according to the *mean square error* criterion, which will be discussed in Section 6.4, model-based estimators are typically preferable to model-free estimators. Thus, if the assumption of a Poisson distribution is correct, $\overline{X}$ is a better estimator of the population variance than the sample variance is, and if the assumption of a uniform$(0, \theta)$ distribution is correct, $X_{(n)}/2$ is a better estimator of the population mean than the sample mean is (at least for large enough n; see Exercise 1 in Section 6.4).

On the other hand, if the model assumption is not correct, model-based estimators can be misleading. The following example illustrates this point by fitting two different models to the same data set and thus obtaining discrepant estimates for a probability and a percentile.

Example 6.2-5

The life times, in hours, of a random sample of 25 electronic components yield sample mean $\overline{X} = 113.5$ hours and sample variance $S^2 = 1205.55$ hours^2. Find model-based estimators of the 95th population percentile of the lifetime distribution, and of the probability that a randomly selected component will last more than 140 hours, under the following two model assumptions:

(a) The distribution of life times is Weibull(α, β). (See Exercise 14 in Section 3.5 for the definition of this distribution.)

(b) The distribution of life times is exponential(λ).

Solution

(a) With the given information, only the method of moments can be used to fit the Weibull(α, β) model. The resulting estimators are $\widehat{\alpha} = 3.634$ and $\widehat{\beta} = 125.892$; see Example 6.3-3. The R commands

```
1-pweibull(140, 3.634, 125.892); qweibull(0.95, 3.634, 125.892)
```

yield 0.230 and 170.264 as estimates of $P(X > 140)$ and $x_{0.05}$, respectively.

(b) For fitting the exponential(λ) distribution, both the method of moments and the method of maximum likelihood yield $\widehat{\lambda} = 1/\overline{X}$ (see Exercise 1 in Section 6.3 and Example 6.3-5). Thus, the fitted model is exponential$(\lambda = 1/113.5)$. The R commands

```
1-pexp(140, 1/113.5); qexp(0.95, 1/113.5)
```

yield 0.291 and 340.016 as estimates of $P(X > 140)$ and $x_{0.05}$, respectively. ■

This example highlights the need for diagnostic checks, such as the Q-Q plot discussed in Section 3.5.2, to help decide whether a stipulated parametric model provides a reasonable fit to the data.

Exercises

1. The data in *OzoneData.txt* contains $n = 14$ ozone measurements (Dobson units) taken from the lower stratosphere, between 9 and 12 miles (15 and 20 km). Compute the sample mean and its estimated standard error.

2. To compare the corrosion-resistance properties of two types of material used in underground pipelines, specimens of both types are buried in soil for a 2-year period and the maximum penetration (in mils) for each specimen is measured. A sample of size $n_1 = 48$ specimens of material type A yielded $\overline{X}_1 = 0.49$ and $S_1 = 0.19$; a sample of size $n_2 = 42$ specimens of material type B gave $\overline{X}_2 = 0.36$ and $S_2 = 0.16$. What is $0.49 - 0.36 = 0.13$ an estimate of? Assuming that the two samples are independent, compute the estimated standard error of $\overline{X}_1 - \overline{X}_2$. (*Hint.* See Example 6.2-3.)

3. In the context of Exercise 2, suppose that the population variance of the maximum penetration is the same for both material types. Call the common population variance σ^2, and show that

$$\widehat{\sigma}^2 = \frac{(n_1 - 1)S_1^2 + (n_2 - 1)S_2^2}{n_1 + n_2 - 2}$$

is an unbiased estimator of σ^2.

4. The financial manager of a department store chain selected a random sample of 200 of its credit card customers and found that 136 had incurred an interest charge during the previous year because of an unpaid balance.

(a) Specify the population parameter of interest in this study, give the empirical estimator for it, and use the information provided to compute the estimate.

(b) Is the estimator in part (a) unbiased?

(c) Compute the estimated standard error of the estimator.

5. In Example 6.3-1 it is shown that if $X_1, \ldots, X_n$ is a random sample from the uniform$(0, \theta)$ distribution, the method of moments estimator of θ is $\widehat{\theta} = 2\overline{X}$. Give the standard error of $\widehat{\theta}$. Is $\widehat{\theta}$ unbiased?

6. To estimate the proportion p_1 of male voters who are in favor of expanding the use of solar energy, take a random sample of size m and set X for the number in favor. To estimate the corresponding proportion p_2 of female voters, take an independent random sample of size n and set Y for the number in favor.

(a) Set $\widehat{p}_1 = X/m$ and $\widehat{p}_2 = Y/n$ and show that $\widehat{p}_1 - \widehat{p}_2$ is an unbiased estimator of $p_1 - p_2$.

(b) Give the standard error and the estimated standard error of $\widehat{p}_1 - \widehat{p}_2$.

(c) The study uses sample sizes of $m = 100$ and $n = 200$, which result in $X = 70$ and $Y = 160$. Compute the estimate of $p_1 - p_2$ and the estimated standard error of the estimator.

7. The fat content measurements of a random sample of 6 jugs of 2% lowfat milk jugs of a certain brand are 2.08, 2.10, 1.81, 1.98, 1.91, 2.06.

(a) Give the model-free estimate of the proportion of milk jugs having a fat content measurement of 2.05 or more.

(b) Assume the fat content measurements are normally distributed, and give the model-based estimate of the same proportion (using $\overline{X}$ and S^2 as estimators of μ and σ^2).

8. The R commands

```
set.seed=1111; m=matrix(runif(20000),
  ncol=10000); mean(apply(m, 2, var));
  mean(apply(m, 2, sd))
```

generate 10,000 samples of size $n = 2$ from the uniform$(0, 1)$ distribution (each column of the matrix m is a sample of size 2), compute the sample variance from each sample, average the 10,000 variances, and do the same for the sample standard deviations.

(a) Compare the average of the 10,000 variances to the population variance $\sigma^2 = 1/12 = 0.0833$; similarly, compare the average of the 10,000 sample standard deviations to the population standard deviation $\sigma = \sqrt{1/12} = 0.2887$. Use the comparisons to conclude that S^2 is unbiased but S is biased.

(b) Repeat the above but use 10,000 samples of size $n = 5$ from the uniform(0, 1). (Use *m=matrix(runif(50000), ncol=10000)*) for generating the random samples.) Use the comparisons to conclude that the bias of S decreases as the sample size increases.

9. Use the R command *set.seed=1111; x=rnorm(50, 11, 4)* to generate a simple random sample of 50 observations from a $N(11, 16)$ population and store it in the R object x.

(a) Give the true (population) values of $P(12 < X \leq 16)$ and of the 15th, 25th, 55th, and 95th percentiles.

(b) Give the empirical/nonparametric estimates of the above population quantities. (*Hint.* The R command *sum(12<x&x<=16)* gives the number of data points that are greater than 12 and less than or equal to 16. For the sample percentiles see the R commands given in (1.7.2).)

(c) Using $\overline{X}$ and S^2 as estimators of the model parameters of $N(\mu, \sigma^2)$, give the model-based estimates of the above population quantities and compare how well the two types of estimates approximate the true population values.

10. Use *cs=read.table("Concr.Strength.1s.Data.txt", header=T); x=cs$Str* to store the data set[2] consisting of 28-day compressive-strength measurements of concrete cylinders using water/cement ratio 0.4 into the R object *x*.

(a) Use the commands given in Section 3.5.2 to produce a normal Q-Q plot for the data. Comment on the appropriateness of the normal model for this data set.

(b) Using $\overline{X}$ and S^2 as estimators of the model parameters of $N(\mu, \sigma^2)$, give model-based estimates of $P(44 < X \leq 46)$, the population median, and the 75th percentile.

(c) Give the empirical, or model-free, estimates of the above population quantities. (*Hint.* The R command *sum(44<x&x<=46)* gives the number of data points that are greater than 44 and less than or equal to 46. For the sample percentiles see the R commands given in (1.7.2).)

(d) Which of the two types of estimates for the above population quantities would you prefer and why?

6.3 Methods for Fitting Models to Data

Model-based estimation of a particular population parameter consists of expressing it in terms of the model parameter(s) θ and plugging the estimator $\widehat{\theta}$ into the expression (see Section 6.2.2). Clearly, the method relies on having an estimator for θ. This section presents three methods for obtaining such estimators.

6.3.1 THE METHOD OF MOMENTS

The method of moments relies on the empirical, or model-free, estimators of population parameter(s), such as the sample mean ($\overline{X}$) or the sample mean and variance ($\overline{X}$ and S^2), and reverses the process of model-based estimation in order to estimate the model parameter(s).

In particular, the method of moments uses the fact that when the population distribution is assumed to be of a particular type, population parameter(s), such as the mean or the mean and variance, can be expressed in terms of the model parameter(s) θ. These expressions can be inverted to express θ in terms of the population mean or the population mean and variance. Plugging the sample mean or the sample mean and variance into these inverted expressions yields the method of moments estimator of θ. A more complete description of moment estimators is given after the next example, which illustrates the above process.

Example 6.3-1

Let $X_1, \ldots, X_n$ be a simple random sample taken from some population. Use the method of moments approach to fit the following models to the data:

(a) The population distribution of the X_i is uniform$(0, \theta)$.

(b) The population distribution of the X_i is uniform(α, β).

Solution

(a) Here we have only one model parameter, so the method of moments starts by expressing the population mean in terms of the model parameter. In this

[2] V. K. Alilou and M. Teshnehlab (2010). Prediction of 28-day compressive strength of concrete on the third day using artificial neural networks. *International Journal of Engineering (IJE)*, 3(6): 521–610.

case, the expression is $\mu = \theta/2$. This expression is then inverted to express θ in terms of μ. In this case, the inverted expression is $\theta = 2\mu$. Finally, the method of moments estimator of θ is obtained by plugging $\overline{X}$ instead of μ into inverted expression: $\widehat{\theta} = 2\overline{X}$.

(b) Here we have two model parameters ($\theta = (\alpha, \beta)$), so the method of moments starts by expressing the population mean and variance in terms of the two model parameters. In this case, the expressions are

$$\mu = \frac{\alpha + \beta}{2} \quad \text{and} \quad \sigma^2 = \frac{(\beta - \alpha)^2}{12}.$$

This expression is then inverted to express α and β in terms of μ and σ^2:

$$\alpha = \mu - \sqrt{3\sigma^2} \quad \text{and} \quad \beta = \mu + \sqrt{3\sigma^2}.$$

Finally, the method of moments estimator of $\theta = (\alpha, \beta)$ is obtained by plugging $\overline{X}$ and S^2 instead of μ and σ^2, respectively, into the inverted expressions:

$$\widehat{\alpha} = \overline{X} - \sqrt{3S^2} \quad \text{and} \quad \widehat{\beta} = \overline{X} + \sqrt{3S^2}.$$

■

The method of moments derives its name from the fact that the expected value of the kth power of a random variable is called its kth **moment**; this is denoted by μ_k:

kth Moment of the Random Variable X

$$\mu_k = E(X^k)$$

In this terminology, the population mean is the first moment and is also denoted by μ_1, while the population variance can be expressed in terms of the first two moments as $\sigma^2 = \mu_2 - \mu_1^2$. If $X_1, \ldots, X_n$ is a sample from a population with a finite kth moment, then the empirical/nonparametric estimator of μ_k is the kth **sample moment**, defined as follows:

kth Sample Moment of the Random Variable X

$$\widehat{\mu}_k = \frac{1}{n}\sum_{i=1}^{n} X_i^k$$

According to the Law of Large Numbers, $\widehat{\mu}_k$ is a consistent estimator of μ_k.

For models with m parameters, method of moments estimators are constructed by (a) expressing the first m population moments in terms of the model parameters, (b) inverting these expressions to obtain expressions of the model parameters in terms of the population moments, and (c) plugging into these inverted expressions the sample moments. Choosing the number of moments in part (a) equal to the number of model parameters assures that the inversion mentioned in part (b) has a unique solution. In this book we will not consider distribution models with more than two model parameters, so in our applications of the method of moments, we will use either only the first moment or the first and the second moments. Equivalently, we will use either only the mean or the mean and the variance as was done in Example 6.3-1.

REMARK 6.3-1 Using the variance instead of the second moment is not exactly the same, because the sample variance is defined by dividing $\sum_i (X_i - \overline{X})^2$ by $n - 1$. Ignoring this (rather insignificant) difference, we will use the variance and the sample variance (instead of the second moment and the second sample moment) when applying the method of moments. ◁

We finish this section with two more examples.

Example 6.3-2

(a) Let $X_1, \ldots, X_n$ be iid Poisson(λ). Find the method of moments estimator of λ. Is it unbiased?

(b) The weekly counts of earthquakes in North America for 30 consecutive weeks are summarized in the following table:

Number of Earthquakes	4	5	6	7	8	9	10	11	12	13	15	16	17
Frequency	1	2	1	5	4	4	1	1	4	1	2	2	2

Assuming that the earthquake counts have the Poisson(λ) distribution, compute the method of moments estimate of λ.

Solution

(a) Because we have only one model parameter, the method of moments starts by expressing the population mean in terms of the model parameter. In this case, the expression is $\mu = \lambda$; see Section 3.4.4. Thus, $\lambda = \mu$ and the method of moments estimator of λ is $\widehat{\lambda} = \overline{X}$. Because $\overline{X}$ is an unbiased estimator of μ, it follows that $\widehat{\lambda}$ is an unbiased estimator of λ.

(b) The average of the given 30 weekly counts of earthquakes is $(1 \times 4 + 2 \times 5 + \cdots + 2 \times 17)/30 = 10.03$. Thus, the method of moments estimate of λ is $\widehat{\lambda} = 10.03$. ■

Example 6.3-3

The life spans, in hours, of a random sample of 25 electronic components yield sample mean $\overline{X} = 113.5$ hours and sample variance $S^2 = 1205.55$ hours^2. Use the method of moments approach to fit the Weibull(α, β) model.

Solution

Because we have two model parameters ($\theta = (\alpha, \beta)$), the method of moments starts by expressing the population mean and variance in terms of the two model parameters. These expressions are (see Exercise 14 in Section 3.5)

$$\mu = \beta\Gamma\left(1 + \frac{1}{\alpha}\right) \quad \text{and} \quad \sigma^2 = \beta^2\left\{\Gamma\left(1 + \frac{2}{\alpha}\right) - \left[\Gamma\left(1 + \frac{1}{\alpha}\right)\right]^2\right\},$$

where Γ is the gamma function (see Exercise 13 in Section 3.5). Note that α enters these equations in a highly non-linear manner, so it is impossible to then invert and find closed-form expressions for α and β in terms of μ and σ^2. As a first step, we replace μ and σ^2 by 113.5 and 1205.55, respectively, solve the first equation with respect to β, and replace β with that solution in the second equation. This results in

$$1205.55 = \left[\frac{113.5}{\Gamma\left(1 + \frac{1}{\alpha}\right)}\right]^2 \left\{\Gamma\left(1 + \frac{2}{\alpha}\right) - \left[\Gamma\left(1 + \frac{1}{\alpha}\right)\right]^2\right\}. \tag{6.3.1}$$

The second step is to solve this equation numerically. This can be done with the function *nleqslv* in the R package *nleqslv*, which should first be installed (*install.packages("nleqslv")*). Then use the following R commands:

```
fn=function(a){(mu/gamma(1+1/a))^2*(gamma(1+2/a)-gamma(1+1/a)
  ^2)-var}
```
this command defines fn as a function of a to be numerically solved

```
library(nleqslv); mu=113.5; var=1205.55
```
this command loads the package *nleqslv* to the current session and sets the values for $\overline{X}$ and S^2

```
nleqslv(13, fn); mu/gamma(1+1/3.634)
```
the first of these commands finds the solution to equation fn(a)=0 (which is $\widehat{\alpha}$) with starting value 13; the second computes $\widehat{\beta}$.

The resulting method of moments estimate for $\theta = (\alpha, \beta)$ is $\widehat{\theta} = (3.634, 125.892)$ (rounded to three decimal places). ■

6.3.2 THE METHOD OF MAXIMUM LIKELIHOOD

The method of maximum likelihood (ML) estimates the parameter θ of a model by addressing the question "what value of the parameter is most likely to have generated the data?" For discrete probability models, the answer to this question is obtained by maximizing, with respect to θ, the probability that a repetition of the experiment will result in the observed data. (A shorter way of saying this is that we "maximize the probability of observing the observed data.") The value of the parameter that maximizes this probability is the **maximum likelihood estimator** (MLE). A more complete description of the method of maximum likelihood is given after the next example, which illustrates the process.

Example 6.3-4

Car manufacturers often advertise damage results from low-impact crash experiments. In an experiment crashing $n = 20$ randomly selected cars of a certain type against a wall at 5 mph, $X = 12$ cars sustain no visible damage. Find the MLE of the probability, p, that a car of this type will sustain no visible damage in such a low-impact crash.

Solution

Intuitively, the value of the parameter p that is most likely to have generated 12 successes in 20 trials is the value that maximizes the probability for observing $X = 12$. Because X has the binomial($n = 20, p$) distribution, this probability is

$$P(X = 12|p) = \binom{20}{12} p^{12}(1-p)^8. \tag{6.3.2}$$

Note that the dependence of the probability on the parameter p is made explicit in the notation. To find the MLE it is more convenient to maximize

$$\log P(X = 12|p) = \log \binom{20}{12} + 12\log(p) + 8\log(1-p) \tag{6.3.3}$$

with respect to p. Note that, because logarithm is a monotone function, maximizing $\log P(X = 12|p)$ is equivalent to maximizing $P(X = 12|p)$. The value of p that maximizes (6.3.3) can be found by setting the first derivative with respect to p equal to zero and solving for p. Doing so yields $\widehat{p} = 12/20$ as the maximum likelihood estimate. In general, the MLE of the binomial probability p is the same as the empirical estimator of p, that is, $\widehat{p} = X/n$. ■

In general, let $x_1, \ldots, x_n$ denote the data, and $f(x|\theta)$ the probability model (PDF or PMF) to be fitted. (As in Example 6.3-4, the dependence of the PDF/PMF on θ is made explicit.) The **likelihood function** is the joint PDF/PMF of the random variables $X_1, \ldots, X_n$ evaluated at $x_1, \ldots, x_n$ and considered as a function of θ. Because the X_i are iid, their joint PDF/PMF is simply the product of their individual PDFs/PMFs:

Definition of the Likelihood Function

$$\text{lik}(\theta) = \prod_{i=1}^{n} f(x_i|\theta) \quad \textbf{(6.3.4)}$$

The value of θ that maximizes the likelihood function is the maximum likelihood estimator $\widehat{\theta}$. Typically, it is more convenient to maximize the logarithm of the likelihood function, which is called the **log-likelihood function**:

Definition of the Log-Likelihood Function

$$\mathcal{L}(\theta) = \sum_{i=1}^{n} \log(f(x_i|\theta)) \quad \textbf{(6.3.5)}$$

In the binomial case of Example 6.3-4, the likelihood function is simply the probability given in (6.3.2) and the log-likelihood function is given in (6.3.3). Two more examples follow.

Example 6.3-5

Let $x_1, \ldots, x_n$ be the waiting times for a random sample of n customers of a certain bank. Use the method of maximum likelihood to fit the exponential(λ) model to this data set.

Solution

The PDF of the exponential(λ) distribution is $f(x|\lambda) = \lambda e^{-\lambda x}$. Thus, the likelihood function is

$$\text{lik}(\lambda) = \lambda e^{-\lambda x_1} \cdots \lambda e^{-\lambda x_n} = \lambda^n e^{-\lambda \sum x_i},$$

and the log-likelihood function is

$$\mathcal{L}(\lambda) = n \log(\lambda) - \lambda \sum_{i=1}^{n} x_i.$$

Setting the first derivative of the log-likelihood function to zero yields the equation

$$\frac{\partial}{\partial \lambda}\left[n \log(\lambda) - \lambda \sum_{i=1}^{n} X_i \right] = \frac{n}{\lambda} - \sum_{i=1}^{n} X_i = 0.$$

Solving this equation with respect to λ yields the MLE $\widehat{\lambda} = 1/\overline{X}$ of λ. ■

The next example demonstrates that the MLE can be very different from the method of moments estimator. It is also an example of a discontinuous likelihood function, which therefore cannot be maximized by differentiation.

Example 6.3-6

(a) Let $X_1, \ldots, X_n$ be iid uniform$(0, \theta)$. Find the maximum likelihood estimator of θ.

(b) The waiting times for a random sample of $n = 10$ passengers of a New York commuter train are: 3.45, 8.63, 8.54, 2.59, 2.56, 4.44, 1.80, 2.80, 7.32, 6.97. Assuming that the waiting times have the uniform$(0, \theta)$ distribution, compute the MLE of θ and the model-based estimate of the population variance.

Solution

(a) Here $f(x|\theta) = 1/\theta$ if $0 < x < \theta$ and 0 otherwise. Thus the likelihood function is

$$\text{lik}(\theta) = \frac{1}{\theta^n} \text{ if } 0 < X_1, \ldots, X_n < \theta \text{ and 0 otherwise.}$$

This likelihood function is maximized by taking θ as small as possible. However, if θ gets smaller than the largest data value, $X_{(n)} = \max\{X_1, \ldots, X_n\}$, then the likelihood function becomes zero. Hence, the MLE is the smallest θ value for which the likelihood function is non-zero, that is, $\widehat{\theta} = X_{(n)}$.

(b) The largest among the given sample of waiting times is $X_{(n)} = 8.63$. Thus, according to the derivation in part (a), the MLE of θ is $\widehat{\theta} = 8.63$. Next, because the variance of the uniform$(0, \theta)$ distribution is $\sigma^2 = \theta^2/12$, the model-based estimate of the population variance is $\widehat{\sigma}^2 = 8.63^2/12 = 6.21$. ■

According to theoretical results, which are beyond the scope of this book, the method of maximum likelihood yields estimators that are optimal, at least when the sample size is large enough, under general regularity conditions. See Exercise 1 in Section 6.4, where the comparison of the methods of moments and ML for fitting the uniform$(0, \theta)$ model confirms the superiority of the MLE in this particular case. Moreover, a function of the MLE, $g(\widehat{\theta})$, is the MLE of $g(\theta)$ and thus its optimal estimator. For example, the estimator $\widehat{\sigma}^2 = X_{(n)}^2/12$, which is the estimator of σ^2 derived in Example 6.3-6, is a function of the MLE and thus it is the MLE, and optimal estimator of σ^2, at least when the sample size is large enough.

6.3.3 THE METHOD OF LEAST SQUARES

The method of *least squares* (LS), which is the most common method for fitting regression models, will be explained here in the context of fitting the simple linear regression model (4.6.4), that is,

$$\mu_{Y|X}(x) = E(Y|X = x) = \alpha_1 + \beta_1 x. \tag{6.3.6}$$

Let $(X_1, Y_1), \ldots, (X_n, Y_n)$ denote a simple random sample from a bivariate population (X, Y) satisfying the simple linear regression model. To explain the method of LS, consider the problem of deciding which of two lines provides a "better" fit to the data. As a first step, we must adopt a principle on whose basis we can judge the quality of a fit. The **principle of least squares** evaluates the quality of a line's fit by the sum of the squared *vertical distances* of each point (X_i, Y_i) from the line. The vertical distance of a point from a line is illustrated in Figure 6-1. Of the two lines in this figure, the line for which this sum of squared vertical distances is smaller is said to provide a better fit to the data.

The best-fitting line according to the principle of least squares is the line that achieves a sum of vertical squared distances smaller than any other line. The best-fitting line will be called the **fitted regression line**. The **least squares estimators** (LSEs) of the intercept and slope of the simple linear regression model (6.3.6) are simply the intercept and slope of the best-fitting line.

The problem of finding the best-fitting line has a surprisingly simple and closed form solution. Since the vertical distance of the point (X_i, Y_i) from a line $a + bx$ is $Y_i - (a + bX_i)$, the method of least squares finds the values $\widehat{\alpha}_1, \widehat{\beta}_1$ that minimize the objective function

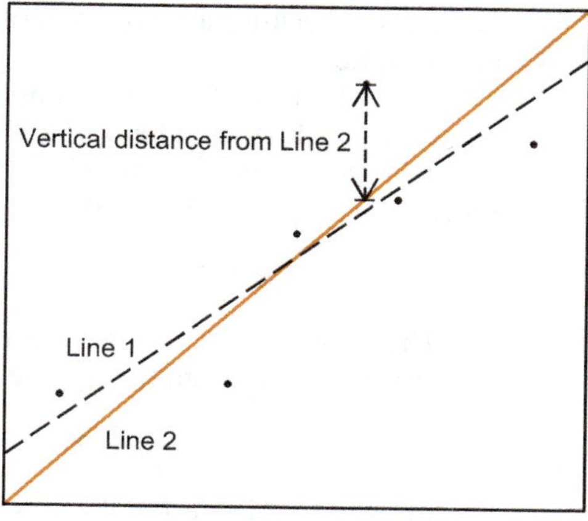

Figure 6-1 Two lines through a data set with an illustration of vertical distance.

$$L(a,b) = \sum_{i=1}^{n}(Y_i - a - bX_i)^2$$

with respect to a, b. This minimization can be carried out by setting the two first partial derivatives to zero. Omitting the details, the LSE of α_1 and β_1 are

Least Squares Estimators of the Slope and Intercept

$$\widehat{\beta}_1 = \frac{n\sum X_iY_i - (\sum X_i)(\sum Y_i)}{n\sum X_i^2 - (\sum X_i)^2}$$

$$\widehat{\alpha}_1 = \overline{Y} - \widehat{\beta}_1\overline{X} \qquad \textbf{(6.3.7)}$$

Thus, the fitted regression line is $\widehat{\mu}_{Y|X}(x) = \widehat{\alpha}_1 + \widehat{\beta}_1 x$. Evaluating the fitted regression line at the X-values of the data set gives the **fitted values**:

$$\widehat{Y}_i = \widehat{\alpha}_1 + \widehat{\beta}_1 X_i, \quad i = 1, \ldots, n.$$

REMARK 6.3-2 The expression $\widehat{\beta}_1 = S_{X,Y}/S_X^2$, where $S_{X,Y}$ is the sample covariance and S_X^2 is the sample variance of the X's, which was derived empirically in Remark 4.6-1, is algebraically equivalent to the expression in (6.3.7). Moreover, under the assumption of normality for the error variable, that is, for the normal simple linear regression model, the maximum likelihood estimators of the slope and intercept coincide with the LSEs. ◁

Example 6.3-7

The summary statistics from $n = 10$ measurements of X = stress applied and Y = time to failure are $\sum_{i=1}^{10} X_i = 200$, $\sum_{i=1}^{10} X_i^2 = 5412.5$, $\sum_{i=1}^{10} Y_i = 484$, and $\sum_{i=1}^{10} X_iY_i = 8407.5$. Find the best-fitting line to this data set.

Solution

According to (6.3.7), the best-fitting line has slope and intercept given by

$$\widehat{\beta}_1 = \frac{10 \times 8407.5 - 200 \times 484}{10 \times 5412.5 - 200^2} = -0.900885, \quad \widehat{\alpha}_1 = \frac{484}{10} - \widehat{\beta}_1\frac{200}{10} = 66.4177.$$

Example 6.3.7 highlights the fact that the best-fitting line can be obtained without having the actual data points. Indeed, as the formulas (6.3.7) suggest, all one needs are the summary statistics as given in Example 6.3-7. This practice, however, should be avoided because a scatterplot of the data may reveal that the linear model is not appropriate.

A third parameter of the simple linear regression model is the conditional variance, σ_ε^2, of Y given the value of X. Recall that σ_ε^2 is also the variance of the intrinsic error variable ε, which appears in the expression (4.6.6) of the simple linear regression model, and rewritten here for convenience:

$$Y = \alpha_1 + \beta_1 X + \varepsilon. \tag{6.3.8}$$

The idea for estimating σ_ε^2 is that, if the true values of α_1 and β_1 were known, then σ_ε^2 would be estimated by the sample variance of

$$\varepsilon_i = Y_i - \alpha_1 - \beta_1 X_i, \quad i = 1, \ldots, n.$$

Of course, α_1 and β_1 are not known and so the intrinsic error variables, ε_i, cannot be computed. But since α_1 and β_1 can be estimated, so can the ε_i:

$$\widehat{\varepsilon}_1 = Y_1 - \widehat{\alpha}_1 - \widehat{\beta}_1 X_1, \ldots, \widehat{\varepsilon}_n = Y_n - \widehat{\alpha}_1 - \widehat{\beta}_1 X_n. \tag{6.3.9}$$

The estimated intrinsic error variables, $\widehat{\varepsilon}_i$, are called **residuals**. The residuals are also expressed in terms of the fitted values as

$$\widehat{\varepsilon}_i = Y_i - \widehat{Y}_i, \quad i = 1, \ldots, n.$$

The residuals and the fitted values are illustrated in Figure 6-2.

Because the computation of residuals requires that two parameters be estimated, which, in statistical jargon, entails the *loss of two degrees of freedom*, we do not use their exact sample variance for estimating σ_ε^2. Instead we use:

Least Squares Estimation of the Intrinsic Error Variance

$$S_\varepsilon^2 = \frac{1}{n-2} \sum_{i=1}^{n} \widehat{\varepsilon}_i^2 \tag{6.3.10}$$

Due to an algebraic identity, which is not derived here, the residuals sum to zero, that is,

$$\sum_{i=1}^{n} \widehat{\varepsilon}_i = \sum_{i=1}^{n} (Y_i - \widehat{Y}_i) = 0.$$

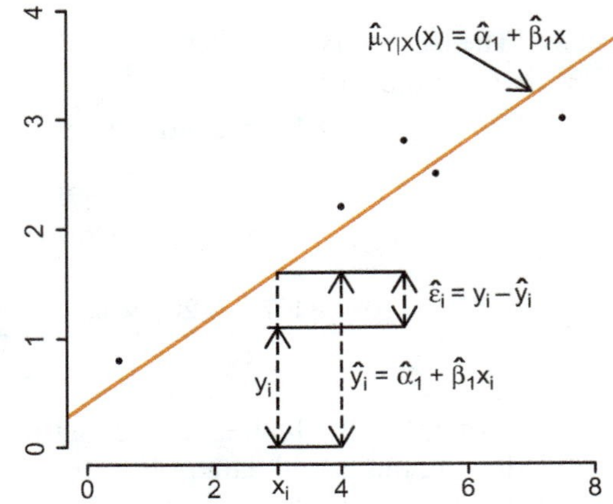

Figure 6-2 Illustration of the fitted regression line, fitted values, and residuals.

Thus, the formula for S_ε^2 in (6.3.10) differs from the sample variance only in that it divides by $n - 2$.

The quantity $\sum_{i=1}^n \widehat{\varepsilon}_i^2$ in (6.3.10) is called the **error sum of squares** and is denoted by **SSE**. Because of the frequent use of this quantity in the chapters that follow, its computational formula is also given here:

$$\text{SSE} = \sum_{i=1}^{n} \widehat{\varepsilon}_i^2 = \sum_{i=1}^{n} Y_i^2 - \widehat{\alpha}_1 \sum_{i=1}^{n} Y_i - \widehat{\beta}_1 \sum_{i=1}^{n} X_i Y_i. \quad \textbf{(6.3.11)}$$

Example 6.3-8

Consider the following data on Y = propagation velocity of an ultrasonic stress wave through a substance and X = tensile strength of substance.

x	12	30	36	40	45	57	62	67	71	78	93	94	100	105
y	3.3	3.2	3.4	3.0	2.8	2.9	2.7	2.6	2.5	2.6	2.2	2.0	2.3	2.1

(a) Use the method of LS to fit the simple linear regression model to this data.

(b) Obtain the error sum of squares and the LSE of the intrinsic error variance.

(c) Compute the fitted value and residual at $X_3 = 36$.

Solution

(a) The scatterplot of the $n = 14$ data points, shown in Figure 6-3, suggests that the assumptions of the simple linear regression model, which are linearity of the regression function and homoscedasticity (i.e., $\text{Var}(Y|X = x)$ is the same for all x), appear to be satisfied for this data set. The summary statistics needed for the LS estimators are

$$\sum_{i=1}^{14} X_i = 890, \quad \sum_{i=1}^{14} Y_i = 37.6, \quad \sum_{i=1}^{14} X_i Y_i = 2234.30,$$

$$\sum_{i=1}^{14} X_i^2 = 67,182, \quad \sum_{i=1}^{14} Y_i^2 = 103.54.$$

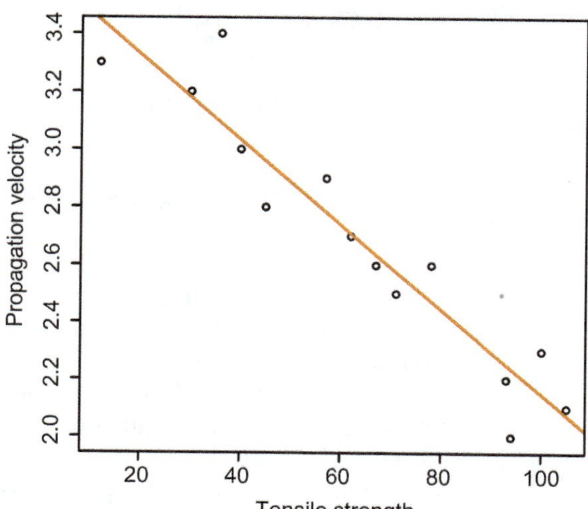

Figure 6-3 Scatterplot for the data of Example 6.3-8.

Plugging these values into formula (6.3.7) we get

$$\widehat{\beta}_1 = \frac{14 \times 2234.3 - 890 \times 37.6}{14 \times 67182 - 890^2} = -0.014711, \quad \widehat{\alpha}_1 = \frac{37.6}{14} - \widehat{\beta}_1 \frac{890}{14} = 3.62091.$$

(b) Using the calculations in part (a) and formula (6.3.11) we have

$$\text{SSE} = 103.54 - \widehat{\alpha}_1(37.6) - \widehat{\beta}_1(2234.30) = 0.26245,$$

so that the LSE of the intrinsic error variance is

$$S_\varepsilon^2 = \frac{1}{n-2}\text{SSE} = \frac{0.26245}{12} = 0.02187.$$

(c) The fitted value and residual at $X_3 = 36$ are

$$\widehat{Y}_3 = \widehat{\alpha}_1 + \widehat{\beta}_1 36 = 3.0913 \quad \text{and} \quad \widehat{\varepsilon}_3 = Y_3 - \widehat{Y}_3 = 3.4 - 3.0913 = 0.3087.$$

The fitted regression line can be used for estimating the expected response at a given value of X, provided the given value of X is within the range of the X-values of the data set. For example, with the data set given in Example 6.3-8, the expected response at $X = 65$, that is, $E(Y|X = 65)$, can be estimated by

$$\widehat{\mu}_{Y|X}(65) = \widehat{\alpha}_1 + \widehat{\beta}_1 65 = 2.6647.$$

On the other hand, it is not appropriate to use the fitted regression line for estimating the expected response at $X = 120$ because the largest X-value is 105. The main reason why it is not appropriate to extrapolate beyond the range of the X-values is that we have no indication that the linear model continues to hold. For example, even though Figure 6-3 suggests that the simple linear regression model is reasonable for this data set, there is no guarantee that the linearity continues to hold for X-values larger than 105 or smaller than 12.

With the X- and Y-values in the R objects x and y, respectively, the R commands for obtaining the LS estimates and other related quantities are as follows:

R Commands for the LS Estimates in Simple Linear Regression

```
lm(y ~ x)$coef # gives α̂1 and β̂1
lm(y ~ x)$fitted # gives the fitted values        (6.3.12)
lm(y ~ x)$resid # gives the residuals
```

Instead of repeating the *lm(y ~ x)* command, it is possible to set all output of this command in the R object *out*, by *out=lm(y ~ x)*, and then use *out$coef*, *out$fitted*, and *out$resid*. It is also possible to obtain specific fitted values or residuals. For example, *out$fitted[3]* and *out$resid[3]* give the third fitted value and residual, respectively, which were calculated in part (c) of Example 6.3-8.

Having issued the command *out=lm(y ~ x)*, the error sum of squares, SSE, and the estimator of the intrinsic error variance, S_ε^2, can be obtained by the following R commands:

```
sum(out$resid**2)
sum(out$resid**2)/out$df.resid
```
(6.3.13)

respectively, where *out$df.resid* gives the value of $n - 2$. Finally, the scatterplot with the fitted regression line shown in Figure 6-3, was generated with the R commands below:

```
plot(x, y, xlab = "Tensile Strength", ylab = "Propagation Velocity");
  abline(out, col = "red")
```

Example 6.3-9

Use R commands and the $n = 153$ measurements (taken in New York from May to September 1973) on solar radiation (lang) and ozone level (ppb) from the R data set *airquality* to complete the following parts.

(a) Use the method of LS to fit the simple linear regression model to this data set.

(b) Construct a scatterplot of the data and comment on whether or not the model assumptions seem to be violated. Comment on the impact of any violations of the model assumptions on the estimators obtained in part (a).

(c) Transform the data by taking the logarithm of both variables, and construct a scatterplot of the transformed data. Comment on whether or not the assumptions of the simple linear regression model appear tenable for the transformed data.

Solution

(a) We first copy the solar radiation and ozone data into the R objects x and y by *x=airquality$Solar.R; y=airquality$Ozone*. The command in the first line of (6.3.12) gives the LSE of the intercept and slope as

$$\widehat{\alpha}_1 = 18.599 \quad \text{and} \quad \widehat{\beta}_1 = 0.127,$$

while the second command in (6.3.13) gives the LSE for σ_ε^2 as $S_\varepsilon^2 = 981.855$.

(b) The scatterplot with the fitted regression line, shown in the left panel of Figure 6-4, suggests that the ozone level increases with the solar radiation level at approximately linear fashion. Thus, the assumption of linearity of the regression function $\mu_{Y|X}(x)$ seems to be, at least approximately, satisfied. On the other hand, the variability in the ozone measurements seems

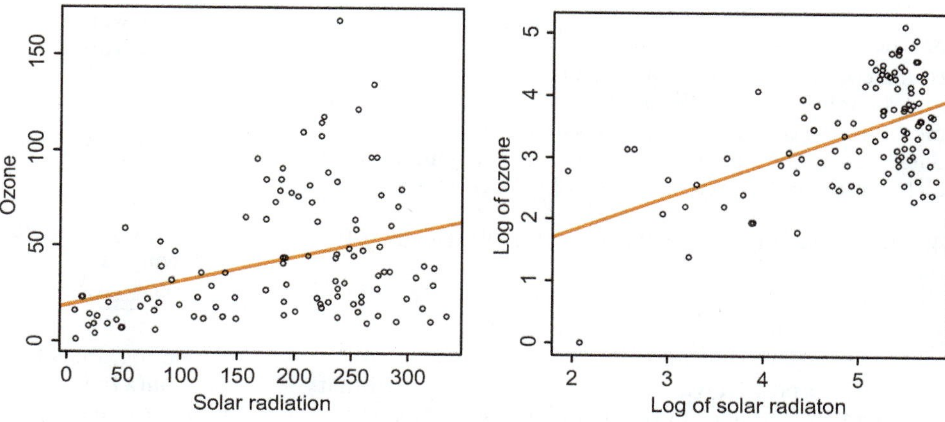

Figure 6-4 Scatterplot for the solar radiation and ozone data in the original scale (left panel) and log scale (right panel).

to increase with the solar radiation level. Thus, the homoscedasticity assumption appears to be violated for this data set. As a consequence, the estimator $S_\varepsilon^2 = 981.855$ does not make sense for this data set. This is because σ_ε^2 is defined to be the conditional variance $\text{Var}(Y|X = x)$ which, in a homoscedastic context, is the same for all x. If homoscedasticity does not hold (so the data are **heteroscedastic**), $\text{Var}(Y|X = x)$ changes with x, as it does for this data set. Hence, the parameter σ_ε^2 is not defined for such data sets.

(c) The scatterplot with the fitted regression line shown in the right panel of Figure 6-4 suggests that the assumptions of linearity and homoscedasticity are satisfied, at least approximately, for the log-transformed data. ■

Exercises

1. Let $X_1, \ldots, X_n$ be iid exponential(λ). Find the method of moments estimator of λ. Is it unbiased?

2. Use *t=read.table("RobotReactTime.txt", header=T); t1=t$Time[t$Robot==1]* to import the data on robot reaction times to simulated malfunctions, and copy the reaction times of Robot 1 into the R object *t1*.

(a) Follow the approach of Example 6.3-3 to fit the Weibull(α, β) model to the data in *t1*. (*Hint.* Use *mean(t1); var(t1)* to compute the sample mean and variance.)

(b) Use Example 6.3-5 to fit the exponential(λ) model to the data in *t1*.

(c) Use each of the fitted models in parts (a) and (b) to give model-based estimates of the 80th population percentile of reaction times, as well as the probability $P(28.15 \leq X \leq 29.75)$, where X denotes the next response time of Robot 1. (*Hint.* See Example 6.2-5.)

(d) Give empirical estimates for the 80th population percentile and the probability $P(28.15 \leq X \leq 29.75)$. (*Hint.* The R command for finding percentiles is given in Section 1.7. The R command *sum(t1>=28.15&t1<=29.75)* counts the number of reaction times that are between 28.15 and 29.75. Finally, *length(t1)* gives the number of observations in *t1*.)

3. The life spans, in hours, of a random sample of 25 electronic components yield sample mean $\overline{X} = 113.5$ hours and sample variance $S^2 = 1205.55$ hours^2. Using the method of moments, fit the gamma(α, β) model to this data. (*Hint.* The mean and variance of the gamma(α, β) distribution are given in Exercise 13 in Section 3.5.)

4. The probability density function of the Rayleigh distribution is

$$f(x) = \frac{x}{\theta^2} e^{-x^2/(2\theta^2)}, \quad x \geq 0,$$

where θ is a positive-valued parameter. It is known that the mean and variance of the Rayleigh distribution are

$$\mu = \theta\sqrt{\frac{\pi}{2}} \quad \text{and} \quad \sigma^2 = \theta^2 \frac{4-\pi}{2}.$$

Let $X_1, \ldots, X_n$ be a random sample from a Rayleigh distribution.

(a) Construct the method of moments estimator of θ. Is it unbiased?

(b) Construct a model-based estimator of the population variance. Is it unbiased?

5. Answer the following questions.

(a) Let $X \sim \text{Bin}(n, p)$. Find the method of moments estimator of p. Is it unbiased?

(b) To determine the probability p that a certain component lasts more than 350 hours in operation, a random sample of 37 components was tested. Of these, 24 lasted more than 350 hours. Compute the method of moments estimate of p.

(c) A system consists of two such components connected in series. Assume the components fail independently of each other. Give the method of moments estimator of the probability the system lasts more than 350 hours in operation, and compute it using the information given in part (b). (*Hint.* Justify that the probability the system lasts more than 350 hours is p^2.)

(d) Is the estimator derived in part (c) unbiased? Justify your answer. (*Hint.* For any random variable X, $E(X^2) = \text{Var}(X) + [E(X)]^2$.)

6. Answer the following questions.

(a) Let $X_1, \ldots, X_n$ be iid Poisson(λ). Find the maximum likelihood estimator of λ.

(b) The numbers of surface imperfections for a random sample of 50 metal plates are summarized in the following table:

Number of Scratches per Item	0	1	2	3	4
Frequency	4	12	11	14	9

Assuming that the imperfection counts have the Poisson(λ) distribution, compute the maximum likelihood estimate of λ.

(c) Give the model-based estimate of the population variance, and compare it with the sample variance. Assuming the Poisson model correctly describes the population distribution, which of the two estimates would you prefer and why?

7. A company manufacturing bike helmets wants to estimate the proportion p of helmets with a certain type of flaw. They decide to keep inspecting helmets until they find $r = 5$ flawed ones. Let X denote the number of helmets that were not flawed among those examined.

(a) Write the log-likelihood function and find the MLE of p.

(b) Find the method of moments estimator of p.

(c) If $X = 47$, give a numerical value to your estimators in (a) and (b).

8. Let $X_1, \ldots, X_n$ be a random sample from the uniform(0, θ) distribution. Use the R command *set.seed(3333); x=runif(20, 0, 10)* to generate a random sample $X_1, \ldots, X_{20}$ from the uniform(0, 10) distribution, and store it into the R object x.

(a) Give the method of moments estimate of θ and the model-based estimate of the population variance σ^2. (*Hint.* See Example 6.3-1; use *mean(x)* to compute the sample mean.)

(b) Use *var(x)* to compute the sample variance S^2. Comment on which of the two estimates (S^2 or the model-based) provides a better approximation to the true value of the population variance σ^2, which is $10^2/12$, for this data set.

9. Plumbing suppliers typically ship packages of plumbing supplies containing many different combinations of item such as pipes, sealants, and drains. Almost invariably a shipment contains one or more incorrectly filled items: a part may be defective, missing, not the type ordered, etc. In this context, the random variable of interest is the proportion P of incorrectly filled items. A family of distributions for modeling the distribution of proportions has PDF

$$f(p|\theta) = \theta p^{\theta-1}, \ 0 < p < 1, \ \theta > 0.$$

It is given that the expected value of a random variable P having this distribution is $E(P) = \theta/(1+\theta)$.

(a) If $P_1, \ldots, P_n$ is a random sample of proportions of incorrectly filled items, find the moments estimator of θ.

(b) In a sample of $n = 5$ shipments, the proportions of incorrectly filled items are 0.05, 0.31, 0.17, 0.23, and 0.08. Give the method of moments estimate of θ.

10. A study was conducted to examine the effects of $NaPO_4$, measured in parts per million (ppm), on the corrosion rate of iron.[3] The summary statistics corresponding to 11 data points, where the $NaPO_4$ concentrations ranged from 2.50 ppm to 55.00 ppm, are as follows: $\sum_{i=1}^n x_i = 263.53$, $\sum_{i=1}^n y_i = 36.66$, $\sum_{i=1}^n x_i y_i = 400.5225$, $\sum_{i=1}^n x_i^2 = 9677.4709$, and $\sum_{i=1}^n y_i^2 = 209.7642$.

(a) Find the estimated regression line.

(b) The engineer in charge of the study wants to estimate the expected corrosion rate for $NaPO_4$ concentrations of 4.5, 34.7, and 62.8 ppm, using a fitted regression line. For each of these concentrations, comment on whether or not it is appropriate to use the fitted regression line. Report the estimates at the concentrations for which the use of the fitted regression line is appropriate.

11. Manatees are large, gentle sea creatures that live along the Florida coast. Many manatees are killed or injured by powerboats. Below are data on powerboat registrations (in thousands) and the number of manatees killed by boats in Florida for four different years between 2001 and 2004:

Number of Boats (thousands)	498	526	559	614
Number of Manatee Deaths	16	25	34	39

Assume the relationship between the number of boats and the number of manatee deaths is linear, and complete the following parts using hand calculations.

(a) Find the estimated regression line and estimate the expected number of manatee deaths in a year with 550 (thousand) powerboat registrations.

(b) Use (6.3.11) to compute the error sum of squares, and give the estimate of the intrinsic error variance.

(c) Compute the four fitted values and the corresponding residuals. Verify that the sum of squared residuals equals the SSE obtained in part (b).

12. Use *sm=read.table("StrengthMoE.txt", header=T)* to read the data on cement's modulus of elasticity and strength into the R data frame *sm*. Copy the data into the R objects x and y by *x=sm$MoE* and *y=sm$Strength*, and use R commands to complete the following.

(a) Construct a scatterplot of the data with the fitted regression line drawn through it. Do the assumptions of the simple linear regression model, which are linearity of the regression function and homoscedasticity, appear to hold for this data set?

(b) Give the LSE for the regression coefficients, and estimate the expected strength at modulus of elasticity $X = 60$.

(c) Give the error sum of squares and the estimator of the intrinsic error variance.

[3] *Sodium Phosphate Hideout Mechanisms: Data and Models for the Solubility and RedoxBehavior of Iron(II) and Iron(III) Sodium-Phosphate Hideout Reaction Products*, EPRI, Palo Alto, CA: 1999. TR-112137.

13. Use *da=read.table("TreeAgeDiamSugarMaple.txt", header=T)* to read the data on the diameter (in millimeters) and age (in years) of $n = 27$ sugar maple trees into the R data frame *da*. Copy the data into the R objects x and y by *x=da$Diamet; y=da$Age*, and use R commands to complete the following.

(a) Use the method of LS to fit the simple linear regression model to this data set.

(b) Construct a scatterplot of the data and comment on whether or not the model assumptions, which are linearity of the regression function of Y on X and homoscedasticity, seem to be violated. Comment on the impact of any violations of the model assumptions on the estimators obtained in part (a).

(c) Transform the data by taking the logarithm (that is, by *x1=log(x); y1=log(y)*), and construct a scatterplot of the transformed data. Comment on whether or not the assumptions of the simple linear regression model seem to be violated for the log-transformed data.

6.4 Comparing Estimators: The MSE Criterion

Given two estimators, $\widehat{\theta}_1, \widehat{\theta}_2$, of the same parameter θ, the estimator of choice is the one that achieves the smaller estimation error. However, the estimation errors, $\widehat{\theta}_1 - \theta$ and $\widehat{\theta}_2 - \theta$, cannot be computed because θ is unknown. Even if θ were known, the estimators are random variables and thus it is possible that for some samples $\widehat{\theta}_1 - \theta$ will be smaller than $\widehat{\theta}_2 - \theta$, while for other samples the opposite will be true. It is thus sensible to look at some type of "average error." Here we will consider the average of the squared error as a criterion for selecting between two estimators.

Definition 6.4-1

The **mean square error** (MSE) of an estimator $\widehat{\theta}$ for the parameter θ is defined to be

$$\text{MSE}\left(\widehat{\theta}\right) = E_\theta\left(\widehat{\theta} - \theta\right)^2$$

The **MSE selection criterion** says that among two estimators, the one with the smaller MSE is preferred.

Again, the correct notation is $\text{MSE}_\theta(\widehat{\theta})$ but we will use $\text{MSE}(\widehat{\theta})$ for simplicity. The following proposition relates the mean square error of an estimator to its variance and bias.

Proposition 6.4-1 If $\widehat{\theta}$ is unbiased for θ then

$$\text{MSE}\left(\widehat{\theta}\right) = \sigma^2_{\widehat{\theta}}.$$

In general,

$$\text{MSE}\left(\widehat{\theta}\right) = \sigma^2_{\widehat{\theta}} + \left[\text{bias}\left(\widehat{\theta}\right)\right]^2.$$

■

For unbiased estimators, the MSE equals the variance. Thus, among two unbiased estimators, the MSE selection criterion selects the one with the smaller standard error. The rationale for this is illustrated in Figure 6-5, which shows the PDFs of two unbiased estimators. Because $\widehat{\theta}_1$ has a smaller standard error than $\widehat{\theta}_2$, the PDF of $\widehat{\theta}_1$ is more concentrated about the true value of θ. Hence, $|\widehat{\theta}_1 - \theta|$ is less likely to take a larger value than $|\widehat{\theta}_2 - \theta|$. For estimators that are not unbiased, however, the MSE selection criterion incorporates both the standard error and the bias for their comparison.

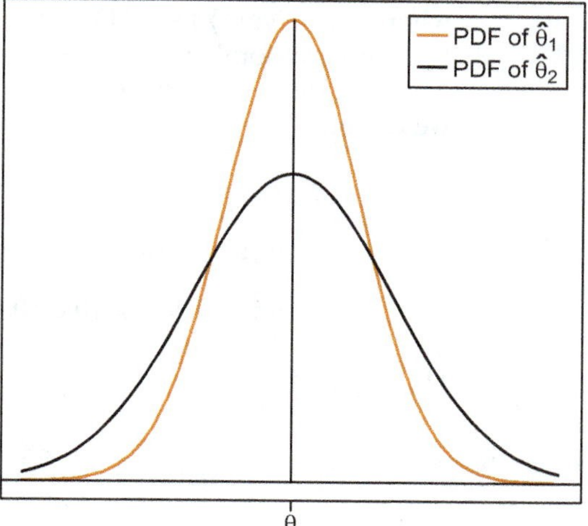

Figure 6-5 PDFs of two unbiased estimators of θ.

Example 6.4-1

Simple random vs stratified sampling. Facilities A and B account for 60% and 40%, respectively, of the production of a certain electronic component. The components from the two facilities are shipped to a packaging location where they are mixed and packaged. A sample of size 100 will be used to estimate the expected life time in the combined population. Use the MSE criterion to decide which of the following two sampling schemes should be adopted: (a) simple random sampling at the packaging location, and (b) stratified random sampling based on a simple random sample of size 60 from facility A and a simple random sample of size 40 from facility B.

Solution

Let Y denote the life time of a randomly selected component from the packaging location (i.e., the combined population), and let X take the value 1 or 2, depending on whether the component was produced in facility A or B, respectively. We are interested in estimating $\mu = E(Y)$. An application of the Law of Total Expectation yields

$$\mu = 0.6E(Y|X = 1) + 0.4E(Y|X = 2) = 0.6\mu_A + 0.4\mu_B, \tag{6.4.1}$$

where μ_A and μ_B are defined implicitly in (6.4.1).

Let $Y_1, \ldots, Y_{100}$, $Y_{A1}, \ldots, Y_{A60}$, $Y_{B1}, \ldots, Y_{B40}$, be simple random samples from the packaging location, facility A, and facility B, respectively, and let $\overline{Y}$, $\overline{Y}_A$, $\overline{Y}_B$ denote the corresponding sample means. Because of (6.4.1), μ can be estimated either as

$$\overline{Y} \quad \text{or as} \quad \overline{Y}_{Str} = 0.6\overline{Y}_A + 0.4\overline{Y}_B.$$

Note that both $\overline{Y}$ and $\overline{Y}_{Str}$ are based on samples of size 100, but the former uses simple random sampling while the latter uses stratified sampling. Because each is an unbiased estimator of μ, its MSE equals its variance. Thus, since $\overline{Y}_A$ and $\overline{Y}_B$ are independent,

$$\begin{aligned} \text{MSE}(\overline{Y}_{Str}) &= 0.6^2\frac{\sigma_A^2}{60} + 0.4^2\frac{\sigma_B^2}{40} \\ &= 0.6\frac{\sigma_A^2}{100} + 0.4\frac{\sigma_B^2}{100}, \end{aligned} \tag{6.4.2}$$

where $\sigma_A^2 = \text{Var}(Y|X = 1)$ and $\sigma_B^2 = \text{Var}(Y|X = 2)$ denote the variance of the life time of a randomly chosen component from facility A and B, respectively.

Next, let $\sigma^2 = \text{Var}(Y)$. Using the Law of Total Expectation and some algebra, we obtain

$$\begin{aligned}\sigma^2 &= E(Y^2) - \mu^2 = 0.6E(Y^2|X = 1) + 0.4E(Y^2|X = 2) - (0.6\mu_A + 0.4\mu_B)^2 \\ &= 0.6\sigma_A^2 + 0.4\sigma_B^2 + 0.6\mu_A^2 + 0.4\mu_B^2 - (0.6\mu_A + 0.4\mu_B)^2 \\ &= 0.6\sigma_A^2 + 0.4\sigma_B^2 + (0.6)(0.4)(\mu_A - \mu_B)^2.\end{aligned}$$

Hence, $\text{MSE}(\overline{Y}) = \text{Var}(\overline{Y}) = \sigma^2/100$ or, in view of the above calculation,

$$\text{MSE}(\overline{Y}) = 0.6\frac{\sigma_A^2}{100} + 0.4\frac{\sigma_B^2}{100} + 0.6 \times 0.4\frac{(\mu_A - \mu_B)^2}{100}. \tag{6.4.3}$$

Comparing (6.4.3) and (6.4.2), it follows that $\text{MSE}(\overline{Y}) \geq \text{MSE}(\overline{Y}_{Str})$, with equality only if $\mu_A = \mu_B$. Hence, stratified sampling is preferable for estimating μ. ■

Exercises

1. Let $X_1, \ldots, X_n$ be a random sample from the uniform(0,θ) distribution, and let $\widehat{\theta}_1 = 2\overline{X}$ and $\widehat{\theta}_2 = X_{(n)}$, that is, the largest order statistic, be estimators for θ. It is given that the mean and variance of $\widehat{\theta}_2$ are

$$E_\theta(\widehat{\theta}_2) = \frac{n}{n+1}\theta \quad \text{and} \quad \text{Var}_\theta(\widehat{\theta}_2) = \frac{n}{(n+1)^2(n+2)}\theta^2.$$

(a) Give an expression for the bias of each of the two estimators. Are they unbiased?

(b) Give an expression for the MSE of each of the two estimators.

(c) Compute the MSE of each of the two estimators for $n = 5$ and true value of θ equal to 10. Which estimator is preferable according to the MSE selection criterion?

2. Let $X_1, \ldots, X_{10}$ be a random sample from a population with mean μ and variance σ^2, and $Y_1, \ldots, Y_{10}$ be a random sample from another population with mean also equal to μ and variance $4\sigma^2$. The two samples are independent.

(a) Show that for any α, $0 \leq \alpha \leq 1$, $\widehat{\mu} = \alpha\overline{X} + (1 - \alpha)\overline{Y}$ is unbiased for μ.

(b) Obtain an expression for the MSE of $\widehat{\mu}$.

(c) Is the estimator $\overline{X}$ preferable over the estimator $0.5\overline{X} + 0.5\overline{Y}$? Justify your answer.

Chapter 7

Confidence and Prediction Intervals

7.1 Introduction to Confidence Intervals

Due to sampling variability (see Section 1.2), point estimators such as the sample mean, median, variance, proportion, and regression parameters approximate corresponding population parameters but, in general, are different from them. This was repeatedly emphasized in Chapter 1 for the point estimators introduced there. Now we can make more precise statements. For example, the Law of Large Numbers (Section 5.2) asserts that $\overline{X}$ approximates the true population mean more accurately as the sample size increases, while the CLT (Section 5.4) allows one to assess the probability that $\overline{X}$ will be within a certain distance from the population mean. This suggests that the practice of reporting only the value of $\overline{X}$ as a point estimate of μ is not as informative as it can be because (a) it does not quantify how accurate the estimator is, and (b) it does not connect the level of accuracy to the sample size. The same comment applies to all point estimators discussed in Chapter 6.

Confidence intervals have been devised to address the lack of information that is inherent in the practice of reporting only a point estimate. A confidence interval is an interval for which we can assert, with a given degree of confidence/certainty, that it includes the true value of the parameter being estimated. The construction of confidence intervals, which is outlined below, relies on the exact distribution of the point estimator, or an approximation to it provided by the Central Limit Theorem. *Prediction intervals*, which are also discussed in this chapter, are similar in spirit to confidence intervals but pertain to future observations as opposed to population parameters.

7.1.1 CONSTRUCTION OF CONFIDENCE INTERVALS

By virtue of the Central Limit Theorem, if the sample size n is large enough, many estimators, $\widehat{\theta}$, are approximately normally distributed with mean equal, or approximately equal, to the true value of the parameter, θ, that is being estimated. Moreover, by virtue of the Law of Large Numbers, the estimated standard error, $S_{\widehat{\theta}}$, is a good approximation to the standard error $\sigma_{\widehat{\theta}}$. Taken together, these facts imply that, if n is large enough,

$$\frac{\widehat{\theta}-\theta}{S_{\widehat{\theta}}} \mathrel{\dot{\sim}} N(0,1). \tag{7.1.1}$$

For example, this is the case for the empirical estimators discussed in Chapters 1 and 4, as well as the moment estimators, least squares estimator, and most maximum likelihood estimators discussed in Chapter 6. In view of the 68-95-99.7% rule of the normal distribution, (7.1.1) implies that the estimation error $|\widehat{\theta} - \theta|$ is less than $2S_{\hat{\theta}}$ approximately 95% of the time. More precisely, an approximate 95% **error bound** in the estimation of θ is

$$|\widehat{\theta} - \theta| \leq 1.96 S_{\hat{\theta}}, \tag{7.1.2}$$

where $1.96 = z_{0.025}$. After some simple algebra, (7.1.2) can be written as

$$\widehat{\theta} - 1.96 S_{\hat{\theta}} \leq \theta \leq \widehat{\theta} + 1.96 S_{\hat{\theta}} \tag{7.1.3}$$

which gives an interval of plausible values for the true value of θ, with degree of plausibility, or **confidence level**, approximately 95%. Such an interval is called a **confidence interval (CI)**. More generally, a $(1 - \alpha)100\%$ error bound,

$$|\widehat{\theta} - \theta| \leq z_{\alpha/2} S_{\hat{\theta}}, \tag{7.1.4}$$

can be written as a CI with confidence level $(1 - \alpha)100\%$, also called $(1 - \alpha)$ 100% CI:

$$\widehat{\theta} - z_{\alpha/2} S_{\hat{\theta}} \leq \theta \leq \widehat{\theta} + z_{\alpha/2} S_{\hat{\theta}}. \tag{7.1.5}$$

Typical values for α are 0.1, 0.05, and 0.01. They correspond to 90%, 95%, and 99% CIs.

7.1.2 Z CONFIDENCE INTERVALS

Confidence intervals that use percentiles from the standard normal distribution, like that in relation (7.1.5), are called ***Z* CIs**, or ***Z* intervals.**

Z intervals for the mean (so $\theta = \mu$) are used only if the population variance is known and either the population is normal or the sample size is at least 30. Because the assumption of a known population variance is not realistic, ***Z*** intervals for the mean are deemphasized in this book. Instead, *Z* intervals will be primarily used for the proportion ($\theta = p$).

7.1.3 THE *T* DISTRIBUTION AND *T* CONFIDENCE INTERVALS

When sampling from normal populations, an estimator $\widehat{\theta}$ of some parameter θ often satisfies, for all sample sizes n,

$$\frac{\widehat{\theta} - \theta}{S_{\widehat{\theta}}} \sim T_\nu, \tag{7.1.6}$$

where $S_{\widehat{\theta}}$ is the estimated standard error of $\widehat{\theta}$, and T_ν stands for the ***T* distribution** with ν **degrees of freedom.** The degrees of freedom, ν, depend on the sample size and the particular estimator $\widehat{\theta}$, and will be given separately for each estimator we consider. The $100(1-\alpha)$-th percentile of the T distribution with ν degrees of freedom will be denoted by $t_{\nu,\alpha}$; see Figure 7-1. The form of the T density will not be used and thus it is not given. The R commands for the T_ν PDF, CDF, and percentiles as well as for simulating random samples are as follows (as usual, both x and s in these commands can be vectors):

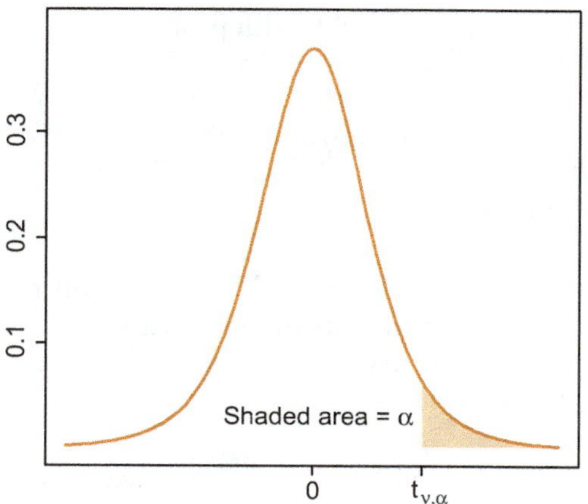

Figure 7-1 Density and percentile of the T_ν distribution.

R Commands for the T_ν Distribution

```
dt(x, ν) # gives the PDF of Tν for x in (−∞, ∞)
pt(x, ν) # gives the CDF of Tν for x in (−∞, ∞)
qt(s, ν) # gives the s100th percentile of Tν for s in (0, 1)
rt(n, ν) # gives a sample of n Tν random variables
```

A T distribution is symmetric and its PDF tends to that of the standard normal as ν tends to infinity. As a consequence, the percentiles $t_{\nu,\alpha}$ approach z_α as ν gets large. For example, the 95th percentile of the T distributions with $\nu = 9, 19, 60$, and 120, are 1.833, 1.729, 1.671, and 1.658, respectively, while $z_{0.05} = 1.645$. The convergence of the T_ν density to that of the standard normal is illustrated in Figure 7-2.

Relation (7.1.6), which also holds approximately when sampling non-normal populations provided $n \geq 30$, leads to the following bound on the error of estimation of θ:

$$|\widehat{\theta} - \theta| \leq t_{\nu,\alpha/2} S_{\widehat{\theta}}, \tag{7.1.7}$$

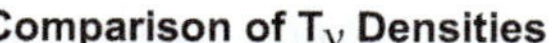

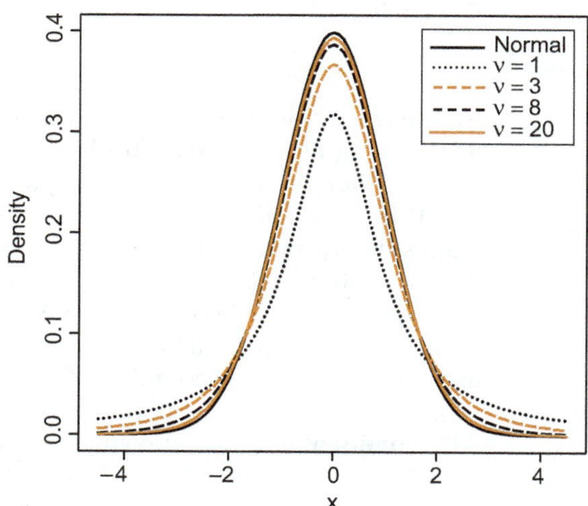

Figure 7-2 T_ν densities converging to the N(0, 1) density as ν gets large.

which holds with probability $1-\alpha$. We note that the probability for this error bound is exact, for all sample sizes n, if the normality assumption holds, whereas it holds with probability approximately $1-\alpha$, provided $n \geq 30$ (according to the rule of thumb introduced in Remark 5.4-1), if the normality assumption does not hold. The error bound in (7.1.7) leads to the following $(1-\alpha)100\%$ CI for θ:

$$(\widehat{\theta} - t_{\nu,\alpha/2}S_{\widehat{\theta}},\ \widehat{\theta} + t_{\nu,\alpha/2}S_{\widehat{\theta}}) \tag{7.1.8}$$

In this book, T intervals will be used for the mean as well as for the regression parameters in the linear regression model. Using short-hand notation, the CI (7.1.8) can also be written as

$$\widehat{\theta} \pm t_{\nu,\alpha/2}S_{\widehat{\theta}}.$$

REMARK 7.1-1 A consequence of the fact that $t_{\nu,\alpha/2}$ approaches $z_{\alpha/2}$ for large ν is that the Z CIs (7.1.5) and the T CIs in (7.1.8) will be almost identical for large sample sizes. ◁

7.1.4 OUTLINE OF THE CHAPTER

In the next section we will discuss the interpretation of CIs. Section 7.3 presents confidence intervals (either Z or T intervals) for population means and proportions, for the regression parameters in the linear regression model, and for the median and other percentiles. A χ^2-type CI for the variance, presented in Section 7.3.5, is valid only under the normality assumption. The aforementioned Z and T intervals are valid also in the non-normal case, provided the sample size is large enough, by virtue of the Central Limit Theorem. The issue of *precision* in estimation, discussed in Section 7.4, considers techniques for manipulating the sample size in order to increase the precision of the estimation of a population mean and proportion. Finally, Section 7.5 discusses the construction of *prediction intervals* under the normality assumption.

7.2 CI Semantics: The Meaning of "Confidence"

A CI can be thought of as a Bernoulli trial: It either contains the true value of the parameter or not. However, the true value of the parameter is unknown and, hence, it is not known whether or not a particular CI contains it. For example, the estimate $\widehat{p} = 0.6$, based on $n = 20$ Bernoulli trials, leads to a 95% CI of

$$(0.39, 0.81) \tag{7.2.1}$$

for the true value of p. (The construction of the CI in (7.2.1) will be explained in the next section.) The interval (0.38,0.82) either contains the true value of p or it does not, and there is no way of knowing what is the case.

Think of flipping a coin, catching it mid-air, and placing it on the table with your hand still covering the coin. Because the coin is covered, it is not known whether it shows heads or tails. Before flipping the coin we say that the probability of heads is 0.5 (assuming the coin is unbiased). When the coin has been flipped there is no more randomness, because the experiment has been performed (the randomness of the experiment lies in the coin flip). Thus, even though the outcome of the experiment is unknown, it is not appropriate to use the word *probability*. Instead, we say we are 50% **confident** that the outcome is heads.

The randomness of the coin flip corresponds to the randomness of collecting the data for the purpose of constructing the CI. Before the data are collected, the CI is a

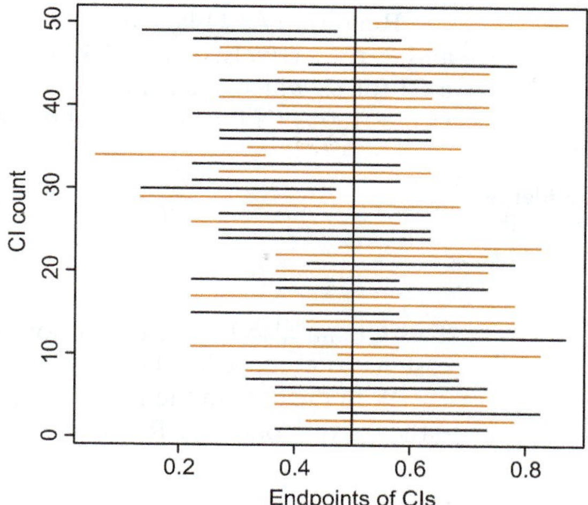

Figure 7-3 Fifty CIs for p.

random interval (see Figure 7-3) and we say that the probability that it will contain the true value of the parameter is, say, 0.95. After the data have been collected, a particular interval is constructed, like that in (7.2.1), and there is no more randomness, even though we do not know whether it contains the true value or not. This corresponds to the coin being covered. Thus we say that we are 95% confident that the interval contains the true value of the parameter.

Figure 7-3 shows 50 90% CIs for p based on $n = 20$ Bernoulli trials with the true value of p being 0.5. Each CI is represented by a horizontal straight line, and the y-axis enumerates the CIs. An interval contains the true value of p if the corresponding line intersects the vertical line at $x = 0.5$. (The advantage of using simulated data is that the true value of the parameter is known!) Because a 90% CI includes the true value of the parameter with probability 0.9, the total number of CIs that intersect the vertical line at $x = 0.5$ is a binomial($n = 50, p = 0.9$) random variable. Thus, we would expect $50 \times 0.9 = 45$ of them to contain the true value of p. In Figure 7-3, 44 of the 50 intervals shown contain 0.5. Being binomial(50, 0.9), the number of CIs that contain the true value p has standard deviation of $\sqrt{50 \times 0.9 \times 0.1} = 2.12$. Thus, even larger deviations from the expected value of 45 would not be unusual.

7.3 Types of Confidence Intervals

7.3.1 T CIs FOR THE MEAN

Let $X_1, \ldots, X_n$ be a simple random sample from a population, and let $\overline{X}$, S^2 denote the sample mean and sample variance, respectively.

Proposition 7.3-1 If the population is normal then,

$$\frac{\overline{X} - \mu}{S/\sqrt{n}} \sim T_{n-1}, \tag{7.3.1}$$

where μ stands for the true value of the population mean. Without the normality assumption, relation (7.3.1) holds approximately provided $n \geq 30$. ■

Relation (7.3.1) is a version of relation (7.1.6) with $\overline{X}$ and μ in the place of $\widehat{\theta}$ and θ, respectively. Thus, (7.3.1) implies a bound for the error in estimating μ and a CI for μ, which are versions of (7.1.7) and (7.1.8), respectively. In particular, the $(1-\alpha)100\%$ CI for the normal mean is

T Confidence Interval for the Population Mean μ

$$\left(\overline{X} - t_{n-1,\,\alpha/2}\frac{S}{\sqrt{n}},\ \overline{X} + t_{n-1,\,\alpha/2}\frac{S}{\sqrt{n}}\right) \tag{7.3.2}$$

This CI can also be used with non-normal populations, provided $n \geq 30$, in which case its confidence level is approximately (i.e., not exactly) $(1-\alpha)100\%$.

With the data in the R object x, the T CI for the mean can also be obtained with either of the following R commands:

R Commands for the $(1-\alpha)100\%$ T CI (7.3.2)

```
confint(lm(x~1), level=1-α)

mean(x)±qt(1-α/2, df=length(x)-1)*sd(x)/sqrt(length(x))
```

In the first command for the CI, the default value for the level is 0.95; thus, *confint(lm(x∼1))* gives a 95% CI. Also, note that the second command for the CI has been given in a condensed form: It has to be used once with the − sign, for the lower endpoint of the interval, and once with the + sign, for the upper endpoint.

Example 7.3-1

The Charpy impact test, developed by French scientist George Charpy, determines the amount of energy, in joules, absorbed by a material during fracture. (Fracture is induced by one blow from a swinging pendulum, under standardized conditions. The test was pivotal in understanding the fracture problems of ships during WWII.) A random sample of $n = 16$ test specimens of a particular metal, yields the following measurements (in kJ):

x | 4.90, 3.38, 3.32, 2.38, 3.14, 2.97, 3.87, 3.39, 2.97, 3.45, 3.35, 4.34, 3.54, 2.46, 4.38, 2.92

Construct a 99% CI for the population mean amount of energy absorbed during fracture.

Solution

Because the sample size is < 30, we must assume that the energy absorption population is normal. The Q-Q plot in Figure 7-4 does not strongly contradict this assumption, so we can proceed with the construction of the CI. (See, however, Section 7.3.4 for an alternative CI that does not require normality or a large sample size.) The given data set yields an average energy absorption of $\overline{X} = 3.42$ kJ, and standard deviation of $S = 0.68$ gr. The degrees of freedom is $\nu = n - 1 = 15$, and, for the desired 99% CI, $\alpha = 0.01$. Using either Table A.4 or the R command *qt(0.995,15)*, we find $t_{n-1,\alpha/2} = t_{15,0.005} = 2.947$. Substituting this information into formula (7.3.2) yields

$$\overline{X} \pm t_{n-1,\,\alpha/2}(S/\sqrt{n}) = 3.42 \pm 2.947(0.68/\sqrt{16}), \quad \text{or } (2.92,\ 3.92).$$

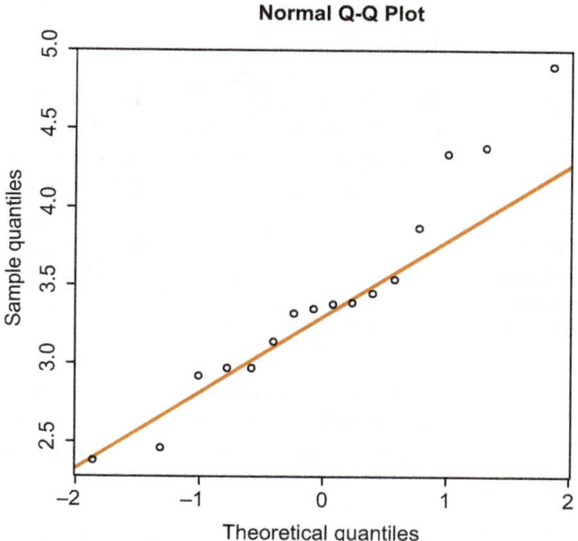

Figure 7-4 Q-Q plot for the data in Example 7.3-1.

Alternatively, with the data in the R object x, the R command *confint(lm(x~1), level=0.99)* returns

```
                 0.5%     99.5%
(Intercept) 2.921772  3.923228
```

which is the same (up to rounding) as the CI obtained by hand calculations. ■

Example 7.3-2

A random sample of $n = 56$ cotton pieces gave average percent elongation of $\overline{X} = 8.17$ and a sample standard deviation of $S = 1.42$. Construct a 95% CI for μ, the population mean percent elongation.

Solution

Because the sample size of $n = 56$ is large enough, the CI (7.3.2) can be used without the assumption of normality. The degrees of freedom is $\nu = 56 - 1 = 55$ and, for the desired 95% CI, $\alpha = 0.05$. Table A.4 does not list the percentiles of $T_{\nu=55}$ but, interpolating between the 97.5th percentiles of $T_{\nu=40}$ and $T_{\nu=80}$, we find that $t_{55,\,\alpha/2} = t_{55,\,0.025} \doteq 2.01$. The R command *qt(0.975, 55)* gives the exact value $t_{55,0.025} = 2.004$. Using the approximate value of the percentile and the given sample information, we obtain

$$\overline{X} \pm t_{55,\,\alpha/2}\frac{S}{\sqrt{n}} = 8.17 \pm 2.01\frac{1.42}{\sqrt{56}} = (7.80,\ 8.54)$$

for the desired 95% CI for μ. ■

7.3.2 Z CIs FOR PROPORTIONS

Let X be the number of successes in n Bernoulli trials, and $\widehat{p} = X/n$ be the sample proportion of successes. By the DeMoivre-Laplace Theorem, and by the fact that $\widehat{p}$ is a consistent estimator of p, the true value of the probability of success, it follows that

$$\frac{\widehat{p}-p}{S_{\widehat{p}}} \stackrel{\cdot}{\sim} N(0,1), \tag{7.3.3}$$

where $S_{\widehat{p}} = \sqrt{\widehat{p}(1-\widehat{p})/n}$. This is a version of (7.1.1) with $\widehat{p}$ and p in the place of $\widehat{\theta}$ and θ, respectively. Thus, (7.3.3) leads to a bound for the error in estimating p and a Z CI for p, which are versions of (7.1.4) and (7.1.5), respectively. In particular, the approximate $(1-\alpha)100\%$ CI for the binomial probability p is

Z Confidence Interval for the Population Proportion p

$$\left(\widehat{p} - z_{\alpha/2}\sqrt{\frac{\widehat{p}(1-\widehat{p})}{n}},\ \widehat{p} + z_{\alpha/2}\sqrt{\frac{\widehat{p}(1-\widehat{p})}{n}}\right) \tag{7.3.4}$$

For the purposes of this book, $(1-\alpha)100\%$ is a reasonable approximation to the true confidence level of the above CI provided there are at least 8 successes and at least 8 failures in our sample of n Bernoulli experiments, that is, provided

Sample Size Requirement for the Z CI (7.3.4)

$$n\widehat{p} \geq 8 \quad \text{and} \quad n(1-\widehat{p}) \geq 8 \tag{7.3.5}$$

Note that the condition (7.3.5) is different from the condition $np \geq 5$ and $n(1-p) \geq 5$ for the application of the DeMoivre-Laplace Theorem. This is due to the fact that p is unknown.

With the value of $\widehat{p}$ in the R object *phat*, the Z CI for p can also be obtained with the following R command:

R Command for the $(1-\alpha)100\%$ Z CI (7.3.4)

```
phat±qnorm(1-α/2)*sqrt(phat*(1-phat)/n)
```

Note that the R command for the CI has been given in a condensed form: It has to be used once with the − sign for the lower endpoint of the interval and once with the + sign for the upper endpoint.

Example 7.3-3

In a low-impact car crash experiment, similar to that described in Example 6.3-4, 18 of 30 cars sustained no visible damage. Construct a 95% CI for the true value of p, the probability that a car of this type will sustain no visible damage in such a low-impact crash.

Solution

The number of successes and the number of failures in this binomial experiment are at least 8, so the sample size requirement (7.3.5) for the CI (7.3.4) holds. Applying the formula with $\widehat{p} = 18/30 = 0.6$, the desired CI is

$$0.6 \pm 1.96\sqrt{\frac{0.6 \times 0.4}{30}} = 0.6 \pm 1.96 \times 0.089 = (0.425, 0.775).$$

Alternatively, the R commands *phat = 0.6; phat-qnorm(0.975)*sqrt(phat*(1-phat)/30); phat+qnorm(0.975)*sqrt(phat*(1-phat)/30)* returns the values 0.4246955 and 0.7753045. These are the same (up to rounding) with the lower and upper endpoint, respectively, of the 95% CI for p obtained by hand calculations. ■

7.3.3 *T* CIs FOR THE REGRESSION PARAMETERS

Let $(X_1, Y_1), \ldots, (X_n, Y_n)$, be a simple random sample from a population of (X, Y) values that satisfy the simple linear regression model. Thus,

$$\mu_{Y|X}(x) = E(Y|X = x) = \alpha_1 + \beta_1 x,$$

and the homoscedasticity assumption holds, that is, the intrinsic error variance $\sigma_\varepsilon^2 = \text{Var}(Y|X = x)$ is the same for all values x of X.

This section presents CIs for the slope, β_1, and the regression line, $\mu_{Y|X}(x)$. The corresponding CI for the intercept is deferred to Exercise 12 because it is a special case of the CI for $\mu_{Y|X}(x)$. This follows from the fact that $\alpha_1 = \mu_{Y|X}(0)$; see also the comment following Proposition 7.3-2 below.

Preliminary Results The formulas for the LSEs $\widehat{\alpha}_1$, $\widehat{\beta}_1$, and S_ε^2, of α_1, β_1, and σ_ε^2, respectively, which were given in Section 6.3.3 of Chapter 6, are restated here for convenient reference:

$$\widehat{\alpha}_1 = \overline{Y} - \widehat{\beta}_1 \overline{X}, \quad \widehat{\beta}_1 = \frac{n \sum X_i Y_i - (\sum X_i)(\sum Y_i)}{n \sum X_i^2 - (\sum X_i)^2}, \quad \text{and} \tag{7.3.6}$$

$$S_\varepsilon^2 = \frac{1}{n-2}\left[\sum_{i=1}^{n} Y_i^2 - \widehat{\alpha}_1 \sum_{i=1}^{n} Y_i - \widehat{\beta}_1 \sum_{i=1}^{n} X_i Y_i\right]. \tag{7.3.7}$$

These estimators are unbiased for their respective parameters. The construction of confidence intervals for β_1 and $\mu_{Y|X}(x)$ is based on the following proposition.

Proposition 7.3-2

1. The estimated standard error of $\widehat{\beta}_1$ is

$$S_{\widehat{\beta}_1} = S_\varepsilon \sqrt{\frac{n}{n \sum X_i^2 - (\sum X_i)^2}}. \tag{7.3.8}$$

2. The estimated standard error of $\widehat{\mu}_{Y|X=x} = \widehat{\alpha}_1 + \widehat{\beta}_1 x$ is

$$S_{\widehat{\mu}_{Y|X}(x)} = S_\varepsilon \sqrt{\frac{1}{n} + \frac{n(x - \overline{X})^2}{n \sum X_i^2 - (\sum X_i)^2}}. \tag{7.3.9}$$

3. Assume now, in addition, that the conditional distribution of Y given $X = x$ is normal or, equivalently, that the intrinsic error variables are normal. Then,

$$\frac{\widehat{\beta}_1 - \beta_1}{S_{\widehat{\beta}_1}} \sim T_{n-2} \quad \text{and} \quad \frac{\widehat{\mu}_{Y|X=x} - \mu_{Y|X=x}}{S_{\widehat{\mu}_{Y|X=x}}} \sim T_{n-2}. \tag{7.3.10}$$

4. Without the normality assumption, relation (7.3.10) holds approximately provided $n \geq 30$. ■

Worth pointing out is the term $(x - \overline{X})^2$ in the expression of $S_{\widehat{\mu}_{Y|X}(x)}$. This term implies that the standard error of $\widehat{\mu}_{Y|X}(x)$ increases with the distance of x from $\overline{X}$; see Figure 7-5 on page 262. Note also that $\widehat{\mu}_{Y|X}(0) = \widehat{\alpha}_1$, which means that the CI for α_1 is the same as that for $\mu_{Y|X}(0)$ given below; see Exercise 12.

CIs for the Slope and the Regression Line The relations in (7.3.10) are versions of (7.1.6) with $\widehat{\theta}$ and θ replaced by each of $\widehat{\beta}_1$ and $\widehat{\mu}_{Y|X}(x)$, and the corresponding parameters they estimate, respectively. Hence, the relations in (7.3.10) imply bounds

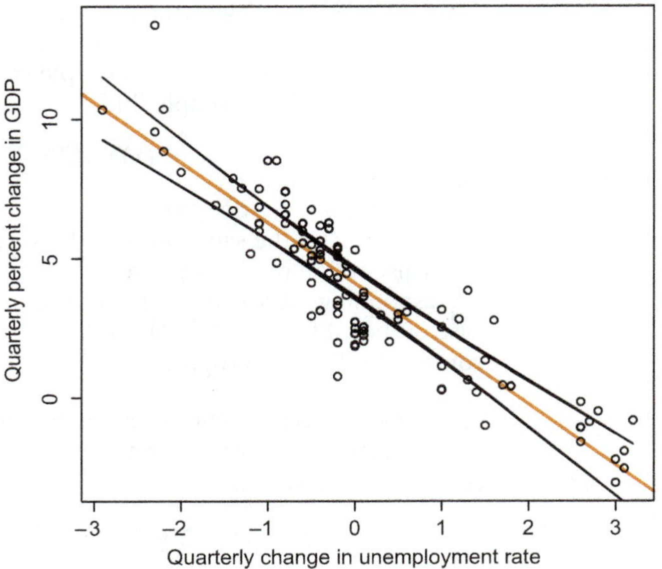

Figure 7-5 Data of Example 7.3-4 with the fitted regression line (in color) and lower and upper limits of the CIs for $\mu_{Y|X}(x)$ (in black).

for the error in estimating β_1 and $\mu_{Y|X}(x)$ and CIs for these parameters, which are versions of (7.1.7) and (7.1.8), respectively. In particular, the $(1-\alpha)100\%$ CI for β_1 is

T Confidence Interval for the Slope of the Regression Line

$$\widehat{\beta}_1 \pm t_{n-2,\,\alpha/2} S_{\widehat{\beta}_1} \tag{7.3.11}$$

where $S_{\widehat{\beta}_1}$ is given in (7.3.8), and the $(1-\alpha)100\%$ CI for the regression line is

T Confidence Interval for the Regression Line $\mu_{Y|X}(x)$

$$\widehat{\mu}_{Y|X}(x) \pm t_{n-2,\,\alpha/2} S_{\widehat{\mu}_{Y|X}(x)} \tag{7.3.12}$$

where $S_{\widehat{\mu}_{Y|X}(x)}$ is given in (7.3.9).

If the normality assumption holds, the confidence level of each of the above CIs is exactly $1-\alpha$ for all n. Without the normality assumption, the confidence level is approximately $1-\alpha$ provided $n \geq 30$.

With the data on the predictor and the response in the R objects x and y, respectively, the T CIs for the regression parameters α_1, β_1 and $\mu_{Y|X}(v)$ (note the use of v instead of the usual x, since x stands for the R object containing the X-values) can be obtained with the following R commands:

R Commands for $(1-\alpha)100\%$ T CIs for α_1, β_1, and $\mu_{Y|X}(v)$

```
out=lm(y~x); t=data.frame(x=v) # defines out, and sets new
  x-value

confint(out, level=1-α) # gives CIs for α1 and β1

predict(out, t, interval="confidence", level=1-α) # gives CI
  for μY|X(v).
```

In the command *t=data.frame(x=v)*, *v* can be either a single value, for example, *t=data.frame(x=4)*, or a set of values, for example, *t=data.frame(x=c(4, 6.3, 7.8))*. In the latter case, the *predict* command will return $(1-\alpha)100\%$ T CIs for $\mu_{Y|X}(4)$, $\mu_{Y|X}(6.3)$, and $\mu_{Y|X}(7.8)$.

Example 7.3-4

Can changes in Gross Domestic Product (GDP) be predicted from changes in the unemployment rate? Quarterly data on changes in unemployment rate (X) and percent change in GDP (Y), from 1949 to 1972, are given in the data file *GdpUemp49-72.txt*.[1] Assuming the simple linear regression model is a reasonable model for the regression function of Y on X, use R commands to complete the following parts:

(a) Estimate the regression function and construct 95% CIs for the slope and the expected percent change in GDP when the unemployment rate increases by 2 percentage points.

(b) Plot the lower and upper limits of the CIs for $\mu_{Y|X}(x)$, for x in the range of X-values, to illustrate the effect of the distance of x from $\overline{X}$ on the width of the CI.

Solution

(a) First, the scatterplot shown in Figure 7-5 suggests that the simple linear regression model is appropriate for this data. With the values of unemployment rate changes in the R object *x*, and those for percent changes in GDP in *y*, the R commands *out=lm(y~x); out$coef; confint(out, level=0.95)* return the following:

```
(Intercept)          x
   4.102769  -2.165556
                  2.5%      97.5%
(Intercept)   3.837120   4.368417
          x  -2.367161  -1.963951
```

Thus, $\widehat{\alpha}_1 = 4.102769$, $\widehat{\beta}_1 = -2.165556$, and the 95% CI for the slope is $(-2.367161, -1.963951)$. Finally, the commands *t=data.frame(x=2); predict(out, t, interval="confidence", level=0.95)* return the following values:

```
       fit         lwr        upr
-0.2283429  -0.6991205  0.2424347
```

So $\widehat{\mu}_{Y|X}(2) = -0.2283429$ and the 95% CI for $\mu_{Y|X}(2)$ is $(-0.6991205, 0.2424347)$.

(b) The commands *plot(x, y, xlab="Quarterly Change in Unemployment Rate", ylab="Quarterly Percent Change in GDP"); abline(lm(y~x), col="red")* produce the scatterplot with the fitted LS line in red. The additional commands *LUL=data.frame(predict(out, interval="confidence", level=0.999)); attach(LUL); lines(x, lwr, col="blue"); lines(x, upr, col="blue")* superimpose the lower and upper limits of the CIs for $\mu_{Y|X}(x)$ for x in the range of X-values. ■

Example 7.3-5

The following summary statistics and least squares estimators were obtained using $n = 14$ data points on Y= propagation velocity of an ultrasonic stress wave through a substance and X= tensile strength of the substance:

[1] Source: http://serc.carleton.edu/sp/library/spreadsheets/examples/41855.html. This data has been used by Miles Cahill, College of the Holy Cross in Worcester, MA, to provide empirical evidence for *Okun's law*.

(1)
$$\sum_{i=1}^{14} X_i = 890, \quad \sum_{i=1}^{14} Y_i = 37.6, \quad \sum_{i=1}^{14} X_i Y_i = 2234.30,$$

$$\sum_{i=1}^{14} X_i^2 = 67,182, \quad \sum_{i=1}^{14} Y_i^2 = 103.54.$$

(2)
$$\widehat{\alpha}_1 = 3.62091, \quad \widehat{\beta}_1 = -0.014711, \quad \text{and} \quad S_\varepsilon^2 = 0.02187.$$

A scatterplot for this data does not indicate violation of the assumptions for the simple linear regression model. Use the information given to complete the following:

(a) Construct a 95% CI for β_1, and use it to produce a 95% CI for $\mu_{Y|X}(66) - \mu_{Y|X}(30)$, the expected difference between propagation velocities at tensile strengths of 66 and 30.

(b) Construct a 95% CI for $\mu_{Y|X}(66)$ and for $\mu_{Y|X}(30)$. Give an explanation for the difference between the widths of these two CIs.

(c) Construct a normal Q-Q plot for the residuals to check the assumption that the intrinsic error variables are normally distributed.

Solution

(a) Plugging these results into formula (7.3.8) we obtain

$$S_{\widehat{\beta}_1} = S_\varepsilon \sqrt{\frac{n}{n \sum X_i^2 - (\sum X_i)^2}} = \sqrt{0.02187} \sqrt{\frac{14}{14 \times 67182 - 890^2}} = 0.001436.$$

Thus, according to (7.3.11) and since $t_{12,\,0.025} = 2.179$, the 95% CI for β_1 is

$$-0.014711 \pm t_{12,\,0.025} 0.001436 = (-0.01784, -0.011582).$$

Next, note that 66 and 30 are within the range of X-values, and thus it is appropriate to consider estimation of the expected response at these covariate values. Note also that $\mu_{Y|X}(66) - \mu_{Y|X}(30) = 36\beta_1$. Since $-0.01784 < \beta_1 < -0.011582$ with 95% confidence, it follows that $-36 \times 0.01784 < 36\beta_1 < -36 \times 0.011582$ also holds with confidence 95%, because the two sets of inequalities are equivalent. Hence, a 95% CI for $\mu_{Y|X}(66) - \mu_{Y|X}(30)$ is

$$(-36 \times 0.01784, -36 \times 0.011582) = (-0.64224, -0.41695).$$

(b) Using the summary statistics given in part (a), formula (7.3.9) gives

$$S_{\widehat{\mu}_{Y|X}(66)} = S_\varepsilon \sqrt{\frac{1}{14} + \frac{14(66 - 63.5714)^2}{14 \times 67182 - 890^2}} = 0.03968 \quad \text{and}$$

$$S_{\widehat{\mu}_{Y|X}(30)} = S_\varepsilon \sqrt{\frac{1}{14} + \frac{14(30 - 63.5714)^2}{14 \times 67182 - 890^2}} = 0.06234.$$

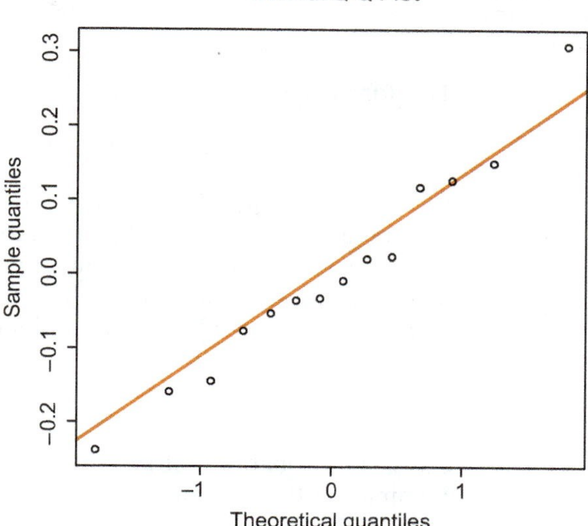

Figure 7-6 Normal Q-Q plot for the LS residuals of Example 7.3-5.

Moreover, $\widehat{\mu}_{Y|X}(66) = 2.65$ and $\widehat{\mu}_{Y|X}(30) = 3.18$. With these calculations, formula (7.3.12), yields the following 95% CIs for $\mu_{Y|X}(66)$ and $\mu_{Y|X}(30)$, respectively:

$$2.65 \pm 2.179 \times 0.03968 = (2.563, 2.736)$$

$$\text{and} \quad 3.18 \pm 2.179 \times 0.06234 = (3.044, 3.315).$$

The width of the CI for $\mu_{Y|X}(66)$ is $2.736 - 2.563 = 0.173$, while that of $\mu_{Y|X}(30)$ is $3.315 - 3.044 = 0.271$. This is explained by the fact that the covariate value of 66 is much closer to $\overline{X} = 63.57$ than 30 is, resulting in $\widehat{\mu}_{Y|X}(66)$ having smaller estimated standard error than $\widehat{\mu}_{Y|X}(30)$.

(c) Since the sample size of this data set is 14, the validity of the CIs constructed in parts (a) and (b) hinges on the assumption that the intrinsic error variables are normal. Of course, the intrinsic error variables, ε_i, are not observable, but their estimated versions, the residuals, are easily available. With the X- and Y-values in the R objects x and y, respectively, the R commands *qqnorm(lm(y~x)$resid); qqline(lm(y~x)$resid, col="red")* produce the Q-Q plot shown in Figure 7-6. The plot does not suggest serious departure from the normality assumption, and thus the above CIs can be reasonably trusted.

7.3.4 THE SIGN CI FOR THE MEDIAN

The sign CI for the median, $\widetilde{\mu}$, is useful because it can be applied also with non-normal data without requiring the sample size to be ≥ 30. To describe it, let $X_1, \ldots, X_n$ be a sample from a continuous population and let $X_{(1)} < \cdots < X_{(n)}$ denote the order statistics. The $(1 - \alpha)100\%$ sign CIs for the median $\widetilde{\mu}$ are based on intervals of the form

$$(X_{(a)}, X_{(n-a+1)}) \tag{7.3.13}$$

for $a < (n+1)/2$ an integer. Each such interval is a CI for the median with confidence level

Confidence Level of the CI (7.3.13) for the Median

$$\left(1 - 2P\left(X_{(a)} > \widetilde{\mu}\right)\right) 100\% \tag{7.3.14}$$

This formula is derived by writing

$$P\left(X_{(a)} \le \widetilde{\mu} \le X_{(n-a+1)}\right) = 1 - 2P\left(X_{(a)} > \widetilde{\mu}\right)$$

since, by symmetry, $P\left(X_{(n-a+1)} < \widetilde{\mu}\right) = P\left(X_{(a)} > \widetilde{\mu}\right)$.

It is a rather surprising fact that, for any integer a, the probability $P\left(X_{(a)} > \widetilde{\mu}\right)$ is the same no matter what continuous population distribution the data has come from. Moreover, this probability can be found by expressing it as a binomial probability. Here is how: Define Bernoulli random variables $Y_1, \ldots, Y_n$ by

$$Y_i = \begin{cases} 1 \text{ if } X_i > \widetilde{\mu} \\ 0 \text{ if } X_i < \widetilde{\mu} \end{cases}.$$

Thus, the probability of success in each Bernoulli trial is 0.5. So $T = \sum_i Y_i$ is a binomial $(n, 0.5)$ random variable. Then the events

$$X_{(a)} > \widetilde{\mu} \quad \text{and} \quad T \ge n - a + 1 \tag{7.3.15}$$

are equivalent. Hence, the level of each CI in (7.3.13) can be found with the use of binomial probabilities as

Confidence Level (7.3.14) in Terms of Binomial Probabilities

$$\left(1 - 2P\left(T \ge n - a + 1\right)\right) 100\% \tag{7.3.16}$$

In other words, (7.3.13) is a $(1 - \alpha)100\%$ CI for $\widetilde{\mu}$, where α can be computed exactly with the R command *2*(1-pbinom(n-a, n, 0.5))*.

Example 7.3-6

Let $X_1, \ldots, X_{25}$ be a sample from a continuous population. Find the confidence level of the following CI for the median:

$$\left(X_{(8)}, X_{(18)}\right).$$

Solution

First note that the CI $(X_{(8)}, X_{(18)})$ is of the form (7.3.13) with $a = 8$. Hence, the confidence level of this interval is computed according to the formula (7.3.16). Using the R command *2*(1-pbinom(25-8, 25, 0.5))*, we find

$$\alpha = 0.0433.$$

Thus, regardless of which continuous distribution the data came from, the confidence level of the CI $\left(X_{(8)}, X_{(18)}\right)$, is $(1 - \alpha)100\% = 95.67\%$. ■

Since the confidence level of each CI in (7.3.13) can be computed, it should be an easy matter to find the one with the desired $(1-\alpha)100\%$ confidence level. This is true except for a slight problem. To see where the problem lies, assume n is large enough for the DeMoivre-Laplace approximation to the probability in (7.3.15). Including continuity correction,

$$P\left(X_{(a)} > \widetilde{\mu}\right) \simeq 1 - \Phi\left(\frac{n - a + 0.5 - 0.5n}{\sqrt{0.25n}}\right).$$

Setting this equal to $\alpha/2$ and solving for a we obtain

$$a = 0.5n + 0.5 - z_{\alpha/2}\sqrt{0.25n}. \tag{7.3.17}$$

The slight problem lies in the fact that the right-hand side of (7.3.17) need not be an integer. Rounding up gives a CI with confidence level less than $(1-\alpha)100\%$, and rounding down results in the confidence level being larger than $(1-\alpha)100\%$. Various interpolation methods exist for constructing a $(1-\alpha)100\%$ sign CI. One such method is employed in the following R command (the package BSDA should first be installed with *install.packages("BSDA")*):

R Commands for the $(1-\alpha)100\%$ Sign CI

```
library(BSDA); SIGN.test(x, alternative="two.sided",
  conf.level = 1-α)
```

where the R object x contains the data.

Example 7.3-7

Using the energy absorption data given in Example 7.3-1, construct a 95% CI for the population median energy absorption.

Solution

With the $n = 16$ data points in the R object x, the R commands *library(BSDA); SIGN.test(x, alternative="two.sided", conf.level=0.95)* generate output the last part of which gives three CIs for the median:

```
                      Conf.Level  L.E.pt  U.E.pt
Lower Achieved CI       0.9232     2.97    3.54
Interpolated CI         0.9500     2.97    3.70
Upper Achieved CI       0.9787     2.97    3.87
```

The 92.32% CI given in the first line corresponds to $a = 5$; thus, its lower endpoint (L.E.pt) is $X_{(5)} = 2.97$, and its upper endpoint (U.E.pt) is $X_{(n-a+1)} = X_{(12)} = 3.54$. Similarly, the 97.87% CI in the third line corresponds to $a = 4$ with lower and upper endpoints of $X_{(4)} = 2.97$ and $X_{(16-4+1)} = 3.87$. Because no CI of the form (7.3.13) has the desired confidence level, the interpolation method used in this package produces the 95% CI given in the second line. For comparison purposes, the R command *confint(lm(x ~1), level=0.95)* returns the CI $(3.06, 3.78)$ for the mean.

7.3.5 χ^2 CIs FOR THE NORMAL VARIANCE AND STANDARD DEVIATION

A special case of the gamma distribution, introduced in Exercise 13 in Section 3.5, is the **χ^2 distribution** with ν degrees of freedom, denoted by χ^2_ν. It corresponds to the gamma parameters $\alpha = \nu/2$ and $\beta = 2$. Thus, the mean and variance of a χ^2_ν random variable are ν and 2ν, respectively; see Exercise 13 in Section 3.5. The prominence of the χ^2 distribution stems from its connection to the standard normal distribution. In particular, if $Z_1, \ldots, Z_\nu$ are ν iid $N(0, 1)$ random variables, then their sum of squares has the χ^2_ν distribution:

$$Z_1^2 + \cdots + Z_\nu^2 \sim \chi^2_\nu.$$

As a consequence it also arises as the distribution of certain other sums of squares that are used in statistics. In particular, if $X_1, \ldots, X_n$ is a random sample and S^2 its sample variance, we have the following result:

Proposition 7.3-3 If the population from which the sample was drawn is normal, then

$$\frac{(n-1)S^2}{\sigma^2} \sim \chi^2_{n-1},$$

where σ^2 stands for the true value of the population variance. ■

This proposition implies that

$$\chi^2_{n-1,\,1-\alpha/2} < \frac{(n-1)S^2}{\sigma^2} < \chi^2_{n-1,\alpha/2} \tag{7.3.18}$$

will be true $(1-\alpha)100\%$ of the time, where $\chi^2_{n-1,\alpha/2}$, $\chi^2_{n-1,\,1-\alpha/2}$ denote percentiles of the χ^2_{n-1} distribution as shown in Figure 7-7. Note that the bounds on the error of estimation of σ^2 by S^2 are given in terms of the ratio S^2/σ^2. After an algebraic manipulation, (7.3.18) yields the following CI:

$(1-\alpha)100\%$ CI for the Normal Variance

$$\frac{(n-1)S^2}{\chi^2_{n-1,\,\alpha/2}} < \sigma^2 < \frac{(n-1)S^2}{\chi^2_{n-1,\,1-\alpha/2}} \tag{7.3.19}$$

Selective percentiles of χ^2 distributions are given in Table A.5, but can also be obtained with the following R command:

R Commands for the χ^2_ν Percentiles

```
qchisq(p, ν) # gives χ²_{ν,1-p}
```

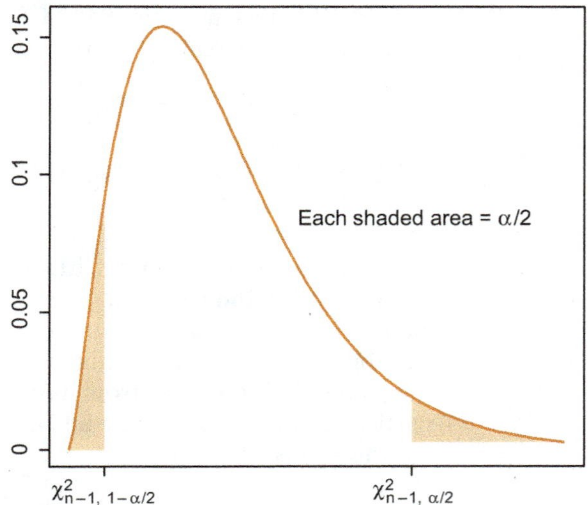

Figure 7-7 Density and percentiles of the χ^2_{n-1} distribution.

Taking the square root of the lower and upper limit of the CI (7.3.19) yields a CI for σ:

$(1-\alpha)100\%$ CI for the Normal Standard Deviation

$$\sqrt{\frac{(n-1)S^2}{\chi^2_{n-1,\,\alpha/2}}} < \sigma < \sqrt{\frac{(n-1)S^2}{\chi^2_{n-1,\,1-\alpha/2}}} \qquad (7.3.20)$$

Unfortunately, the CIs (7.3.19) and (7.3.20) are valid only when the sample has been drawn from a normal population, regardless of the sample size. Thus, they should only be used when normality appears to be a reasonable model for the population distribution.

Example 7.3-8

An optical firm purchases glass to be ground into lenses. As it is important that the various pieces of glass have nearly the same index of refraction, the firm is interested in controlling the variability. A simple random sample of size $n = 20$ measurements yields $S^2 = (1.2)10^{-4}$. From previous experience, it is known that the normal distribution is a reasonable model for the population of these measurements. Find a 95% CI for σ.

Solution

With $n - 1 = 19$ degrees of freedom, Table A.5 gives $\chi^2_{0.975,\,19} = 8.906$ and $\chi^2_{0.025,\,19} = 32.852$. The same values (up to rounding) are obtained with the R commands *qchisq(0.025, 19)* and *qchisq(0.975, 19)*, respectively. Thus, according to (7.3.20), a 95% CI for σ is

$$\sqrt{\frac{(19)(1.2\times 10^{-4})}{32.852}} < \sigma < \sqrt{\frac{(19)(1.2\times 10^{-4})}{8.906}},$$

or $0.0083 < \sigma < 0.0160$. ■

Exercises

1. A question relating to a study of the echolocation system for bats is how far apart the bat and an insect are when the bat first senses the insect. The technical problems for measuring this are complex and so only $n = 11$ data points were obtained (in dm):

x | 57.16, 48.42, 46.84, 19.62, 41.72, 36.75, 62.69, 48.82, 36.86, 50.59, 47.53

It is given that $\overline{X} = 45.18$ and $S = 11.48$.

(a) Construct by hand a 95% CI for the mean distance μ. What assumptions, if any, are needed for the validity of this CI?

(b) Copy the data into the R object x, that is, by *x=c(57.16, 48.42, 46.84, 19.62, 41.72, 36.75, 62.69, 48.82, 36.86, 50.59, 47.53)*, and do the normal Q-Q plot for this data. Comment on the appropriateness of the CI in part (a).

2. Analysis of the venom of 7 eight-day-old worker bees yielded the following observations on histamine content in nanograms: 649, 832, 418, 530, 384, 899, 755.

(a) Construct by hand a 90% CI for the true mean histamine content for all worker bees of this age. What assumptions, if any, are needed for the validity of the CI?

(b) The true mean histamine content will be in the CI you constructed in part (a) with probability 90%. True or false?

3. For a random sample of 50 measurements of the breaking strength of cotton threads, $\overline{X} = 210$ grams and $S = 18$ grams.

(a) Obtain an 80% CI for the true mean breaking strength. What assumptions, if any, are needed for the validity of the CI?

(b) Would the 90% CI be wider than the 80% CI constructed in part (a)?

(c) A classmate offers the following interpretation of the CI you obtained in part (a): We are confident that 80% of all breaking strength measurements of cotton threads will be within the calculated CI. Is this interpretation correct?

4. Refer to Exercises 1 and 2.

(a) For the data in Exercise 1, find the confidence level of the CIs (36.86, 50.59) and (36.75, 57.16) for the population median distance.

(b) For the data in Exercise 2, find the confidence level of the CI (418, 832) for the population median histamine content.

5. The data file *OzoneData.txt* contains $n = 14$ ozone measurements (Dobson units) taken from the lower stratosphere, between 9 and 12 miles (15 and 20 km). Import the data into the R data frame *oz*, copy the data into the R object x by *x=oz$OzoneData*, and use R commands to complete the following parts.

(a) Do the normal Q-Q plot for this data, and comment on the appropriateness of the normal distribution as a model for the population distribution of ozone measurements.

(b) Construct a 90% CI for the mean ozone level.

(c) Construct a 90% CI for the median ozone level.

(d) Compare the lengths of the two CIs and comment on the appropriateness of using each of them.

6. The data file *SolarIntensAuData.txt* contains $n = 40$ solar intensity measurements (watts/m^2) on different days at a location in southern Australia. Import the data into the R data frame *si*, copy the data into the R object x by *x=si$SI*, and use R commands to complete the following parts.

(a) Construct a 95% CI for the mean solar intensity, and state any assumptions needed for the validity of the CI.

(b) Construct a 95% CI for the median solar intensity, and state any assumptions needed for the validity of the CI.

(c) It is suggested that 95% of the (conceptual) population of solar intensity measurements taken in the same location on similar days of the year will lie within the CI in part (a). Is this a correct interpretation of the CI? What about a similar interpretation for the CI in part (b)?

7. Fifty newly manufactured items are examined and the number of scratches per item are recorded. The resulting frequencies of the number of scratches is:

Number of scratches per item	0	1	2	3	4
Observed frequency	4	12	11	14	9

Assume that the number of scratches per item is a Poisson(λ) random variable.

(a) Construct a 95% CI for λ. (*Hint.* $\lambda = \mu$.)

(b) Construct a 95% CI for σ, the standard deviation of the number of scratches per item. (*Hint.* In the Poisson model, $\sigma^2 = \lambda$.)

8. Copy the Old Faithful geyser's eruption durations data into the R object *ed* with the command *ed=faithful$eruptions*, and use R commands to complete the following parts.

(a) Construct a 95% CI for the mean eruption duration.

(b) Construct a 95% CI for the median eruption duration.

(c) Construct a 95% CI for the probability that an eruption duration will last more than 4.42 min. (*Hint.* The sample proportion, $\widehat{p}$, can be found with the R command *sum(ed>4.42)/length(ed)*.)

9. In making plans for an executive traveler's club, an airline would like to estimate the proportion of its current customers who would qualify for membership. A random sample of 500 customers yielded 40 who would qualify.

(a) Construct a 95% CI for the population proportion, p, of customers who qualify.

(b) What assumptions, if any, are needed for the validity of the above CI?

10. A health magazine conducted a survey on the drinking habits of young adult (ages 21–35) US citizens. On the question "Do you drink beer, wine, or hard liquor each week?" 985 of the 1516 adults interviewed responded "yes."

(a) Find a 95% confidence interval for the proportion, p, of young adult US citizens who drink beer, wine, or hard liquor on a weekly basis.

(b) The true proportion of young adults who drink on a weekly basis lies in the interval obtained in part (a) with probability 0.95. True or false?

11. To determine the probability that a certain component lasts more than 350 hours in operation, a random sample of 37 components was tested. Of these, 24 lasted longer than 350 hours.

(a) Construct a 95% CI for the probability, p, that a randomly selected component lasts more than 350 hours.

(b) A system consists of two such components connected in series. Thus, the system operates if and only if both components operate properly. Construct a 95% CI for the probability that the system lasts more than 350 hours. You can assume that the life spans of the two components in the system are independent. (*Hint.* Express the probability that the system lasts more than 350 hours in terms of p.)

12. In the parametrization $\mu_{Y|X}(x) = \alpha_1 + \beta_1 x$ of the simple linear regression model, $\mu_{Y|X}(0) = \alpha_1$ and $\widehat{\mu}_{Y|X}(0) = \widehat{\alpha}_1$. Use this fact and the formula for the $(1-\alpha)100\%$ CI for $\mu_{Y|X}(x)$ to give the formula for the $(1-\alpha)100\%$ CI for α_1.

13. A study was conducted to determine the relation between the (easier to measure) conductivity (μS/cm) of surface water and water in the sediment at the bank of a river.[2] The summary statistics for 10 pairs of surface (X) and sediment (Y) conductivity measurements are $\sum X_i = 3728$, $\sum Y_i = 5421$, $\sum X_i^2 = 1816016$, $\sum Y_i^2 = 3343359$, and $\sum X_i Y_i = 2418968$, with the X-values ranging from 220 to 800. Assume the regression function of Y on X is linear, and that $\text{Var}(Y|X = x)$ is for all x.

(a) Give the LSEs for α_1, β_1, and σ_ε^2.

(b) Construct a 95% CI for the true slope of the regression line. What additional assumptions, if any, are needed for the validity of T CIs for this data set?

(c) Construct a 95% CI for the expected sediment conductivity when the surface conductivity is 500. Repeat the same for surface conductivity of 900. Comment on the appropriateness of the two CIs.

14. A study examined the effect of varying the water/cement ratio (X) on the strength (Y) of concrete that has been aged 28 days.[3] Use *CS=read.table("WaCeRat28DayS.txt", header=T)* to read the $n = 13$ pairs of measurements into the R data frame *CS*. Copy the data into the R objects x and y by *x=CS$X* and *y=CS$Y*, and use R commands to complete the following parts.

(a) Make a scatterplot of the data and comment on whether the simple linear regression model assumptions, which are linearity of the regression function of Y on X and homoscedasticity (i.e., $\text{Var}(Y|X = x)$ for all x), appear to be satisfied for this data set.

(b) Give the LSEs for α_1, β_1, and σ_ε. (*Hint.* Use (6.3.12) and (6.3.13).)

(c) Give 90% CIs for β_1 and for the mean difference of the strengths at water/cement ratios 1.55 and 1.35. (*Hint.* The mean difference of the strengths at water/cement ratios 1.55 and 1.35 is $\mu_{Y|X}(1.55) - \mu_{Y|X}(1.35) = 0.2\beta_1$).

(d) Give 90% CIs for the mean strength at water/cement ratios 1.35, 1.45, and 1.55.

(e) Since the sample size of this data set is 13, the validity of the CIs in parts (b) and (c) requires the assumption of normality for the intrinsic error variables. Construct a normal Q-Q plot of the residuals and comment on whether or not this assumption appears to be violated.

15. Copy the $n = 153$ daily measurements of temperature (^oF) and wind speed (mph), taken in New York from May to September 1973 and included in the R data set *airquality*, by *x=airquality$Temp; y=airquality$Wind*, and use R commands to complete the following parts.

(a) Contruct a scatterplot of the data and comment on whether or not the linearity and homoscedasticity assumptions of the simple linear regression model seem to be violated.

(b) Give 95% CIs for β_1 and for the expected wind speed on an 80^oF day.

16. Let $X_1, \ldots, X_{30}$ be a random sample from a continuous population. It has been decided to use $(X_{(10)}, X_{(21)})$ as a CI for the median. Find the confidence level of this CI.

17. Use formula (7.3.16) to confirm that in a sample of size $n = 16$ from a continuous population, $a = 4$ and $a = 5$ yield confidence levels as reported in Example 7.3-7.

18. An important quality characteristic of the lapping process that is used to grind certain silicon wafers is the population standard deviation, σ, of the thickness of die pieces sliced out from the wafers. If the thickness of 15 dice cut from such wafers have sample standard deviation of 0.64 μm, construct a 95% confidence interval for σ. What assumption is needed for the validity of this CI?

19. Kingsford's regular charcoal with hickory is available in 15.7-lb bags. Long-standing uniformity standards require the standard deviation of weights not to exceed 0.1 lb. The quality control team uses daily samples of 35 bags to see if there is evidence that the standard deviation is within the required limit. A particular day's sample yields $S = 0.117$. Make a 95% CI for σ. Does the traditional value of 0.1 lie within the CI?

20. Use *rt=read.table("RobotReactTime.txt", header=T); t2=rt$Time[rt$Robot==2]* to import the robot reaction times data set into the R data frame *rt* and to copy the 22 reaction times of Robot 2 into the R object *t2*. Construct a 95% CI for the population variance of reaction times of Robot 2, using R commands to compute the sample variance of the data and to find the needed percentiles of the χ^2 distribution.

[2] M. Latif and E. Licek (2004). Toxicity assessment of wastewaters, river waters, and sediments in Austria using cost-effective microbiotests. *Environmental Toxicology*, 19(4): 302–308.

[3] V. K. Alilou and M. Teshnehlab (2010). Prediction of 28-day compressive strength of concrete on the third day using artificial neural networks. *International Journal of Engineering (IJE)*, 3(6): 521–670.

7.4 The Issue of Precision

Precision in the estimation of a parameter θ is quantified by the size of the bound of the error of estimation $|\widehat{\theta} - \theta|$, or, equivalently, by the length of the CI for θ, which is twice the size of the error bound. A shorter error bound, or shorter CI, implies more precise estimation.

The bounds on the error of estimation of a population mean μ and population proportion p are of the form

$$
\left.\begin{aligned}
\left|\overline{X} - \mu\right| &\le z_{\alpha/2}\frac{\sigma}{\sqrt{n}} \quad \text{(known } \sigma\text{; normal case or } n > 30\text{)} \\
\left|\overline{X} - \mu\right| &\le t_{n-1,\,\alpha/2}\frac{S}{\sqrt{n}} \quad \text{(unknown } \sigma\text{; normal case or } n > 30\text{)} \\
|\widehat{p} - p| &\le z_{\alpha/2}\sqrt{\frac{\widehat{p}(1-\widehat{p})}{n}} \quad (n\widehat{p} \ge 8, n(1-\widehat{p}) \ge 8).
\end{aligned}\right\} \qquad \textbf{(7.4.1)}
$$

The above expressions suggest that, for a given σ or p, the size of the error bound (or length of CI) depends on the sample size n and the choice of α: For a fixed α, a larger sample size yields a smaller error bound, and thus more precise estimation. For a fixed n, a 90% CI is shorter than a 95% CI, which is shorter than a 99% CI. This is so because

$$z_{0.05} = 1.645 < z_{0.025} = 1.96 < z_{0.005} = 2.575,$$

and similar inequalities hold for the t critical values.

The increase in the length of the CI with the level of confidence is to be expected. Indeed, we are more confident that the wider CI will contain the true value of the parameter. However, we rarely want to reduce the length of the CI by decreasing the level of confidence. Hence, more precise estimation is achieved by selecting a larger sample size.

In principle, the problem of selecting the sample size needed to achieve a desired level of precision has a straightforward solution. For example, suppose we are sampling a normal population with known variance. Then the sample size needed to achieve length L for the $(1-\alpha)100\%$ CI for the mean is found by solving

$$2z_{\alpha/2}\frac{\sigma}{\sqrt{n}} = L,$$

for n. The solution is

$$n = \left(2z_{\alpha/2}\frac{\sigma}{L}\right)^2. \qquad \textbf{(7.4.2)}$$

More likely than not, the solution will not be an integer, in which case the recommended procedure is to round up. The practice of rounding up guarantees that the desired precision level will be more than met.

The main obstacle to getting a completely satisfactory solution to the problem of sample size selection is twofold: (a) the true value of the variance is rarely known, and (b) the estimated standard error, which enters the second and third error bounds in (7.4.1), is unknown prior to the data collection. Thus, sample size determinations must rely on some preliminary approximation of, or bound to, the standard error. Methods for such approximations are discussed separately for the mean and the proportion.

Sample Size Determination for μ The most common method for sample size determination for μ is to use the sample standard deviation, S_{pr}, from a preliminary/pilot sample of size n_{pr}. Then the sample size needed to achieve length L for the $(1-\alpha)100\%$ T CI for the mean is found by solving for n, and rounding the solution up, giving either

$$2t_{n_{pr}-1,\,\alpha/2}\frac{S_{pr}}{\sqrt{n}} = L \quad \text{or} \quad 2z_{\alpha/2}\frac{S_{pr}}{\sqrt{n}} = L,$$

depending on whether or not S_{pr} came from a sample of known sample size. Thus, the recommended sample size is the rounded-up value of

$$n = \left(2t_{n_{pr}-1,\,\alpha/2}\frac{S_{pr}}{L}\right)^2 \quad \text{or} \quad n = \left(2z_{\alpha/2}\frac{S_{pr}}{L}\right)^2, \tag{7.4.3}$$

depending on whether or not n_{pr} is known. Note that the second formula in (7.4.3) is formula (7.4.2) with S_{pr} replacing the unknown σ.

When a pilot sample is not available, but the likely range of values the variable can take can be guessed, the sample size is determined by the second formula in (7.4.3) with

$$S_{pr} = \frac{\text{range}}{3.5} \quad \text{or} \quad S_{pr} = \frac{\text{range}}{4}.$$

These approximations are justified by the relations

$$\sigma = \frac{B-A}{\sqrt{12}} = \frac{B-A}{3.464} \quad \text{and} \quad \sigma = \frac{z_{0.025}-z_{0.975}}{3.92},$$

which hold for the uniform(A, B), and the normal$(0, 1)$ distributions, respectively.

Sample Size Determination for p The two most commonly used methods for sample size determination for p correspond to whether or not a preliminary estimator $\widehat{p}_{pr}$ of p (obtained, e.g., from a pilot sample) is available.

When a preliminary estimator $\widehat{p}_{pr}$ exists, the sample size needed to achieve a desired length L for the CI for p is found by solving for n and rounding the solution up:

$$2z_{\alpha/2}\sqrt{\frac{\widehat{p}_{pr}(1-\widehat{p}_{pr})}{n}} = L.$$

Thus, the required sample size is the rounded up value of

$$n = \frac{4z_{\alpha/2}^2\widehat{p}_{pr}(1-\widehat{p}_{pr})}{L^2}. \tag{7.4.4}$$

When no preliminary estimator for p exists, we use (7.4.4) with $\widehat{p}_{pr} = 0.5$. The rationale for doing so is seen by noting that the value of $\widehat{p}_{pr}\ (1-\widehat{p}_{pr})$ is largest when $\widehat{p}_{pr} = 0.5$. Hence, the calculated sample size will be at least as large as needed for meeting the precision specification. With $\widehat{p}_{pr} = 0.5$, (7.4.4) becomes

$$n = \frac{z_{\alpha/2}^2}{L^2}. \tag{7.4.5}$$

REMARK 7.4-1 Sample size determination for μ by either of the formulas in (7.4.3) is not completely satisfactory because the standard deviation of the final sample, upon which the CI will be calculated, will be different from S_{pr}. Thus it is possible that the desired level of precision objective may not be met. Typically, the desired level of precision is achieved after some trial-and-error iteration. A similar comment applies for the sample size determination for p by the formula (7.4.4). ◁

Having installed the package BSDA with *install.packages("BSDA")*, the following R commands can be used for sample size calculation.

R Commands for Sample Size Calculation for CIs

```
library(BSDA) # makes package available

nsize(b=L/2, sigma=Spr, conf.level=1-α, type="mu") # gives
  sample size for μ

nsize(b=L/2, p=p̂, conf.level=1-α, type="pi") # gives sample
  size for p
```

In the commands for sample size determination, b stands for the desired bound on the error of estimation, which is half the desired length for the CI. Note also that the command for sample size determination for μ does not allow input of the size of the preliminary sample. Hence, it always uses the second formula in (7.4.3).

Example 7.4-1

The estimation of a new operating system's mean response time to an editing command should have an error bound of 5 milliseconds with 95% confidence. Experience with other operating systems suggests that $S_{pr} = 25$ is a reasonable approximation to the population standard deviation. What sample size n should be used?

Solution

The second formula in (7.4.3) with $\alpha = 0.05$ (so $z_{\alpha/2} = 1.96$) and $L = 10$ gives

$$n = \left(2 \times 1.96 \times \frac{25}{10}\right)^2 = 96.04,$$

which is rounded up to $n = 97$. The same answer is found with the R command *nsize(b=5, sigma=25, conf.level=0.95, type="mu")*.

Example 7.4-2

A new method of pre-coating fittings used in oil, brake, and other fluid systems in heavy-duty trucks is being studied for possible adoption. In this context, the proportion of such fittings that leak must be determined *to within* 0.01 with 95% confidence.

(a) What sample size is needed if a preliminary sample gave $\widehat{p}_{pr} = 0.9$?

(b) What sample size is needed if there is no prior information regarding the true value of p?

Solution

"To within 0.01" is another way of saying that the 95% bound on the error of estimation should be 0.01, or the desired CI should have length $L = 0.02$.

(a) Since we have preliminary information, we use (7.4.4):

$$n = \frac{4(1.96)^2(0.9)(0.1)}{0.02^2} = 3457.44.$$

This is rounded up to 3458. The R command *nsize(b=0.01, p=0.9, conf.level=0.95, type="pi")* gives the same answer.

(b) If no prior information is available, we use (7.4.5):

$$n = \frac{1.96^2}{0.02^2} = 9604.$$

The same answer is found with the R command *nsize(b=0.01, p=0.5, conf.level=0.95, type="pi")*. ■

Exercises

1. The estimation of the average shrinkage percentage of plastic clay should have an error bound of 0.2 with 98% confidence. A pilot sample of $n_{pr} = 50$ gave $S_{pr} = 1.2$. Use either hand calculations or an R command to determine the sample size that should be used.

2. A pilot sample of 50 measurements of the breaking strength of cotton threads gave $S_{pr} = 18$ grams. Use either hand calculations or an R command to determine the sample size needed to obtain a 90% CI of length 4.

3. A food processing company, considering the marketing of a new product, is interested in the proportion p of consumers that would try the new product. In a pilot sample of 40 randomly chosen consumers, 9 said that they would purchase the new product and give it a try. Use either hand calculations or R commands to answer the following.

(a) What sample size is needed for the 90% CI for p to have length 0.1?

(b) What would your answer be if no information from a pilot study were available?

4. A pilot study of the electro-mechanical protection devices used in electrical power systems showed that of 193 devices that failed when tested, 75 failed due to mechanical parts failures. Use either hand calculations or R commands to answer the following.

(a) How large a sample is required to estimate p to within 0.03 with 95% confidence?

(b) What would your answer be if no information from a pilot study were available?

7.5 Prediction Intervals

7.5.1 BASIC CONCEPTS

The meaning of the word *prediction* is related to, but is distinct from, the word *estimation*. The latter is used for a population or model parameter, while the former is used for a future observation. For a concrete example, suppose a person eating a hot dog wonders about the amount of fat in the hot dog he or she is eating. This is different from the question "what is the expected (mean) amount of fat in a hot dog of the type being eaten?" To further emphasize the difference between the two, suppose that the mean amount of fat is known to be 20 grams. Even so, the amount of fat in the particular hot dog being eaten is unknown, simply because it is a random variable.

The basic result in prediction is given in the next proposition.

Proposition 7.5-1 According to the mean square error (MSE) criterion, the best predictor of a random variable Y is its mean value μ_Y.

Proof The prediction error of a predictor P_Y of Y is $Y - P_Y$. According to the MSE criterion, the best predictor is the one that achieves the smallest MSE, where the MSE of a predictor P_Y of Y is $E[(Y - P_Y)^2]$. Adding and subtracting μ_Y, the MSE of a predictor P_Y can be written as

$$\begin{aligned} E[(Y - P_Y)^2] &= E[(Y - \mu_Y + \mu_Y - P_Y)^2] \\ &= E[(Y - \mu_Y)^2] + 2E[(Y - \mu_Y)(\mu_Y - P_Y)] + E[(\mu_Y - P_Y)^2] \\ &= \sigma_Y^2 + 0 + (\mu_Y - P_Y)^2, \end{aligned} \quad \textbf{(7.5.1)}$$

where the last equality holds because $(\mu_Y - P_Y)$ is a constant and, hence, $E[(\mu_Y - P_Y)^2] = (\mu_Y - P_Y)^2$ and $E[(Y - \mu_Y)(\mu_Y - P_Y)] = (\mu_Y - P_Y)E[(Y - \mu_Y)] = 0$. Relation (7.5.1) implies that the smallest MSE any predictor of Y can achieve is σ_Y^2, and $P_Y = \mu_Y$ achieves this minimum value. ■

Proposition 7.5-1 also implies that if prior information is available, the best predictor of Y is the conditional expectation of Y given the available information. In particular, given the information that a predictor variable X takes the value x, the best predictor of Y is $E(Y|X = x) = \mu_{Y|X}(x)$.

The confidence interval analogue in prediction is called the **prediction interval (PI)**. Roughly speaking, the $(1 - \alpha)100\%$ PI for a future observation Y is

$$(y_{(1-\alpha/2)}, y_{\alpha/2}),$$

where y_α denotes the $(1 - \alpha)100$th percentile of the distribution (or the conditional, given $X = x$, distribution) of Y. Because of this, construction of a PI requires knowledge of the distribution (or the conditional, given $X = x$, distribution) of Y. This is a sharp difference between CIs and PIs.

If we assume that $Y \sim N(\mu_Y, \sigma_Y^2)$, and if μ_Y and σ_Y^2 are known, the $(1-\alpha)100\%$ PI for a future observation Y is

$$(\mu_Y - z_{\alpha/2}\sigma_Y, \mu_Y + z_{\alpha/2}\sigma_Y). \quad \textbf{(7.5.2)}$$

This PI is illustrated in Figure 7-8.

Similarly, if we assume the normal linear regression model, that is, $Y|X = x \sim N(\alpha_1 + \beta_1 x, \sigma_\varepsilon^2)$, and if all population parameters are known, the $(1-\alpha)100\%$ PI for a future observation Y, taken when the predictor variable has the value $X = x$, is

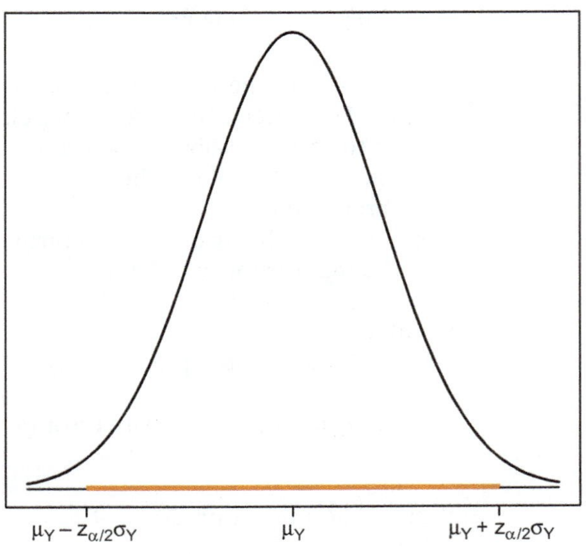

Figure 7-8 The $(1 - \alpha)100\%$ PI for a future observation Y having the normal distribution when μ_Y and σ_Y^2 are known.

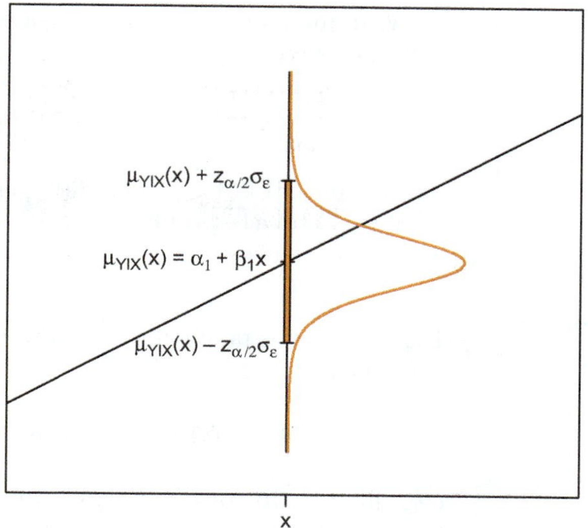

Figure 7-9 Prediction interval for a future observation Y at $X = x$ under the normal simple linear regression model with all model parameters known.

$$(\alpha_1 + \beta_1 x - z_{\alpha/2}\sigma_\varepsilon,\ \alpha_1 + \beta_1 x + z_{\alpha/2}\sigma_\varepsilon). \tag{7.5.3}$$

This PI is illustrated in Figure 7-9.

The model parameters μ_Y and σ_Y (and, in a regression context, α_1, β_1, and σ_ε) are, of course, unknown and need to be estimated. However, simple substitution of estimators in the PIs (7.5.2) and (7.5.3) does not suffice. This is because the variability of the estimators must be taken into consideration. This is described next.

7.5.2 PREDICTION OF A NORMAL RANDOM VARIABLE

Let $Y_1, \ldots, Y_n$ be a sample from a normal(μ_Y, σ_Y^2) population, and let $\overline{Y}$, S^2 denote the sample mean and variance. By Proposition 7.5-1, the best predictor of a future observation Y is μ_Y. Since $\overline{Y}$ estimates the unknown μ_Y (in fact, $\overline{Y}$ is the best estimator of μ_Y under the normal model), we use $\overline{Y}$ as a prediction of the value of Y. If Y is independent from the sample (which is typically the case), the prediction error $Y - \overline{Y}$ has variance

$$\sigma_Y^2 + \frac{\sigma_Y^2}{n} = \sigma_Y^2\left(1 + \frac{1}{n}\right).$$

(Compare the above with the variance of the prediction error $Y - \mu_Y$, which is σ_Y^2.) Thus, replacing μ_Y by $\overline{Y}$ in (7.5.2) requires σ_Y to be replaced by $\sigma_Y\sqrt{1 + 1/n}$. Finally, replacing σ_Y by the sample standard deviation S requires that $z_{\alpha/2}$ in (7.5.2) be replaced by $t_{n-1,\,\alpha/2}$. With these substitutions, (7.5.2) transforms into the following $(1 - \alpha)100\%$ PI:

$(1 - \alpha)100\%$ PI for a Future Observation Y

$$\overline{Y} \pm t_{n-1,\,\alpha/2} S\sqrt{1 + \frac{1}{n}} \tag{7.5.4}$$

Note that for a large enough sample size, the PI in (7.5.4) is nearly the same as the PI in (7.5.2). This is because if n is large enough, $\overline{Y} \simeq \mu_Y$, $S \simeq \sigma_Y$ (both by the Law of Large Numbers), $t_{n-1,\,\alpha/2} \simeq z_{\alpha/2}$, and $1 + 1/n \simeq 1$.

With the data set in the R object y, the PI in (7.5.4) can be obtained with the following R command:

R Command for the $(1-\alpha)100\%$ PI in (7.5.4)

```
predict(lm(y~1), data.frame(1), interval="predict",
  level=1-α)
```
(7.5.5)

Example 7.5-1

The following is fat content measurements (in g) from a sample of size $n = 10$ hot dogs of a certain type:

$$y \mid 24.21, 20.15, 14.70, 24.38, 17.02, 25.03, 26.47, 20.74, 26.92, 19.38$$

Use this data to construct a 95% PI for the fat content of the next hot dog to be sampled. What assumptions are needed for the validity of this PI?

Solution

The sample mean and standard deviation from this data is $\overline{Y} = 21.9$ and $S = 4.13$. Using this, and the percentile value $t_{9,0.025} = 2.262$, formula (7.5.4) yields the 95% PI

$$\overline{Y} \pm t_{9,0.025}\, S\sqrt{1+\frac{1}{n}} = (12.09, 31.71).$$

Copying the data into the R object y, that is, by *y=c(24.21, 20.15, 14.70, 24.38, 17.02, 25.03, 26.47, 20.74, 26.92, 19.38)*, the R command *predict(lm(y~1), data.frame(1), interval="predict", level=0.95)* yields the same PI. The constructed PI is valid under the assumption that the 10 measurements have come from a normal population. The normal Q-Q plot for this data (not shown here) suggests that this assumption may indeed be (at least approximately) correct.

7.5.3 PREDICTION IN NORMAL SIMPLE LINEAR REGRESSION

Let $(X_1, Y_1), \ldots, (X_n, Y_n)$ be a sample from a population of (X, Y) values, and let $\widehat{\alpha}_1$, $\widehat{\beta}_1$, and S_ε^2 denote the LSEs of α_1, β_1, and σ_ε^2, respectively. By Proposition 7.5-1, the best predictor of a future observation Y at $X = x$ is $\mu_{Y|X}(x) = \alpha_1 + \beta_1 x$. Assume now that the population of (X, Y) values satisfies the assumptions of the normal simple linear regression model. Since $\widehat{\alpha}_1$ and $\widehat{\beta}_1$ estimate α_1 and β_1, respectively (in fact, they are the best estimators under the normal simple linear regression model), we use $\widehat{\mu}_{Y|X}(x) = \widehat{\alpha}_1 + \widehat{\beta}_1 x$ to predict the value of Y.

Using arguments similar to those leading to the PI (7.5.4), it follows that replacing the unknown model parameters in the formula (7.5.3) by their estimators requires additional substitutions that transform the formula into the following $100(1-\alpha)\%$ PI for a future observation Y at $X = x$:

$100(1-\alpha)\%$ PI for an Observation Y at $X = x$

$$\widehat{\mu}_{Y|X=x} \pm t_{n-2,\,\alpha/2} S_\varepsilon \sqrt{1+\frac{1}{n}+\frac{n(x-\overline{X})^2}{n\sum X_i^2-(\sum X_i)^2}} \tag{7.5.6}$$

With the data on the predictor and the response variables in the R objects x and y, respectively, the PI in (7.5.6) for a new observation Y at X=v (note the use of v instead of the x used in (7.5.6), since now x stands for the R object containing the X-values) can be obtained with the following R commands:

R Command for the $(1-\alpha)100\%$ PI in (7.5.7)

```
predict(lm(y~x), data.frame(x=v), interval="predict",
  level=1-α)
```

(7.5.7)

The v in *data.frame(x=v)* of the above command can be either a single value, for example, *data.frame(x=4)*, or a set of values, for example, *data.frame(x=c(4, 6.3, 7.8))*. In the latter case, the command will return $(1-\alpha)100\%$ PIs for future observations at $X = 4$, $X = 6.3$, and $X = 7.8$.

Example 7.5-2

Data on rainfall volume (X) and soil runoff volume (Y) can be found in *SoilRunOffData.txt*.[4] Make a prediction for the volume of soil runoff at the next rainfall of volume $X = 62$, and construct a 95% PI for the volume of soil runoff at $X = 62$. What assumptions are needed for the validity of the prediction? What assumptions are needed for the validity of the PI?

Solution

The $n = 15$ data points give the following summary statistics: $\sum_i X_i = 798$, $\sum_i X_i^2 = 63040$, $\sum_i Y_i = 643$, $\sum_i Y_i^2 = 41999$, and $\sum_i X_i Y_i = 51232$. On the basis of these summary statistics, the LSEs of α_1, β_1, and σ_ε are found to be (see formulas (7.3.6) and (7.3.7)) $\widehat{\alpha}_1 = -1.128$, $\widehat{\beta}_1 = 0.827$, and $S_\varepsilon = 5.24$. The soil runoff at the next rainfall of volume $X = 62$ is predicted to be

$$\widehat{\mu}_{Y|X=62} = -1.128 + 0.827 \times 62 = 50.15.$$

Using the above calculations and the percentile value $t_{13,\,0.25} = 2.16$, formula (7.5.6) yields the following 95% PI for a Y measurement to be made at $X = 62$:

$$\widehat{\mu}_{Y|X=62} \pm t_{13,\,0.025}(5.24)\sqrt{1 + \frac{1}{15} + \frac{15(62 - 798/15)^2}{15 \times 63,040 - 798^2}} = (38.43, 61.86).$$

With the data on the predictor and the response variables in the R objects x and y, respectively, the R command *predict(lm(y~x), data.frame(x=62), interval="predict", level=0.95)* yields the same PI. For the validity of the prediction, the assumption of linearity of the regression function of Y on X must be satisfied. The validity of the PI requires that all the assumptions of the normal simple linear regression model, that is, the additional assumptions of homoscedasticity and normality of the intrinsic error variables, be satisfied. The scatterplot in the left panel of Figure 7-10 suggests that the first two assumptions are reasonable for this data set, while the normal Q-Q plot for the residuals in the right panel does not suggest gross violation of the normality assumption.

[4] M. E. Barrett et al. (1995). Characterization of Highway Runoff in Austin, Texas Area, Center for Research in Water Resources, University of Texas at Austin, Tech. Rep.# CRWR 263.

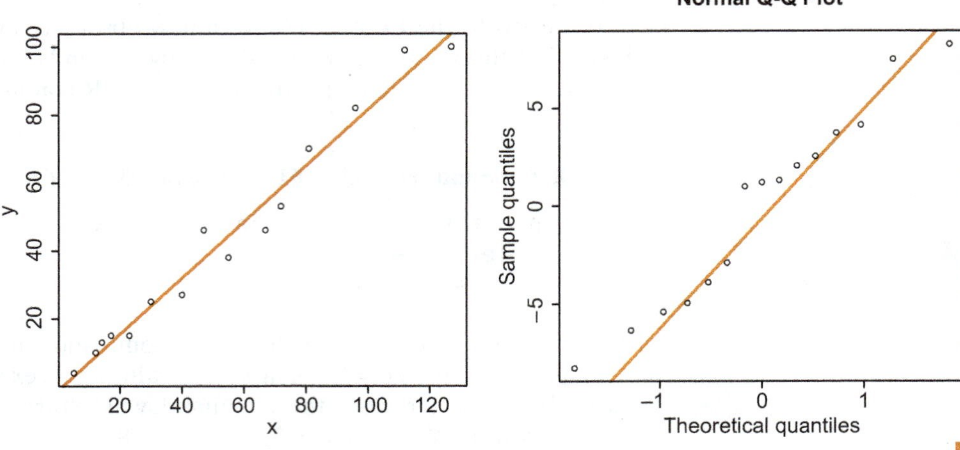

Figure 7-10 Scatterplot (left panel) and residual Q-Q plot (right panel) for the data of Example 7.5-2.

Exercises

1. A random sample of 16 chocolate chip cookies made by a machine yields an average weight of 3.1 oz and a standard deviation of 0.3 oz. Construct an interval that contains the weight of the next chocolate chip cookie to be sampled with confidence 90%, and state any assumptions that are needed for its validity.

2. Soil heat flux is used extensively in agrometeorological studies, as it relates to the amount of energy stored in the soil as a function of time. A particular application is in the prevention of frost damage to orchards. Heat flux measurements of eight plots covered with coal dust yielded $\overline{X} = 30.79$ and $S = 6.53$. A local farmer using this method of frost prevention wants to use this information for predicting the heat flux of his coal-dust-covered plot.

(a) Construct a 90% prediction interval and state what assumptions, if any, are needed for its validity.

(b) Construct a 90% CI for the mean heat flux, and compare the lengths of the confidence and prediction intervals.

3. Use *cs=read.table("Concr.Strength.1s.Data.txt", header =T)* to read into the R data frame *cs* data on 28-day compressive strength measurements of concrete cylinders using water/cement ratio 0.4,[5] and the command *y=cs$Str* to copy the $n = 32$ data points into the R object *y*. Use R commands to complete the following parts.

(a) Construct a 95% PI for the compressive strength of the next concrete specimen.

(b) Construct a normal Q-Q plot for the data. Comment on whether or not the plot appears to contradict the normality assumption.

4. The article *Effects of Bike Lanes on Driver and Bicyclist Behavior* reports data from a study on X = distance between a cyclist and the roadway center line, and Y = the separation distance between the cyclist and a passing car (both determined by photography).[6] The data, in feet, from $n = 10$ streets with bike lanes yield the summary statistics $\sum_i X_i = 154.2$, $\sum_i Y_i = 80$, $\sum_i X_i^2 = 2452.18$, $\sum_i Y_i^2 = 675.16$, $\sum_i X_i Y_i = 1282.74$, with the X-values ranging from 12.8 to 20.8 feet. Fitting the simple linear regression model with the method of LS yields the estimates $\widehat{\alpha}_1 = -2.1825$, $\widehat{\beta}_1 = 0.6603$, and $S_\varepsilon^2 = 0.3389$ for α_1, β_1, and σ_ε^2, respectively.

(a) Construct a 90% prediction interval for the separation distance between the next cyclist, whose distance from the roadway center line is 15 feet, and a passing car.

(b) A state-level committee charged with investigating the safety of existing bike lanes requests the civil engineering department to provide a 90% PI for the separation distance when the cyclist is 12 feet from the center line. What should the civil engineering department's response be?

5. Fifty black bears, captured in the period September to November, were anesthetized and their bodies were measured and weighed.[7] Use *bd=read.table("BearsData.txt", header=T); x=bd$Chest.G; y=bd$Weight* to import the

[5] V. K. Alilou and M. Teshnehlab (2010). Prediction of 28-day compressive strength of concrete on the third day using artificial neural networks. *International Journal of Engineering (IJE)*, 3(6): 521–670.

[6] B. J. Kroll and M. R. Ramey (1977). Effects of bike lanes an driver and bicyclist behavior, *Transportation Eng. J.*, 103(2): 243–256.

[7] This data set is a subset of a data set contributed to Minitab by Gary Alt.

data into the R data frame *bd* and to copy the chest girth and weight measurements into the R objects x and y, respectively. Use R commands to complete the following parts.

(a) Using only the weight measurements, that is, the data in the R object y, make a prediction for the weight of the next black bear that will be captured during the same time period and construct a 95% PI.
  (i) What assumptions are needed for the validity of your prediction?
  (ii) What assumptions are needed for the validity of the PI?

(b) Using both the chest girth and the weight measurements, make a prediction for the weight of the next bear that will be captured during the same time period if its chest girth measures 40 cm. Construct also a 95% PI.
  (i) What assumptions are needed for the validity of your prediction?
  (ii) What assumptions are needed for the validity of the PI?

(c) Comment on the lengths of the PIs constructed in parts (a) and (b).

Chapter 8

Testing of Hypotheses

8.1 Introduction

In many situations, mainly involving decision making, investigators are called upon to decide whether a statement regarding the value of a parameter is true. For example, the decision to implement a new design for a product may rest on whether the mean value of some quality characteristic of the product exceeds a certain threshold. A statement regarding the value of a parameter of interest is called a *hypothesis*. This chapter deals with the statistical procedures for *testing hypotheses*, that is, for deciding whether a certain hypothesis regarding the true value of a parameter θ is supported by the data.

Confidence intervals, which provide a set of plausible (i.e., compatible with the data) values for the true value of θ, can be used for hypothesis testing. For example, consider testing a hypothesis of the form

$$H_0 : \theta = \theta_0, \tag{8.1.1}$$

where θ_0 is a specified value. A sensible way of testing this hypothesis is to construct a CI for θ and check whether or not the specified value θ_0 belongs in the CI. If it does then the hypothesis H_0 is compatible with the data and cannot be refuted/rejected, while if it does not belong in the CI then H_0 is not supported by the data and is rejected. For a concrete example of this testing procedure, suppose it is hypothesized that the population mean value is 9.8. Thus, $\theta = \mu$, $\theta_0 = 9.8$, and

$$H_0 : \mu = 9.8.$$

Suppose further that the data yield a 95% CI of $(9.3, 9.9)$ for the true value of μ. Since 9.8 belongs in the 95% CI, we conclude that H_0 is not refuted by the data or, in proper statistical parlance, that H_0 is not rejected at *level of significance* $\alpha = 0.05$.

Even though there is a close connection between CIs and hypothesis testing, there are a number of specific issues that arise in hypothesis testing and deserve separate treatment. These issues are listed below.

1. **The null hypothesis and the alternative hypothesis.** In every hypothesis-testing situation, there is a null and an alternative hypothesis. Typically, the statement of the alternative hypothesis is the complement of the statement of the null hypothesis. For example, the alternative to the null hypothesis $H_0 : \theta = \theta_0$ is

$$H_a : \theta \neq \theta_0.$$

This alternative hypothesis is called *two-sided*. Other common null hypotheses are

$$H_0 : \theta \le \theta_0 \quad \text{or} \quad H_0 : \theta \ge \theta_0, \tag{8.1.2}$$

with corresponding alternative hypotheses

$$H_a : \theta > \theta_0 \quad \text{or} \quad H_a : \theta < \theta_0. \tag{8.1.3}$$

The alternative hypotheses in (8.1.3) are called *one-sided*. Testing procedures do not treat the null and the alternative hypotheses equally. Basically, test procedures treat the null hypothesis in a manner similar to the manner in which the presumption of innocence is treated in a court of law. Thus, the null hypothesis is not rejected unless there is strong evidence (beyond "reasonable doubt," in legalistic terms) against it. Perhaps the most important learning objective of this chapter is the designation of the null and alternative hypotheses in a given testing situation.

2. **Rejection rules.** The intuitive, CI-based procedure described above for rejecting the null hypothesis in relation (8.1.1) is not suitable for testing the one-sided null hypotheses in (8.1.2). While it is possible to define *one-sided* CIs and base test procedures for one-sided hypotheses on them, there are more informative ways of reporting the outcome of a test procedure (see issue 4 below). Moreover, there exist test procedures that are quite distinct from the CI-based procedure. For these reasons, the rejection rules presented in this chapter do not make explicit reference to CIs.
3. **Sample size determination.** This issue involves considerations that are quite distinct from the considerations for determining the sample size required to achieve a desired level of precision in CI construction.
4. **Reporting the outcome.** The practice of reporting the outcome of a test procedure as "H_0 is rejected" or "H_0 is not rejected" fails to convey all available information regarding the strength (or lack thereof) of the evidence against H_0. Full information is conveyed by also reporting the so-called *p-value*.

In this chapter we will learn how to deal with these issues for testing hypotheses about a population mean, median, variance, and proportion, and we will learn about regression parameters in the simple linear regression (SLR) model.

8.2 Setting Up a Test Procedure

8.2.1 THE NULL AND ALTERNATIVE HYPOTHESES

The hypothesis testing problems we will consider take the form of deciding between two competing hypotheses, the **null hypothesis**, denoted by H_0, and the **alternative hypothesis**, denoted by H_a. Proper designation of H_0 and H_a is very important, because test procedures do not treat the two hypotheses symmetrically. In particular, test procedures are designed to favor the null hypothesis, so that H_0 will not be rejected unless the data present strong evidence against it. To draw an analogy, test procedures treat a null hypothesis like the presumption of innocence is treated in a court of law, where the accused is presumed innocent unless proven guilty.

An immediate implication of this is that when H_0 is not rejected, it cannot be claimed that it is true—one can only say that the evidence in the data is not strong enough to reject it. This is best demonstrated with the CI-based test procedure that rejects $H_0 : \theta = \theta_0$ if θ_0 does not belong in the CI for θ. Since any value in the CI is a plausible (given the data set) candidate for the true parameter value, it is evident that by not rejecting $H_0 : \theta = \theta_0$, we have not proved that H_0 is true. For example, if $H_0 : \mu = 9.8$ and the 95% CI for μ is $(9.3, 9.9)$, then $H_0 : \mu = 9.8$ is not rejected at level $\alpha = 0.05$; on the other hand $H_0 : \mu = 9.4$ is not rejected either, so that by not rejecting $H_0 : \mu = 9.8$, we have not proved that $H_0 : \mu = 9.8$ is true. The court of law analogy to this is that when an accused is acquitted, his or her innocence has not been established. A test procedure provides statistical proof only when H_0 is rejected. In that case it can be claimed that the alternative has been proved (in the statistical sense) at **level of significance** α. The level of significance quantifies the *reasonable doubt* we are willing to accept when rejecting a null hypothesis; see Section 8.2.2 for a more precise definition.

The above discussion leads to the following rule for designating H_0 and H_a:

> **Rule for Designating H_0 and H_a**
>
> The statement for which the investigator seeks evidence, or statistical proof, is designated as H_a. The complementary statement is designated as H_0.

It is important to note that, as suggested by (8.1.1) and (8.1.2), the equality sign ($=$, $\geq$, or $\leq$) is always part of the statement of H_0.

Example 8.2-1

Designate H_0 and H_a for each of the following testing situations.

(a) A trucking firm suspects that a tire manufacturer's claim that certain tires last at least 28,000 miles, on average, is faulty. The firm intends to initiate a study, involving data collection and hypothesis testing, to provide evidence supporting this suspicion.

(b) A tire manufacturing firm wants to claim that certain tires last, on average, more than 28,000 miles. The firm intends to initiate a study, involving data collection and hypothesis testing, to support the validity of the claim.

Solution

(a) Let μ denote the mean life span of the tires in question. The trucking firm seeks evidence that the claim made by the tire manufacturer is wrong, that is, it seeks evidence supporting the statement that $\mu < 28{,}000$. According to the rule, this statement is designated as H_a and the complementary statement is designated as H_0. Thus, the hypotheses to be tested are $H_0 : \mu \geq 28{,}000$ vs $H_a : \mu < 28{,}000$.

(b) The manufacturing firm seeks evidence in support of the claim that is about to be made, that is, that $\mu > 28{,}000$. According to the rule, this statement is designated as H_a and the complementary statement is designated as H_0. Thus, the hypotheses to be tested are $H_0 : \mu \leq 28{,}000$ vs $H_a : \mu > 28{,}000$. ■

8.2.2 TEST STATISTICS AND REJECTION RULES

A test procedure is specified in terms of a **test statistic** and a **rejection rule** (RR). The test statistic for testing a null hypothesis H_0 about a parameter θ can be based on a point estimator $\widehat{\theta}$ of θ. (Other types of test statistics will be seen in Sections 8.3.4 and 8.3.5.) The rejection rule prescribes when H_0 is to be rejected. Basically, H_0 is rejected when the test statistic takes a value of such magnitude (too large, or too small, or either, depending on the alternative hypothesis) that the value is unlikely if H_0 were true.

Consider, for example, the hypothesis-testing problem in part (a) of Example 8.2-1. Where the null hypothesis is $H_0 : \mu \geq 28{,}000$. Let $\overline{X}$ denote the average tread life span of a random sample of n tires. Being an estimator of μ, $\overline{X}$ is unlikely to take values that are much smaller than 28,000 if H_0 were true. Thus, the rejection region is of the form $\overline{X} \leq C_1$ for some constant smaller than 28,000. For example, the rejection region can be $\overline{X} \leq 27,000$, $\overline{X} \leq 26,000$, etc. Similarly, the null hypothesis $H_0 : \mu \leq 28,000$ in part (b) of Example 8.2-1 is rejected if the test statistic $\overline{X}$ takes a value so large that it is unlikely to happen if H_0 were true. Thus, the rejection region is of the form $\overline{X} \geq C_2$, where C_2 can be 29,000, 30,000, or some other constant larger than 28,000. Finally, for testing $H_0 : \mu = 28{,}000$ vs $H_a : \mu \neq 28,000$, the rejection region is of the form $\overline{X} \leq C_3$ or $\overline{X} \geq C_4$, where, for the CI-based procedure, C_3, C_4 are the endpoints of the CI.

But how exactly are the values of the constants C_1, C_2, C_3, and C_4 to be selected? The answer to this question rests on the **level of significance**, which is defined as

Definition of Level of Significance

The level of significance is the (largest) probability of incorrectly rejecting H_0, or, in other words, the (largest) probability of rejecting H_0 when H_0 is true.

As already mentioned, the level of significance specifies the risk (or, in legalistic terms, the "reasonable doubt") we are willing to accept for being wrong when concluding that H_0 is false. It turns out that the constants C_1 and C_2 can be determined by specifying the level of significance. The way this works is demonstrated in the following example.

Example 8.2-2

Consider the two testing problems in parts (a) and (b) of Example 8.2-1, and suppose that the tread life spans are approximately normally distributed and the population variance σ^2 is known. Let $\overline{X} \leq C_1$ and $\overline{X} \geq C_2$ be the rejection regions for the testing problems in parts (a) and (b), respectively. Determine the values of C_1 and C_2 so the tests have level of significance $\alpha = 0.05$.

Solution

Consider first the testing problem in part (a) where $H_0 : \mu \geq 28,000$. The requirement that the level of significance, that is, the probability of incorrectly rejecting H_0, is no more than 0.05 can be expressed, in mathematical notation, as

$$P(\overline{X} \leq C_1) \leq 0.05 \quad \text{if } H_0 \text{ is true.} \tag{8.2.1}$$

Clearly, the probability $P(\overline{X} \leq C_1)$ depends on the actual value of μ, which, when H_0 is true, can be any number $\geq$ 28,000. Thus the task is to choose C_1 so that the above probability does not exceed 0.05 no matter what value, in the range specified by H_0 (i.e., $\geq$ 28,000) μ takes. Because the event $\overline{X} \leq C_1$ specifies "small" values for $\overline{X}$, its probability is largest when $\mu =$ 28,000, that is, when μ takes the smallest value specified by H_0. Thus, the requirement (8.2.1) will be satisfied if C_1 is chosen so that when $\mu =$ 28,000, $P(\overline{X} \leq C_1) = 0.05$. This is achieved by choosing C_1 to be the 5th percentile of the distribution of $\overline{X}$ when $\mu = 28{,}000$, that is, $C_1 = 28{,}000 - z_{0.05}\sigma/\sqrt{n}$ (recall that σ is assumed known). This yields a rejection region of the form

$$\overline{X} \leq 28{,}000 - z_{0.05}\sigma/\sqrt{n}. \tag{8.2.2}$$

Similarly, the constant C_2 for testing the null hypothesis $H_0 : \mu \leq 28{,}000$ of part (b), is determined from the requirement that the probability of incorrectly rejecting H_0 is no more than 0.05 or, in mathematical notation, from

$$P(\overline{X} \geq C_2) \leq 0.05 \quad \text{if } H_0 \text{ is true.} \tag{8.2.3}$$

Again, it can be argued that, over the range of μ values specified by H_0 (i.e., $\mu \leq 28{,}000$), the probability $P(\overline{X} \geq C_2)$ is largest when $\mu = 28{,}000$. Thus, the requirement (8.2.3) will be satisfied if C_2 is chosen so that when $\mu = 28{,}000$, $P(\overline{X} \geq C_2) = 0.05$. This is achieved by choosing C_2 to be the 95th percentile of the distribution of $\overline{X}$ when $\mu = 28{,}000$. This yields a rejection region of the form

$$\overline{X} \geq 28{,}000 + z_{0.05}\sigma/\sqrt{n}. \tag{8.2.4}$$

■

The next example expresses the CI-based RR for testing $H_0 : \mu = \mu_0$, where μ_0 is a specified value, in terms of a test statistic, and shows that the level of significance of the test is related to the confidence level of the CI.

Example 8.2-3

Let $X_1, \ldots, X_n$ be a simple random sample from a normal(μ, σ^2) population, and consider testing the hypothesis $H_0 : \mu = \mu_0$ versus $H_a : \mu \neq \mu_0$ by rejecting H_0 if μ_0 does not lie inside the $(1-\alpha)100\%$ T CI for μ. Express this RR in terms of a test statistic, and show that the level of significance is α.

Solution

Since the $(1-\alpha)100\%$ CI for μ is $\overline{X} \pm t_{n-1,\,\alpha/2}S/\sqrt{n}$, H_0 is rejected if

$$\mu_0 \leq \overline{X} - t_{n-1,\,\alpha/2}S/\sqrt{n} \quad \text{or} \quad \mu_0 \geq \overline{X} + t_{n-1,\,\alpha/2}S/\sqrt{n}.$$

After an algebraic manipulation, the two inequalities can be rewritten as

$$\frac{\overline{X} - \mu_0}{S/\sqrt{n}} \geq t_{n-1,\,\alpha/2} \quad \text{or} \quad \frac{\overline{X} - \mu_0}{S/\sqrt{n}} \leq -t_{n-1,\,\alpha/2}.$$

Setting

$$T_{H_0} = \frac{\overline{X} - \mu_0}{S/\sqrt{n}}$$

for the test statistic, the CI-based RR can be expressed as

$$\left|T_{H_0}\right| \geq t_{n-1,\,\alpha/2}. \tag{8.2.5}$$

Finally, since $T_{H_0} \sim T_{n-1}$ when H_0 is true (see Proposition 7.3-1), it follows that the level of significance, that is, the (largest) probability of incorrectly rejecting H_0, is α.

■

Note that the test statistic T_{H_0} in Example 8.2-3 is a *standardized version*[1] of $\overline{X}$. In general, RRs for testing hypotheses about a parameter θ are expressed more concisely in terms of the standardized $\widehat{\theta}$. For example, using the standardized test statistic

$$Z_{H_0} = \frac{\overline{X} - \mu_0}{\sigma/\sqrt{n}} \tag{8.2.6}$$

with $\mu_0 = 28{,}000$, the RRs (8.2.2) and (8.2.4) can be written as

$$Z_{H_0} \le -z_{0.05} \quad \text{and} \quad Z_{H_0} \ge z_{0.05}, \tag{8.2.7}$$

respectively. For this reason, the test statistics based on $\widehat{\theta}$ will always be given in terms of the standardized $\widehat{\theta}$.

8.2.3 *Z* TESTS AND *T* TESTS

The description of the test procedures will be simplified by adopting the convention of always stating the null hypothesis about a parameter θ as

$$H_0 : \theta = \theta_0, \tag{8.2.8}$$

where θ_0 is a specified value, regardless of whether the alternative hypothesis is of the form $H_a : \theta < \theta_0$ or $H_a : \theta > \theta_0$ or $H_a : \theta \neq \theta_0$. This convention, done for reasons of convenience and simplicity in notation, should cause no confusion since the actual null hypothesis being tested is always the complementary version of a given alternative hypothesis.

As with confidence intervals, we have Z tests and T tests. The rejection rule in relation (8.2.7) is an example of a Z test, while that in (8.2.5) is an example of a T test. In general, tests that use percentiles from the standard normal distribution for the specification of the rejection rule are called Z tests, while tests that use percentiles from the T distribution are called T tests. As with CIs, there are also other types of tests which, typically but not always, are named after the distribution whose percentiles are used to specify the rejection rules. These will be discussed in the next section.

Like the Z intervals, Z tests for the mean are used only if the population variance is known and either the population is normal or the sample size is large enough (≥ 30). In this case, the test statistic for testing $H_0 : \mu = \mu_0$, where μ_0 is a specified value, against any of the common alternatives, is Z_{H_0} as given in (8.2.6). The subscript H_0 serves as a reminder that Z_{H_0} has the standard normal distribution (exactly, if sampling from a normal population; otherwise approximately, if $n \ge 30$) only if H_0 is true.

Because the assumption of a known population variance is not realistic, Z tests for the mean are deemphasized in this book. Instead, Z tests will be used primarily for the proportion and the median. The Z test statistic for testing $H_0 : p = p_0$, where p_0 is a specified value, against any of the common alternatives, is

Z Test Statistic for $H_0 : p = p_0$

$$Z_{H_0} = \frac{\widehat{p} - p_0}{\sqrt{p_0(1 - p_0)/n}} \tag{8.2.9}$$

[1] The standardized version of an estimator $\widehat{\theta}$ is obtained by subtracting its expected value and dividing by its standard error or estimated standard error.

By the DeMoivre-Laplace Theorem, if n is large enough ($np_0 \geq 5$ and $n(1-p_0) \geq 5$, for our purposes), Z_{H_0} has approximately the standard normal distribution only if $H_0 : p = p_0$ is true.

Like the T intervals, T tests will be used for the mean and the regression parameters (slope, intercept, and regression line). Let θ denote the true value of any of these parameters, and let $\widehat{\theta}$ and $S_{\widehat{\theta}}$ denote its estimator and the estimator's estimated standard error, respectively. As mentioned in Section 7.1.3, when sampling from normal populations,

$$\frac{\widehat{\theta} - \theta}{S_{\widehat{\theta}}} \sim T_\nu \qquad \textbf{(8.2.10)}$$

is true for all sample sizes n. Moreover, (8.2.10) is approximately true if $n \geq 30$ without the normality assumption. (Recall that ν is $n-1$ when $\theta = \mu$ and is $n-2$ when θ stands for any of the simple linear regression parameters.) Because of this, the test statistic for testing $H_0 : \theta = \theta_0$ is

T Test Statistic for $H_0 : \theta = \theta_0$ when θ is either μ, α_1, β_1, or $\mu_{Y|X=x}$

$$T_{H_0} = \frac{\widehat{\theta} - \theta_0}{S_{\widehat{\theta}}} \qquad \textbf{(8.2.11)}$$

According to (8.2.10), T_{H_0} has a T_ν distribution only if $H_0 : \theta = \theta_0$ is true.

8.2.4 P-VALUES

Reporting the outcome of a test procedure as only a rejection of a null hypothesis does not convey the full information contained in the data regarding the strength of evidence against it. To illustrate the type of information that is not conveyed, consider the following testing situation.

Example 8.2-4

It is suspected that a machine, used for filling plastic bottles with a net volume of 16.0 oz, on average, does not perform according to specifications. An engineer will collect 15 measurements and will reset the machine if there is evidence that the mean fill volume is different from 16 oz. The resulting data, given in *FillVolumes.txt*, yield $\overline{X} = 16.0367$ and $S = 0.0551$. Test the hypothesis $H_0 : \mu = 16$ vs $H_a : \mu \neq 16$ at level of significance $\alpha = 0.05$.

Solution

The value of the test statistic is

$$T_{H_0} = \frac{16.0367 - 16}{0.0551/\sqrt{15}} = 2.58.$$

According to the CI-based RR given in (8.2.5), H_0 will be rejected at level $\alpha = 0.05$ if $|T_{H_0}| \geq t_{14,0.025}$. Since $t_{14,0.025} = 2.145$, H_0 is rejected at level of significance $\alpha = 0.05$. ■

Knowing only the outcome of the test procedure in Example 8.2-4, we do not have a full appreciation of the strength of evidence against H_0. For example, we cannot know if H_0 would have been rejected had the level of significance been $\alpha =$ 0.01. Indeed, we cannot know whether or not $|T_{H_0}| \geq t_{14,0.005} = 2.977$ if it is only known that $|T_{H_0}| \geq 2.145$. With the data of this particular example, H_0 is not rejected at level of significance $\alpha = 0.01$.

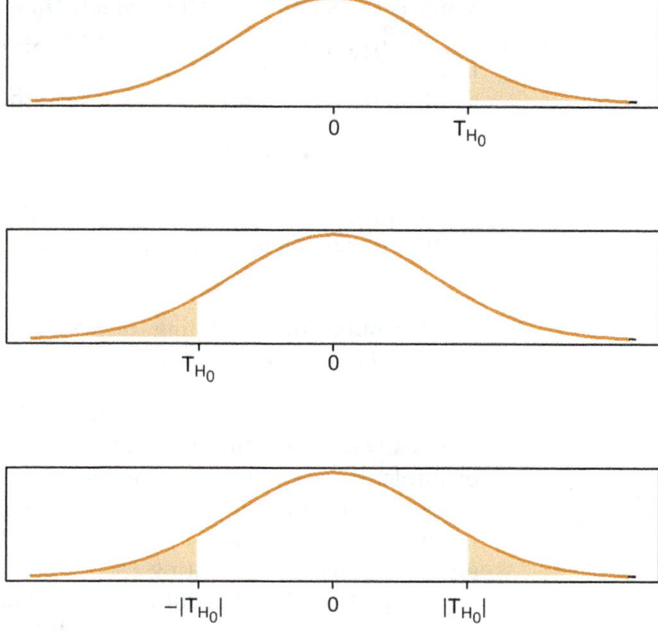

Figure 8-1 P-values of the T test for $H_a : \mu > \mu_0$ (top panel), $H_a : \mu < \mu_0$ (middle panel), and $H_a : \mu \neq \mu_0$ (bottom panel). In all panels the PDF is that of the T_{n-1} distribution.

We can convey the full information about the strength of evidence against the null hypothesis by reporting the ***p*-value**. Basically, the p-value is the (largest) probability, computed under H_0, of the test statistic taking a value more "extreme" (i.e., smaller or larger, depending on the alternative hypothesis) than the value we computed from the data. For a T test, with T_{H_0} denoting the value of the test statistic computed from the data, the p-values for the different alternative hypotheses are illustrated in Figure 8-1.

The relation of the p-value to the strength of evidence is inverse: The smaller the p-value, the stronger the evidence against H_0. The formal definition of the p-value is often given as follows.

Definition 8.2-1

In any hypothesis-testing problem, the ***p*-value** is the smallest level of significance at which H_0 would be rejected for a given data set.

The practical consequence of this definition is that the rejection rule for any hypothesis-testing problem can be stated in terms of the p-value as follows:

Rejection Rule in Terms of the *p*-Value

$$\text{If the } p\text{-value} \leq \alpha \Rightarrow \text{reject } H_0 \text{ at level } \alpha \tag{8.2.12}$$

The computation of the p-value is illustrated in the following example.

Example 8.2-5

Consider testing the hypothesis $H_0 : \mu = 16$ vs $H_a : \mu \neq 16$ using the sample of 15 observations given in Example 8.2-4. Find the p-value.

Solution

From Example 8.2-4 and the discussion following it, we have that the value of the test statistic is $T_{H_0} = 2.58$ and H_0 is rejected at level 0.05 but is not rejected at level 0.01.

What is the smallest level at which H_0 is rejected? Since the RR is $|T_{H_0}| \geq t_{14,\alpha/2}$, the smallest level at which H_0 is rejected is found by solving $2.58 = t_{14,\alpha/2}$ for α. Letting G_ν denote the cumulative distribution function of the T_ν distribution, the solution to this equation, which is also the p-value, is

$$p\text{-value} = 2(1 - G_{14}(2.58)) = 2 \times 0.0109 = 0.0218.$$

Alternatively (and more simply), since $T_{H_0} = 2.58$, the shaded area shown in the bottom panel of Figure 8-1 is $2(1 - G_{14}(2.58)) = 0.0218$, as computed above. ■

Formulas for calculating the p-value will be given for each of the test procedures discussed in the following section.

Statistical Significance vs Practical Significance The extent to which the alternative physically differs from the null hypothesis is referred to as *practical significance*. For example, suppose that gasoline additive A increases the mileage of a certain type of car by an average of $\mu_A = 2$ miles per gallon (mpg), while additive B results in an average increase of $\mu_B = 4$ mpg. Then, from the practical point of view, the alternative μ_B (when testing $H_0 : \mu \leq 0$ vs $H_a : \mu > 0$) is more significant (meaning more significantly different from H_0) than μ_A.

As already mentioned, the smaller the p-value the stronger the evidence against the null hypothesis, or, in statistical parlance, the higher the *statistical significance* of the alternative. Is it true that high statistical significance implies high practical significance? The correct answer is: Not necessarily! To see why, let's focus on the T test statistic T_{H_0}. As suggested by the bottom panel of Figure 8-1, the larger the value of $|T_{H_0}|$ the smaller the p-value when testing against the two-sided alternative. (Similar statements, accounting also for the sign of T_{H_0}, hold for the other alternative hypotheses.) $|T_{H_0}|$ is large if its numerator, $|\overline{X} - \mu_0|$, is large, or its denominator, $S/\sqrt{n}$, is small, or both. The practical significance of an alternative affects the value of the numerator, but not that of the denominator. For example, $\overline{X}$ tends to be larger under the alternative $\mu_B = 4$ than under the alternative $\mu_A = 2$, but the alternative has no effect on the sample size or the sample variance (at least in homoscedastic settings). Hence, since the practical significance of the alternative is not the only factor affecting the value of T_{H_0}, it follows that high statistical significance does not necessarily imply high practical significance.

For a numerical demonstration of the above discussion, we resort to simulated data. The R commands *n1=10; n2=100; set.seed(333); x1=rnorm(n1, 4, 5); x2=rnorm(n2, 2, 5)* generate two data sets, stored in objects *x1* and *x2*, of sizes 10 and 100, respectively. Thus, both population distributions are normal with $\sigma^2 = 25$, but for the data in *x1* the population mean is $\mu_1 = 4$, while for *x2* it is $\mu_2 = 2$. The additional commands *T1=mean(x1)/sqrt(var(x1)/n1); T2=mean(x2)/sqrt(var(x2)/n2); 1-pt(T1, n1-1); 1-pt(T2, n2-1)* compute the test statistics and yield p-values of 0.03 and 9.57×10^{-5} for samples *x1* and *x2*, respectively. Thus, there is stronger evidence (a smaller p-value) that $\mu_2 = 2$ is different from zero, even though $\mu_1 = 4$ is bigger.

Exercises

1. Researchers are exploring alternative methods for preventing frost damage to orchards. It is known that the mean soil heat flux for plots covered only with grass is 29 units. An alternative method is to use coal dust cover. Due to the additional cost of covering plots with coal dust, this method will not be recommended unless there is significant evidence that coal dust cover raises the mean soil heat flux by more than 2 units. Formulate this decision problem as a hypothesis-testing problem by

(a) stating the null and alternative hypotheses, and

(b) stating what action should be taken if the null hypothesis is rejected.

2. In 10-mph crash tests, 25% of a certain type of automobile sustain no visible damage. A modified bumper design has been proposed in an effort to increase this percentage. Let p denote the probability that a car with the modified bumper design sustains no visible damage in a 10-mph crash test. Due to cost considerations, the new design will not be implemented unless there is significant evidence that the new bumper design improves the crash test results. Formulate this decision problem as a hypothesis-testing problem by

(a) stating the null and alternative hypotheses, and

(b) stating what action should be taken if the null hypothesis is rejected.

3. The CEO of a car manufacturer is considering the adoption of a new type of grille guard for the upcoming line of SUVs. If μ_0 is the average protection index of the current grille guard and μ is the corresponding average for the new grille guard, the manufacturer wants to test the null hypothesis $H_0 : \mu = \mu_0$ against a suitable alternative.

(a) What should H_a be if the CEO wants to adopt it (because, by using the latest innovation from material science, it is lighter and thus will not affect the mileage) unless there is evidence that it has a lower protection index?

(b) What should H_a be if the CEO does not want to adopt it (because the new material is more expensive and he does not really care for the new design) unless there is evidence that it has a higher protection index?

(c) For each of the two cases above, state whether or not the CEO should adopt the new grille guard if the null hypothesis is rejected.

4. An appliance manufacturer is considering the purchase of a new machine for cutting sheet metal parts. If μ_0 is the average number of metal parts cut per hour by her old machine and μ is the corresponding average for the new machine, the manufacturer wants to test the null hypothesis $H_0 : \mu = \mu_0$ against a suitable alternative.

(a) What should H_a be if she does not want to buy the new machine unless there is evidence it is more productive than the old one?

(b) What should H_a be if she wants to buy the new machine (which has additional improved features) unless there is evidence it is less productive than the old one?

(c) For each of the two cases above, state whether she should buy the new machine if the null hypothesis is rejected.

5. In making plans for an executive traveler's club, an airline would like to estimate the proportion of its current customers who would qualify for membership. A random sample of 500 customers yielded 40 who would qualify.

(a) The airline wants to proceed with the establishment of the executive traveler's club unless there is evidence that less that 5% of its customers qualify. State the null and alternative hypotheses.

(b) State the action the airline should take if the null hypothesis is rejected.

6. Consider the hypothesis-testing problem in part (b) of Example 8.2-1. Thus, the null hypothesis is $H_0 : \mu \le 28,000$. Let $\overline{X}$ denote the average tread life span of a random sample of n tires.

(a) The rule for rejecting this null hypothesis should be of the form $\overline{X} \ge C$, for some constant $C > 28{,}000$. True or false?

(b) Use the rationale of Example 8.2-2 to find C when the level of significance is chosen as $\alpha = 0.05$, assuming that the population variance σ^2 is known.

(c) Express the RR in terms of the standardized $\overline{X}$.

7. In a simple linear regression context, consider testing a hypothesis regarding the expected response at a value $X = x$ of the covariate.

(a) If the hypothesis is of the form $H_0 : \mu_{Y|X}(x) = \mu_{Y|X}(x)_0$ versus $H_a : \mu_{Y|X}(x) > \mu_{Y|X}(x)_0$, where $\mu_{Y|X}(x)_0$ is a specified value, state the form of the RR. (*Hint.* It is either of the form $\widehat{\mu}_{Y|X}(x) \ge C$ or of the form $\widehat{\mu}_{Y|X}(x) \le C$ for some constant C.)

(b) Use the rationale of Example 8.2-2 to determine the constant C from the requirement that the level of significance is $\alpha = 0.05$.

8. A coal mining company suspects that certain detonators used with explosives do not meet the requirement that at least 90% will ignite. To investigate this suspicion, a random sample of n detonators is selected and tested. Let X denote the number of those that ignite.

(a) State the null and alternative hypotheses.

(b) Give the standardized test statistic Z_{H_0} in terms of $\widehat{p} = X/n$.

(c) The RR will be of the form $Z_{H_0} \ge C$ for some constant C. True or false?

9. Working as in Example 8.2-3, give the CI-based rejection rule for testing $H_0 : \theta = \theta_0$ versus $H_a : \theta \ne \theta_0$ in terms of a test statistic when

(a) θ stands for the regression slope β_1.

(b) θ stands for the expected response $\mu_{Y|X}(x)$ at a given value x of the covariate.

10. Suppose a tire manufacturer wants to claim that the average tread life of a certain type of tire is more than 28,000 miles. To gain empirical evidence that this claim is true, a study is initiated to test $H_0 : \mu = 28{,}000$ vs H_a :

$\mu > 28{,}000$. The tread life spans of a random sample of $n = 25$ yields sample mean of $\overline{X} = 28{,}640$. Assume the tread life spans are normally distributed and the standard deviation is known to be $\sigma = 900$.

(a) Find the p-value. (*Hint.* Use the RR in (8.2.7) and argue as in Example 8.2-5.)

(b) Should the null hypothesis be rejected at level of significance 0.05?

11. In the context of Exercise 2, suppose it is decided to test $H_0 : p = 0.25$ vs $H_a : p > 0.25$. In $n = 50$ independent crashes of car prototypes with the new bumper, $X = 8$ result in no visible damage. Let Z_{H_0} be the test statistic given in (8.2.9).

(a) Draw by hand a figure similar to those in Figure 8-1, using Z_{H_0} instead of T_{H_0}, and shade the area that corresponds to the p-value. The PDF you drew is that of what distribution?

(b) Use the figure you drew to compute the p-value. Should the H_0 be rejected at level of significance $\alpha = 0.05$?

8.3 Types of Tests

8.3.1 *T* TESTS FOR THE MEAN

Let $X_1, \ldots, X_n$ be a simple random sample from a population, and let $\overline{X}$, S^2 denote the sample mean and sample variance, respectively. The T test procedures for testing $H_0 : \mu = \mu_0$, where μ_0 is a given value, against the various alternative hypotheses, as well as formulas for the p-value, are given below:

The T Test Procedures for $H_0 : \mu = \mu_0$

(1) *Assumptions:* The population is normal, or $n \geq 30$

(2) *Test Statistic:* $T_{H_0} = \dfrac{\overline{X} - \mu_0}{S/\sqrt{n}}$

(3) *Rejection Rules for the Different H_a:*

H_a	RR at Level α
$\mu > \mu_0$	$T_{H_0} > t_{n-1,\alpha}$
$\mu < \mu_0$	$T_{H_0} < -t_{n-1,\alpha}$
$\mu \neq \mu_0$	$\lvert T_{H_0}\rvert > t_{n-1,\alpha/2}$

(8.3.1)

(4) *Formulas for the p-Value:*

$$p\text{-value} = \begin{cases} 1 - G_{n-1}(T_{H_0}) & \text{for } H_a : \mu > \mu_0 \\ G_{n-1}(T_{H_0}) & \text{for } H_a : \mu < \mu_0 \\ 2\left[1 - G_{n-1}(|T_{H_0}|)\right] & \text{for } H_a : \mu \neq \mu_0 \end{cases}$$

where G_{n-1} is the CDF of the T_{n-1} distribution

With the data set in the R object x, the R command for computing the above T test statistic and p-value is as follows:

R Commands for the T Test Statistic and p-Values in (8.3.1)

```
t.test(x, mu=μ0, alternative="greater") # for testing
  against Ha : μ > μ0

t.test(x, mu=μ0, alternative="less") # for testing against
  Ha : μ < μ0

t.test(x, mu=μ0, alternative="two.sided") # for testing
  against Ha : μ ≠ μ0
```

These R commands will also return a CI for μ (one-sided CIs for *alternative="greater"* and *alternative="less"* and the usual two-sided CI for *alternative="two.sided"*). The default confidence level is 95%, but adding *conf.level=1-α* in any of these commands, for example, *t.test(x, mu=μ_0, alternative="two.sided", conf.level = 1-α)*, gives a $(1-\alpha)100\%$ CI.

Example 8.3-1

The proposed federal health standard for the maximum acceptable level of exposure to microwave radiation in the US[2] is an average of 10 W/cm^2. It is suspected that a radar installation used for air traffic control may be pushing the average level of radiation above the safe limit. A random sample of $n = 25$ measurements, taken at different points around the installation and at different times of the day, is given in *ExRadiationTestData.txt.*

(a) Specify the alternative hypothesis. What are the requirements for the validity of the T test in this case?

(b) Compute the test statistic and the corresponding p-value. Do the data provide strong evidence in favor of the suspicion? Use the p-value to test $H_0 : \mu = 10$ at level of significance $\alpha = 0.05$.

(c) Conduct the above test using the rejection rules given in (8.3.1) with $\alpha = 0.05$.

Solution

(a) To gain evidence in support of the suspicion, the alternative hypothesis is specified as $H_a : \mu > 10$. Since the sample size is less than 30, the validity of the test requires the data to have come from a normal population. A normal Q-Q plot (not shown here) suggests that normality is a reasonable assumption for this data set.

(b) The data yield $\overline{X} = 10.6$ and $S = 2.0$. Hence, the test statistic is

$$T_{H_0} = \frac{\overline{X} - \mu_0}{s/\sqrt{n}} = \frac{10.6 - 10}{2/\sqrt{25}} = 1.5.$$

Table A.4 is not detailed enough for exact computation of the p-value. The table gives 1.318 and 1.711 as the 90th and 95th percentiles of the T_{24} distribution. Hence, the p-value, which is $1 - G_{24}(1.5)$, lies between 0.05 and 0.1. The exact p-value, which can be found with the R command *1-pt(1.5, 24)*, is 0.0733. With the data copied in the R object x, the R command *t.test(x, alternative="greater", mu=10)* returns the same values (up to round-off error) for the test statistic and the p-value. The evidence in favor of the suspicion (and hence against H_0) suggested by the p-value is only moderately strong. In particular, H_0 cannot be rejected at level of significance $\alpha = 0.05$ (since $0.0733 > 0.05$) but it would have been rejected at $\alpha = 0.1$.

(c) According to the RR given in (8.3.1), H_0 is rejected in favor of $H_a : \mu > 10$ if $T_{H_0} > t_{24,0.05} = 1.711$. Since 1.5 is not greater than 1.711, H_0 is not rejected. ■

REMARK 8.3-1 The result in Example 8.3-1, namely, that $\overline{X} = 10.6$ but $H_0 : \mu = 10$, is not rejected at level 0.05 in favor of $H_a : \mu > 10$, serves to highlight the fact that test procedures do not treat the two hypotheses evenly. ◁

[2] S. Henry (1978). Microwave radiation: level of acceptable exposure subject of wide disagreement. *Can. Med. Assoc. J.*, 119(4): 367–368.

8.3.2 Z TESTS FOR PROPORTIONS

Let X denote the number of successes in n Bernoulli trials, and let $\widehat{p} = X/n$ denote the sample proportion. The Z test procedure for testing $H_0 : p = p_0$, where p_0 is a specified value, against the various alternative hypotheses, as well as formulas for the p-value, are given below:

The Z Test Procedures for $H_0 : p = p_0$

(1) *Condition:* $np_0 \geq 5$ and $n(1 - p_0) \geq 5$

(2) *Test Statistic:* $Z_{H_0} = \dfrac{\widehat{p} - p_0}{\sqrt{p_0(1 - p_0)/n}}$

(3) *Rejection Rules for the Different H_a:*

H_a	RR at Level α
$p > p_0$	$Z_{H_0} \geq z_\alpha$
$p < p_0$	$Z_{H_0} \leq -z_\alpha$
$p \neq p_0$	$\lvert Z_{H_0}\rvert \geq z_{\alpha/2}$

(8.3.2)

(4) *Formulas for the p-Value:*

$$p\text{-value} = \begin{cases} 1 - \Phi(Z_{H_0}) & \text{for } H_a : p > p_0 \\ \Phi(Z_{H_0}) & \text{for } H_a : p < p_0 \\ 2[1 - \Phi(|Z_{H_0}|)] & \text{for } H_a : p \neq p_0 \end{cases}$$

With the number of successes and the number of trials in the R objects x, n, respectively, the R commands *1-pnorm((x/n-p$_0$)/sqrt(p$_0^*$(1-p$_0$)/n))*, *pnorm((x/n-p$_0$)/sqrt(p$_0^*$(1-p$_0$)/n))*, and *2*(1-pnorm(abs((x/n-p$_0$)/sqrt(p$_0$*(1-p$_0$)/n))))* compute the p-values for $H_a : p > p_0$, $H_a : p < p_0$, and $H_a : p \neq p_0$, respectively.

Example 8.3-2

It is thought that more than 70% of all faults in transmission lines are caused by lightning. In a random sample of 200 faults from a large data base, 151 are due to lightning. Does the data provide strong evidence in support of this contention? Test at level of significance $\alpha = 0.01$, and report the p-value.

Solution

To assess the evidence in favor of this contention, the alternative hypothesis is specified as $H_a : p > 0.7$. Since $200(0.7) \geq 5$ and $200(0.3) \geq 5$, the condition needed for the validity of the test procedure in (8.3.2) is satisfied. From the data given we have $\widehat{p} = 151/200 = 0.755$, and thus the test statistic is

$$Z_{H_0} = \frac{\widehat{p} - 0.7}{\sqrt{(0.7)(0.3)/200}} = 1.697.$$

According to the RR given in (8.3.2), H_0 is rejected in favor of $H_a : p > 0.7$ at level of significance $\alpha = 0.01$ if $Z_{H_0} > z_{0.01} = 2.33$. Since $1.697 \not> 2.33$, H_0 is not rejected. Next, since this is an upper tail test, the p-value is $1 - \Phi(1.697)$. Using Table A.3, we find

$$p\text{-value} = 1 - \Phi(1.697) \simeq 1 - \Phi(1.7) = 1 - 0.9554 = 0.0446.$$

The R command *1-pnorm((151/200-0.7)/sqrt(0.7*0.3/200))* returns 0.0448 for the p-value. The additional information conveyed by reporting the p-value is that the data does provide strong evidence against H_0, and, hence, in support of the contention that more than 70% of all faults in transmission lines are caused by lightning. In fact, had the level of significance been set at 0.05, H_0 would have been rejected because $0.0446 < 0.05$. ■

REMARK 8.3-2 With the number of successes and the number of trials in the R objects x, n, respectively, the R command *prop.test(x, n, p=p*$_0$*, alternative="two.sided", conf.level=1-α)* gives the p-value for $H_a : p \neq p_0$, and a $(1-\alpha)100\%$ CI for p. This p-value and CI are based on different formulas than the ones given, and thus will not be used. ◁

REMARK 8.3-3 The R commands *1-pbinom(x-1, n, p*$_0$*)* and *pbinom(x, n, p*$_0$*)* give exact p-values, valid for any n, for $H_a : p > p_0$ and $H_a : p < p_0$, respectively. For example, the command *1-pbinom(150, 200, 0.7)* returns 0.0506 for the exact p-value in Example 8.3-2. ◁

8.3.3 *T* TESTS ABOUT THE REGRESSION PARAMETERS

Let $(X_1, Y_1), \ldots, (X_n, Y_n)$ come from the simple linear regression model

$$Y_i = \alpha_1 + \beta_1 X_i + \varepsilon_i, \quad i = 1, \ldots, n, \tag{8.3.3}$$

where the intrinsic error variables have variance $\mathrm{Var}(\varepsilon_i) = \sigma_\varepsilon^2$. The LSEs of α_1, β_1, and σ_ε^2, which were given in Section 6.3.3 of Chapter 6, are restated here for convenient reference:

$$\widehat{\alpha}_1 = \overline{Y} - \widehat{\beta}_1 \overline{X}, \quad \widehat{\beta}_1 = \frac{n\sum X_i Y_i - (\sum X_i)(\sum Y_i)}{n\sum X_i^2 - (\sum X_i)^2}, \quad \text{and} \tag{8.3.4}$$

$$S_\varepsilon^2 = \frac{1}{n-2}\left[\sum_{i=1}^n Y_i^2 - \widehat{\alpha}_1 \sum_{i=1}^n Y_i - \widehat{\beta}_1 \sum_{i=1}^n X_i Y_i\right] \tag{8.3.5}$$

Tests about the Regression Slope and the Regression Line The T test procedure for testing $H_0 : \beta_1 = \beta_{1,0}$, where $\beta_{1,0}$ is a specified value, against the various alternatives, as well as formulas for the p-value, are given below:

The T Test Procedures for $H_0 : \beta_1 = \beta_{1,0}$

(1) *Assumptions:* Either the ε_i in (8.3.3) are normal or $n \geq 30$

(2) *Test Statistic:* $T_{H_0} = \dfrac{\widehat{\beta}_1 - \beta_{1,0}}{S_{\widehat{\beta}_1}}$,

$$\text{where } S_{\widehat{\beta}_1} = \sqrt{\frac{S_\varepsilon^2}{\sum X_i^2 - \frac{1}{n}(\sum X_i)^2}} \text{ and } S_\varepsilon^2 \text{ is given in (8.3.5)} \tag{8.3.6}$$

(3) *Rejection Rules for the Different H_a:*

H_a	RR at Level α
$\beta_1 > \beta_{1,0}$	$T_{H_0} > t_{n-2,\alpha}$
$\beta_1 < \beta_{1,0}$	$T_{H_0} < -t_{n-2,\alpha}$
$\beta_1 \neq \beta_{1,0}$	$\lvert T_{H_0}\rvert > t_{n-2,\alpha/2}$

(4) *Formulas for the p-Value:*

$$p\text{-value} = \begin{cases} 1 - G_{n-2}(T_{H_0}) & \text{for } H_a: \beta_1 > \beta_{1,0} \\ G_{n-2}(T_{H_0}) & \text{for } H_a: \beta_1 < \beta_{1,0} \\ 2[1 - G_{n-2}(|T_{H_0}|)] & \text{for } H_a: \beta_1 \neq \beta_{1,0}, \end{cases}$$

where G_{n-2} is the CDF of the T_{n-2} distribution

The most common testing problem is $H_0 : \beta_1 = 0$ vs $H_a : \beta_1 \neq 0$. This is called the **model utility test**, a terminology justified by the fact that, if $H_0 : \beta_1 = 0$ is true, then the regression model has no utility, that is, that X has no predictive value for Y.

The T procedures for testing hypotheses about the regression line and formulas for computing the p-value for each hypothesis are given in the display (8.3.7). The R commands for testing in regression are given and demonstrated in Example 8.3-3.

The T Test Procedures for $H_0 : \mu_{Y|X}(x) = \mu_{Y|X}(x)_0$

(1) *Assumptions:* Either the ε_i in (8.3.3) are normal or $n \geq 30$
(2) *Condition:* The value x lies in the range of X-values

(3) *Test Statistic:* $T_{H_0} = \dfrac{\widehat{\mu}_{Y|X}(x) - \mu_{Y|X}(x)_0}{S_{\widehat{\mu}_{Y|X}(x)}}$, where

$$S_{\widehat{\mu}_{Y|X}(x)} = S_\varepsilon \sqrt{\frac{1}{n} + \frac{n(x - \overline{X})^2}{n\sum X_i^2 - (\sum X_i)^2}}$$ and S_ε^2 is given in (8.3.5).

(4) *Rejection Rules for the Different H_a:*

H_a	RR at Level α
$\mu_{Y\|X}(x) > \mu_{Y\|X}(x)_0$	$T_{H_0} > t_{\alpha,n-2}$
$\mu_{Y\|X}(x) < \mu_{Y\|X}(x)_0$	$T_{H_0} < -t_{\alpha,n-2}$
$\mu_{Y\|X}(x) \neq \mu_{Y\|X}(x)_0$	$\|T_{H_0}\| > t_{\alpha/2,n-2}$

(8.3.7)

(5) *Formulas for the p-Value*:

$$p\text{-value} = \begin{cases} 1 - G_{n-2}(T_{H_0}) & \text{for } H_a: \mu_{Y|X}(x) > \mu_{Y|X}(x)_0 \\ G_{n-2}(T_{H_0}) & \text{for } H_a: \mu_{Y|X}(x) < \mu_{Y|X}(x)_0 \\ 2[1 - G_{n-2}(|T_{H_0}|)] & \text{for } H_a: \mu_{Y|X}(x) \neq \mu_{Y|X}(x)_0, \end{cases}$$

where G_{n-2} is the CDF of the T_{n-2} distribution

Example 8.3-3

Measurements along the river Ijse[3] on temperature (^oC) and dissolved oxygen (mg/L), taken from March 1991 to December 1997, can be found in *OxygenTempData.txt*.

(a) The question of scientific interest is whether temperature (X) can be used for predicting the amount of dissolved oxygen (Y).

(i) Does the normal simple linear regression model seem appropriate for this data?

(ii) Using the summary statistics $\sum_i X_i = 632.3$, $\sum_i X_i^2 = 7697.05$, $\sum_i Y_i = 537.1$, $\sum_i Y_i^2 = 5064.73$, and $\sum_i X_i Y_i = 5471.55$, test the question of scientific interest at level of significance $\alpha = 0.01$ and report the p-value.

[3] River Ijse is a tributary of the river Dijle, Belgium. Data from VMM (Flemish Environmental Agency) compiled by G. Wyseure.

(b) Two questions of secondary scientific interest have to do with the rate of change in oxygen as temperature changes and with the average level of oxygen at $10^o C$. The corresponding testing problems are: (i) $H_0 : \beta_1 = -0.25$ vs $H_a : \beta_1 < -0.25$ and (ii) $H_0 : \mu_{Y|X}(10) = 9$ vs $H_a : \mu_{Y|X}(10) \neq 9$. Test the first at level $\alpha = 0.05$, the second at level $\alpha = 0.1$, and report the p-value for both.

Solution

(a) The scatterplot shown in the first panel of Figure 8-2 suggests that the assumptions of the simple linear regression model, which are linearity of the regression function of Y on X and homoscedasticity, are, at least approximately, satisfied. However, the normal Q-Q plot for the residuals suggests that the normality assumption is not realistic. Since the sample size is $n = 59$, the T test procedures for the regression coefficients and the regression line can still be applied. The question of whether temperature can be used for predicting the amount of dissolved oxygen can be answered by performing the model utility test: If $H_0 : \beta_1 = 0$ is rejected, we conclude that temperature can indeed be used for the stated purpose. Using the summary statistics and the formulas for $\widehat{\beta}_1$ and its standard error given in (8.3.4) and (8.3.6), we obtain

$$\widehat{\beta}_1 = \frac{59 \times 5471.55 - 632.3 \times 537.1}{59 \times 7697.05 - 632.3^2} = -0.309,$$

$$\widehat{\alpha}_1 = \frac{537.1}{59} + 0.309\frac{632.3}{59} = 12.415,$$

$$S_\varepsilon^2 = \frac{1}{57}(5064.73 - 12.415 \times 537.1 + 0.309 \times 5471.55) = 1.532, \quad \text{and}$$

$$S_{\widehat{\beta}_1} = \sqrt{\frac{1.532}{7697.05 - 632.3^2/59}} = 0.0408$$

Thus, the T statistic and p-value for the model utility test are

$$T_{H_0} = \frac{-0.309}{0.0408} = -7.573,$$

$$p\text{-value} = 2(1 - G_{57}(7.573)) = 3.52 \times 10^{-10}, \tag{8.3.8}$$

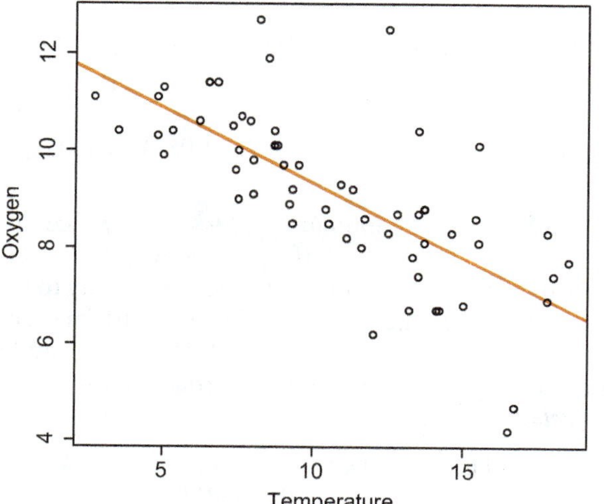

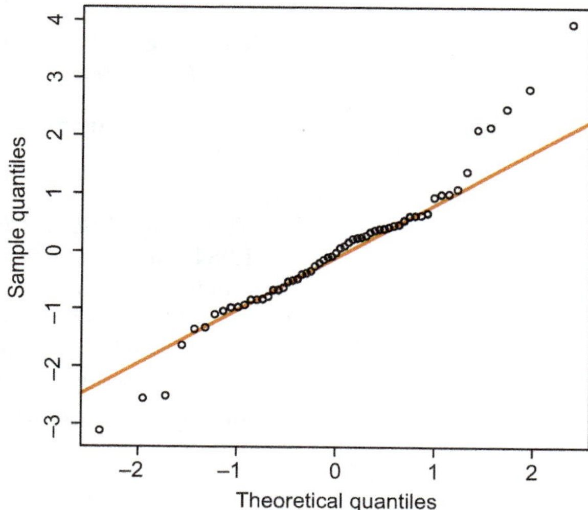

Figure 8-2 Scatterplot (left panel) and residual Q-Q plot (right panel) for the data of Example 8.3-3.

where the R command *2*(1-pt(7.573, 57))* was used to get the above p-value. Since $|T_{H_0}| = 7.573 > t_{57,0.005} = 2.6649$ (or, equivalently, since the p-value $= 3.52 \times 10^{-10} < 0.01$), $H_0 : \beta_1 = 0$ is rejected at level of significance 0.01. Using *to=read.table("OxygenTempData.txt", header=T); x=to$T; y=to$DO* to copy the data into the R objects x and y, the R commands *out=lm(y~x); summary(out)* produce output, part of which is

```
Coefficients:
              Estimate  Std. Error  t value    Pr(>|t|)
(Intercept)    12.4152      0.4661   26.639     < 2e-16
x              -0.3090      0.0408   -7.573    3.52e-10
--
Residual standard error: 1.238 on 57 degrees of freedom
```

The column headed "Estimate" gives the LSEs $\widehat{\alpha}_1$ and $\widehat{\beta}_1$, and the next column gives their standard errors. The column headed "t value" gives the ratios of each estimate over its standard error. These ratios are the T statistics for testing $H_a : \alpha_1 \neq 0$ and $H_a : \beta_1 \neq 0$. The final column gives the p-values for these tests. In particular, the T statistic and p-value for the model utility test are -7.573 and 3.52e-10, as found in (8.3-3). The last line of the displayed output gives 1.238 for the value of S_ε, in agreement with the value of S_ε^2 found by hand calculations ($1.238^2 = 1.532$).

(b) According to (8.3.6), the test statistic and p-value for testing $H_0 : \beta_1 = -0.25$ vs $H_a : \beta_1 < -0.25$ are

$$T_{H_0} = \frac{-0.309 + 0.25}{0.0408} = -1.4461, \quad p\text{-value} = G_{57}(-1.4461) = 0.0768,$$

where the value 0.0768 was obtained with the R command *pt(-1.4461, 57)*. Since $T_{H_0} = -1.4461 \not< t_{57,0.05} = -1.672$ (or, equivalently, since the p-value $= 0.0768 > 0.05$), the null hypothesis is not rejected.

Finally, using the LSEs obtained in part (a), we have $\widehat{\mu}_{Y|X}(10) = 12.415 - 0.309 \times 10 = 9.325$. Using the value of S_ε found in part (a) and the formula for the standard error of $\widehat{\mu}_{Y|X}(x)$ given in (8.3.7), we obtain

$$S_{\widehat{\mu}_{Y|X}(10)} = 1.238\sqrt{\frac{1}{59} + \frac{59(10 - 10.717)^2}{59 \times 7697.05 - 632.3^2}} = 0.1638$$

Thus, the test statistic and p-value for testing $H_0 : \mu_{Y|X}(10) = 9$ vs $H_a : \mu_{Y|X}(10) \neq 9$ are

$$T_{H_0} = \frac{9.325 - 9}{0.1638} = 1.984, \quad p\text{-value} = 2(1 - G_{57}(1.984)) = 0.052,$$

where $G_{57}(1.984)$ was found by the R command *pt(1.984, 57)*. Since $|T_{H_0}| = 1.984 > t_{57,0.05} = 1.672$ (or, equivalently, since the p-value $= 0.052 < 0.1$), the null hypothesis is rejected at level of significance 0.1. The easiest way to test this hypothesis with R commands is to make a 90% CI for $\mu_{Y|X}(10)$ and check if the value 9 belongs in the CI (i.e., use the CI-based test). Having created the R object *out* in part (a), the additional R command *predict(out, data.frame(x=10), interval="confidence", level=.9)* returns

```
       fit       lwr       upr
1 9.324943   9.05103  9.598855
```

for the fitted value ($\widehat{\mu}_{Y|X}(10)$) and the lower and upper endpoints of the 90% CI. Since the value 9 is not included in the CI, the null hypothesis is rejected at level of significance 0.1. The calculation of the p-value with R still requires that the value of the test statistic be computed. The additional R commands *fit=predict(out, data.frame(x=10), interval="confidence", 1); fit$se.fit* return the same value for $S_{\widehat{\mu}_{Y|X}(10)}$ that was obtained previously by hand calculations. Using the value of $\widehat{\mu}_{Y|X}(10)$ that was also given with the output for the 90% CI, the computation of the test statistic and p-value proceeds with the same calculations as done above. ■

8.3.4 THE ANOVA *F* TEST IN REGRESSION

ANOVA, an acronym for Analysis Of Variance, is a very useful and generally applicable approach to hypothesis testing. All ANOVA F tests involve the **F distribution.** F distributions are positively skewed distributions that are characterized by two degrees of freedom, the *numerator degrees of freedom*, denoted by ν_1, and the *denominator degrees of freedom*, denoted by ν_2. If W_1 and W_2 are two independent random variables having χ^2 distributions (see Section 7.3.5) with degrees of freedom ν_1 and ν_2, respectively, then

$$F = \frac{W_1/\nu_1}{W_2/\nu_2} \tag{8.3.9}$$

has the F distribution with ν_1 and ν_2 degrees of freedom. This is denoted by writing $F \sim F_{\nu_1,\nu_2}$. The notation F_{ν_1,ν_2} is also used for the cumulative distribution function of the F_{ν_1,ν_2} distribution; thus, if $F \sim F_{\nu_1,\nu_2}$, $P(F \le x) = F_{\nu_1,\nu_2}(x)$. The $(1-\alpha)$100th percentile of the F_{ν_1,ν_2} distribution is denoted by $F_{\nu_1,\nu_2,\alpha}$. Selected percentiles of F distributions are given in Table A.6, but can also be obtained with the following R command:

R Commands for the F_{ν_1,ν_2} Percentiles and Cumulative Probabilities

```
qf(1-α, ν1, ν2) # gives Fν1,ν2,α

pf(x, ν1, ν2) # gives Fν1,ν2(x)
```

In the regression context, the ANOVA F test is an alternative (but equivalent) way of conducting the model utility test, that is, testing $H_0: \beta_1 = 0$ vs $H_a: \beta_1 \ne 0$. The ANOVA methodology is based on a decomposition of the *total variability* (to be defined below) into components. Variability in ANOVA is represented by the so-called **sums of squares**, or **SS**. The total variability, or **total SS**, abbreviated by **SST**, is defined as

$$\text{SST} = \sum_{i=1}^{n}(Y_i - \overline{Y})^2 = \sum_{i=1}^{n} Y_i^2 - \frac{1}{n}\left(\sum_{i=1}^{n} Y_i\right)^2. \tag{8.3.10}$$

Note that SST is related to the sample variance, S_Y^2, of the Y values by SST $= (n-1)S_Y^2$.

SST is decomposed into the **regression SS (SSR)**, which represents the *variability explained by the regression model*, and the **error SS (SSE)**, which represents the *variability due to the intrinsic scatter*:

Decomposition of the Total SS in Simple Linear Regression

$$\text{SST} = \text{SSR} + \text{SSE} \tag{8.3.11}$$

SSE has already been introduced, as the sum of squared residuals, in Chapter 6 in connection with the estimator S_ε^2 of the intrinsic error variance σ_ε^2. A computational formula for SSE is given in (6.3.11), and is restated below for convenient reference:

$$\text{SSE} = \sum_{i=1}^{n} Y_i^2 - \widehat{\alpha}_1 \sum_{i=1}^{n} Y_i - \widehat{\beta}_1 \sum_{i=1}^{n} X_i Y_i. \tag{8.3.12}$$

The regression SS is computed from (8.3.11) by subtraction: SSR = SST − SSE.

The proportion of the total variability that is explained by the regression model, that is,

$$R^2 = \frac{\text{SSR}}{\text{SST}}, \tag{8.3.13}$$

is called the **coefficient of determination**. R^2, also denoted by "R-squared" in the output of software packages, is widely used in practice as a measure of the utility of the regression model: The larger R^2 is, the better the predictive power of the regression model is. This interpretation of R^2 is also justified by the fact that $R^2 = \text{SSR}/\text{SST}$ equals the square of Pearson's correlation coefficient.

The ANOVA F test rejects $H_0 : \beta_1 = 0$ if SSR is large compared to SSE. For a proper comparison of SSR and SSE, however, they must be divided by their respective *degrees of freedom*, which are as follows:

Degrees of Freedom for SST, SSE, SSR

$$\text{DF}_{\text{SST}} = n - 1, \quad \text{DF}_{\text{SSE}} = n - 2, \quad \text{DF}_{\text{SSR}} = 1 \tag{8.3.14}$$

Dividing each of the SSE and SSR by their degrees of freedom we obtain, respectively, the **mean squares for error** (**MSE**) and the **mean squares for regression** (**MSR**):

Mean Squares for Error and Regression

$$\text{MSE} = \frac{\text{SSE}}{n-2}, \quad \text{MSR} = \frac{\text{SSR}}{1} \tag{8.3.15}$$

From the computational formula (8.3.5), it can be seen that $S_\varepsilon^2 = \text{SSE}/(n-2)$. Thus, MSE is just another notation for the estimator S_ε^2 of σ_ε^2.

The ANOVA F test statistic for the model utility test is

***F* Test Statistic for the Model Utility Test**

$$F = \frac{\text{MSR}}{\text{MSE}} \tag{8.3.16}$$

The sums of squares, the mean squares, and the F statistic are summarized in an organized fashion in the so-called **ANOVA table** as follows:

Source	df	SS	MS	F
Regression	1	SSR	MSR= $\frac{\text{SSR}}{1}$	$F = \frac{\text{MSR}}{\text{MSE}}$
Error	$n-2$	SSE	MSE= $\frac{\text{SSE}}{n-2}$	
Total	$n-1$	SST		

(8.3.17)

In the normal simple linear regression model, the exact null distribution of the F statistic (i.e., its distribution when $H_0: \beta_1 = 0$ is true) is F with numerator degrees of freedom 1 and denominator degrees of freedom $n-2$:

Exact Null Distribution of the ANOVA *F* Statistic in the Normal Simple Linear Regression Model

$$F = \frac{\text{MSR}}{\text{MSE}} \sim F_{1,n-2} \tag{8.3.18}$$

Without the normality assumption, $F_{1,n-2}$ is the approximate null distribution of the F statistic, provided $n \geq 30$. On the basis of (8.3.18), the null hypothesis $H_0: \beta_1 = 0$ is rejected in favor of $H_a: \beta_1 \neq 0$, at level of significance α, if $F > F_{1,n-2,\alpha}$.

The ANOVA F test procedure for $H_0: \beta_1 = 0$ vs $H_a: \beta_1 \neq 0$, including the p-value, and the R command for constructing the ANOVA table are given below:

The *F* Test Procedure for the Model Utility Test

(1) *Assumptions:* $Y_i = \alpha_1 + \beta_1 X_i + \epsilon_i, \quad i = 1, \ldots, n$, where either the iid ϵ_is are normal or $n \geq 30$

(2) *Test Statistic:* $F = \dfrac{\text{MSR}}{\text{MSE}}$,

where MSR and MSE are defined in (8.3.15)

(3) *Rejection Rule at Level* α: $F > F_{1,n-2,\alpha}$,
where $F_{1,n-2,\alpha}$ is the $100(1-\alpha)$th percentile of the $F_{1,n-2}$ distribution.

(4) *Formula for the p-Value:* p-value $= 1 - F_{1,n-2}(F)$,
where $F_{1,n-2}$ is the CDF of the $F_{1,n-2}$ distribution.

(8.3.19)

R Commands for Constructing the ANOVA Table

```
out=lm(y~x); anova(out)
```

(8.3.20)

Example 8.3-4

The $n = 59$ measurements of temperature and dissolved oxygen of Example 8.3-3 yield MSE $= 1.5329$ and MSR $= 87.924$.

(a) Using the given information, construct the ANOVA table by hand calculations and report the percent of the total variability explained by the regression model.

(b) Conduct the ANOVA F test for the model utility test at level $\alpha = 0.1$ and report its p-value.

(c) Confirm the hand calculations in parts (a) and (b) by constructing the ANOVA table with an R command.

Solution

(a) According to (8.3.14), or (8.3.17), the degrees of freedom in the second column of the ANOVA table are 1, $n-2 = 57$, and $n-1 = 58$. Since the SSs equal the MSs times the corresponding degrees of freedom (see (8.3.15) or (8.3.17)), the entries in the third column of the ANOVA table are SSR $= 87.924$, SSE $= 1.5329 \times 57 = 87.375$ and SST $=$ SSR$+$SSE $= 175.299$. Thus, the F statistic in the fifth column is MSR/MSE $= 57.358$. With these calculations, the ANOVA table shown below:

Source	df	SS	MS	F
Regression	1	87.924	87.924	57.358
Error	57	87.375	1.5329	
Total	58	175.299		

Finally, the percent of the total variability explained by the regression model is $R^2 = (87.924/175.299) = 0.5016$.

(b) According to the rejection rule given in (8.3.19), $H_0 : \beta_1 = 0$ is rejected in favor of $H_a : \beta_1 \neq 0$ at level α if $F > F_{1,57,0.1}$. From Table A.6 it can be seen that $F_{1,100,0.1} = 2.76 < F_{1,57,0.1} < F_{1,50,0.1} = 2.81$, while the R command *qf(.9, 1, 57)* returns $F_{1,57,0.1} = 2.796$. Since $F = 57.358$, the null hypothesis is rejected at level 0.1. The R command *1-pf(57.358, 1, 57)* returns 3.522×10^{-10} for the p-value.

(c) The R command *out=lm(y~x); anova(out)* returns the following ANOVA table:

```
Analysis of Variance Table
Response: y
          Df  Sum Sq  Mean Sq  F value     Pr(>F)
x          1  87.924   87.924   57.358  3.522e-10
Residuals 57  87.375    1.533
```

Note that the lines labeled "x" and "Residuals" correspond to the lines labeled "Regression" and "Error", respectively, of the ANOVA table in part (a). Also note that the R output does not include the last line (labeled "Total"). With these clarifications, it is seen that the ANOVA table generated by R confirms the hand calculations done in parts (a) and (b). ■

REMARK 8.3-4 The T test statistic for the model utility test and the ANOVA F test statistic for the same test are related by

$$T^2_{H_0} = F.$$

This rather surprising algebraic identity will not be proved here. As a check of this identity, the square of the T statistic computed in part (a) of Example 8.3-3 is $T^2_{H_0} = (-7.573)^2 = 57.35$, while the F statistic computed in Example 8.3-4 is $F = 57.358$. The two values represent the same quantity but have different round-off errors. This identity, together with the relationship

$$t^2_{\nu,\alpha/2} = F_{1,\nu,\alpha}$$

between the quantiles of the T_ν and $F_{1,\nu}$ distributions, proves that the ANOVA F test and the T test are two equivalent ways of conducting the model utility test. ◁

8.3.5 THE SIGN TEST FOR THE MEDIAN

When the sample size is small and the sample $X_1, \ldots, X_n$ has come from a non-normal population, T tests for the mean are not valid. In such cases, the *sign test* for the median, $\tilde{\mu}$, is useful as it can be applied with non-normal data without requiring the sample size to be ≥ 30. Of course, conclusions reached about the median do not translate into conclusions about the mean (unless the population distribution is known to be symmetric). However, the median is meaningful in its own right.

The sign test procedure is based on the fact that, in continuous populations, the probability that an observation is larger than $\widetilde{\mu}$ is 0.5. This fact helps convert a null hypothesis of the form $H_0 : \widetilde{\mu} = \widetilde{\mu}_0$ into a null hypothesis of the form $H_0 : p = 0.5$, where p is the probability that an observation is larger than $\widetilde{\mu}_0$. To see how alternative hypotheses about the median are converted, suppose that $\widetilde{\mu}_0 = 3$ but the true value of the median is $\widetilde{\mu} > 3$, for example, $\widetilde{\mu} = 5$. In this case, the probability that an observation is larger than 3 is larger than 0.5. Thus, the alternative hypothesis $H_a : \widetilde{\mu} > \widetilde{\mu}_0$ is converted into $H_a : p > 0.5$. Similarly, the alternative hypothesis $H_a : \widetilde{\mu} < \widetilde{\mu}_0$ is converted into $H_a : p < 0.5$, while the alternative hypothesis $H_a : \widetilde{\mu} \neq \widetilde{\mu}_0$ is converted into $H_a : p \neq 0.5$.

With slight modifications to the above argument, hypotheses about other percentiles can also be converted into hypotheses about probabilities. See Exercise 13.

The steps for carrying out the sign test procedure for testing hypotheses about the median are displayed in (8.3.21).

Example 8.3-5

Elevated blood pressure among infants is thought to be a risk factor for hypertensive disease later in life.[4] However, because blood pressure is rarely measured on children under the age of three, there is little understanding of what blood pressure levels should be considered elevated. Systolic blood pressure (SBP) measurements from a sample of 36 infants can be found in *InfantSBP.txt*. Do the data suggest that the median is greater than 94? Test at $\alpha = 0.05$ and report the p-value.

The Sign Test Procedure for $H_0 : \widetilde{\mu} = \widetilde{\mu}_0$

(1) *Assumptions:* $X_1, \ldots, X_n$ has come from a continuous population and $n \geq 10$

(2) *Converted Null Hypothesis:* $H_0 : p = 0.5$, where p is the probability that an observation is $> \widetilde{\mu}_0$

(3) *The Sign Statistic:* $Y = \#$ of observations that are $> \widetilde{\mu}_0$

(4) *Converted Alternative Hypotheses*:

H_a for $\widetilde{\mu}$		H_a for p
$\widetilde{\mu} > \widetilde{\mu}_0$	Converts to	$p > 0.5$
$\widetilde{\mu} < \widetilde{\mu}_0$	Converts to	$p < 0.5$
$\widetilde{\mu} \neq \widetilde{\mu}_0$	Converts to	$p \neq 0.5$

(5) *Test Statistic:* $Z_{H_0} = \dfrac{\widehat{p} - 0.5}{0.5/\sqrt{n}}$, where $\widehat{p} = \dfrac{Y}{n}$ **(8.3.21)**

(6) *Rejection Rules for the Different H_a*:

H_a	RR at Level α
$\widetilde{\mu} > \widetilde{\mu}_0$	$Z_{H_0} \geq z_\alpha$
$\widetilde{\mu} < \widetilde{\mu}_0$	$Z_{H_0} \leq -z_\alpha$
$\widetilde{\mu} \neq \widetilde{\mu}_0$	$\lvert Z_{H_0} \rvert \geq z_{\alpha/2}$

(7) *Formulas for the p-Value:*

$$p\text{-value} = \begin{cases} 1 - \Phi(Z_{H_0}) & \text{for } H_a : \widetilde{\mu} > \widetilde{\mu}_0 \\ \Phi(Z_{H_0}) & \text{for } H_a : \widetilde{\mu} < \widetilde{\mu}_0 \\ 2[1 - \Phi(|Z_{H_0}|)] & \text{for } H_a : \widetilde{\mu} \neq \widetilde{\mu}_0 \end{cases}$$

[4] Andrea F. Duncan et al. (2008). Interrater reliability and effect of state on blood pressure measurements in infants 1 to 3 years of age, *Pediatrics*, 122, e590–e594; http://www.pediatrics.org/cgi/content/full/122/3/e590.

Solution

The null and alternative hypotheses are $H_0 : \widetilde{\mu} = 94$ vs $H_a : \widetilde{\mu} > 94$. The assumptions on the population and sample size are satisfied and thus the sign test can be applied. With the data copied into the R object x, the R command *sum(x > 94)* returns 22 for the value of the sign statistic Y. Thus, $\widehat{p} = Y/n = 22/36 = 0.611$. The converted hypothesis-testing problem is $H_0 : p = 0.5$ vs $H_a : p > 0.5$ and the Z_{H_0} test statistic is

$$Z_{H_0} = \frac{0.611 - 0.5}{0.5/\sqrt{36}} = 1.332.$$

Since $1.332 < z_{0.05} = 1.645$, there is not enough evidence to conclude, at level 0.05, that the median is larger than 94. The p-value is $1 - \Phi(1.332) = 0.091$. ■

8.3.6 χ^2 TESTS FOR A NORMAL VARIANCE

As with the CIs of Section 7.3.5, hypothesis testing for a normal variance will be based on Proposition 7.3-3, according to which $(n-1)S^2/\sigma^2 \sim \chi^2_{n-1}$. This leads to the test procedure and formulas for the p-value detailed below:

The χ^2 Test Procedures for $H_0 : \sigma^2 = \sigma_0^2$

(1) *Assumptions:* $X_1, \ldots, X_n$ has come from a normal distribution

(2) *Test Statistic:* $\chi^2_{H_0} = \dfrac{(n-1)S^2}{\sigma_0^2}$

(3) *Rejection Rules for the Different H_a:*

H_a	RR at Level α
$\sigma^2 > \sigma_0^2$	$\chi^2_{H_0} > \chi^2_{n-1,\alpha}$
$\sigma^2 < \sigma_0^2$	$\chi^2_{H_0} < \chi^2_{n-1,1-\alpha}$
$\sigma^2 \neq \sigma_0^2$	$\chi^2_{H_0} > \chi^2_{n-1,\alpha/2}$ or $\chi^2_{H_0} < \chi^2_{n-1,1-\alpha/2}$

(8.3.22)

(4) *Formulas for the p-Value:*

$$p\text{-value} = \begin{cases} 1 - \Psi_{n-1}(\chi^2_{H_0}) & \text{for } H_a : \sigma^2 > \sigma_0^2 \\ \Psi_{n-1}(\chi^2_{H_0}) & \text{for } H_a : \sigma^2 < \sigma_0^2 \\ 2\min\left\{\Psi_{n-1}(\chi^2_{H_0}), 1 - \Psi_{n-1}(\chi^2_{H_0})\right\} & \text{for } H_a : \sigma^2 \neq \sigma_0^2, \end{cases}$$

where Ψ_{n-1} is the CDF of the χ^2_{n-1} distribution

Example 8.3-6

An optical firm is considering the use of a particular type of glass for making lenses. As it is important that the various pieces of glass have nearly the same index of refraction, this type of glass will be used if there is evidence that the standard deviation of the refraction index is less than 0.015. The refraction index measurements for a simple random sample of $n = 20$ glass specimens yields $S = 0.01095$. With this information, should this glass type be used? Test at $\alpha = 0.05$ and report the p-value.

Solution

The testing problem for answering this question is $H_0 : \sigma^2 = 0.015^2$ vs $H_a : \sigma^2 < 0.015^2$. Assume that the refraction measurements follow the normal distribution. Then, according to the rejection rules in (8.3.22), H_0 will be rejected at $\alpha = 0.05$ if

$$\chi^2_{H_0} = \frac{(n-1)S^2}{\sigma_0^2} = \frac{19 \times 0.01095^2}{0.015^2} = 10.125$$

is less than $\chi^2_{19,0.95}$, which, from Table A.5, is seen to be 10.117. Since $10.125 \not< 10.117$, the null hypothesis is not rejected, and this type of glass will not be adopted for use. Working only with Table A.5, we see that the value 10.125 is sandwiched between the 5th and 10th percentiles (i.e., between 10.117 and 11.651) of the χ^2_{19} distribution. From this, and the formulas for the p-value given in (8.3.22), we can say that the p-value is between 0.05 and 0.1. The R command *pchisq(10.125, 19)* returns 0.0502 for the exact p-value. ■

Exercises

1. In the context of Exercise 4 in Section 8.2, suppose that the old machine achieves $\mu_0 = 9.5$ cuts per hour. The appliance manufacturer decides to test $H_0 : \mu = 9.5$ against $H_a : \mu > 9.5$ at level $\alpha = 0.05$. She gets the machine manufacturer to lend her a new machine and she measures the number of cuts made by the machine in 50 one-hour time periods. The summary statistics for this random sample are $\overline{X} = 9.8$ and $S = 1.095$.

(a) Carry out the test at level $\alpha = 0.05$, and report whether or not H_0 should be rejected.

(b) State any assumptions needed for the validity of the test procedure.

2. Studies have shown that people who work with benzene longer than five years have 20 times the incidence of leukemia than the general population. As a result, OSHA (Occupational Safety and Health Administration) has set a time-weighted average permissible exposure limit of 1 ppm. A steel manufacturing plant, which exposes its workers to benzene daily, is under investigation for possible violations of the permissible exposure limit. Thirty-six measurements, taken over a period of 1.5 months, yielded $\overline{X} = 2.1$ ppm, $S = 4.1$ ppm. Complete the following parts to determine if there is sufficient evidence to conclude that the steel manufacturing plant is in violation of the OSHA exposure limit.

(a) State the null and alternative hypotheses.

(b) Carry out the test at $\alpha = 0.05$ and state your conclusion. What assumptions, if any, are needed for the validity of this test?

(c) Give the p-value (i) approximately using Table A.4 and (ii) exactly using R.

3. To investigate the corrosion-resistance properties of a certain type of steel conduit, 16 specimens are buried in soil for a 2-year period. The maximum penetration (in mils) for each specimen is then measured, yielding a sample average penetration of $\overline{X} = 52.7$ and a sample standard deviation of $S = 4.8$. The conduits will be used unless there is strong evidence that the (population) mean penetration exceeds 50 mils.

(a) State the null and alternative hypotheses.

(b) Carry out the test at level $\alpha = 0.1$ and state your conclusion. What assumptions, if any, are needed for the validity of this test?

(c) Give the p-value (i) approximately using Table A.4 and (ii) exactly using R.

4. Researchers are exploring alternative methods for preventing frost damage to orchards. It is known that the mean soil heat flux for plots covered only with grass is $\mu_0 = 29$ units. An alternative method is to use coal dust cover.

(a) Heat flux measurements of 8 plots covered with coal dust yielded $\overline{X} = 30.79$ and $S = 6.53$. Test the hypothesis $H_0 : \mu = 29$ vs $H_a : \mu > 29$ at $\alpha = 0.05$, and report the p-value (either exactly, using R, or approximately, using Table A.4).

(b) What assumption(s) underlie the validity of the above test?

5. In the airline executive traveler's club setting of Exercise 5 in Section 8.2, a random sample of 500 customers yielded 40 who would qualify.

(a) Test the hypothesis $H_0 : p = 0.05$ vs $H_a : p < 0.05$ at $\alpha = 0.01$, and state what action the airline should take.

(b) Compute the p-value using either Table A.3 or R.

6. KitchenAid will discontinue the bisque color for its dishwashers, due to reports suggesting it is not popular west of the Mississippi, unless more than 30% of its customers in states east of the Mississippi prefer it. As part of the decision process, a random sample of 500 customers east of the Mississippi is selected and their preferences are recorded.

(a) State the null and alternative hypotheses.

(b) Of the 500 interviewed, 185 said they prefer the bisque color. Carry out the test at level $\alpha = 0.05$ and state what action KitchenAid should take.

(c) Compute the p-value, and use the rejection rule (8.2.12) to conduct the above test at $\alpha = 0.01$.

7. A food processing company is considering the marketing of a new product. The marketing would be profitable if more than 20% of consumers would be willing to try this new product. Among 42 randomly chosen consumers, 9 said that they would purchase the new product and give it a try.

(a) Set up the appropriate null and alternative hypotheses.

(b) Carry out a test at level $\alpha = 0.01$ and report the p-value. Is there evidence that the marketing would be profitable?

8. An experiment examined the effect of temperature on the strength of new concrete. After curing for several days at $20^o C$, specimens were exposed to temperatures of $-10^o C$, $-5^o C$, $0^o C$, $10^o C$, or $20^o C$ for 28 days, at which time their strengths were determined. The data can be found in *Temp28DayStrength.txt*.

(a) Import the data on X = exposure temperature and Y = 28-day strength into the R objects x and y, respectively, and use R commands to fit the simple linear regression model. Give the estimated regression line.

(b) Use R commands to construct a scatterplot of the data and a normal Q-Q plot of the residuals. Do these plots suggest any possible violations of the normal simple linear regression model assumptions?

(c) Using output from an R command, give SST, SSE, and SSR. On the basis of these SSs, what percent of the total variability is explained by the regression model?

(d) Give the value of the F statistic, and use it to carry out the model utility test at level of significance 0.05.

(e) Because this concrete is used in structures located in cold climates, there is concern that the decrease in temperature would weaken the concrete. State the relevant H_0 and H_a for assessing the evidence in support of this concern. Carry out the test for the hypotheses specified at level $\alpha = 0.05$, and state your conclusion.

9. In the context of a study on the relationship between exercise intensity and energy consumption, the percentage of maximal heart rate reserve (X) and the percentage of maximal oxygen consumption (Y) were measured for 26 male adults during steady states of exercise on a treadmill.[5] Use the data set given in *HeartRateOxygCons.txt* and R commands to complete the following parts.

(a) Construct a scatterplot of the data and a normal Q-Q plot of the residuals. Do these plots suggest any possible violations of the normal simple linear regression model assumptions?

(b) Give the ANOVA table. What proportion of the total variability in oxygen consumption is explained by the regression model?

(c) Give the fitted regression line, and the estimated change in the average oxygen consumption when the percentage of maximal heart rate reserve increases by 10 points?

(d) Is there evidence that oxygen consumption increases by more than 10 points when the percentage of maximal heart rate reserve increases by 10 points?

(i) Formulate the null and alternative hypotheses for answering this question. (*Hint.* Express the hypotheses in terms of the slope.)

(ii) Test the hypotheses at level 0.05 and report the p-value.

10. A historic (circa 1920) data set on the relationship between car speed (X) and stopping distance (Y) is given in the R data frame *cars*. Use the R output given below to answer the following questions.

```
             Estimate  Std. Error  t value  Pr(>|t|)
(Intercept)  -17.5791      6.7584   -2.601    0.0123
x              3.9324      0.4155    9.464  1.49e-12
Multiple R-squared: 0.6511
```

```
Analysis of Variance Table
           Df  Sum Sq  Mean Sq  F value     Pr(>F)
x           1   21186                    1.490e-12
Residuals  48   11354
```

(a) What is the sample size in this study?

(b) Give the estimate of the standard deviation of the intrinsic error.

(c) Give the values of the T test statistics for testing $H_0: \alpha_1 = 0$ vs $H_a: \alpha_1 \neq 0$ and $H_0: \beta_1 = 0$ vs $H_a: \beta_1 \neq 0$. What are the corresponding p-values?

(d) Fill in the missing entries in the ANOVA table.

(e) What proportion of the total variability of the stopping distance is explained by the regression model?

11. It is claimed that the median increase in home owners' taxes in a certain county is \$300. A random sample of 20 home owners gives the following tax-increase data (arranged from smallest to largest): 137, 143, 176, 188, 195, 209, 211, 228, 233, 241, 260, 279, 285, 296, 312, 329, 342, 357, 412, 517.

(a) Does the data present strong enough evidence to conclude that the claim is false? State the null and alternative hypotheses and test at level $\alpha = 0.05$.

(b) Does the data present strong enough evidence to conclude that the median increase is less than 300? Test at level $\alpha = 0.05$.

[5] T. Bernard et al. (1997). Relationships between oxygen consumption and heart rate in transitory and steady states of exercise and during recovery: Influence of type of exercise. *Eur. J. Appl.*, 75: 170–176.

12. Use *rt=read.table("RobotReactTime.txt", header=T); r2=rt$Time[rt$Robot==2]* to copy the reaction times of robot 2 into the R object *r2*.

(a) It is desired to test either $H_0 : \widetilde{\mu} = 28$ vs $H_a\ \widetilde{\mu} > 28$ or $H_0 : \mu = 28$ vs $H_a\ \mu > 28$. Use a normal Q-Q plot to decide which of the two hypotheses tests can be used with this data set. Justify your answer.

(b) Test $H_0 : \widetilde{\mu} = 28$ vs $H_a\ \widetilde{\mu} > 28$ at level 0.05 and report the p-value.

13. Sign test for percentiles. The sign test procedure can also be applied to test a null hypothesis $H_0 : x_\pi = \eta_0$ for the $(1-\pi)100$th percentile, x_π, by converting it to the null hypothesis $H_0 : p = \pi$, where p is the probability that an observation is larger than η_0. The alternative hypotheses $H_a : x_\pi > \eta_0$, $H_a : x_\pi < \eta_0$, and $H_a : x_\pi \neq \eta_0$ convert to $H_a : p > \pi$, $H_a : p < \pi$, and $H_a : p \neq \pi$, respectively. Let

$$Y = \#\text{ of observations that are } > \eta_0,$$

$$Z_{H_0} = \frac{\widehat{p} - \pi}{\sqrt{\pi(1-\pi)/n}},$$

where $\widehat{p} = Y/n$, be the sign statistic and test statistic, respectively, and suppose $n \times \pi \geq 5$ and $n \times (1-\pi) \geq 5$. With these changes, the rejection rules and formulas for the p-value given in (8.3.2) apply for testing $H_0 : x_\pi = \eta_0$ against the various alternatives.

(a) Use *v=read.table("ProductionVol.txt", header=T); x=v$V* to copy a sample of 16 hourly outputs of a production facility into the R object *x*, and the R command *sum(x>250)* to find the number of observations that are greater than 250. Is there sufficient evidence to accept the alternative hypothesis that the 25th percentile is smaller than 250? Test at level 0.05.

(b) Use the data given in Example 8.3-5 to test, at level 0.05, the alternative hypothesis that the 75th percentile of the systolic blood pressure is greater than 104.

14. A tire manufacturer will adopt a new tire design unless there is evidence that the standard deviation of tread life span of the new tires is more than 2.5 thousand miles.

(a) Specify the null and alternative hypotheses.

(b) Use *t=read.table("TireLifeTimes.txt", header=T); var(t$x)* to compute the sample variance of the tread life spans of a sample of 20 new tires. Test the hypotheses specified in part (a) at level $\alpha = 0.025$. On the basis of this test, will the new design be adopted?

(c) Construct a normal Q-Q plot and comment on any concerns that ought to be raised regarding the procedure used to decide on the adoption of the new tire design.

15. The standard deviation of a random sample of 36 chocolate chip cookie weights is 0.25 oz. Test the hypothesis $H_0 : \sigma = 0.2$ vs $H_a : \sigma \neq 0.2$ at $\alpha = 0.05$, and report the p-value. State any assumptions that are needed to justify your results.

8.4 Precision in Hypothesis Testing

8.4.1 TYPE I AND TYPE II ERRORS

Because of sampling variability, it is possible that the test statistic will take a value in the rejection region when the null hypothesis is true. Similarly, it is possible that the test statistic will not take a value in the rejection region when the alternative hypothesis is true. For example, in the testing problem

$$H_0 : \mu = 28{,}000 \text{ vs } H_a : \mu < 28{,}000$$

with rejection region $\overline{X} < 27{,}000$, it is possible that $\overline{X} < 27{,}000$ (which would lead the investigator to reject H_0) even when the true value of the mean is $\mu = 28{,}500$; similarly, it is possible that $\overline{X} > 27{,}000$ (which would lead the investigator not to reject H_0) even when the true value of the mean is $\mu = 26{,}000$. **Type I error** is committed when the null hypothesis is rejected when in fact it is true. **Type II error** is committed when the null hypothesis is not rejected when in fact it is false. These two types of error are illustrated in the following table.

		Truth	
		H_0	H_a
Outcome of Test	H_0	Correct decision	Type II
	H_a	Type I	Correct decision

The next two examples illustrate the calculation of the probabilities for committing type I and type II errors in the context of testing a hypothesis for a binomial proportion.

Example 8.4-1

A coal mining company suspects that certain detonators used with explosives do not meet the requirement that at least 90% will ignite. To test the hypothesis

$$H_0 : p = 0.9 \text{ vs } H_a : p < 0.9,$$

a random sample of $n = 20$ detonators is selected and tested. Let X denote the number of those that ignite. For each of the two rejection rules

$$\text{Rule 1: } X \leq 16 \quad \text{and} \quad \text{Rule 2: } X \leq 17,$$

(a) calculate the probability of type I error, and

(b) calculate the probability of type II error when the true value of p is 0.8.

Solution

Using Table A.1, the probability of committing type I error with Rule 1 is

$$\begin{aligned} P(\text{type I error}) &= P(H_0 \text{ is rejected when it is true}) \\ &= P(X \leq 16 \mid p = 0.9, n = 20) = 0.133. \end{aligned}$$

Thus, there is a 13.3% chance that H_0 will be rejected when it is true. Now suppose that $p = 0.8$, so H_a is true. Using Table A.1, the probability of type II error with Rule 1 is

$$\begin{aligned} P(\text{type II error when } p = 0.8) &= P(H_0 \text{ is not rejected when } p = 0.8) \\ &= P(X > 16 \mid p = 0.8, n = 20) = 1 - 0.589 = 0.411. \end{aligned}$$

Using the same calculations for Rule 2 we obtain

$$P(\text{type I error}) = P(X \leq 17 \mid p = 0.9, n = 20) = 0.323, \text{ and}$$

$$P(\text{type II error when } p = 0.8) = P(X > 17 \mid p = 0.8, n = 20) = 0.206.$$

The calculations in the above example demonstrate the very important fact that it is not possible to reduce the probabilities of both types of errors simply by changing the rejection rule. This is due to the fact that the events involved in the calculation of the two probabilities are complementary. Thus, shrinking the rejection region (which results in decreased probability of type I error), expands its complement (thereby increasing the probability of type II error). Hence, because type I error is deemed more important, all test procedures given earlier in this chapter were constructed so the (maximum) probability of type I error does not exceed the level of significance α, ignoring the probability of type II error.

However, the issue of type II error arises naturally in many hypothesis-testing situations. For example, in Example 8.3-1 it was seen that $H_0 : \mu = 10$ was not rejected in favor of $H_a : \mu > 10$ even though the point estimate, $\overline{X} = 10.6$, suggested that H_a was true. While this is a manifestation of the fact that test procedures favor H_0, situations like this raise questions regarding the performance characteristics of the test procedure. For example, in the context of the aforementioned example, it would be of interest to know what the probability of type II error is when the true value of the population mean is 10.5 or 11. The next example demonstrates the calculation of the probability of type II error in the simple setting of sampling from a normal population with known variance.

Example 8.4-2

A proposed change to the tire design is justifiable only if the average life span of tires with the new design exceeds 20,000 miles. The life spans of a random sample of $n = 16$ tires with the new design will be used for the decision. It is known that the life spans are normally distributed with $\sigma = 1{,}500$. Compute the probability of type II error of the size $\alpha = 0.01$ test at $\mu = 21{,}000$.

Solution

Since σ is known, we will use $Z_{H_0} = \left(\overline{X} - \mu_0\right)/(\sigma/\sqrt{n})$ as the test statistic and, since $H_a : \mu > 20{,}000$, the appropriate rejection region is $Z_{H_0} > z_\alpha$. Thus,

$$
\begin{aligned}
\beta(21,000) &= P(\text{type II error} \mid \mu = 21{,}000) \\
&= P\left(\frac{\overline{X} - \mu_0}{\sigma/\sqrt{n}} < z_\alpha \;\middle|\; \mu = 21{,}000\right) \\
&= P\left(\overline{X} < \mu_0 + z_\alpha \frac{\sigma}{\sqrt{n}} \;\middle|\; \mu = 21{,}000\right) \\
&= \Phi\left(\frac{\mu_0 - 21,000}{\sigma/\sqrt{n}} + z_\alpha\right).
\end{aligned}
$$

With $\mu_0 = 20{,}000$, $\sigma = 1{,}500$, and $\alpha = 0.01$ (thus $z_\alpha = 2.33$), we obtain $\beta(21{,}000) = \Phi(-0.34) = 0.3669$, so the power (defined below) at $\mu = 21{,}000$ is $1 - 0.3669 = 0.6331$.

When testing hypotheses about a parameter θ, the probability of committing a type II error when $\theta = \theta_a$, where θ_a belongs in the domain of the alternative hypothesis H_a, is denoted by $\beta(\theta_a)$. That is,

$$
\beta(\theta_a) = P(\text{type II error when } \theta = \theta_a). \qquad \textbf{(8.4.1)}
$$

One minus the probability of type II error, evaluated at θ_a, is called the **power** of the test procedure at θ_a. That is

$$
\text{Power at } \theta_a = 1 - \beta(\theta_a). \qquad \textbf{(8.4.2)}
$$

Thus, power is the probability of rejecting H_0 when the alternative is true.

Precision in hypothesis testing is quantified by the power of the test procedure. Plotting the power as a function of θ gives a visual impression of the *efficiency* of the test procedure. In general, the power increases as the alternative value of θ moves farther away from the null hypothesis; it also increases as the sample size increases. These facts are illustrated in Figure 8-3.

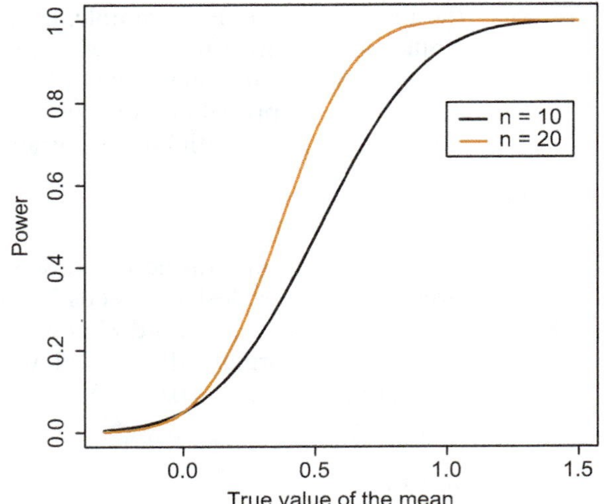

Figure 8-3 Power functions for $H_0 : \mu \le 0$ vs $H_a : \mu > 0$.

8.4.2 POWER AND SAMPLE SIZE CALCULATIONS

Since the level of significance remains fixed for each testing problem, a precision objective, that is, a specified level of power at a given alternative value of the parameter, is achieved by increasing the sample size.

However, calculation of the power of T tests involves the *non-central T* distribution and will not be presented. Instead, power and sample-size calculations will be done using R. Commands for power and sample-size calculation when testing for a proportion are also given. The R commands for power calculations are shown below; those for sample-size determination follow Example 8.4-3.

R Commands for Power Calculations for Testing $H_0 : \mu = \mu_0$ and $H_0 : p = p_0$

```
library(pwr) # Loads the package pwr

pwr.t.test(n, (μa − μ0)/sd, α, power=NULL, "one.sample",
  c("greater", "less", "two.sided")) # Gives the power at
  μa

h=2*asin(sqrt(pa))-2*asin(sqrt(p0)); pwr.p.test(h, n, α,
  power=NULL, c("greater", "less", "two.sided")) # Gives the
  power at pa
```

REMARK 8.4-1 The above commands require installation of the package *pwr*, which can be done with the command *install.packages("pwr")*. ◁

REMARK 8.4-2 Underlying the R command for the power of the test for a proportion is an approximation formula, so the power value returned differs somewhat from the exact power. For example, noting that the level of significance of the test that corresponds to Rule 1 of Example 8.4-1 is 0.133, the command *h=2*asin(sqrt(0.8))-2*asin(sqrt(0.9)); pwr.p.test(h, 20, 0.133, alternative="less")* returns 0.562 for the power at $p_a = 0.8$. This is slightly smaller than the exact value of the power found in Example 8.4-1. ◁

Example 8.4-3

(a) For the testing problem of Example 8.3-1, find the power at $\mu_a = 11$.

(b) For the testing problem of Example 8.3-2, find the power at $p = 0.8$.

Solution

(a) Example 8.3-1 tests $H_0: \mu = 10$ vs $H_a: \mu > 10$ at level $\alpha = 0.05$ using a sample of size 25, which results in $S = 2$. The R command *pwr.t.test(25, (11-10)/2, 0.05, power=NULL, "one.sample", "greater")* returns a power of 0.78.

(b) Example 8.3-2 tests $H_0: p = 0.7$ vs $H_a: p > 0.7$ at level of significance $\alpha = 0.01$ using a sample of size 200. The R command *h=2*asin(sqrt(0.8))-2*asin(sqrt(0.7)); pwr.p.test(h, 200, 0.01,power=NULL,"greater")* returns a power of 0.83. ■

With the package *pwr* installed (see Remark 8.4-1), the R commands for determining the sample size needed to achieve a desired level of power, at a specified μ_a or p_a, when testing $H_0: \mu = \mu_0$ or $H_0: p = p_0$, respectively, are shown below:

R Commands for Sample Size Determination

```
library(pwr) # Loads the package pwr

pwr.t.test(n=NULL, (μa − μ0)/Spr, α, 1-β(μa), "one.sample", alter-
  native=c("greater", "less", "two.sided")) # Gives the n
  needed for power 1-β(μa) at μa

h=2*asin(sqrt(pa))-2*asin(sqrt(p0)); pwr.p.test(h, n=NULL, α,
  1-β(pa), c("greater", "less", "two.sided")) # Gives the n
  needed for power 1-β(pa) at pa
```

The quantity S_{pr}, which is used in the command for sample size determination for the T test for the mean, is a preliminary estimator of the standard deviation. See Section 7.4 for a discussion of methods for obtaining such a preliminary estimator.

Example 8.4-4

(a) For the testing problem of Example 8.3-1, find the sample size needed to achieve power of 0.9 at $\mu_a = 11$.

(b) For the testing problem of Example 8.3-2, find the sample size needed to achieve power of 0.95 at $p = 0.8$.

Solution

(a) Example 8.3-1 tests $H_0: \mu = 10$ vs $H_a: \mu > 10$ at level $\alpha = 0.05$. A sample of size 25 gives a preliminary estimator $S_{pr} = 2$ of the standard deviation. The R command *pwr.t.test(n=NULL, (11-10)/2, 0.05, 0.9, "one.sample", alternative="greater")* returns a sample size of 35.65, which is rounded up to 36.

(b) Example 8.3-2 tests $H_0: p = 0.7$ vs $H_a: p > 0.7$ at level of significance $\alpha = 0.01$ using a sample of size 200. The R command *h=2*asin(sqrt(0.8))-2*asin(sqrt(0.7)); pwr.p.test(h, n=NULL, 0.01, 0.95, alternative="greater")* returns a sample size of 293.04, which is rounded up to 294. ■

Exercises

1. (a) When a null hypothesis is rejected, there is risk of committing which type of error?

(b) When a null hypothesis is not rejected, there is risk of committing which type of error?

2. To investigate the flame resistance of material type A, used in children's pajamas, 85 specimens of the material were subjected to high temperatures and 28 of those specimens ignited. Consider testing $H_0 : p_A = 0.3$ vs $H_a : p_A \neq 0.3$, where p_A is the probability that a specimen exposed to such high temperatures will ignite. Specify true or false for each of the following:

(a) The probability of type I error is defined as the probability of concluding that $p_A \neq 0.3$ when in fact $p_A = 0.3$.

(b) The probability of type I error is the same thing as the level of significance.

(c) The probability of type II error is defined as the probability of concluding that $p_A \neq 0.3$ when in fact $p_A = 0.3$.

3. In 10-mph crash tests, 25% of a certain type of automobile sustain no visible damage. A modified bumper design has been proposed in an effort to increase this percentage. Let p denote the proportion of all cars with this new bumper that sustain no visible damage in 10-mph crash tests. The hypothesis to be tested is $H_0 : p = 0.25$ vs $H_a : p > 0.25$. The test will be based on an experiment involving $n = 20$ independent crashes of car prototypes with the new bumper. Let X denote the number of crashes resulting in no visible damage, and consider the test procedure that rejects H_0 if $X \geq 8$.

(a) Use the binomial table to find the probability of type I error.

(b) Use the binomial table to find the power at $p = 0.3$.

(c) Using R, find the probability of type I error and the power at $p = 0.3$, when $n = 50$ and rejection region $X \geq 17$. Compare the level and power achieved by the two different sample sizes.

4. Use R commands and the information given in Exercise 2 in Section 8.3, that is, $n = 36$, $S = 4.1$, $H_0 : \mu = 1$ vs $H_a : \mu > 1$, and $\alpha = 0.05$, to complete the following.

(a) Find the probability of type II error when the true concentration is 2 ppm.

(b) OSHA would like the probability of type II error not to exceed 1% when the true concentration is 2 ppm, while keeping the level of significance at 0.05. What sample size should be used?

5. Acid baths are used to clean impurities from the surfaces of metal bars used in laboratory experiments. For effective cleaning, the average acidity of the solution should be 8.5. Average acidity in excess of 8.65 may damage the plated surface of the bars. Before delivering a batch of acid bath solution, the chemical company will make several acidity measurements and test $H_0 : \mu = 8.5$ against $H_a : \mu > 8.5$ at level $\alpha = 0.05$. If the null hypothesis is rejected, the batch will not be delivered. A preliminary sample gave sample standard deviation of 0.4. Use R commands to determine the sample size needed to satisfy the laboratory's requirement that the probability of delivering a batch of acidity 8.65 should not exceed 0.05.

6. Use R commands and the information given in Exercise 7 in Section 8.3, that is, $n = 42$, $H_0 : p = 0.2$ vs $H_a : p > 0.2$, and $\alpha = 0.01$, to complete the following.

(a) Find the probability of type II error at $p_a = 0.25$.

(b) What sample size should be used to achieve power of 0.3 at $p_a = 0.25$ while keeping the level of significance at 0.01?

7. Ues R commands and the information given in Exercise 5 in Section 8.3, that is, $n = 500$, $H_0 : p = 0.05$ vs $H_a : p < 0.05$, and $\alpha = 0.01$, to complete the following.

(a) Find the probability of type II error at $p_a = 0.04$.

(b) What sample size should be used to achieve power of 0.5 at $p_a = 0.04$ while keeping the level of significance at 0.01?

Comparing Two Populations

Chapter

9.1 Introduction

In this chapter we use confidence intervals and hypothesis testing for comparing particular aspects of two populations. Specifically, we will develop procedures for comparing two population means, two population proportions, and two population variances.

Studies aimed at comparing two populations are the simplest kind of comparative studies, as mentioned in Section 1.8. These two populations are also called treatments, or factor levels. In order to establish causation, that is, to claim that a statistically significant comparison is due to a difference in the treatments, the allocation of experimental units to the treatments must be done in a randomized fashion, in other words, a statistical experiment must be performed. In all that follows, we assume that the data have been collected from a statistical experiment.

Section 9.2 presents CIs and tests for the difference of two means and two proportions. Section 9.3 presents a procedure for testing the equality of two populations based on the *ranks* of the data. The rank-based test is the recommended procedure for small sample sizes from non-normal populations, but it is also a useful procedure regardless of the sample sizes. Section 9.4 presents two test procedures for comparing the variances of two populations. The procedures in Sections 9.2, 9.3, and 9.4 are based on two independent samples, one from each population. Section 9.5 presents CIs and tests for the difference of two means and of two proportions based on non-independent samples, where the dependence in the two samples is caused by a process of *pairing* observations.

9.2 Two-Sample Tests and CIs for Means

Let μ_1, σ_1^2 denote the mean and variance of population 1, and μ_2, σ_2^2 denote the mean and variance of population 2. If the two populations are Bernoulli, then $\mu_i = p_i$ and $\sigma_i^2 = p_i(1-p_i)$, where p_i is the probability of success in a random selection from population i, $i = 1, 2$. The comparison of the two populations will be based on a simple random sample from each of the two populations. Let

$$X_{i1}, X_{i2}, \ldots, X_{in_i}, \quad i = 1, 2, \tag{9.2.1}$$

denote the two samples. Thus, the sample from population 1 has size n_1, with observations denoted by $X_{11}, \ldots, X_{1n_1}$, and the sample from population 2 has size n_2 with observations denoted by $X_{21}, \ldots, X_{2n_2}$. The two samples are assumed independent. Let

$$\overline{X}_i = \frac{1}{n_i}\sum_{j=1}^{n_i} X_{ij}, \quad S_i^2 = \frac{1}{n_i - 1}\sum_{j=1}^{n_i}\left(X_{ij} - \overline{X}_i\right)^2, \tag{9.2.2}$$

be the sample mean and sample variance from the ith sample, $i = 1, 2$. When the populations are Bernoulli

$$\overline{X}_i = \widehat{p}_i \quad \text{and} \quad S_i^2 = \frac{n_i}{n_i - 1}\widehat{p}_i(1 - \widehat{p}_i),$$

though $\widehat{p}_i(1 - \widehat{p}_i)$ is typically used as an estimator of $\sigma_i^2 = p_i(1 - p_i)$. Moreover, typically, only the proportion, $\widehat{p}_i$, or the number of successes, $n_i\widehat{p}_i$, in the sample from population i, $i = 1, 2$, is given when sampling Bernoulli populations.

9.2.1 SOME BASIC RESULTS

The following proposition collects the results that are used for the construction of CIs and tests for the difference of two means or two proportions.

Proposition 9.2-1

1. If both populations are normal with the same variance, that is, $\sigma_1^2 = \sigma_2^2$, then, for any sample sizes,

$$\frac{\overline{X}_1 - \overline{X}_2 - (\mu_1 - \mu_2)}{\sqrt{S_p^2\left(\frac{1}{n_1} + \frac{1}{n_2}\right)}} \sim T_{n_1+n_2-2}, \tag{9.2.3}$$

where, letting S_1^2, S_2^2 be the two sample variances defined in (9.2.2),

$$S_p^2 = \frac{(n_1 - 1)S_1^2 + (n_2 - 1)S_2^2}{n_1 + n_2 - 2} \tag{9.2.4}$$

is the **pooled estimator** of the common variance.

2. If the populations are non-normal with the same variance, then, provided the sample sizes are large enough ($n_1 \geq 30$, $n_2 \geq 30$), (9.2.3) holds approximately.
3. If both populations are normal with possibly unequal variances, that is, $\sigma_1^2 \neq \sigma_2^2$, then

$$\frac{\overline{X}_1 - \overline{X}_2 - (\mu_1 - \mu_2)}{\sqrt{\frac{S_1^2}{n_1} + \frac{S_2^2}{n_2}}} \stackrel{\cdot}{\sim} T_\nu, \quad \text{where} \quad \nu = \left[\frac{\left(\frac{S_1^2}{n_1} + \frac{S_2^2}{n_2}\right)^2}{\frac{(S_1^2/n_1)^2}{n_1-1} + \frac{(S_2^2/n_2)^2}{n_2-1}}\right], \tag{9.2.5}$$

holds for any sample sizes, where S_1^2, S_2^2 are the two sample variances defined in (9.2.2), and brackets around a number x, $[x]$, denote the integer part of x (i.e., x rounded down to its nearest integer). The approximate distribution in (9.2.5) is called the **Smith-Satterthwaite** approximation.

4. If the populations are non-normal, then, provided the sample sizes are large enough ($n_1 \geq 30$, $n_2 \geq 30$), the Smith-Satterthwaite approximation (9.2.5) can be used.
5. When the sample sizes are large enough ($n_1\widehat{p}_1 \geq 8$, $n_1(1-\widehat{p}_1) \geq 8$, and $n_2\widehat{p}_2 \geq 8$, $n_2(1 - \widehat{p}_2) \geq 8$, for our purposes),

$$\frac{\widehat{p}_1 - \widehat{p}_2 - (p_1 - p_2)}{\widehat{\sigma}_{\widehat{p}_1 - \widehat{p}_2}} \stackrel{\cdot}{\sim} N(0, 1), \tag{9.2.6}$$

where

$$\widehat{\sigma}_{\widehat{p}_1 - \widehat{p}_2} = \sqrt{\frac{\widehat{p}_1(1 - \widehat{p}_1)}{n_1} + \frac{\widehat{p}_2(1 - \widehat{p}_2)}{n_2}}. \tag{9.2.7}$$

■

9.2.2 CONFIDENCE INTERVALS

CIs for $\mu_1 - \mu_2$: Equal Variances Let S_p^2 be the pooled estimator of the common variance defined in (9.2.4). Relation (9.2.3) yields the following $(1-\alpha)100\%$ confidence interval for $\mu_1 - \mu_2$.

T CI for $\mu_1 - \mu_2$ for Equal Variances

$$\overline{X}_1 - \overline{X}_2 \pm t_{n_1+n_2-2,\alpha/2}\sqrt{S_p^2\left(\frac{1}{n_1} + \frac{1}{n_2}\right)} \tag{9.2.8}$$

The R command for this CI is given in Section 9.2.3.

If the two populations are normal (and have equal variances), (9.2.8) is an exact $(1 - \alpha)100\%$ CI for $\mu_1 - \mu_2$ with any sample sizes. If the populations are not normal, and $\sigma_1^2 = \sigma_2^2$, (9.2.8) is an approximate $(1 - \alpha)100\%$ CI for $\mu_1 - \mu_2$ provided $n_1, n_2 \geq 30$.

Section 9.4 discusses procedures for testing the assumption of equal population variances, which is needed for the CI (9.2.8). However, the following rule of thumb, based on the ratio of the larger sample variance ($\max\{S_1^2, S_2^2\}$) to the smaller one ($\min\{S_1^2, S_2^2\}$), can be used as a rough check for $\sigma_1^2 \simeq \sigma_2^2$, that is, that σ_1^2, σ_2^2 are approximately equal:

Rule of Thumb for Checking If $\sigma_1^2 \simeq \sigma_2^2$

$$\frac{\max\{S_1^2, S_2^2\}}{\min\{S_1^2, S_2^2\}} < \begin{cases} 5 & \text{if } n_1, n_2 \simeq 7 \\ 3 & \text{if } n_1, n_2 \simeq 15 \\ 2 & \text{if } n_1, n_2 \simeq 30 \end{cases} \tag{9.2.9}$$

Example 9.2-1

To compare two catalysts in terms of the mean yield of a chemical process, $n_1 = 8$ chemical processes are performed with catalyst A, and $n_2 = 8$ are performed with catalyst B. From catalyst A we obtain $\overline{X}_1 = 92.255$, $S_1 = 2.39$. From catalyst B we obtain $\overline{X}_2 = 92.733$, $S_2 = 2.98$. Construct a 95% CI for the contrast $\mu_1 - \mu_2$. What assumptions, if any, are needed for the validity of the CI?

Solution

Since the ratio of the larger sample variance to the smaller one is $S_2^2/S_1^2 = 1.55$, the assumption of equal variances appears to be approximately satisfied, according to the rule of thumb (9.2.9). The pooled estimator of the common standard deviation is

$$S_p = \sqrt{\frac{(8-1)S_1^2 + (8-1)S_2^2}{8+8-2}} = \sqrt{7.30} = 2.7.$$

According to (9.2.8), the requested 95% CI for $\mu_1 - \mu_2$ is

$$\overline{X}_1 - \overline{X}_2 \pm t_{14,\,0.025}\sqrt{S_p^2\left(\frac{1}{8}+\frac{1}{8}\right)} = -0.478 \pm 2.145 \times 1.351 = (-3.376,\ 2.420).$$

Because both sample sizes are less than 30, the validity of the above CI requires that both populations are normal. ■

Example 9.2-2

A manufacturer of video display units wants to compare two micro circuit designs in terms of the current flow they produce. Using Design 1, $n_1 = 35$ current flow measurements give $\overline{X}_1 = 24.2$ and $S_1 = \sqrt{10.0}$ amps. Using Design 2, $n_2 = 40$ measurements give $\overline{X}_2 = 23.9$ and $S_2 = \sqrt{14.3}$ amps. Construct a 90% CI for $\mu_1 - \mu_2$, and state any assumptions that are needed for its validity.

Solution

Since the ratio of the larger sample variance to the smaller one is $S_2^2/S_1^2 = 1.43$, the rule of thumb (9.2.9) suggests that the data do not contradict the assumption that $\sigma_1^2 = \sigma_2^2$. The pooled estimator of the common standard deviation is

$$S_p = \sqrt{\frac{(35-1)S_1^2 + (40-1)S_2^2}{35+40-2}} = \sqrt{12.3} = 3.51.$$

Since both sample sizes are greater than 30, we do not need the normality assumption. The requested 90% CI is

$$\overline{X}_1 - \overline{X}_2 \pm t_{35+40-2,\,0.05}\sqrt{S_p^2\left(\frac{1}{35}+\frac{1}{40}\right)} = 0.3 \pm 1.666 \times 0.812 = (-1.05,\ 1.65).$$ ■

CIs for $\mu_1 - \mu_2$: Possibly Unequal Variances Regardless of whether or not the two population variances are equal, the Smith-Satterthwaite approximation (9.2.5) yields the following $(1-\alpha)100\%$ confidence interval for $\mu_1 - \mu_2$:

T CI for $\mu_1 - \mu_2$ for Unequal Variances

$$\overline{X}_1 - \overline{X}_2 \pm t_{\nu,\alpha/2}\sqrt{\frac{S_1^2}{n_1}+\frac{S_2^2}{n_2}} \qquad (9.2.10)$$

where the degrees of freedom ν is given in (9.2.5). The R command for this CI is given in Section 9.2.3.

If the two populations are either normal or the sample sizes are large enough ($n_1 \geq 30$, $n_2 \geq 30$), the Smith-Satterthwaite CI (9.2.10) is an approximate $(1-\alpha)100\%$ CI for $\mu_1 - \mu_2$. Because the CI (9.2.10) does not require any assumptions when the two sample sizes are at least 30, it is the default procedure in R and other statistical software. However, the CIs (9.2.8) tend to be shorter, and thus should be preferred if the assumption $\sigma_1^2 = \sigma_2^2$ appears tenable.

Example 9.2-3

A random sample of $n_1 = 32$ specimens of cold-rolled steel give average strength $\overline{X}_1 = 29.80$ ksi, and sample standard deviation of $S_1 = 4.00$ ksi. A random sample of $n_2 = 35$ specimens of two-sided galvanized steel give average strength $\overline{X}_2 = 34.70$ ksi, and $S_2 = 6.74$ ksi. Construct a 99% CI for $\mu_1 - \mu_2$, and state any assumptions that are needed for its validity.

Solution
Because both sample sizes are greater than 30, the Smith-Satterthwaite CI (9.2.10) requires no assumptions for its validity (except finite variances). Here

$$\nu = \left[\frac{\left(\frac{16}{32} + \frac{45.4276}{35}\right)^2}{\frac{(16/32)^2}{31} + \frac{(45.4276/35)^2}{34}} \right] = [56.11] = 56,$$

and $t_{\nu,\alpha/2} = t_{56,0.005} = 2.6665$. Thus, the desired CI is

$$\overline{X}_1 - \overline{X}_2 \pm t_{\nu,\alpha/2}\sqrt{\frac{S_1^2}{n_1} + \frac{S_2^2}{n_2}} = -4.9 \pm 3.5754 = (-8.475, -1.325).$$

Confidence Intervals for $p_1 - p_2$ Relation (9.2.6) yields the following approximate $(1-\alpha)100\%$ confidence interval for $p_1 - p_2$, provided $n_1\widehat{p}_1 \geq 8$, $n_1(1-\widehat{p}_1) \geq 8$, and $n_2\widehat{p}_2 \geq 8$, $n_2(1-\widehat{p}_2) \geq 8$:

$(1-\alpha)100\%$ Z CI for $p_1 - p_2$

$$\widehat{p}_1 - \widehat{p}_2 \pm z_{\alpha/2}\widehat{\sigma}_{\widehat{p}_1-\widehat{p}_2} \qquad \textbf{(9.2.11)}$$

where $\widehat{\sigma}_{\widehat{p}_1-\widehat{p}_2}$ is given in (9.2.7). The R command for this CI is given in Section 9.2.3.

Example 9.2-4

A certain type of tractor is being assembled at two locations, L_1 and L_2. An investigation into the proportion of tractors requiring extensive adjustments after assembly finds that in random samples of 200 tractors from L_1 and 400 from L_2, the number requiring extensive adjustments were 16 and 14, respectively. Construct a 99% CI for $p_1 - p_2$, the difference of the two proportions.

Solution
Here $\widehat{p}_1 = 16/200 = 0.08$ and $\widehat{p}_2 = 14/400 = 0.035$, and the conditions needed for the CI (9.2.11) are satisfied. Moreover, $\alpha = 0.01$, so that $z_{\alpha/2} = z_{0.005} = 2.575$. Thus the 99% CI is

$$0.08 - 0.035 \pm 2.575\sqrt{\frac{(0.08)(0.92)}{200} + \frac{(0.035)(0.965)}{400}} = (-0.01,\ 0.10).$$

9.2.3 HYPOTHESIS TESTING

In comparing two population means, the null hypothesis of interest can be put in the form

$$H_0: \mu_1 - \mu_2 = \Delta_0, \qquad \textbf{(9.2.12)}$$

where the constant Δ_0 is specified in the context of a particular application. Note that if $\Delta_0 = 0$, the null hypothesis claims that $\mu_1 = \mu_2$. The null hypothesis in (9.2.12) is tested against one of the alternatives

$$H_a: \mu_1 - \mu_2 > \Delta_0 \quad \text{or} \quad H_a: \mu_1 - \mu_2 < \Delta_0 \quad \text{or} \quad H_a: \mu_1 - \mu_2 \neq \Delta_0. \qquad \textbf{(9.2.13)}$$

In comparing two population proportions, the null and alternative hypotheses are as shown above, with p_1 and p_2 replacing μ_1 and μ_2, respectively.

Tests about $\mu_1 - \mu_2$ As with confidence intervals, there is one test procedure for when the two population variances can be assumed equal and a different procedure that can be applied regardless of whether $\sigma_1^2 = \sigma_2^2$ or not. The procedure that requires the assumption of equal variances deserves its place in the statistical toolbox as it can yield higher power.

When $\sigma_1^2 = \sigma_2^2$ can be assumed, the statistic for testing the null hypothesis (9.2.12) is (the superscript *EV* stands for *Equal Variance*)

$$T_{H_0}^{EV} = \frac{\overline{X}_1 - \overline{X}_2 - \Delta_0}{\sqrt{S_p^2\left(\frac{1}{n_1} + \frac{1}{n_2}\right)}}, \tag{9.2.14}$$

where S_p^2 is the pooled variance given in (9.2.4). An alternative test statistic, which does not rely on the assumption that $\sigma_1^2 = \sigma_2^2$ (and thus is to be preferred, especially when this assumption appears to be violated), is (the superscript *SS* stands for *Smith-Satterthwaite*)

$$T_{H_0}^{SS} = \frac{\overline{X}_1 - \overline{X}_2 - \Delta_0}{\sqrt{\frac{S_1^2}{n_1} + \frac{S_2^2}{n_2}}}. \tag{9.2.15}$$

If the H_0 is true, that is, if $\mu_1 - \mu_2 = \Delta_0$, relations (9.2.3) and (9.2.5) of Proposition 9.2-1 imply

$$T_{H_0}^{EV} \sim T_{n_1+n_2-2} \quad \text{and} \quad T_{H_0}^{SS} \stackrel{\cdot}{\sim} T_\nu, \quad \text{where } \nu = \left[\frac{\left(\frac{S_1^2}{n_1} + \frac{S_2^2}{n_2}\right)^2}{\frac{(S_1^2/n_1)^2}{n_1-1} + \frac{(S_2^2/n_2)^2}{n_2-1}}\right] \tag{9.2.16}$$

provided that either both populations are normal or that $n_1 \geq 30$ and $n_2 \geq 30$. The rules for rejecting (9.2.12) in favor of the various alternatives listed in (9.2.13), as well as formulas for the *p*-value, are given in (9.2.17).

With the data from populations 1 and 2 in the R objects *x1* and *x2*, respectively, the R commands for computing $T_{H_0}^{SS}$, the degrees of freedom ν, and *p*-value, are:

R Commands for Testing $H_0 : \mu_1 - \mu_2 = \Delta_0$ without Assuming $\sigma_1^2 = \sigma_2^2$

```
t.test(x1, x2, mu=Δ0, alternative="greater")
  # if Ha : μ1−μ2 > Δ0

t.test(x1, x2, mu=Δ0, alternative="less")
  # if Ha : μ1−μ2 < Δ0

t.test(x1, x2, mu=Δ0, alternative="two.sided")
  # if Ha : μ1−μ2 ≠ Δ0
```

For the test procedure that assumes $\sigma_1^2 = \sigma_2^2$, include *var.equal=T* in any of the above R commands. For example

```
t.test(x1, x2, var.equal=T)
```

tests $H_0: \mu_1 - \mu_2 = 0$ (which is the default value of Δ_0) against a two-sided alternative (which is the default alternative) using the test statistic $T_{H_0}^{EV}$. Occasionally, the data file has both samples stacked in a single data column, with a second column indicating the sample index of each observation in the data column. If the data and sample index columns are in the R objects x and s, respectively, use *x~s* instead of *x1, x2* in any of the above commands. These R commands will also return a CI for $\mu_1 - \mu_2$ (one-sided CIs if *alternative="greater"* or *"less"*, and the CIs of Section 9.2.2 if *alternative="two.sided"*). The default confidence level is 95%, but a $(1-\alpha)100\%$ CI can be obtained by adding *conf.level=1-α* in any of these commands; for example, *t.test(x1, x2, conf.level=0.99)* gives the 99% CI (9.2.10).

The T Test Procedures for $H_0: \mu_1 - \mu_2 = \Delta_0$

(1) *Assumptions:*
(a) $X_{11}, \ldots, X_{1n_1}$, $X_{21}, \ldots, X_{2n_2}$ are independent simple random samples
(b) Either the populations are normal or n_1 and $n_2 \geq 30$

(2) *Test Statistic:*

$$T_{H_0} = \begin{cases} T_{H_0}^{EV} \text{ if } \sigma_1^2 = \sigma_2^2 \text{ is assumed} \\ T_{H_0}^{SS} \text{ regardless of whether or not } \sigma_1^2 = \sigma_2^2 \end{cases}$$

where $T_{H_0}^{EV}$ and $T_{H_0}^{SS}$ are defined in (9.2.14) and (9.2.15)

(3) *Rejection Rules for the Different H_a:*

H_a	RR at Level α
$\mu_1 - \mu_2 > \Delta_0$	$T_{H_0} > t_{\alpha,df}$
$\mu_1 - \mu_2 < \Delta_0$	$T_{H_0} < -t_{\alpha,df}$
$\mu_1 - \mu_2 \neq \Delta_0$	$\lvert T_{H_0}\rvert > t_{\alpha/2,df}$

(9.2.17)

where $df = n_1 + n_2 - 2$ if $T_{H_0} = T_{H_0}^{EV}$, or else $df = \nu$, where ν is given in (9.2.16), if $T_{H_0} = T_{H_0}^{SS}$

(4) *Formulas for the p-Value:*

$$p\text{-value} = \begin{cases} 1 - G_{df}(T_{H_0}) & \text{for } H_a: \mu_1 - \mu_2 > \Delta_0 \\ G_{df}(T_{H_0}) & \text{for } H_a: \mu_1 - \mu_2 < \Delta_0 \\ 2[1 - G_{df}(|T_{H_0}|)] & \text{for } H_a: \mu_1 - \mu_2 \neq \Delta_0 \end{cases}$$

where G_{df} is the cumulative distribution function of T_{df} and the degrees of freedom df are as described in (3)

Example 9.2-5

Consider the data from the experiment of Example 9.2-3, which compares the strengths of two types of steel. Thus, the $n_1 = 32$ cold-rolled steel observations yield $\overline{X}_1 = 29.8$, $S_1 = 4.0$, and the $n_2 = 35$ two-sided galvanized steel observations yield $\overline{X}_2 = 34.7$, $S_2 = 6.74$. Are the mean strengths of the two types of steel different?

(a) Test at level $\alpha = 0.01$ and compute the p-value.

(b) Use *ss=read.table("SteelStrengthData.txt", header=T)* to read the data into the R data frame *ss*, then use R commands to conduct the test of part (a) and to construct a 99% CI.

Solution

(a) Here we want to test $H_0 : \mu_1 - \mu_2 = 0$ vs $H_a : \mu_1 - \mu_2 \neq 0$; thus, $\Delta_0 = 0$. The ratio of the larger to the smaller sample variance is $6.74^2/4.0^2 = 2.84$. Since both sample sizes are over 30, the rule of thumb (9.2.9) suggests that the assumption of equal population variances is not plausible. Hence, the test statistic should be

$$T_{H_0} = T_{H_0}^{SS} = \frac{29.8 - 34.7}{\sqrt{\frac{16}{32} + \frac{45.4276}{35}}} = -3.654.$$

In Example 9.2-3 it was found that $\nu = 56$, and $t_{56,0.005} = 2.6665$. Since $|-3.654| > 2.6665$, H_0 is rejected at level 0.01. Using the formula given in (9.2.17), the p-value is $2[1 - G_\nu(|T_{H_0}|)] = 2[1 - G_{56}(3.654)] = 0.00057$.

(b) The command *t.test(Value~Sample, conf.level=0.99, data=ss)* returns $T_{H_0} = -3.6543$, $\nu = 56$ (rounded down from 56.108), and a p-value of 0.000569, which match the hand calculations done in part (a). It also returns a 99% CI of (−8.475447, −1.324785), which matches the 99% CI obtained by hand calculations in Example 9.2-3. ■

Example 9.2-6

Consider the data from the experiment of Example 9.2-2, which compares the current flow of two micro circuit designs. Thus, the $n_1 = 35$ measurements using Design 1 give $\overline{X}_1 = 24.2$ amps, $S_1 = \sqrt{10}$ amps, and the $n_2 = 40$ measurements using Design 2 give $\overline{X}_2 = 23.9$ amps and $S_2 = \sqrt{14.3}$ amps. Is the mean current flow with Design 1 (statistically) significantly bigger than that of Design 2?

(a) Test at level $\alpha = 0.1$ and compute the p-value.

(b) Use *cf=read.table("MicroCirCurFlo.txt", header=T)* to read the data into the R data frame *cf*, then use R commands to conduct the test of part (a) and to construct a 90% CI.

Solution

(a) Here we want to test $H_0 : \mu_1 - \mu_2 = 0$ vs $H_a : \mu_1 - \mu_2 > 0$; thus, $\Delta_0 = 0$. The ratio of the larger to the smaller sample variance is $S_2^2/S_1^2 = 1.43$, so that, according to the rule of thumb (9.2.9), the assumption of equal population variances is not contradicted by the data. Hence, the test statistic that requires the assumption of equal population variances can be used. Moreover, since both sample sizes are greater than 30, no additional assumptions are required. From Example 9.2-2 we have that the pooled estimator of the variance is $S_p^2 = 12.3$. Thus, the test statistic is

$$T_{H_0} = T_{H_0}^{EV} = \frac{24.2 - 23.9}{\sqrt{12.3(1/35 + 1/40)}} = 0.369.$$

Since $0.369 \not> t_{35+40-2,0.1} = 1.293$, H_0 is not rejected at level 0.1. Using the formula given in (9.2.17), the p-value is $1 - G_{73}(0.369) = 0.357$.

(b) The command *t.test(CurFlo~Design, var.equal=T, alternative="greater", data=cf)* returns a T statistic, degrees of freedom, and p-value of 0.3692, 73, and 0.3565, respectively, in agreement with the hand calculations in part (a). However, the CI it returns is one-sided (in addition to having the default level of 95%). To get the CI (9.2.8) at the 90% level, the *t.test* command needs to be re-issued with the (default) alternative *"two.sided"*: *t.test(CurFlo~Design, var.equal=T, conf.level=0.9, data=cf)*. Doing so returns the 90% CI (−1.052603, 1.652032), in agreement with the CI obtained in Example 9.2-2. ■

Tests about $p_1 - p_2$ Given sample proportions $\widehat{p}_1, \widehat{p}_2$, the test statistic for testing $H_0: p_1 - p_2 = \Delta_0$, in favor of the various alternative hypotheses, depends on whether Δ_0 is zero or not. If $\Delta_0 \neq 0$, the test statistic is

$$Z_{H_0}^{P_1 P_2} = \frac{\widehat{p}_1 - \widehat{p}_2 - \Delta_0}{\sqrt{\dfrac{\widehat{p}_1(1-\widehat{p}_1)}{n_1} + \dfrac{\widehat{p}_2(1-\widehat{p}_2)}{n_2}}}. \tag{9.2.18}$$

If $\Delta_0 = 0$, the test statistic is

$$Z_{H_0}^{P} = \frac{\widehat{p}_1 - \widehat{p}_2}{\sqrt{\widehat{p}(1-\widehat{p})\left(\dfrac{1}{n_1} + \dfrac{1}{n_2}\right)}}, \quad \text{where } \widehat{p} = \frac{n_1\widehat{p}_1 + n_2\widehat{p}_2}{n_1 + n_2}. \tag{9.2.19}$$

The estimator $\widehat{p}$ of the common value p, under $H_0: p_1 = p_2$, of p_1 and p_2 is called the **pooled estimator** of p.

If the null hypothesis $H_0: p_1 - p_2 = \Delta_0$ for $\Delta_0 \neq 0$ is true, $Z_{H_0}^{P_1 P_2}$ has, approximately, a N(0, 1) distribution provided $n_1\widehat{p}_1 \geq 8$, $n_1(1-\widehat{p}_1) \geq 8$, and $n_2\widehat{p}_2 \geq 8$, $n_2(1-\widehat{p}_2) \geq 8$. Similarly, under the same sample size conditions, if $H_0: p_1 - p_2 = 0$ is true, $Z_{H_0}^{P}$ has, approximately, a N(0, 1) distribution. These facts lead to the following test procedures and formulas for the p-value when testing for the difference of two proportions.

The Z Test Procedures for $H_0: p_1 - p_2 = \Delta_0$

(1) *Assumptions:*
(a) $\widehat{p}_1, \widehat{p}_2$, are independent
(b) $n_i\widehat{p}_i \geq 8$, $n_i(1-\widehat{p}_i) \geq 8$, $i = 1, 2$

(2) *Test Statistic:*

$$Z_{H_0} = \begin{cases} Z_{H_0}^{P_1 P_2} & \text{if } \Delta_0 \neq 0 \\ Z_{H_0}^{P} & \text{if } \Delta_0 = 0 \end{cases}$$

where $Z_{H_0}^{P_1 P_2}$ and $Z_{H_0}^{P}$ are given in (9.2.18) and (9.2.19), respectively

(3) *Rejection Rules for the Different H_a:* (9.2.20)

H_a	RR at Level α
$p_1 - p_2 > \Delta_0$	$Z_{H_0} > z_\alpha$
$p_1 - p_2 < \Delta_0$	$Z_{H_0} < -z_\alpha$
$p_1 - p_2 \neq \Delta_0$	$\lvert Z_{H_0}\rvert > z_{\alpha/2}$

(4) *Formulas for the p-Value:*

$$p\text{-value} = \begin{cases} 1 - \Phi(Z_{H_0}) & \text{for } H_a: p_1 - p_2 > \Delta_0 \\ \Phi(Z_{H_0}) & \text{for } H_a: p_1 - p_2 < \Delta_0 \\ 2[1 - \Phi(\lvert Z_{H_0}\rvert)] & \text{for } H_a: p_1 - p_2 \neq \Delta_0 \end{cases}$$

where Φ is the CDF of the standard normal distribution

Let x_1, x_2 denote the number of successes from populations 1 and 2, respectively. The following R commands return $(Z_{H_0}^{P})^2$, that is, the square of statistic in (9.2.20) for testing $H_0: p_1 = p_2$, and the p-value:

R Commands for the Z Procedure for Testing $H_0 : p_1 - p_2 = 0$

```
x = c(x1, x2); n = c(n1, n2)

prop.test(x, n, alternative="greater", correct=F)
  # if Ha : p1 - p2 > 0

prop.test(x, n, alternative="less", correct=F)
  # if Ha : p1 - p2 < 0

prop.test(x, n, alternative="two.sided", correct=F)
  # if Ha : p1 - p2 ≠ 0
```

Omitting the *correct=F* part in the commands (or using the default *correct=T* instead) gives a version of the procedure in (9.2.20) with continuity correction. The default alternative is *"two.sided"*. These R commands will also return a CI for $p_1 - p_2$ (one-sided CIs if the alternative is specified as *"greater"* or *"less"*, and the CIs of Section 9.2.2 if *"two.sided"* is used). The default confidence level is 95%, but a $(1-\alpha)100\%$ CI can be obtained by adding *conf.level=1-*α in any of these commands; for example, *prop.test(c(*x_1, x_2*), c(*n_1, n_2*), correct=F, conf.level=0.99)* gives the 99% CI (9.2.11).

Example 9.2-7

During an investigation into the flame-resistance properties of material types A and B, which are being considered for use in children's pajamas, 85 randomly selected specimens of material type A and 100 randomly selected specimens of material type B were subjected to high temperatures. Twenty-eight of the specimens of type A material and 20 of the specimens of type B material ignited. Material type A will be used unless there is evidence that its probability of ignition exceeds that of material B by more than 0.04. On the basis of these data, should material A be used? Test at level $\alpha = 0.05$ and compute the p-value.

Solution

Let p_1, p_2 denote the ignition probabilities for material types A, B, respectively. To answer the question posed, we need to test $H_0 : p_1 - p_2 = 0.04$ vs $H_a : p_1 - p_2 > 0.04$. Here $n_1 = 85$, $n_2 = 100$, $\widehat{p}_1 = 0.3294$, and $\widehat{p}_2 = 0.2$, so the sample size requirements are satisfied. Moreover, the description of the experiment suggests that the assumption of independence of $\widehat{p}_1$ and $\widehat{p}_2$ is satisfied. The value of the test statistic is

$$Z_{H_0} = Z_{H_0}^{P_1 P_2} = \frac{\widehat{p}_1 - \widehat{p}_2 - 0.04}{\sqrt{\dfrac{\widehat{p}_1(1-\widehat{p}_1)}{n_1} + \dfrac{\widehat{p}_2(1-\widehat{p}_2)}{n_2}}} = 1.38.$$

Since $1.38 \not> z_{0.05} = 1.645$, the null hypothesis is not rejected. Hence, material type A should be used. The p-value is $1 - \Phi(1.38) = 0.084$; hence, H_0 would have been rejected had the test been conducted at $\alpha = 0.1$. ■

Example 9.2-8

Consider the manufacturing of tractors using two different assembly lines, as described in Example 9.2-4. Let p_1 denote the proportion of tractors coming out of assembly line L_1 that require adjustments, and let p_2 be the corresponding proportion for assembly line L_2.

(a) Test $H_0 : p_1 = p_2$ against $H_a : p_1 > p_2$ at level of significance at $\alpha = 0.01$, and compute the p-value.

(b) Use R commands to conduct the test of part (a) and to construct a 99% CI for $p_1 - p_2$.

Solution

(a) Here $n_1 = 200$, $n_2 = 400$, $\widehat{p}_1 = 0.08$, and $\widehat{p}_2 = 0.035$. Thus, the sample size conditions are satisfied. The pooled estimate of the common value (under H_0) of the two probabilities is $\widehat{p} = 0.05$. The test statistic is

$$Z_{H_0} = Z_{H_0}^P = \frac{0.08 - 0.035}{\sqrt{(0.05)(0.95)(1/200 + 1/400)}} = 2.384.$$

Since $2.384 > z_{0.01} = 2.33$, H_0 is rejected. The p-value is $1 - \Phi(2.384) = 0.0086$.

(b) The command *prop.test(c(16, 14), c(200, 400), correct=F, conf.level=0.99)* returns 5.6842 for the value of the squared test statistic, in agreement with $2.384^2 = 5.683$ obtained in part (a), and $(-0.00979, 0.09979)$ for the 99% CI for $p_1 - p_2$, in agreement with the 99% CI of $(-0.01, 0.10)$ obtained in Example 9.2-4 by hand calculations. ■

Exercises

1. An article reports on a study regarding the effect of thickness in fatigue crack growth in aluminum alloy 2024-T351.[1] Two groups of specimens were created, one with a thickness of 3 mm and the other with a thickness of 15 mm. Each specimen had an initial crack length of 15 mm. The same cyclic loading was applied to all specimens, and the number of cycles it took to reach a final crack length of 25 mm was recorded. Suppose that for the group having a thickness of 3 mm, a sample of size 36 gave $\overline{X}_1 = 160{,}592$ and $S_1 = 3{,}954$, and for the group having a thickness of 15 mm, a sample of size 42 gave $\overline{X}_2 = 159{,}778$ and $S_2 = 15{,}533$. The scientific question is whether or not thickness affects fatigue crack growth.

(a) State the null and alternative hypotheses. Is the test statistic in (9.2.14) appropriate for this data? Justify your answer.

(b) State which statistic you will use, test at level $\alpha = 0.05$, and compute the p-value. What assumptions, if any, are needed for the validity of this test procedure?

(c) Construct a 95% CI for the difference in the two means. Explain how the testing problem in (a) can be conducted in terms of the CI, and check if the test result remains the same.

2. To compare the corrosion-resistance properties of two types of material used in underground pipe lines, specimens of both types are buried in soil for a 2-year period and the maximum penetration (in mils) for each specimen is measured. A sample of size 42 specimens of material type A yielded $\overline{X}_1 = 0.49$ and $S_1 = 0.19$, and a sample of size 42 specimens of material type B gave $\overline{X}_2 = 0.36$ and $S_2 = 0.16$. Is there evidence that the average penetration for material A exceeds that of material B by more than 0.1?

(a) State the null and alternative hypotheses. Is the test statistic in (9.2.14) appropriate for this data? Justify your answer.

(b) State which statistic you will use, test at level $\alpha = 0.05$, and compute the p-value. What assumptions, if any, are needed for the validity of this test procedure?

(c) Construct a 95% CI for the difference in the two means.

3. A company is investigating how long it takes its drivers to deliver goods from its factory to a port for export. Records reveal that with a standard driving route, the last 48 delivery times have sample mean 432.7 minutes and sample standard deviation 20.38 minutes. A new driving route is proposed and this has been tried 34 times with sample mean 403.5 minutes and sample standard deviation 15.62 minutes. Is this sufficient evidence for the company to conclude, at $\alpha = 0.05$, that the new route is faster than the standard one?

(a) State the null and alternative hypotheses. Is the test statistic in (9.2.14) appropriate for this data? Justify your answer.

---

[1] J. Dominguez, J. Zapatero, and J. Pascual (1997). Effect of load histories on scatter fatigue crack growth in aluminum alloy 2024-T351. *Engineering Fracture Mechanics*, 56(1): 65–76.

(b) State which statistic you will use, test at level $\alpha = 0.05$, and compute the p-value. What assumptions, if any, are needed for the validity of this test procedure?

(c) Construct a 99% CI for the difference in the two means.

(d) Use *dd=read.table("DriveDurat.txt", header=T)* to import the data set into the R data frame *dd*, then use R commands to perform the test and construct the CI specified in parts (b) and (c).

4. After curing for several days at 20^oC, concrete specimens were exposed to temperatures of either -8^oC or 15^oC for 28 days, at which time their strengths were determined. The $n_1 = 9$ strength measurements at -8^oC resulted in $\overline{X}_1 = 62.01$ and $S_1 = 3.14$, and the $n_2 = 9$ strength measurements at 15^oC resulted in $\overline{X}_2 = 67.38$ and $S_2 = 4.92$. Is there evidence that temperature has an effect on the strength of new concrete?

(a) State the null and alternative hypotheses. Is the test statistic in (9.2.14) appropriate for this data? Justify your answer.

(b) State which statistic you will use, test at level $\alpha = 0.1$, and compute the p-value. What assumptions, if any, are needed for the validity of this test procedure?

(c) Construct a 90% CI for the difference in the two means.

(d) Use *cs = read.table("Concr.Strength.2s.Data.txt", header = T)* to import the data set into the R data frame *cs*, then use R commands to perform the test and construct the CI specified in parts (b) and (c).

5. Wrought aluminum alloy 7075-T6 is commonly used in applications such as ski poles, aircraft structures, and other highly stressed structural applications, where very high strength and good corrosion resistance are needed. A laboratory conducts an experiment to determine if the ultimate tensile strength (UTS) of holed specimens of 7075-T6 wrought aluminum is, on average, more than 126 units greater than that of notched specimens. Random samples of 15 specimens of each type give $\overline{X}_1 = 557.47$, $S_1^2 = 52.12$, $\overline{X}_2 = 421.40$, $S_2^2 = 25.83$.

(a) State the null and alternative hypotheses. Is the test statistic in (9.2.14) appropriate for this data? Justify your answer.

(b) State which statistic you will use, test at level $\alpha = 0.05$, and compute the p-value. What assumptions, if any, are needed for the validity of this test procedure?

(c) Construct a 95% CI for the difference in the two means.

(d) Use *uts=read.table("HoledNotchedUTS.txt", header=T)* to import the data set into the R data frame *uts*, then use R commands to test, at level $\alpha = 0.05$, the hypotheses stated in part (a) in two ways, once using the statistic in (9.2.14), and once using the statistic in (9.2.15). Report the p-values from each procedure.

6. A facility for bottling soft drinks uses two fill and seal machines. As part of quality control, data are periodically collected to test if the fill weight is the same for the two machines. A particular data collection of 12 fill weights from each machine yields sample mean and sample variance of $\overline{X}_1 = 966.75$, $S_1^2 = 29.30$ from the first and $\overline{X}_2 = 962.33$, $S_2^2 = 26.24$ from the second.

(a) State the null and alternative hypotheses. Is the test statistic in (9.2.14) appropriate for this data? Justify your answer.

(b) Test, at level $\alpha = 0.05$, the hypotheses stated in part (a) in two ways, once using the statistic in (9.2.14), and once using the statistic in (9.2.15). Report the p-values from each procedure. What assumptions, if any, are needed for the validity of this test procedure?

(c) Construct a 95% CI for the difference in the two means. Explain how the testing problem in (a) can be conducted in terms of the CI, and check if the test result remains the same.

7. *Earnings management* refers to a wide array of accounting techniques that can help earnings per share (EPS) meet analyst expectations. Because reported EPS in the United States are rounded to the nearest cent, earnings of 13.4 cents are rounded down to 13 cents while earnings of 13.5 cents are rounded up to 14 cents. Thus, under-representation of the number four in the first post-decimal digit of EPS data, termed *quadrophobia*,[2] suggests a particular form of earnings management, that is, that managers of publicly traded firms want to increase their reported earnings by one cent. In a typical year (1994), 692 out of 9,396 EPS reports by firms with analyst coverage had the number four in the first post-decimal digit. The corresponding number for the 13,985 EPS reports by firms with no analyst coverage is 1,182. Do the data suggest a significantly different level of quadrophobia for the two types of firms?

(a) State the null and alternative hypotheses, carry out the test at level 0.01, and report the p-value.

(b) Construct a 99% CI for $p_1 - p_2$.

(c) Repeat parts (a) and (b) using R commands.

8. An article[3] reported results of arthroscopic meniscal repair with an absorbable screw. For tears greater

[2] Joseph Grundfest and Nadya Malenko (2009). Quadrophobia: Strategic Rounding of EPS Data, available at http://ssrn.com/abstract=1474668.

[3] M. E. Hantes, E. S. Kotsovolos, D. S. Mastrokalos, J. Ammenwerth, and H. H. Paessler (2005). Anthroscopic meniscal repair with an absorbable screw: results and surgical technique, *Knee Surgery, Sports Traumatology, Arthroscopy*, 13: 273–279.

than 25 millimeters, 10 of 18 repairs were successful, while for tears less than 25 millimeters, 22 of 30 were successful.

(a) Is there evidence that the success rates for the two types of tears are different? Test at $\alpha = 0.1$ and report the p-value.

(b) Construct a 90% confidence interval for $p_1 - p_2$.

(c) Repeat parts (a) and (b) using R commands.

9. A tracking device, used for enabling robots to home in on a beacon that produces an audio signal, is said to be fine-tuned if the probability of correct identification of the direction of the beacon is the same for each side (left and right) of the tracking device. Out of 100 signals from the right, the device identifies the direction correctly 85 times. Out of 100 signals from the left, the device identifies the direction correctly 87 times.

(a) State the null and alternative hypotheses, carry out the test at level 0.01, and report the p-value.

(b) Construct a 99% CI for $p_1 - p_2$.

(c) Repeat parts (a) and (b) using R commands.

10. In 85 10-mph crash tests with type A cars, 19 sustained no visible damage. For type B cars, 22 out of 85 sustained no visible damage. Is this evidence sufficient to claim that type B cars do better in 10-mph crash tests than type A cars?

(a) State the null and alternative hypotheses, carry out the test at level 0.05, and report the p-value.

(b) Construct a 90% CI for $p_1 - p_2$.

(c) Repeat parts (a) and (b) using R commands.

9.3 The Rank-Sum Test Procedure

The procedures described in Section 9.2 for inference about $\mu_1 - \mu_2$ require either the normality assumption or large sample sizes. An alternative procedure, which can be used with both small and large sample sizes regardless of whether or not the normality assumption is tenable, is the *Mann-Whitney-Wilcoxon rank-sum test*, also referred to as *rank-sum test* or *MWW test* for short.

Underlying the versatility of the rank-sum test is the amazing fact that, for any given sample sizes n_1 and n_2, the exact null distribution of the statistic is the same no matter what the continuous population distribution is. Moreover, if the population distribution is discrete, the null distribution of the statistic can be well approximated with much smaller sample sizes. The popularity of the rank-sum test is also due to its desirable power properties (i.e., low probability of type II error), especially if the two population distributions are heavy tailed, or skewed.

The null hypothesis tested by the rank-sum procedure is

$$H_0^F : F_1 = F_2, \tag{9.3.1}$$

where F_1, F_2 denote the two population cumulative distribution functions. However, the rank-sum test is widely interpreted to be a test for equality of the medians, that is, $H_0 : \widetilde{\mu}_1 = \widetilde{\mu}_2$. For this reason, the different alternative hypotheses are stated in terms of the medians in (9.3.5). Remark 9.3-1(a) gives a different view of the alternative hypotheses. The rank-sum test procedure can also be adapted for testing $H_0 : \widetilde{\mu}_1 - \widetilde{\mu}_2 = \Delta_0$, for some constant Δ_0, against the different alternatives; see Remark 9.3-1(d) and the R commands following it. Finally, it is possible to construct a confidence interval for the difference $\widetilde{\mu}_1 - \widetilde{\mu}_2$ of the medians (or, more precisely, the median of the distribution of $X_1 - X_2$). While description of the confidence interval is beyond the scope of this book, the R command for the rank-sum test can produce it.

Implementation of the rank-sum test procedure begins by *ranking* the data, a process that consists of the following steps:

- Combine the observations, $X_{11}, \ldots, X_{1n_1}$ and $X_{21}, \ldots, X_{2n_2}$, from the two samples into an overall set of $N = n_1 + n_2$ observations.
- Arrange the combined set of observations from smallest to largest.

- For each observation X_{ij}, define its **rank** R_{ij} to be the position that X_{ij} occupies in this ordered arrangement.

In Table 9.1, 0.03, −1.42, −0.25 are the $n_1 = 3$ observations from population 1, and −0.77, −2.93, 0.48, −2.38 are the $n_2 = 4$ observations from population 2. The combined data set, the ordered observations, and the rank of each observation are displayed in the table. For example, $R_{11} = 6$ because 0.03 occupies the sixth position in the ordered arrangement of the combined set of observations. If some observations share the same value, that is, if there are *tied* observations, then they cannot be assigned ranks as described above because they cannot be arranged from smallest to largest in a unique way. For example, if X_{13} were −0.77 (instead of −0.25) then observations X_{13} and X_{21} would be tied, so the combined set of observations could also be arranged from smallest to largest as

$$X_{22} \quad X_{24} \quad X_{12} \quad X_{13} \quad X_{21} \quad X_{11} \quad X_{23}.$$

In either case, the pair of tied observations, that is, X_{13} and X_{21}, would occupy the pair of ranks 4 and 5 but it is not clear which observation should be ranked 4 and which should be ranked 5. The solution is to assign **mid-ranks** to tied observations, which is the average of the ranks they occupy. In particular, X_{13} and X_{21} would both receive the mid-rank of 4.5 and all other observations would maintain their previously assigned ranks. The R command *rank(x)* returns the ranks (and mid-ranks if there are tied observations) for each number in the object x. For example, *rank(c(0.03, -1.42, -0.77, -0.77, -2.93, 0.48, -2.38))* returns

6.0 3.0 4.5 4.5 1.0 7.0 2.0.

As the name of the procedure suggests, the rank-sum test is based on the sum of the ranks. But the sum of the ranks of all $N = n_1 + n_2$ in the combined set is always

$$1 + 2 + \cdots + N = \frac{N(N+1)}{2} \tag{9.3.2}$$

and thus does not provide any information regarding the validity of the null hypothesis (9.3.1). On the other hand, the sum of the ranks of the observations of each sample separately is quite informative. For example, if the sum of the ranks of the observations from sample 1 is "large," it implies that the corresponding sum from sample 2 is "small" (since, by (9.3.2), the sum of the two is $N(N+1)/2$), and both imply that the observations from sample 1 tend to be larger than those from

Table 9-1 Illustration of the ranking process

Original Data						
X_{11}	X_{12}	X_{13}	X_{21}	X_{22}	X_{23}	X_{24}
0.03	−1.42	−0.25	−0.77	−2.93	0.48	−2.38
Ordered Observations						
X_{22}	X_{24}	X_{12}	X_{21}	X_{13}	X_{11}	X_{23}
−2.93	−2.38	−1.42	−0.77	−0.25	0.03	0.48
Ranks of the Data						
R_{11}	R_{12}	R_{13}	R_{21}	R_{22}	R_{23}	R_{24}
6	3	5	4	1	7	2

sample 2. The rank-sum statistic is typically taken to be the sum of the ranks of the observations in the first sample:

$$W_1 = R_{11} + \cdots + R_{1n_1} \tag{9.3.3}$$

If the null hypothesis (9.3.1) is true, the distribution of W_1 is the same no matter what the continuous population distribution is. On the basis of this null distribution, it can be determined if W_1 is "large" enough, or "small" enough, to reject the null hypothesis. Before statistical software became widely available, the null distribution of W_1, for data without ties, was given in tables for each combination of values of n_1 and n_2. If both n_1 and n_2 are > 8, the distribution of the standardized W_1 (i.e., $(W_1 - E(W_1))/\sigma_{W_1}$) is well approximated by the standard normal distribution, even for data with ties. Formulas for the standardized W_1, for data with and without ties, are given in Remark 9.3-1(b), but both of these formulas are included in the unified formula given in (9.3.5). Set

$$\overline{R}_1 = \frac{1}{n_1}\sum_{j=1}^{n_1} R_{1j}, \quad \overline{R}_2 = \frac{1}{n_2}\sum_{j=1}^{n_2} R_{2j}, \quad \text{and}$$
$$S_R^2 = \frac{1}{N-1}\sum_{i=1}^{2}\sum_{j=1}^{n_i}\left(R_{ij} - \frac{N+1}{2}\right)^2. \tag{9.3.4}$$

With this notation the rank-sum test procedures are as follows:

The Rank-Sum Test Procedures for $H_0 : \widetilde{\mu}_1 = \widetilde{\mu}_2$

(1) *Assumptions:* $X_{11}, \ldots, X_{1n_1}, X_{21}, \ldots, X_{2n_2}$, are independent simple random samples and $n_1, n_2 > 8$

(2) *Test Statistic:* $Z_{H_0} = \dfrac{\overline{R}_1 - \overline{R}_2}{\sqrt{S_R^2\left(\dfrac{1}{n_1} + \dfrac{1}{n_2}\right)}}$,

where $\overline{R}_1$, $\overline{R}_2$, and S_R^2 are given in (9.3.4)

(3) *Rejection Rules for the Different H_a:*

H_a	RR at Level α
$\widetilde{\mu}_1 - \widetilde{\mu}_2 > 0$	$Z_{H_0} > z_\alpha$
$\widetilde{\mu}_1 - \widetilde{\mu}_2 < 0$	$Z_{H_0} < -z_\alpha$
$\widetilde{\mu}_1 - \widetilde{\mu}_2 \neq 0$	$\lvert Z_{H_0}\rvert > z_{\alpha/2}$

(9.3.5)

(4) *Formulas for the p-Value:*

$$p\text{-value} = \begin{cases} 1 - \Phi(Z_{H_0}) & \text{for } H_a : \widetilde{\mu}_1 - \widetilde{\mu}_2 > 0 \\ \Phi(Z_{H_0}) & \text{for } H_a : \widetilde{\mu}_1 - \widetilde{\mu}_2 < 0 \\ 2[1 - \Phi(|Z_{H_0}|)] & \text{for } H_a : \widetilde{\mu}_1 - \widetilde{\mu}_2 \neq 0 \end{cases}$$

where Φ is the cumulative distribution function of the standard normal distribution

REMARK 9.3-1

(a) The alternative $\widetilde{\mu}_1 - \widetilde{\mu}_2 > 0$ is also commonly stated as $P(X_1 > X_2) > 0.5$. Expressed this way, the alternative means that an observation from population 1 is more likely to be larger than an observation from population 2 than vice-versa. The left-sided and two-sided alternatives can also be expressed as $P(X_1 > X_2) < 0.5$ and $P(X_1 > X_2) \neq 0.5$, respectively, with similar interpretations.

(b) In the case of no ties, an alternative (and easier to compute) form of the standardized rank-sum statistic is

$$Z_{H_0} = \sqrt{\frac{12}{n_1 n_2(N+1)}}\left(W_1 - n_1\frac{N+1}{2}\right).$$

In the case of ties, an alternative form of the standardized rank-sum statistic is

$$Z_{H_0} = \left[\frac{n_1 n_2(N+1)}{12} - \frac{n_1 n_2 \sum_k d_k(d_k^2 - 1)}{12N(N-1)}\right]^{-1/2}\left(W_1 - n_1\frac{N+1}{2}\right),$$

where the summation is over all groups of tied observations in the combined sample, and d_k is the number of tied observations at the kth group.

(c) The *Mann-Whitney* form of the rank-sum statistic is computed as the number of pairs (X_{1i}, X_{2j}) for which $X_{1i} > X_{2j}$; pairs for which $X_{1i} = X_{2j}$ count as 0.5. While this number, called W in the R output, is different from the rank-sum statistic W_1 given in (9.3.3), the two are equivalent in the sense that they lead to the same inference, that is, same p-value and CIs.

(d) The rank-sum test procedure is easily adapted for testing $H_0 : \widetilde{\mu}_1 - \widetilde{\mu}_2 = \Delta_0$ against the various alternatives. To do so, modify sample 1 by subtracting Δ_0 from each observation in sample 1, that is, form $X_{11} - \Delta_0, \ldots, X_{1n_1} - \Delta_0$, and apply the procedure (9.3.5) to the modified sample 1 and sample 2. ◁

With the data from populations 1 and 2 in the R objects *x1* and *x2*, respectively, the following R commands return the Mann-Whitney form of the rank-sum statistic (see Remark (9.3-1)(c)), and the p-value for testing the null hypothesis in (9.3.1) against the different alternatives.

R Commands for the Rank-Sum Test for $H_0 : \widetilde{\mu} - \widetilde{\mu}_2 = \Delta_0$

```
wilcox.test(x1, x2, mu=Δ0, alternative="greater")
  # if Ha : μ̃1 − μ̃2 > Δ0

wilcox.test(x1, x2, mu=Δ0, alternative="less")
  # if Ha : μ̃1 − μ̃2 < Δ0

wilcox.test(x1, x2, mu=Δ0, alternative="two.sided")
  # if Ha : μ̃1 − μ̃2 ≠ Δ0
```

In the above commands, the default value of μ is zero, and the default alternative is *"two.sided"*. Thus, *wilcox.test(x1, x2)* tests $H_0^F : F_1 = F_2$ (or $H_0 : \widetilde{\mu}_1 = \widetilde{\mu}_2$) against $H_a : \widetilde{\mu}_1 \neq \widetilde{\mu}_2$. Occasionally, the data file has both samples stacked in a single data column, with a second column indicating the sample index of each observation in the data column. If the data and sample index columns are in the R objects x and s, respectively, use $x \sim s$ instead of $x1, x2$ in any of the above commands. A $(1-\alpha)100\%$ CI for $\widetilde{\mu}_1 - \widetilde{\mu}_2$ (one-sided if the alternative is specified as *"greater"* or *"less"*, and two-sided if the default alternative is chosen) can be obtained by adding *conf.int=T, conf.level=1-α* in any of the commands. For example *wilcox.test(x1, x2, conf.int=T)* will also return a 95% CI, while *wilcox.test(x1, x2, conf.int=T, conf.level=0.9)* will return a 90% CI. (If there are ties in the data, R gives a warning that the p-value and the level of the CI are not exact.)

Example 9.3-1

The sputum histamine levels (in μg/g) from a sample of size 9 allergic individuals and 13 non-allergic individuals[4] are as follows:

Allergic	67.7, 39.6, 1,651.0, 100.0, 65.9, 1,112.0, 31.0, 102.4, 64.7
Non-Allergic	34.3, 27.3, 35.4, 48.1, 5.2, 29.1, 4.7, 41.7, 48.0, 6.6, 18.9, 32.4, 45.5

Is there a difference between the two populations? Test at level $\alpha = 0.01$, and use R commands to construct a 95% CI for the median of the difference between the histamine levels of an allergic and non-allergic individual.

Solution

The data set for allergic individuals contains a couple of huge outliers, so the normality assumption is untenable. Because both sample sizes are larger than 8, we can use the test procedure in (9.3.5). Moreover, since there are no ties, the simpler formula for Z_{H_0} given in Remark 9.3-1(b) can be used. The ranks of the observations of sample 1 (allergic individuals) are $R_{11} = 18, R_{12} = 11, R_{13} = 22, R_{14} = 19, R_{15} = 17, R_{16} = 21, R_{17} = 7, R_{18} = 20, R_{19} = 16$. Thus $W_1 = \sum_j R_{1j} = 151$ and the test statistic

$$Z_{H_0} = \frac{151 - 9(23)/2}{\sqrt{9(13)(23)/12}} = 3.17$$

yields a p-value of $2[1 - \Phi(3.17)] = 0.0015$. Thus H_0 is rejected in favor of $H_a : \tilde{\mu}_1 \neq \tilde{\mu}_2$. For the sake of illustration, we recalculate Z_{H_0} using the unified formula given in (9.3.5). This is most easily done using R commands. With the data from sample 1 and 2 in the R objects *x1* and *x2*, respectively, the commands

```
n1=9; n2=13; N=n1+n2; x=c(x1, x2); r=rank(x)
w1=sum(r[1:n1]); w2=sum(r[n1+1:n2])
s2r=sum((r-(N+1)/2)^2)/(N-1); z=(w1/n1-w2/n2)
  /sqrt(s2r*(1/n1+1/n2)); z
```

return the same value for Z_{H_0}. Finally, the R command *wilcox.test(x1, x2, conf.int=T)* returns an exact p-value of 0.000772, and a 95% CI for the median of the difference between the histamine levels of an allergic and a non-allergic individual of (22.2, 95.8).

Exercises

1. An article reports on a cloud seeding experiment conducted to determine whether cloud seeding with silver nitrate increases rainfall.[5] Out of 52 clouds, 26 were randomly selected for seeding, with the remaining 26 serving as controls. The rainfall measurements, in acre-feet, are given in *CloudSeedingData.txt*. State the null and the alternative hypotheses and use R commands to (i) carry out the test at $\alpha = 0.05$, and (ii) construct a 95% CI for the median of the difference in rainfall between a seeded and an unseeded cloud.

2. Six water samples taken from the eastern part of a lake and seven taken from the western part are subjected to a chemical analysis to determine the percent content of a certain pollutant. The data are as shown below:

Eastern Part	1.88 2.60 1.38 4.41 1.87 2.89
Western Part	1.70 3.84 1.13 4.97 0.86 1.93 3.36

Is the pollutant concentration on the two sides of the lake significantly different at $\alpha = 0.1$?

[4] S. K. Hong, P. Cerretelli, J. C. Cruz, and H. Rahn (1969). Mechanics of respiration during submersion in water, *J. Appl. Physiol.*, 27(4): 535–538.

[5] J. Simpson, A. Olsen, and J. C. Eden (1975). A Bayesian analysis of a multiplicative treatment effect in weather modification, *Technometrics*, 17: 161–166.

(a) State the appropriate null and alternative hypotheses. Is the test procedure (9.3.5) recommended for this data set? Justify your answer.

(b) Use R commands to (i) carry out the test, and (ii) construct a 95% CI for the median of the difference between a measurement from the eastern part of the lake and one from the western part.

3. An article reports on a study using high temperature strain gages to measure the total strain amplitude ($\Delta\varepsilon_m/2$) of different types of cast iron for use in disc brakes.[6] The results for spheroidal graphite (SG) and compacted graphite (CG), multiplied by 10,000, are given below:

SG	105	77	52	27	22	17	12	14	65
CG	90	50	30	20	14	10	60	24	76

Are the total amplitude strain properties of the different types of cast iron significantly different at level of significance $\alpha = 0.05$?

(a) State the appropriate null and alternative hypotheses, conduct the rank-sum procedure in (9.3.5), and compute the p-value. (You may use the R commands given in Example 9.3-1 instead of hand calculations.)

(b) Use R commands to (i) carry out the test on the basis of the exact p-value, and (ii) construct a 95% CI for the median of the difference between a measurement from SG and one from CG cast iron.

4. One of the variables measured during automobile driver-side crash tests with dummies is the left femur load (the femur is the largest and strongest bone in the human body, situated between the pelvis and the knee). The data file *FemurLoads.txt* gives left femur load measurements for vehicles of 2800 lb (type 1 vehicles) and 3200 lb (type 2 vehicles).[7] Are the Femur loads for the two types of cars significantly different at level 0.1?

(a) Construct a boxplot for each of the two data sets, and comment on whether or not the normality assumption is tenable.

(b) State the appropriate null and alternative hypotheses, then use R commands to (i) carry out the test, and (ii) construct a 95% CI for the median of the difference between a femur load measurement from a type 1 vehicle and one from a type 2 vehicle.

5. Consider the data set and testing problem given in Exercise 3 in Section 9.2, and compare the p-values obtained from the T test and the MWW rank-sum test. Next, compare the 90% CI obtained through the rank-sum procedure and with the 90% T CI. Use R commands to obtain all p-values and CIs.

9.4 Comparing Two Variances

Let $X_{11}, \ldots, X_{1n_1}$ and $X_{21}, \ldots, X_{2n_2}$ denote the samples from populations 1 and 2, respectively. We will present two procedures for testing the equality of the two variances, Levene's test and the F test. The former is a more generally applicable test procedure, while the latter requires the normality assumption.

9.4.1 LEVENE'S TEST

Levene's test (also called the Brown-Forsythe test) is based on the idea that if $\sigma_1^2 = \sigma_2^2$, then the induced samples

$$V_{1j} = |X_{1j} - \widetilde{X}_1|, \quad j = 1, \ldots, n_1, \quad \text{and} \quad V_{2j} = |X_{2j} - \widetilde{X}_2|, \quad j = 1, \ldots, n_2,$$

where $\widetilde{X}_i$ is the sample median of $X_{i1}, \ldots, X_{in_i}$ for $i = 1, 2$, have equal population means and variances. Moreover, if $\sigma_1^2 > \sigma_2^2$, then the population mean μ_{V_1} of the V_1 sample will be larger than the population mean μ_{V_2} of the V_2 sample. Thus, testing $H_0 : \sigma_1^2 = \sigma_2^2$ versus $H_a : \sigma_1^2 > \sigma_2^2$ or $H_a : \sigma_1^2 < \sigma_2^2$ or $H_a : \sigma_1^2 \neq \sigma_2^2$ can be performed by testing the hypothesis $H_0^V : \mu_{V_1} = \mu_{V_2}$ versus

$$H_a^V : \mu_{V_1} > \mu_{V_2} \text{ or } H_a^V : \mu_{V_1} < \mu_{V_2} \text{ or } H_a^V : \mu_{V_1} \neq \mu_{V_2},$$

[6] F. Sherratt and J. B. Sturgeon (1981). Improving the thermal fatigue resistance of brake discs, *Materials, Experimentation and Design in Fatigue: Proceedings of Fatigue 1981*: 60–71.

[7] Data from the National Transportation Safety Administration, reported in http://lib.stat.cmu.edu/DASL.

respectively, using the two-sample T test with pooled variance, that is, the procedure (9.2.17) based on the statistic $T_{H_0}^{EV}$ given in (9.2.14), using the two V samples.

The R function *levene.test*, which is available in the R package *lawstat* (to install it use *install.packages("lawstat")*), performs Levene's test for testing $H_0 : \sigma_1^2 = \sigma_2^2$ versus $H_a : \sigma_1^2 \neq \sigma_2^2$. With the two samples in the R objects *x1, x2*, the R commands

```
library(lawstat)
x=c(x1, x2); ind=c(rep(1, length(x1)), rep(2, length(x2)));     (9.4.1)
  levene.test(x, ind)
```

return the square of the test statistic $T_{H_0}^{EV}$ evaluated on the two induced V samples and the p-value for the two-sided alternative $H_a : \sigma_1^2 \neq \sigma_2^2$. Alternatively, the R commands

```
v1=abs(x1-median(x1)); v2=abs(x2-median(x2));
  t.test(v1, v2, var.equal=T)                                   (9.4.2)
```

return the value of $T_{H_0}^{EV}$ evaluated on the two induced V samples and the same p-value, that is, for the two-sided alternative, as the *levene.test* function. Moreover, by adding *alternative="greater"* or *"less"* one obtains the p-value for one-sided alternatives.

Example 9.4-1

Numerous studies have shown that cigarette smokers have a lower plasma concentration of ascorbic acid (vitamin C) than nonsmokers. Given the health benefits of ascorbic acid, there is also interest in comparing the variability of the concentration in the two groups. The following data represent the plasma ascorbic acid concentration measurements (μmol/l) of five randomly selected smokers and nonsmokers:

Nonsmokers	41.48	41.71	41.98	41.68	41.18
Smokers	40.42	40.68	40.51	40.73	40.91

Test the hypothesis $H_0 : \sigma_1^2 = \sigma_2^2$ versus $H_a : \sigma_1^2 \neq \sigma_2^2$ at $\alpha = 0.05$.

Solution

Here the two medians are $\widetilde{X}_1 = 41.68, \widetilde{X}_2 = 40.68$. Subtracting them from their corresponding sample values, and taking the absolute values, we obtain the two V samples:

V_1 Values for Nonsmokers	0.20	0.03	0.30	0.00	0.50
V_2 Values for Smokers	0.26	0.00	0.17	0.05	0.23

The two-sample test statistic $T_{H_0}^{EV}$ evaluated on the two induced V samples takes a value of 0.61. With 8 degrees of freedom, this corresponds to a p-value of 0.558. Thus, there is not enough evidence to reject the null hypothesis of equality of the two population variances. Instead of using hand calculations, the two samples can be imported into R with the commands *x1=c(41.48, 41.71, 41.98, 41.68, 41.18); x2=c(40.42, 40.68, 40.51, 40.73, 40.91)* and then we can use either of the commands in (9.4.1) or (9.4.2) to get the same p-value. ■

9.4.2 THE F TEST UNDER NORMALITY

The F test for equality of two population variances derives its name from the class of F distributions, which was introduced in Section 8.3.4. When the two samples have been drawn from normal populations, the exact distribution of the ratio of the two sample variances is a multiple of an F distribution. This fact, which is the basis for the F test, is stated precisely in the following theorem.

Theorem 9.4-1

Let $X_{11}, \ldots, X_{1n_1}$ and $X_{21}, \ldots, X_{2n_2}$ be two independent random samples from normal populations with variances σ_1^2 and σ_2^2, respectively, and let S_1^2 and S_2^2 denote the two sample variances. Then

$$\frac{S_1^2/\sigma_1^2}{S_2^2/\sigma_2^2} \sim F_{n_1-1,\, n_2-1},$$

that is, the ratio has an F distribution with $\nu_1 = n_1 - 1$ and $\nu_2 = n_2 - 1$ degrees of freedom.

This result suggests that $H_0 : \sigma_1^2 = \sigma_2^2$ can be tested using the statistic

$$F_{H_0} = \frac{S_1^2}{S_2^2}. \tag{9.4.3}$$

Indeed, if $H_0 : \sigma_1^2 = \sigma_2^2$ is true then $F_{H_0} \sim F_{\nu_1,\, \nu_2}$, but if $\sigma_1^2 > \sigma_2^2$ then F_{H_0} will tend to take a larger value than would have been anticipated under the null hypothesis. Similarly, if $\sigma_1^2 < \sigma_2^2$ then F_{H_0} will tend to take a smaller value (or, equivalently, $1/F_{H_0} = S_2^2/S_1^2$ would tend to take a larger value) than would have been anticipated under the null hypothesis. This leads to the following rejection rules and formulas for the p-value when testing $H_0 : \sigma_1^2 = \sigma_2^2$ against the different alternatives.

The F Test Procedures for $H_0 : \sigma_1^2 = \sigma_2^2$

(1) *Assumption:* $X_{11}, \ldots, X_{1n_1}$ and $X_{21}, \ldots, X_{2n_2}$ are independent samples from normal populations

(2) *Test Statistic:* $F_{H_0} = \dfrac{S_1^2}{S_2^2}$.

(3) *Rejection Rules for the Different H_a:*

H_a	RR at Level α
$\sigma_1^2 > \sigma_2^2$	$F_{H_0} > F_{n_1-1,n_2-1;\,\alpha}$
$\sigma_1^2 < \sigma_2^2$	$\dfrac{1}{F_{H_0}} > F_{n_2-1,n_1-1;\,\alpha}$
$\sigma_1^2 \neq \sigma_2^2$	either $F_{H_0} > F_{n_1-1,n_2-1;\,\alpha/2}$ or $\dfrac{1}{F_{H_0}} > F_{n_2-1,n_1-1;\,\alpha/2}$

(9.4.4)

where $F_{\nu_1,\nu_2;\alpha}$ denotes the $(1-\alpha)100$th percentile of the $F_{\nu_1,\,\nu_2}$ distribution

(4) *Formulas for the p-Value*:

$$p\text{-value} = \begin{cases} p_1 = 1 - F_{n_1-1,\, n_2-1}(F_{H_0}) & \text{for } H_a: \sigma_1^2 > \sigma_2^2 \\ p_2 = 1 - F_{n_2-1,\, n_1-1}(1/F_{H_0}) & \text{for } H_a: \sigma_1^2 < \sigma_2^2 \\ 2[\min(p_1, p_2)] & \text{for } H_a: \sigma_1^2 \neq \sigma_2^2 \end{cases}$$

where F_{ν_1,ν_2} denotes the CDF of the F_{ν_1,ν_2} distribution

Percentiles $F_{\nu_1,\nu_2;\alpha}$ can be obtained with the R command *qf(1-α, ν_1, ν_2)*. Moreover, *pf(x, ν_1, ν_2)* gives $F_{\nu_1,\nu_2}(x)$, that is, the CDF F_{ν_1,ν_2} evaluated at x.

With the data from populations 1 and 2 in the R objects *x1* and *x2*, respectively, the following R commands return the F test statistic and the p-value for testing $H_0: \sigma_1^2 = \sigma_2^2$ against the different alternatives.

R Commands for the F Test for $H_0: \sigma_1^2 = \sigma_2^2$

```
var.test(x1, x2, alternative="greater") # if Ha : σ1² > σ2²
var.test(x1, x2, alternative="less") # if Ha : σ1² < σ2²
var.test(x1, x2, alternative="two.sided") # if Ha : σ1² ≠ σ2²
```

In the above commands the default alternative is *"two.sided"*. Occasionally, the data file has both samples stacked in a single data column, with a second column indicating the sample index of each observation in the data column. If the data and sample index columns are in the R objects x and s, respectively, use *x~s* instead of *x1, x2* in any of the above commands. The above commands will also return a 95% CI for σ_1^2/σ_2^2 (one-sided if the alternative is specified as *"greater"* or *"less"*, and two-sided if the default alternative is chosen). A $(1-\alpha)100\%$ CI can be obtained by adding *conf.level=1-α* in any of the commands. For example *var.test(x1, x2, conf.level=0.9)* will return a 90% CI.

Example 9.4-2

Consider the data and testing problem described in Example 9.4-1, and assume that the underlying populations are normal.

(a) Test $H_0: \sigma_1^2 = \sigma_2^2$ versus $H_a: \sigma_1^2 \neq \sigma_2^2$ using the F test procedure (9.4.4).

(b) Implement the F test procedure using R commands, and report the 95% CI for σ_1^2/σ_2^2.

Solution

(a) The test statistic is

$$F_{H_0} = \frac{0.08838}{0.03685} = 2.40.$$

The value 2.4 corresponds to the 79th percentile of the F distribution with $\nu_1 = 4$ and $\nu_2 = 4$ degrees of freedom. Thus, noting also that $1 - F_{4,4}(2.4) < 1 - F_{4,4}(1/2.4)$, the p-value is $2(1 - 0.79) = 0.42$.

(b) The commands *x1=c(41.48, 41.71, 41.98,41.68, 41.18); x2=c(40.42, 40.68, 40.51, 40.73, 40.91); var.test(x1, x2)* return 2.3984 and 0.4176 for the F statistic and p-value, respectively. Rounded to two decimal places, these values match those obtained in part (a). In addition, the above commands return (0.25, 23.04) as a 95% CI for σ_1^2/σ_2^2.

Exercises

1. Consider the information given in Exercise 3 in Section 9.3, and apply Levene's test to test the equality of the two variances against the two-sided alternative at level $\alpha = 0.05$.

2. Consider the information given in Exercise 1 in Section 9.2, and apply the F test to test for the equality of the two variances against the two-sided alternative at level $\alpha = 0.05$. What assumptions are needed for the validity of this test?

3. Consider the information given in Exercise 3 in Section 9.2, and apply the F test to test for the equality of the two variances against the two-sided alternative at level $\alpha = 0.05$. What assumptions are needed for the validity of this test?

9.5 Paired Data

9.5.1 DEFINITION AND EXAMPLES OF PAIRED DATA

Paired data arise from an alternative sampling design used for the comparison of two population means. In particular, such data arise whenever each of n randomly chosen experimental units (subjects or objects) yields two measurements, one from each of the two populations whose means are to be compared. This section develops CIs and test procedures for the comparison of two means using paired data. The following examples highlight two contexts where such data arise.

Example 9.5-1

A certain lake has been designated for pollution clean-up. One way to assess the effectiveness of the clean-up measures is to randomly select a number n of locations from which water samples are taken and analyzed both before and after the clean-up. The n randomly chosen locations are the experimental units, each of which yields two measurements. This results in paired data. Another way of designing the comparative study is to select a random sample of n_1 locations from which water samples are taken to assess the water quality before the clean-up measures, and a different random sample of n_2 locations which will serve to assess the water quality after the clean-up measures. The second sampling design will result in two independent samples, one from each population. ■

Example 9.5-2

Two different types of materials for making soles for children's shoes are to be compared for durability. One way of designing this comparative experiment is to make n pairs of shoes where one (either the left or the right) is randomly selected to be made with material A, and the other with material B. Then a random sample of n children is selected and each is fitted with such a pair of shoes. After a certain amount of time, the shoes are evaluated for wear and tear. In this example, the n children in the sample are the subjects, and the two treatments are the two types of material. For each subject there will be two measurements: the quantification of wear and tear in the shoe made with material A and the corresponding quantification in the shoe made with material B. This results in paired data. Another way of designing the comparative study is to select a random sample of n_1 children who are each fitted with shoes made with material A and a random sample of n_2 children who are each fitted with shoes made with material B. This will result in two independent samples, one from each population. ■

Example 9.5-3

Two different methods for determining the percentage of iron in ore samples are to be compared. One way of designing this comparative study is to obtain n ore samples and subject each of them to the two different methods for determining the iron content. In this example, the n ore samples are the objects, and the two methods are the treatments. For each ore sample there will be two measurements, resulting in paired data. Another way of designing the comparative study is to obtain n_1 ore samples to be evaluated with method 1, and, independently, obtain a different set of n_2 ore samples to be evaluated with method 2. This will result in two independent samples, one from each population. ■

From the above examples it follows that designs involving paired data have the potential to eliminate a large part of the uncontrolled variability. In Example 9.5-2, a large part of the uncontrolled variability is due to children having different weights, walking in different terrains, etc. Fitting the children with pairs of shoes where one shoe is made with material A and the other with material B eliminates this source of uncontrolled variability. Similarly, in Example 9.5-3, though the ore samples come from the same area, individual ore samples may differ in iron content due to natural variability. Subjecting the same ore samples to each method eliminates this source of uncontrolled variability. Elimination of uncontrolled variability means that we can achieve a more accurate comparison with smaller sample sizes. Thus, comparative studies should be designed to yield paired data whenever it is reasonable to expect that such a design can eliminate a large part of the uncontrolled variability. A popular class of designs yielding paired data are the so-called **before-after** designs that are used to evaluate the effectiveness of a treatment or program. See Example 9.5-1 for an application of such a design in the evaluation of the effectiveness of a clean-up program in reducing pollution. Other applications include the evaluation of the effectiveness of a new diet in reducing weight, and the effectiveness of a political speech in changing public opinion on a certain matter.

Because the two samples are derived from the same set of experimental units, they are not independent. As a consequence, the procedures described in Sections 9.2 and 9.3 for constructing CIs and tests for the difference of two population means and two proportions, which assume that the two samples are independent, cannot be used. The adaptation of these procedures to paired data is discussed in the next sections. An alternative procedure, the *signed-rank test*, which is applicable without the normality assumption even with a small sample size, will also be discussed.

9.5.2 THE PAIRED DATA T TEST

Let X_{1i} denote the observation on experimental unit i receiving treatment 1, and X_{2i} denote the observation on experimental unit i receiving treatment 2. Thus, experimental unit i contributes the pair of observations (X_{1i}, X_{2i}) to the data set, which can be put in the form

$$(X_{11}, X_{21}), \ldots, (X_{1n}, X_{2n}).$$

Because they are associated with the same experimental unit, X_{1i} and X_{2i} are not independent. This, in turn, implies that the sample averages $\overline{X}_1$ and $\overline{X}_2$ are not independent. Hence, the formula for the estimated standard error of the sample contrast $\overline{X}_1 - \overline{X}_2$, that is,

$$\widehat{\sigma}_{\overline{X}_1 - \overline{X}_2} = \sqrt{\frac{S_1^2}{n_1} + \frac{S_2^2}{n_2}},$$

where S_1^2, S_2^2 are the sample variances from populations 1, 2, respectively, does not apply because the standard error also involves the covariance of $\overline{X}_1$ and $\overline{X}_2$; see Proposition 4.4-4. We will now describe a way of estimating the standard error of the sample contrast without estimating the covariance of $\overline{X}_1$ and $\overline{X}_2$.

Let $D_i = X_{1i} - X_{2i}$ denote the difference of the two observations on the ith unit, $i = 1, \ldots, n$. For example, if each of 12 ore samples are analyzed by both methods in the context of Example 9.5-3, the paired data and differences D_i are shown below:

Ore Sample	Method A	Method B	D
1	38.25	38.27	−0.02
2	31.68	31.71	−0.03
⋮	⋮	⋮	⋮
12	30.76	30.79	−0.03

Set $\overline{D}$, S_D^2 for the sample average and sample variance, respectively, of the D_i, that is,

$$\overline{D} = \frac{1}{n}\sum_{i=1}^{n} D_i, \quad S_D^2 = \frac{1}{n-1}\sum_{i=1}^{n}\left(D_i - \overline{D}\right)^2. \qquad \textbf{(9.5.1)}$$

Thus, the estimated standard error of $\overline{D}$ is $\widehat{\sigma}_{\overline{D}} = S_D/\sqrt{n}$. However,

$$\overline{D} = \overline{X}_1 - \overline{X}_2.$$

Hence, the standard error of $\overline{X}_1 - \overline{X}_2$ is the same as the standard error of $\overline{D}$, that is,

$$\widehat{\sigma}_{\overline{X}_1-\overline{X}_2} = \widehat{\sigma}_{\overline{D}} = \frac{S_D}{\sqrt{n}}. \qquad \textbf{(9.5.2)}$$

Moreover, the population mean of the differences D_i is

$$\mu_D = E(D_i) = E(X_{1i}) - E(X_{2i}) = \mu_1 - \mu_2,$$

which means that a CI for $\mu_1 - \mu_2$ is the same as a CI for μ_D, which can be constructed as described in Chapter 7. Similarly, $H_0 : \mu_1 - \mu_2 = \Delta_0$ is true if and only if $\tilde{H}_0 : \mu_D = \Delta_0$, which can be tested with the procedures described in Chapter 8. This leads to the following *paired T test* procedures and confidence intervals for $\mu_1 - \mu_2$.

The Paired T Test Procedures and CIs

(1) *Assumptions:* The differences $D_i = X_{1i} - X_{2i}$, $i = 1, \ldots, n$, are independent, and either normal or $n \geq 30$

(2) $(1-\alpha)100\%$ *CI for* $\mu_1 - \mu_2$: $\overline{D} \pm t_{n-1,\alpha/2}\sqrt{\dfrac{S_D^2}{n}}$, where S_D^2 is given in (9.5.1)

(3) *Test Statistic for* $H_0: \mu_1 - \mu_2 = \Delta_0$: $T_{H_0} = \dfrac{\overline{D} - \Delta_0}{\sqrt{S_D^2/n}}$,

where S_D^2 is given in (9.5.1)

(4) *Rejection Rules for the Different* H_a:

$$
\begin{array}{ll}
\underline{H_a} & \underline{\text{RR at Level } \alpha} \\
\mu_1 - \mu_2 > \Delta_0 & T_{H_0} > t_{n-1,\alpha} \\
\mu_1 - \mu_2 < \Delta_0 & T_{H_0} < -t_{n-1,\alpha} \\
\mu_1 - \mu_2 \neq \Delta_0 & |T_{H_0}| > t_{n-1,\alpha/2}
\end{array}
\qquad (9.5.3)
$$

(5) *Formulas for the p-Value*:

$$
p\text{-value} = \begin{cases} 1 - G_{n-1}(T_{H_0}) & \text{for } H_a: \mu_1 - \mu_2 > \Delta_0 \\ G_{n-1}(T_{H_0}) & \text{for } H_a: \mu_1 - \mu_2 < \Delta_0 \\ 2[1 - G_{n-1}(|T_{H_0}|)] & \text{for } H_a: \mu_1 - \mu_2 \neq \Delta_0 \end{cases}
$$

where G_{n-1} is the CDF of T_{n-1}

With the data from populations 1 and 2 in the R objects *x1* and *x2*, respectively, the R commands for computing the paired T statistic, T_{H_0}, and p-value, are

R Commands for the Paired T Test for $H_0: \mu_1 - \mu_2 = \Delta_0$

```
t.test(x1, x2, mu=Δ0, paired=T, alternative="greater")
  # if Ha : μ1 − μ2 > Δ0

t.test(x1, x2, mu=Δ0, paired=T, alternative="less")
  # if Ha : μ1 − μ2 < Δ0

t.test(x1, x2, mu=Δ0, paired=T, alternative="two.sided")
  # if Ha : μ1 − μ2 ≠ Δ0
```

Occasionally, the data file has both samples stacked in a single data column, with a second column indicating the sample index of each observation in the data column. If the data and sample index columns are in the R objects x and s, respectively, use *x~s* instead of *x1, x2* in any of the above commands. These R commands will also return a CI for $\mu_1 - \mu_2$ (one-sided CIs if *alternative="greater"* or *"less"*, and the CIs shown in (9.5.3) if *alternative="two.sided"*). The default confidence level is 95%, but a $(1-\alpha)100\%$ CI can be obtained by adding *conf.level=1-α* in any of these commands; for example, *t.test(x1, x2, conf.level = 0.99)* gives the 99% CI shown in (9.5.3).

Example 9.5-4

Consider the study for comparing two methods for determining the iron content in ore samples described in Example 9.5-3. A total of 12 ore samples are analyzed by both methods, producing the paired data that yield $\overline{D} = -0.0167$ and $S_D = 0.02645$. Is there evidence that method B gives a higher average percentage than method A? Test at $\alpha = 0.05$, compute the p-value, and state any needed assumptions.

Solution

The appropriate null and alternative hypotheses are $H_0: \mu_1 - \mu_2 = 0$ and $H_a: \mu_1 - \mu_2 < 0$. Because the sample size is small, we must assume normality. The test statistic is

$$T_{H_0} = \frac{\overline{D}}{S_D/\sqrt{n}} = \frac{-0.0167}{0.02645/\sqrt{12}} = -2.1872.$$

Since $T_{H_0} < -t_{11,0.05} = -1.796$, H_0 is rejected. According to the formula in (9.5.3), the p-value is $G_{11}(-2.1865) = 0.026$. ■

Example 9.5-5

Changes in the turbidity of a body of water are used by environmental or soil engineers as an indication that surrounding land may be unstable, which allows sediments to be pulled into the water. Turbidity measurements using the Wagner test, from 10 locations around a lake taken both before and after a land-stabilization project, can be found in *Turbidity.txt*. Is there evidence that the land-stabilizing measures have reduced the turbidity?

(a) Using R commands test the appropriate hypothesis at level $\alpha = 0.01$, give the p-value, and construct a 99% CI. What assumptions, if any, are needed for the validity of this procedure?

(b) Conduct the test and CI assuming that the two samples are independent and compare the results with those obtained in part (a). Use the default statistic that does not assume equal variances.

Solution

(a) Having read the data into the R data frame *tb*, the command *t.test(tb$Before, tb$After, paired=T, alternative="greater")* returns the paired T statistic, degrees of freedom, and p-value of 8.7606, 9, and 5.32×10^{-06}, respectively. The p-value is much smaller than the chosen level of significance, and thus the null hypothesis is rejected. Because of the small sample size, the normality assumption is needed for the validity of the procedure. Repeating the command with the default alternative and confidence level specification of 0.99, that is, *t.test(tb$Before, tb$After, paired=T, conf.level=0.99)*, returns a 99% CI of (1.076, 2.344).

(b) Without "*paired=T*", the first of the two R commands in part (a) returns the two-sample T statistic, degrees of freedom, and p-value of 0.956, 17.839, and 0.1759, respectively. Now the p-value is large enough that the null hypothesis cannot be rejected even at level 0.1. Next, the second of the two commands in (a) without "*paired=T*" returns a 99% CI of (−3.444, 6.864). It is seen that the p-value is radically different from that obtained in part (a) and leads to a different hypothesis-testing decision. Similarly, the 99% CI is radically different from that obtained in part (a). This highlights the need to apply the appropriate procedure to each data set. ■

9.5.3 THE PAIRED T TEST FOR PROPORTIONS

When the paired observations consist of Bernoulli variables, each pair (X_{1j}, X_{2j}) can be either $(1, 1)$ or $(1, 0)$ or $(0, 1)$ or $(0, 0)$. To illustrate, consider the before-after example where a random sample of n voters are asked, both before and after a presidential speech, whether or not they support a certain policy. Thus, X_{1j} takes the value 1 if the jth voter is in favor of the policy before the President's speech and 0 otherwise, while $X_{2j} = 1$ or 0 according to whether or not the same voter is in favor of the policy after the speech. Since $E(X_{1j}) = p_1$, which is the probability that

$X_{1j} = 1$, and $E(X_{2j}) = p_2$, which is the probability that $X_{2j} = 1$, the paired T test procedure and CIs of (9.5.3) pertain to $p_1 - p_2$. Note that

$$\overline{X}_1 = \widehat{p}_1, \quad \overline{X}_2 = \widehat{p}_2, \quad \overline{D} = \widehat{p}_1 - \widehat{p}_2. \tag{9.5.4}$$

Also, as before, S_D^2/n, where S_D^2 is given in (9.5.1) correctly estimates the variance of $\widehat{p}_1 - \widehat{p}_2$.

In spite of these similarities to the general case, two issues deserve special mention in the case of Bernoulli variables. The first has to do with a different sample size requirement; this is stated in (9.5.9). The second issue is computational. This issue arises because, in the case of Bernoulli variables, it is customary to present the data in the form of a summary table like the following

		After	
		1	0
Before	1	Y_1	Y_2
	0	Y_3	Y_4

(9.5.5)

where "Before" refers to the first coordinate (observation) of a pair and "After" refers to the second. Thus, Y_1 is the number of pairs (X_{1j}, X_{2j}) whose coordinates are both 1, that is, the number of $(1, 1)$ pairs, Y_2 is the number of $(1, 0)$ pairs, and so forth. Clearly, $Y_1 + Y_2 + Y_3 + Y_4 = n$. To construct CIs for $p_1 - p_2$ and to compute the test statistic in (9.5.3) note first that

$$\frac{Y_1 + Y_2}{n} = \widehat{p}_1, \quad \frac{Y_1 + Y_3}{n} = \widehat{p}_2. \tag{9.5.6}$$

Thus, from (9.5.4) and (9.5.6) it follows that

$$\overline{D} = \frac{Y_2 - Y_3}{n}. \tag{9.5.7}$$

Expressing S_D^2 in terms of the Y_i's requires some algebra. Omitting the details, we have

$$\frac{n-1}{n} S_D^2 = \frac{1}{n} \sum_{i=1}^{n} (D_i - \overline{D})^2 = \widehat{q}_2 + \widehat{q}_3 - (\widehat{q}_2 - \widehat{q}_3)^2, \tag{9.5.8}$$

where $\widehat{q}_2 = Y_2/n$, $\widehat{q}_3 = Y_3/n$. Paired data CIs and tests for $p_1 - p_2$ are conducted according

Computation of T_{H_0} and Sample Size Requirement for the Procedures in (9.5.3) with Bernoulli Variables

(1) *Sample Size Requirement:* The numbers n_{10} and n_{01} of $(1, 0)$ and $(0, 1)$ pairs, respectively, must satisfy $n_{10} + n_{01} \geq 16$

(2) *Test Statistic for $H_0 : p_1 - p_2 = \Delta_0$:*

$$T_{H_0} = \frac{(\widehat{q}_2 - \widehat{q}_3) - \Delta_0}{\sqrt{(\widehat{q}_2 + \widehat{q}_3 - (\widehat{q}_2 - \widehat{q}_3)^2)/(n-1)}}, \tag{9.5.9}$$

where $\widehat{q}_2 = Y_2/n$, $\widehat{q}_3 = Y_3/n$ with the Y_k depicted in (9.5.5)

to (9.5.3) by replacing $\overline{D}$ and S_D^2 with the expressions in (9.5.7) and (9.5.8). The resulting form of the paired data test statistic and new sample size requirement are given in (9.5.9).

McNemar's Test For testing $H_0 : p_1 - p_2 = 0$, a variation of the paired T test statistic for proportions bears the name of *McNemar's test*. The variation consists of omitting the term $(\widehat{q}_2 - \widehat{q}_3)^2$ in the denominator of (9.5.9) and using n instead of $n - 1$. Thus, McNemar's test statistic is

McNemar's Statistic

$$MN = \frac{(\widehat{q}_2 - \widehat{q}_3)}{\sqrt{(\widehat{q}_2 + \widehat{q}_3)/n}} = \frac{Y_2 - Y_3}{\sqrt{Y_2 + Y_3}}. \tag{9.5.10}$$

Omission of the term $(\widehat{q}_2 - \widehat{q}_3)^2$ is justified on the grounds that, under the null hypothesis, $P((X_{1j}, X_{2j}) = (1,0)) = P((X_{1j}, X_{2j}) = (0,1))$; that is, the quantity estimated by $\widehat{q}_2 - \widehat{q}_3$ is zero. Use of such a variance estimator, however, restricts McNemar's procedure to testing only for the equality of two proportions, that is, only in the case where $\Delta_0 = 0$.

Example 9.5-6

To assess the effectiveness of a political speech in changing public opinion on a proposed reform, a random sample of $n = 300$ voters was asked, both before and after the speech, if they support the reform. The before-after data are given in the table below:

		After	
		Yes	No
Before	Yes	80	100
	No	10	110

Was the political speech effective in changing public opinion? Test at $\alpha = 0.05$.

Solution

According to (9.5.10) McNemar's test statistic is

$$MN = \frac{90}{\sqrt{110}} = 8.58,$$

and according to (9.5.9) the paired T test statistic is

$$T_{H_0} = \frac{0.3}{\sqrt{(11/30 - 0.3^2)/299}} = 9.86.$$

Because of the large sample size we use $z_{\alpha/2} = z_{0.025} = 1.96$ as the critical point. Since both 8.58 and 9.86 are greater than 1.96, we conclude that the political speech was effective in changing public opinion.

9.5.4 THE WILCOXON SIGNED-RANK TEST

The *signed-rank test* is a one-sample procedure that can be used even with small sample sizes provided the population distribution is continuous and symmetric. Because it requires the sample to have come from a symmetric population, the test pertains to both the mean and the median. If the population distribution is known to be symmetric, the signed-rank test is preferable to (i.e., is more powerful than) the sign test discussed in Section 8.3.5. The assumption of symmetry, however, cannot be tested effectively with small sample sizes, which is why the signed-rank test was not discussed in Chapter 8.

The relevance of the signed-rank test to paired data (X_{1i}, X_{2i}), $i = 1, \ldots, n$, stems from the fact that, if X_{1i}, X_{2i} have the same marginal distributions, that is, $H_0^F : F_1 = F_2$ is true, the distribution of the differences $D_i = X_{1i} - X_{2i}$ is symmetric about zero. Thus, the crucial requirement of symmetry is automatically met for the differences, and the signed-rank procedure can be used for testing $H_0 : \mu_D = 0$.

The construction of the signed-rank statistic, denoted by S_+, involves ranking the absolute values of the differences D_i and taking the sum of the ranks of the positive D_i's. To appreciate the relevance of S_+ for testing $H_0 : \mu_D = 0$, note that if μ_D were larger than 0, then more differences D_i will tend to be positive and also the positive differences will tend to be larger than the absolute value of the negative differences. Thus, S_+ will tend to take larger values if the alternative $H_a : \mu > 0$ is true. Similarly, S_+ will tend to take smaller values if the alternative $H_a : \mu < 0$ is true. To test $H_a : \mu = \Delta_0$ against the different alternatives, Δ_0 is subtracted from the D_i. A detailed description of the signed-rank test procedure is given in the display (9.5.11) below.

The signed-rank statistic shares the property of the rank-sum statistic that its exact null distribution does not depend on the continuous distribution of the D_i. Thus, it can be carried out even with small sample sizes with the use of tables or a software package.

With the data from populations 1 and 2 in the R objects *x1* and *x2*, respectively, the R command *wilcox.test*, given in Section 9.3, with the *paired=T* specification, returns the signed-rank statistic and the p-value for testing the null hypothesis $H_0 : \mu_D = \Delta_0$ against the different alternatives. The *paired=T* specification should be omitted if only the differences (*d=x1-x2*) are used as input. Moreover, a $(1-\alpha)100\%$ CI for μ_D (one-sided if the alternative is specified as *"greater"* or *"less"* and two-sided if the default alternative is chosen) can be obtained by adding *conf.int=T, conf.level=1-α*. The procedure for hand calculation of the signed-rank test statistic and p-value is given in (9.5.11).

The Signed-Rank Test Procedures for $H_0 : \mu_D = \Delta_0$

(1) *Assumptions:* The pairs (X_{1i}, X_{2i}), $i = 1, \ldots, n$, are iid with continuous marginal distributions; $n \geq 10$

(2) *Construction of the Signed-Rank Statistic S_+:*
(a) Rank the absolute differences $|D_1 - \Delta_0|, \ldots, |D_n - \Delta_0|$ from smallest to largest. Let R_i denote the rank of $|D_i - \Delta_0|$
(b) Assign to R_i the sign of $D_i - \Delta_0$ **(9.5.11)**
(c) Let S_+ be the sum of the ranks R_i with positive sign

(3) *Test Statistic:* $$Z_{H_0} = \frac{\left(S_+ - \frac{n(n+1)}{4}\right)}{\sqrt{n(n+1)(2n+1)/24}}$$

(4) *Rejection Rules for the Different H_a:*

H_a	RR at Level α
$\mu_D > \Delta_0$	$Z_{H_0} > z_\alpha$
$\mu_D < \Delta_0$	$Z_{H_0} < -z_\alpha$
$\mu_D \neq \Delta_0$	$\|Z_{H_0}\| > z_{\alpha/2}$

(5) *Formulas for the p-Value:*

$$p\text{-value} = \begin{cases} 1 - \Phi(Z_{H_0}) & \text{for } H_a: \mu_D > 0 \\ \Phi(Z_{H_0}) & \text{for } H_a: \mu_D < 0 \\ 2[1 - \Phi(|Z_{H_0}|)] & \text{for } H_a: \mu_D \neq 0 \end{cases}$$

where Φ is the CDF of the standard normal distribution

Example 9.5-7

The octane ratings of 12 gasoline blends were determined by two standard methods. The 12 differences D_i are given in the first row of the table below.

D_i	2.1	3.5	1.7	0.2	−0.6	2.2	2.5	2.8	2.3	6.5	−4.6	1.6
$\|D_i\|$	2.1	3.5	1.7	0.2	0.6	2.2	2.5	2.8	2.3	6.5	4.6	1.6
R_i	5	10	4	1	2	6	8	9	7	12	11	3
Signed R_i	5	10	4	1	−2	6	8	9	7	12	−11	3

Is there evidence, at level $\alpha = 0.1$, that the rating produced by the two methods differs?

(a) Using hand calculations, carry out the test procedure specified in (9.5.11) and report the p-value.

(b) Use R commands to obtain the p-value and to construct a 90% CI for μ_D.

Solution

(a) Rows 2, 3, and 4 of the above table are formed according to the steps in part (2) of (9.5.11). From this we have $S_+ = 5+10+4+1+6+8+9+7+12+3 = 65$. Thus

$$Z_{H_0} = \frac{65 - (156/4)}{\sqrt{12 \times 13 \times 25/24}} = \frac{65 - 39}{\sqrt{162.5}} = 2.04.$$

The p-value equals $2[1 - \Phi(2.04)] = 0.04$ and thus, H_0 is rejected at level $\alpha = 0.1$.

(b) With the differences entered in the R object d, that is, with *d=c(2.1, 3.5, 1.6, 0.2, -0.6, 2.2, 2.5, 2.8, 2.3, 6.5, -4.6, 1.6)*, the R command *wilcox.test(d, conf.int=T, conf.level=0.9)* returns a p-value of 0.045 and a 90% CI of (0.8, 2.8). ■

Exercises

1. A study was conducted to see whether two types of cars, A and B, took the same time to parallel park. Seven drivers were randomly obtained and the time required for each of them to parallel park each of the 2 cars was measured. The results are listed in the following table.

	Driver						
Car	1	2	3	4	5	6	7
A	19.0	21.8	16.8	24.2	22.0	34.7	23.8
B	17.8	20.2	16.2	41.4	21.4	28.4	22.7

Is there evidence that the time required to parallel park the two types of car are, on average, different?

(a) Test at $\alpha = 0.05$ using the paired T test. What assumptions are needed for the validity of this test? Comment on the appropriateness of the assumptions for this data set.

(b) Test at $\alpha = 0.05$ using the signed-rank test.

2. Two different analytical tests can be used to determine the impurity levels in steel alloys. The first test is known to perform very well but the second is cheaper. A specialty steel manufacturer will adopt the second method unless there is evidence that it gives significantly different results than the first. Eight steel specimens are cut in half and one half is randomly assigned to one test and the other half to the other test. The results are shown in the following table.

Specimen	Test 1	Test 2
1	1.2	1.4
2	1.3	1.7
3	1.5	1.5
4	1.4	1.3
5	1.7	2.0
6	1.8	2.1
7	1.4	1.7
8	1.3	1.6

(a) State the appropriate null and alternative hypotheses.

(b) Carry out the paired data T test at level $\alpha = 0.05$, and state whether H_0 is rejected or not. Should the specialty steel manufacturer adopt the second method?

(c) Construct a 95% CI for the mean difference.

3. The percent of soil passing through a sieve is one of several soil properties studied by organizations such as the National Highway Institute. A particular experiement considered the percent of soil passing through a 3/8-inch sieve for soil taken from two separate locations. It is known that the percent of soil passing through the sieve is affected by weather conditions. One measurement on soil taken from each location was made on each of 32 different days. The data are available in *SoilDataNhi.txt*. Is there evidence that the average percent of soil passing through the sieve is different for the two locations? Use R commands to complete the following parts.

(a) Enter the data in R and use the paired T procedure to conduct the test at level $\alpha = 0.05$, and construct a 95% CI for the difference of the population means.

(b) Use the signed-rank procedure to repeat part (a). Are the p-values and CIs from the two procedures similar?

(c) Conduct the test and CI using the *t.test* and *wilcox.test* commands without the *paired=T* specification. How do the new p-values and CIs compare to those obtained in parts (a) and (b)?

4. Two brands of motorcycle tires are to be compared for durability. Eight motorcycles are selected at random and one tire from each brand is randomly assigned (front or back) on each motorcycle. The motorcycles are then run until the tires wear out. The data, in km, are given in *McycleTiresLifeT.txt*. Use either hand calculations or R commands to complete the following parts.

(a) State the null and alternative hypotheses, then use the paired T test procedure to test the hypothesis at level $\alpha = 0.05$ and to construct a 90% CI. What assumptions are needed for its validity?

(b) Conduct, at level $\alpha = 0.05$, the signed-rank test.

5. During the evaluation of two speech recognition algorithms, A_1 and A_2, each is presented with the same sequence, $u_1, \ldots, u_n$, of labeled utterances for recognition. The $\{u_i\}$ are assumed to be a random sample from some population of utterances. Each algorithm makes a decision about the label of each u_i which is either correct or incorrect. The data are shown below:

		A_2	
		Correct	Incorrect
A_1	Correct	1325	3
	Incorrect	13	59

Is there evidence that the two algorithms have different error rates? Test at $\alpha = 0.05$ using both the paired T test procedure and McNemar's test procedure.

6. To assess a possible change in voter attitude toward gun control legislation between April and June, a random sample of $n = 260$ was interviewed in April and in June. The resulting responses are summarized in the following table:

		June	
		No	Yes
April	No	85	62
	Yes	95	18

Is there evidence that there was a change in voter attitude? Test at $\alpha = 0.05$ using both the paired T test procedure and McNemar's test procedure.

Chapter 10

COMPARING $k > 2$ POPULATIONS

10.1 Introduction

This chapter considers the comparison of several population means or proportions. The basic hypothesis-testing problem in the comparison of $k > 2$ means is

$$H_0 : \mu_1 = \cdots = \mu_k \text{ vs } H_a : H_0 \text{ is not true.} \tag{10.1.1}$$

The alternative H_a is equivalent to the statement that at least one pair of mean values differs. When the null hypothesis is rejected, the additional question—which pairs of means differ—arises naturally. Addressing this additional question leads to the subjects of simultaneous confidence intervals (simultaneous CIs or SCIs) and *multiple comparisons*. Moreover, the primary focus of comparative experiments is often the comparison of specific contrasts instead of the general testing problem (10.1.1); see Example 1.8-3 for different types of contrasts that might be of interest. Such specialized comparisons can also be carried out in the framework of multiple comparisons and simultaneous CIs. Similar comments apply for the comparison of $k > 2$ population proportions.

Section 10.2 deals with different methods for testing the basic hypothesis (10.1.1). Section 10.2.1 describes the analysis of variance (ANOVA) approach. Testing for a specific contrast and testing for the assumptions needed for the validity of the ANOVA procedure are also considered. Section 10.2.2 describes the Kruskal-Wallis test procedure, which extends the rank-sum test to the present setting of comparing $k > 2$ populations. Section 10.2.3 describes the chi-square test for the equality of $k > 2$ proportions, including its contingency table formulation. Section 10.3 introduces the concepts of simultaneous CIs and multiple comparisons. There are several methods for performing these procedures, but this chapter (and book) presents *Bonferroni's* and Tukey's methods, including Tukey's method on the ranks. The procedures in Sections 10.2 and 10.3 require that the samples from the k populations are collected independently. A generalization of the paired data design, which is discussed in Section 9.5, is the *randomized block design*. The k samples arising from *randomized block designs* are not independent for the same reason that paired data are not independent. Section 10.4 considers methods for testing the hypothesis (10.1.1) when the $k > 2$ samples arise from a randomized block design. These methods include the ANOVA procedure (Section 10.4.2) and Friedman's test and the ANOVA procedure on the ranks (Section 10.4.3). Finally, the Bonferroni and Tukey methods for simultaneous CIs and multiple comparisons are described in Section 10.4.4.

10.2 Types of k-Sample Tests

10.2.1 THE ANOVA F TEST FOR MEANS

The comparison of the k populations means, $\mu_1, \ldots, \mu_k$, will be based on k simple random samples, which are drawn independently, one from each of the populations. Let

$$X_{i1}, X_{i2}, \ldots, X_{in_i}, \quad i = 1, \ldots, k, \tag{10.2.1}$$

denote the k samples. Thus, the random sample from population 1 has size n_1 with observations denoted by $X_{11}, \ldots, X_{1n_1}$, the random sample from population 2 has size n_2 with observations denoted by $X_{21}, \ldots, X_{2n_2}$, and so forth. This is the simplest type of comparative study for which it is common to write a **statistical model** for the data. The model can be written as

$$X_{ij} = \mu_i + \epsilon_{ij} \quad \text{or} \quad X_{ij} = \mu + \alpha_i + \epsilon_{ij}, \tag{10.2.2}$$

where $\mu = \frac{1}{k}\sum_{i=1}^{k} \mu_i$, $\alpha_i = \mu_i - \mu$ are the population (or treatment) effects defined in (1.8.3), and the intrinsic error variables ϵ_{ij} are independent with zero mean and variance σ_ϵ^2. Note that the definition of the treatment effects implies that they satisfy the condition

$$\sum_{i=1}^{k} \alpha_i = 0. \tag{10.2.3}$$

The first expression in (10.2.2) is called the *mean-plus-error* form of the model, while the second is the *treatment effects* form. The null hypothesis of equality of the k means in (10.1.1) is often expressed in terms of the treatment effects as

$$H_0 : \alpha_1 = \alpha_2 = \cdots = \alpha_k = 0. \tag{10.2.4}$$

As mentioned in Section 8.3.4, the ANOVA methodology is based on a decomposition of the *total variability* into the variability due to the differences in the population means (called *between groups variability*) and the variability due to the intrinsic error (called *within groups variability*). As in the regression setting, variability is represented by the sums of squares, or SS. To define them, set

$$\overline{X}_i = \frac{1}{n_i}\sum_{j=1}^{n_i} X_{ij} \quad \text{and} \quad \overline{X} = \frac{1}{N}\sum_{i=1}^{k}\sum_{j=1}^{n_i} X_{ij} = \frac{1}{N}\sum_{i=1}^{k} n_i \overline{X}_i, \tag{10.2.5}$$

where $N = n_1 + \cdots + n_k$, for the sample mean from the ith sample and the average of all observations. The *treatment SS* and the *error SS* are defined as follows:

Treatment Sum of Squares

$$\text{SSTr} = \sum_{i=1}^{k} n_i (\overline{X}_i - \overline{X})^2 \tag{10.2.6}$$

Error Sum of Squares

$$\text{SSE} = \sum_{i=1}^{k}\sum_{j=1}^{n_i} (X_{ij} - \overline{X}_i)^2 \tag{10.2.7}$$

It can be shown that SSTr and SSE decompose the total sum of squares, which is defined as $\text{SST} = \sum_{i=1}^{k}\sum_{j=1}^{n_i}(X_{ij} - \overline{X})^2$. That is,

$$\text{SST} = \text{SSTr} + \text{SSE}.$$

This is similar to the decomposition of the total sum of squares we saw in (8.3.11) for the construction of the F statistic in the context of simple linear regression.

The ANOVA F test rejects H_0 in (10.2.4) if SSTr is large compared to SSE. For a proper comparison of SSTr and SSE, however, they must be divided by their respective degrees of freedom, which are (recall that $N = n_1 + \cdots + n_k$) given below:

Degrees of Freedom for SSTr and SSE

$$\text{DF}_{SSTr} = k - 1, \quad \text{DF}_{SSE} = N - k \tag{10.2.8}$$

Dividing each of the SSTr and SSE by their degrees of freedom, we obtain the *mean squares for treatment*, or MSTr, and the *mean squares for error*, or MSE, shown below:

Mean Squares for Treatment and Error

$$\text{MSTr} = \frac{\text{SSTr}}{k-1}, \quad \text{MSE} = \frac{\text{SSE}}{N-k} \tag{10.2.9}$$

The k-sample version of the pooled variance we saw in (9.2.4), denoted also by S_p^2, is

$$S_p^2 = \frac{(n_1 - 1)S_1^2 + \cdots + (n_k - 1)S_k^2}{n_1 + \cdots + n_k - k}, \tag{10.2.10}$$

where S_i^2 is the sample variance from the ith sample. It is not difficult to verify that

$$\text{MSE} = S_p^2. \tag{10.2.11}$$

The ANOVA F statistic, F_{H_0}, is computed as the ratio of MSTr/MSE. Since MSE can be computed from the sample variances and MSTr can be computed from the sample means, it follows that F_{H_0} can be computed from the sample means and the sample variances of the k samples. Under the assumptions of normality and homoscedasticity (i.e., $\sigma_1^2 = \cdots = \sigma_k^2$), its exact distribution is known to be $F_{k-1,\,N-k}$, where $N = n_1 + \cdots + n_k$. (In the notation introduced in Section 8.3.4, $F_{\nu_1,\,\nu_2}$ stands both for the F distribution with numerator degrees of freedom ν_1 and denominator degrees of freedom ν_2 and for its CDF.) The discussion of this paragraph is summarized in the following display:

ANOVA *F* Statistic and Its Null Distribution for Homoscedastic Normal Data

$$F_{H_0} = \frac{\text{MSTr}}{\text{MSE}} \sim F_{k-1,\,N-k} \tag{10.2.12}$$

Under the alternative hypothesis, F_{H_0} tends to take larger values so the p-value is computed from the formula $1 - F_{k-1,N-k}(F_{H_0})$. The F statistic F_{H_0}, its p-value, and all quantities required for their computation are shown in an ANOVA table in the form shown below:

Source	DF	SS	MS	F	P
Treatment	$k-1$	SSTr	$\text{MSTr} = \dfrac{\text{SSTr}}{k-1}$	$F_{H_0} = \dfrac{\text{MSTr}}{\text{MSE}}$	$1 - F_{k-1,N-k}(F_{H_0})$
Error	$N-k$	SSE	$\text{MSE} = \dfrac{\text{SSE}}{N-k}$		
Total	$N-1$	SST			

With the k samples stacked in a single data column, called *Value*, and a second column, called *Sample*, indicating the sample membership index of each observation in the data column, after the data are read into the R data frame *df*, the following R command gives the ANOVA table including the value of the F statistic and the p-value.

R Commands for the ANOVA F Test for $H_0 : \mu_1 = \cdots = \mu_k$

```
fit=aov(Value~as.factor(Sample), data=df); anova(fit)
```

The *as.factor* designation of the sample membership column *Sample* is only needed if the column is numeric, but it can always be used. In the first of the two examples below, the sample membership column is not numeric and the *as.factor* designation can be omitted. The two commands can be combined into one: *anova(aov(Value~as.factor(Sample), data=df))*.

The following display summarizes all assumptions and formulas needed for implementing the ANOVA F test procedure discussed in this section.

The F Procedure for Testing the Equality of k Means

(1) *Assumptions:*
  (a) For each $i = 1, \ldots, k$, $X_{i1}, \ldots, X_{in_i}$ is a simple random sample from the ith population
  (b) The k samples are independent
  (c) The k variances are equal: $\sigma_1^2 = \cdots = \sigma_k^2$
  (d) The k populations are normal, or $n_i \geq 30$ for all i

(2) *Construction of the F Test Statistic:*
  (a) Compute the sample mean and sample variance of each of the k samples
  (b) Compute SSTr according to (10.2.6), and MSTr as MSTr=SSTr/DF_{SSTr}, where the degrees of freedom are given in (10.2.8) **(10.2.13)**
  (c) Compute MSE from (10.2.10) and (10.2.11)
  (d) Compute $F_{H_0} = \text{MSTr}/\text{MSE}$

(3) *Rejection Region:*
  $F_{H_0} > F_{k-1,N-k,\alpha}$, where $F_{k-1,N-k,\alpha}$ is the $(1-\alpha)100$th percentile of the $F_{k-1,N-k}$ distribution

(4) *p-Value:*
  $p\text{-value} = 1 - F_{k-1,N-k}(F_{H_0})$

Example 10.2-1

To compare three different mixtures of methacrylic acid and ethyl acrylate for stain/soil release effectiveness, 5 cotton fabric specimens are treated with each mixture and tested. The data are given in *FabricSoiling.txt*. Do the three mixtures differ in the stain/soil release effectiveness? Test at level of significance $\alpha = 0.05$ and report the p-value. What assumptions, if any, are needed?

Solution

Let μ_1, μ_2, μ_3 denote the average stain/soil release for the three different mixtures. We want to test $H_0 : \mu_1 = \mu_2 = \mu_3$ vs $H_a : H_0$ is false at $\alpha = 0.05$. Because of the small sample sizes, we need to assume that the three populations are normal and homoscedastic (i.e., have equal variances); methods for testing these assumptions are described below. Following the steps for computing the F statistic given in the display (10.2.13), we first compute the three sample means and sample variances:

$$\overline{X}_1 = 0.918, \qquad \overline{X}_2 = 0.794, \qquad \overline{X}_3 = 0.938$$
$$S_1^2 = 0.04822, \qquad S_2^2 = 0.00893, \qquad S_3^2 = 0.03537.$$

This gives an overall mean of $\overline{X} = 0.883$ and pooled sample variance of $S_p^2 = \text{MSE} = 0.0308$. Next, using formula (10.2.6) we obtain $\text{SSTr} = 0.0608$ and $\text{MSTr} = \text{SSTr}/2 = 0.0304$. Thus, $F_{H_0} = 0.0304/0.0308 = 0.99$. The calculations are summarized in the ANOVA table below. From Table A.6, we find $F_{2,12,0.05} = 3.89$. Since $0.98 \not> 3.89$, the null hypothesis is not rejected. The R command *1-pf(0.99, 2, 12)* returns a p-value of 0.4.

Source	DF	SS	MS	F	P
Treatment	$k-1=2$	0.0608	0.0304	0.99	0.4
Error	$N-k=12$	0.3701	0.0308		
Total	$N-1=14$	0.4309			

With the data read into the R data fame *df*, the R commands

```
anova(aov(Value~Sample, data=df))
```

returns the same ANOVA table. ■

Testing for a Particular Contrast Instead of the basic, overall hypothesis of equality of all k means, interest often lies in testing for specialized contrasts. The concept of a contrast was discussed in Section 1.8; see Example 1.8-3 for some examples of contrasts that arise in the comparison of k means. In general, a contrast of the k means, $\mu_1, \ldots, \mu_k$, is any linear combination

$$\theta = c_1\mu_1 + \cdots + c_k\mu_k \tag{10.2.14}$$

of these means whose coefficients sum to zero, that is, $c_1 + \cdots + c_k = 0$. For example, the two sets of coefficients

$$(1, -1, 0, \ldots, 0) \text{ and } \left(1, -\frac{1}{k-1}, \ldots, -\frac{1}{k-1}\right)$$

define the contrasts $\mu_1 - \mu_2$ and $\mu_1 - (\mu_2 + \cdots + \mu_k)/(k-1)$, respectively. The first contrast compares only the first two means, while the second compares the first with the average of all other means. The basic, or overall, null hypothesis $H_0 : \mu_1 = \cdots = \mu_k$ implies that all contrasts are zero. [To see this note that, under H_0, any contrast of the form (10.2.14) can be written as $\mu(c_1 + \cdots + c_k)$, where μ denotes the common value of $\mu_1, \ldots, \mu_k$; hence, by the definition of a contrast, it is zero.] However, testing for individual contrasts is more powerful than the overall ANOVA F test. For example, if $\mu_2 = \cdots = \mu_k$ but μ_1 differs from the rest, a test is much more likely to reject the null hypothesis that specifies that the contrast

$$\mu_1 - (\mu_2 + \cdots + \mu_k)/(k-1)$$

is zero than the overall null hypothesis. This is demonstrated in Example 10.2-2.

The one-sample T test procedure of Chapter 8 and the one-sample T CI of Chapter 7 can be used for testing hypotheses and for constructing CIs for the contrast θ in (10.2.14). In particular, the estimator of θ and its estimated standard error are

$$\hat{\theta} = c_1\overline{X}_1 + \cdots + c_k\overline{X}_k \quad \text{and} \quad \hat{\sigma}_{\hat{\theta}} = \sqrt{S_p^2\left(\frac{c_1^2}{n_1} + \cdots + \frac{c_k^2}{n_k}\right)}, \tag{10.2.15}$$

where S_p^2 is the pooled variance given in (10.2.10). If the populations are normal,

$$\frac{\hat{\theta} - \theta}{\hat{\sigma}_{\hat{\theta}}} \sim T_{N-k},$$

and the same relationship holds approximately if the sample sizes are large. It follows that a $(1-\alpha)100\%$ CI for θ is given by (7.1.8), with $\hat{\theta}$, $\hat{\sigma}_{\hat{\theta}}$ given by (10.2.15), and $\nu = N - k$. Also, the test statistic for $H_0 : \theta = 0$ is

$$T_{H_0} = \frac{\hat{\theta}}{\hat{\sigma}_{\hat{\theta}}}, \tag{10.2.16}$$

and the p-value for testing vs $H_a : \theta \neq 0$ is given by $2(1 - G_{N-k}(|T_{H_0}|))$.

Example 10.2-2

A quantification of coastal water quality converts measurements on several pollutants to a water quality index with values from 1 to 10. An investigation into the after-clean-up water quality of a lake focuses on five areas encompassing the two beaches on the eastern shore and the three beaches on the western shore. Water quality index values are obtained from 12 water samples from each beach. The data are found in *WaterQualityIndex.txt*. One objective of the study is the comparison of the water quality on the eastern and western shores of the lake.

(a) Identify the relevant contrast, θ, for the study's objective, and use the data to (i) carry out the test of $H_0 : \theta = 0$ vs $H_a : \theta \neq 0$ at level of significance $\alpha = 0.05$, and (ii) to construct a 95% CI for θ.

(b) Test, at $\alpha = 0.05$, the overall hypothesis that the average pollution index is the same for all five beaches.

Solution

(a) Let μ_1, μ_2 denote the average pollution index at the two beaches in the eastern shore, and μ_3, μ_4, μ_5 denote the average pollution index at the three beaches in the western shore. The hypothesis-testing problem relevant to the stated objective is $H_0 : \theta = 0$ vs $H_a : \theta \neq 0$, where the contrast θ is defined as

$$\theta = \frac{\mu_1 + \mu_2}{2} - \frac{\mu_3 + \mu_4 + \mu_5}{3},$$

so the corresponding set of coefficients is $c_1 = 1/2,\ c_2 = 1/2,\ c_3 = -1/3,\ c_4 = -1/3,\ c_5 = -1/3$. From the data we obtain

$$\overline{X}_1 = 8.01,\quad \overline{X}_2 = 8.33,\quad \overline{X}_3 = 8.60,\quad \overline{X}_4 = 8.63,\quad \overline{X}_5 = 8.62$$
$$S_1^2 = 0.43,\quad S_2^2 = 0.63,\quad S_3^2 = 0.38,\quad S_4^2 = 0.18,\quad S_5^2 = 0.56.$$

This gives (note that when the sample sizes are equal, the pooled sample variance, S_p^2, is the average of the sample variances)

$$\hat{\theta} = \frac{\overline{X}_1 + \overline{X}_2}{2} - \frac{\overline{X}_3 + \overline{X}_4 + \overline{X}_5}{3} = -0.45 \quad \text{and} \quad S_p^2 = \frac{S_1^2 + \cdots + S_5^2}{5} = 0.44,$$

and $\hat{\sigma}_{\hat{\theta}} = \sqrt{S_p^2\left[\frac{1}{4}\left(\frac{1}{n_1}+\frac{1}{n_2}\right)+\frac{1}{9}\left(\frac{1}{n_3}+\frac{1}{n_4}+\frac{1}{n_5}\right)\right]} = 0.17$, using the fact that all $n_i = 12$. The degrees of freedom are $\nu = N - k = 60 - 5 = 55$. Thus, the 95% CI is

$$\hat{\theta} \pm t_{55,0.025}\sigma_{\hat{\theta}} = (-0.45 \pm 0.17 \times 2.00) = (-0.79, -0.11),$$

where the value of $t_{\nu,0.025}$, with $\nu = 55$, was obtained from the command *qt(0.975, 55)*. Note that the CI does not include zero, suggesting that the contrast is significantly different from zero at $\alpha = 0.05$. Next, the test statistic and p-value are

$$T_{H_0} = \frac{\hat{\theta}}{\hat{\sigma}_{\hat{\theta}}} = \frac{-0.45}{0.17} = -2.65, \quad p\text{-value} = 2(1 - G_{55}(2.65)) = 0.01.$$

Again we see that the null hypothesis $H_0 : \theta = 0$ is rejected in favor of $H_a : \theta \neq 0$ at level $\alpha = 0.05$, since the p-value is less than 0.05.

The above calculations can easily be done with R commands. First, read the data into the R data frame *wq*, attach it with *attach(wq)*, and then use the following commands:

```
sm=by(Index, Beach, mean); svar=by(Index, Beach, var)
t=(sm[1]+sm[2])/2-(sm[3]+sm[4]+sm[5])/3
st=sqrt(mean(svar)*((1/4)*(2/12)+(1/9)*(3/12)))
t-qt(0.975, 55)*st; t+qt(0.975, 55)*st; TS=t/st
```

The commands in the first three lines compute the five sample means, the sample variances, the contrast, and the standard deviation of the contrast, which are stored in the R objects *t* and *st*, respectively. The commands in the last line compute the lower and upper endpoint of the 95% CI for the contrast, and store the value of the test statistic T_{H_0} in the R object *TS*. With three decimals, the obtained value of T_{H_0} is -2.555. Finally, the command *2*(1-pt(abs(TS), 55))* returns 0.013 for the p-value.

(b) The R command *anova(aov(Index~as.factor(Beach)))* returns the ANOVA table

	Df	Sum Sq	Mean Sq	F value	Pr(>F)
as.factor(Beach)	4	3.496	0.87400	1.9921	0.1085
Residuals	55	24.130	0.43873		

Since the p-value is greater than $\alpha = 0.05$, the null hypothesis $H_0 : \mu_1 = \cdots = \mu_5$ is not rejected. Note that the outcome of the test for the overall null hypothesis of equality of all means would suggest that the contrast θ in part (a) is also not significantly different from zero. However, the test procedure applied in part (a), which targets that specific contrast, rejects the null hypothesis that $\theta = 0$. ■

Testing the Validity of the Assumptions The validity of the ANOVA F test procedure rests on the assumptions stated in (10.2.13). The assumptions that the data constitute simple random samples from their respective populations and that the k samples are independent are best checked by reviewing the data collection protocol. Here we will discuss ways of checking the assumption of equal variances and, in case the sample sizes are less than 30, the normality assumption.

When dealing with comparative studies involving more than two populations, the group sample sizes are rarely 30 or more. Because the normality assumption cannot be checked reliably with small sample sizes, it is customary to perform a single normality test for all samples. Combining the raw observations from each population, however, is not recommended because differences in the population means and variances render the normality test invalid. For this reason, the normality test is applied on the residuals, after the homoscedasticity assumption has been judged tenable. The details of testing the homoscedasticity and normality assumptions are given next.

Note that the variance of the ith population equals the variance of each of the intrinsic error variables ϵ_{ij}, $j = 1, \ldots, n_i$, in the model (10.2.2). Moreover, since the intrinsic error variables have zero mean,

$$\mathrm{Var}(X_{ij}) = E(\epsilon_{ij}^2).$$

It follows that the equal variance assumption can be tested by testing that the k samples

$$\epsilon_{ij}^2, \quad j = 1, \ldots, n_i, \quad i = 1, \ldots, k, \tag{10.2.17}$$

have equal means. Thus, if the intrinsic error variables were observed, we would perform the ANOVA F test on the k samples in (10.2.17). Since the intrinsic error variables are not observed, we use the squared *residuals* obtained by fitting the model (10.2.2):

$$\widehat{\epsilon}_{ij}^2 = \left(X_{ij} - \overline{X}_i\right)^2, \quad j = 1, \ldots, n_i, \quad i = 1, \ldots, k. \tag{10.2.18}$$

If the p-value resulting from performing the ANOVA F test on the squared residuals in (10.2.18) is greater than 0.1, conclude that the assumption of equal variances is approximately satisfied.

REMARK 10.2-1 Levene's test for testing the equality of two variances (see Section 9.4) can be extended to testing the equality of k variances by performing the ANOVA F test on the k samples $|X_{ij} - \widetilde{X}_i|$, $j = 1, \ldots, n_i$, $i = 1, \ldots, k$, where $\widetilde{X}_i$ is the sample median of the ith sample. However, the described procedure is simpler to implement in R. ◁

Next, noting that the normality assumption is satisfied if the intrinsic error variables ϵ_{ij} in the model (10.2.2) are normal, the normality assumption can be checked by performing the *Shapiro-Wilk* normality test on the residuals. If the p-value is greater than 0.1, it can be concluded that the normality assumption is approximately satisfied. With the data read into the data frame *df*, the first column of which, named *Value*, contains the stacked data, and the second column, named *Sample*, specifies the treatment level for each observation, the following R commands perform the described test for equal variances and the Shapiro-Wilk normality test:

R Command for Testing the Assumption $\sigma_1^2 = \cdots = \sigma_k^2$

```
anova(aov(resid(aov(df$Value~df$Sample))**2~df$Sample))
```

(10.2.19)

R Command for Testing the Normality Assumption

```
shapiro.test(resid(aov(df$Value~df$Sample)))
```

(10.2.20)

Residual plots can shed light on the nature of assumptions violations. With the object *fit* defined as *fit=aov(df$Value~df$Sample)*, the commands

```
plot(fit, which=1)
plot(fit, which=2)
```

(10.2.21)

will display the residuals by groups (labeled by the fitted, i.e., the $\widehat{\mu}_i = \overline{X}_i$, values), and produce a Q-Q plot for the combined residuals, respectively. A boxplot of the combined residuals can also be informative.

Example 10.2-3

In the context of the water quality measurements of Example 10.2-2, test the validity of the assumptions of equal variances and normality.

Solution

Use *wq=read.table("WaterQualityIndex.txt", header=T)* to import the data into the data frame *wq*, and set the R object *Sample* by *Sample=as.factor(wq$Beach)*. The R commands *fit=aov(wq$Index~Sample); anova(aov(resid(fit)**2~Sample))* return a p-value of 0.166. Thus, it is concluded that the assumption of equal variances is approximately satisfied. Plotting the residuals against the fitted values through the command *plot(fit, which=1)*, shown in Figure 10-1, confirms that the variability within each group is approximately the same. Next, the R command

```
shapiro.test(resid(aov(wq$Index~Sample)))
```

returns a p-value of 0.427. Thus, it is concluded that the normality assumption is approximately satisfied. Finally, the R commands

```
Resid=resid(aov(wq$Index~Sample)); boxplot(Resid)
qqnorm(Resid); qqline(Resid, col=2)
```

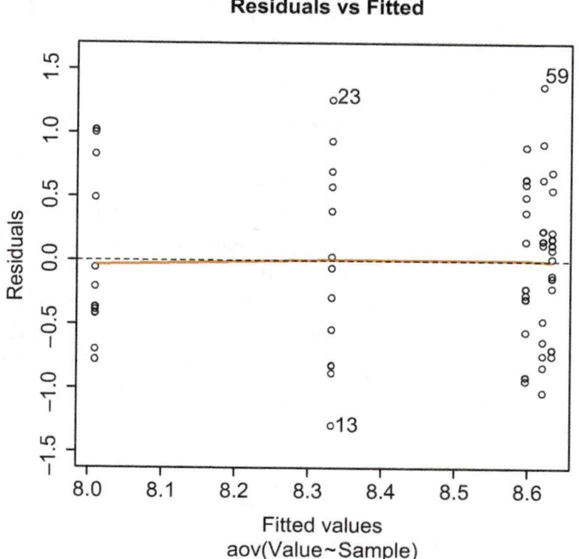

Figure 10-1 Plotting residuals by groups in Example 10.2-3.

produce the boxplot and Q-Q plot of the residuals, shown in Figure 10-2. (The Q-Q plot could also have been produced by *plot(fit, which=2)*.) These plots also suggest that the normality assumption is approximately satisfied, in agreement with the Shapiro-Wilk test p-value.

10.2.2 THE KRUSKAL-WALLIS TEST

This section describes the *Kruskal-Wallis* test, a rank-based procedure for testing the hypothesis of equality of k populations. Like the two-sample rank-sum test, which it generalizes to k samples, the Kruskal-Wallis test can be used with both small and large sample sizes, regardless of the normality assumption. The popularity of this procedure is also due to its relatively high power (or low probability of type II error), especially if the two population distributions are heavy tailed, or skewed.

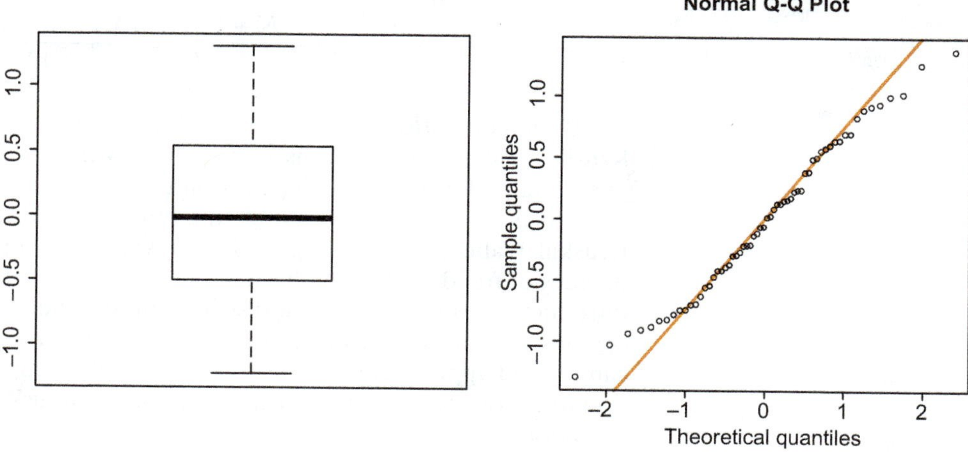

Figure 10-2 Boxplot and Q-Q plot of the residuals in Example 10.2-3.

The null hypothesis tested by the Kruskal-Wallis procedure is

$$H_0^F : F_1 = \cdots = F_k, \tag{10.2.22}$$

where $F_1, \ldots, F_k$ are the cumulative distribution functions of the k populations. Note that if H_0^F is true, then so is the hypothesis of equality of the k population means, $H_0 : \mu_1 = \cdots = \mu_k$.

The calculation of the rank test statistic begins by combining the observations from the k samples, $X_{i1}, \ldots, X_{in_i}$, $i = 1, \ldots, k$, into an overall set of $N = n_1 + \cdots + n_k$ observations, arranging them from smallest to largest, and assigning to each observation X_{ij} its rank (or mid-rank). The process of combining the $k = 2$ samples and ranking the combined set of observations was explained in detail in Section 9.3 in the context of the rank-sum test. With more than two samples, the process of ranking the data works in the same way. Let R_{ij} denote the rank, or mid-rank, of observation X_{ij}, and set

$$\overline{R}_i = n_i^{-1} \sum_{j=1}^{n_i} R_{ij}, \quad S_{KW}^2 = \frac{1}{N-1} \sum_{i=1}^{k} \sum_{j=1}^{n_i} \left(R_{ij} - \frac{N+1}{2} \right)^2, \tag{10.2.23}$$

for the average rank in group i and the sample variance of the ranks of the combined sample. (Recall that the average of the ranks is $(N+1)/2$.) With this notation, the Kruskal-Wallis test statistic is

Kruskal-Wallis Test Statistic

$$KW = \frac{1}{S_{KW}^2} \sum_{i=1}^{k} n_i \left(\overline{R}_i - \frac{N+1}{2} \right)^2 \tag{10.2.24}$$

If there are no tied observations, the Kruskal-Wallis test statistic has the following simpler form:

Kruskal-Wallis Statistic for Data with No Ties

$$KW = \frac{12}{N(N+1)} \sum_{i=1}^{k} n_i \left(\overline{R}_i - \frac{N+1}{2} \right)^2 \tag{10.2.25}$$

The equivalence of the formulas (10.2.24) and (10.2.25) in the case of no ties follows from the identity

$$\sum_{i=1}^{N} \left(i - \frac{N+1}{2} \right)^2 = \frac{(N-1)N(N+1)}{12}.$$

If the populations are continuous (so no ties), the exact null distribution of the Kruskal-Wallis test statistic is known even with very small sample sizes. However, this exact null distribution depends on the group sample sizes, so it requires extensive tables for its presentation. On the other hand, the exact null distribution of the Kruskal-Wallis statistic is approximated well by a chi-square distribution with $k - 1$ degrees of freedom even with small (≥ 8) sample sizes; see Section 7.3.5 for a quick introduction to the chi-square distribution and the R command for finding its percentiles. For this reason, and also because the exact p-value of the Kruskal-Wallis test can be well approximated in R for even smaller sample sizes (see Example 10.2-4), no tables for the exact null distribution of the Kruskal-Wallis statistic are given in this book.

The rule for rejecting $H_0^F : F_1 = \cdots = F_k$ using the approximate null distribution of the statistic and the formula for the approximate p-value follow:

Kruskal-Wallis Rejection Rule and p-Value when $n_i \geq 8$

$$KW > \chi^2_{k-1,\,\alpha}, \quad p\text{-value} = 1 - \Psi_{k-1}(KW) \tag{10.2.26}$$

where Ψ_ν denotes the CDF of the χ^2_ν distribution.

With the data from the k samples stacked in a single data column, and a second column indicating the sample index of each observation in the data column, the R command for computing the Kruskal-Wallis test statistic and p-value is:

R Command for the Kruskal-Wallis Test

```
kruskal.test(x~s)
```

where x and s are the R objects containing the data column and the sample index column, respectively.

Example 10.2-4

Organizations such as the American Society for Testing Materials are interested in the flammability properties of clothing textiles. A particular study performed a standard flammability test on six pieces from each of three types of fabric used in children's clothing. The response variable is the length of the burn mark made when the fabric piece is exposed to a flame in a specified way. Using the Kruskal-Wallis procedure on the data from this study (*Flammability.txt*), test the hypothesis that the flammability of the three materials is the same, at level of significance $\alpha = 0.1$, and compute the p-value.

Solution

The ranks and rank averages of the flammability data are

	Ranks						$\overline{R}_i$
Material 1	1	18	8	15.5	10	12	10.75
Material 2	4	3	7	5	6	9	5.667
Material 3	2	17	15.5	14	13	11	12.08

Note that observation 2.07 in group 1 is tied with an observation in group 3. Thus, we will use the more general expression of the Kruskal-Wallis statistic given in (10.2.24). The sample variance of the combined set of the $N = 6 + 6 + 6 = 18$ ranks is $S^2_{KW} = 28.47$. Next, using the average ranks $\overline{R}_i$ we have

$$\sum_{i=1}^{3} n_i \left(\overline{R}_i - \frac{N+1}{2}\right)^2 = 6(10.75 - 9.5)^2 + 6(5.667 - 9.5)^2 + 6(12.08 - 9.5)^2 = 137.465.$$

Thus, the Kruskal-Wallis statistic is

$$KW = 137.465/28.47 = 4.83.$$

The 90th percentile of the chi-square distribution with 2 degrees of freedom is $\chi^2_2(0.1) = 4.605$. Since $4.83 > 4.605$, the null hypothesis is rejected at

$\alpha = 0.1$. Using the formula for the p-value in (10.2.26) and the R command *1-pchisq(4.83, 2)*, we find a p-value of 0.089. Reading the flammability data into the R data frame *fl*, the R commands *x=fl$BurnL; s=fl$Material; kruskal.test(x~s)* give the same values for the test statistic and p-value. Because the sample sizes are $n_i = 6$, that is, < 8, a more accurate p-value (computed through *Monte-Carlo resampling*) can be obtained from the R commands *library(coin); kw=kruskal_test(x~factor(s), distribution=approximate(B=9999)); pvalue(kw)*. These commands return a p-value of 0.086. Thus, the original conclusion of rejecting the null hypothesis at $\alpha = 0.1$ holds.

10.2.3 THE CHI-SQUARE TEST FOR PROPORTIONS

If we want to compare different types of cars in terms of the proportion that sustain no damage in 5-mph crash tests, or if we want to compare the proportion of defective products of a given production process in different weeks, we end up wanting to test

$$H_0 : p_1 = p_2 = \cdots = p_k = p \text{ versus } H_a : H_0 \text{ is not true.} \tag{10.2.27}$$

Since the probability p of a 1 (i.e., of a "success") in a Bernoulli experiment is also the mean value of the Bernoulli random variable, this testing problem is a special case of testing for the equality of k means. Moreover, because the variance of the ith population is $\sigma_i^2 = p_i(1 - p_i)$, the homoscedasticity (i.e., equal variances) assumption is automatically satisfied under the null hypothesis. However, testing (10.2.27) is typically not performed with the ANOVA F test procedure.

Let the k sample proportions and the overall, or pooled, proportion be denoted by

$$\widehat{p}_1, \ldots, \widehat{p}_k, \quad \text{and} \quad \widehat{p} = \sum_{i=1}^{k} \frac{n_i}{N} \widehat{p}_i, \tag{10.2.28}$$

respectively. The test statistic for testing (10.2.27) is

$$Q_{H_0} = \sum_{i=1}^{k} \frac{n_i (\widehat{p}_i - \widehat{p})^2}{\widehat{p}(1 - \widehat{p})}. \tag{10.2.29}$$

There are similarities and differences between Q_{H_0} and the ANOVA F statistic F_{H_0}. First, the denominator $\widehat{p}(1 - \widehat{p})$ of Q_{H_0} is the variance corresponding to the pooled proportion instead of the pooled variance. Next, note that substituting the sample means $\overline{X}_i$ by $\widehat{p}_i$ and the overall sample mean $\overline{X}$ by $\widehat{p}$ in the formula for SSTr in (10.2.6) gives

$$\text{SSTr} = \sum_{i=1}^{k} n_i (\widehat{p}_i - \widehat{p})^2 .$$

Thus, the numerator of Q_{H_0} is SSTr, not MSTr. Because of this, the null distribution of Q_{H_0} is (approximately) chi-square with $k - 1$ degrees of freedom. Thus, the rule for rejecting H_0 at level of significance α and the formula for the p-value follow:

Chi-Square Test and p-Value for (10.2.27), with $n_i \geq 8$

$$Q_{H_0} > \chi^2_{k-1,\alpha}, \quad p\text{-value} = 1 - \Psi_{k-1}(Q_{H_0}) \tag{10.2.30}$$

where Ψ_ν denotes the CDF of the χ^2_ν distribution.

The Contingency Table Form of Q_{H_0} Even though the expression of Q_{H_0} in (10.2.29) is quite simple, an equivalent form of it is more common. To describe this equivalent form, we need the following additional notation:

$$O_{1i} = n_i\widehat{p}_i, \qquad O_{2i} = n_i(1-\widehat{p}_i) \tag{10.2.31}$$

$$E_{1i} = n_i\widehat{p}, \qquad E_{2i} = n_i(1-\widehat{p}), \tag{10.2.32}$$

where $\widehat{p}$ is the overall sample proportion defined in (10.2.28). Thus, O_{1i} is the observed number of successes (or 1's) in the ith group, O_{2i} is the observed number of failures (or 0's) in the ith group, and E_{1i}, E_{2i} are the corresponding expected numbers under the null hypothesis that the probability of 1 is the same in all groups. With this notation, the alternative form for Q_{H_0}, called the *contingency table* form, is

$$Q_{H_0} = \sum_{i=1}^{k}\sum_{\ell=1}^{2}\frac{(O_{\ell i}-E_{\ell i})^2}{E_{\ell i}}. \tag{10.2.33}$$

The equivalence of the two expressions for Q_{H_0} (10.2.29) and (10.2.33) is easily seen by noting the following easy algebraic identities:

$$(O_{1i}-E_{1i})^2 = (O_{2i}-E_{2i})^2 = n_i^2(\widehat{p}_i-\widehat{p})^2$$

$$\frac{1}{E_{1i}}+\frac{1}{E_{2i}} = \frac{E_{1i}+E_{2i}}{E_{1i}E_{2i}} = \frac{1}{n_i\widehat{p}(1-\widehat{p})}.$$

Thus, for each i,

$$\sum_{\ell=1}^{2}\frac{(O_{\ell i}-E_{\ell i})^2}{E_{\ell i}} = n_i^2(\widehat{p}_i-\widehat{p})^2\left(\frac{1}{E_{1i}}+\frac{1}{E_{2i}}\right) = \frac{n_i(\widehat{p}_i-\widehat{p})^2}{\widehat{p}(1-\widehat{p})},$$

which shows the equivalence of the expressions (10.2.29) and (10.2.33).

Using the notation O_{1i}, O_{2i}, for $i = 1, \ldots, k$, defined in (10.2.31), the R commands for computing the chi-square statistic (10.2.33) and the corresponding p-value follow:

R Commands for the Chi-Square Test for $H_0: p_1 = \cdots = p_k$

```
table=matrix(c(O11, O21, ..., O1k, O2k), nrow=2)
chisq.test(table)
```

Example 10.2-5

A commercial airline is considering four different designs of the control panel for the new generation of airplanes. To see if the designs have an effect on the pilot's response time to emergency displays, emergency conditions were simulated and the response times of pilots were recorded. The sample sizes, n_i, and number of times, O_{1i}, that the response times were below 3 seconds for the four designs are as follows: $n_1 = 45$, $O_{11} = 29$; $n_2 = 50$, $O_{12} = 42$; $n_3 = 55$, $O_{13} = 28$; $n_4 = 50$, $O_{14} = 24$. Perform the test at level of significance $\alpha = 0.05$.

Solution

Let p_i denote the probability that the response time of a pilot to a simulated emergency condition with control panel design i is below 3 seconds. The airline wants to test $H_0 : p_1 = p_2 = p_3 = p_4$ vs $H_a : H_0$ is not true, at level of significance $\alpha = 0.05$. Thus, the rejection region is $Q_{H_0} > \chi^2_3(0.05) = 7.815$. With the data given,

$$\widehat{p} = \frac{O_{11} + O_{12} + O_{13} + O_{14}}{n_1 + n_2 + n_3 + n_4} = \frac{123}{200} = 0.615,$$

so the common denominator of the terms in the expression (10.2.29) for Q_{H_0} is $\widehat{p}(1 - \widehat{p}) = 0.2368$. Continuing with the expression (10.2.29) for Q_{H_0}, and using $\widehat{p}_i = O_{1i}/n_i$, we obtain

$$\begin{aligned} Q_{H_0} &= \frac{45(0.6444 - 0.615)^2}{0.2368} + \frac{50(0.84 - 0.615)^2}{0.2368} \\ &\quad + \frac{55(0.5091 - 0.615)^2}{0.2368} + \frac{50(0.48 - 0.615)^2}{0.2368} \\ &= 0.1643 + 10.6894 + 2.6048 + 3.8482 = 17.307. \end{aligned}$$

Since $17.307 > 7.815$, the null hypothesis is rejected. The R commands *table=matrix(c(29, 16, 42, 8, 28, 27, 24, 26), nrow=2); chisq.test(table)* return the same value for the test statistic and a p-value of 0.0006, which also leads to the null hypothesis being rejected. ■

Exercises

1. In a study aimed at comparing the average tread lives of four types of truck tires, 28 trucks were randomly divided into four groups of seven trucks. Each group of seven trucks was equipped with tires from one of the four types. The data in *TireLife1Way.txt* consist of the average tread lives of the four tires of each truck.

(a) Are there any differences among the four types of tires? State the relevant null and alternative hypotheses for answering this question. Use either hand calculations or R commands to conduct the ANOVA F test at level of significance $\alpha = 0.1$, and state any assumptions that are needed for the validity of this test procedure. (*Hint.* If you choose hand calculations, you may use the following summary statistics: $\overline{X}_1 = 40.069$, $\overline{X}_2 = 40.499$, $\overline{X}_3 = 40.7$, $\overline{X}_4 = 41.28$, $S_1^2 = 0.9438$, $S_2^2 = 0.7687$, $S_3^2 = 0.7937$, and $S_4^2 = 0.9500$.)

(b) It is known that tire types 1 and 2 are brand A tires, while tire types 3 and 4 are brand B tires. Of interest is to compare the two brands of tires.

(i) Write the relevant contrast, and the null and alternative hypotheses to be tested for this comparison.

(ii) Use either hand calculations or R commands similar to those given in Example 10.2-2 to conduct the test at $\alpha = 0.1$ and to construct a 90% CI.

(iii) Is the outcome of the test for the specialized contrast in agreement with the outcome of the test for the overall null hypothesis in part (a)? If not, provide an explanation.

2. In the context of Example 10.2-4, where three types of fabric are tested for their flammability, materials 1 and 3 have been in use for some time and are known to possess similar flammability properties. Material 2 has been recently proposed as an alternative. Of primary interest in the study is the comparison of the joint population of materials 1 and 3 with that of material 2. Import the data into the R data frame *fl* and complete the following parts.

(a) Specify the contrast θ of interest, and use hand calculations to test $H_0 : \theta = 0$ vs $H_a : \theta \neq 0$ at level of significance $\alpha = 0.05$ using the procedure described in Section 10.2.1. (*Hint.* You may use *attach(fl); sm=by(BurnL, Material, mean); sv=by(BurnL, Material, var)* to obtain the sample means and sample variances.)

(b) Combine the burn-length measurements from materials 1 and 3 into one data set, and use the rank-sum procedure for testing that the combined populations of materials 1 and 3 is the same as that of material 2. (*Hint.* Use the R commands *x=c(BurnL[(Material==1)], BurnL[(Material==3)]); y=BurnL[(Material==2)] ; wilcox.test(x, y).*)

3. Four different concentrations of ethanol are compared at level $\alpha = 0.05$ for their effect on sleep time. Each concentration was given to a sample of 5 rats and the REM (rapid eye movement) sleep time for each rat was recorded (*SleepRem.txt*). Do the four concentrations differ in terms of their effect on REM sleep time?

(a) State the relevant null and alternative hypotheses for answering this question, and use hand calculations to conduct the ANOVA F test at level of significance 0.05. State any assumptions needed for the validity of this test procedure. (*Hint.* You may use the summary statistics $\overline{X}_1 = 79.28$, $\overline{X}_2 = 61.54$, $\overline{X}_3 = 47.92$, $\overline{X}_4 = 32.76$, and MSE $= S_p^2 = 92.95$.)

(b) Import the data into the R data frame *sl*, and use the R command *anova(aov(sl\$values~sl\$ind))* to conduct the ANOVA F test. Give the ANOVA table, stating the p-value, and the outcome of the test at level of significance 0.05.

(c) Use R commands to test the assumptions of equal variances and normality. Report the p-values from the two tests and the conclusions reached. Next, construct a boxplot and the normal Q-Q plot for the residuals, and comment on the validity of the normality assumption on the basis of these plots.

4. Consider the setting and data of Exercise 3.

(a) Use hand calculations to conduct the Kruskal-Wallis test at level $\alpha = 0.05$, stating any assumptions needed for its validity. (*Hint.* There are no ties in this data set. You may use *attach(sl); ranks=rank(values); rms=by(ranks, ind, mean)* to compute the rank averages.)

(b) Use the R command *kruskal.test(sl\$values~sl\$ind)* to conduct the Kruskal-Wallis test. Report the value of the test statistic, the p-value, and whether or not the null hypothesis is rejected at level $\alpha = 0.05$.

5. As part of a study on the rate of combustion of artificial graphite in humid air flow, researchers conducted an experiment to investigate oxygen diffusivity through a water vapor mixture. An experiment was conducted with mole fraction of water at levels $MF_1 = 0.002$, $MF_2 = 0.02$, and $MF_3 = 0.08$. Nine measurements at each of the three mole fraction levels were taken. The total sum of squares of the resulting data is SST $= 24.858$, and the treatment sum of squares is given in the partly filled out ANOVA table below.

	Df	Sum Sq	Mean Sq	F value
Treatment		0.019		
Residuals				

(a) State the null and alternative hypotheses, and use the information given to complete the ANOVA table.

(b) Use the value of the F statistic to test the hypothesis at level $\alpha = 0.05$.

(c) Compute the p-value and use it to decide whether or not the null hypothesis is rejected at level $\alpha = 0.05$. (*Hint.* You may use R to compute the p-value. See Example 10.2-1.)

6. Porous carbon materials are used commercially in several industrial applications, including gas separation, membrane separation, and fuel cell applications. For the purpose of gas separation, the pore size is important. To compare the mean pore size of carbon made at temperatures (in °C) of 300, 400, 500, and 600, an experiment uses 5 measurements at each temperature setting (*PorousCarbon.txt*). Is there any difference in the average pore size of carbon made at the different temperatures?

(a) State the relevant null and alternative hypotheses for answering this question, and use hand calculations to conduct the ANOVA F test at level $\alpha = 0.05$, stating any assumptions needed for its validity. (*Hint.* You may use the following summary statistics: $\overline{X}_1 = 7.43$, $\overline{X}_2 = 7.24$, $\overline{X}_3 = 6.66$, $\overline{X}_4 = 6.24$, $S_1^2 = 0.2245$, $S_2^2 = 0.143$, $S_3^2 = 0.083$, $S_4^2 = 0.068$.)

(b) Use R commands to import the data into the R data frame *pc* and to conduct the ANOVA F test. Report the value of the test statistic, the p-value, and whether or not the null hypothesis is rejected at level $\alpha = 0.05$.

(c) Use R commands to test the assumptions of equal variances and normality. Report the p-values from the two tests and the conclusions reached. Next, construct a boxplot and the normal Q-Q plot for the residuals from fitting the model (10.2.2), and comment on the validity of the normality assumption on the basis of these plots.

7. Consider the setting and data of Exercise 6.

(a) Use hand calculations to conduct the Kruskal-Wallis test at level $\alpha = 0.05$, stating any assumptions needed for its validity. (*Hint.* This data set has ties. You may use *attach(pc); ranks=rank(values); vranks=var(ranks); rms= by(ranks, temp, mean)* to compute S_{KW}^2 and the rank averages.)

(b) Use R commands to conduct the Kruskal-Wallis test. Report the value of the test statistic, the p-value, and whether or not the null hypothesis is rejected at level $\alpha = 0.05$.

8. Compression testing of shipping containers aims at determining if the container will survive the compression loads expected during distribution. Two common types of compression testers are fixed platen and floating platen. The two methods were considered in a study, using different types of corrugated fiberboard containers.[1] Suppose that $n_1 = 36$, $n_2 = 49$, and $n_3 = 42$ fixed-platen strength measurements for types 1, 2, and 3 of corrugated containers, respectively, yielded the following summary statistics: $\overline{X}_1 = 754$, $\overline{X}_2 = 769$, $\overline{X}_3 = 776$, $S_1 = 16$, $S_2 = 27$, and $S_3 = 38$. Is there evidence that the three types of corrugated fiberboard containers differ in terms of their average strength using fixed-platen testers? State the null and alternative hypotheses relevant for answering this question, and use the ANOVA procedure for conducting the test at level $\alpha = 0.05$. State any assumptions needed for the validity of the ANOVA test procedure.

9. An article reports on a study where fatigue tests were performed by subjecting the threaded connection of large diameter pipes to constant amplitude stress of either L1 = 10 ksi, L2 = 12.5 ksi, L3 = 15 ksi, L4 = 18 ksi, or L5 = 22 ksi.[2] The measured fatigue lives, in number of cycles to failure (in units of 10,000 cycles), are given in *FlexFatig.txt*. Read the data into the data frame *ff* and complete the following.

(a) Use R commands to test the assumptions of equal variances and normality. Report the p-values from the two tests and the conclusions reached. Next, construct a boxplot and the normal Q-Q plot for the residuals from fitting the model (10.2.2), and comment on the validity of the normality assumption on the basis of these plots.

(b) Taking into consideration the conclusions reached in part (a), which procedure for testing whether the stress level impacts the fatigue life would you recommend? Justify your recommendation.

(c) Carry out the test procedure you recommended in part (b) at level of significance 0.01. Report the value of the test statistic and the p-value.

10. The flame resistance of three materials used in children's pajamas was tested by subjecting specimens of the materials to high temperatures. Out of 111 specimens of material A, 37 ignited. Out of 85 specimens of material B, 28 ignited. Out of 100 specimens of material C, 21 ignited. Test the hypothesis that the probability of ignition is the same for all three materials versus the alternative that this hypothesis is false at $\alpha = 0.05$.

11. A certain brand of tractor is assembled in five different locations. To see if the proportion of tractors that require warranty repair work is the same for all locations, a random sample of 50 tractors from each location is selected and followed up for the duration of the warranty period. The numbers requiring warranty repair work are 18 for location A, 8 for location B, 21 for location C, 16 for location D, and 13 for location E. Is there evidence that the five population proportions differ at level of significance $\alpha = 0.05$? State the null and alternative hypotheses relevant for answering this question, and use an appropriate method to conduct the test. Report the method used, the value of the test statistic, the test outcome, and any assumptions needed for the validity of the test.

12. Wind-born debris (from roofs, passing trucks, insects, or birds) can wreak havoc on architectural glass in the upper stories of a building. A paper reports the results of an experiment where 10 configurations of glass were subjected to a 2-gram steel ball projectile traveling under 5 impact velocity ranges.[3] Here we report the results for configurations 1, 2, 3, and 5. Out of 105 inner glass ply breaks (IPBs) of configuration 1, 91 were at impact velocity of 139 ft/s or less. For configurations 2, 3, and 5 the results were 128 out of 148, 46 out of 87, and 62 out of 93. Test the hypothesis that the five population proportions are the same at $\alpha = 0.05$.

10.3 Simultaneous CIs and Multiple Comparisons

Because the null hypothesis $H_0 : \mu_1 = \mu_2 = \cdots = \mu_k$ is tested against the alternative $H_a : H_0$ is false, it follows that when H_0 is rejected it is not clear which μ_i's are significantly different. It would seem that this question can be addressed quite simply by conducting individual tests of $H_0 : \mu_i - \mu_j = 0$ vs $H_a : \mu_i - \mu_j \neq 0$ for each pair of means μ_i, μ_j. If a test rejects the null hypothesis, the corresponding means

[1] S. P. Singh, G. Burgess, and M. Langlois (1992). Compression of single-wall corrugated shipping containers using fixed and floating test platens, *J. Testing and Evaluation*, 20(4): 318–320.

[2] A. H. Varma, A. K. Salecha, B. Wallace, and B. W. Russell (2002). Flexural fatigue behavior of threaded connections for large diameter pipes, *Experimental Mechanics*, 42: 1–7.

[3] N. Kaiser, R. Behr, J. Minor, L. Dharani, F. Ji, and P. Kremer (2000). Impact resistance of laminated glass using "sacrificial ply" design concept, *Journal of Architectural Engineering*, 6(1): 24–34.

are declared significantly different. With some fine-tuning, this approach leads to a correct **multiple comparisons** method.

The individual tests mentioned above can be conducted through any of the test procedures we have described; for example, the two-sample T test procedure, or the rank-sum procedure, or the Z procedure for the comparison of two proportions in the case of Bernoulli populations. Alternatively, the tests can be conducted by constructing individual CIs for each contrast $\mu_i - \mu_j$ and checking whether or not each CI contains zero. If one of these confidence intervals does not contain zero, then the means involved in the corresponding contrast are declared significantly different.

The reason the above simple procedure needs fine-tuning has to do with the **overall**, or **experiment-wise error rate**, which is the probability of at least one pair of means being declared different when in fact all means are equal.

To appreciate the experiment-wise error rate, suppose that the problem involves the comparison of $k = 5$ population means. If the overall null hypothesis $H_0 : \mu_1 = \cdots = \mu_5$ is rejected, then, to determine which pairs of means are significantly different, $(1 - \alpha)100\%$ CIs for all 10 pairwise differences,

$$\mu_1 - \mu_2, \ldots, \mu_1 - \mu_5, \mu_2 - \mu_3, \ldots, \mu_2 - \mu_5, \ldots, \mu_4 - \mu_5, \qquad \textbf{(10.3.1)}$$

must be made. (Equivalently, we could test $H_0 : \mu_i - \mu_j = 0$ vs $H_a : \mu_i - \mu_j \neq 0$ for each of the above 10 contrasts at level of significance α, but we will focus the discussion on the CIs approach.) Assume for the moment that the confidence intervals for the 10 contrasts in (10.3.1) are independent. (They are not independent because the confidence intervals for, say, $\mu_1 - \mu_2$ and $\mu_1 - \mu_3$ both involve the sample $X_{11}, \ldots, X_{1n_1}$ from population 1.) In that case, and if all means are the same (so all contrasts are zero), the probability that each interval contains zero is $1-\alpha$ and thus, by the assumed independence, the probability that all 10 confidence intervals contain zero is $(1 - \alpha)^{10}$. Thus, the experiment-wise error rate is $1 - (1 - \alpha)^{10}$. If $\alpha = 0.05$, then the experiment-wise error rate is

$$1 - (1 - 0.05)^{10} = 0.401. \qquad \textbf{(10.3.2)}$$

It turns out that, in spite of the dependence of the confidence intervals, the above calculation gives a fairly close approximation to the true experiment-wise error rate. Thus, the chances are approximately 40% that at least one of contrasts will be declared significantly different from zero when using traditional 95% confidence intervals.

Confidence intervals that control the experiment-wise error rate at a desired level α will be called $(1 - \alpha)100\%$ **simultaneous confidence intervals**. We will see two methods of fine-tuning the naive procedure of using traditional confidence intervals. One method, which gives an upper bound on the experiment-wise error rate, is based on *Bonferroni's inequality*. The other, *Tukey*'s procedure, gives the exact experiment-wise error rate if the samples come from normal homoscedastic populations, but can also be used as a good approximation with large samples from other homoscedastic populations. Moreover, Tukey's method can be applied on the ranks, with smaller sample sizes, when sampling from skewed distributions. On the other hand, Tukey's method applies only to the set of all pairwise comparisons, while Bonferonni's method can also be used for multiple comparisons and simultaneous CIs of the particular specialized contrasts that might be of interest.

10.3.1 BONFERRONI MULTIPLE COMPARISONS AND SIMULTANEOUS CIs

The idea behind Bonferroni's intervals is to adjust the level of the traditional confidence intervals in order to achieve the desired experiment-wise error rate. As mentioned in connection to the calculation in (10.3.2) above, due to the dependence among the confidence intervals it is not possible to know the exact experiment-wise error rate when performing a total of m confidence intervals. However, *Bonferroni's inequality* asserts that, when each of m confidence intervals are performed at level α, then the probability that at least one does not contain the true value of the parameter, that is, the experiment-wise error rate, is no greater than $m\alpha$. Similarly, if each of m pairwise tests are performed at level α, the experiment-wise level of significance (i.e., the probability of rejecting at least one of the m null hypotheses when all are true), is no greater than $m\alpha$.

The above discussion leads to the following procedure for constructing Bonferroni simultaneous CIs and multiple comparisons:

1. **Bonferroni Simultaneous CIs:**
   For each of the m contrasts construct a $(1-\alpha/m)100\%$ CI. This set of m CIs are the $(1-\alpha)100\%$ Bonferroni simultaneous CIs for the m contrasts.
2. **Bonferroni Multiple Comparisons:**
   (a) *Multiple Comparisons through Simultaneous CIs*: If any of the m $(1-\alpha)100\%$ Bonferroni simultaneous CIs does not contain zero, the corresponding contrast is declared significantly different from 0 at experiment-wise level α.
   (b) *Multiple Comparisons through Testing:* For each of m contrasts, test the null hypothesis that the contrast is zero vs the alternative that it is not zero at level of significance α/m. If any of the m tests rejects the null hypothesis, the corresponding contrast is declared significantly different from zero at experiment-wise level of significance α.

Example 10.3-1

In the context of Example 10.2-5, use multiple comparisons based on Bonferroni simultaneous CIs to determine which pairs of panel designs differ, at experiment-wise level of significance $\alpha = 0.05$, in terms of their effect on the pilot's reaction time.

Solution

Since there are four panel designs, there are $m = \binom{4}{2} = 6$ possible pairwise contrasts:

$$p_1 - p_2,\ p_1 - p_3,\ p_1 - p_4,\ p_2 - p_3,\ p_2 - p_4,\ p_3 - p_4.$$

Recall that the sample sizes and number of successes (where "success" is a response time below 3 seconds) for the four panel designs are: $n_1 = 45$, $O_{11} = 29$; $n_2 = 50$, $O_{12} = 42$; $n_3 = 55$, $O_{13} = 28$; $n_4 = 50$, $O_{14} = 24$. The 95% Bonferroni simultaneous CIs for these contrasts consist of individual $(1 - 0.05/6)100\% = 99.17\%$ CIs for each of them. The following R commands give the CIs for the six contrasts:

```
k=4; alpha=0.05/(k*(k-1)/2); o=c(29, 42, 28, 24);
  n=c(45,50,55,50)

for(i in 1:(k-1)){for(j in (i+1):k){print(prop.test(c(o[i],
  o[j]), c(n[i], n[j]), conf.level=1-alpha,
  correct=F)$conf.int)}}
```

The CIs for the six contrasts are shown in the second column of the following table:

Contrast	Individual 99.17% CI	Contains Zero?
$p_1 - p_2$	(−0.428, 0.0371)	Yes
$p_1 - p_3$	(−0.124, 0.394)	Yes
$p_1 - p_4$	(−0.100, 0.429)	Yes
$p_2 - p_3$	(0.107, 0.555)	No
$p_2 - p_4$	(0.129, 0.591)	No
$p_3 - p_4$	(−0.229, 0.287)	Yes

To conduct multiple comparisons based on these intervals, we check for any that do not contain zero. The answers are given in the third column of the above table. Thus, p_2 is significantly different, at experiment-wise level $\alpha = 0.05$, from p_3 and p_4. All other contrasts are not significantly different from zero.

The results from a multiple comparisons procedure can be summarized by listing the estimates of the parameters being compared in increasing order, and joining each pair that is not significantly different by an underline. For example, the results from the Bonferroni multiple comparisons procedure in Example 10.3-1 can be displayed as

$\widehat{p}_4$	$\widehat{p}_3$	$\widehat{p}_1$	$\widehat{p}_2$
0.48	0.51	0.64	0.84

Example 10.3-2 Iron concentration measurements from four ore formations are given in *FeData.txt*. Use Bonferroni multiple comparisons, based on rank-sum tests, to determine which pairs of ore formations differ, at experiment-wise level of significance $\alpha = 0.05$, in terms of iron concentration.

Solution

The four ore formations yield $m = 6$ possible pairwise median contrasts:

$$\widetilde{\mu}_1 - \widetilde{\mu}_2,\ \widetilde{\mu}_1 - \widetilde{\mu}_3,\ \widetilde{\mu}_1 - \widetilde{\mu}_4,\ \widetilde{\mu}_2 - \widetilde{\mu}_3,\ \widetilde{\mu}_2 - \widetilde{\mu}_4,\ \widetilde{\mu}_3 - \widetilde{\mu}_4.$$

The hypothesis that each of the above contrasts is zero vs the two-sided alternative will be tested at an individual level of $\alpha/6 = 0.0083$. With the data imported into the R data frame *fe*, the R commands *f1=fe$conc[fe$ind=="V1"]; f2=fe$conc[fe$ind=="V2"]; f3=fe$conc[fe$ind=="V3"]; f4=fe$conc[fe$ind=="V4"]* assign the concentration samples from the four formations into the R objects *f1,...,f4*. Using the command *wilcox.test* for each pair of samples (i.e., *wilcox.test(f1, f2)*, for testing $H_0 : \tilde{\mu}_1 - \tilde{\mu}_2 = 0$ vs $H_a : \tilde{\mu}_1 - \tilde{\mu}_2 \neq 0$, and similarly for the other contrasts), we obtain the following table of p-values and their comparisons with $\alpha/6 = 0.0083$:

Contrast	p-Value for $H_0 : \tilde{\mu}_i - \tilde{\mu}_j = 0$	Less than 0.0083?
$\tilde{\mu}_1 - \tilde{\mu}_2$	0.1402	No
$\tilde{\mu}_1 - \tilde{\mu}_3$	0.0172	No
$\tilde{\mu}_1 - \tilde{\mu}_4$	0.0013	Yes
$\tilde{\mu}_2 - \tilde{\mu}_3$	0.0036	Yes
$\tilde{\mu}_2 - \tilde{\mu}_4$	0.0017	Yes
$\tilde{\mu}_3 - \tilde{\mu}_4$	0.0256	No

These results are summarized as

$\tilde{X}_2$	$\tilde{X}_1$	$\tilde{X}_3$	$\tilde{X}_4$
25.05	27.65	30.10	34.20

Thus, at experiment-wise level of significance $\alpha = 0.05$, ore formation 1 is not significantly different from formations 2 and 3, nor is formation 3 from 4. All other pairs of formations are significantly different. ■

10.3.2 TUKEY'S MULTIPLE COMPARISONS AND SIMULTANEOUS CIs

Tukey's simultaneous CIs are appropriate under normality and homoscedasticity, and apply to all $m = k(k-1)/2$ pairwise contrasts $\mu_i - \mu_j$. If group sample sizes are all large (≥ 30), they are approximately valid without the normality assumption, though the homoscedasticity assumption is still needed.

Tukey's intervals are based on the so-called studentized range distribution which is characterized by a numerator degrees of freedom and a denominator degrees of freedom. The numerator degrees of freedom equals the number of means, k, that are being compared. The denominator degrees of freedom equals the degrees of freedom of the MSE in the ANOVA table, which is $N - k$, where $N = n_1 + \cdots + n_k$. The 90th and 95th percentiles of the studentized range distribution are given in Table A.7, where the denominator degrees of freedom, $N - k$, is denoted by ν.

With $Q_{\alpha,k,N-k}$ denoting the upper-tail α critical value of the studentized range distribution with k and $\nu = N - k$ degrees of freedom, selected from Table A.7, Tukey's simultaneous CIs and multiple comparisons are as follows:

1. **Tukey's Simultaneous CIs:** The $(1-\alpha)100\%$ Tukey's simultaneous CIs for all contrasts $\mu_i - \mu_j$, $i \neq j$, are constructed as

$$\overline{X}_i - \overline{X}_j \pm Q_{\alpha,k,N-k}\sqrt{\frac{S_p^2}{2}\left(\frac{1}{n_i} + \frac{1}{n_j}\right)}, \qquad \textbf{(10.3.3)}$$

   where $S_p^2 = \text{MSE}$ is the pooled sample variance given in (10.2.10).

2. **Tukey's Multiple Comparisons at Level α:** If for a pair (i,j), $i \neq j$, the interval (10.3.3) does not contain zero, it is concluded that μ_i and μ_j differ significantly at level α.

Example 10.3-3

Four different concentrations of ethanol are compared at level $\alpha = 0.05$ for their effect on sleep time. Each concentration was given to a sample of 5 rats and the

REM sleep time for each rat was recorded. Use the resulting data set (*SleepRem.txt*) to construct Tukey's 95% simultaneous CIs and apply the Tukey multiple comparisons method to identify the pairs of concentrations that are significantly different at experiment-wise level of significance 0.05.

Solution

Because the sample sizes are small, we assume that the four populations are normal, with the same variance. Importing the data into the R data frame *sl*, the R command *by(sl$values, sl$ind, mean)* returns the four sample means as $\overline{X}_1 = 79.28$, $\overline{X}_2 = 61.54$, $\overline{X}_3 = 47.92$, and $\overline{X}_4 = 32.76$. The additional command *anova(aov(sl$values~sl$ind))* returns MSE $= S_p^2 = 92.95$ (it also returns a p-value of 8.322×10^{-6}, so that the null hypothesis of equality of the four means is rejected at $\alpha = 0.05$). Next, from Table A.7 we find that the 95th percentile of the studentized range distribution with $k = 4$ and $N - k = 16$ degrees of freedom is $Q_{0.05,4,16} = 4.05$. With the above information, the computed 95% Tukey's simultaneous CIs, using the formula (10.3.3), are given in the second column of the following table:

Contrast	Simultaneous 95% CI	Contains Zero?
$\mu_1 - \mu_2$	(0.28, 35.20)	No
$\mu_1 - \mu_3$	(13.90, 48.82)	No
$\mu_1 - \mu_4$	(29.06, 63.98)	No
$\mu_2 - \mu_3$	(-3.84, 31.08)	Yes
$\mu_2 - \mu_4$	(11.32, 46.24)	No
$\mu_3 - \mu_4$	(-2.30, 32.62)	Yes

Checking whether or not each CI contains zero results in the third column of the table. Thus, all pairs of means except for the pairs (μ_2, μ_3) and (μ_3, μ_4) are significant at experiment-wise level of significance 0.05. The results are summarized as

$\overline{X}_4$	$\overline{X}_3$	$\overline{X}_2$	$\overline{X}_1$
32.76	47.92	61.54	79.28

With the observations stacked in the R object y, and the sample membership of each observation contained in the R object s, the R commands for constructing Tukey's simultaneous CIs, including a plot for their visual display, are

R Commands for Tukey's $(1-\alpha)$100% Simultaneous CIs

```
TukeyHSD(aov(y~s), conf.level=1-α)

plot(TukeyHSD(aov(y~s), conf.level=1-α))
```

The default value of α is 0.05. For example, the command *TukeyHSD(aov(sl$values~sl$ind))* gives the 95% Tukey's simultaneous CIs shown in the solution of Example 10.3-3 (except for reversing the sign of the contrasts, i.e., $\mu_2 - \mu_1$ instead of $\mu_1 - \mu_2$, etc). The additional command *plot(TukeyHSD(aov(sl$values~sl$ind)))* produces the plot in Figure 10-3. The vertical line at zero in Figure 10-3 helps identify the CIs that contain zero, which are the CIs for $\mu_4 - \mu_3$ and $\mu_3 - \mu_2$. This is in agreement with the findings of Example 10.3-3.

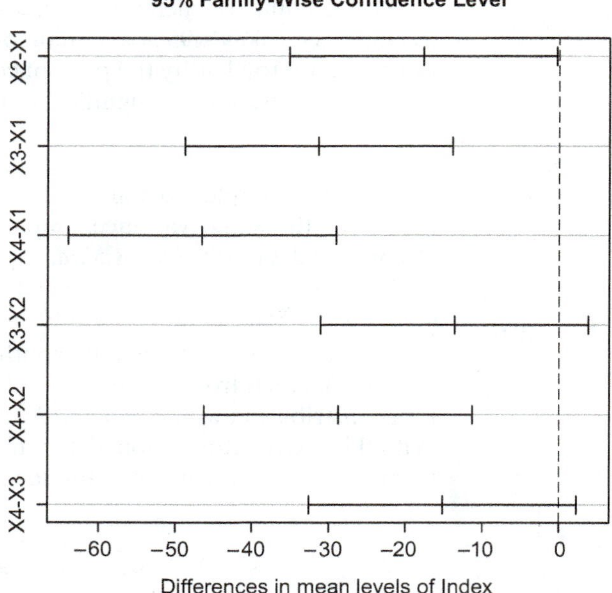

Figure 10-3 Visual display of Tukey's simultaneous CIs for Example 10.3-3.

10.3.3 TUKEY'S MULTIPLE COMPARISONS ON THE RANKS

This procedure consists of first combining the observations from the k samples into an overall set of $N = n_1 + \ldots + n_k$ observations, sorting them from smallest to largest, and assigning ranks or mid-ranks to each observation. See Section 10.2.2 for the details of this ranking step. As a second step, arrange the ranks of the N observations into k rank-samples, so that the ith rank-sample contains the ranks of the observations in the ith sample. Finally, apply Tukey's multiple comparisons procedure on the k rank-samples.

Tukey's multiple comparisons procedure on the ranks is recommended whenever the sample sizes are small and/or non-normal. It should be kept in mind, however, that the resulting simultaneous CIs are not CIs for the contrasts of the medians or the means of the original observations. Instead, the results of this multiple comparisons procedure are interpretable as multiple comparisons for pairs of the k populations the observations came from, or as multiple comparisons for all pairs of median contrasts $\widetilde{\mu}_i - \widetilde{\mu}_j$, for $i \neq j$.

Example 10.3-4

Consider the setting of Example 10.2-4, where three types of fabric are tested for their flammability, and use Tukey's multiple comparisons method on the ranks to identify which materials differ in terms of flammability at experiment-wise level of significance $\alpha = 0.1$.

Solution

With the flammability data read into the R data frame *fl*, we can use the following R commands:

```
r=rank(fl$BurnL); s=as.factor(fl$Material)
plot(TukeyHSD(aov(r~s), conf.level=0.9))
```

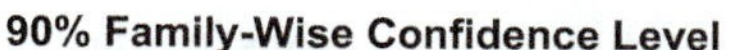

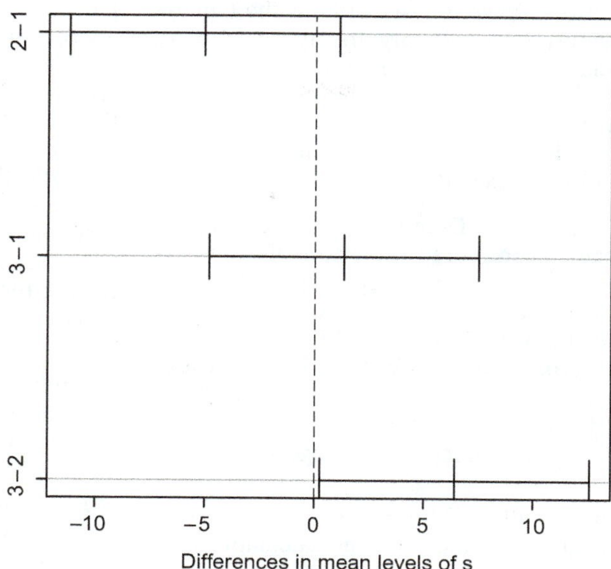

Figure 10-4 Tukey's simultaneous CIs on the ranks for Example 10-4.

These commands generate the plot shown in Figure 10-4 Thus, only materials 2 and 3 are significantly different at experiment-wise level of significance $\alpha = 0.1$. The results are summarized as

$\overline{R}_3$	$\overline{R}_1$	$\overline{R}_2$
12.083	10.750	5.667

Exercises

1. An article reports on a study using high temperature strain gages to measure the total strain amplitude of three different types of cast iron [spheroidal graphite (S), compacted graphite (C), and gray (G)] for use in disc brakes.[4] Nine measurements from each type were taken and the results (multiplied by 10,000) are in *FatigueThermal.txt*. The scientific question is whether the total strain amplitude properties of the different types of cast iron differ.

(a) State the relevant null and alternative hypotheses for answering this question.

(b) Import the data into the R data frame *tf* and use the R command *kruskal.test(tf$values~tf$ind)* to conduct the the Kruskal-Wallis procedure for testing the equality of the three populations. State the *p*-value and whether or not the null hypothesis is rejected at level 0.05.

(c) On the basis of the test outcome in part (b), is there a need to conduct multiple comparisons to determine which pairs of populations differ at experiment-wise level of significance 0.05? Justify your answer.

2. Consider the data in Exercise 8 in Section 10.2 on fixed-platen compression strengths of three types of corrugated fiberboard containers.

(a) One of the assumptions needed for the validity of the ANOVA *F* test is homoscedasticity, or equal variances. Since only summary statistics are given, the homoscedasticity assumption can be tested by performing all pairwise tests for the three population variances and use the Bonferroni multiple comparisons procedure. In particular, apply the *F* test for the equality of two variances, given in Section 9.4, to test $H_{10} : \sigma_1^2 = \sigma_2^2$, $H_{20} : \sigma_1^2 = \sigma_3^2$, and $H_{30} : \sigma_2^2 = \sigma_3^2$ versus the corresponding two-sided alternative hypotheses, and apply the Bonferroni method at experiment-wise level of significance 0.1. Does the homoscedasticity assumption appear to be violated?

[4] "Improving the thermal fatigue resistance of brake discs" in *Materials, Experimentation and Design in Fatigue*. Eds. F. Sherrat and J. B. Sturgeon. 1982. 60–71.

(b) Use the Bonferroni multiple comparisons method, with pairwise two-sample T tests without the equal-variance assumption, to identify the types of corrugated containers whose population means differ significantly at experiment-wise level of significance $\alpha = 0.05$. Summarize your results by arranging the means in increasing order and underlining the ones that do not differ significantly.

3. A study conducted at Delphi Energy & Engine Management Systems considered the effect of blow-off pressure during the manufacture of spark plugs on the spark plug resistance. The resistance measurements of 150 spark plugs manufactured at each of three different blow-off pressures (10, 12, and 15 psi) yield the following summary statistics: $\overline{X}_1 = 5.365$, $\overline{X}_2 = 5.415$, $\overline{X}_3 = 5.883$, $S_1^2 = 2.241$, $S_2^2 = 1.438$, and $S_3^2 = 1.065$. Is there evidence that the blow-off pressure affects the average spark plug resistance?

(a) State the relevant null and alternative hypotheses for answering this question, and use hand calculations to conduct the ANOVA F test at level of significance 0.05. State any assumptions needed for the validity of this test procedure.

(b) Use the Bonferroni multiple comparisons method, with pair wise two-sample T tests using the equal-variance assumption, to identify the groups whose population means differ significantly at experiment-wise level of significance $\alpha = 0.05$. Summarize your results by arranging the means in increasing order and underlining the ones that do not differ significantly.

4. Consider the setting and data of Exercise 6 in Section 10.2.

(a) Use R commands or hand calculations to compute Tukey's 95% simultaneous CIs, and perform Tukey's multiple comparisons at experiment-wise level of significance $\alpha = 0.05$. (*Hint.* You may use the summary statistics given in Exercise 6 in Section 10.2.)

(b) Use R commands or hand calculations to perform Tukey's multiple comparisons, procedure on the ranks at experiment-wise level $\alpha = 0.05$. (*Hint.* For hand calculations you may use the rank summary statistics obtained by commands given in Exercise 7 in Section 10.2.)

(c) For each analysis performed in parts (a) and (b), summarize your results by arranging the means in increasing order and underlining the ones that do not differ significantly.

5. Records from an honors statistics class, Experimental Design for Engineers, indicated that each professor had adopted one of three different teaching methods: (A) use of a textbook as the main source of teaching material, (B) use of a textbook combined with computer activities, and (C) use of specially designed instructional notes together with computer activities. It was decided to compare the three teaching methods by randomly dividing 24 students into three groups of eight, with each group receiving one of the three teaching methods. A common exam was administered at the end of the study. The scores for the three groups (*GradesTeachMeth.txt*) will be used to compare the pedagogical effectiveness of the three methods.

(a) State the null and alternative hypotheses relevant for this comparison, and use the ANOVA procedure to test at level $\alpha = 0.05$. What assumptions, if any, are needed for the validity of the test?

(b) Construct Tukey's 95% simultaneous CIs for all pairwise contrasts, and use them to conduct Tukey's multiple comparisons procedure at experiment-wise level of significance $\alpha = 0.05$ to determine which teaching methods are (statistically) significantly different.

(c) Use R commands to test the validity of the needed assumptions. Report the p-values for each test and the conclusion reached. According to the conclusion reached, are the procedures in parts (a) and (b) valid?

6. Consider the data in Exercise 5.

(a) Conduct the Kruskal-Wallis test at level $\alpha = 0.05$ and state what assumptions, if any, are needed for its validity.

(b) Use the Bonferroni multiple comparisons method with the rank-sum test to identify the groups whose population means differ significantly at experiment-wise level of $\alpha = 0.05$.

(c) Repeat part (b) using Tukey's multiple comparisons procedure on the ranks.

7. For the data in Exercise 10 in Section 10.2, construct Bonrerroni 95% simultaneous CIs for all pairwise differences of proportions, and use them to identify the pairs of proportions that differ at experiment-wise level of significance 0.05.

8. For the data in Exercise 12 in Section 10.2, use the Bonferroni multiple comparisons method to identify the groups whose population proportions differ significantly at experiment-wise level of significance $\alpha = 0.05$.

10.4 Randomized Block Designs

A *randomized block design* generalizes the paired data design, which we saw in Section 9.5, to the comparison of $k > 2$ populations. Analogously with the paired data design, the k samples generated from a randomized block design are not independent. Thus, the methods for comparing $k > 2$ populations discussed in the previous sections, which are all based on the assumption that the k samples are independent, are not valid.

A randomized block design arises when a random sample of n individuals (subjects or objects) receives each of the k treatments that are to be compared. Because k observations, one for each treatment, are obtained from the same subject or object, the k samples are not independent. The subjects or objects on which observations are made are called **blocks**. A block design is called **randomized** if the order in which the k treatments are applied is randomized for each block. The term **randomized complete block design** is also used to emphasize the fact that each block receives all k treatments. For our purposes, the term randomized block design refers to a randomized complete block design.

The following examples highlight two contexts where such data arise.

Example 10.4-1

Four different types of truck tires, A, B, C, and D, are to be compared for durability. One way of designing this comparative experiment is to select a random sample of n trucks and fit each of them with one tire of each type. The locations (front left, front right, rear left, and rear right) where each tire is fitted are selected at random for each truck. After a pre-specified number of miles on the road, the tires are evaluated for wear and tear. In this example, the sample of n trucks are the blocks, and the four populations/treatments to be compared correspond to the four tire types. From each block four measurements are made, which are quantifications of wear and tear of each tire. Because of the specific way that a truck affects the wear and tear of its tires (load, road conditions, driver, etc.), the four measurements from each block cannot be assumed independent. However, measurements from different trucks can be assumed independent. Another design for this comparative study is to use tire type A on a random sample of n_1 trucks, fit a different sample of n_2 trucks with tire type B, a different sample of n_3 trucks with tire type C, and a different sample of n_4 trucks with tire type D. From each truck the average wear and tear of its four tires is recorded, resulting in four independent samples.

Example 10.4-2

Three different methods for determining the percentage of iron in ore samples are to be compared. A randomized block design for this comparative study consists of obtaining n ore samples and subjecting each of them to the three different methods for determining its iron content. The order in which the three methods are applied is randomized for each ore sample. In this example, the n ore samples are the blocks, and the populations that are compared correspond to the three different methods. For each ore sample there will be three measurements that are dependent, because they depend on the ore sample's true iron content. Another design for this comparative study is to use different ore samples for each method, resulting in three independent samples.

From the above two examples it follows that a randomized block design eliminates a lot of uncontrolled variability in the measurements. In Example 10.4-1, the randomized block design eliminates the uncontrolled variability caused by the trucks having different loads and traveling different routes at different speeds with different drivers. Similarly, in Example 10.4-2, the randomized block design eliminates the uncontrolled variability caused by the different iron content of the various ore samples.

10.4.1 THE STATISTICAL MODEL AND HYPOTHESIS

The data notation uses the subscript i to represent the treatment and the subscript j to represent the block. Thus, X_{ij} denotes the jth observation ($j = 1, \ldots, n$) from treatment i ($i = 1, \ldots, k$). Figure 10-5 shows the data from a randomized block design with three treatments. The k observations within each block j, that is

$$(X_{1j}, X_{2j}, \ldots, X_{kj}), \quad j = 1, \ldots, n, \qquad \textbf{(10.4.1)}$$

are, in general, correlated, but observations in different blocks (different columns in Figure 10-5) are assumed to be independent.

The set of observations for treatment i is assumed to be a random sample from the population corresponding to treatment i. Letting μ_i denote the mean of observations coming from treatment i, that is, $E(X_{ij}) = \mu_i$, decompose these means into an overall mean and treatment effect as in (10.2.2):

$$\mu_i = \mu + \alpha_i,$$

where $\mu = \frac{1}{k}\sum_{i=1}^{k} \mu_i$, and $\alpha_i = \mu_i - \mu$. Thus, the null hypothesis of equality of the treatment means, $H_0 : \mu_1 = \cdots = \mu_k$, can also be written as

$$H_0 : \alpha_1 = \cdots = \alpha_k = 0. \qquad \textbf{(10.4.2)}$$

Note, however, that the statistical model (10.2.2) does not apply to data X_{ij} from a randomized block design, because it does not account for the dependence of the k observations within each block. To account for the dependence within the observations of each k-tuple in (10.4.1), model (10.2.2) is modified by including a **random effect** for each block j, denoted by b_j:

$$X_{ij} = \mu + \alpha_i + b_j + \epsilon_{ij}. \qquad \textbf{(10.4.3)}$$

The random effects $b_j, j = 1, \ldots, n$, are assumed to be iid, with

$$E(b_j) = 0 \quad \text{and} \quad \mathrm{Var}(b_j) = \sigma_b^2.$$

As in (10.2.2), the intrinsic error variables ϵ_{ij} are assumed to be uncorrelated (also independent for different blocks) with zero mean and variance σ_ϵ^2. Moreover, the intrinsic error variables are uncorrelated from the random effects. Model (10.4.3), and properties of covariance, yield the following expressions for the variance of an observation X_{ij} and the covariance between any two observations, $X_{i_1 j}$ and $X_{i_2 j}$, with $i_1 \neq i_2$, from block j:

$$\mathrm{Var}(X_{ij}) = \sigma_b^2 + \sigma_\epsilon^2, \qquad \mathrm{Cov}\left(X_{i_1 j}, X_{i_2 j}\right) = \sigma_b^2. \qquad \textbf{(10.4.4)}$$

Figure 10-5 Data display for a randomized block design.

	Blocks				
Treatments	1	2	3	$\cdots$	n
1	X_{11}	X_{12}	X_{13}	$\cdots$	X_{1n}
2	X_{21}	X_{22}	X_{23}	$\cdots$	X_{2n}
3	X_{31}	X_{32}	X_{33}	$\cdots$	X_{3n}s

Thus, according to model (10.4.3), all pairs of observations within each block are assumed to have equal covariance. This assumption is made tenable by the fact that the treatments are randomized within each block.

10.4.2 THE ANOVA *F* TEST

As mentioned in Sections 8.3.4 and 10.2.1, the ANOVA methodology is based on a decomposition of the total variability, represented by the total sums of squares (SST), into different components. In the context of a randomized block design, SST is decomposed into the sum of squares due to differences in the treatments (SSTr), due to differences in the blocks (SSB), and due to the intrinsic error variable (SSE). To define these sums of squares, set

$$\overline{X}_{i\cdot} = \frac{1}{n}\sum_{j=1}^{n} X_{ij}, \quad \overline{X}_{\cdot j} = \frac{1}{k}\sum_{i=1}^{k} X_{ij}, \quad \text{and} \quad \overline{X}_{\cdot\cdot} = \frac{1}{kn}\sum_{i=1}^{k}\sum_{j=1}^{n} X_{ij}.$$

This is similar to the dot and bar notation introduced in Figure 1-20. The *treatment SS*, the *block SS*, and the *error SS* are defined as follows:

Treatment Sum of Squares

$$\text{SSTr} = \sum_{i=1}^{k} n\left(\overline{X}_{i\cdot} - \overline{X}_{\cdot\cdot}\right)^2 \tag{10.4.5}$$

Block Sum of Squares

$$\text{SSB} = \sum_{j=1}^{n} k\left(\overline{X}_{\cdot j} - \overline{X}_{\cdot\cdot}\right)^2 \tag{10.4.6}$$

Error Sum of Squares

$$\text{SSE} = \sum_{i=1}^{k}\sum_{j=1}^{n}\left(X_{ij} - \overline{X}_{i\cdot} - \overline{X}_{\cdot j} + \overline{X}_{\cdot\cdot}\right)^2 \tag{10.4.7}$$

It can be shown that SSTr, SSB and SSE decompose the total sum of squares, which is defined as $\text{SST} = \sum_{i=1}^{k}\sum_{j=1}^{n}\left(X_{ij} - \overline{X}_{\cdot\cdot}\right)^2$. That is,

$$\text{SST} = \text{SSTr} + \text{SSB} + \text{SSE}. \tag{10.4.8}$$

The ANOVA *F* test rejects the hypothesis of equality of the k means (or, of no main treatment effects, see (10.4.2)), if SSTr is large compared to SSE. Though not of primary concern in the present context of randomized block designs, the hypothesis of no significant block effects can be rejected if SSB is large compared to SSE.

REMARK 10.4-1 The F test for the hypothesis of no main treatment effects remains the same even if the block effects are fixed instead of random; that is, even under the model

$$X_{ij} = \mu + \alpha_i + \beta_j + \epsilon_{ij}, \quad \sum_i \alpha_i = 0, \quad \sum_j \beta_j = 0, \quad E(\epsilon_{ij}) = 0, \tag{10.4.9}$$

which is the model for a two-factor design without interaction (see Section 1.8.4). ◁

For a proper comparison of SSTr and SSB with SSE, these sums of squares must be divided by their respective degrees of freedom, which are:

Degrees of Freedom for SSTr, SSB, and SSE

$$\text{DF}_{SSTr} = k - 1, \quad \text{DF}_{SSB} = n - 1, \quad \text{DF}_{SSE} = (k-1)(n-1) \qquad \textbf{(10.4.10)}$$

Dividing each of SSTr, SSB, and SSE by their degrees of freedom, we obtain the corresponding *mean squares*:

Mean Squares for Treatment, Block, and Error

$$\text{MSTr} = \frac{\text{SSTr}}{k-1}, \quad \text{MSB} = \frac{\text{SSB}}{n-1}, \quad \text{MSE} = \frac{\text{SSE}}{(k-1)(n-1)} \qquad \textbf{(10.4.11)}$$

The ANOVA F statistic is computed as the ratio MSTr/MSE. If the random effects and the error terms in the model (10.4.3) have a normal distribution, the exact null distribution of the F statistic, that is, its distribution when $H_0 : \mu_1 = \cdots = \mu_k$ is true, is known to be F with $k-1$ and $(k-1)(n-1)$ degrees of freedom:

Null Distribution of the ANOVA F-Statistic under Normality

$$F_{H_0}^{Tr} = \frac{\text{MSTr}}{\text{MSE}} \sim F_{k-1,\,(k-1)(n-1)} \qquad \textbf{(10.4.12)}$$

This null distribution of $F_{H_0}^{Tr}$ is approximately correct without the normality assumption provided $n \geq 30$.

Under the alternative hypothesis $H_a : H_0$ is not true, the ANOVA F statistic tends to take larger values. Thus, $H_0 : \mu_1 = \cdots = \mu_k$ is rejected at level of significance α if

ANOVA Region for Rejecting $H_0 : \mu_1 = \cdots = \mu_k$ at Level α

$$F_{H_0}^{Tr} > F_{k-1,\,(k-1)(n-1),\alpha} \qquad \textbf{(10.4.13)}$$

Moreover, the p-value is computed as $1 - F_{k-1,\,(k-1)(n-1)}(F_{H_0}^{Tr})$, where $F_{k-1,\,(k-1)(n-1)}$ denotes the CDF of the $F_{k-1,\,(k-1)(n-1)}$ distribution. The F statistic $F_{H_0}^{Tr}$, its p-value, and all quantities required for their computation are shown in an ANOVA table in the form summarized below:

Source	DF	SS	MS	F	P
Treatment	$k-1$	SSTr	MSTr	$F_{H_0}^{Tr} = \frac{\text{MSTr}}{\text{MSE}}$	$1 - F_{k-1,\,(k-1)(n-1)}(F_{H_0}^{Tr})$
Blocks	$n-1$	SSB	MSB	$F_{H_0}^{Bl} = \frac{\text{MSB}}{\text{MSE}}$	$1 - F_{n-1,\,(k-1)(n-1)}(F_{H_0}^{Bl})$
Error	$(n-1)(k-1)$	SSE	MSE		

Note that this ANOVA table also shows the test statistic $F_{H_0}^{Bl}$ and the corresponding p-value that can be used for testing the hypothesis of no block effects. This hypothesis, however, will not typically concern us.

Example 10.4-3

A random sample of 36 Napa Valley visitors tested and rated four wine varieties on a scale of 1–10. For impartiality purposes, the wines were identified only by numbers 1–4. The order in which each of the four wines were presented to each visitor was randomized. The average rating for each wine, and overall average rating, are $\overline{X}_{1\cdot} = 8.97$, $\overline{X}_{2\cdot} = 9.04$, $\overline{X}_{3\cdot} = 8.36$, $\overline{X}_{4\cdot} = 8.31$, and $\overline{X}_{\cdot\cdot} = 8.67$. Moreover, it is given that the sum of squares due to the visitors (blocks) is SSB $= 11.38$ and the total sum of squares is SST $= 65.497$. With the information given, construct the ANOVA table. Is there a significant difference in the rating of the four wines? Test at $\alpha = 0.05$.

Solution

In this example, the visitors constitute the blocks and the wines constitute the "treatments" (in the sense that the visitors are "treated" with different wines). The MS, F, and P columns of the ANOVA table follow from the first two columns. The entries in the degrees of freedom column are 3, 35, and 105. The information on the average rating of each wine, and formula (10.4.5) give

$$\text{SSTr} = 36\left[(8.97 - 8.67)^2 + (9.04 - 8.67)^2 + (8.36 - 8.67)^2 + (8.31 - 8.67)^2\right] = 16.29.$$

The information given about the values SSB and SST, together with formula (10.4.8) give

$$\text{SSE} = \text{SST} - \text{SSTr} - \text{SSB} = 65.497 - 16.29 - 11.38 = 37.827.$$

The resulting ANOVA table is

Source	df	SS	MS	F	P
Wines	3	16.29	5.43	15.08	3.13×10^{-8}
Visitors	35	11.38	0.325	0.9	0.63
Error	105	37.827	0.36		

The p-values in the last column were found with the R commands *1-pf(15.08, 3, 105)* and *1-pf(0.9, 35, 105)*. Since the p-value for the wine effect is very small, it is concluded that the wines differ significantly (at $\alpha = 0.05$, as well as at any of the common levels of significance) in terms of their average rating. ■

With the data stacked in the single column in the R object *values*, and additional columns in the R objects *treatment* and *block* indicating the treatment and block, respectively, for each observation, the R command for generating the randomized block design ANOVA table is

R Command for the Randomized Block Design

```
summary(aov(values~treatment+block))
```

As an example, we import the wine tasting data of Example 10.4-3 into the R data frame *wt* by *wt=read.table("NapaValleyWT.txt", header=T)*. (Note that the observations for each treatment, i.e., the ratings for each wine, are given in the columns of this data file, whereas in Figure 10-5 they are depicted as rows.) The R commands

```
st=stack(wt); wine=st$ind; visitor=as.factor(rep(1:36,4))
summary(aov(st$values~wine+visitor))
```

will generate the ANOVA table of Example 10.4-3 (up to round-off errors).

Testing the Validity of the Assumptions The ANOVA F test procedure requires that the intrinsic error variables ϵ_{ij} in the model (10.4.3) are homoscedastic and, if $n < 30$, normally distributed. The validity of these assumptions can be tested from the residuals obtained from fitting the model (10.4.3) in a manner similar to that described in Section 10.2.1. In particular, with the data stacked in the single column in the R object *values*, and additional columns in the R objects *trt* and *blk* indicating the treatment and block, respectively, for each observation, the R commands for testing the validity of the assumptions are:

R Command for Testing the Homoscedasticity Assumption

```
anova(aov(resid(aov(values~trt+blk))**2~trt+blk))
```
(10.4.14)

R Command for Testing the Normality Assumption

```
shapiro.test(resid(aov(values~trt+blk)))
```
(10.4.15)

As an example, using the R objects previously defined for the wine tasting data of Example 10.4-3, the R command:

```
anova(aov(resid(aov(st$values~wine+visitor))**2~wine+visitor))
```

produces p-values of 0.7906 and 0.4673 for the wine and visitor effects. These p-values suggest that the residual variance is not significantly different for different wines or visitors, and thus the homoscedasticity assumption is approximately valid. However, the additional command

```
shapiro.test(resid(aov(st$values~wine+visitor)))
```

returns a p-value of 0.006, suggesting that the normality assumption is violated. The residual boxplot and Q-Q plot shown in Figure 10-6 shed some insight into the nature of the violation. The boxplot does not suggest any violations in terms of lack of symmetry or existence of outliers, but the Q-Q plot suggests that the data come from a distribution with lighter tails than the normal. However, the sample size in

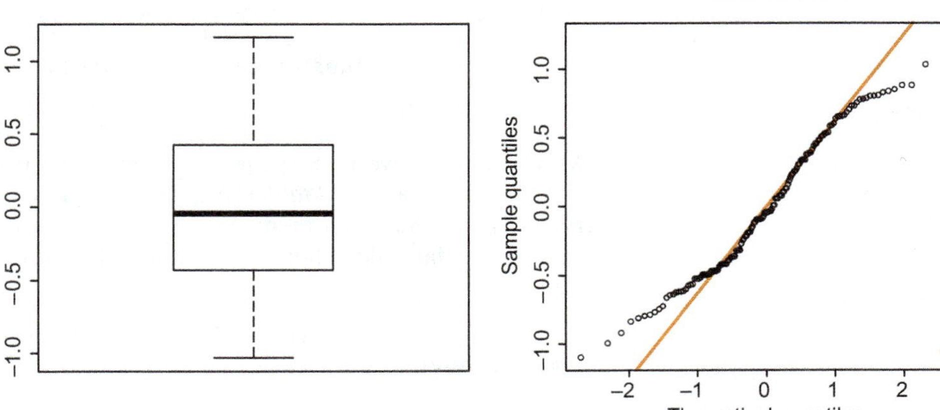

Figure 10-6 Boxplot and Q-Q plot of the residuals in Example 10.4-3.

this study is greater than 30, so the ANOVA F test procedure is approximately valid even under violations of the normality assumption.

10.4.3 FRIEDMAN'S TEST AND F TEST ON THE RANKS

When the data appear to contradict the normality assumption, Friedman's test and the F test on the ranks of the observations are alternative procedures for testing the equality of the distributions of k treatment populations.

Let $F_1, \ldots, F_k$ denote the cumulative distribution functions of the k populations, and $\widetilde{\mu}_1, \ldots, \widetilde{\mu}_k$ be their respective medians. The two procedures described in this section are widely interpreted as tests for

$$H_0 : \widetilde{\mu}_1 = \cdots = \widetilde{\mu}_k. \tag{10.4.16}$$

F Test on the Ranks As the name of this procedure suggests, it consists of applying the ANOVA F test on the ranks of the observations. In particular, the procedure consists of the following steps:

1. Combine the data from the k samples, $X_{i1}, \ldots, X_{in}$, $i = 1, \ldots, k$, into an overall set of $n \times k$ observations.
2. Assign ranks, or mid-ranks, to the combined set of observations as discussed in Sections 9.3 and 10.2.2. Let R_{ij} denote the (mid-)rank of observation X_{ij}.
3. Compute rank sums of squares SSTr_R, SSB_R, and SSE_R from the formulas (10.4.5), (10.4.6), and (10.4.7), respectively, using R_{ij} instead of X_{ij}, and compute the mean rank sums of squares MSTr_R, MSB_R, and MSE_R, by dividing the rank sums of squares by their respective degrees of freedom, as given in (10.4.10).
4. Compute the F statistic on the ranks by $FR^{Tr}_{H_0} = \mathrm{MSTr}_R/\mathrm{MSE}_R$, and the p-value by $1 - F_{k-1,(k-1)(n-1)}(FR^{Tr}_{H_0})$. The null hypothesis (10.4.16) is rejected at level α if $FR^{Tr}_{H_0} > F_{k-1,(k-1)(n-1),\alpha}$ or if the p-value is less than α.

This test procedure is approximately valid for $n \geq 8$.

Example 10.4-4

For the wine tasting data set of Example 10.4-3, the summary statistics on the ranks are $\overline{R}_{1\cdot} = 90.93$, $\overline{R}_{2\cdot} = 94.89$, $\overline{R}_{3\cdot} = 52.97$, $\overline{R}_{4\cdot} = 51.21$, and $\overline{R}_{\cdot\cdot} = 72.50$. Moreover, it is given that the rank sum of squares due to the visitors (blocks) is $\mathrm{SSB}_R = 41{,}843$ and the rank total sum of squares is $\mathrm{SST}_R = 248203$. With the information given, calculate the F statistic on the ranks and test the hypothesis of no difference in the ratings of the four wines at $\alpha = 0.05$.

Solution

Plugging the given summary statistics into the formula (10.4.5) for the treatment sum of squares, we have

$$\begin{aligned}\mathrm{SSTr}_R &= 36\left[(90.93 - 72.5)^2 + (94.89 - 72.5)^2 + (52.97 - 72.5)^2 + (51.21 - 72.5)^2\right] \\ &= 60{,}323.83.\end{aligned}$$

The information about the values SSB_R and SST_R together with formula (10.4.8) give

$$\mathrm{SSE}_R = \mathrm{SST}_R - \mathrm{SSTr}_R - \mathrm{SSB}_R = 248{,}203.00 - 41{,}843.00 - 60{,}323.83 = 146{,}036.17.$$

Thus, $\text{MSTr}_R = 60{,}323.83/3 = 20{,}107.94$, $\text{MSE}_R = 146{,}036.2/105 = 1390.82$,

$$FR_{H_0}^{Tr} = \frac{20{,}107.94}{1390.82} = 14.46, \text{ and } p\text{-value} = 1 - F_{3,105}(14.46) = 5.93 \times 10^{-8}.$$

Since the p-value is smaller than 0.05 (or, since $14.46 > F_{3,105,0.05} = 2.69$), it is concluded that the wines differ significantly at $\alpha = 0.05$ in terms of their median rating. ■

With the wine tasting data read into the R data frame *wt* and stacked into the data frame *st* (by *st=stack(wt)*), the R commands

```
ranks=rank(st$values); wine=st$ind;
  visitor=as.factor(rep(1:36,4))
summary(aov(ranks~wine+visitor))
```

will generate the sums of squares and F statistic of Example 10.4-4 (up to round-off errors).

Friedman's Test The main difference between Friedman's test and the ANOVA F test on the ranks is that in the former test each observation is ranked among the k observations of its own block. These *within block* ranks are denoted by r_{ij}. There are three additional differences. First, the rank mean error sum of squares is computed differently and is denoted by MSE_r^*. Second, the test statistic is the ratio of $\text{SSTr}_r/\text{MSE}_r^*$. Finally, the null distribution of Friedman's test statistic is approximately χ_{k-1}^2 (chi-square with $k-1$ degrees of freedom).

The steps for computing Friedman's test statistic and its p-value follow:

1. Let r_{ij} denote the (mid-)rank of observation X_{ij} among the observations in the jth block, that is, among $X_{1j}, \ldots, X_{kj}$.
2. Compute SSTr_r from (10.4.5), using r_{ij} instead of X_{ij}, and compute MSE_r^* as

$$\text{MSE}_r^* = \frac{1}{n(k-1)} \sum_{i=1}^{k} \sum_{j=1}^{n} (r_{ij} - \bar{r}_{..})^2 \qquad \textbf{(10.4.17)}$$

3. Compute Friedman's statistic by $Q_{H_0} = \text{SSTr}_r/\text{MSE}_r^*$ and the p-value by $1 - \Psi_{k-1}(Q_{H_0})$, where Ψ_ν denotes the CDF of the χ_ν^2 distribution. H_0 in (10.4.16) is rejected at level α if $Q_{H_0} > \Psi_{k-1,\alpha}$ or if the p-value is less than α.

Friedman's test procedure is approximately valid for $n > 15$ and $k > 4$.

Example 10.4-5

For the wine tasting data set of Example 10.4-3, the summary statistics on the within-block ranks r_{ij} are $\bar{r}_{1.} = 3.11$, $\bar{r}_{2.} = 3.14$, $\bar{r}_{3.} = 1.92$, $\bar{r}_{4.} = 1.83$, and $\bar{r}_{..} = 2.5$. Moreover, the sample variance of these ranks is 1.248. With the information given, calculate Friedman's test statistic and p-value, and test the hypothesis of no difference in the ratings of the four wines at $\alpha = 0.05$.

Solution

Plugging the given summary statistics into the formula (10.4.5) for the treatment sum of squares, we have

$$\text{SSTr}_r = 36\left[(3.11-2.5)^2 + (3.14-2.5)^2 + (1.92-2.5)^2 + (1.83-2.5)^2\right] = 56.41.$$

Formula (10.4.17) implies that MSE_r^* can be computed from the sample variance, S_r^2, of the within-block ranks as

$$\text{MSE}_r^* = \frac{1}{n(k-1)}(nk-1)S_r^2. \qquad \textbf{(10.4.18)}$$

Thus, using the information that $S_r^2 = 1.248$,

$$\text{MSE}_r^* = \frac{(36 \times 4 - 1)1.248}{36 \times 3} = 1.65 \quad \text{and} \quad Q_{H_0} = \frac{56.41}{1.65} = 34.19.$$

The command *1-pchisq(34.19, 3)* returns a p-value of $1 - \Psi_3(34.19) = 1.8 \times 10^{-7}$. Moreover, from Table A.5 we have $\Psi^2_{3,\,0.05} = 7.815$. Since $Q_{H_0} = 34.19 > 7.815$, and also since the p-value is less than 0.05, it is concluded that the wines differ significantly at $\alpha = 0.05$ in terms of their median rating. ■

With the data read into the R data frame *wt*, and stacked into the data frame *st* (by *st=stack(wt)*), the R command for performing Friedman's test is

R Command for Friedman's Test

```
friedman.test(st$values, st$ind, as.factor(rep(1:n, k)))
```

In particular, with the wine tasting data read into the R data frame *wt* and stacked into the data frame *st*, the command *friedman.test(st$values, st$ind, as.factor(rep(1:36, 4)))* gives the value of Friedman's test statistic and p-value (up to round-off errors) that was found in Example 10.4-5.

REMARK 10.4-2 Though Friedman's test is the more commonly used rank statistic for randomized block designs, it is generally less powerful than the ANOVA F test on the ranks. ◁

10.4.4 MULTIPLE COMPARISONS

As discussed in Section 10.3, when the null hypothesis of equality of the k means/medians is rejected, the further question arises as to which of the pairwise contrasts is different from zero at a given level of significance α. This section presents the Bonferroni and Tukey multiple comparisons procedures as they apply to data from randomized block designs.

Bonferroni Multiple Comparisons and Simultaneous CIs The procedure for constructing Bonferroni simultaneous CIs and multiple comparisons remains as described in Section 10.3.1. The only difference now is that the simultaneous CIs for the pairwise differences of means take the form of the paired T CI of Section 9.5. Similarly, if the multiple comparisons are to be done through pairwise testing, the only difference is that we use either the paired T test of Section 9.5 or the signed-rank test of Section 9.5.4. These Bonferroni multiple comparison procedures, through simultaneous CIs, and through pairwise testing, are demonstrated in the following example.

Example 10.4-6

In the context of the wine tasting data of Example 10.4-3, apply the following Bonferroni multiple comparisons procedures to identify which of the $\binom{4}{2} = 6$ pairs of wines are significantly different at experiment-wise level of significance $\alpha = 0.05$.

(a) Construct 95% Bonferroni simultaneous CIs, and perform multiple comparisons based on them.

(b) Perform the Bonferroni multiple comparisons procedure through pairwise testing using the Wilcoxon's signed-rank procedure.

(c) Perform the Bonferroni multiple comparisons procedure through pairwise testing using the paired T test procedure.

Solution

Since four wines were found to be significantly different in terms of their average, or median, ratings in Examples 10.4-3, 10.4-4, and 10.4-5, application of a multiple comparisons procedure is warranted.

(a) Because there are $m = 6$ 95% simultaneous CIs to be constructed, the Bonferroni method constructs each CI at confidence level of $(1 - 0.05/6)100\% = 99.167\%$. The six simultaneous CIs are displayed in the following table:

Comparison	99.17% CI	Includes 0?
$\mu_1 - \mu_2$	(−0.470, 0.326)	Yes
$\mu_1 - \mu_3$	(0.237, 0.974)	No
$\mu_1 - \mu_4$	(0.278, 1.044)	No
$\mu_2 - \mu_3$	(0.295, 1.061)	No
$\mu_2 - \mu_4$	(0.320, 1.147)	No
$\mu_3 - \mu_4$	(−0.370, 0.481)	Yes

(10.4.19)

For example, recalling that there are $n = 36$ blocks, the first interval is constructed as

$$\overline{X}_{1\cdot} - \overline{X}_{2\cdot} \pm t_{35,\,0.05/12} S_{1,2}/6$$

where 0.05/12 is half of 0.05/6, and $S_{1,2}$ is the standard deviation of the differences $X_{1j} - X_{2j}$, $j = 1, \ldots, 36$. Alternatively, with the data read into the R data frame *wt*, this first interval can be constructed with the R command *t.test(wt$W1,wt$W2, paired=T, conf.level=0.99167)*, and similarly for the others. Using these CIs, Bonferroni 95% multiple comparisons are performed by checking which of them includes zero. If an interval does not include zero, the corresponding comparison is declared significant at experiment-wise level of 0.05. The results, given in the last column of the table in (10.4.19), mean that each of the wines 1 and 2 is significantly different from wines 3 and 4, but wine 1 is not significantly different from wine 2, and wine 3 is not significantly different from wine 4.

(b) The p-values resulting from the signed-rank test applied on the data from each of the $m = 6$ pairs of wine ratings are given in the table in (10.4.20). Because the desired experiment-wise level of significance is 0.05, each p-value is compared to $0.05/6 = 0.00833$. Comparisons with p-values less than 0.00833 are declared significantly different.

Comparison	p-Value	Less than 0.0083?
1 vs 2	0.688	No
1 vs 3	0.000	Yes
1 vs 4	0.000	Yes
2 vs 3	0.000	Yes
2 vs 4	0.000	Yes
3 vs 4	0.712	No

(10.4.20)

According to the last column of the table in (10.4.20), each of the wines 1 and 2 is significantly different from wines 3 and 4, but wine 1 is not significantly different from wine 2, and wine 3 is not significantly different from wine 4. This conclusion is in agreement with the conclusion reached in part (a).

(c) The p-values resulting from the paired T test applied on the data from each of the $m = 6$ pairs of wine ratings are: 0.615, 5.376×10^{-5}, 2.728×10^{-5}, 1.877×10^{-5}, 1.799×10^{-5}, and 0.717. Arranging these p-values in the second column of a table like the one in (10.4.20) and comparing them with 0.00833 result in exactly the same conclusion that was reached in parts (a) and (b).

The multiple comparison results from all three methods are summarized as

$\overline{X}_4$	$\overline{X}_3$	$\overline{X}_1$	$\overline{X}_2$
8.31	8.36	8.97	9.04

Tukey's Multiple Comparisons and Simultaneous CIs Tukey's method is appropriate under the normality assumption of Section 10.4.2 or if the number of blocks is large (≥ 30).

The procedures for simultaneous CIs of all pairwise contrasts and for multiple comparisons, are similar to those described in Section 10.3.2, but we use the MSE for the randomized block designs and now the denominator degrees of freedom of the studentized range distribution is $(k-1)(n-1)$. More precisely, the procedures are as follows:

1. **Tukey's Simultaneous CIs:** The $(1-\alpha)100\%$ Tukey's simultaneous CIs for all contrasts $\mu_i - \mu_j$, $i \neq j$, are constructed as

$$\overline{X}_{i\cdot} - \overline{X}_{j\cdot} \pm Q_{\alpha,k,(k-1)(n-1)}\sqrt{\frac{\text{MSE}}{n}} \qquad \textbf{(10.4.21)}$$

where MSE is given in (10.4.11).

2. **Tukey's Multiple Comparisons at Level α:** If for a pair (i,j), $i \neq j$, the interval (10.4.21) does not contain zero, it is concluded that μ_i and μ_j differ significantly at level α.

Example 10.4-7

In the context of the wine tasting data of Example 10.4-3, apply Tukey's method to construct 95% simultaneous CIs and multiple comparisons to identify which of the $\binom{4}{2} = 6$ pairs of wines are significantly different at experiment-wise level of significance $\alpha = 0.05$.

Solution

Formula (10.4.21) yields the following 95% simultaneous CIs for the six pairwise differences of means.

Comparison	95% Tukey's SCI	Includes 0?
$\mu_1 - \mu_2$	(−0.442, 0.297)	Yes
$\mu_1 - \mu_3$	(0.236, 0.975)	No
$\mu_1 - \mu_4$	(0.292, 1.030)	No
$\mu_2 - \mu_3$	(0.308, 1.047)	No
$\mu_2 - \mu_4$	(0.364, 1.103)	No
$\mu_3 - \mu_4$	(−0.314, 0.425)	Yes

(10.4.22)

By checking which of these intervals include zero, the third column of the table in (10.4.22) yields multiple comparison results that are in agreement with those of Example 10.4-6. ■

REMARK 10.4-3 Note that the widths of the CIs in (10.4.22) tend to be smaller than the Bonferroni CIs of Example 10.4-6(a). This is due, in part, to the fact that Tukey's intervals use the assumption of equal variances, whereas the Bonferroni intervals do not. ◁

With the data stacked in the single column in the R object *values*, and additional columns in the R objects *treatment* and *block* indicating the treatment and block, respectively, for each observation, the R commands for Tukey's simultaneous CIs are:

R Commands for Tukey's $(1-\alpha)100\%$ Simultaneous CIs

```
TukeyHSD(aov(values~treatment+block), "treatment",
  conf.level=1-α)

plot(TukeyHSD(aov(values~treatment+block), "treatment",
  conf.level=1-α))
```

The default value of α is 0.05. As an example, with the wine tasting data of Example 10.4-3 imported into the R data frame *wt*, the R commands

```
st=stack(wt); wine=st$ind; visitor=as.factor(rep(1:36, 4))
TukeyHSD(aov(st$values~wine+visitor),"wine")
```

will generate the table of 95% simultaneous CIs of Example 10.4-7 (except for reversing the sign of the contrasts, i.e., $\mu_2 - \mu_1$ instead of $\mu_1 - \mu_2$, etc., and up to round-off errors). The additional command *plot(TukeyHSD(aov(st$values~ wine+visitor),"wine"))* produces the plot of Figure 10-7, which is an effective visual display of the multiple comparisons results.

Tukey's Multiple Comparisons on the Ranks This procedure consists of first combining all observations into an overall set of $N = n \times k$ observations, sorting them from smallest to largest, and assigning ranks or mid-ranks to each observation. This ranking process is described in Section 10.2.2. Then the data are replaced by their ranks, and Tukey's multiple comparisons procedure is applied on them. The application of Tukey's method on the (mid-)ranks of the combined data is approximately valid for $n \geq 8$, and is the recommended procedure with non-normal data.

As explained in Section 10.3.3, Tukey's simultaneous CIs are not relevant to the contrasts $\mu_i - \mu_j$, but the resulting multiple comparisons procedure are interpretable as multiple comparisons for pairs of the k populations the observations came from.

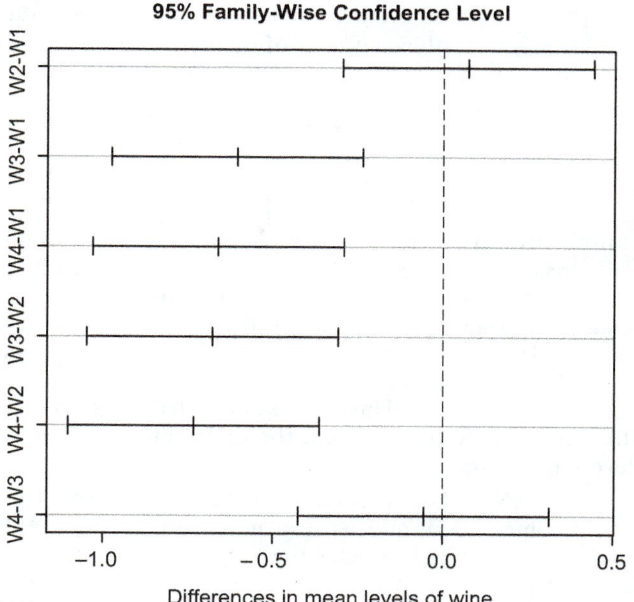

Figure 10-7 Visual display of Tukey's simultaneous CIs for Example 10.4-7.

In particular, the results of Tukey's multiple comparisons procedure on the ranks will be interpreted as multiple comparisons for all pairs of median contrasts $\widetilde{\mu}_i - \widetilde{\mu}_j$.

Example 10.4-8

In the context of the wine tasting data of Example 10.4-3, perform multiple comparisons at experiment-wise level of significance $\alpha = 0.05$ using Tukey's method on the ranks.

Solution

With the wine tasting data imported into the R data frame *wt*, the R commands

```
st=stack(wt); wine=st$ind; visitor=as.factor(rep(1:36, 4))
TukeyHSD(aov(rank(st$values)~wine+visitor),"wine")
```

give the 95% Tukey simultaneous CIs on the ranks shown in the table below.

Comparison	95% Tukey's SCI	Includes 0?
1 vs 2	(−26.91, 18.99)	Yes
1 vs 3	(15.01, 60.91)	No
1 vs 4	(16.77, 62.67)	No
2 vs 3	(18.97, 64.87)	No
2 vs 4	(20.73, 66.63)	No
3 vs 4	(−21.18, 24.71)	Yes

As already mentioned, these CIs do not pertain to contrasts of the means or medians of the wine ratings. However, the multiple comparison procedure resulting from checking whether or not each interval includes zero does pertain to the contrasts of the medians. It is seen that the multiple comparisons results that follow from the third column of the above table are in agreement with those of Examples 10.4-6 and 10.4-7. ■

Exercises

1. The data for the rate of combustion in humid air flow study of Exercise 5 in Section 10.2 can be found in *CombustRate.txt*. In the context of that exercise, an engineer with statistical training observes that there is a certain pattern in the data and inquires about other experimental conditions. It turned out that different rows corresponded to different temperatures. Having this additional information, the engineer decides to treat the different temperature levels as random blocks.

(a) Explain why the ANOVA F procedure of Section 10.2.1 is not recommended, and write a model for the observations X_{ij} that incorporates the additional information.

(b) Using parameters of the model you wrote in part (a), state the null and alternative hypotheses relevant for deciding if variation in the mole fraction of water affects the mean diffusivity.

(c) Use R commands to construct the ANOVA table for the model in part (a), and use the p-value to conduct the test at level $\alpha = 0.01$.

(d) Compare the p-value obtained in (c) with that obtained in Exercise 5 in Section 10.2, and give a brief explanation for the observed difference.

(e) Use R commands to construct Tukey's 99% simultaneous CIs and multiple comparisons.

2. A service center for electronic equipment is interested in investigating possible differences in service times of the three types disk drives that it regularly services. Each of the three technicians currently employed was randomly assigned to one repair of each type of drive and the repair times were recorded. The results are shown in the table below:

	Technician		
Drive	1	2	3
1	44.8	33.4	45.2
2	47.8	61.2	60.8
3	73.4	71.2	64.6

(a) Write a model for the observations. Do the random blocks correspond to the drives or the technicians? (*Hint.* The technicians currently employed can be viewed as a random sample from the population of available technicians.)

(b) Complete the ANOVA table and report the p-value for testing the null hypothesis that the average service time of the three drives is the same.

(c) Compute Friedman's test statistic for the hypothesis in (b) and report the p-value.

3. A commercial airline is considering four different designs of the control panel for the new generation of airplanes. To see if the designs have an effect on the pilot's response time to emergency displays, emergency conditions were simulated and the response times, in seconds, of 8 pilots were recorded. The same 8 pilots were used for all four designs. The order in which the designs were evaluated was randomized for each pilot. The data are given in *PilotReacTimes.txt.*

(a) Write a model for the observations X_{ij} and specify which parameters represent the treatment effects and which represent the random block effects.

(b) Using parameters of the model you wrote in part (a), state the null and alternative hypotheses relevant for deciding if the designs differ in terms of the pilot's mean response time.

(c) Using R commands, perform the tests for checking the assumptions of homoscedasticity and normality of the intrinsic error variables of the model in (a). Report the p-values of the tests, and construct the residual boxplot and Q-Q plot. Comment on the validity of the two assumptions on the basis of the p-values and the plots.

(d) Regardless of the conclusions reached in part (c), carry out the ANOVA F procedure at level of significance $\alpha = 0.01$ for testing the hypothesis in part (b), and report the p-value.

4. In the context of Exercise 3, use R commands to construct Tukey's 99% simultaneous CIs, including the plot that visually displays them, and perform multiple comparisons at experiment-wise level 0.01 to determine which pairs of designs differ significantly in terms of the pilot's response time. Summarize the conclusions by arranging the sample means in increasing order and underlining the pairs of means that are not significantly different.

5. An experiment was performed to determine the effect of four different chemicals on the strength of a fabric. Five fabric samples were selected and each chemical was tested once in random order on each fabric sample. The total sum of squares for this data is SST = 8.4455. Some additional summary statistics are given in the partially filled ANOVA table below:

	Df	Sum Sq	Mean Sq	F value
treatment				
block		5.4530		
Residuals		0.5110		

(a) Which of the two factors in this study, that is, which of the factors "chemical" and "fabric," is the blocking factor?

(b) Complete the ANOVA table and test the null hypothesis at level $\alpha = 0.05$ that the four chemicals do not differ in terms of the mean strength of the fabric. Report the p-value.

6. The data file *FabricStrengRbd.txt* contains the fabric strength data of Exercise 5. Import the data into the R object *fs*, and use the R commands

```
ranks=rank(fs$streng);anova(aov(ranks~
  fs$chemical+fs$fabric))
```

to construct the ANOVA table on the ranks.

(a) Report the p-value for testing the null hypothesis that the four chemicals do not differ in terms of the mean strength of the fabric. Should the hypothesis be rejected at level $\alpha = 0.01$?

(b) Conduct Tukey's multiple comparisons on the ranks, and report which pairs of chemicals differ at experiment-wise level of significance $\alpha = 0.01$.

7. A study was conducted to see whether three cars, A, B, and C, took the same time to parallel park. A random sample of seven drivers was obtained and the time required for each of them to parallel park each of the three cars was measured. The results are listed in the table below.

	Driver						
Car	1	2	3	4	5	6	7
A	19.0	21.8	16.8	24.2	22.0	34.7	23.8
B	17.8	20.2	16.2	41.4	21.4	28.4	22.7
C	21.3	22.5	17.6	38.1	25.8	39.4	23.9

Is there evidence that the time required to parallel park the three types of car are, on average, different? You may use hand calculations or R commands to complete the following parts.

(a) Construct Bonferroni's 95% simultaneous CIs and perform the corresponding multiple comparisons.

(b) Perform Bonferroni's multiple comparisons at experiment-wise error rate of 0.05, using the signed-rank test.

Chapter 11

MULTIFACTOR EXPERIMENTS

11.1 Introduction

A statistical experiment involving several factors is called a *factorial experiment* if all factor-level combinations are considered, that is, if data are collected from all factor-level combinations.

Section 1.8.4 introduced the important concepts of *main effects* and *interaction* in the context of a two-factor factorial experiment. It was seen that in additive designs (i.e., designs where there is no interaction among the factors) the different levels of a factor can be compared in terms of the main effect, and the "best" factor-level combination is the one that corresponds to the "best" level of each factor. However, in non-additive designs (i.e., when there is interaction among the factors) the comparison of different levels of a factor is more complicated. The same is true for designs with more than two factors. For this reason, the analysis of data from a factorial experiment often begins by determining whether or not the design is additive. The interaction plot we saw in Section 1.8.4 is useful in this regard but needs to be followed up by a formal test.

In this chapter we will discuss the ANOVA F procedure for testing the null hypothesis that a two-factor design is additive versus the alternative that states the opposite. ANOVA F tests for the main effects will also be presented. The concepts of main effects and interactions will then be extended to a three-factor design, and the ANOVA F procedures for testing them will be presented. Finally, the last section presents a special class of experimental designs, 2^r factorial and fractional factorial designs, which are used extensively in quality improvement programs.

11.2 Two-Factor Designs

11.2.1 *F* TESTS FOR MAIN EFFECTS AND INTERACTIONS

We begin by reviewing some of the notation and terminology introduced in Section 1.8.4, but we advise the reader to go back and review that section before proceeding with this section.

A design where factor A, the *row factor*, has a levels and factor B, the *column factor* has b levels is referred to as an $a \times b$ design; a 2×2 design is also referred to as a 2^2 design. Let μ_{ij} denote the mean value of an observation taken at factor-level combination (i, j), that is, when the level of factor A is i and the level of factor B is j. The set of mean values μ_{ij}, $i = 1, \ldots, a$, $j = 1, \ldots, b$, can be decomposed as

Decomposition of Means in a Two-Factor Design

$$\mu_{ij} = \mu + \alpha_i + \beta_j + \gamma_{ij} \tag{11.2.1}$$

where $\mu = \overline{\mu}_{..}$ is the average of all the μ_{ij}, and α_i, β_j, and γ_{ij} are the main row effects, main column effects, and interaction effects, defined in (1.8.4) and (1.8.6), respectively. Note that the definition of the main effects and interactions implies that they satisfy the following conditions:

$$\sum_{i=1}^{a} \alpha_i = 0, \quad \sum_{j=1}^{b} \beta_j = 0, \quad \sum_{i=1}^{a} \gamma_{ij} = 0, \quad \sum_{j=1}^{b} \gamma_{ij} = 0. \tag{11.2.2}$$

It can be shown that the α_i, β_j, and γ_{ij} defined in (1.8.4) are the only sets of numbers that satisfy both the decomposition (11.2.1) and the conditions (11.2.2).

From each factor-level combination (i, j) we observe a simple random sample

$$X_{ijk}, \quad i = 1, \ldots, a, \quad j = 1, \ldots, b, \quad k = 1, \ldots, n_{ij}.$$

Note that the first two indices refer to the levels of factors A and B, and the third index enumerates the observations within each factor-level combination; see Figure 1-19 for an illustration of the arrangement of the observations within each factor-level combination. When all group sample sizes are equal, that is, $n_{ij} = n$, for all i, j, we say that the design is **balanced**.

The statistical model for the data can be written as

$$X_{ijk} = \mu_{ij} + \epsilon_{ijk} \quad \text{or} \quad X_{ijk} = \mu + \alpha_i + \beta_j + \gamma_{ij} + \epsilon_{ijk}, \tag{11.2.3}$$

where the intrinsic error variables ϵ_{ijk} are assumed to be independent with zero mean and common variance σ_ϵ^2 (*homoscedasticity* assumption). The first expression in (11.2.3) is called the *mean-plus-error* form of the model, while the second is the *treatment-effects* form. The null hypotheses of no interaction effect and no main factor effects are

No Interaction Effects

$$H_0^{AB} : \gamma_{11} = \cdots = \gamma_{ab} = 0 \tag{11.2.4}$$

No Main Row Effects

$$H_0^{A} : \alpha_1 = \cdots = \alpha_a = 0 \tag{11.2.5}$$

No Main Column Effects

$$H_0^{B} : \beta_1 = \cdots = \beta_b = 0 \tag{11.2.6}$$

Testing for these hypotheses will be based on the ANOVA methodology, which was used in Sections 8.3.4 and 10.2.1 in the context of the simple linear regression model and testing for the equality of k means, respectively. For technical reasons, the formulas for the decomposition of the total sum of squares and the resulting F statistics will be given only for the case of a balanced design, that is, $n_{ij} = n$ for all i, j.

For data denoted by multiple indices, the *dot and bar* notation (already partly used in Section 1.8.4) is very useful for denoting summations and averages over some of the indices. According to this notation, replacing an index by a dot means summation over that index, while a dot and a bar means averaging over that index. For example, the sample mean from factor-level combination, or group, (i, j) is denoted by

$$\overline{X}_{ij\cdot} = \frac{1}{n}\sum_{k=1}^{n} X_{ijk}. \tag{11.2.7}$$

For simplicity in notation, however, we write $\overline{X}_{ij}$ instead of $\overline{X}_{ij\cdot}$. Moreover, set

$$\overline{X}_{i\cdot} = \frac{1}{b}\sum_{j=1}^{b}\overline{X}_{ij}, \quad \overline{X}_{\cdot j} = \frac{1}{a}\sum_{i=1}^{a}\overline{X}_{ij}, \quad \overline{X}_{\cdot\cdot} = \frac{1}{a}\frac{1}{b}\sum_{i=1}^{a}\sum_{j=1}^{b}\overline{X}_{ij}. \tag{11.2.8}$$

The total sum of squares, and its degrees of freedom, are defined as the numerator of the sample variance of all X_{ijk} and their total number minus one, respectively:

$$\text{SST} = \sum_{i=1}^{a}\sum_{j=1}^{b}\sum_{k=1}^{n}\left(X_{ijk} - \overline{X}_{\cdot\cdot}\right)^2, \quad DF_{SST} = abn - 1. \tag{11.2.9}$$

The total sum of squares is decomposed as

$$\text{SST} = \text{SSA} + \text{SSB} + \text{SSAB} + \text{SSE}, \tag{11.2.10}$$

where the sum of squares due to the main row effects, SSA, the main column effects, SSB, the interaction effects, SSAB, and the error term, SSE, are defined as

$$\text{SSA} = bn\sum_{i=1}^{a}\left(\overline{X}_{i\cdot} - \overline{X}_{\cdot\cdot}\right)^2, \quad \text{SSB} = an\sum_{j=1}^{b}\left(\overline{X}_{\cdot j} - \overline{X}_{\cdot\cdot}\right)^2, \tag{11.2.11}$$

$$\text{SSAB} = n\sum_{i=1}^{a}\sum_{j=1}^{b}\left(\overline{X}_{ij} - \overline{X}_{i\cdot} - \overline{X}_{\cdot j} + \overline{X}_{\cdot\cdot}\right)^2, \quad \text{and} \tag{11.2.12}$$

$$\text{SSE} = \sum_{i=1}^{a}\sum_{j=1}^{b}\sum_{k=1}^{n}\left(X_{ijk} - \overline{X}_{ij}\right)^2. \tag{11.2.13}$$

It is instructive to note that

$$\text{SSA} = bn\sum_{i=1}^{a}\widehat{\alpha}_i^2, \quad \text{SSB} = an\sum_{j=1}^{b}\widehat{\beta}_j^2, \quad \text{and} \quad \text{SSAB} = n\sum_{i=1}^{a}\sum_{j=1}^{b}\widehat{\gamma}_{ij}^2, \tag{11.2.14}$$

where $\widehat{\alpha}_i$, $\widehat{\beta}_j$, and $\widehat{\gamma}_{ij}$ are the estimated effects defined in (1.8.8) and (1.8.9). The corresponding decomposition of the total degrees of freedom, DF_{SST}, is

$$\begin{aligned} abn - 1 &= (a-1) + (b-1) + (a-1)(b-1) + ab(n-1) \\ &= DF_{SSA} + DF_{SSB} + DF_{SSAB} + DF_{SSE}, \end{aligned} \tag{11.2.15}$$

where DF_{SSA}, DF_{SSB}, DF_{SSAB}, and DF_{SSE} are defined implicitly in (11.2.15). The *mean squares* are obtained by dividing the sums of squares by their degrees of freedom:

$$\text{MSA} = \frac{\text{SSA}}{a-1}, \ \text{MSB} = \frac{\text{SSB}}{b-1}, \ \text{MSAB} = \frac{\text{SSAB}}{(a-1)(b-1)}, \ \text{MSE} = \frac{\text{SSE}}{ab(n-1)}.$$

It is not difficult to verify that the pooled sample variance, S_p^2, which, since all sample sizes are equal, is just the average of the ab sample variances, equals MSE. See also (10.2.11) for the one-factor analogue of this result.

The F statistics for testing H_0^{AB}, H_0^A, and H_0^B and their null distributions are:

F-Statistic for H_0^{AB}

$$F_{H_0}^{AB} = \frac{\text{MSAB}}{\text{MSE}} \sim F_{(a-1)(b-1),\, ab(n-1)} \tag{11.2.16}$$

F-Statistic for H_0^A

$$F_{H_0}^{A} = \frac{\text{MSA}}{\text{MSE}} \sim F_{a-1,\, ab(n-1)} \tag{11.2.17}$$

F-Statistic for H_0^B

$$F_{H_0}^{B} = \frac{\text{MSB}}{\text{MSE}} \sim F_{b-1,\, ab(n-1)} \tag{11.2.18}$$

Recall that the notation $F_{\nu_1,\,\nu_2}$ is used to denote both the F distribution with numerator and denominator degrees of freedom ν_1 and ν_2, respectively, and its CDF. The above computations are summarized in an ANOVA table in the form summarized below:

Source	Df	SS	MS	F	p-Value
Main Effects					
A	$a-1$	SSA	MSA	$F_{H_0}^A$	$1 - F_{a-1,ab(n-1)}(F_{H_0}^A)$
B	$b-1$	SSB	MSB	$F_{H_0}^B$	$1 - F_{b-1,ab(n-1)}(F_{H_0}^B)$
Interactions					
AB	$(a-1)(b-1)$	SSAB	MSAB	$F_{H_0}^{AB}$	$1 - F_{(a-1)(b-1),ab(n-1)}(F_{H_0}^{AB})$
Error	$ab(n-1)$	SSE	MSE		
Total	$abn-1$	SST			

It is typically recommended that the test for interaction be performed first. If H_0^{AB} is rejected we may conclude that both factors influence the response *even if H_0^A and H_0^B* are not rejected. This is because the main effects are *average* effects, where the averaging is over the levels of the other factor. For example, if interaction is present it may be that at level $j = 1$ of factor B, level $i = 1$ of factor A results in higher mean than level $i = 2$, while the opposite is true for level $j = 2$ of factor B. The phenomenon of insignificant main effects in the presence of significant interaction effects is often referred to as the *masking* of main effects due to interaction.

When one of the hypotheses is rejected, simultaneous CIs and multiple comparisons can be performed to determine which pairs of parameters are significantly different. For example, if H_0^A is rejected, simultaneous CIs and multiple comparisons can determine which pairs of contrasts of the row main effects are significantly different, that is, which differences $\alpha_i - \alpha_j$ are significantly different from zero. Instead of giving formulas, we rely on R output for conducting such multiple comparisons.

With the response variable and the levels of the two factors read into the R objects y, A, and B, respectively, the following R commands for constructing the ANOVA table and Tukey's simultaneous CIs apply also for **unbalanced** designs, that is, when the sample sizes n_{ij} are not equal:

R Commands for the ANOVA F Tests and Tukey's $(1 - \alpha)100\%$ Simultaneous CIs

```
fit=aov(y~A*B); anova(fit); TukeyHSD(fit, conf.level=1-α)
```

The following display summarizes all assumptions and formulas needed for implementing the ANOVA F test procedure discussed in this section.

ANOVA F Tests in Two-Factor Designs

(1) *Assumptions:*
(a) $X_{ij1}, \ldots, X_{ijn}$ is a simple random sample from the (i, j)th population, $i = 1, \ldots, a, j = 1, \ldots, b$
(b) The ab samples are independent
(c) The ab variances are equal (homoscedasticity)
(d) The ab populations are normal or $n \geq 30$

(2) *Construction of the F Test Statistics:*
(a) Compute the sample variance of each of the ab samples, and average them to form $S_p^2 = \text{MSE}$.
(b) Compute $\overline{X}_{ij}, \overline{X}_{i\cdot}, \overline{X}_{\cdot j}, \overline{X}_{\cdot\cdot}$, as shown in (11.2.7) and (11.2.8), and use them to compute SSA, SSB and SSAB as shown in (11.2.11) and (11.2.12)
(c) Compute $F_{H_0}^{AB}, F_{H_0}^{A}$ and $F_{H_0}^{B}$ by following the steps leading to (11.2.16), (11.2.17), and (11.2.18)

(3) *Rejection Regions and p-Values:* **(11.2.19)**

Hypothesis	Rejection Region
H_0^{AB}	$F_{H_0}^{AB} > F_{(a-1)(b-1),ab(n-1),\alpha}$
H_0^{A}	$F_{H_0}^{A} > F_{a-1,ab(n-1),\alpha}$
H_0^{B}	$F_{H_0}^{B} > F_{b-1,ab(n-1),\alpha}$

where $F_{\nu_1,\nu_2,\alpha}$ is the $(1 - \alpha)100$th percentile of the F_{ν_1,ν_2} distribution

Hypothesis	p-Value
H_0^{AB}	$1 - F_{(a-1)(b-1),ab(n-1)}(F_{H_0}^{AB})$
H_0^{A}	$1 - F_{a-1,ab(n-1)}(F_{H_0}^{A})$
H_0^{B}	$1 - F_{b-1,ab(n-1)}(F_{H_0}^{B})$

Example 11.2-1

Data were collected[1] on the amount of rainfall, in inches, in select target areas of Tasmania with and without cloud seeding during the different seasons. The sample means and sample variances of the $n = 8$ measurements from each factor-level combination are given in the table below.

[1] A. J. Miller et al. (1979). Analyzing the results of a cloud-seeding experiment in Tasmania, *Communications in Statistics—Theory & Methods*, A8(10): 1017–1047.

		Season							
		Winter		Spring		Summer		Autumn	
		Mean	Var	Mean	Var	Mean	Var	Mean	Var
Seeded	No	2.649	1.286	1.909	1.465	0.929	0.259	2.509	1.797
	Yes	2.192	3.094	2.015	3.141	1.127	1.123	2.379	2.961

Carry out the tests for no interaction and no main effects at level of significance 0.05.

Solution

The sample versions of the main effects and interactions were computed in Example 1.8-12, and the interaction plot is shown in Figure 1-21. Here we will determine whether the differences in the sample versions are big enough to declare the population quantities significantly different. (The assumptions needed for the validity of the tests will be checked in Example 11.2-3.) Averaging the eight sample variances yields

$$\text{MSE} = 1.891.$$

Let "Seeded" be factor A, with levels "No" ($i = 1$) and "Yes" ($i = 2$), and "Season" be factor B, with levels "Winter," "Spring," "Summer," "Autumn" corresponding to $j = 1, 2, 3, 4$, respectively. Averaging the means overall and by rows and columns yields

$$\bar{x}_{\cdot\cdot} = 1.964, \quad \bar{x}_{1\cdot} = 1.999, \quad \bar{x}_{2\cdot} = 1.928$$
$$\bar{x}_{\cdot 1} = 2.421, \quad \bar{x}_{\cdot 2} = 1.962, \quad \bar{x}_{\cdot 3} = 1.028, \quad \bar{x}_{\cdot 4} = 2.444.$$

Using formulas (11.2.11) and (11.2.12), we obtain SSA $= 0.079$, SSB $= 21.033$, SSAB $= 1.024$. Dividing these by the corresponding degrees of freedom, which are 1, 3, and 3, respectively, we obtain

$$\text{MSA} = 0.079, \quad \text{MSB} = 7.011, \quad \text{MSAB} = 0.341.$$

Using also the previously obtained MSE, the test statistics are

$$F_{H_0}^{AB} = \frac{0.341}{1.891} = 0.180, \quad F_{H_0}^{A} = \frac{0.079}{1.891} = 0.042, \quad F_{H_0}^{B} = \frac{7.011}{1.891} = 3.708.$$

The corresponding p-values are

$$1 - F_{3,56}(0.180) = 0.909, \quad 1 - F_{1,56}(0.042) = 0.838, \quad 1 - F_{3,56}(3.708) = 0.017.$$

Thus, only the levels of factor B (the seasons) are significantly different at level 0.05 in terms of the amount of rainfall produced. ■

Example 11.2-2

The data file *CloudSeed2w.txt* contains the rainfall measurements of the previous example. Use R commands to construct the ANOVA table and to perform Tukey's 95% simultaneous CIs and multiple comparisons for determining the pairs of seasons that are significantly different at experiment-wise error rate of 0.05.

Solution

The commands

```
cs=read.table("CloudSeed2w.txt", header=T)
y=cs$rain; A=cs$seeded; B=cs$season
fit=aov(y~A*B); anova(fit); TukeyHSD(fit)
```

import the data into the R data frame *cs*; define the objects *y*, *A*, and *B* as columns containing the rainfall data, the level of factor *A*, and the level of factor *B*, respectively; and perform the analysis of variance including Tukey's 95% simultaneous CIs. The resulting ANOVA table is shown below:

	Df	Sum Sq	Mean Sq	F value	Pr(>F)
A	1	0.079	0.0791	0.0418	0.83868
B	3	21.033	7.0108	3.7079	0.01667
A:B	3	1.024	0.3414	0.1806	0.90914
Residuals	56	105.885	1.8908		

The *p*-values shown in the ANOVA table are the same (up to round-off errors) with those obtained in the previous example, and thus the conclusion that only seasons are significantly different at level 0.05 remains. It follows that multiple comparisons are needed only to determine which seasons are different from each other. (Note that factor *A* has only two levels, so there would be no need for multiple comparisons even if H_0^A had been rejected.) The command *TukeyHSD(fit)* gives 95% simultaneous CIs for the contrasts of the form $\alpha_{i_1} - \alpha_{i_2}$ (so an ordinary 95% CI if factor *A* has only two levels), a separate set of 95% simultaneous CIs for contrasts of the form $\beta_{j_1} - \beta_{j_2}$, and a separate set of 95% simultaneous CIs for contrasts of the form $\mu_{i_1 j_1} - \mu_{i_2 j_2}$ which, however, are typically of no interest. The 95% simultaneous CIs for contrasts of the form $\beta_{j_1} - \beta_{j_2}$ (the seasons' contrasts) are given in the following table. From this table it is seen that only the simultaneous CIs for the Summer-Autumn and Winter-Summer contrasts do not contain zero, so these are the only contrasts that are significantly different from zero at experiment-wise level of significance 0.05.

Tukey multiple comparisons of means
95% family-wise confidence level

	diff	lwr	upr	p adj
Spring-Autumn	-0.481875	-1.7691698	0.8054198	0.7550218
Summer-Autumn	-1.415625	-2.7029198	-0.1283302	0.0257497
Winter-Autumn	-0.023125	-1.3104198	1.2641698	0.9999609
Summer-Spring	-0.933750	-2.2210448	0.3535448	0.2311927
Winter-Spring	0.458750	-0.8285448	1.7460448	0.7815268
Winter-Summer	1.392500	0.1052052	2.6797948	0.0291300

11.2.2 TESTING THE VALIDITY OF ASSUMPTIONS

The validity of the ANOVA *F* tests for interaction and main effects rests on the assumptions stated in (11.2.19). The assumptions that the data constitute simple random samples from their respective populations, that is, factor-level combinations, and the *ab* samples being independent are best checked by reviewing the data collection protocol. Here we will discuss ways of checking the assumption of equal variances and, in case the sample sizes are less than 30, the normality assumption.

As remarked in Section 10.2.1, in comparative studies involving more than two populations, the group sample sizes are rarely 30 or more. Because the normality assumption cannot be checked reliably with the small sample sizes, it is customary to perform a single normality test on the combined set of residuals after the homoscedasticity assumption has been judged tenable.

As explained in Section 10.2.1, the homoscedasticity assumption can be checked by performing the two-factor ANOVA F test on the squared residuals obtained from fitting the model in (11.2.3). If the resulting p-values for no main effects and no interactions are all greater than 0.05, conclude that the assumption of equal variances is approximately satisfied. Once the homoscedasticity assumption has been judged tenable, the normality assumption can be checked by performing the Shapiro-Wilk normality test on the residuals. Again, if the p-value is greater than 0.05 it can be concluded that the normality assumption is approximately satisfied. Residual plots can shed light on the nature of assumptions violations.

With the R object *fit* containing the output from fitting the model (11.2.3), that is, by using the command *fit=aov(y~A*B)*, the R commands for obtaining the p-values from the two-factor ANOVA F tests on the squared residuals and from the Shapiro-Wilk normality test on the residuals are:

R Command for Testing the Homoscedasticity Assumption

```
anova(aov(resid(fit)**2~A*B))
```
(11.2.20)

R Command for Testing the Normality Assumption

```
shapiro.test(resid(fit))
```
(11.2.21)

Further, the commands

```
plot(fit, which=1)
plot(fit, which=2)
```
(11.2.22)

will display the residuals by groups (labeled by the fitted, or the $\widehat{\mu}_{ij}$, values) and produce a Q-Q plot for the combined residuals, respectively.

Example 11.2-3

Check whether the data of Examples 11.2-1 and 11.2-2 satisfy the homoscedasticity and normality assumptions.

Solution

Using the R object *fit* generated in Example 11.2-2, the R command shown in (11.2.20) generates the following ANOVA table:

```
          Df   Sum Sq  Mean Sq  F value   Pr(>F)
A          1   23.264  23.2643   4.1624  0.04606
B          3   23.736   7.9121   1.4156  0.24781
A:B        3    1.788   0.5960   0.1066  0.95588
Residuals 56  312.996   5.5892
```

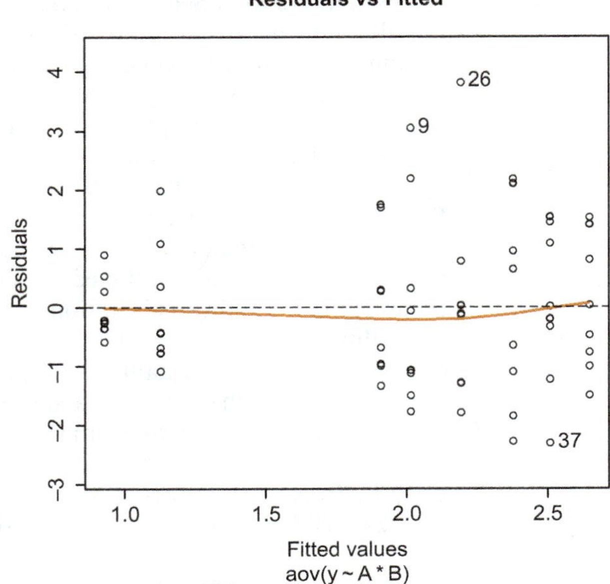

Figure 11-1 Residuals plotted against the fitted values for Example 11.2-3.

Since the p-value corresponding to factor A is less than 0.05, we conclude that the two levels of factor A (seeded and not seeded) have different effect on the residual variance. It follows that the homoscedasticity assumption does not appear tenable and, hence, it does not make sense to test for the normality assumption because heteroscedasticity renders the test invalid. The command *plot(fit, which=1)* generates the plot in Figure 11-1, which shows that the variability of the residuals varies significantly among the different groups. In such cases, a *weighted analysis* is recommended; see Sections 12.4.1 and 12.4.2.

11.2.3 ONE OBSERVATION PER CELL

In some cases, factorial experiments are designed with only one observation per cell. In such cases, the previous analysis cannot be carried out. The main reason for this is that MSE cannot be formed. Recall that MSE is obtained by pooling the sample variances from all ab samples. Since the calculation of the sample variance requires at least two observations, the sample variances, and hence MSE, cannot be formed. In technical parlance, interactions are *confounded* with the error term, or there are as many parameters (which are the ab group means μ_{ij}) as there are observations, and thus *there are no degrees of freedom left* for estimating the error variance. The usual remedy is to model the ab group means using fewer parameters, and the most common way of doing so is to assume an additive model, that is, all $\gamma_{ij} = 0$:

$$\mu_{ij} = \mu + \alpha_i + \beta_j, \tag{11.2.23}$$

where the main effects satisfy the usual conditions $\sum_{i=1}^{a} \alpha_i = 0 = \sum_{j=1}^{b} \beta_j$. Note that, under the additive model, the ab group means are given in terms of $1+(a-1)+(b-1) = a+b-1$ parameters, so there are $ab-a-b+1 = (a-1)(b-1)$ degrees of freedom left for estimating the error variance. (Essentially, what was the interaction sum of squares now becomes the error sum of squares.)

The procedures for generating the ANOVA table, for testing $H_0^A : \alpha_1 = \cdots = \alpha_a = 0$ and $H_0^B : \beta_1 = \cdots = \beta_b = 0$, and for constructing Tukey's simultaneous CIs and multiple comparisons, under the additive model (11.2.23), are the same as the procedures described in Section 10.4.2 for testing for no treatment (and for no block) effect in a randomized block design. For convenient reference, however, we restate here the R commands for constructing the ANOVA table and for performing Tukey's multiple comparisons. With the response variable and the levels of the two factors read into the R objects *y*, *A*, and *B*, respectively, these commands are:

```
fit=aov(y~A+B); anova(fit); TukeyHSD(fit)
```
(11.2.24)

Analyzing data under an incorrect model, however, can lead to misleading results. In particular, analyzing data under the additive model (11.2.23) when, in fact, the model is non-additive can mask significant main effects. For this reason, the experimenter should examine the data for indications of non-additivity. The interaction plot we saw in Chapter 1 is a useful graphical tool. A formal test for interaction, called *Tukey's one degree of freedom* test, is also possible. The procedure uses the squared fitted values as an added covariate in the model (11.2.23). The p-value for testing the significance of this covariate is the p-value for Tukey's one degree of freedom test for interaction.

To describe the implementation of Tukey's test, let *fit* be the object from fitting the additive model, for example, as in (11.2.24), and use the R commands

```
fitteds=(fitted(fit))**2; anova(aov(y~A+B+fitteds))
```
(11.2.25)

These commands generate an ANOVA table from fitting the additive model that includes the two factors and the covariate *fitteds*, which is the square of the fitted values. The p-value for Tukey's test for interaction can be read at the end of the row that corresponds to the covariate *fitteds* in the ANOVA table. If this p-value is less than 0.05, the assumption of additivity is not tenable. Conclude that both factors influence the response.

The following example uses simulated data to illustrate the implementation of Tukey's one degree of freedom test for interaction, and to highlight the masking of main effects when the additive model is erroneously assumed.

Example 11.2-4

Use R commands to generate data from an additive, and also from a non-additive, 3×3 design with one observation per cell and non-zero main effects.

(a) Apply Tukey's one degree of freedom test for interaction to both data sets, stating whether or not the additivity assumption is tenable.

(b) Regardless of the outcome of Tukey's tests in part (a), assume the additive model to analyze both data sets, stating whether or not the main effects for factors A and B are significantly different from zero at level of significance $\alpha = 0.05$.

Solution

The commands

```
S=expand.grid(a=c(-1, 0, 1), b=c(-1, 0, 1));
  y1=2+S$a+S$b+rnorm(9, 0, 0.5)
```

generate data according to the additive model (11.2.23) with $\alpha_1 = -1$, $\alpha_2 = 0$, $\alpha_3=1$, $\beta_1 = -1$, $\beta_2 = 0$, $\beta_3 = 1$, and normal errors with zero mean and standard deviation

0.5. The additional command *y2=2+S$a+S$b+S$a*S$b+rnorm(9, 0, 0.5)* generates data from a non-additive model having the same main effects for factors A and B and the same type of error variable. (Note that the seed for random number generation was not set. Thus, the numerical results in parts (a) and (b) are not reproducible.)

(a) To perform Tukey's test for additivity for the first data set, use

```
A=as.factor(S$a); B=as.factor(S$b); fit=aov(y1~A+B)
fitteds=(fitted(fit))**2; anova(aov(y1~A+B+fitteds))
```

These commands generate the following ANOVA table:

	Df	Sum Sq	Mean Sq	F value	Pr(>F)
A	2	4.3357	2.16786	16.9961	0.02309
B	2	4.7662	2.38310	18.6836	0.02026
fitteds	1	0.0074	0.00737	0.0578	0.82548
Residuals	3	0.3827	0.12755		

The p-value for Tukey's test for additivity, given at the end of the row that corresponds to "fitteds," is 0.82548. Since this is greater than 0.05, we conclude that the additivity assumption is tenable, in agreement with the fact that the data are generated from an additive design. Repeating the same commands for the second data set (i.e., replacing $y1$ by $y2$) yields the ANOVA table below:

	Df	Sum Sq	Mean Sq	F value	Pr(>F)
A	2	2.378	1.1892	23.493	0.0147028
B	2	53.154	26.5772	525.031	0.0001521
fitteds	1	5.012	5.0123	99.017	0.0021594
Residuals	3	0.152	0.0506		

This time, the p-value for Tukey's test for additivity is 0.0021594. Since this is less than 0.05, we conclude that the additivity assumption is not tenable, in agreement with the fact that the data are generated from a non-additive design. The interaction plots for the two data sets are displayed in Figure 11-2.

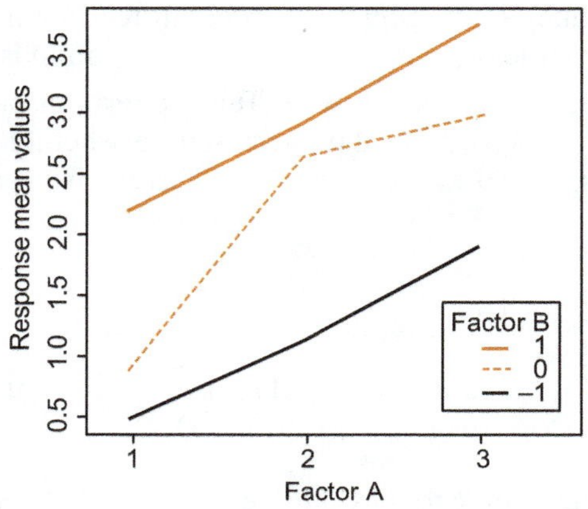

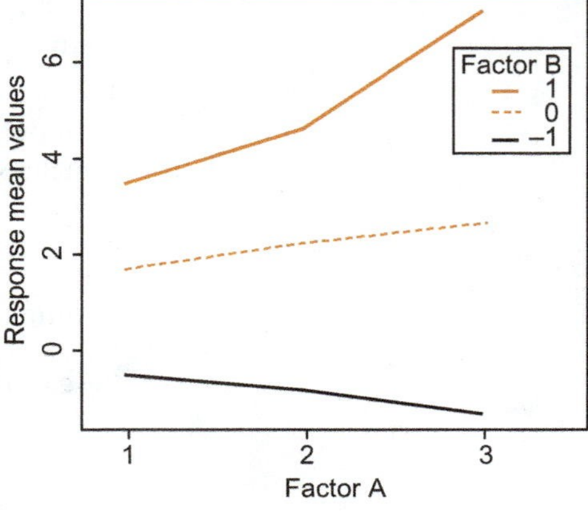

Figure 11-2 Interaction plots for the additive (left panel) and non-additive (right panel) designs of Example 11.2-4.

(b) The command *anova(aov(y1~A+B))* generates the ANOVA table below:

	Df	Sum Sq	Mean Sq	F value	Pr(>F)
A	2	4.3357	2.16786	22.233	0.006812
B	2	4.7662	2.38310	24.440	0.005722
Residuals	4	0.3900	0.09751		

The p-values for the main effects of both factors are quite small, suggesting that the main effects are non-zero in agreement with the fact that the main effects are truly non-zero. Performing the same analysis on the second data set (that is, *anova(aov(y2~A+B))*) generates the ANOVA table shown here:

	Df	Sum Sq	Mean Sq	F value	Pr(>F)
A	2	2.378	1.1892	0.9212	0.468761
B	2	53.154	26.5772	20.5859	0.007841
Residuals	4	5.164	1.2910		

The conclusion from this analysis is that only the main effects of factor B are significantly different from zero. This, of course, is not correct since both factors have non-zero main effects. The incorrect conclusion was reached because the data analysis assumes (incorrectly in this case) that the model is additive. The interaction between the factors A and B has masked the effect of factor A.

Exercises

1. An experiment studying the effect of growth hormone and sex steroid on the change in body mass fat in men resulted in the data shown in *GroHormSexSter.txt* (P, p for placebo, and T, t for treatment).[2] This is an unbalanced design with sample sizes $n_{11} = 17$, $n_{12} = 21$, $n_{21} = 17$, $n_{22} = 19$, where level 1 is placebo and factor A is growth hormone. Use R commands to complete the following parts.

(a) Test for no interaction and for no main effects for each of the two factors. Report the three test statistics and the corresponding p-values, and state whether each of the three hypotheses is rejected at level of significance 0.05.

(b) Generate the interaction plot and comment on its interpretation. Is this in agreement with the formal F test?

(c) Generate residual plots for checking the homoscedasticity and normality assumptions. What do you conclude from these plots?

(d) Conduct formal tests for the homoscedasticity and normality assumptions, and state your conclusions.

2. A cellphone's SAR (Specific Absorption Rate) is a measure of the amount of radio frequency (RF) energy absorbed by the body when using the handset. For a phone to receive FCC certification, its maximum SAR level must be 1.6 watts per kilogram (W/kg); the level is the same in Canada, while in Europe it is capped at 2 W/kg. All cellphones emit RF energy and the SAR varies by handset model. Moreover, in weak signal areas the handset generates more radiation in order to connect to the tower. Simulated data comparing the radiation emitted by five different types of cellphones, at three signal levels, are given in *CellPhoneRL.txt*. Use R commands to complete the following parts.

(a) Test for no interaction and no main effects. Report the three test statistics and the corresponding p-values, and state whether each of the three hypotheses is rejected at level of significance 0.01.

(b) Perform multiple comparisons to determine which pairs of cellphone types are significantly different at level 0.01 in terms of their main effects. Do the same for the three signal levels.

(c) Conduct formal tests for the homoscedasticity and normality assumptions, and state your conclusions.

(d) Generate the interaction plot and the residual plots for checking the homoscedasticity and normality assumptions. What do you conclude from these plots?

3. The data file *Alertness.txt* contains data from a study on the effect on alertness of two doses of a medication on male and female subjects. This is a 2 × 2 design with four

[2] Marc Blackman et al. (2002). Growth hormone and sex steroid administration in healthy aged women and men, *Journal of the American Medical Association*, 288(18): 2282–2292.

replications. The question of interest is whether changing the dose changes the average alertness of male and female subjects by the same amount.

(a) Express in words the null hypothesis that should be tested for answering the question of interest (use statistical terminology).

(b) Use either R commands or hand calculations to test the hypothesis in part (a).

4. The data file *AdhesHumTemp.txt* contains simulated data from a study conducted to investigate the effect of temperature and humidity on the force required to separate an adhesive product from a certain material. Two temperature settings (20°, 30°) and four humidity settings (20%, 40%, 60%, 80%) are considered, and three measurements are made at each temperature-humidity combination. Let μ_{ij} denote the mean separating force at temperature-humidity combination (i,j).

(a) The experimenter wants to know if the difference in the mean separating forces between the two temperature levels is the same at each humidity level. In statistical notation, of interest is whether

$$\mu_{11} - \mu_{21} = \mu_{12} - \mu_{22} = \mu_{13} - \mu_{23} = \mu_{14} - \mu_{24}$$

is false. Which is the relevant hypothesis to be tested?

(b) The experimenter wants to know if the average (over the two temperature settings) of the mean separating forces is the same for each humidity setting. In statistical notation, of interest is whether

$$\frac{\mu_{11} + \mu_{21}}{2} = \frac{\mu_{12} + \mu_{22}}{2} = \frac{\mu_{13} + \mu_{23}}{2} = \frac{\mu_{14} + \mu_{24}}{2}$$

is false. Which is the relevant hypothesis to be tested?

(c) Test for no interaction and for no main effects for each of the two factors. Report the three test statistics and the corresponding p-values, and state whether each of the three hypotheses is rejected at level of significance 0.05.

(d) Is it appropriate to conduct multiple comparisons for determining which of the humidity level main effects are different from each other? If so, perform the multiple comparisons at level of siginficance 0.01, and state your conclusions.

5. The data file *AdLocNews.txt* contains the number of inquiries regarding ads placed in a local newspaper. The ads are categorized according to the day of the week and section of the newspaper in which they appeared. Use R commands to complete the following:

(a) Construct plots to help assess the validity of the homoscedasticity and normality assumptions. What is your impression?

(b) Perform formal tests for the validity of the homoscedasticity and normality assumptions. State your conclusions.

(c) Perform multiple comparisons to determine which days are significantly different and which newspaper sections are significantly different, both at experiment-wise error rate $\alpha = 0.01$, in terms of the average number of inquiries received.

6. A large research project studied the physical properties of wood materials constructed by bonding together small flakes of wood. Three different species of trees (aspen, birch, and maple) were used, and the flakes were made in two different sizes (0.015 inches by 2 inches and 0.025 inches by 2 inches). One of the physical properties measured was the tension modulus of elasticity in the direction perpendicular to the alignment of the flakes in pounds per square inch. There are three observations per cell. With the response values and the corresponding levels of the factors "tree species" and "flake size" in the R objects *y, A, B*, respectively, the command *anova(aov(y~A*B))* produces an ANOVA table, which is shown partially filled below:

	DF	SS	MS	F	P
Species	2				0.016
Size	1	3308			
Interaction			20854		0.224
Error		147138			
Total	17	338164			

(a) Fill out the remaining entries in the ANOVA table. (*Hint.* You need to use R to fill out the missing p-value. The command is of the form $1\text{-}pf(F^B_{H_0}, \nu_1, \nu_2)$.)

(b) Let Y_{ijk} be the kth observation in factor-level combination (i,j). Write the statistical model for Y_{ijk} that was used to produce the given ANOVA table, and state the assumptions needed for the validity of the statistical analysis.

(c) Write the null and alternative hypotheses for testing whether an additive model is appropriate for this data, and test this hypothesis at level $\alpha = 0.05$.

(d) Using the p-values from the filled out ANOVA table, test each of the hypotheses H_0^A and H_0^B at level $\alpha = 0.05$. (Recall A = "tree species" and B = "flake size".)

7. When an additive model is considered for a balanced design, the decomposition of the total sum of squares in (11.2.10) reduces to SST = SSA + SSB + SSE, where SST and its degrees of freedom are still given by (11.2.9), and SSA, SSB, and their degrees of freedom are still given by (11.2.11) and (11.2.15), respectively. Use this information to construct the ANOVA table from fitting the additive model to the data in Exercise 6 and to test the hypotheses H_0^A and H_0^B at level $\alpha = 0.05$ using the additive model.

8. It is known that the life span of a particular type of root system is influenced by the amount of watering it receives and its depth. An experiment is designed to study the effect of three watering regimens ("W1," "W2," "W3"), and their possible interaction with the depth factor. The depth variable was categorized as "D1" (< 4 cm), "D2" (between 4 and 6 cm), and "D3" (> 6 cm). The life spans of five root systems were recorded for each factor-level combination. The logarithms of the life spans yield the following cell means and variances.

		Depth					
		D1		D2		D3	
		Mean	Var	Mean	Var	Mean	Var
Watering	W1	2.94	2.10	3.38	6.62	5.30	3.77
	W2	3.03	1.18	5.24	0.15	6.45	2.97
	W3	4.78	2.12	5.34	0.41	6.61	2.40

Compute the ANOVA table, and carry out the tests for no interaction and no main effects at level of significance 0.01.

9. The data file *InsectTrap.txt* contains the average number of insects trapped for three kinds of traps used in five periods.[3] Read the data into the data frame *df*, set *y=df$catch; A=as.factor(df$period); B=as.factor(df$trap)*, and use R commands to complete the following.

(a) Construct the ANOVA *F* table, and give the *p*-values for testing for no main row and no main column effects in an additive design. Are the null hypotheses rejected at level 0.05?

(b) Generate the interaction plot. Does it appear that the two factors interact? Perform Tukey's one degree of freedom test for additivity at level 0.05, and state your conclusion.

(c) As an overall conclusion, do the factors "period" and "trap" make a difference in terms of the number of insects caught?

10. Consider the cellphone radiation data of Exercise 2, but use one observation per cell. (This serves to highlight the additional power gained by using several observations per factor-level combination.) Having imported the data into the data frame df, use $y = df\$y[1:15]$, $S = df\$S[1{:}15]$, $C = df\$C[1:15]$, a one-observation-per-cell data set with the response in *y* and the levels of the two factors in *S* and *C*.

(a) Construct the ANOVA F table, and give the p-values for testing for no main row and no main column effects in an additive design. Are the null hypotheses rejected at level 0.01?

(b) Perform multiple comparisons to determine which pairs of cellphone types and/or which pairs of signal levels are significantly different at level 0.01.

11. Data from an article investigating the effect of auxin-cytokinin interaction on the organogenesis of haploid geranium callus can be found in *AuxinKinetinWeight.txt*.[4] Read the data into the data frame *Ac*, and use *attach(Ac); A=as.factor(Auxin); B=as.factor(Kinetin); y=Weight* to copy the response variable and the levels of factors "Auxin" and "Kinetin" into the R objects *y, A, B*, respectively.

(a) Construct the ANOVA *F* table, and give the *p*-values for testing for no main row and no main column effects in an additive design. Are the null hypotheses rejected at level 0.01?

(b) Perform multiple comparisons to determine which pairs of the Auxin factor levels are significantly different at level 0.01.

(c) Perform Tukey's one degree of freedom test for additivity and state your conclusion.

12. A soil scientist is considering the effect of soil pH level (factor *A*) on the breakdown of a pesticide residue. Two pH levels are considered in the study. Because pesticide residue breakdown is also affected by soil temperature (factor *B*), four different temperatures are included in the study. The eight observations are given in the following table.

	Temp *A*	Temp *B*	Temp *C*	Temp *D*
pH I	$X_{11} = 108$	$X_{12} = 103$	$X_{13} = 101$	$X_{14} = 100$
pH II	$X_{21} = 111$	$X_{22} = 104$	$X_{23} = 100$	$X_{24} = 98$

(a) Construct the ANOVA table given in Section 10.4.2. (*Hint.* Use formulas (10.4.5) and (10.4.6) for SSA and SSB, respectively, keeping in mind that, in the notation of Section 10.4.2, k is the number of row levels and n is the number of column levels. Thus, for this data set, $k = 2$ and $n = 4$.)

(b) Test each of the hypotheses H_0^A and H_0^B at level $\alpha = 0.05$.

11.3 Three-Factor Designs

This section extends the two-factor model, the concepts of main effects and interactions, and the corresponding test procedures to designs with three factors. For

[3] George Snedecor and William Cochran (1989). *Statistical Methods*, 8th ed. Ames: Iowa State University Press.

[4] M. M. El-Nil, A. C. Hildebrandt, and R. F. Evert (1976). Effect of auxin-cytokinin interaction on organogenesis in haploid callus of *Pelargonium hortorum, In Vitro*, 12(8): 602–604.

example, the model for a three-factor design involves the *three-factor interaction*, which generalizes the concept of two-factor interactions. As with two-factor designs, formulas for the decomposition of the total sum of squares, and the resulting F statistics, will be given only in the case of balanced designs, that is, when all cell sample sizes are equal.

In designs with three or more factors, it is quite common to assume that some of the effects (typically higher order interactions) are zero and to fit a *reduced model* that does not include these effects. The decomposition of the total sums of squares corresponding to the reduced model essentially involves combining the sums of squares of the omitted effects with the error sum of squares. This is similar to using the interaction sum of squares as the error sum of squares in the two-factor design with one observation per cell; see also Exercise in Section 11.2.

Designs with more than three factors involve even higher order interaction effects. However, the interpretation of higher order interaction effects, the formulas for the decomposition of the total sum of squares and of the F test statistics (both for the full and reduced models), as well as the R commands for fitting both full and reduced models, are quite similar and will not be discussed in detail.

Statistical Model for a Three-Factor Design A design where factor A has a levels, factor B has b levels, and factor C has c levels is referred to as an $a \times b \times c$ design; a $2 \times 2 \times 2$ design is also referred to as a 2^3 design. Let μ_{ijk} denote the mean value of an observation taken at factor level combination (i, j, k), that is, when the level of factor A is i, the level of factor B is j, and the level of factor C is k. The set of mean values μ_{ijk}, $i = 1, \ldots, a$, $j = 1, \ldots, b$, $k = 1, \ldots, c$, can be decomposed as

Decomposition of Means in a Three-Factor Design

$$\mu_{ijk} = \mu + \alpha_i + \beta_j + \gamma_k + (\alpha\beta)_{ij} + (\alpha\gamma)_{ik} + (\beta\gamma)_{jk} + (\alpha\beta\gamma)_{ijk}. \tag{11.3.1}$$

The decomposition (11.3.1) builds on the decomposition (11.2.1) for two factor designs, in the sense that μ, α_i, β_j and $(\alpha\beta)_{ij}$ decompose the average $\overline{\mu}_{ij.}$ of μ_{ijk} over k:

$$\overline{\mu}_{ij.} = \mu + \alpha_i + \beta_j + (\alpha\beta)_{ij}.$$

Thus they are given by

$$\mu = \overline{\mu}_{...}, \quad \alpha_i = \overline{\mu}_{i..} - \mu, \quad \beta_j = \overline{\mu}_{.j.} - \mu, \quad (\alpha\beta)_{ij} = \overline{\mu}_{ij.} - \overline{\mu}_{i..} - \overline{\mu}_{.j.} + \mu. \tag{11.3.2}$$

Similarly, μ, α_i, γ_k, and $(\alpha\gamma)_{ik}$ decompose $\overline{\mu}_{i.k}$, and so forth. It follows that the main effects and interaction of the decomposition (11.3.1) satisfy the zero-sum conditions given in (11.2.2). The only really new component in the decomposition (11.3.1) is the three-factor interaction term $(\alpha\beta\gamma)_{ijk}$, which is defined from (11.3.1) by subtraction. It can be shown that three-factor interaction terms also satisfy zero-sum conditions:

$$\sum_{i=1}^{a} (\alpha\beta\gamma)_{ijk} = \sum_{j=1}^{b} (\alpha\beta\gamma)_{ijk} = \sum_{k=1}^{c} (\alpha\beta\gamma)_{ijk} = 0. \tag{11.3.3}$$

The interpretation of the three-factor interaction terms becomes clear upon examining the interaction term in a two-factor decomposition of μ_{ijk} obtained by holding the level of one of the factors fixed. For example, the interaction term in the decomposition

$$\mu_{ijk} = \mu^k + \alpha_i^k + \beta_j^k + (\alpha\beta)_{ij}^k \tag{11.3.4}$$

of μ_{ijk}, $i = 1, \ldots, a$, $j = 1, \ldots, b$, which is obtained by holding the level of factor C fixed at k, is given by (see Exercise 6)

$$(\alpha\beta)_{ij}^{k} = (\alpha\beta)_{ij} + (\alpha\beta\gamma)_{ijk}. \tag{11.3.5}$$

It follows that the three-factor interaction terms capture the change in two-factor interactions as the level of the remaining factor changes. Hence, when the null hypothesis of no third order interaction effects is rejected, it may be concluded that second order interaction effects exist (i.e., the $(\alpha\beta)_{ij}^{k}$ are not zero) even if the average second order interaction effects (i.e., $(\alpha\beta)_{ij}$) are not significantly different from zero.

From each factor-level combination (i, j, k) we observe a simple random sample of size n: $X_{ijk\ell}$, $\ell = 1, \ldots n$. The mean-plus-error form of the statistical model for the data is

$$X_{ijk\ell} = \mu_{ijk} + \epsilon_{ijk\ell}. \tag{11.3.6}$$

Replacing the means μ_{ijk} by their decomposition (11.3.1), the mean-plus-error form of the model becomes the treatment-effects form of the model. As always, the cell means μ_{ijk} are estimated by the corresponding cell averages, that is,

$$\widehat{\mu}_{ijk} = \overline{X}_{ijk} = \frac{1}{n}\sum_{\ell=1}^{n} X_{ijk\ell}.$$

(Note that, for simplicity, we use $\overline{X}_{ijk}$ instead of $\overline{X}_{ijk\cdot}$.) The parameters in the treatment-effects form of the model are estimated by replacing μ_{ijk} by $\overline{X}_{ijk}$ in expressions like (11.3.2). Thus, $\widehat{\mu} = \overline{X}_{\cdots}$, and

Estimators of Main Effects	$\widehat{\alpha}_i = \overline{X}_{i\cdot\cdot} - \overline{X}_{\cdots}, \quad \widehat{\beta}_j = \overline{X}_{\cdot j\cdot} - \overline{X}_{\cdots}, \quad \widehat{\gamma}_k = \overline{X}_{\cdot\cdot k} - \overline{X}_{\cdots}$	
Estimators of Two-Factor Interactions	$\widehat{(\alpha\beta)}_{ij} = \overline{X}_{ij\cdot} - \overline{X}_{i\cdot\cdot} - \overline{X}_{\cdot j\cdot} + \overline{X}_{\cdots}, \quad \widehat{(\alpha\gamma)}_{ik} = \overline{X}_{i\cdot k} - \overline{X}_{i\cdot\cdot} - \overline{X}_{\cdot\cdot k} + \overline{X}_{\cdots},$ $\widehat{(\beta\gamma)}_{jk} = \overline{X}_{\cdot jk} - \overline{X}_{\cdot j\cdot} - \overline{X}_{\cdot\cdot k} + \overline{X}_{\cdots}$	**(11.3.7)**
Estimators of Three-Factor Interactions	$\widehat{(\alpha\beta\gamma)}_{ijk} = \overline{X}_{ijk} - \overline{X}_{\cdots} - \widehat{\alpha}_i - \widehat{\beta}_j - \widehat{\gamma}_k - \widehat{(\alpha\beta)}_{ij} - \widehat{(\alpha\gamma)}_{ik} - \widehat{(\beta\gamma)}_{jk}$	

The relevant sums of squares and their degrees of freedom are

Sums of Squares	Degrees of Freedom	
$\text{SST} = \sum_{i=1}^{a}\sum_{j=1}^{b}\sum_{k=1}^{c}\sum_{\ell=1}^{n}(X_{ijk\ell} - \overline{X}_{\cdots})^2$	$abcn - 1$	
$\text{SSA} = bcn\sum_{i=1}^{a}\widehat{\alpha}_i^2$	$a - 1$	
$\text{SSAB} = cn\sum_{i=1}^{a}\sum_{j=1}^{b}\widehat{(\alpha\beta)}_{ij}^2$	$(a-1)(b-1)$	**(11.3.8)**
$\text{SSABC} = n\sum_{i=1}^{a}\sum_{j=1}^{b}\sum_{k=1}^{c}\widehat{(\alpha\beta\gamma)}_{ijk}^2$	$(a-1)(b-1)(c-1)$	
$\text{SSE} = \sum_{i=1}^{a}\sum_{j=1}^{b}\sum_{k=1}^{c}\sum_{\ell=1}^{n}(X_{ijk\ell} - \overline{X}_{ijk})^2$	$abc(n-1)$	

with the sums of squares for the other main effects and two-factor interactions defined symmetrically. In total there are eight sums of squares for main effects and interactions, and their sum equals SST. Similarly, the degrees of freedom for SST is the sum of the degrees of freedom for all other sums of squares.

As usual, the mean squares are defined as the sums of squares divided by their degrees of freedom. For example, MSA = SSA/$(a-1)$, and so forth. The ANOVA F test procedures for the hypotheses of no main effects and no interactions are

Hypothesis	F Test Statistic	Rejection Rule at Level α
H_0^{ABC}: all $(\alpha\beta\gamma)_{ijk} = 0$	$F_{H_0}^{ABC} = \frac{\text{MSABC}}{\text{MSE}}$	$F_{H_0}^{ABC} > F_{(a-1)(b-1)(c-1),abc(n-1),\alpha}$
H_0^{AB}: all $(\alpha\beta)_{ij} = 0$	$F_{H_0}^{AB} = \frac{\text{MSAB}}{\text{MSE}}$	$F_{H_0}^{AB} > F_{(a-1)(b-1),abc(n-1),\alpha}$
H_0^{A}: all $\alpha_i = 0$	$F_{H_0}^{A} = \frac{\text{MSA}}{\text{MSE}}$	$F_{H_0}^{A} > F_{a-1,abc(n-1),\alpha}$

and similarly for testing for the other main effects and two-factor interactions. These test procedures are valid under the assumptions of homoscedasticity (i.e., same population variance for all abc factor-level combinations) and normality.

Hand calculations will only be demonstrated in 2^3 designs. For such designs, there is a specialized method for efficient hand calculations, which is described in the following example.

Example 11.3-1

Hand calculations in a 2^3 design. Surface roughness is of interest in many manufacturing processes. A paper[5] considers the effect of several factors including *tip radius* (TR), *surface autocorrelation length* (SAL), and *height distribution* (HD) on surface roughness, on the nanometer scale, by the atomic force microscope (AFM). The cell means, based on two simulated replications (i.e., two observations per cell), are given in Table 11-1. Use this information to compute the main effects and interactions and the corresponding sums of squares.

Solution

Let factors A, B, and C be TR, SAL, and HD, respectively. Let $\widehat{\mu}$ denote the average of all eight observations in the table. According to the formulas in (11.3.7), $\widehat{\alpha}_1$ is the average of the four observations in the first line of the table minus $\widehat{\mu}$, that is,

Table 11-1 Cell means in the 2^3 design of Example 11.3-1

		HD			
		1		2	
		SAL		SAL	
		1	2	1	2
TR	1	$\bar{x}_{111} = 6.0$	$\bar{x}_{121} = 8.55$	$\bar{x}_{112} = 8.0$	$\bar{x}_{122} = 7.0$
	2	$\bar{x}_{211} = 8.9$	$\bar{x}_{221} = 9.50$	$\bar{x}_{212} = 11.5$	$\bar{x}_{222} = 13.5$

[5] Y. Chen and W. Huang (2004). Numerical simulation of the geometrical factors affecting surface roughness measurements by AFM, *Measurement Science and Technology*, 15(10): 2005–2010.

$$\widehat{\alpha}_1 = \frac{\overline{x}_{111} + \overline{x}_{121} + \overline{x}_{112} + \overline{x}_{122}}{4} - \frac{\overline{x}_{111} + \overline{x}_{121} + \overline{x}_{112} + \overline{x}_{122} + \overline{x}_{211} + \overline{x}_{221} + \overline{x}_{212} + \overline{x}_{222}}{8}$$
$$= \frac{\overline{x}_{111} + \overline{x}_{121} + \overline{x}_{112} + \overline{x}_{122} - \overline{x}_{211} - \overline{x}_{221} - \overline{x}_{212} - \overline{x}_{222}}{8} = -1.73125. \quad \textbf{(11.3.9)}$$

Similarly, $\widehat{\beta}_1$ is the average of the four observations in the two columns where SAL is 1 (the first and third columns) minus $\widehat{\mu}$, and $\widehat{\gamma}_1$ is the average of the four observations where HD is 1 (the first two columns) minus $\widehat{\mu}$. Working as above, we arrive at the expressions

$$\widehat{\beta}_1 = \frac{\overline{x}_{111} + \overline{x}_{211} + \overline{x}_{112} + \overline{x}_{212} - \overline{x}_{121} - \overline{x}_{221} - \overline{x}_{122} - \overline{x}_{222}}{8} = -0.51875$$
$$\widehat{\gamma}_1 = \frac{\overline{x}_{111} + \overline{x}_{211} + \overline{x}_{121} + \overline{x}_{221} - \overline{x}_{112} - \overline{x}_{212} - \overline{x}_{122} - \overline{x}_{222}}{8} = -0.88125.$$

Because of the zero-sum conditions, $\widehat{\alpha}_2 = -\widehat{\alpha}_1$, $\widehat{\beta}_2 = -\widehat{\beta}_1$, and $\widehat{\gamma}_2 = -\widehat{\gamma}_1$, they do not need to be calculated. For the same reason, only one from each of the three types of two-factor interactions needs to be calculated, and only one three-factor interaction.

Before computing estimators of the remaining effects, we describe an organized way for calculating these effects. This method consists of creating a column of cell means and a column of pluses and minuses for each effect, as shown in Table 11-2. The numerator of each effect estimator is formed by combining the cell means according to the signs of the corresponding column. Since the denominator is always 8, the numerators suffice to determine the estimators. For example, it is easily checked that the numerator in the expression for $\widehat{\alpha}_1$ given in (11.3.9) is

$$\overline{x}_{111} - \overline{x}_{211} + \overline{x}_{121} - \overline{x}_{221} + \overline{x}_{112} - \overline{x}_{212} + \overline{x}_{122} - \overline{x}_{222}.$$

The column of cell means in Table 11-2 is formed by stacking the columns in the cell-means of Table 11-1. Except for the μ column, the columns for each effect consist of four + signs and four − signs. Thus, in the terminology introduced in Section 1.8, the estimators of all main effects and interactions are *contrasts*. (Because the estimator of the overall mean μ, represented by the μ column in Table 11-2, is not a contrast, it is often not considered to be one of the *effects*.)

Table 11-2 Signs for estimating the effects in a 2^3 design

Factor-Level Combination	Cell Means	μ	α_1	β_1	γ_1	$(\alpha\beta)_{11}$	$(\alpha\gamma)_{11}$	$(\beta\gamma)_{11}$	$(\alpha\beta\gamma)_{111}$
1	$\overline{x}_{111}$	+	+	+	+	+	+	+	+
a	$\overline{x}_{211}$	+	−	+	+	−	−	+	−
b	$\overline{x}_{121}$	+	+	−	+	−	+	−	−
ab	$\overline{x}_{221}$	+	−	−	+	+	−	−	+
c	$\overline{x}_{112}$	+	+	+	−	+	−	−	−
ac	$\overline{x}_{212}$	+	−	+	−	−	+	−	+
bc	$\overline{x}_{122}$	+	+	−	−	−	−	+	+
abc	$\overline{x}_{222}$	+	−	−	−	+	+	+	−

Different columns have very distinct patterns of +'s and −'s. In the α_1 column, +'s and −'s alternate; in the β_1 column, two +'s; alternate with two −'s; and in the γ_1 column, four +'s are followed by four −'s. The interaction columns are formed by the products (in the row-wise sense) of the respective main effects columns. For example, the $(\alpha\beta)_{11}$ column is the product of the α_1 column times the β_1 column, and so forth.

The first column of Table 11-2 contains codes for each cell (factor-level combination). According to this code (which is commonly used in all 2^r designs), level 1 is the default level of a factor. The letters a, b, and c, respectively, are used to denote that factors A, B, and C are at level 2. Thus, the first cell, which corresponds to level 1 for all three factors, is the default cell and is coded by 1. Thus, the letter a in the second row of the table denotes factor-level combination (2, 1, 1), the letters ab in the fourth row of the table correspond to cell (2, 2, 1), and so forth. It is worth noting that the codes in the last four rows are formed by adding c to the codes of the first four rows.

Using this method, the remaining effects are computed as follows (note that 6.0, 8.9, 8.55, 9.5, 8.0, 11.5, 7.0, 13.5 are the cell means in the order shown in the second column of Table 11-2):

$$\widehat{(\alpha\beta)}_{11} = \frac{6.0 - 8.9 - 8.55 + 9.5 + 8.0 - 11.5 - 7.0 + 13.5}{8} = 0.13125$$

$$\widehat{(\alpha\gamma)}_{11} = \frac{6.0 - 8.9 + 8.55 - 9.5 - 8.0 + 11.5 - 7.0 + 13.5}{8} = 0.76875$$

$$\widehat{(\beta\gamma)}_{11} = \frac{6.0 + 8.9 - 8.55 - 9.5 - 8.0 - 11.5 + 7.0 + 13.5}{8} = -0.26875$$

$$\widehat{(\alpha\beta\gamma)}_{111} = \frac{6.0 - 8.9 - 8.55 + 9.5 - 8.0 + 11.5 + 7.0 - 13.5}{8} = -0.61875.$$

Finally, the sums of squares for the main effects and interactions can be obtained from the formulas (11.3.8). Since $a = b = c = 2$, the zero-sum conditions imply that each sum of squares equals 16 times the square of the corresponding effect computed above. For example (note also that $n = 2$), $\text{SSA} = 2 \cdot 2 \cdot 2 \cdot (\widehat{\alpha}_1^2 + \widehat{\alpha}_2^2) = 16\widehat{\alpha}_1^2$. Thus, rounding the numerical value of each effect to three decimal places,

$$\text{SSA} = 16(1.731^2) = 47.956, \quad \text{SSB} = 16(0.519^2) = 4.306,$$

$$\text{SSC} = 16(0.881^2) = 12.426$$

$$\text{SSAB} = 16(0.131^2) = 0.276, \quad \text{SSAC} = 16(0.769^2) = 9.456$$

$$\text{SSBC} = 16(0.269^2) = 1.156, \quad \text{SSABC} = 16(0.619^2) = 6.126.$$

It should be noted that the +'s and −'s in Table 11-2 can (and are) also used to represent the factor levels. A "+" in the α_1 column represents level 1 for factor A and a "−" represents level 2. Similarly, the +'s and −'s in the β_1 column represent levels 1 and 2, respectively, for factor B, and those in the γ_1 column represent the levels of factor C. The sequence of three +'s and/or −'s from the α_1, β_1, and γ_1 columns reveals the factor-level combination in each row.

With the response variable and the levels of the three factors read into the R objects y, A, B, and C, respectively, the R commands for fitting different versions of the three-factor model are:

R Commands for Fitting the Full and Reduced Models

```
out=aov(y~A*B*C); anova(out) # for fitting the full model
out=aov(y~A+B+C); anova(out) # for fitting the additive
  model
out=aov(y~A*B+A*C+B*C); anova(out) # for fitting the model
  without ABC interactions
```

To clarify the syntax of the R commands, we note that a longer version of the command *anova(aov(y ~ A*B*C))* is *anova(aov(y~A+B+C+A*B+A*C+B*C+A*B*C))*. Thus, *anova(aov(y~A*C+B*C))* fits the model without AB and ABC interactions, while *anova(aov(y~A+B*C))* fits the model without AB, AC, and ABC interactions.

Example 11.3-2

A paper reports on a study sponsored by CIFOR (Center for International Forestry Research) to evaluate the effectiveness of monitoring methods related to water and soil management.[6] Part of the study considered soil runoff data from two catchment areas (areas number 37 and 92) using runoff plots classified as "undisturbed/control" and "harvested." The runoff volume was calculated at each rainfall event, with the amount of rainfall serving as an additional factor at three levels (3.5–10 mm, 10–20 mm, and > 20 mm). The data, consisting of four measurements per factor-level combination, is in *SoilRunoff3w.txt*. Use R commands to complete the following parts using level of significance of $\alpha = 0.05$.

(a) Construct the ANOVA table corresponding to fitting the full model and report which of the null hypotheses are significant.

(b) Construct the ANOVA table corresponding to fitting the additive model and report which of the null hypotheses are significant.

(c) Construct the ANOVA table corresponding to fitting the model with no three-factor interactions, and report which of the null hypotheses are significant.

(d) Using the residuals from fitting the full model, check the homoscedasticity and normality assumptions.

Solution

(a) With the data read into the R data frame *Data*, the commands *attach(Data); out=aov(y~Rain*Log*Catch); anova(out)* produce the ANOVA table below:

	Df	Sum Sq	Mean Sq	F value	Pr(>F)
Rain	2	0.1409	0.0704	13.396	4.48e-05
Log	1	0.0630	0.0630	11.984	0.001
Catch	1	0.0018	0.0018	0.349	0.558
Rain:Log	2	0.0799	0.0400	7.603	0.002
Rain:Catch	2	0.0088	0.0044	0.834	0.442
Log:Catch	1	0.0067	0.0067	1.265	0.268
Rain:Log:Catch	2	0.0032	0.0016	0.301	0.742
Residuals	36	0.1893	0.0053		

[6] Herlina Hartanto et al. (17 July 2003). Factors affecting runoff and soil erosion: Plot-level soil loss monitoring for assessing sustainability of forest management, *Forest Ecology and Management*, 180(13): 361–374. The data used are based on the information provided in Tables 2 and 3.

It follows that the main effects of the factors "Rain" and "Log" (for "logging") are significant at level 0.05, and so are their interaction effects. All other effects are not significant; thus, the "Catch" factor has no impact on the average (over the other factors) mean runoff volume, does not interact with the other two factors, and has no impact on the interaction of the factors "Rain" and "Log" (i.e., no significant three-factor interaction).

(b) The command *anova(aov(y~Rain+Log+Catch))* produces the following ANOVA table:

	Df	Sum Sq	Mean Sq	F value	Pr(>F)
Rain	2	0.1409	0.0704	10.5224	0.0002
Log	1	0.0630	0.0630	9.4132	0.004
Catch	1	0.0018	0.0018	0.2746	0.603
Residuals	43	0.2878	0.0067		

It follows that the main effects of the factors "Rain" and "Log" are significant at level 0.05, but the main effects of the factor "Catch" are not significantly different from zero. It is worth noting that the error, or residual, degrees of freedom in this table is the sum of the error degrees of freedom plus the degrees of freedom of all two- and three-factor interactions in the ANOVA table of part (a), that is, $43 = 2 + 2 + 1 + 2 + 36$. Similarly for the sums of squares: $0.2878 = 0.0799 + 0.0088 + 0.0067 + 0.0032 + 0.1893$, up to rounding.

(c) The commands *out1=aov(y~Rain*Log+Rain*Catch+Log*Catch); anova (out1)* produce an ANOVA table, shown below, according to which the main effects of the factors "Rain" and "Log" are significant at level 0.05, and so are the interaction effects of these two factors. All other effects are not significant, implying that the "Catch" factor has no impact on the mean runoff volume.

	Df	Sum Sq	Mean Sq	F value	Pr(>F)
Rain	2	0.1409	0.0704	13.9072	2.94e-05
Log	1	0.0630	0.0630	12.4412	0.001
Catch	1	0.0018	0.0018	0.3629	0.550
Rain:Log	2	0.0799	0.0400	7.8933	0.001
Rain:Catch	2	0.0088	0.0044	0.8663	0.429
Log:Catch	1	0.0067	0.0067	1.3133	0.259
Residuals	38	0.1924	0.0051		

Again, we point out that the error, or residual, degrees of freedom in this table is the sum of the error degrees of freedom plus those of the three-factor interactions in the ANOVA table of part (a), that is, $38 = 2 + 36$. Similarly for the sums of squares: $0.1924 = 0.0032 + 0.1893$, up to rounding.

(d) To test for homoscedasticity, use *anova(aov(resid(out)**2~Rain*Log*Catch))*. The resulting output suggests that only the main rain effect is significant (p-value of 0.026). As mentioned in Example 11.2-3, a weighted analysis (see Sections 12.4.1, 12.4.2) is recommended. Heteroscedasticity affects the validity of the normality test but, for illustration purposes, the command *shapiro.test(resid(out))* yields a p-value of 0.032. The commands *plot(out, which=1)* and *plot(out, which=2)* produce plots (not shown here) that add insight into the nature of violations of these assumptions. ■

When the assumptions underlying the validity of the ANOVA F tests appear to be violated, as in Example 11.3-2, a Q-Q plot of the estimated main effects and interactions can add credibility to the significant outcomes of the F tests. Because of the zero-sum constraints the effects satisfy, only a linearly independent subset of them is plotted. The rationale behind this plot is that if an effect is zero, the estimated effect has expected value zero, while estimated effects corresponding to non-zero effects have either negative or positive expected value. Thus, estimated effects that appear as outliers, that is, either underneath the line on the left side of the plot or above the line on the right side of the plot, most likely correspond to non-zero effects. With the R object *out* containing the results of fitting the full three-factor model, the R command *eff=effects(out)* generates a length N vector (where $N = abcn$ is the total number of observations) whose first element is the overall mean $\widehat{\mu}$, the next seven elements are the main effects and interactions (multiplied by $\sqrt{N}$), and the remaining $abc(n-1)$ elements are uncorrelated single-degree-of-freedom values which, in technical terms, span the residual space. However, the effectiveness of the Q-Q plot in helping to discern significant effects can be enhanced by including all values of *eff* except for the first (i.e., except for the $\widehat{\mu}$ value). Thus, it is recommended that the Q-Q plot produced by the following R commands be examined:

```
eff=effects(out); qqnorm(eff[-1]); qqline(eff[-1], col="red")
```

As an example, Figure 11-3 was generated by the above command with *out* containing the results of fitting the full three-factor model to the data of Example 11.3-2. The outlier on the left side of Figure 11-3, with value of −0.361, corresponds to one of the main effects of the "Rain" factor; the two smaller outliers on the right side of the plot, with values of 0.251 and 0.279, correspond to the main effect of the "Log" factor and one of the "Rain*Log" interactions. This is in agreement with the formal

Figure 11-3 Q-Q plot of the estimated main effects and interactions of Example 11.3-2.

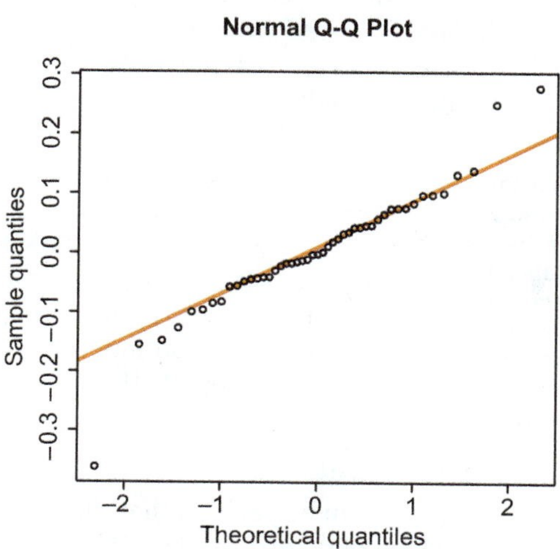

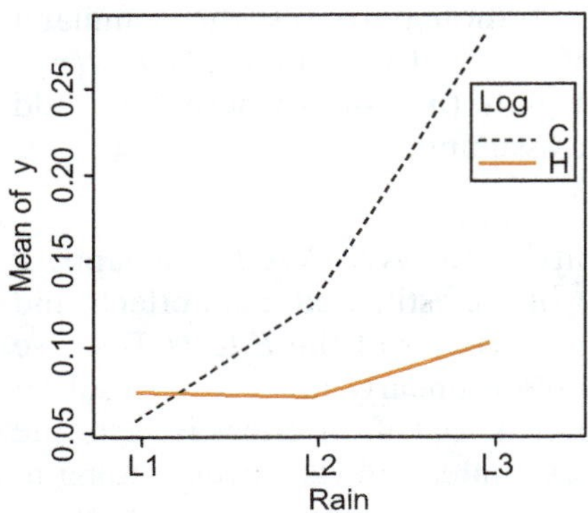

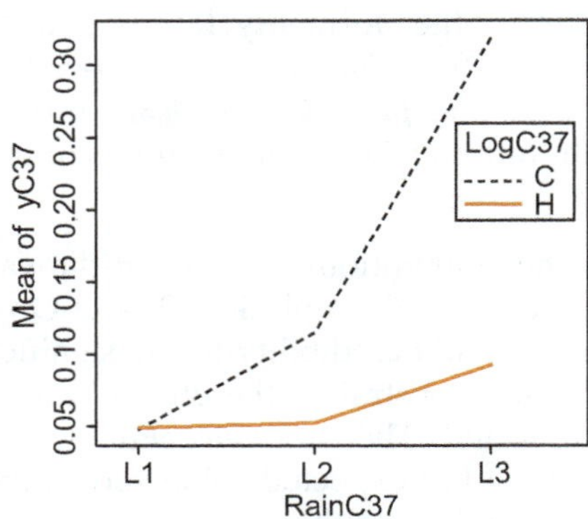

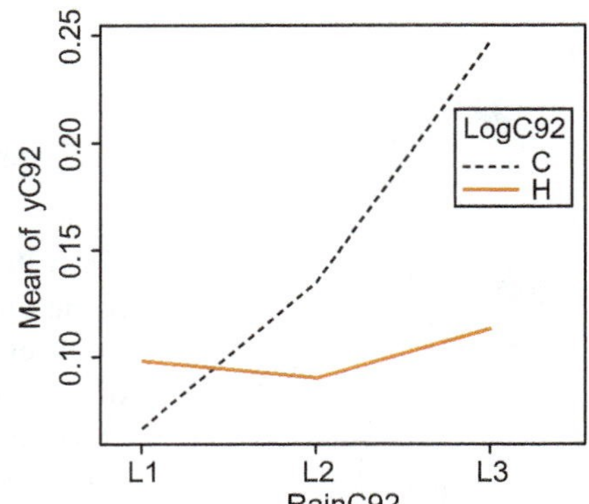

Figure 11-4 "Rain*Log" interaction plots for Example 11.3-2: Average interaction over the levels of "Catch" (left panel), for catchment 37 (middle panel), and for catchment 92 (right panel).

analysis, suggesting that the main effects of the factors "Rain" and "Log," as well as their interactions, are indeed significant.

Figure 11-4 shows three "Rain*Log" interaction plots. The first uses all data points but ignores the factor "Catch." The other two are "Rain*Log" interaction plots within each level of the "Catch" factor. Since the three-factor interactions are not significantly different from zero, the similarity of the three plots illustrates the given interpretation of the three-factor interaction (based on relation (11.3.4)), namely, that it captures the change in two-factor interactions as the level of the third factor changes. The middle plot in Figure 11-4 was produced by the R command

```
interaction.plot(Rain[Catch=="C37"],
  Log[Catch=="C37"],  y[Catch=="C37"], col=c(2,3))
```

As pointed out in Exercise 7 in Section 11.2 and highlighted in parts (b) and (c) of Example 11.3-2, when a reduced model (i.e., when some of the effects are omitted) is fitted, the sums of squares of the remaining effects in the reduced model remain the same as when fitting the full model, while the error sums of squares in the reduced model equals the sums of squares corresponding to all omitted effects plus the full model error sum. This fact, which is true *only for balanced designs*, allows us to form the ANOVA tables for reduced models from that of the full-model ANOVA table. This is important enough to be highlighted:

Rules for Forming ANOVA Tables for Reduced Models from the Full-Model ANOVA Table in *Balanced* Designs

(a) The sums of squares of the effects in the reduced model, and their degrees of freedom, remain the same as when fitting the full model.

(b) The reduced-model SSE equals the full-model SSE plus the sums of squares of all omitted effects. Similarly, the reduced-model error degrees of freedom equals the full-model error degrees of freedom plus the degrees of freedom of all omitted effects.

It can be verified that the ANOVA tables in parts (b) and (c) of Example 11.3-2, which were obtained by R commands, could also have been derived by applying the above rules to the ANOVA table in part (a).

Exercises

1. A number of studies have considered the effectiveness of membrane filtration in removing or rejecting organic micropollutants. Membranes act as nano-engineered sieves and thus the rejection rate depends on the molecular weight of the pollutant or solute. In addition, the rejection rate depends on the membrane–solute electrostatic and hydrophobic interactions and several other factors.[7,8] The file *H2Ofiltration3w.txt* contains simulated data of rejection rates for uncharged solutes in a $2 \times 3 \times 2$ design with five replications. Factor A is "Molecule Size" (MS) at two levels, above or below the membrane's molecular weight cut off (MWCO). Factor B is "Solute Hydrophobicity" (SH) at three levels, $\log K_{ow} < 1$, $1 < \log K_{ow} < 3$, and $\log K_{ow} > 3$, where $\log K_{ow}$ is the logarithm of the octanol-water partition coefficient. Factor C is "Membrane Hydrophobicity" (MH) at two levels, $\theta < 45^o$ and $\theta \geq 45^o$, where θ is the contact angle with a water droplet on the membrane surface. Use R commands to complete the following.

(a) Write the full model, construct the corresponding ANOVA table, and report which of the main effects and interactions are significant at level 0.05.

(b) Write the model without the three-factor interaction, construct the corresponding ANOVA table, and report which of the main effects and interactions are significant at level 0.05.

(c) Using residuals from fitting the full model, conduct the homoscedasticity and normality tests and state your conclusion.

(d) The square root arcsine transformation is often used to transform the rate or proportion data so the normality and homoscedasticity assumptions are approximately satisfied. With the response variable in the R object *y*, use the command *yt=sqrt(asin(y))* to transform the data. Fit the full model to the transformed data and test for homoscedasticity and normality. State your conclusion.

2. In a study of the effectiveness of three types of home insulation methods in cold weather, an experiment considered the energy consumption, in kilowatt hours (kWh) per 1000 square feet, at two outside temperature levels (15°–$20^o F$ and 25°–$30^o F$), using two different thermostat settings (68°–$69^o F$ and 70°–$71^o F$). The three replications at each factor-level combination were obtained from homes using similar types of electric heat pumps. The simulated data are available in *ElectrCons3w.txt*. Use R commands to complete the following.

(a) Write the full model, construct the corresponding ANOVA table, and report which of the main effects and interactions are significant at level 0.05.

(b) Using residuals from fitting the full model, conduct the homoscedasticity and normality tests and state your conclusion.

(c) Construct three interaction plots for the factors "insulation type" and "outside temperature" similar to those in Figure 11-4 (i.e., an overall plot and a plot for each of the two levels of the factor "thermostat setting"). Comment on the agreement of these plots with the test results in part (a).

(d) Write the model without the $(\alpha\beta)$, $(\alpha\gamma)$, and $(\alpha\beta\gamma)$ interactions. Use the ANOVA table from part (a) to construct the ANOVA table corresponding to the model without the aforementioned interactions and three-factor interaction, and to test for the significance of the remaining effects at level 0.05.

3. Line yield and defect density are very important variables in the semiconductor industry as they are directly correlated with production cost and quality. A small pilot study considered the effect of three factors, "PMOS transistor threshold voltage," "polysilicon sheet resistance," and "N-sheet contact chain resistance," on the defect density in batches of wafers (determined by synchrotron X-ray topographs and etch pit micrographs). The design considered two levels for each factor with one replication. Using the system of coding the factor-level combinations explained in Example 11.3-1, the data are

Treatment Code							
1	*a*	*b*	*ab*	*c*	*ac*	*bc*	*abc*
11	10	8	16	12	15	18	19

(a) Which are the observations denoted by x_{221}, x_{112}, and x_{122}?

(b) Construct a table of signs similar to Table 11-2.

(c) Use the table of signs constructed in part (b) to estimate the main effects and interactions.

(d) Compute the sum of squares for each effect.

(e) Assuming all interactions are zero, test for the significance of the main effects at level 0.05.

(f) Enter the estimates of the seven main and interaction effects in the R object *eff* and use an R command to produce a probability plot of the effects. Comment on the assumption that the three interaction effects are zero.

---

[7] C. Bellonal et al. (2004). Factors affecting the rejection of organic solutes during NF/RO treatment: A literature review, *Water Research*, 38(12): 2795–2809.

[8] Arne Roel Dirk Verliefde (2008). *Rejection of organic micropollutants by high pressure membranes (NF/RO)*. Water Management Academic Press.

4. The data shown in Table 11-3 are from an experiment investigating the compressive strength of two different mixtures at two temperatures and two aging periods.

Table 11-3 Observations in the 2^3 design of Exercise 4

		Mixture			
		1		2	
		Age		Age	
		1	2	1	2
Temperature	1	459, 401	458, 385	475, 524	473, 408
	2	468, 532	466, 543	479, 542	479, 544

(a) Compute a table of signs for estimating the main and interaction effects, similar to Table 11-2. (Include the first column of treatment codes.)

(b) Use the table of signs constructed in part (a) to estimate the main effects and interactions.

(c) Compute the sum of squares for each effect.

(d) You are given that MSE $= 2096.1$. Test for the significance of the main and interaction effects at level 0.05.

5. Assume the 2^3 design of Example 11.3-1 has only one replication, that is, $n = 1$, with the observations as given in Table 11-1. Use hand calculations to complete the following.

(a) The estimated main and interaction effects computed in Example 11.3-1 remain the same. True or false? If you answer "false," recompute all effects.

(b) The sums of squares for the main and interaction effects computed in Example 11.3-1 remain the same. True or false? If you answer "false," recompute the sums of squares for all effects.

(c) Assume that the three-factor interactions are zero and conduct tests for the significance of all other effects. Report the values of the F test statistics, the p-values, and whether or not each hypothesis is rejected at level 0.05.

6. Let μ, α_i, $(\alpha\beta)_{ij}$, $(\alpha\beta\gamma)_{ijk}$, etc., be the terms in the decomposition (11.3.1), and let

$$\mu^k = \overline{\mu}_{\cdot\cdot k}, \quad \alpha_i^k = \overline{\mu}_{i\cdot k} - \mu^k,$$

$$\beta_j^k = \overline{\mu}_{\cdot jk} - \mu^k, \quad (\alpha\beta)_{ij}^k = \mu_{ijk} - \mu^k - \alpha_i^k - \beta_j^k,$$

be the terms in the decomposition (11.3.4) when the level of factor C is held fixed at k. Verify the following relations between the terms of the two decompositions.

(a) $\mu^k = \gamma_k + \mu$.

(b) $\alpha_i^k = \alpha_i + (\alpha\gamma)_{ik}$.

(c) $\beta_j^k = \beta_j + (\beta\gamma)_{jk}$.

(d) Relation (11.3.5), that is, $(\alpha\beta)_{ij}^k = (\alpha\beta)_{ij} + (\alpha\beta\gamma)_{ijk}$.

11.4 2^r Factorial Experiments

When many factors can potentially influence a response, and each *run* of the experiment (i.e., taking an observation/measurement under each experimental condition) is time consuming, pilot/screening experiments using 2^r designs are very common for two main reasons. First, such designs use the smallest number of runs for studying r factors in a complete factorial design. Second, a 2^r design is an efficient tool for identifying the relevant factors to be further studied, especially if the two levels for each of the p factors are carefully selected.

As the number of factors increases, so does the number of effects that must be estimated. In particular, a 2^r design has $2^r - 1$ effects which, for $r > 3$, include interactions, of order higher than three. For example, a 2^4 design has 4 main effects, 6 two-factor interactions, 4 three-factor interactions, and 1 four-factor interaction. However, the computation of the effects and the statistical analysis for 2^r designs with $r > 3$ parallels that for three factors. Thus, this section focuses on some additional issues that arise in the analysis of 2^r designs with $r > 3$. These issues arise because of the need to incorporate as many factors as possible, and, due to limitations placed by cost and time considerations, 2^r designs with $r > 3$ are often *unreplicated*. (The term *unreplicated* means there is only $n = 1$ observation per experimental condition or factor-level combination.)

The first such issue has to do with the fact that MSE cannot be formed (the error variance cannot be estimated) in unreplicated designs. As we did with unreplicated two- and three-factor designs, this issue is dealt with by assuming that higher order

interactions are zero. The *sparsity of effects principle*, according to which multifactor systems are usually driven by main effects and low-order interactions, underlies this approach. In addition, as explained in the previous section, a Q-Q plot of the effects can help confirm the appropriateness of applying this principle in each particular case. However, even unreplicated multifactor factorial experiments require considerable effort. Another issue has to do with finding efficient ways of conducting either the full factorial design, or a fraction of it, while preserving the ability to test for main effects and lower order interactions. This is addressed in the following two sections.

11.4.1 BLOCKING AND CONFOUNDING

Due to the time and space requirements in running multifactor experiments, it is often impossible to keep the experimental settings homogeneous for all experimental runs. The different settings may correspond to incidental factors (whose effects are not being investigated), including different laboratories, different work crews, and different time periods over which the scheduled experimental runs are conducted. The different settings under which the experimental runs are performed are called **blocks**. The presence of blocks results in additional effects, called *block effects*. In the terminology introduced in Chapter 1, blocks may act as lurking variables and confound certain effects of the factors being investigated.

It turns out that, with judicious allocation of the scheduled runs in the different blocks, it is possible to confound the block effects only with higher order interactions. To see how this works, consider a 2^2 design and two blocks. Assuming that the blocks do not interact with the factors, the blocks contribute only main effects, say θ_1 and $\theta_2 = -\theta_1$. Thus, the model for an observation in factor-level combination (i, j) is

$$Y_{ij} = \mu + \alpha_i + \beta_j + (\alpha\beta)_{ij} + \theta_1 + \epsilon_{ij} \quad \text{or} \qquad Y_{ij} = \mu + \alpha_i + \beta_j + (\alpha\beta)_{ij} + \theta_2 + \epsilon_{ij}, \tag{11.4.1}$$

depending on whether the experimental condition (i, j) has been allocated in block 1 or block 2. Suppose that experimental conditions $(1, 1)$, $(2, 2)$ are allocated to block 1, and $(1, 2)$, $(2, 1)$ are allocated to block 2. Using (11.4.1), it is seen that the block effect cancels from the estimators of α_1 and β_1. For example, $\widehat{\alpha}_1$ is the contrast

$$\overline{Y}_{i\cdot} - \overline{Y}_{\cdot\cdot} = \frac{Y_{11} - Y_{21} + Y_{12} - Y_{22}}{4}$$

and θ_1 is eliminated as Y_{11} and Y_{22} enter in opposite signs; similarly, θ_2 is eliminated as Y_{21} and Y_{12} enter in opposite signs. In the same manner, the block effects cancel out in the contrast estimating β_1. However, the block effects do not cancel out in the contrast

$$\frac{Y_{11} - Y_{21} - Y_{12} + Y_{22}}{4}$$

and thus, instead of estimating $(\alpha\beta)_{11}$ it estimates $(\alpha\beta)_{11} + \theta_1$. It follows that the main effects can be estimated as usual but the interaction effect is confounded with the block effect.

Using the letter codes for the experimental conditions introduced in Example 11.3-1 (i.e., *1*, *a*, *b*, and *ab* for $(1, 1)$, $(2, 1)$, $(1, 2)$ and $(2, 2)$, respectively), the allocation of experimental conditions to blocks so that the block effect is confounded only with the interaction effect is shown in Figure 11-5. Note that experimental conditions with codes sharing an even number of letters with *ab*, such as 1 (zero counts as even) and *ab*, go in one block, and experimental conditions having an odd number

Figure 11-5 Treatment allocation in a 2^2 design to confound the AB interaction.

Block 1	Block 2
1, *ab*	*a*, *b*

Figure 11-6 Treatment allocation in a 2^3 design to confound the ABC interaction.

Block 1	Block 2
1, ab, ac, bc	a, b, c, abc

of letters in common with *ab* go to the other block. In fact, this is a rule of general applicability (but always under the assumption that blocks do not interact with the factors). Thus, in a 2^3 design the allocation of factor-level combinations to two blocks so that the block effect is confounded only with the three-factor interactions, experimental conditions with codes sharing an even number of letters with *abc* go in one block, and experimental conditions having an odd number of letters in common with *abc* go to the other block; see Figure 11-6

The R package *conf.design* has a function that shows the allocation of factor-level combinations to blocks so that only the specified effect is confounded with the block effect. As usual, the command *install.packages("conf.design")* installs the package and *library(conf.design)* makes it available in the current session. Then the R command shown below produces the allocation of treatments to two blocks that results in the block effect being confounded only with the highest order interaction:

R Command for Simple Confounding in 2^r Designs

```
conf.design(c(1,...,1), p=2,
  treatment.names=c("A1",...,"Ar"))
```
(11.4.2)

In this command, the length of the vector of 1's (i.e., the c(1,. . .,1) part of the command) is r, that is, the number of factors in the study, the fact that this vector consists of 1's specifies that the highest order interaction is to be confounded with the block effect, and the *p=2* part of the command informs R that each factor has two levels. As an example, the R command *conf.design(c(1, 1), p=2, treatment.names=c("A", "B"))* returns

```
   Blocks  A  B
1       0  0  0
2       0  1  1
3       1  1  0
4       1  0  1
```

which is the allocation of experimental conditions to blocks given in Figure 11-5. (Note that in the R output, the blocks are numbered by 0 and 1 and the two levels of factors A and B are also numbered by 0 and 1. Thus, in the letter coding of the experimental conditions introduced in Example 11.3-1, the experimental condition (0, 0) of the first line is denoted by 1, the condition (1, 1) of the second line is denoted by *ab* and so forth.) As an additional example, the R command *conf.design(c(1, 1, 1), p=2, treatment.names=c("A", "B", "C"))* returns

```
  Blocks A B C
1      0 0 0 0
2      0 1 1 0
3      0 1 0 1
4      0 0 1 1
5      1 1 0 0
6      1 0 1 0
7      1 0 0 1
8      1 1 1 1
```

which is the allocation of experimental conditions to blocks given in Figure 11-6.

It is also possible to designate a lower order interaction, or even a main effect, to be confounded with the block effect (though, typically, this is not desirable). For example, the R commands

```
conf.design(c(1, 0), p=2)
conf.design(c(0, 1, 1), p=2)
```
(11.4.3)

give the allocations of treatments to blocks that result in the block effect to be confounded with the main effect of factor A in a 2^2 design and with the BC interaction in a 2^3 design, respectively.

Recall that to test for main effects in an unreplicated 2^2 design, the interaction effect is assumed zero, and the interaction sum of squares is used as the error sum of squares. Thus, if an unreplicated 2^2 design is run in two blocks, with the block effect confounding the interaction effect, it is not possible to test for main effects. Similarly, if an unreplicated 2^3 design is run in two blocks, with the block effect confounding the three-factor interaction effect, the only way to test for the main effects is by assuming that at least one of the two-factor interactions is zero. Rules similar to those given at the end of Section 11.3 for forming ANOVA tables for reduced models apply here, too. For replicated 2^r designs that are run in two blocks, the error sum of squares can be formed without assuming any effects to be zero. The command for fitting a 2^3 design that has been run in two blocks, with the block effect confounded with the ABC interaction term, is demonstrated in the following example.

Example 11.4-1

Due to time considerations, an accelerated life testing experiment is conducted in two labs. There are two levels for each of the three factors involved and the experimental runs are allocated in the two labs so that the lab effect is confounded only with the three-factor interaction effect; see Figure 11-6. The experiment is replicated four times. The data are given in *ALT2cbBlockRepl.txt*. Use R commands to test for significance of the main effects and two-factor interactions.

Solution

With the data set read into the data frame *dat* the command

```
anova(aov(LIFE~Block+A*B*C, data=dat))
```
(11.4.4)

produces the ANOVA table below:

```
          Df   Sum Sq   Mean Sq   F value   Pr(>F)
Block      1   1176.1    1176.1    1.4362    0.242
A          1    378.1     378.1    0.4617    0.503
B          1   1035.1    1035.1    1.2640    0.272
C          1  11628.1   11628.1   14.1994    0.001
A:B        1    406.1     406.1    0.4959    0.488
A:C        1   2278.1    2278.1    2.7819    0.108
B:C        1  16471.1   16471.1   20.1133   0.0001
Residuals 24  19654.0     818.9
```

It follows that the main effect of factor C and the BC two-factor interaction are significant, but all other effects are not. Note that the ANOVA table does not include a line for the ABC interaction effect since this effect is confounded with the main Block effect. Note also that the syntax of the R command in (11.4.4) serves as a reminder of the assumption of no interaction between the block and the factors. ■

In general, it is possible to allocate the runs of a 2^r design in 2^p blocks. In that case, $2^p - 1$ effects will be confounded with block effects. However, it is only possible to hand pick p effects to be confounded with block effects. The p chosen effects are called the *defining* effects. The remaining $2^p - 1 - p$ effects confounded with block effects are **generalized interactions** formed from the p defining effects. To explain the notion of generalized interaction, consider a 2^3 design whose runs are allocated into $2^2 = 4$ blocks. Then, $2^2 - 1 = 3$ effects will be confounded with the block effects and we can hand pick only two of them. Suppose the two defining events we choose are the AB and AC interaction effects. Their generalized interaction is obtained by writing the two defining effects side by side and canceling any letters common to both, that is, $ABAC = BC$. Thus, the three interaction effects that will be confounded with block effects are the three two-factor interactions. Note that had the defining effects been chosen as AC and ABC, their generalized interaction is $ACABC = B$, which shows that a generalized interaction can be a main effect. For $p > 2$, the generalized interactions are formed by any two, any three, etc., of the effects in the defining set. (Note that $2^p - 1 - p = \binom{p}{2} + \cdots + \binom{p}{p}$; see the Binomial Theorem in Exercise 18 in Section 2.3.)

The command *conf.design* we saw before can also be used to get the allocation of experimental runs of a 2^r design into 2^p blocks by specifying the p defining effects. Each of the p defining events is denoted by a vector of 0's and 1's of length r, as described in (11.4.2) and further illustrated in (11.4.3), and the p 0-1 vectors are arranged into a matrix. For example, to get the allocation of the runs of a 2^3 design into $2^2 = 4$ blocks, the R commands

```
G=rbind(c(1, 0, 1), c(1, 1, 1)); conf.design(G, p=2)     (11.4.5)
```

produce the following output (we could have also specified *treatment.names=c("A", "B", "C")* in the above *conf.design* command):

```
        Blocks  T1  T2  T3
1           00   0   0   0
2           00   1   0   1
3           01   0   1   0
4           01   1   1   1
5           10   1   1   0
6           10   0   1   1
7           11   1   0   0
8           11   0   0   1
```

which shows the allocation of runs into four blocks that result in the main B effect and the AC and ABC interactions being confounded with the block effects.

11.4.2 FRACTIONAL FACTORIAL DESIGNS

Another way of dealing with the space and time requirements in running multifactor factorial experiments is to conduct only a fraction of the 2^r runs, that is, take measurements at only a fraction of the 2^r factor-level combinations. An experiment where only a fraction of the factor-level combinations is run is called a *fractional factorial experiment*. A *half-replicate* of a 2^r design involves 2^{r-1} runs, a *quarter-replicate* involves 2^{r-2} runs, and so forth. We write 2^{r-p} to denote a $1/2^p$ replicate of a 2^r factorial design.

The success of fractional factorial designs rests on the sparsity of effects principle, that is, the assumption that higher order interactions are negligible. Under this assumption, a fraction of the runs (experimental conditions) can be carefully selected so as to ensure that the main effects and low-order interactions can be estimated.

To see how this works, consider a 2^3 design where all interactions are assumed to be zero. In this case, a half-replicate that allows the estimation of the main effects consists of the runs that correspond to the lines with a "+" in the $(\alpha\beta\gamma)_{111}$ column of Table 11-2. These are shown in Table 11-4. Note that except for the μ and $(\alpha\beta\gamma)_{111}$ columns, all other columns in Table 11-4 have two pluses and two minuses and thus define contrasts in the cell means. Note also that the contrast defined by each of the main effects columns is identical to that defined by a two-factor interaction column. For example, the α_1 column is identical to the $(\beta\gamma)_{11}$ column, and so forth. Effects with identical columns of pluses and minuses are confounded (the term *aliased* is also used in this context). To illustrate this, consider the contrast defined by the α_1 column (which is the same as the $(\beta\gamma)_{11}$ column), that is,

$$\frac{\overline{x}_{111} - \overline{x}_{221} - \overline{x}_{212} + \overline{x}_{122}}{4}. \tag{11.4.6}$$

Table 11-4 Estimating effects in a 2^{3-1} fractional factorial design

Factor-Level Combination	Cell Means	μ	α_1	β_1	γ_1	$(\alpha\beta)_{11}$	$(\alpha\gamma)_{11}$	$(\beta\gamma)_{11}$	$(\alpha\beta\gamma)_{111}$
1	$\overline{x}_{111}$	+	+	+	+	+	+	+	+
ab	$\overline{x}_{221}$	+	–	–	+	+	–	–	+
ac	$\overline{x}_{212}$	+	–	+	–	–	+	–	+
bc	$\overline{x}_{122}$	+	+	–	–	–	–	+	+

According to the model (11.3.1), each cell sample mean $\bar{x}_{ijk}$ estimates $\mu_{ijk} = \alpha_i + \beta_j + \gamma_k + (\alpha\beta)_{ij} + (\alpha\gamma)_{ik} + (\beta\gamma)_{jk} + (\alpha\beta\gamma)_{ijk}$. Thus, the contrast in (11.4.6) estimates

$$\begin{aligned}
&\frac{1}{4}[\alpha_1 + \beta_1 + \gamma_1 + (\alpha\beta)_{11} + (\alpha\gamma)_{11} + (\beta\gamma)_{11} + (\alpha\beta\gamma)_{111}] \\
&-\frac{1}{4}[-\alpha_1 - \beta_1 + \gamma_1 + (\alpha\beta)_{11} - (\alpha\gamma)_{11} - (\beta\gamma)_{11} + (\alpha\beta\gamma)_{111}] \\
&-\frac{1}{4}[-\alpha_1 + \beta_1 - \gamma_1 - (\alpha\beta)_{11} + (\alpha\gamma)_{11} - (\beta\gamma)_{11} + (\alpha\beta\gamma)_{111}] \\
&+\frac{1}{4}[\alpha_1 - \beta_1 - \gamma_1 - (\alpha\beta)_{11} - (\alpha\gamma)_{11} + (\beta\gamma)_{11} + (\alpha\beta\gamma)_{111}] \\
&= \alpha_1 + (\beta\gamma)_{11}, \qquad\qquad \text{(11.4.7)}
\end{aligned}$$

where we also used the fact that, because each factor has two levels, the zero-sum constraints, which the effects satisfy, imply that each main effect and interaction can be expressed in terms of the corresponding effect with all indices equal to 1. For example, $(\alpha\beta\gamma)_{211} = -(\alpha\beta\gamma)_{111} = (\alpha\beta\gamma)_{222}$ and so forth. Relation (11.4.7) shows that α_1 is confounded, or aliased, with $(\beta\gamma)_{11}$, implying that α_1 cannot be estimated unless $(\beta\gamma)_{11} = 0$. Similarly, it can be seen that β_1 is confounded with $(\alpha\gamma)_{11}$ and γ_1 is confounded with $(\alpha\beta)_{11}$; see Exercise 8.

The pairs of confounded effects in 2^{r-1} designs are called **alias pairs**. Thus, denoting the main effects and interactions by capital letters and their products, respectively, the alias pairs in the 2^{3-1} design of Table 11-4 are

$$[A, BC], [B, AC], \text{ and } [C, AB].$$

Note that each main effect is confounded with its generalized interaction with the three-factor interaction effect. The 2^{3-1} design that is complimentary to that of Table 11-4, that is, the one that consists of the runs that correspond to the lines with a "−" in the $(\alpha\beta\gamma)_{111}$ column of Table 11-2, has the same set of alias pairs; see Exercise 9.

For the purpose of designating a 2^{3-1} design, a shorter version of Table 11-4 suffices. This shorter version displays only the columns α_1, β_1 and γ_1, with labels A, B, and C, respectively. Thus, the 2^{3-1} design of Table 11-4 is completely specified by Table 11-5, each row of which specifies the treatment combination using the convention that a "+" denotes a factor's level 1 and a "−" denotes level 2. Note that the columns A and B represent the factor-level combinations of a 2^2 design, and the C column is the (row-wise) product of columns A and B. As will be seen shortly, this is a special case of a generally applicable rule.

Table 11-5 Concise description of the 2^{3-1} design of Table 11-4

A	B	C
+	+	+
−	−	+
−	+	−
+	−	−

Half-replicates of 2^r fractional factorial designs can be constructed in a manner similar to that for constructing the 2^{3-1} design. The first step is to select an effect, called the **generator** effect. The generator effect, which will be non-estimable, is typically chosen to be the highest order interaction effect. Hence, the discussion will be concentrated on this choice. (In technical terms, choosing the highest order interaction to be the generator effect results in the desirable highest *resolution* design.) As a second step, form a table of pluses and minuses for estimating the main effects and the highest order interaction (the columns for the other interactions are not needed) of a design with r factors. This can be done by extending the pattern described for Table 11-2 to more than three factors. Then the 2^{r-1} fractional factorial design consists of the runs corresponding to the "+" entries in the highest order interaction column.

Instead of writing all 2^r runs and then selecting those with "+" entries in the highest order interaction column, it is more expedient to construct directly the runs of the 2^{r-1} design in the form of Table 11-5. This can be accomplished by first writing down the factor-level combinations for a full factorial design with $r-1$ factors in the form of $r-1$ columns of pluses and minuses, with "+" denoting level 1 and "−" denoting level 2 as usual. Then, add an rth column formed by the (row-wise) product of all $r-1$ columns. The r columns of pluses and minuses thus formed specify the 2^{r-1} runs of the same half-replicate of a 2^r design described in the previous paragraph.

As with the 2^{3-1} design, the constructed 2^{r-1} design has the property that each effect is aliased with its generalized interaction with the highest order interaction effect.

Example 11.4-2

Construct a 2^{4-1} design by selecting $ABCD$ for the generator effect, and give the set of aliased pairs.

Solution

For instructional purposes, we will derive the 2^{4-1} fractional factorial design by both ways described above. The first approach uses the table on the left side of Table 11-6, which shows the pluses and minuses columns for the main effects and the four-factor interaction of a 2^4 factorial design. According to this approach, the factor-level combinations of the four factors in the half-replicate correspond to the rows where there is a "+" in the $ABCD$ column. The second (and more expedient) approach constructs directly the factor-level combinations of the 2^{4-1} fractional factorial design by adding a column D to the table of factor-level combinations of a 2^3 factorial design. The column D is constructed as the (row-wise) product of the columns A, B, and C, and this shorter version of the half-replicate is shown on the right side of Table 11-6. It is easily checked that each level combination of the A, B, C, and D factors shown on the right side of Table 11-6 is a level combination with a "+" in the

Table 11-6 Two ways of constructing a 2^{4-1} design

2^4 Factorial Design: Main Effects and 4-Factor Interaction					2^3 Factorial Design: Main Effects and Their Product			
A	*B*	*C*	*D*	*ABCD*	*A*	*B*	*C*	*D*(=*ABC*)
+	+	+	+	+	+	+	+	+
−	+	+	+	−	−	+	+	−
+	−	+	+	−	+	−	+	−
−	−	+	+	+	−	−	+	+
+	+	−	+	−	+	+	−	−
−	+	−	+	+	−	+	−	+
+	−	−	+	+	+	−	−	+
−	−	−	+	−	−	−	−	−
+	+	+	−	−				
−	+	+	−	+				
+	−	+	−	+				
−	−	+	−	−				
+	+	−	−	+				
−	+	−	−	−				
+	−	−	−	−				
−	−	−	−	+				

$ABCD$ column on the left side of the table (though not listed in the same order). The set of $(2^4 - 2)/2 = 7$ alias pairs is

$$[A, BCD], [B, ACD], [C, ABD], [D, ABC],$$
$$[AB, CD], [AC, BD], [AD, BC].$$

It is worth noting that the factor D, whose column was added on the right side of Table 11-6, is aliased with ABC. This observation justifies setting the D column as the product of A, B, and C. ■

To construct a quarter-replicate of a 2^r factorial design, two generator effects must be selected. The runs where both generator effects have a "+" define a 2^{r-2} design. (The other three quarter-replicates consist of the runs with signs "+, −", "−, +", and "−, −", respectively.) The generator effects and their generalized interaction will be the non-estimable effects. The expedient way of forming the 2^{r-2} fractional factorial design is to first construct a table of signs representing the factor-level combinations of a factorial design consisting of the first $r - 2$ factors. The columns for the two missing factors are the interaction terms that are aliased with each of them.

Example 11.4-3

Construct a 2^{5-2} fractional factorial design using ABD and BCE as the generator effects.

Solution

Note that the main effects corresponding to the factors denoted by D and E are aliased with the AB and BC interaction effects, respectively. The first step is to construct the 2^3 design for factors A, B, and C. Then add two columns by forming the AB and BC products, and label these columns as D and E, respectively. The result is shown in Table 11-7. ■

It should be clear that the expedient way of constructing a 2^{r-2} fractional factorial design requires that the main effects of the factors denoted by the last two letters are aliased with interactions of some of the factors denoted by the first $r - 2$ letters.

In 2^{r-2} designs, each effect is aliased with its generalized interactions with each of the three non-estimable effects (i.e., the two generator effects and their generalized interaction). In the 2^{5-2} design of Example 11.4-3, the

Table 11-7 The 2^{5-2} fractional factorial design of Example 11.4-3

A	B	C	D(=AB)	E(=BC)
+	+	+	+	+
−	+	+	−	+
+	−	+	−	−
−	−	+	+	−
+	+	−	+	−
−	+	−	−	−
+	−	−	−	+
−	−	−	+	+

$(2^5 - 4)/4 = 7$ aliased groups of four are (note that the third non-estimable effect is $(ABD)(BCE){=}ACDE$):

$$\left.\begin{array}{l}[A, BD, ABCE, CDE], [B, AD, CE, ABCDE], [C, ABCD, BE, ADE],\\ \qquad [D, AB, BCDE, ACE], [E, ABDE, BC, ACD],\\ \qquad [AC, CBD, ABE, DE], [AE, BDE, ABC, CD]\end{array}\right\} \quad \textbf{(11.4.8)}$$

The method for constructing half- and quarter-replicates of a 2^r factorial design can be extended for constructing $1/2^p$ replicates for any p. For details, and additional applications to real-life engineering experiments see, for example, the book by Montgomery.[9]

Sums of squares in 2^{r-p} fractional factorial designs can be computed by formulas analogous to those in (11.3.8), except that there is one sum of squares for each class of aliased effects. A method similar to that used in Example 11.3-1 can be used for their computation. As a first step in this computation, the table of r columns of $+$'s and $-$'s, which describes the treatment combinations of the design, is expanded with a column on the left containing the observations at each treatment combination and with additional columns on the right, one for each class of aliased effects. The sum of squares corresponding to each class of aliased effects columns is computed as

$$\mathrm{SS}_{\mathrm{effect}} = \frac{\left(\mathrm{Sum}\{(\mathrm{Observations})(\mathrm{Effect\ Column})\}\right)^2}{2^{r-p}} \quad \textbf{(11.4.9)}$$

The somewhat imprecise notation of (11.4.9) will be clarified in Example 11.4-4. These sums of squares for classes of aliased effects have one degree of freedom. Because fractional factorial experiments are typically unreplicated, the significance of effects cannot be tested unless some aliased pairs of higher order interactions are assumed zero.

Example 11.4-4

The data in *Ffd.2.5-2.txt* are from an experiment using a 2^{5-2} design with treatment combinations given in Example 11.4-3.

(a) Compute the sums of squares for each class of aliased effects using (11.4.9).

(b) Assuming that one class of aliased effects is zero (or negligible), test for the significance of the remaining (classes of aliased) effects.

Solution

(a) The effects denoted by A, B, C, D, E, AC, and CD can be used to represent each of the seven classes of aliased effects given in (11.4.8). Thus, as a first step, Table 11-7 is expanded to include the observations (y) column on the left and the AC and CD columns on the right. Table 11-8 shows the expanded table. Next, using (11.4.9) we have

$$\mathrm{SSA} = \frac{(12.1 - 4.4 + 5.8 - 5.1 + 12.7 - 12.3 + 5.7 - 7.0)^2}{2^{5-2}} = \frac{56.25}{8} = 7.031.$$

The remaining sums of squares can be computed by applying the same formula.

[9] Douglas C. Montgomery (2005). *Design and Analysis of Experiments*, 6th ed. John Wiley & Sons.

Table 11-8 Expansion of Table 11-7 for calculating the sums of squares

y	*A*	*B*	*C*	*D*	*E*	*AC*	*CD*
12.1	+	+	+	+	+	+	+
4.4	−	+	+	−	+	−	−
5.8	+	−	+	−	−	+	−
5.1	−	−	+	+	−	−	+
12.7	+	+	−	+	−	−	−
12.3	−	+	−	−	−	+	+
5.7	+	−	−	−	+	−	+
7.0	−	−	−	+	+	+	−

The results are summarized in the sum of squares column (Sum Sq) of the following ANOVA table.

	Df	Sum Sq	Mean Sq	F value	Pr(>F)
A	1	7.031	7.031	2.002	0.392
B	1	40.051	40.051	11.407	0.183
C	1	13.261	13.261	3.777	0.303
D	1	9.461	9.461	2.695	0.348
E	1	5.611	5.611	1.598	0.426
A:C	1	10.811	10.811	3.079	0.330
C:D	1	3.511	3.511		

(b) This data set has no replications ($n = 1$) and, as already mentioned, the error sum of squares cannot be formed and thus tests cannot be conducted. Since the sum of squares of the (alias class of the) effect *CD* is the smallest (smaller by more than a factor of 10 from the largest SS), we may tentatively assume that it is negligible. Treating the SS for *CD* as error sum of squares with one degree of freedom, we can compute the F statistics and the corresponding p-values shown in the above ANOVA table. All p-values are greater than 0.05, so none of the effects is statistically significant.

Having read the data into the data frame *df*, the sum of squares in the ANOVA table of Example 11.4-4 can also be obtained with the R command

```
anova(aov(y~A*B*C*D*E, data=df))
```

This command does not produce any F statistics or p-values because, as mentioned above, there are no degrees of freedom left for the error SS. To require R to treat the *CD* interaction as error term, use the command

```
anova(aov(y~A+B+C+D+E+A*C, data=df))
```

This command produces the entire ANOVA table of Example 11.4-4, except that the "C:D" row is labeled "Residuals." Finally, we note that it is possible to assume that a main effect is negligible. The command for fitting the model assuming the *E* and *CD* effects are negligible is

```
anova(aov(y~A+B+C+D+A*C, data=df))
```

This command gives error sum of squares that is equal to the sum of squares of effects *E* and *CD* and has two degrees of freedom.

Exercises

1. Verify that the treatment allocation to blocks described in Figure 11-6 confounds only the three-factor interactions with the block effect. (*Hint.* The contrasts estimating the different effects are given in Example 11.3-1.)

2. Verify the following:

(a) Use the first of the two commands given in (11.4.3) to verify that the block effect is confounded only with the main effect of factor A.

(b) Use the second of the two commands given in (11.4.3) to verify that the block effect is confounded only with the BC interaction effect.

3. A 2^5 design will be run in four blocks. Let the five factors be denoted by A, B, C, D, and E.

(a) Find the generalized interaction of the defining effects ABC and CDE.

(b) Find the generalized interaction of the defining effects BCD and CDE.

(c) For each of the two cases above, use R commands to find the allocation of runs into the four blocks so the block effects are confounded only with the defining effects and their generalized interaction.

4. A 2^5 design will be run in eight blocks. Let the five factors be denoted by A, B, C, D, and E.

(a) What is the total number of effects that must be confounded with the block effects?

(b) Construct the set of confounded effects from the defining effects ABC, BCD, and CDE.

(c) Use R commands to find the allocation of runs into the eight blocks so the block effects are confounded only with the defining effects and their generalized interaction.

5. A paper presents a study investigating the effects of feed rate (factor A at levels 50 and 30), spindle speed (factor B at levels 1500 and 2500), depth of cut (factor C at levels 0.06 and 0.08), and the operating chamber temperature on surface roughness.[10] In the data shown in *SurfRoughOptim.txt*, temperature is used as a blocking variable so that it is confounded with the ABC interaction effect. Use R commands to complete the following.

(a) Read the data into the data frame *sr* and introduce a blocking variable indicating the splitting of the experimental conditions into the two blocks. (*Hint.* The eight entries of the "block" variable will be either 1 or 2 depending on whether the corresponding line in the data set is from block 1 or 2. [The block numbering does not matter.] The "block" variable can be introduced into the data frame *sr* by the command *sr$block=c(1, 2, . . .)*.)

(b) Test for the significance of the main factor effects and their interactions at level 0.05.

6. In the context of Exercise 5, the experiment accounted for the possible influence of the blocking factor "tool inserts" at two levels, with the eight runs allocated into four blocks in such a way that the interactions AB, AC, and their generalized interaction, which is BC, are confounded with the block effects. Repeat parts (a) and (b) of Exercise 5.

7. Arc welding is one of several fusion processes for joining metals. By intermixing the melted metal between two parts, a metallurgical bond is created resulting in desirable strength properties. A study investigated the effect on the strength of welded material (factor A at levels SS41 and SB35), thickness of welded material (factor B at levels 8 mm and 12 mm), angle of welding device (factor C at levels 70° and 60°), and current (factor D at levels 150 A and 130 A). The simulated data are in *ArcWeld.txt*. Four blocks are used in this experiment, and the defining effects for confounding were AB and CD. Use R commands to complete the following.

(a) Find the allocation of runs into the four blocks.

(b) Read the data into the data frame *aw* and introduce a blocking variable indicating the splitting of the experimental conditions into the four blocks identified in part (a). (*Hint.* See the hint in part (a) of Exercise 5 for how to introduce the blocking variable into the data frame *aw*.)

(c) Test for the significance of the main factor effects and their interactions at level 0.01.

8. Consider the 2^{3-1} design shown in Table 11-4, and verify that β_1 is confounded with $(\alpha\gamma)_{11}$ and γ_1 is confounded with $(\alpha\beta)_{11}$.

9. Verify that the set of alias pairs in the 2^{3-1} design

Factor-Level Combination	Cell Means	μ	α_1	β_1	γ_1	$(\alpha\beta)_{11}$	$(\alpha\gamma)_{11}$	$(\beta\gamma)_{11}$	$(\alpha\beta\gamma)_{111}$
a	$\overline{x}_{211}$	+	−	+	+	−	−	+	−
b	$\overline{x}_{121}$	+	+	−	+	−	+	−	−
c	$\overline{x}_{112}$	+	+	+	−	+	−	−	−
abc	$\overline{x}_{222}$	+	−	−	−	+	+	+	−

are the same as those for its complimentary 2^{3-1} design given in Table 11-4. (*Hint.* Write the contrast corresponding to each main effect column, as was done in (11.4.6), and find what it estimates, as was done in (11.4.7).)

[10] J. Z. Zhang et al. (2007). Surface roughness optimization in an end-milling operation using the Taguchi design method, *Journal of Materials Processing Technology*, 184(13): 233–239.

10. Answer the following questions.

(a) Construct a 2^{5-1} design, using the five-factor interaction as the generator effect, and give the set of $(2^5 - 2)/2 = 15$ aliased pairs.

(b) Construct a 2^{6-2} fractional factorial design using $ABCE$ and $BCDF$ as the generator effects. Give the third non-estimable effect and the set of $(2^6 - 4)/4 = 15$ groups of four aliased effects.

(c) Construct a 2^{7-2} fractional factorial design using $ABCDF$ and $BCDEG$ as the generator effects, and give the third non-estimable effect. How many groups of four aliased effects are there? (You need not list them.)

11. Photomasks are used to generate various design patterns in the fabrication of liquid crystal displays (LCDs). A paper[11] reports on a study aimed at optimizing process parameters for laser micro-engraving iron oxide coated glass. The effect of five process parameters was explored (letter designation in parentheses): beam expansion ratio (A), focal length (B), average laser power (C), pulse repetition rate (D), and engraving speed (E). The primary response is the engraving linewidth. Data from an unreplicated 2^{5-1} design, using the five-factor interaction as generator, are in *Ffd.2.5-1.1r.txt*.

(a) Give the set of aliased pairs.

(b) Use R commands to compute the sums of squares of (the classes of) aliased effects. Is it possible to test for the significance of effects?

(c) Use the AC interaction term as error, and test for the significance of the effects at level 0.05. (*Hint.* With the data read into the data frame *df*, the R command for omitting the AC interaction term is *anova(aov(y~A+B+C+D+E+A*B+A*D+A*E+B*C+B*D+B*E+C*D+C*E+D*E, data=df)).*)

[11] Y. H. Chen et al. (1996). Application of Taguchi method in the optimization of laser micro-engraving of photomasks, *International Journal of Materials and Product Technology*, 11(3/4).

Chapter 12

Polynomial and Multiple Regression

12.1 Introduction

The simple linear regression model, introduced in Section 4.6.2, specifies that the regression function, $\mu_{Y|X}(x) = E(Y|X = x)$, of a response variable Y is a linear function of the predictor/explanatory variable X. Many applications, however, require more general regression models either because the regression function is not linear or because there are more than one predictor variables. Including polynomial terms in the regression model is a common way to achieve a better approximation to the true regression function. This leads to the term **polynomial regression**. Incorporating several predictor variables, with or without additional polynomial terms, leads to the term **multiple regression**. This chapter discusses estimation, testing, and prediction in the context of polynomial and multiple regression models. Additional issues such as *weighted least squares*, *variable selection*, and *multi-collinearity* are discussed.

The multiple regression model allows categorical covariates and, as such, it can be (and is) used to model factorial designs. In addition to providing a new perspective on factorial designs, and a unified data analytic methodology, this provides a natural way of handling heteroscedastic factorial designs.

A suitable transformation of the response variable, the predictor variables, or the regression function often leads to a better approximation of the true regression function by the multiple regression model. One such transformation, commonly used for Bernoulli response variables, leads to **logistic regression**, which is also briefly discussed.

12.2 The Multiple Linear Regression Model

The multiple linear regression (MLR) model specifies that the conditional expectation, $E(Y|X_1 = x_1, \ldots, X_k = x_k) = \mu_{Y|X_1,\ldots,X_k}(x_1, \ldots, x_k)$, of a response variable Y given the values of k predictor variables, $X_1, \ldots, X_k$, is a linear function of the predictors' values:

$$\mu_{Y|X_1,\ldots,X_k}(x_1, \ldots, x_k) = \beta_0 + \beta_1 x_1 + \cdots + \beta_k x_k. \tag{12.2.1}$$

Equation (12.2.1) describes a hyperplane in the $(k+1)$-dimensional space of the response and predictor variables. For example, for $k = 2$ predictors, relation (12.2.1), that is,

$$\mu_{Y|X_1,\,X_2}(x_1, x_2) = \beta_0 + \beta_1 x_1 + \beta_2 x_2,$$

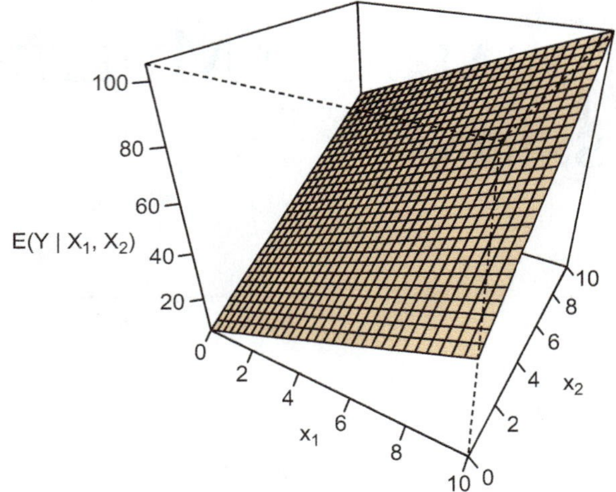

Figure 12-1 The regression plane for the model $\mu_{Y|X_1,X_2}(x_1, x_2) = 5 + 4x_1 + 6x_2$.

describes a plane in the 3D space. Figure 12-1 shows such a plane for a specific choice of the regression parameters β_0, β_1, and β_2. For notational simplicity, the intercept term in the MLR model will always be denoted by β_0 regardless of whether or not the predictor variables have been centered (see (4.6.4) and (4.6.5)). The regression parameters β_j have a clear physical interpretation. Noting that $\beta_j, j \geq 1$, is the partial derivative of $\mu_{Y|X_1,\ldots,X_k}(x_1, \ldots, x_k)$ with respect to x_j, it follows that β_j represents the change in the regression function per unit change in x_j when all other predictors are held fixed. Moreover, if the predictor variables have been centered, the intercept parameter β_0 is the marginal mean of Y; see Proposition 4.6-1.

The multiple linear regression model is most commonly written as an equation relating the response variable Y to the explanatory/predictor variables $X_1, \ldots, X_k$ and an intrinsic error variable as

$$Y = \beta_0 + \beta_1 X_1 + \cdots + \beta_k X_k + \varepsilon, \qquad \textbf{(12.2.2)}$$

with the intrinsic error variable ε being implicitly defined by (12.2.2). The basic properties of the intrinsic error variable are analogous to those summarized in Proposition 4.6-2 for the simple linear regression model. Namely, ε has zero mean, is uncorrelated from the predictor variables $X_1, \ldots, X_k$, and its variance σ_ε^2 is the conditional variance of Y given the values of $X_1, \ldots, X_k$. As in the simple linear regression model, it will be assumed that the intrinsic error variance does not depend on (i.e., does not change with) the values of the predictor variables (homoscedasticity assumption); see, however, Section 12.4.2.

The Polynomial Regression Model When there is only one predictor variable, but either theoretical reasoning or a scatterplot of the data suggests that the true regression function has one or more peaks or valleys (i.e., local maxima or minima), the simple linear regression model is not appropriate. In such cases incorporation of polynomial terms in the model for the regression function becomes necessary. A kth-degree polynomial may accommodate up to $k - 1$ peaks and valleys. Figure 12-2 shows a regression function modeled as a third-degree polynomial.

In general, the kth-degree polynomial regression model is

$$Y = \beta_0 + \beta_1 X + \beta_2 X^2 + \cdots + \beta_k X^k + \varepsilon. \qquad \textbf{(12.2.3)}$$

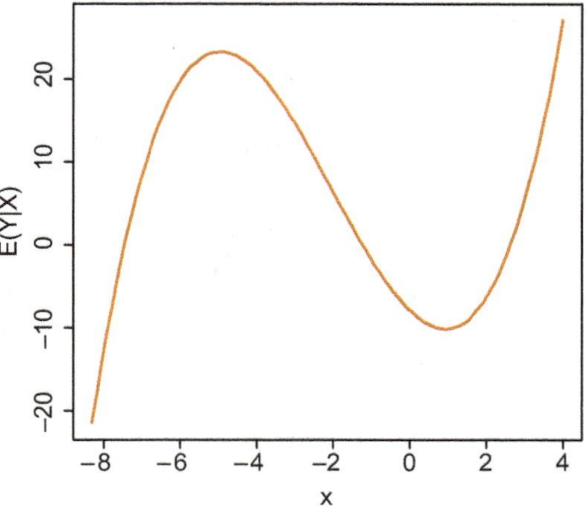

Figure 12-2 Polynomial regression model $\mu_{Y|X}(x) = -8 - 4.5x + 2x^2 + 0.33x^3$.

A comparison of the MLR model (12.2.2) with the polynomial regression model (12.2.3) reveals that the polynomial regression model can also be written as an MLR model:

$$Y = \beta_0 + \beta_1 X_1 + \beta_2 X_2 + \cdots + \beta_k X_k + \varepsilon, \quad \text{where } X_i = X^i \quad \text{for } i = 1, \ldots, k. \tag{12.2.4}$$

Thus, procedures for estimating the parameters of an MLR model apply for polynomial regression models as well. However, the interpretation of the parameters is different. For example, the coefficient β_1 in (12.2.3) is the derivative of the regression function at $x = 0$, and thus it represents the *rate* of change of the regression function at $x = 0$. Similarly, β_2 in (12.2.3) is one-half of the second derivative of the regression function at $x = 0$, and so forth. Of course, if zero is not in the range of the covariate values, knowing the rate of change of the regression function at $x = 0$ has no practical value. This is one of the reasons why centering the covariate in a polynomial regression model is recommended: If the covariate is centered, β_1 represents the rate of change of the regression function at the average x-value. The second reason for centering is grounded in empirical evidence suggesting that coefficients of polynomial regression models are more accurately estimated if the covariate has been centered. Lastly, even when the predictor variable in a polynomial regression model is centered, the intercept β_0 is not the marginal expected value of Y. To see this, consider the quadratic regression model with centered covariate:

$$Y = \beta_0 + \beta_1(X - \mu_X) + \beta_2(X - \mu_X)^2 + \varepsilon, \tag{12.2.5}$$

and recall that the use of a capital letter for the covariate indicates it is considered a random variable. Taking the expected value on both sides of (12.2.5), it follows that

$$E(Y) = \beta_0 + \beta_2 \text{Var}(X).$$

In practice, the covariate is centered by its sample mean but the interpretation of coefficients is similar.

Polynomial Models with Interaction Terms When there are more than one predictors it is usually a good idea to include **interaction** effects in the model. The concept of interaction in multiple regression is similar to the concept of interaction in two-way factorial designs (see Section 1.8.4 and Chapter 11). In particular, we say that predictor variables X_1 and X_2 interact if the effect of changing the level of one variable (X_1, say) depends on the level of the other variable (X_2). The most common

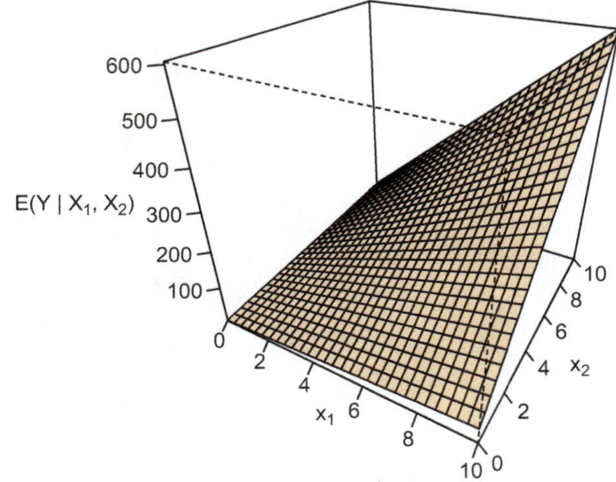

Figure 12-3 The regression plane for the model $\mu_{Y|X_1,X_2}(x_1, x_2) = 5 + 4x_1 + 6x_2 + 5x_1x_2$.

way of modeling the interaction effect of two variables is by their cross-product. For example, the regression function of Y on X_1 and X_2 can be modeled as

$$\mu_{Y|X_1,X_2}(x_1, x_2) = \beta_0 + \beta_1 x_1 + \beta_2 x_2 + \beta_3 x_1 x_2. \tag{12.2.6}$$

Figure 12-3 shows the plot of the regression function (12.2.6) with a particular choice of the regression coefficients. Because of the interaction term, this regression function is not a plane, as the one in Figure 12-1 is. Figure 12-3 illustrates the fact that the expected change in Y when x_1 is changed (say, by one unit) is a function of x_2.

Finally, the modeling flexibility can be enhanced by combining polynomial terms with interaction terms. For example, the regression function of Y on X_1 and X_2 may be modeled as a second-degree polynomial with interaction as

$$\mu_{Y|X_1,X_2}(x_1, x_2) = \beta_0 + \beta_1 x_1 + \beta_2 x_2 + \beta_3 x_1^2 + \beta_4 x_2^2 + \beta_5 x_1 x_2. \tag{12.2.7}$$

Figure 12-4 shows the plot of the regression function (12.2.7) with a particular choice of the regression coefficients. By varying the values of the regression coefficients in

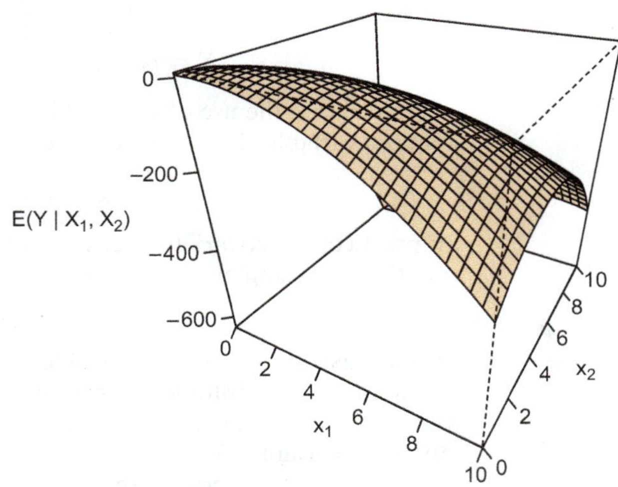

Figure 12-4 The regression plane for the model $\mu_{Y|X_1,X_2}(x_1, x_2) = 5 + 4x_1 + 6x_2 - 3.5x_1^2 - 7x_2^2 + 5x_1x_2$.

(12.2.7), the second-degree polynomial model with interaction can assume a wide variety of shapes.

Regression models with interaction terms can also be put in the form of the MLR model (12.2.2). For example, the regression function in (12.2.7) can be written as

$$\mu_{Y|X_1,X_2}(x_1,x_2) = \beta_0 + \beta_1 x_1 + \beta_2 x_2 + \beta_3 x_3 + \beta_4 x_4 + \beta_5 x_5, \tag{12.2.8}$$

where $x_3 = x_1^2$, $x_4 = x_2^2$, and $x_5 = x_1 x_2$. It follows that procedures for estimating the parameters of an MLR model apply for regression models with polynomial and interaction terms as well.

Transformations Occasionally, empirical evidence or science suggests a non-linear (and non-polynomial) relationship between the response and predictor variables. In such cases it may be possible that a suitable transformation of the response, or of the covariate, or both, will restore the suitability of a multiple linear or polynomial regression model. For example, the *exponential growth model*, for example, for cell counts, specifies a relationship of the form

$$E(Y_t) = \alpha_0 e^{\beta_1 t}, \quad t \geq 0,$$

for the expected number of cells at time t. Taking logs, the relationship becomes linear: $\log(E(Y_t)) = \log(\alpha_0) + \beta_1 t$. Assuming further that the intrinsic error term for Y_t is *multiplicative*, that is, $Y_t = \alpha_0 e^{\beta_1 t} U_t$, the transformed variable $\widetilde{Y}_t = \log(Y_t)$ follows the simple linear regression model

$$\widetilde{Y}_t = \beta_0 + \beta_1 t + \varepsilon_t, \quad \text{where } \beta_0 = \log(\alpha_0) \quad \text{and} \quad \varepsilon_t = \log U_t. \tag{12.2.9}$$

Note that for model (12.2.9) to be a *normal* simple linear regression model, U_t must have the log-normal distribution (see Exercise 12 in Section 3.5).

As another example, consider the non-linear model $\mu_{Y|X}(x) = \alpha_0 x^{\beta_1}$. Assuming again a multiplicative error term for Y, the transformed variables $\widetilde{Y} = \log(Y)$ and $\widetilde{X} = \log(X)$ follow the simple linear regression model

$$\widetilde{Y} = \beta_0 + \beta_1 \widetilde{X} + \varepsilon, \quad \text{where } \beta_0 = \log \alpha_0.$$

Models fitted on transformed variables can be used for prediction in a straightforward manner. For example, if fitting model (12.2.9) to data yields least squares estimators $\widehat{\beta}_0$ and $\widehat{\beta}_1$, the number of cells at time t_0 is predicted by $\widehat{Y}_{t_0} = e^{\widehat{\beta}_0 + \widehat{\beta}_1 t_0}$.

Exercises

1. A response variable is related to two predictors through the multiple linear regression model $Y = 3.6 + 2.7X_1 + 0.9X_2 + \varepsilon$.

(a) Give $\mu_{Y|X_1,X_2}(12, 25) = E(Y|X_1 = 12, X_2 = 25)$.

(b) If $E(X_1) = 10$ and $E(X_2) = 18$, find the (marginal) expected value of Y.

(c) What is the expected change in Y when X_1 increases by one unit while X_2 remains fixed?

(d) The model can be expressed equivalently in terms of the centered variables as $Y = \beta_0 + \beta_1(X_1 - \mu_{X_1}) + \beta_2(X_2 - \mu_{X_2}) + \varepsilon$. Using the information about the expected values of X_1 and X_2 given in part (b), give the values of β_0, β_1, and β_2.

2. A response variable is related to two predictors through the multiple regression model with an interaction term $Y = 3.6 + 2.7X_1 + 0.9X_2 + 1.5X_1X_2 + \varepsilon$.

(a) If $E(X_1) = 10$, $E(X_2) = 18$, and $\text{Cov}(X_1, X_2) = 80$, find the (marginal) expected value of Y. (*Hint.* $\text{Cov}(X_1, X_2) = E(X_1X_2) - E(X_1)E(X_2)$.)

(b) Express the model in terms of the centered variables as $Y = \beta_0 + \beta_1(X_1 - \mu_{X_1}) + \beta_2(X_2 - \mu_{X_2}) + \beta_3(X_1 - \mu_{X_1})(X_2 - \mu_{X_2}) + \varepsilon$. Justify that $\beta_3 = 1.5$. Next, show that $\beta_0 = E(Y) - \beta_3 \text{Cov}(X_1, X_2)$, and use part (a) to find its value.

3. A response variable is related to a predictor variable through the quadratic regression model $\mu_{Y|X}(x) = -8.5 - 3.2x + 0.7x^2$.

(a) Give the rate of change of the regression function at $x = 0$, 2, and 3.

(b) Express the model in terms of the centered variables as $\mu_{Y|X}(x) = \beta_0 + \beta_1(x - \mu_X) + \beta_2(x - \mu_X)^2$. If $\mu_X = 2$, give the values of β_0, β_1, and β_2. (*Hint.* Match the coefficients of the different powers of x.)

4. The data set *Leinhardt* in the R package *car* has data on per capita income and infant mortality rate from several countries. Use the R commands *install.packages("car")* to install the package, the command *library(car)* to make the package available to the current R session, and the command *x=Leinhardt$income; y=Leinhardt$infant* to set the income values and mortality rates in the R objects x and y, respectively.

(a) Construct the (x, y) and (log(x), log(y)) scatterplots. Which scatterplot suggests a linear relationship? [Note: Cases with missing values are omitted when plotting.]

(b) Fit the simple linear regression model to the set of variables whose scatterplot suggests a linear relationship and give the estimated regression line.

(c) Use the estimated regression line from part (b) to predict the infant mortality rate of a country with per capita income of 1400.

5. The file *BacteriaDeath.txt* has simulated data on bacteria deaths over time. Import the time and bacteria counts into the R objects t and y, respectively.

(a) Construct the (t, y) and (t, log(y)) scatterplots. Which scatterplot suggests a linear relationship?

(b) Construct a predictive equation for the bacteria count Y at time t.

6. The data set in *WindSpeed.txt* has 25 measurements of current output, produced by a wind mill, and wind speed (in miles per hour).[1] Import the wind speed and output into the R objects x and y, respectively.

(a) Construct the (x, y) and $(1/x, y)$ scatterplots. Which scatterplot suggests a linear relationship?

(b) Fit the simple linear regression model to the set of variables whose scatterplot suggests a linear relationship and give the estimated regression line.

(c) Use the estimated regression line from part (b) to predict the current produced at a wind speed of 8 miles per hour.

12.3 Estimation, Testing, and Prediction

12.3.1 THE LEAST SQUARES ESTIMATORS

The method of least squares, introduced in Section 6.3.3, can also be used to fit a multiple linear regression model to the data. To describe it in this context, let

$$(y_i, x_{i1}, x_{i2}, \ldots, x_{ik}), \qquad i = 1, \ldots, n,$$

be $n > k$ observations on the k covariates/predictor variables and the response variable. Typically, the data are presented in table form as:

y	x_1	x_2	$\cdots$	x_k
y_1	x_{11}	x_{12}	$\cdots$	x_{1k}
y_2	x_{21}	x_{22}	$\cdots$	x_{2k}
$\vdots$	$\vdots$	$\vdots$	$\vdots$	$\vdots$
y_n	x_{n1}	x_{n2}	$\cdots$	x_{nk}

Each row of data is assumed to satisfy the model equation

$$y_i = \beta_0 + \beta_1 x_{i1} + \beta_2 x_{i2} + \cdots + \beta_k x_{ik} + \varepsilon_i, \quad i = 1, \ldots, n, \tag{12.3.1}$$

according to (12.2.2). The **least squares estimates**, $\widehat{\beta}_0, \widehat{\beta}_1, \ldots, \widehat{\beta}_k$, of the regression coefficients are obtained by minimizing the objective function

[1] D. C. Montgomery, E. A. Peck, and G. G. Vining (2012). *Introduction to Linear Regression Analysis*, 5th ed. Hoboken: Wiley & Sons. Table 5.5.

$$L(b_0, b_1, \ldots, b_k) = \sum_{i=1}^{n} (y_i - b_0 - b_1 x_{i1} - b_2 x_{i2} - \cdots - b_k x_{ik})^2 \qquad \textbf{(12.3.2)}$$

with respect to $b_0, b_1, \ldots, b_k$. Equivalently, the least squares estimates are obtained as the solution to the following system of equations known as the **normal equations**:

$$\frac{\partial L}{\partial b_0} = -2 \sum_{i=1}^{n} \left(y_i - b_0 - \sum_{j=1}^{k} b_j x_{ij} \right) = 0$$

$$\frac{\partial L}{\partial b_j} = -2 \sum_{i=1}^{n} \left(y_i - b_0 - \sum_{j=1}^{k} b_j x_{ij} \right) x_{ij} = 0, \quad j = 1, \ldots, k.$$

The normal equations simplify to

$$nb_0 + b_1 \sum_{i=1}^{n} x_{i1} + b_2 \sum_{i=1}^{n} x_{i2} + \cdots + b_k \sum_{i=1}^{n} x_{ik} = \sum_{i=1}^{n} y_i$$

$$b_0 \sum_{i=1}^{n} x_{i1} + b_1 \sum_{i=1}^{n} x_{i1}^2 + b_2 \sum_{i=1}^{n} x_{i1} x_{i2} + \cdots + b_k \sum_{i=1}^{n} x_{i1} x_{ik} = \sum_{i=1}^{n} x_{i1} y_i$$

$$\vdots \quad \vdots \quad \vdots$$

$$b_0 \sum_{i=1}^{n} x_{ik} + b_1 \sum_{i=1}^{n} x_{ik} x_{i1} + b_2 \sum_{i=1}^{n} x_{ik} x_{i2} + \cdots + b_k \sum_{i=1}^{n} x_{ik}^2 = \sum_{i=1}^{n} x_{ik} y_i.$$

As an example, the normal equations for the simple linear regression model are:

$$nb_0 + b_1 \sum_{i=1}^{n} x_{i1} = \sum_{i=1}^{n} y_i$$

$$b_0 \sum_{i=1}^{n} x_{i1} + b_1 \sum_{i=1}^{n} x_{i1}^2 = \sum_{i=1}^{n} x_{i1} y_i.$$

Solving them yields the least squares estimators as shown in (6.3.7). (Note that in (6.3.7), x_{i1} is denoted by x_i and the intercept estimator is denoted by $\widehat{\alpha}_1$ instead of $\widehat{\beta}_0$.) The solution of the normal equations for the general MLR model, that is, for $k > 1$ predictors, is most conveniently given in matrix notation and will be discussed later.

According to the terminology introduced in Section 6.3.3 for the simple linear regression model, the least squares estimators specify the *estimated regression function*, the *fitted (or predicted) values*, and the *residuals* (or *estimated errors*):

Estimated Regression Function

$$\widehat{\mu}_{Y|X_1,\ldots,X_k}(x_1, \ldots, x_k) = \widehat{\beta}_0 + \widehat{\beta}_1 x_1 + \cdots + \widehat{\beta}_k x_k$$

Fitted Values

$$\widehat{y}_i = \widehat{\beta}_0 + \widehat{\beta}_1 x_{i1} + \cdots + \widehat{\beta}_k x_{ik}, \quad 1 \le i \le n \qquad \textbf{(12.3.3)}$$

Residuals

$$e_i = y_i - \widehat{y}_i, \quad 1 \le i \le n$$

Moreover, the sum of the squared residuals gives the *error sum of squares*, *SSE*, whose degrees of freedom is $\text{DF}_{SSE} = n - k - 1$, that is, the number of observations minus the total number of regression coefficients being estimated. Dividing the SSE by DF_{SSE} gives the *MSE*, which is an unbiased estimator of the intrinsic error variance σ_ε^2:

Error Sum of Squares

Mean Square Error

$$\text{SSE} = \sum_{i=1}^{n} e_i^2 = \sum_{i=1}^{n} (y_i - \widehat{y}_i)^2$$

$$\text{MSE} = \frac{\text{SSE}}{\text{DF}_{SSE}} = \frac{\text{SSE}}{n-k-1} = \widehat{\sigma}_\varepsilon^2 \qquad \textbf{(12.3.4)}$$

As in the simple linear regression model, SSE expresses the variability in the response variable that is not explained by the regression model (*unexplained variability*). Subtracting SSE from the total variability (which, as always, is the numerator of the sample variance of the response variable) gives the *regression sum of squares, SSR*. SSR expresses the variability in the response variable that is explained by the regression model (*explained variability*). The proportion of the total variability explained by the model, SSR/SST or 1−SSE/SST, is called the **coefficient of multiple determination** and is denoted by R^2. The following display summarizes these quantities:

Total Sum of Squares

Regression Sum of Squares

Coefficient of Multiple Determination

$$\text{SST} = \sum_{i=1}^{n} (y_i - \overline{y}_i)^2$$

$$\text{SSR} = \text{SST} - \text{SSE} \qquad \textbf{(12.3.5)}$$

$$R^2 = 1 - \frac{\text{SSE}}{\text{SST}} = \frac{\text{SSR}}{\text{SST}}$$

The notion of correlation between X and Y can be extended to the notion of **multiple correlation** between $X_1, \ldots, X_k$ and Y. The *multiple correlation coefficient* between the response variable and the predictor variables is computed as the Pearson's linear correlation coefficient between the pairs $(y_i, \widehat{y}_i)$, $i = 1, \ldots, n$. It can be shown that the positive square root of R^2 equals the multiple correlation coefficient between the response and the predictors.

Since a main objective of regression analysis is to reduce the unexplained variability, it should be that the higher the R^2, the more successful the model. It turns out, however, that R^2 can be inflated by including additional covariates, even if the additional covariates have no predictive power. For example, inclusion of polynomial and interaction terms in the regression model will increase R^2. At the same time, polynomial and interaction terms complicate the interpretation of the effect each covariate has. According to the **principle of parsimony**, models should be as simple as possible. One way of balancing the loss of interpretability due to the inclusion of additional parameters, against the gain in R^2, that such inclusion entails, is to use the **adjusted coefficient of multiple determination** denoted by $R^2(\text{adj})$:

Adjusted Coefficient of Multiple Determination

$$R^2(\text{adj}) = 1 - \frac{\text{MSE}}{\text{MST}} = \frac{(n-1)R^2 - k}{n-1-k} \qquad \textbf{(12.3.6)}$$

R Commands for Regression Analysis Let the R objects *y, x1, x2, x3* contain the values of the response variable Y and those of the three covariates,

Table 12-1 R commands for fitting multiple linear and polynomial regression models

R Command	Model
`fit=lm(y~x1+ x2+x3)`	Fits the basic model $Y = \beta_0 + \beta_1 X_1 + \beta_2 X_2 + \beta_3 X_3 + \varepsilon$
`fit=lm(y~(x1+x2+x3)^2)`	Fits the basic model plus all pairwise interaction terms
`fit=lm(y~x1+x2+x3+x1:x2+x1:x3+x2:x3)`	Fits the basic model plus all pairwise interaction terms
`fit=lm(y~x1*x2*x3)`	Fits the basic model plus all interactions up to order three
`fit=lm(y~(x1+x2+x3)^3)`	Fits the basic model plus all interactions up to order three
`fit=lm(y~x1+I(x1^2)+x2+x3)`	Fits the basic model plus a quadratic term in X_1
`fit=lm(y~poly(x1, 2, raw=T)+x2+x3)`	Fits the basic model plus a quadratic term in X_1
`fit=lm(y~poly(x1, 3, raw=T)+x2+x3)`	Fits the basic model plus quadratic and cubic terms in X_1

X_1, X_2, X_3, respectively. The R commands in Table 12.1 fit the basic MLR model $Y = \beta_0 + \beta_1 X_1 + \beta_2 X_2 + \beta_3 X_3 + \varepsilon$ and certain variations of it that include polynomial and interaction terms. The commands extend to regression models with more than three predictors in a straightforward manner. Regardless of the model to be fitted, it is always a good idea to first center the variables:

Centering the Predictors Prior to Fitting the Model

```
x1=x1-mean(x1); x2=x2-mean(x2); x3=x3-mean(x3)
```

Table 12-1 shows that some models can be specified in more than one way. For example, the model with all pairwise interactions can be specified by the command in either the second or third row of Table 12-1, as well as by the command

```
fit=lm(y~x1*x2*x3-x1:x2:x3)
```

Similarly, the model with all interaction terms, including the third order interaction, can be specified by the command in either the fourth or fifth row of Table 12-1, as well as by the command *fit=lm(y~(x1+x2+x3)^2+x1:x2:x3)*. Finally, the commands in the last three rows of Table 12-1 show a couple of options for including polynomial terms in the model.

REMARK 12.3-1 Using simply *poly(x1, 3)* instead of *poly(x1, 3, raw=T)*, gives what is known as **orthogonal polynomial** terms. Orthogonal polynomials result in "cleaner" inference procedures, and are preferable for testing the significance of higher order polynomial terms. On the other hand, their construction is complicated and thus it is difficult to use the fitted model for prediction. ◁

Adding the R command *fit*, or *coef(fit)*, at the end of any of the Table 12-1 commands will give the least squares estimates of the regression coefficients. Adding *summary(fit)* gives, in addition, standard errors of the estimated coefficients, the residual standard error (i.e., $\widehat{\sigma}_\varepsilon$) together with the error degrees of freedom, R^2,

R^2(adj), and p-values for hypothesis-testing problems that will be discussed in the next section.

Example 12.3-1

The data[2] in *Temp.Long.Lat.txt* give the average (over the years 1931 to 1960) daily minimum January temperature in degrees Fahrenheit with the latitude and longitude of 56 US cities. Let Y, X_1, X_2 denote the temperature and the centered latitude and longitude variables, respectively.

(a) Using R commands to fit the following models, report the least squares estimates of the regression parameters, the estimated intrinsic error variance, as well as R^2 and R^2(adj) for each of the models:
(1) the basic multiple regression model $Y = \beta_0 + \beta_1 X_1 + \beta_2 X_2 + \varepsilon$,
(2) the model $Y = \beta_0 + \beta_1 X_1 + \beta_2 X_2 + \beta_3 X_1 X_2 + \varepsilon$, which adds the interaction term to the basic model, and
(3) the model $Y = \beta_0 + \beta_1 X_1 + \beta_{1,2} X_1^2 + \beta_2 X_2 + \beta_{2,2} X_2^2 + \beta_3 X_1 X_2 + \varepsilon$, which includes second-degree polynomial terms in X_1 and X_2 and their interaction. (Note that indexing coefficients of polynomial terms by double subscripts, as done above, is not very common.)

(b) Use each model's estimated intrinsic variance and the corresponding degrees of freedom to find each model's SSE and (their common) SST.

(c) The average minimum January temperature in Mobile, AL, and its latitude and longitude are 44, 31.2, 88.5. For each of the three models, give the fitted value and the residual for Mobile, AL.

Solution

(a) Use *df=read.table("Temp.Long.Lat.txt", header=T)* to import the data into the R data frame *df*, and *y=df$JanTemp; x1=df$Lat-mean(df$Lat); x2=df$Long-mean(df$Long)* to set the response variable and the centered covariates into the R objects *y, x1, x2*, respectively. Then the commands

```
fit1=lm(y~x1+x2); fit2=lm(y~x1*x2);
fit3=lm(y~poly(x1, 2, raw=T)+poly(x2, 2, raw=T)+x1:x2)
summary(fit1); summary(fit2); summary(fit3)
```

fit the three models and produce the desired information, which is summarized in the table below:

Model	Intercept	X_1	X_1^2	X_2	X_2^2	X_1X_2	$\widehat{\sigma}_\varepsilon$	R^2	R^2(adj)
(1)	26.52	−2.16		0.13			6.935	0.741	0.731
(2)	26.03	−2.23		0.034		0.04	6.247	0.794	0.782
(3)	21.31	−2.61	−0.01	−0.18	0.02	0.04	4.08	0.916	0.907

(b) According to (12.3.4), SSE $= \widehat{\sigma}_\varepsilon^2(n - k - 1)$, where n is the sample size and k is the number of regression coefficients excluding the intercept. Since in this data set $n = 56$, and $\widehat{\sigma}_\varepsilon$ and R^2 are reported in part (a), we have

[2] J. L. Peixoto (1990). A property of well-formulated polynomial regression models. *American Statistician*, 44: 26–30.

Model	k	$\text{DF}_{SSE} = n - k - 1$	$\widehat{\sigma}_\varepsilon^2$	SSE
(1)	2	53	48.09	2548.99
(2)	3	52	39.02	2029.3
(3)	5	50	16.61	830.69

Note the decrease in SSE as interaction and polynomial terms are added to the model. The SST, which is the numerator of the sample variance of the response variable, is most accurately computed with the command *var(y)*55*; its value is 9845.98. It can also be computed through the formula

$$\text{SST} = \frac{\text{SSE}}{1 - R^2},$$

which follows from (12.3.5). For each of the three models, this formula gives 9845.48, 9846.19, and 9842.28, respectively, due to different round-off errors.

(c) Since Mobile, AL, is listed first in the data set, the corresponding response and covariate values are subscripted by 1. The centered covariate values are $x_{11} = 31.2 - 38.97 = -7.77$, $x_{12} = 88.5 - 90.96 = -2.46$, where 38.97 and 90.96 are the average latitude and longitude in the data set. According to the formulas in (12.3.3), and the MLR model representation of polynomial regression models given, for example, in (12.2.8), we have

Model	$\widehat{y}_1$	$e_1 = y_1 - \widehat{y}_1$
(1)	$25.52 - 2.16x_{11} + 0.13x_{12} = 43$	1.00
(2)	$26.03 - 2.23x_{11} + 0.034x_{12} + 0.04x_{11}x_{12} = 44.1$	−0.10
(3)	$21.31 - 2.61x_{11} - 0.01x_{11}^2 - 0.18x_{12} + 0.02x_{12}^2 + 0.04x_{11}x_{12} = 42.29$	1.71

The above fitted values and residuals can also be obtained with the R commands

```
fitted(fit1)[1]; fitted(fit2)[1]; fitted(fit3)[1]
resid(fit1)[1];resid(fit2)[1];resid(fit3)[1],
```

respectively. Given the different outcomes produced by the different models, the question as to which model should be preferred arises. This question can be addressed through hypothesis testing and *variable selection*; see Sections 12.3.3 and 12.4.3.

Matrix Notation* The MLR model equations for the data in (12.3.1) can be written in matrix notation as

$$\mathbf{y} = \mathbf{X}\boldsymbol{\beta} + \boldsymbol{\epsilon} \tag{12.3.7}$$

where

$$\mathbf{y} = \begin{pmatrix} y_1 \\ y_2 \\ \vdots \\ y_n \end{pmatrix}, \mathbf{X} = \begin{pmatrix} 1 & x_{11} & x_{12} & \cdots & x_{1k} \\ 1 & x_{21} & x_{22} & \cdots & x_{2k} \\ \vdots & \vdots & \vdots & \vdots & \\ 1 & x_{n1} & x_{n2} & \cdots & x_{nk} \end{pmatrix}, \boldsymbol{\beta} = \begin{pmatrix} \beta_0 \\ \beta_1 \\ \vdots \\ \beta_k \end{pmatrix}, \boldsymbol{\epsilon} = \begin{pmatrix} \epsilon_1 \\ \epsilon_2 \\ \vdots \\ \epsilon_n \end{pmatrix}.$$

*This section may be skipped at first reading.

The above matrix **X** is known as the **design matrix**. Using matrix notation, with the convention that ′ denotes transpose, the objective function (12.3.2) can be written as

$$L(\mathbf{b}) = (\mathbf{y} - \mathbf{X}\mathbf{b})'(\mathbf{y} - \mathbf{X}\mathbf{b}) = \mathbf{y}'\mathbf{y} - 2\mathbf{b}'\mathbf{X}'\mathbf{y} + \mathbf{b}'\mathbf{X}'\mathbf{X}\mathbf{b}. \quad (12.3.8)$$

The least squares estimator, $\widehat{\boldsymbol{\beta}}$, that minimizes the objective function satisfies

$$\left.\frac{\partial L(\mathbf{b})}{\partial \mathbf{b}}\right|_{\widehat{\boldsymbol{\beta}}} = -2\mathbf{X}'\mathbf{y} + 2\mathbf{X}'\mathbf{X}\widehat{\boldsymbol{\beta}} = \mathbf{0},$$

or, equivalently, $\widehat{\boldsymbol{\beta}}$ satisfies the matrix version of the normal equations

$$\mathbf{X}'\mathbf{X}\widehat{\boldsymbol{\beta}} = \mathbf{X}'\mathbf{y}. \quad (12.3.9)$$

Multiplying both sides of (12.3.9) by the inverse of $\mathbf{X}'\mathbf{X}$ we obtain the closed form solution to the normal equations:

Least Squares Estimators of the Regression Coefficients

$$\widehat{\boldsymbol{\beta}} = (\mathbf{X}'\mathbf{X})^{-1}\mathbf{X}'\mathbf{y} \quad (12.3.10)$$

The matrix form of the fitted values and residuals is

Fitted Values and Residuals in Matrix Notation

$$\widehat{\mathbf{y}} = \mathbf{X}\widehat{\boldsymbol{\beta}} \quad \text{and} \quad \mathbf{e} = \mathbf{y} - \widehat{\mathbf{y}} \quad (12.3.11)$$

Example 12.3-2

Consider the first 8 of the 56 data points of the temperature-latitude-longitude data set of Example 12.3-1.

(a) Give the design matrices for models (1) and (2) specified in Example 12.3-1.

(b) Write the normal equations for model (1) and obtain the least squares estimators.

(c) Give the estimated regression function, fitted values, and residuals for model (1).

Solution

(a) The values of the covariates in the first 8 lines of the data set are 31.2, 32.9, 33.6, 35.4, 34.3, 38.4, 40.7, 41.7 with sample mean 36.025 for latitude, and 88.5, 86.8, 112.5, 92.8, 118.7, 123.0, 105.3, 73.4 with sample mean 100.125 for longitude. The design matrices $\mathbf{X}$ and $\widetilde{\mathbf{X}}$ for models (1) and (2), respectively, are

$$\mathbf{X} = \begin{pmatrix} 1 & -4.825 & -11.625 \\ 1 & -3.125 & -13.325 \\ 1 & -2.425 & 12.375 \\ 1 & -0.625 & -7.325 \\ 1 & -1.725 & 18.575 \\ 1 & 2.375 & 22.875 \\ 1 & 4.675 & 5.175 \\ 1 & 5.675 & -26.725 \end{pmatrix}, \quad \widetilde{\mathbf{X}} = \begin{pmatrix} 1 & -4.825 & -11.625 & 56.09 \\ 1 & -3.125 & -13.325 & 41.64 \\ 1 & -2.425 & 12.375 & -30.01 \\ 1 & -0.625 & -7.325 & 4.58 \\ 1 & -1.725 & 18.575 & -32.04 \\ 1 & 2.375 & 22.875 & 54.33 \\ 1 & 4.675 & 5.175 & 24.19 \\ 1 & 5.675 & -26.725 & -151.66 \end{pmatrix}.$$

Thus, $\mathbf{X}$ consists of a column of 1's, a column for the centered latitude values, and a column for the centered longitude values, while $\widetilde{\mathbf{X}}$ has an additional column with entries the products of the centered latitude and longitude values.

(b) The matrix $\mathbf{X}'\mathbf{X}$ and its inverse are

$$\mathbf{X}'\mathbf{X} = \begin{pmatrix} 8 & 0 & 0 \\ 0 & 101.99 & -32.88 \\ 0 & -32.88 & 2128.79 \end{pmatrix}, \quad (\mathbf{X}'\mathbf{X})^{-1} = \begin{pmatrix} 0.125 & 0 & 0 \\ 0 & 0.010 & 0.0002 \\ 0 & 0.0002 & 0.0005 \end{pmatrix}.$$

Moreover, the vector of responses is $\mathbf{y} = (44, 38, 35, 31, 47, 42, 15, 22)'$ and $\mathbf{X}'\mathbf{y} = (274,\ -221.65,\ 511.65)'$. Using the above, the normal equations (12.3.9) take the form

$$\begin{aligned} 8\widehat{\beta}_0 &= 274 \\ 101.99\widehat{\beta}_1 - 32.88\widehat{\beta}_2 &= -221.65 \\ -32.88\widehat{\beta}_1 + 2128.79\widehat{\beta}_2 &= 511.65, \end{aligned}$$

and their solution (12.3.10) is

$$\widehat{\boldsymbol{\beta}} = (\mathbf{X}'\mathbf{X})^{-1}\mathbf{X}'\mathbf{y} = (34.25,\ -2.1061,\ 0.2078)'.$$

Thus, $\widehat{\beta}_0 = 34.25$, $\widehat{\beta}_1 = -2.1061$, and $\widehat{\beta}_2 = 0.2078$.

(c) The estimated regression function is

$$\mu_{Y|X_1,X_2}(x_1, x_2) = 34.25 - 2.1061(x_1 - 36.025) + 0.2078(x_2 - 100.125),$$

the fitted or predicted values $\widehat{y}_i = \widehat{\beta}_0 + \widehat{\beta}_1(x_{i1} - 36.025) + \widehat{\beta}_2(x_{i2} - 100.125)$ are

$$\begin{aligned} &\widehat{y}_1 = 41.996, \quad \widehat{y}_2 = 38.062, \quad \widehat{y}_3 = 41.929, \quad \widehat{y}_4 = 34.044, \\ &\widehat{y}_5 = 41.743, \quad \widehat{y}_6 = 34.002, \quad \widehat{y}_7 = 25.479, \quad \widehat{y}_8 = 16.744, \end{aligned}$$

and the residuals $e_i = y_i - \widehat{y}_i$ are

$$\begin{aligned} &e_1 = 2.004, \quad e_2 = -0.063, \quad e_3 = -6.929, \quad e_4 = -3.044 \\ &e_5 = 5.257, \quad e_6 = 7.998, \quad e_7 = -10.479, \quad e_8 = 5.256. \end{aligned}$$

The closed form expression (12.3.10) of the least squares estimators can be used to prove that $\widehat{\boldsymbol{\beta}}$ is an unbiased estimator of the vector of regression coefficients and to obtain a closed form expression for the so-called *variance-covariance* matrix of $\widehat{\boldsymbol{\beta}}$:

$$E\left(\widehat{\boldsymbol{\beta}}\right) = \boldsymbol{\beta} \quad \text{and} \quad \text{Var}(\widehat{\boldsymbol{\beta}}) = \sigma_\varepsilon^2(\mathbf{X}'\mathbf{X})^{-1}. \tag{12.3.12}$$

In particular, the variance of $\widehat{\beta}_j$, the jth component of $\widehat{\boldsymbol{\beta}}$, is

$$\text{Var}(\widehat{\beta}_j) = \sigma_\varepsilon^2 C_j, \quad j = 0, 1, \ldots, k, \tag{12.3.13}$$

where C_j is the jth diagonal element of $(\mathbf{X}'\mathbf{X})^{-1}$. Moreover, (12.3.12) implies that the estimated regression function, $\widehat{\mu}_{Y|X_1,\ldots,X_k}(x_1, \ldots, x_k)$, see (12.3.3), is an unbiased estimator of the true regression function, and its variance is

$$\text{Var}(\widehat{\mu}_{Y|X_1,\ldots,X_k}(x_1, \ldots, x_k)) = \sigma_\varepsilon^2\left[(1, x_1, \ldots, x_k)(\mathbf{X}'\mathbf{X})^{-1}(1, x_1, \ldots, x_k)'\right]. \tag{12.3.14}$$

In particular, the variance of the ith fitted value $\widehat{y}_i$, see (12.3.3), is

$$\text{Var}(\widehat{y}_i) = \sigma_\varepsilon^2 h_i, \quad \text{where } h_i \text{ is the } i\text{th diagonal element of } \mathbf{X}(\mathbf{X}'\mathbf{X})^{-1}\mathbf{X}'. \tag{12.3.15}$$

12.3.2 MODEL UTILITY TEST

As in the simple linear regression model we saw in Section 8.3.3, the model utility test is used to confirm the usefulness of the multiple or polynomial regression model for predicting (i.e., for reducing the variability of) the response variable. (Throughout, the expression "reduction of the variability" is to be interpreted as a decrease of SSE.) Though not as informative in a multiple regression context (see Section 12.3.3 for more interesting hypotheses) its p-value is routinely reported in applications. Formally, the model utility test is interpreted as a test for the hypothesis

$$H_0 : \beta_1 = \cdots = \beta_k = 0 \text{ versus } H_a : H_0 \text{ is not true.} \tag{12.3.16}$$

The F test statistic for this hypothesis is

Test Statistic for the Model Utility Test

$$F_{H_0} = \frac{\text{MSR}}{\text{MSE}} \tag{12.3.17}$$

where MSE is given in (12.3.4) and MSR = SSR/k, where SSR is given in (12.3.5). Under the assumption that the intrinsic error variance is normal—the *normality assumption*—and that its variance does not depend on the covariate values—the *homoscedasticity assumption*—(already implicitly made), the null distribution of the test statistic is F with k and $n-k-1$ degrees of freedom ($F_{H_0} \sim F_{k,\, n-k-1}$). Under the alternative hypothesis in (12.3.16), F tends to take larger values, leading to the rejection rule and p-value shown below:

Rejection Region and *p*-Value for the Model Utility Test

$$F_{H_0} > F_{k,n-k-1;\alpha} \qquad p\text{-value} = 1 - F_{k,n-k-1}(F_{H_0}) \tag{12.3.18}$$

Since the F statistic (12.3.17) is used for testing if the reduction in variability is significant, it makes sense that it can be expressed in terms of R^2. This alternative expression of the F statistic is given in Exercise 6.

The model utility test is easily implemented in R. Let *fit* be the object generated from the command *fit=lm(y~(model specification))*; see Table 12-1 in Section 12.3.1. Part of the information contained in the output from the R command *summary(fit)* are the value of the F test statistic (12.3.17) and the corresponding p-value.

Standardized and Studentized Residuals The validity of the homoscedasticity and normality assumptions can be ascertained through residual plots and formal tests. Because the variance of the residuals depends on the covariate values (even if the homoscedasticity assumption holds!), it is recommended that they first be divided by an estimate of their standard deviation before applying the diagnostic plots and formal tests to them. The *standardized* and the *studentized* residuals are defined, respectively, as

$$r_i = \frac{e_i}{\widehat{\sigma}_\varepsilon \sqrt{1-h_i}} \quad \text{and} \quad \widetilde{r}_i = \frac{e_i}{\widehat{\sigma}_\varepsilon^{(-i)} \sqrt{1-h_i}}, \tag{12.3.19}$$

where h_i is defined in (12.3.15), $\widehat{\sigma}_\varepsilon$ is the square root of the MSE defined in (12.3.4), and $\widehat{\sigma}_\varepsilon^{(-i)}$ is the square root of the MSE computed without the ith data point. If *fit* is the object generated from the command *fit=lm(y~(model specification))*, the R commands

```
resid(fit); rstandard(fit)
```

give the residuals and standardized residuals, respectively. The studentized residuals are available through *library(MASS); studres(fit)*.

The diagnostic plots and formal tests are illustrated in the following example.

Example 12.3-3

For each of models (1) and (3) mentioned in Example 12.3-1 for the temperature-latitude-longitude data set, do the following.

(a) Carry out the model utility test at level $\alpha = 0.01$ and report the p-value.

(b) Use diagnostic plots and formal tests to test the assumptions of homoscedasticity and normality of the intrinsic error variable.

Solution

(a) From Example 12.3-1 we have that for both models SST = 9845.98, and the SSEs for models (1) and (3), respectively, are 2548.99 and 830.69. Thus, the SSRs are 9845.98 − 2548.99 = 7296.99 and 9845.98 − 830.69 = 9015.29 for models (1) and (3), respectively. By (12.3.17), the F statistics for models (1) and (3), respectively, are

$$F_{H_0}^{(1)} = \frac{7296.99/2}{2548.99/53} = 75.86, \quad F_{H_0}^{(3)} = \frac{9015.29/5}{830.69/50} = 108.53.$$

By (12.3.18), the corresponding p-values are $1 - F_{2,53}(75.86) = 2.22 \times 10^{-16}$ and $1 - F_{5,50}(108.53) = 0$, found by the commands *1-pf(75.86, 2, 53); 1-pf(108.53, 5, 50)*. The commands *summary(fit1); summary(fit3)*, used in Example 12.3-1, give "F-statistic: 75.88 on 2 and 53 DF, p-value: 2.792e-16" for model (1), and "F-statistic: 108.5 on 5 and 50 DF, p-value: < 2.2e-16" for model (3). Since both p-values are less than 0.01, the model utility test rejects the null hypothesis for

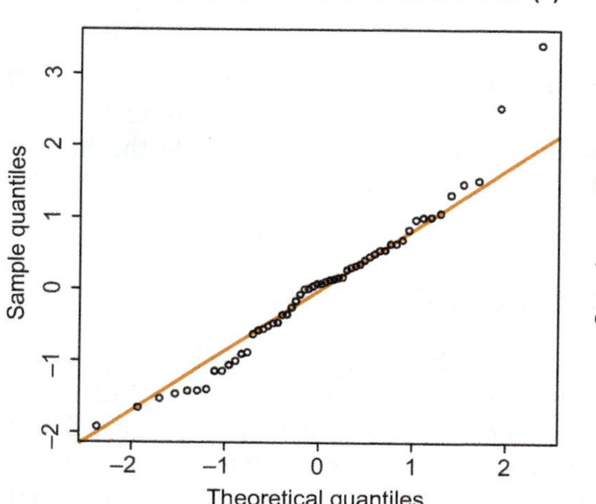

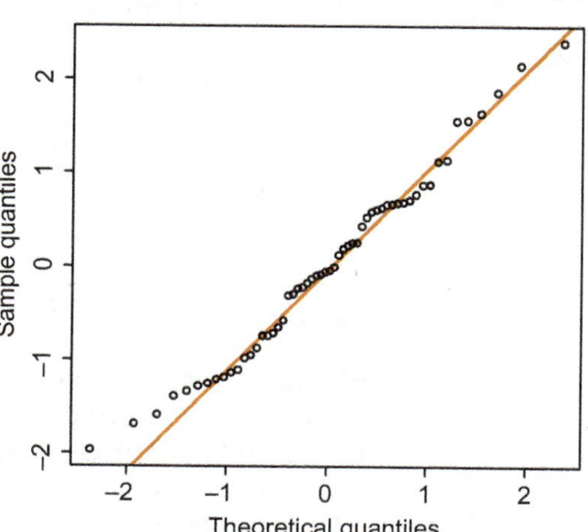

Figure 12-5 Normal Q-Q plots for the standardized residuals of models (1) and (3).

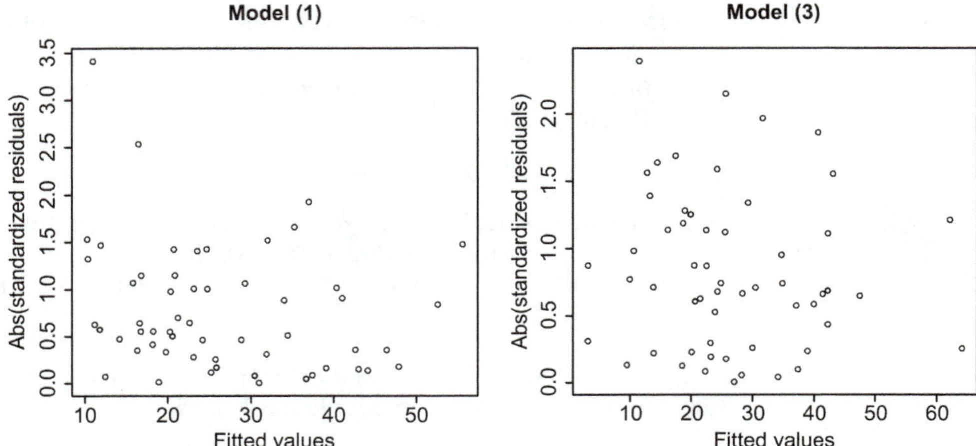

Figure 12-6 Plots of $|r_i|$ vs $\widehat{y}_i$ for models (1) and (3).

both models. Thus, both models achieve a significant reduction in the variability of the response variable, that is, SSR is a significant portion of SST for both models.

(b) With the objects *fit1* and *fit3* as produced in Example 12.3-1, the commands

```
shapiro.test(rstandard(fit1)); shapiro.test(rstandard(fit3))
```

return p-values 0.08 and 0.51, respectively. This suggests that the normality of the residuals in model (1) is suspect (probably due to the fact that model (1) does not include important predictors), while for model (3) the normality assumption appears to be satisfied. The Q-Q plot for the standardized residuals of model (1), shown in the left panel of Figure 12-5, reveals a couple of larger than expected observations contributing to the smaller p-value for model (1).

Plotting the standardized residuals, or their absolute values, against the fitted values helps reveal possible heteroscedasticity for the model (1) residuals; see Figure 12-6. To assign p-values with such plots one can try to fit a model with fitted values serving as the covariate and the absolute value of the standardized residuals serving as the response variable. The commands

```
r1=lm(abs(rstandard(fit1))~poly(fitted(fit1), 2));
  summary(r1)
r3=lm(abs(rstandard(fit3))~poly(fitted(fit3), 2));
  summary(r3)
```

give "F-statistic: 3.234 on 2 and 53 DF, p-value: 0.04731" for model (1), and "F-statistic: 0.2825 on 2 and 53 DF, p-value: 0.755" for model (3). This suggests that the model (1) residuals are heteroscedastic, while the homoscedasticity assumption is not contradicted for the model (3) residuals. ■

12.3.3 TESTING THE SIGNIFICANCE OF REGRESSION COEFFICIENTS

The most common inference question in multiple/polynomial regression is whether a variable, or a group of variables, contributes to significant *additional* decrease of variability, that is, significant *additional* decrease of the error sum of squares, given all other variables in the model. According to the aforementioned (see Section 12.3.1) principle of parsimony, if a variable or group of variables do not contribute to a significant decrease of variability they should be omitted from the model. Groups of variables that are commonly tested for their significance include all higher order polynomial terms, or all interaction terms, etc.

Formally, testing the significance of a variable or group of variables amounts to testing the null hypothesis that the coefficient or group of coefficients of the variables in question are zero, versus the alternative, which states that the null hypothesis is false.

Under the null hypothesis $H_0^j : \beta_j = 0$, and the additional assumptions of normality and homoscedasticity, the ratio of $\widehat{\beta}_j$ to its standard error has a T_ν distribution, where $\nu = \mathrm{DF}_{SSE} = n - k - 1$, that is,

$$T_{H_0^j} = \frac{\widehat{\beta}_j}{\widehat{\sigma}_{\widehat{\beta}_j}} \sim T_{n-k-1}, \quad \text{where } \widehat{\sigma}_{\widehat{\beta}_j} = \widehat{\sigma}_\varepsilon \sqrt{C_j} \text{ with } C_j \text{ given in (12.3.13).} \qquad \textbf{(12.3.20)}$$

The rejection rule and p-value for testing $H_0^j : \beta_j = 0$ vs $H_a^j : H_0^j$ is false, are given next.

Procedure for Testing $H_0^j : \beta_j = 0$ vs $H_a^j : H_0^j$ Is False

$$\text{Test statistic: } T_{H_0^j} \text{ given in (12.3.20)}$$
$$\text{Rejection region: } |T_{H_0^j}| \geq t_{n-k-1;\alpha/2}, \qquad \textbf{(12.3.21)}$$
$$p\text{-value} = 2(1 - G_{n-k-1}(T_{H_0^j}))$$

where G_ν denotes the cumulative distribution function of the T_ν distribution. A scatterplot of the residuals from the null hypothesis model (i.e., the model without the variable being tested), against the variable being tested, provides a useful visual impression of its potential for additional significant reduction of the error sum of squares. See Example 12.3-4 for an illustration of such a plot. Finally, using (12.3.20), testing H_0^j versus one-sided alternative hypotheses and CIs for β_j are constructed as usual; see Section 12.3.4 for a discussion of CIs.

The standard errors $\widehat{\sigma}_{\widehat{\beta}_j}$, the ratios $\widehat{\beta}_j / \widehat{\sigma}_{\widehat{\beta}_j}$, and the p-values for testing each H_0^j : $\beta_j = 0$ vs $H_a^j : H_0^j$ is false, $j = 0, 1, \ldots, k$, are all part of the output of the R commands

```
fit=lm(y~(model specification)); summary(fit).
```

See Example 12.3-4 for an illustration.

To describe the procedure for testing for the significance of a group of variables, let the **full model** and the **reduced model** refer to the model that includes the variables being tested and the one without the variables being tested, respectively. Thus, if out of the k variables $X_1, \ldots, X_k$ (some of which may correspond to polynomial and/or interaction terms) we want to test the significance of the last ℓ of them, that is, of $X_{k-\ell+1}, \ldots, X_k$, for $\ell < k$, the full and reduced models are

$$Y = \beta_0 + \beta_1 X_1 + \cdots + \beta_k X_k + \varepsilon \quad \text{and} \quad Y = \beta_0 + \beta_1 X_1 + \cdots + \beta_{k-\ell} X_{k-\ell} + \varepsilon,$$

respectively. Testing for the reduced model versus the full model is equivalent to testing

$$H_0 : \beta_{k-\ell+1} = \cdots = \beta_k = 0 \text{ vs } H_a : H_0 \text{ is not true.} \qquad \textbf{(12.3.22)}$$

Let SSE_f and SSE_r denote the error sums of squares resulting from fitting the full and reduced models, respectively. If SSE_f is significantly smaller than SSE_r, that is, if $X_{k-\ell+1}, \ldots, X_k$ contribute significantly in reducing the variability, then these variables should be kept in the model. The formal test procedure is

Test Procedure for the Hypotheses in (12.3.22)

$$\text{Test Statistic: } F_{H_0} = \frac{(\text{SSE}_r - \text{SSE}_f)/\ell}{\text{SSE}_f/(n-k-1)}$$
$$\text{Rejection Region: } F_{H_0} \geq F_{\ell,n-k-1;\alpha}, \quad p\text{-value} = 1 - F_{\ell,n-k-1}(F_{H_0}) \qquad (12.3.23)$$

When $\ell = 1$, that is, when testing for the significance of a single covariate, the test procedures (12.3.21) and (12.3.23) give identical p-values. With *fitF* and *fitR* being the R objects generated from fitting the full and reduced models, respectively, that is, *fitF=lm(y~(full model)); fitR=lm(y~(reduced model))*, the R command

```
anova(fitF, fitR)
```

gives the value of the F statistic in (12.3.23) and the corresponding p-value.

Example 12.3-4

For the temperature-latitude-longitude data set of Example 12.3-1, let X_1 and X_2 denote the centered latitude and longitude variables, and use R commands to complete the following parts.

(a) Fit the model that includes second order polynomial terms in X_1 and X_2 and X_1X_2 (the X_1X_2 interaction term), and report the p-values from the T tests for individually testing the significance of each of the five covariates, that is, X_1, X_1^2, X_2, X_2^2 and $X_1\ X_2$. Which of the coefficients are significantly different from zero at level $\alpha = 0.01$?

(b) Perform the procedure (12.3.23) for testing the significance of the X_1X_2 interaction term. Confirm that the p-value of the F test is identical to the p-value of the corresponding T test in part (a). Construct a scatterplot of the reduced model residuals against X_1X_2, and comment on whether or not the visual impression it conveys about the significance of the X_1X_2 interaction term for additional reduction of the error variability is consistent with the p-value.

(c) Perform the procedure (12.3.23) for testing the significance of th X_1^2 term, and confirm that the p-value of the F test is identical to the p-value of the corresponding T test in part (a). Construct a scatterplot of the reduced model residuals against X_1^2, and comment on whether or not the visual impression it conveys about the significance of the X_1^2 term for additional reduction of the error variability is consistent with the p-value.

(d) Perform the procedure (12.3.23) for testing the joint (i.e., as a group) significance of the two quadratic terms, X_1^2, X_2^2, in the model also containing X_1, X_2, X_1X_2.

Solution

(a) The commands for fitting the model with the second order polynomials and interaction term, and for producing the T test statistics and p-values for the significance of each regression coefficient are:

```
fitF=lm(y~poly(x1, 2, raw=T)+poly(x2, 2, raw=T)+x1:x2);
  summary(fitF)
```

The outcome it produces is

	Estimate	Std. Error	t value	Pr(> |t|)
(Intercept)	21.307601	0.908516	23.453	< 2e-16
poly(x1, 2, raw=T)1	-2.613500	0.134604	-19.416	< 2e-16
poly(x1, 2, raw=T)2	-0.011399	0.019880	-0.573	0.568968
poly(x2, 2, raw=T)1	-0.177479	0.048412	-3.666	0.000596
poly(x2, 2, raw=T)2	0.023011	0.002715	8.474	3.10e-11
x1:x2	0.041201	0.009392	4.387	5.93e-05

The p-values from the T test for each term are given in the last column of the above table. Five of the six p-values are less than 0.01. Thus, all regression coefficients except the one for X_1^2 are significantly different from zero at level $\alpha = 0.01$.

(b) To perform the F test for the significance of X_1X_2 interaction term, the model fitted in part (a) is designated as the "full model" and the model resulting from omitting X_1X_2 interaction is designated as the "reduced model." With the object *fitF* as generated in part (a), the commands

```
fitR=lm(y~poly(x1, 2, raw=T)+poly(x2, 2, raw=T));
  anova(fitR, fitF)
```

produce the output

```
Model 1: y~poly(x1, 2, raw=T)+poly(x2, 2, raw=T)+x1:x2
Model 2: y~poly(x1, 2, raw=T)+poly(x2, 2, raw=T)
```

	Res.Df	RSS	Df	Sum of Sq	F	Pr(>F)
1	51	1150.48				
2	50	830.72	1	319.76	19.246	5.931e-05

Thus, the value of the F statistic is $F_{H_0} = 19.246$, which corresponds to a p-value of 5.93×10^{-5}. This p-value is identical to the p-value for the interaction term reported in part (a); note also that $19.246 = 4.387^2$. The command

```
plot(x1*x2, resid(fitR), main="Reduced Model
  Residuals vs x1*x2")
```

produces the scatterplot in the left panel of Figure 12-7. The plot conveys the impression of a linear trend between the reduced model and the product X_1X_2. This visual impression suggests that the interaction term contributes to a significant additional reduction of the reduced model SSE, which is consistent with the small p-value of the F test.

(c) To perform the F test for the significance of X_1^2, the model fitted in part (a) is designated as the "full model" and the model resulting from omitting the X_1^2 term is designated as the "reduced model." With the object *fitF* as generated in part (a), the commands (note that the first command redefines the object *fitR*)

```
fitR=lm(y~x1+poly(x2, 2, raw=T)+x1:x2); anova(fitR, fitF)
```

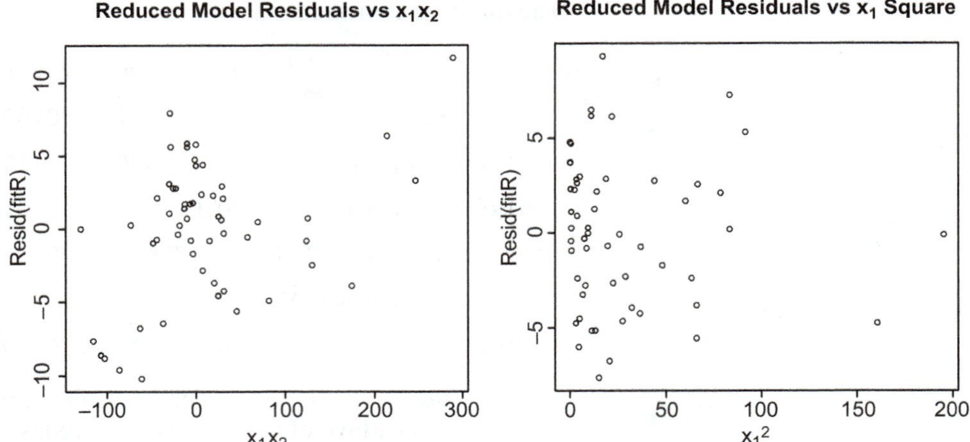

Figure 12-7 Reduced model residual plots for assessing the significance of the X_1X_2 interaction term (left panel) and the X_1^2 term (right panel).

produce the output

```
Model 1: y~poly(x1, 2, raw=T)+poly(x2, 2, raw=T)+x1:x2
Model 2: y~x1+poly(x2, 2, raw=T)+x1:x2
   Res.Df     RSS  Df  Sum of Sq       F   Pr(>F)
1      51  836.18
2      50  830.72   1     5.4619  0.3287    0.569
```

Thus, the value of the F statistic is $F_{H_0} = 0.3287$ with a corresponding p-value of 0.569. This p-value is identical (up to round-off) to the p-value for the X_1^2 term reported in part (a); note also that $0.3287 = (-0.573)^2$. The command

```
plot(x1^2, resid(fitR), main="Reduced Model
  Residuals vs x1 Square")
```

produces the scatterplot in the right panel of Figure 12-7. The plot does not convey the impression of any trend between the reduced model residuals and X_1^2. This visual impression suggests that X_1^2 does not contribute to a significant additional reduction of the reduced model SSE, which is consistent with the large p-value of the F test.

(d) To perform the F test for the joint significance of X_1^2 and X_2^2, the model fitted in part (a) is designated as the "full model" and the model resulting from omitting the X_1^2 and X_2^2 terms is designated as the "reduced model." With the object *fitF* as generated in part (a), the commands (note that *fitR* is being redefined again)

```
fitR=lm(y~x1+x2+x1:x2); anova(fitF, fitR)
```

produce output, similar in layout to the output in parts (b) and (c) but not shown here, giving $F_{H_0} = 36.065$ with a corresponding p-value of 2.012×10^{-10}. This suggests that the two quadratic terms as a group contribute to a significant additional reduction of the reduced model SSE, without specifying the contribution of each individual quadratic term. ■

Scatterplot Matrices Scatterplot matrices were used in Section 1.5.2 as a tool for identifying the single most useful predictor for a response variable. When used for identifying *groups* of important predictors, as is often done in multiple regression contexts, one should not be surprised if formal tests contradict what a scatterplot matrix suggests. Indeed, it is possible for a variable to be a significant predictor when taken individually (which is what the scatterplot shows), but an insignificant predictor when other variables are taken into consideration. The data in Exercise 2 offer such an example. Surprisingly, the opposite can also occur. Namely, an individual scatterplot between a covariate and a response variable may not show any relationship, or show a negative association, but, when other predictors are taken into consideration, the covariate becomes significant, or the nature of the association reverses. The data in Exercise 12 in Section 12.4 offer such an example.

12.3.4 CONFIDENCE INTERVALS AND PREDICTION

Confidence intervals for the regression parameters β_j and the regression function $\mu_{Y|X_1,\ldots,X_k}(x_1,\ldots,x_k)$, as well as prediction intervals for a future observation, are as follows:

1. With $\widehat{\sigma}_{\widehat{\beta}_j}$ given in (12.3.20), a $100(1-\alpha)\%$ CI for β_j is $\widehat{\beta}_j \pm t_{n-k-1;\alpha/2}\widehat{\sigma}_{\widehat{\beta}_j}$.
2. A $100(1-\alpha)\%$ CI for $\mu_{Y|X_1,\ldots,X_k}(\mathbf{x})$, where $\mathbf{x} = (x_1,\ldots,x_k)$, is

$$\widehat{\mu}_{Y|X_1,\ldots,X_k}(\mathbf{x}) \pm t_{n-k-1;\alpha/2}\widehat{\sigma}_{\widehat{\mu}_{Y|X_1,\ldots,X_k}(\mathbf{x})},$$

   where $\widehat{\sigma}_{\widehat{\mu}_{Y|X_1,\ldots,X_k}(\mathbf{x})} = \widehat{\sigma}_\varepsilon^2(1,x_1,\ldots,x_k)(\mathbf{X}'\mathbf{X})^{-1}(1,x_1,\ldots,x_k)'$; see (12.3.14).
3. A $100(1-\alpha)\%$ PI for a future observation y to be taken at $\mathbf{x} = (x_1,\ldots,x_k)$ is

$$\widehat{\mu}_{Y|X_1,\ldots,X_k}(\mathbf{x}) \pm t_{n-k-1;\alpha/2}\sqrt{\widehat{\sigma}_\varepsilon^2 + \widehat{\sigma}^2_{\widehat{\mu}_{Y|X_1,\ldots,X_k}(\mathbf{x})}}.$$

The matrix operations required for the computation of the standard errors in these formulas make hand calculations impractical. Their implementation in R is very convenient and is described next.

With *fit* being the object generated from fitting the multiple/polynomial regression model, that is, *fit=lm(y~(model specification))*, the R command

```
confint(fit, level=1-α)
```

gives $(1-\alpha)100\%$ CIs for all $k+1$ regression coefficients. (By default, *confint(fit)* gives a 95% CI for all coefficients.)

As in the simple linear regression model, the function *predict( )* can be used to make both CIs for the regression function, $\mu_{Y|X_1,\ldots,X_k}(x_1,\ldots,x_k)$ at specified covariate values $(x_1,\ldots,x_k)$, and prediction intervals for a future value of the response variable taken at covariate values $(x_1,\ldots,x_k)$. To make a CI for the regression function use the option *interval="confidence"*. To make a PI use the option *interval="prediction"*. As always, the default level is 95%, but the option *level=*1-α can be added for other confidence levels. Specification of the covariate values $(x_1,\ldots,x_k)$ is done through the *data.frame( )* option; see Example 12.3-5, parts (b) and (c), for an illustration.

Example 12.3-5

For the data of Example 12.3-1, use R commands to fit the model that includes second order polynomial terms in latitude and longitude, as well as the latitude-longitude interaction, and to complete the following parts.

(a) Construct 90% CIs for each of the six regression coefficients.

(b) Construct a 90% CI for the mean average daily minimum temperature for a typical month of January at latitude and longitude of 35 and 110 degrees, respectively.

(c) Construct a 90% PI for the average daily minimum temperature of next January at latitude and longitude of 35 and 110 degrees, respectively.

Solution

As in Example 12.3-1, set the values of the response variable and those of the centered covariates into the R objects *y, x1, x2*, respectively, and use *fit=lm(y~poly(x1, 2, raw=T)+poly(x2, 2, raw=T)+x1:x2)* to fit the model with second order polynomial terms and interaction.

(a) The command *confint(fit, level=0.9)* generates the output

```
                              5%          95%
(Intercept)          19.78501346  22.83018765
poly(x1, 2, raw=T)1  -2.83908334  -2.38791667
poly(x1, 2, raw=T)2  -0.04471617   0.02191884
poly(x2, 2, raw=T)1  -0.25861371  -0.09634472
poly(x2, 2, raw=T)2   0.01846049   0.02756225
x1:x2                 0.02546145   0.05694009
```

Thus, a 90% CI for the coefficient of X_1 is $(-2.839, -2.388)$, suggesting that this coefficient is significantly different from zero; a 90% CI for the coefficient of X_1^2 is $(-0.0447, 0.0219)$, suggesting that this coefficient is not significantly different from zero, and so forth.

(b) The command for constructing a 90% CI for the regression line at latitude and longitude of 35 and 110 degrees, is

```
predict(fit, data.frame(x1=35-38.97, x2=110-90.96),
  interval="confidence", level=0.9).
```

Note that in this command, the sample mean of latitude has been subtracted from 35 and the sample mean of longitude has been subtracted from 110. This is necessary because the model was fitted using the centered latitude and longitude values. The output produced by this command is

```
       fit       lwr      upr
1 33.35018  31.00375  35.6966
```

Thus, the average daily minimum temperature for a typical month of January at latitude and longitude of 35 and 110 is estimated to be $33.3^o F$, and a 90% CI for it is (31.0, 35.7).

(c) The command for constructing a 90% PI for the regression line at latitude and longitude of 35 and 110 degrees, is

```
predict(fit, data.frame(x1=35-38.97, x2=110-90.96),
  interval="prediction", level=0.9).
```

The output produced by this command is

```
       fit      lwr      upr
1 33.35018 26.12731 40.57304
```

Thus, the average daily minimum temperature next January at latitude and longitude of 35 and 110 degrees is predicted to be 33.3^oF, and a 90% PI for it is (26.1, 40.6). ■

Exercises

1. An article reports on a study of methyl tertiary-butyl ether (MTBE) in the vicinity of two gas stations, one urban and one roadside, equipped with stage I vapor recovery systems.[3] The data set in *GasStatPoll.txt* contains the MTBE concentration measurements along with the covariates "Gas Sales," "Wind Speed," and "Temperature" from eight days in May–June and October. Use R commands to complete the following.

(a) Fit an MLR model, with no polynomial or interaction terms, for predicting concentration on the basis of the three covariates. Report the estimated regression model, R^2, and the p-value for the model utility test. Is the model useful for predicting MTBE concentrations?

(b) Give the fitted value and the residual corresponding to the first observation in the data set. (*Hint.* Use *fitted(fit); resid(fit)*.)

(c) Use diagnostic plots and formal tests to check the assumptions of normality and heteroscedasticity. (See Example 12.3-3, part (b).)

(d) Use the T test procedure to test the significance of each predictor at level of significance 0.05.

(e) Construct 95% CIs for the regression coefficients.

2. The R data set *stackloss*[4] contains data from 21 days of operation of a plant whose processes include the oxidation of ammonia (NH_3) to nitric acid (HNO_3). The nitric-oxide wastes produced are absorbed in a countercurrent absorption tower. There are three predictor variables: Air.Flow represents the rate of operation of the plant; Water.Temp is the temperature of cooling water circulated through coils in the absorption tower; and Acid.Conc. is the concentration of the acid circulating (given as the actual percentage minus 50 then times 10; that is, 89 corresponds to 58.9 percent acid). The dependent variable, stack.loss, is 10 times the percentage of the ingoing ammonia to the plant that escapes from the absorption tower unabsorbed; that is, an (inverse) measure of the overall efficiency of the plant. Use *y=stackloss$stack.loss; x1=stackloss$Air.Flow; x2=stackloss$Water.Temp; x3=stackloss$Acid.Conc.* (do not omit the period!) to set the dependent variable and the predictors in the R objects *y, x1, x2, x3*, respectively, and *x1=x1-mean(x1); x2=x2-mean(x2); x3=x3-mean(x3)* to center the predictor variables. Then use R commands to complete the following.

(a) Fit an MLR model, with no polynomial or interaction terms, for predicting stackloss on the basis of the three covariates. Report the estimated regression model, the adjusted R^2, and the p-value for the model utility test. Is the model useful for predicting stackloss?

(b) Report the p-value for *x3*; is it a useful predictor in the model? Fit the MLR model using only *x1* and *x2*, with no polynomial or interaction terms. Report the adjusted R^2 value and compare it with the adjusted R^2 value of part (a). Justify the comparison in terms of the p-value for *x3*.

(c) Using the model with only *x1* and *x2*, give a 95% CI for the expected stackloss at water temperature 20 and acid concentration 85. (*Hint.* The values 20 and 85 are not centered!)

(d) Fit the MLR model based on second order polynomials for *x1* and *x2*, as well as their interaction. Test the joint (i.e., as a group) significance of the two quadratic terms and the X_1X_2 interaction, using $\alpha = 0.05$.

(e) Use *pairs(stackloss)* to produce a scatterplot matrix for the data. Does "Acid.Conc." appear correlated with "stack.loss"? What explanation do you give for the high p-value shown for this variable reported in part (b)?

3. The R data set *state.x77*, collected by the US Bureau of the Census in the 1970s, has the population, per capita income, illiteracy, life expectancy, murder rate, percent high school graduation, mean number of frost days (defined as days with minimum temperature below freezing in the capital or a large city for years 1931–1960),

[3] Vainiotalo et al. (1998). MTBE concentrations in ambient air in the vicinity of service stations. *Atmospheric Environment*, 32(20): 3503–3509.

[4] K. A. Brownlee (1960, 2nd ed. 1965). *Statistical Theory and Methodology in Science and Engineering*. New York: Wiley, pp 491–500.

and land area in square miles for each of the 50 states. It can be imported into the R data frame *st* by *data(state); st=data.frame(state.x77, row.names=state.abb, check.names=T)*, or by *st = read.table("State.txt", header=T)*. We will consider life expectancy to be the response variable and the other seven as predictor variables. Use R commands to complete the following.

(a) Use *h1=lm(Life.Exp~Population+Income+Illiteracy+Murder+HS.Grad+Frost+Area, data=st); summary(h1)* to fit an MLR model, with no polynomial or interaction terms, for predicting life expectancy on the basis of the seven predictors. Report the estimated regression model, the R^2adj, and the p-value for the model utility test. Is the model useful for predicting life expectancy?

(b) Test the joint significance of the variables "Income," "Illiteracy," and "Area" at level 0.05. (*Hint.* The reduced model is most conveniently fitted through the function *update*. The R command for this is *h2=update(h1, .~. -Income-Illiteracy-Area)*. According to the syntax of *update( )*, a dot means "same." So the above update command is read as follows: "Update *h1* using the same response variables and the same predictor variables, except remove (minus) "Income," "Illiteracy," and "Area.")

(c) Compare the R^2 and R^2adj values for the full and reduced models. Is the difference in R^2 consistent with the p-value in part (b)? Explain. Why is the difference in R^2adj bigger?

(d) Using standardized residuals from the reduced model, test the assumptions of normality and homoscedasticity, both graphically and with formal tests.

(e) Using the reduced model, give the fitted value for the state of California. (Note that California is listed fifth in the data set.) Next, give a prediction for the life expectancy in the state of California with the murder rate reduced to 5. Finally, give a 95% prediction interval for life expectancy with the murder rate reduced to 5.

4. The file *HardwoodTensileStr.txt* has data on hardwood concentration and tensile strength.[5] Import the data into the R data frame *hc* and use *x=hc$Concentration; x=x-mean(x); y=hc$Strength* to set the centered predictor and the response variable in the R objects x and y, respectively.

(a) Use *hc3=lm(y~x+I(x^2)+I(x^3)); summary(hc3)* to fit a third order polynomial model to this data. Report the adjusted R^2 and comment on the significance of the model utility test and the regression coefficients, using level of significance 0.01.

(b) Use *plot(x,y); lines(x,fitted(hc3), col="red")* to produce a scatterplot of the data with the fitted curve superimposed. Are you satisfied with the fit provided by the third order polynomial model?

(c) The plot of the residuals vs the fitted values (*plot(hc3, which=1)*—try it!) suggests that the fit can be improved. Use *hc5=lm(y~x+I(x^2)+I(x^3)+I(x^4)+I(x^5)); summary(hc5)* to fit a fifth order polynomial model to this data. Report the adjusted R^2 and comment on the significance of the regression coefficients, using level of significance 0.01.

(d) Omit the polynomial term with the largest p-value from the fit in part (c) and fit the model with all other terms. Comment on the significance of the regression coefficients, using $\alpha = 0.01$. Compare the adjusted R^2 to that of part (a). Finally, construct a scatterplot of the data with the fitted curve from the final model superimposed. Compare this scatterplot to the one of part (b).

5. The data in *EmployPostRecess.txt* has the number of employees of a particular company during 11 post-recession quarters. Import the data into the R data frame *pr* and use *x=pr$Quarter; xc=x-mean(x); y=pr$Population* to set the centered predictor and the response variable in the R objects x and y, respectively.

(a) Use *pr3=lm(y~x+I(x^2)+I(x^3))* to fit a third-degree polynomial to this data set. Report R^2 and the adjusted R^2 and comment on the significance of the model utility test at level of significance 0.01.

(b) Test the joint significance of the quadratic and cubic terms at $\alpha = 0.01$ by fitting a suitably reduced model.

(c) Use *pr8=lm(y~poly(x, 8, raw=T))* to fit an eighth-degree polynomial, and use it and the fit of part (a) to test the joint significance of the polynomial terms of orders four through eight, at level of significance 0.01. Next, report R^2 and the adjusted R^2 from fit *pr8*, and compare with those from fit *pr3* of part (a).

(d) Use *plot(x, y); lines(x, fitted(pr3))* to superimpose the fit *pr3* on the scatterplot of the data. Construct a different plot superimposing the fit *pr8* to the data. Finally, fit a tenth-degree polynomial using *pr10=lm(y~poly(x, 10, raw=T))* and superimpose its fit on the scatterplot of the data. What do you notice?

6. Show that the F statistic for the model utility test given in (12.3.17) can be expressed in terms of R^2 as

$$F = \frac{R^2/k}{(1 - R^2)/(n - k - 1)}.$$

[5] D. C. Montgomery, E. A. Peck, and G. G. Vining (2012). *Introduction to Linear Regression Analysis*, 5th ed. Hoboken: Wiley & Sons. Table 7.1.

12.4 Additional Topics

12.4.1 WEIGHTED LEAST SQUARES

When the intrinsic error variable is homoscedastic, that is, its variance is constant throughout the range of values of the predictor variables, the least squares estimators have smaller variance than any other linear unbiased estimator. Moreover, if the sample size is sufficiently large, the tests of hypotheses and confidence intervals discussed in Section 12.3 do not require the normality assumption. This is no longer the case if the intrinsic error variable is heteroscedastic, due to the fact that the formulas for the standard errors of the regression coefficients are valid only under homoscedasticity. Under heteroscedasticity, however, the formulas for the standard errors of the regression coefficients are not valid. Thus, while the estimators remain consistent, the CIs are not valid; see Exercise 1.

The basic reason underlying this breakdown of the least squares method is that all data points are weighted equally in the objective function that is minimized. Under heteroscedasticity, however, observations with higher variance are less precise and should be weighted less. The **weighted least squares estimators**, $\widehat{\beta}_0^w, \widehat{\beta}_1^w, \ldots, \widehat{\beta}_k^w$ are obtained by minimizing the objective function

$$L_w(b_0, b_1, \ldots, b_k) = \sum_{i=1}^{n} w_i \left(y_i - b_0 - \sum_{j=1}^{k} b_j x_{ij} \right)^2, \tag{12.4.1}$$

where the w_i's are weights that decrease with increasing intrinsic error variance. Let σ_i^2 denote the intrinsic error variance at the ith data point, that is,

$$\sigma_i^2 = \text{Var}(\varepsilon_i) = \text{Var}(Y_i | X_1 = x_{i1}, \ldots, X_k = x_{ik}). \tag{12.4.2}$$

Choosing the weights w_i to be inversely proportional to σ_i^2, that is, $w_i = 1/\sigma_i^2$, yields optimal estimators of the regression parameters. The R command for fitting a regression model with the method of weighted least squares is

```
fitw=lm(y~(model specification), weights=c(w1,...,wn))
```

The main disadvantage of weighted least squares lies in the fact that the intrinsic error variances σ_i^2, defined in (12.4.2), and hence the optimal weights $w_i = 1/\sigma_i^2$, are unknown. In the absence of a "Regression Oracle" (see Figure 12-8), these weights will have to be estimated. Keeping in mind that inferences about coefficients may not be valid for small sample sizes when the weights are estimated from the data, even under normality, a simple method for estimating the weights is as follows:

1. Use *fit=lm(y~(model specification))* to fit the regression model using ordinary least squares.
2. Use *abse=abs(resid(fit)); yhat=fitted(fit)* to set the absolute values of the residuals in the object *abse* and the fitted values in the object *yhat*.
3. Use *efit=lm(abse~poly(yhat,2)); shat=fitted(efit)* to regress *abse* on *yhat* and to set the resulting fitted values in *shat*.
4. Use *w=1/shat^2; fitw=lm(y~(model specification), weights=w)* to obtain the vector of weights and fit the regression model using weighted least squares.

Figure 12-8 Statistician (right) getting the heteroscedastic intrinsic error variances from the Regression Oracle (left).

Example 12.4-1

The data file *AgeBlPrHeter.txt* contains data on age and diastolic blood pressure of 54 adult male subjects. The scatterplot shown in Figure 12-9 suggests a heteroscedastic linear relation between the two variables. Use R commands to estimate the heteroscedastic intrinsic error variances and to construct a 95% CI for the regression slope using a weighted least squares analysis.

Solution

Import the age and blood pressure values into the R objects x and y, respectively, and use *fit=lm(y~x); abse=abs(resid(fit)); yhat=fitted(fit); efit=lm(abse~yhat); w=1/fitted(efit)^2* to generate the weights for the weighted least squares analysis. The 95% CI for the regression slope, generated by the commands

```
fitw=lm(y~x, weights=w); confint(fitw),
```

is (0.439, 0.755). As a comparison, the corresponding CI obtained by the command *confint(fit)* is (0.385, 0.773), which is a little wider than the weighted least squares CI. More important, the actual confidence level of the interval (0.385, 0.773) is less than 95% due to the fact that, under heteroscedasticity, the ordinary least squares analysis underestimates the standard error of the slope; see Exercise 1. ■

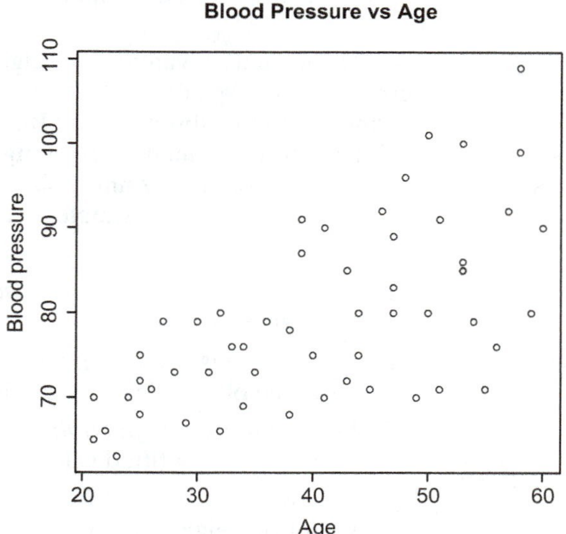

Figure 12-9 Scatterplot showing linear trend with heteroscedasticity.

12.4.2 APPLICATIONS TO FACTORIAL DESIGNS

The traditional applications of regression analysis involve quantitative predictor variables. Categorical predictors, however, such as type of root system, coastal or urban locations, gender, type of filter or material, education or income category, strength of opinion, and so forth, can also be incorporated into regression models by assigning numerical codes to the different categories. In particular, factorial designs can be represented as multiple regression models and can be analyzed through the multiple regression methodology. The main advantage of using the multiple regression formulation for analyzing factorial designs is that R and other software packages make the option of weighted least squares available in their multiple regression menus.

For the multiple regression analysis to yield the information obtained from the factorial design analysis presented in Chapters 10 and 11, each main effect and interaction term must correspond to one of the coefficients in the multiple regression model. This is possible with a suitable numerical coding of the levels, the so-called **contrast coding.** For a factor with k levels, contrast coding uses the following $k-1$ **indicator variables**:

$$X_i = \begin{cases} 1 \text{ if the observation comes from level } i \\ -1 \text{ if the observation comes from level } k, \quad i = 1, \ldots, k-1 \\ 0 \text{ otherwise.} \end{cases} \tag{12.4.3}$$

Taken jointly, these $k-1$ indicator variables provide the convention for representing the k factor levels shown in Table 12.2. This indicator variable representation of the k levels of a factor can be used to cast factorial designs into the multiple regression format. The details for one- and two-way layouts are given in the following paragraphs.

Regression Formulation of the k-Sample Problem Let $Y_{i1}, \ldots, Y_{in_i}$, $i = 1, \ldots, k$, denote the k samples. According to model (10.2.2),

$$Y_{ij} = \mu_i + \epsilon_{ij} \quad \text{or} \quad Y_{ij} = \mu + \alpha_i + \epsilon_{ij}, \tag{12.4.4}$$

where the treatment effects α_i satisfy the condition (10.2.3). A basic difference between the data representation in regression models and in factorial design models is the single index enumeration of the observations in regression models versus the multiple indices used in the factorial design representation. Regression models avoid multiple indices by giving explicitly the covariate values corresponding to each value of the response variable.

To cast data in (12.4.4) into the multiple linear regression format, set $n = n_1 + \cdots + n_k$, let $Y_1, \ldots, Y_n$ be a single index enumeration of the Y_{ij}'s, and let

Table 12-2 Indicator variable coding of k factor levels

	Indicator Variable				
Factor Level	X_1	X_2	X_3	$\cdots$	X_{k-1}
1	1	0	0	$\cdots$	0
2	0	1	0	$\cdots$	0
$\vdots$	$\vdots$	$\vdots$	$\vdots$	$\vdots$	
$k-1$	0	0	0	$\cdots$	1
k	-1	-1	-1	$\cdots$	-1

Table 12-3 Cell means in (12.4.4) in terms of the coefficients in (12.4.6)

Cell	Cell Mean	Value of Predictors $(x_1, \ldots, x_{k-1})$	Regression Function $\mu_{Y\mid X_1,\ldots,X_{k-1}}(x_1, \ldots, x_{k-1})$
1	$\mu_1 = \mu + \alpha_1$	$(1, 0, \ldots, 0)$	$\beta_0 + \beta_1$
2	$\mu_2 = \mu + \alpha_2$	$(0, 1, \ldots, 0)$	$\beta_0 + \beta_2$
$\vdots$	$\vdots$	$\vdots$	$\vdots$
$k-1$	$\mu_{k-1} = \mu + \alpha_{k-1}$	$(0, 0, \ldots, 1)$	$\beta_0 + \beta_{k-1}$
k	$\mu_k = \mu + \alpha_k$	$(-1, -1, \ldots, -1)$	$\beta_0 - \beta_1 - \cdots - \beta_{k-1}$

$X_{j1}, \ldots, X_{j,k-1}$, $j = 1, \ldots, n$, be $k-1$ indicator variables designating, in the coding convention of Table 12.2, the level each Y_j came from. Thus, the regression format representation (i.e., response and covariate values) of the data in (12.4.4) is

$$(Y_j, X_{j1}, \ldots, X_{j,k-1}), \quad j = 1, \ldots, n. \tag{12.4.5}$$

To see to what extent parameters of the multiple linear regression model

$$Y_j = \beta_0 + \beta_1 X_{j1} + \cdots + \beta_{k-1} X_{j,k-1} + \varepsilon_j, \quad j = 1, \ldots, n, \tag{12.4.6}$$

correspond to those of (12.4.4), we express the cell means μ_i in terms of the regression coefficients. These expressions, given in Table 12.3, yield the following correspondence between the parameters of model (12.4.4) with those of model (12.4.6):

$$\mu_0 = \beta_0, \quad \alpha_1 = \beta_1, \ldots, \quad \alpha_{k-1} = \beta_{k-1}, \quad \alpha_k = -\beta_1 - \cdots - \beta_{k-1}. \tag{12.4.7}$$

Note that the last equation in (12.4.7) is consistent with the restriction (10.2.3), that is, $\alpha_1 + \cdots + \alpha_k = 0$, satisfied by the parameters of (12.4.4).

The p-value from the F test for the hypothesis $H_0 : \alpha_1 = \cdots = \alpha_k$, in the context of the ANOVA model (12.4.4), equals the p-value for the model utility test obtained from an unweighted least squares analysis of the MLR model (12.4.6). Rephrasing this in terms of R commands, the p-value of the F test obtained from *anova(aov(y~A))*, where A is the non-numeric (*as.factor*) column designating the factor level of each observation, equals the p-value for the model utility test obtained from *fit=lm(y~X_1+$\cdots$+X_{k-1}); summary(fit)*. Moreover, the MLR model (12.4.6) permits a weighted least squares analysis through the R commands

R Commands for Weighted Least Squares Analysis of One-Factor Designs

```
w=c(rep(1/S_1^2, n_1), ..., rep(1/S_k^2,n_k));
  lm(y~X_1+···+X_{k-1}, weights=w)
```
(12.4.8)

where $S_1^2, \ldots, S_k^2$ denote the sample variances from each group.

Regression Formulation of Two-Factor Designs Let Y_{ijk}, $k = 1, \ldots, n_{ij}$, denote the observations taken from factor-level combination (i, j), for $i = 1, \ldots, a$, $j = 1, \ldots, b$. According to the statistical model (11.2.3) for an $a \times b$ factorial design,

$$Y_{ijk} = \mu_{ij} + \epsilon_{ijk} \quad \text{or} \quad Y_{ijk} = \mu + \alpha_i + \beta_j + \gamma_{ij} + \epsilon_{ijk}, \tag{12.4.9}$$

where the main effects and interaction satisfy the conditions (11.2.2).

To cast model (12.4.9) into the multiple linear regression format, set $n = \sum_{i=1}^{a}\sum_{j=1}^{b} n_{ij}$, and let $Y_1, \ldots, Y_n$ be a single index enumeration of the Y_{ijk}'s. To indicate which factor level combination an observation Y_k comes from we need to introduce two sets of indicator variables,

$$X_{k1}^{A}, \ldots, X_{k,a-1}^{A} \quad \text{and} \quad X_{k1}^{B}, \ldots, X_{k,b-1}^{B}.$$

In the coding convention of Table 12.2, $X_{k1}^{A}, \ldots, X_{k,a-1}^{A}$ and $X_{k1}^{B}, \ldots, X_{k,b-1}^{B}$ indicate the level of factor A and level of factor B, respectively, that Y_k came from. Thus, the regression format representation of the data in (12.4.9) is

$$(Y_k, X_{k1}^{A}, \ldots, X_{k,a-1}^{A}, X_{k1}^{B}, \ldots, X_{k,b-1}^{B}), \quad k = 1, \ldots, n. \tag{12.4.10}$$

The regression model for the additive two-factor design, $Y_{ijk} = \mu + \alpha_i + \beta_j + \epsilon_{ijk}$, is

$$Y_k = \beta_0 + \beta_1^{A} X_{k1}^{A} + \cdots + \beta_{a-1}^{A} X_{k,a-1}^{A} + \beta_1^{B} X_{k1}^{B} + \cdots + \beta_{b-1}^{B} X_{k,b-1}^{B} + \varepsilon_k. \tag{12.4.11}$$

In Exercise 4 you are asked to show that $\beta_i^{A} = \alpha_i$, $i = 1, \ldots, a-1$, with $\alpha_a = -\beta_1^{A} - \cdots - \beta_{a-1}^{A}$, and a similar correspondence between the β_j's and β_j^{B}'s. The regression model for the non-additive two-factor design, shown in (12.4.9), includes, in addition, all interaction terms between the X_i^{A} and X_j^{B} covariates. Thus, in shorthand notation, it is written as

$$Y_k = \beta_0 + (\text{terms } \beta_i^{A} X_{ki}^{A}) + (\text{terms } \beta_j^{B} X_{kj}^{B}) + (\text{terms } \beta_{ij}^{AB} X_{ki}^{A} X_{kj}^{B}) + \varepsilon_k, \tag{12.4.12}$$

where i goes from 1 to $a-1$ and j goes from 1 to $b-1$. For example, if observation Y_k comes from cell (1, 1), the factor A indicator variables take values $X_{k1}^{A} = 1, X_{k2}^{A} = 0, \ldots, X_{k,a-1}^{A} = 0$, and the factor B indicator variables take values $X_{k1}^{B} = 1, X_{k2}^{B} = 0, \ldots, X_{k,b-1}^{B} = 0$. Thus, the only non-zero of the $\beta_i^{A} X_{ki}^{A}$ terms is β_1^{A}, the only non-zero of the $\beta_j^{B} X_{kj}^{B}$ terms is β_1^{B}, and the only non-zero of the $\beta_{ij}^{AB} X_{ki}^{A} X_{kj}^{B}$ terms is β_{11}^{AB}. It follows that if observation Y_k comes from cell (1, 1), (12.4.12) becomes

$$Y_k = \beta_0 + \beta_1^{A} + \beta_1^{B} + \beta_{11}^{AB} + \varepsilon_k.$$

Working similarly with other cells yields the following correspondence between the parameters of model (12.4.9) and those of model (12.4.12)

$$\mu = \beta_0, \quad \alpha_i = \beta_i^{A}, \quad \beta_j = \beta_j^{B}, \quad \gamma_{ij} = \beta_{ij}^{AB}. \tag{12.4.13}$$

For balanced two-factor designs, that is, when all sample sizes n_{ij} are equal, the p-value from the F test for the hypothesis $H_0 : \alpha_1 = \cdots = \alpha_a$ of no main factor A effects, which was discussed in Chapter 11, equals the p-value for testing the significance of the *group* of variables $X_1^{A}, \ldots, X_{a-1}^{A}$, that is, for testing the hypothesis $H_0 : \beta_1^{A} = \cdots = \beta_{a-1}^{A} = 0$, through an unweighted least squares analysis of the MLR model (12.4.12). Similarly the p-values for the hypotheses of no main factor B effects and no interaction, which are obtained through the *aov(y~A*B)* command, are equal to the p-values for testing the significance of the group of indicator variables for factor B and the group interaction variables, respectively, through an unweighted LS analysis of (12.4.12). Moreover, the MLR model (12.4.12) permits a weighted LS analysis. The R commands for doing this are similar to the commands in (12.4.8) and are demonstrated in Example 12.4-2.

For unbalanced designs, the equivalence of the p-values obtained through *aov* and the unweighted *lm* continues to hold for additive designs and for the interaction effects in non-additive designs. Thus, for a weighted least squares analysis of an unbalanced two-factor design, one should test for the significance of the interaction variables in the MLR model (12.4.12). In the case of a significant outcome, the main effects may also be declared significant. In the case of a non-significant outcome, the main effects can be tested under the additive model (12.4.11). The benefit of using weighted LS analysis is more pronounced in unbalanced designs.

Example 12.4-2

The file *H2Ofiltration3w.txt* contains data from a $2 \times 2 \times 3$ design with five replications; see Exercise 1 in Section 11.3 for a description. Use a weighted least squares analysis to test for main effects of factors A, B, and C, as well as their two- and three-way interactions.

Solution

Use *ne=read.table("H2Ofiltration3w.txt", header=T)* to read the data into the R data frame *ne* and *Y=ne$y; A=ne$MS; B=ne$SH; C=ne$MH* to set the response variable and the levels of the three factors in the R objects *Y, A, B*, and *C*, respectively. The MLR model representation of the data requires only one indicator variable for each of the factors A and C, say xA and xC, and two indicator variables to represent the three levels of factor B, say *xB1, xB2*. To set the values of the indicator variables xA and *xB1*, use the following commands. The values of xC and *xB2* are set similarly.

```
xA=rep(0, length(Y)); xB1=xA
xA[which(A=="A1")]=1; xA[which(A=="A2")]=-1
xB1[which(B=="B1")]=1; xB1[which(B=="B2")]=0;
  xB1[which(B=="B3")]=-1
```

The commands

```
vm=tapply(Y, ne[, c(1, 2, 3)], var);
  s2=rep(as.vector(vm), 5); w=1/s2
```

compute the matrix of cell variances and define the weights for the weighted LS methodology. (The data must be listed as in the given file for this to work.) Finally, use

```
summary(lm(Y~xA*xB1*xC+xA*xB2*xC, weights=w))
```

to produce the weighted LS analysis (regression coefficients with p-values, etc.). Two separate p-values are produced for the main factor B effect, one for each of the two indicator variables. The usual p-value for the main factor B effect is obtained by testing for the joint significance of *xB1* and *xB2*. Because the design is balanced (and also because the factors have a small number of levels), the results of the weighted LS analysis do not differ much from those of the unweighted analysis. For example, the p-values for the main factor B effects with unweighted and weighted analysis are 0.08 and 0.07, respectively. ■

12.4.3 VARIABLE SELECTION

The discussion so far has assumed that all predictors are included in the model. As data collection technologies improve, it becomes easier to collect data on a large number of covariates that may influence a particular response variable. Inadvertently, some (occasionally even most!) of the variables collected are not useful predictors. The question then becomes how to identify the subset of useful predictors in order to build a parsimonious model. The process of identifying a subset of useful predictors is called *variable selection*. There are two basic classes of variable selection procedures, *criterion-based procedures*, also called *best subset procedures*, and *stepwise procedures*.

Criterion-based, or best subset, procedures consider all subsets of predictors that can be formed from the available predictors and fit all possible models. Thus, if the number of available predictors is k, there are $2^{k+1}-1$ possible models to be fitted (this includes also models without the intercept term). Each model fitted is assigned a score according to a criterion for evaluating the quality of a fitted model, and the model with the best score is selected as the model of choice. The most common criteria used for this purpose are listed below:

1. **Adjusted R^2 criterion.** It selects the model with the largest adjusted R^2.
2. **Akaike information criterion (AIC).** It selects the model with the smallest AIC value. Loosely speaking, a model's AIC value is a relative measure of the information lost when the model is used to describe reality. The AIC for a multiple regression model with p parameters, intercept included, and error sum of squares SSE is (up to an additive constant) computed as

$$\text{AIC} = n\log\left(\frac{\text{SSE}}{n}\right) + 2p.$$

   For small sample sizes, a corrected version of AIC, $\text{AIC}_c = \text{AIC} + 2p(p+1)/(n-p-1)$, is recommended.
3. **Bayes information criterion (BIC).** It selects the model with the smallest BIC value. For a model with p parameters, intercept included, and error sum of squares SSE, its BIC value is computed as

$$\text{BIC} = n\log\left(\frac{\text{SSE}}{n}\right) + p\log n.$$

4. **Mallow's C_p criterion.** It selects the model with the smallest C_p value. For a model with p parameters, intercept included, whose error sum of squares is SSE, its C_p value is computed as

$$C_p = \frac{\text{SSE}}{\text{MSE}_{k+1}} + 2p - n,$$

   where MSE_{k+1} stands for the mean square error of the full model, that is, the model with all k predictor variables (so $k+1$ parameters, counting the intercept).
5. **Predicted residual sum of squares (PRESS) criterion.** This selects the model with the smallest value of the PRESS statistic. To describe the computation of the PRESS statistic for a particular model involving the covariates $X_1, \ldots, X_p$, let $\widehat{\beta}_{0,-i}, \widehat{\beta}_{1,-i}, \ldots, \widehat{\beta}_{p,-i}$ be the least squares estimators of the regression coefficients obtained by fitting the model after removing the ith data point, and let $\widehat{y}_{i,-i} = \widehat{\beta}_{0,-i} + \widehat{\beta}_{1,-i}x_{i1} + \cdots + \widehat{\beta}_{p,-i}x_{ip}$ denote the corresponding predicted value at the ith data point. Then the PRESS statistic for that model is computed as

$$\text{PRESS} = \sum_{i=1}^{n}\left(y_i - \widehat{y}_{i,-i}\right)^2.$$

Thus, all model selection criteria, except for PRESS, use a "penalized" score based on SSE, with the "penalty" increasing as the number of parameters in the model increases. The degree to which a small SSE value is "penalized" is most easily compared for the AIC and BIC selection criteria: BIC penalizes more heavily, especially for larger sample sizes, and thus it will select a more parsimonious model. Mallow's C_p has the property that for the full model, that is, the model with all k predictors and the intercept term, $C_{k+1} = k + 1$. If a p-parameter model provides a good fit, then its C_p value should be close to p; otherwise the C_p value is much bigger than p. It is usual to plot C_p vs p and select the model with the smallest p for which the point (p, C_p) is close to, or under, the diagonal. Note also that C_p orders the models in exactly the same way as AIC. The PRESS criterion, which is not based on a penalized SSE score, is designed to select models with the best predictive properties.

Stepwise procedures adopt a sequential approach to model building. At each step, these procedures use p-values obtained from testing the significance of each predictor to decide whether that predictor should be dropped or added to the model. The main algorithms for such stepwise model building are the following:

1. **Backward elimination**. This algorithm consists of the following steps:
   (a) Fit the full model.
   (b) If the p-values resulting from testing the significance of each predictor in the model are all less than a preselected critical value α_{cr}, stop. Otherwise proceed to step (c).
   (c) Eliminate the predictor with the largest p-value and fit the model with all remaining predictors. Go to step (b).

   The critical value α_{cr}, which is called "p-to-remove," need not be 0.05. Most typically, it is chosen in the range of 0.1–0.2.
2. **Forward selection**. Basically, this reverses the backward elimination algorithm:
   (a) Start with no predictors in the model.
   (b) Add each of the remaining predictors to the model, one at a time and compute each predictor's p-value. If all p-values are larger than a preselected critical value α_{cr}, stop. Otherwise proceed to step (c).
   (c) Add to the model the predictor with the smallest p-value. Go to step (b).

   In the context of forward selection, the critical value α_{cr} is called "p-to-enter."
3. **Stepwise regression**. This algorithm is a combination of backward elimination and forward selection, and there are at least a couple of versions for implementing it. One version starts with no predictors in the model, as in forward selection, but at each stage a predictor may be added or removed. For example, suppose the first predictor to be added, in a forward selection manner, is x_1. Continuing as in forward selection, suppose the second variable to be added is x_2. At this stage, variable x_1 is re-evaluated, as in backward elimination, as to whether or not it should be retained in a model that includes x_2. The process continues until no variables are added or removed. Another version starts with all predictors in the model, as in backward elimination, but at each stage a previously eliminated predictor may be re-entered in the model.

Model Selection in R The principal functions for best subset model selection are *regsubsets* and *leaps* in the R package *leaps*. These functions order the models according to each of the criteria: adjusted R^2, BIC (available in the *regsubsets* function), and C_p criterion (available in the *leaps* function). The model ordering (and the corresponding value of the criterion used for each model) can be displayed in plots. Ordering according to the AIC criterion is not displayed explicitly (and therefore the AIC value for each model is not produced automatically) because it is the same as the ordering according to the C_p criterion. However, the AIC for any particular model can be obtained with the R command

```
fit=lm(y~(model specification)); AIC(fit).
```

(Using *AIC(fit, k=log(length(y)))* produces the model's BIC.) Finally, the function *press* in the R package *DAAG* computes the PRESS criterion. It is invoked simply as *press(fit)*.

Stepwise procedures can also be performed with the *regsubsets* function, but the *step* function in the *stats* package, which is described next, is just as convenient. The argument *direction* can be used to specify backward elimination, forward selection, or stepwise regression. The default setting is *direction="both"* for stepwise regression; use *direction="backward"* for backward elimination and *direction="forward"* for forward selection. The use of these R commands is illustrated in the next example.

Example 12.4-3

An article[6] reports data on six attributes/characteristics and a performance measure (based on a benchmark mix relative to IBM 370/158-3) of 209 CPUs. The data are available in the R package *MASS* under the name *cpus*. (For a description of the six attributes used, type *library(MASS); ?cpus*.) Use these data to build a parsimonious model for predicting a CPU's relative performance from its attributes.

Solution

Use *install.packages("leaps")* to install the package *leaps*, and then load it to the current R session by *library(leaps)*. A basic invocation of *regsubsets* is

```
vs.out=regsubsets(perf~syct+mmin+mmax+cach
  +chmin+chmax, nbest=3, data=cpus)
```
(12.4.14)

The outcome object *vs.out* contains information on the best three models for each model size, that is, the best three models with one variable, the best three models with two variables, and so forth; by default, all models include the intercept term. (Note that, because ranking models of the same size is identical for all criteria [except for PRESS], no particular criterion is specified in the *regsubsets* command.) In particular, the command in (12.4.14) contains information on 16 models (the full model and the three best for each size of one through five). The command *summary(vs.out)* will display these 16 models. This output is not shown here because, typically, we are not concerned about which models are the best for each size. Instead, we are concerned with the overall ranking of these 16 models, and the overall best model, according to each of the criteria. Using Mallow's C_p criterion, the overall ranking is given, in the form of the plot in Figure 12-10, by the command

R Command Generating the Plot of Figure 12-10, Where *vs.out* comes from (12.4.14)

```
plot(vs.out, scale="Cp")
```
(12.4.15)

Admittedly, such a plot seems strange when first encountered. Each *row* in the plot represents a model, with the white spots representing the variables that are not included in the model. The best model is at the top of the figure (darkest shading)

[6] P. Ein-Dor and J. Feldmesser (1987). Attributes of the performance of central processing units: A relative performance prediction model. *Comm. ACM*, 30: 308–317.

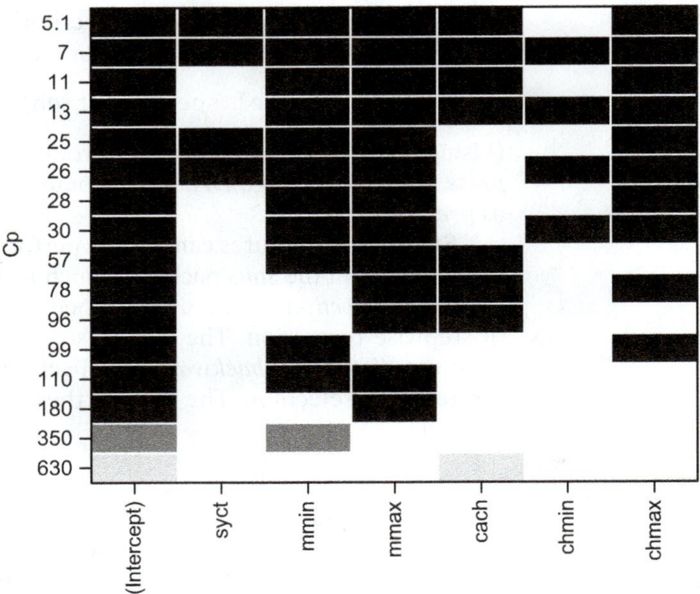

Figure 12-10 Ordering of 16 models, identified by *regsubsets*, using Mallow's C_p criterion.

with C_p value of 5.1; it leaves out the variable "chmin." The second best model, with C_p value of 7, is the full model with all six predictor variables. The third best model, with C_p value of 11, leaves out "syct" and "chmin," and so forth. Note that the three worst models all use only one predictor variable.

Recall that the ordering of the models shown in Figure 12-10 is identical to the ordering produced by the AIC criterion. Plots showing the rankings of the 16 models according to the adjusted R^2 criterion and the BIC criterion are produced by

R Commands for Ranking Models According to R^2 Adj and BIC, Where *vs.out* Comes from (12.4.14)

```
plot(vs.out, scale="adjr2")
plot(vs.out, scale="bic")
```
(12.4.16)

It should be kept in mind that all criteria are computed from data and, hence, are subject to variability. If the criterion value of the best model does not differ by much from that of the second best or the third best, etc., claims about having identified the "best" model should be made with caution. The AIC can give the relative probability that a model that was not ranked best is indeed the best (in terms of minimizing the information lost), as follows: Let AIC_{min} be the minimum AIC score among the models considered, and let AIC_j be the AIC score of the jth model considered. Then, the relative probability that the jth model is the best is

$$e^{D_j/2}, \quad \text{where } D_j = \mathrm{AIC}_{min} - \mathrm{AIC}_j. \qquad \textbf{(12.4.17)}$$

In this example, the AIC scores of the top five models are 2311.479, 2313.375, 2317.471, 2319.272, and 2330.943. Thus the relative probability that each of the models ranked second through fifth is actually the best is 0.39, 0.050, 0.020, and 5.93×10^{-5}, respectively.

The best-fitting model according to either the C_p or the adjusted R^2 criterion can also be obtained through the *leaps* function (in the *leaps* package). The commands for the C_p criterion are

R Commands for the Best-Fitting Model According to C_p

```
fit=lm(perf~syct+mmin+mmax+cach+chmin+chmax, data=cpus)
X=model.matrix(fit)[, -1]
cp.leaps=leaps(X, cpus$perf, nbest=3, method="Cp")
cp.leaps$which[which(cp.leaps$Cp==min(cp.leaps$Cp)),]
```
(12.4.18)

The output produced is

```
   1     2     3     4     5     6
TRUE  TRUE  TRUE  TRUE FALSE  TRUE
```

This output specifies as best the model that includes all variables except the fifth, which is "chmin." Note that this is the same model identified as best in Figure 12-10. The p vs C_p scatterplot of Figure 12-11, with the diagonal line superimposed, is produced with the following additional commands:

R Commands for the Plot of Figure 12-11

```
plot(cp.leaps$size, cp.leaps$Cp, pch=23,
  bg="orange", cex=3)
abline(0, 1)
```
(12.4.19)

Replacing the third and fourth lines of the commands in (12.4.18) by *adjr2.leaps=leaps(X, cpus$perf, nbest=3, method="adjr2")* and *adjr2.leaps$which [which(adjr2.leaps$adjr2==min(adjr2.leaps$adjr2)),]*, respectively, gives the best model according to the adjusted R^2 criterion. For this data set, the adjusted R^2 criterion gives as best the same model identified above. The BIC criterion has not been implemented in the *leaps* function. Using *method="r2"* will give the model with the highest R^2 value, though this is not an appropriate model ranking criterion.

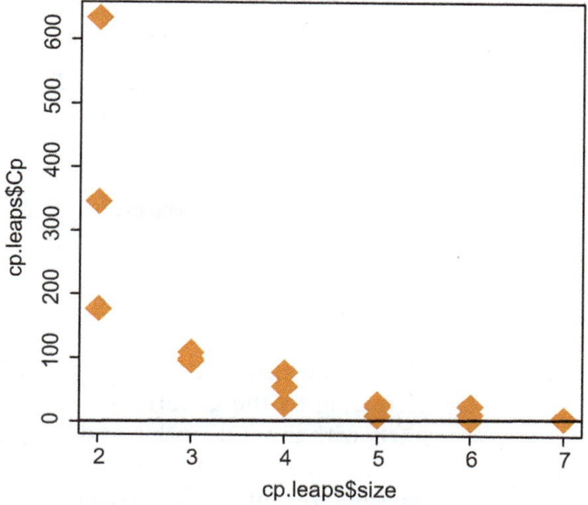

Figure 12-11 Model size vs Mallow's C_p plot for the top 16 models (up to 3 for each size).

With *fit* defined from *fit=lm(perf~syct+mmin+mmax+cach+chmin+chmax, data=cpus)*, the following R commands perform stepwise variable selection (both, backward or forward).

R Commands for Stepwise Variable Selection

```
step(lm(perf~1, data=cpus), list(upper=~1+syct+mmin+
  mmax+cach+chmin+chmax), direction ="forward")
step(fit, direction="backward")
step(fit, direction="both")
```
(12.4.20)

These commands give the steps leading to the final model. As an illustration, the output produced by the commands for stepwise regression (both directions) is shown below. Note that the AIC values given in this output correspond to what is known as *generalized* AIC, and are different from the usual AIC values given in connection with (12.4.17).

```
Start: AIC=1718.26
perf~syct+mmin+mmax+cach+chmin+chmax
         Df  Sum of Sq     RSS     AIC
- chmin   1        358  727360  1716.4
<none>                  727002  1718.3
 - syct   1      27995  754997  1724.2
 - cach   1      75962  802964  1737.0
- chmax   1     163396  890398  1758.6
 - mmin   1     252211  979213  1778.5
 - mmax   1     271147  998149  1782.5

Step: AIC=1716.36
perf~syct+mmin+mmax+cach+chmax
         Df  Sum of Sq     RSS     AIC
<none>                  727360  1716.4
+ chmin   1        358  727002  1718.3
 - syct   1      28353  755713  1722.3
 - cach   1      78670  806030  1735.8
- chmax   1     177174  904534  1759.9
 - mmin   1     258289  985649  1777.9
 - mmax   1     270827  998187  1780.5
```

The model displayed in the second and final step is the same as the best model according to the criterion-based procedures. In this example, stepwise regression starts with the full model, and, because the best model includes all variables but one, it stops after only two steps. The backward elimination procedure produces exactly the same output, while forward selection takes several steps to reach the final (best) model; see Exercise 7. To also obtain the statistical analysis for the final model, use *summary(step(fit, direction="both or forward or backward"))*. ■

REMARK 12.4-1 The default rule for entering or deleting variables in the stepwise algorithm of the function *step* is unclear. However, the function *update*, described in Exercise 3 in Section 12.3, can be used for implementing backward elimination with any chosen p-to-remove; see Exercise 8. ◁

12.4.4 INFLUENTIAL OBSERVATIONS

It is possible for a single observation to have a great influence on the results of a regression analysis, in general, and on the variable selection process in particular.

The **influence** of the ith observation, $y_i, x_{i1}, \ldots, x_{ik}$, is defined in terms of how much the predicted, or fitted, values would differ if the ith observation were omitted from the data. It is quantified by *Cook's distance* (also referred to as *Cook's D*), which is defined as follows:

$$D_i = \frac{\sum_{j=1}^{n}(\widehat{y}_j - \widehat{y}_{j,-i})^2}{p\text{MSE}}, \tag{12.4.21}$$

where p is the number of fitted parameters (so $p = k + 1$ if an intercept is also fitted), $\widehat{y}_j$ is the jth fitted value (see (12.3.3)) and $\widehat{y}_{j,-i}$ is the jth fitted value when the ith observation has been omitted from the data. A common rule of thumb is that an observation with a value of Cook's D greater than 1.0 has too much influence.

REMARK 12.4-2 A related concept is **leverage**. The leverage of the ith observation is defined to be the ith diagonal element of $\mathbf{X}(\mathbf{X}'\mathbf{X})^{-1}\mathbf{X}'$, which has been denoted by h_i in (12.3.15). It quantifies how far $(x_{i1}, \ldots, x_{ik})$ is from the rest of the covariates' values. Put in different words, h_i quantifies the extent to which $(x_{i1}, \ldots, x_{ik})$ is an *outlier* with respect to the other covariates' values. The ith observation is defined to be a high-leverage point if $h_i > 3p/n$, that is, if h_i is larger than three times the average leverage.[7] High-leverage observations have the potential of being influential, but they need not be. An influential observation, however, will typically have high leverage. On the other hand, a high-leverage observation with a large (in absolute value) studentized residual (see (12.3.19)) will be an influential observation. ◁

As mentioned, an influential observation has the potential of changing the outcome of a variable selection algorithm. Therefore, it is recommended that the variable selection algorithm be reapplied to the data set with the influential observation(s) removed. If the same model is selected both with and without the influential observation(s) in the data set, then all is well and no further action is needed. In the opposite case, that is, if the algorithm selects a different model when the influential observation(s) are removed, one should check the "validity" of the influential observation. For example, it is possible that there has been a recording error, or the observation(s) came from experimental unit(s) erroneously included in the sample. If the observations pass the validity check, it is probably a good idea to use the model that includes all predictors included in the two selected models.

---

[7] It can be shown that $\sum_{i=1}^{n} h_i = p$, where p is the number of fitted parameters.

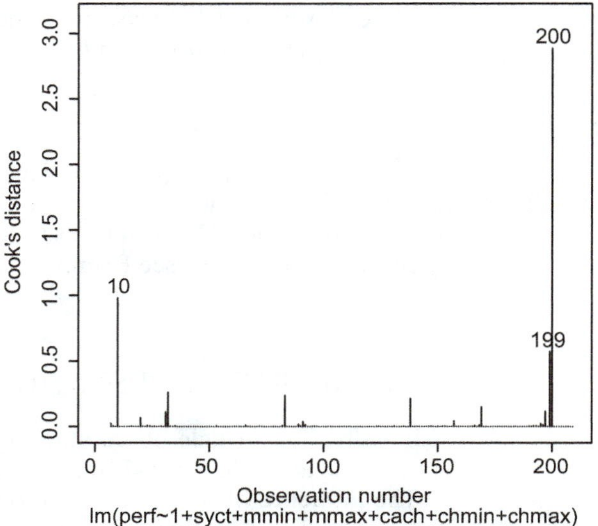

Figure 12-12 Cook's D for the observations in the data set of Example 12.4-3.

With *fit* generated from *fit=lm(perf~syct+mmin+mmax+cach+chmin+chmax, data=cpus)*, the R command

```
plot(fit, which=4)
```

generates the plot in Figure 12-12, which, according to the rule of thumb, identifies the observations numbered 10 and 200 in the data set *cpus* as influential.[8]

To see if the presence of these observations has an effect on the outcome of the variable selection algorithms, use the R command

```
cpus1=cpus[-c(10, 200),]
```
(12.4.22)

to define the data frame *cpus1* which includes all observations in *cpus* except for the 10th and 200th. Applying to *cpus1* the variable selection algorithms used in Example 12.4-3, we see that the stepwise procedures as well as the best subset procedure with criterion "adjr2" select the full model, while the best subset with the "bic" criterion selects the model without the variables "syct" and "chmin." Only best subset with C_p selects the same model as before. On the basis of these results, the final recommendation is to use the full model.

12.4.5 MULTICOLLINEARITY

The predictors in multiple regression data sets often exhibit strong linear dependencies. The stronger the interdependence among the predictors, the larger the *condition number*[9] of the normal equations (12.3.9).

Put in different words, strong interdependency among the predictors makes the inverse $(\mathbf{X}'\mathbf{X})^{-1}$ unstable, where $\mathbf{X}$ is the design matrix defined in (12.3.7). As a consequence, the least squares estimators, which are given in terms of this inverse matrix (see (12.3.10)), are imprecise. Moreover, the variance of each $\widehat{\beta}_j$ is given in

[8] The command *cooks.distance(fit)[c(10, 199, 200)]* returns Cook's D values of 0.98, 0.57, and 2.89 for observations numbered 10, 199, and 200, respectively.

[9] The condition number of a linear equation $\mathbf{Ax} = \mathbf{b}$ captures the rate at which the solution, $\mathbf{x}$, will change with respect to a change in $\mathbf{b}$. The larger the condition number, the more sensitive the solution to changes in $\mathbf{b}$.

terms of C_j, the jth diagonal element of this inverse matrix; see (12.3.13). It can be shown that C_j can be written as

$$C_j = \frac{1}{1 - R_j^2}, \tag{12.4.23}$$

where R_j^2 is the coefficient of multiple determination resulting from regressing the jth predictor, X_j, on the remaining predictors. Clearly, if the linear dependence of X_j on the remaining $k - 1$ predictors is strong, R_j^2 will be close to one, and this results in an "inflation" of the variance of $\widehat{\beta}_j$ by the quantity C_j. For this reason, C_j is called the **variance inflation factor** for β_j. The data is said to exhibit **multicollinearity** if the variance inflation factor for any β_j exceeds 4 or 5, and multicollinearity is called severe if a variance inflation factor exceeds 10.

Under multicollinearity, p-values and confidence intervals for individual parameters cannot be trusted. Nevertheless, even though individual parameters are imprecisely estimated, the significance of the fitted model may still be tested by the model utility test, and its usefulness for prediction may still be judged by the coefficient of multiple determination. An interesting side effect of (but also indication for) multicollinearity is the phenomenon of a significant model utility test when all predictors are not significant; see Exercise 13.

The R package *car* has the function *vif* for computing the variance inflation factor for each regression coefficient. The R commands for it are shown below:

R Commands for Computing the Variance Inflation Factors

```
library(car); vif(lm(y~(model specification)))
```
(12.4.24)

As an example, use *df=read.table("Temp.Long.Lat.txt", header=T); y=df$JanTemp; x1=df$Lat-mean(df$Lat); x2=df$Long-mean(df$Long)* to read the temperature values and the centered latitude and longitude values into the R objects *y, x1, x2*, respectively. Then the command

```
library(car); vif(lm(y~x1+x2+I(x1^2)+I(x2^2)+x1:x2))
```

generates the following output:

```
   x1     x2  I(x1^2)  I(x2^2)  x1:x2
1.735  1.738    1.958    1.764  1.963
```

All inflation factors are less than 2, so multicollinearity is not an issue for this data set.

12.4.6 LOGISTIC REGRESSION

As mentioned in Section 12.2, there are cases where the regression function is not a linear function of the predictor, but linearity can be restored through a suitable transformation. This section deals with a particular case where the simple (or multiple) linear regression model is not appropriate, namely, when the response variable Y is Bernoulli.

Experiments with a binary response variable are encountered frequently in reliability studies investigating the impact of certain variables, such as stress level, on the probability of failure of a product, or the probability of no major issues within a product's warranty period; they are also quite common in medical research and other fields of science.[10]

To see why the simple linear regression model is not appropriate when Y is Bernoulli, recall that $E(Y) = p$, where p is the probability of success. Thus, when the probability of success depends on a covariate x, the regression function, which is

$$\mu_{Y|X}(x) = E(Y|X = x) = p(x),$$

is constrained to take values between 0 and 1. On the other hand, the regression function implied by the linear regression model, that is, $\mu_{Y|X}(x) = \beta_0 + \beta_1 x$, does not conform to the same constraint as $p(x)$. The MLR model is also not appropriate for the same reason.

Continuing with one covariate for simplicity, the **logistic regression model** assumes that the **logit** transformation of $p(x)$, which is defined as

$$\text{logit}(p(x)) = \log\left(\frac{p(x)}{1 - p(x)}\right), \qquad \textbf{(12.4.25)}$$

is a linear function of x. That is, the logistic regression model assumes that

$$\text{logit}(p(x)) = \beta_0 + \beta_1 x. \qquad \textbf{(12.4.26)}$$

Simple algebra shows that the logistic regression model is equivalently written as

$$p(x) = \frac{e^{\beta_0 + \beta_1 x}}{1 + e^{\beta_0 + \beta_1 x}}. \qquad \textbf{(12.4.27)}$$

The expression on the right-hand side of (12.4.27) is the **logistic function** of $\beta_0 + \beta_1 x$.[11] This reveals the origin of the term *logistic regression*. Two logistic regression functions, one with positive and one with negative slope parameter (β_1), are shown in Figure 12-13.

If p is the probability of success (or, in general, of an event E), $p/(1-p)$ is called the *odds ratio*, and is used as an alternative quantification of the likelihood of success through the expression "the odds of success are $p/(1-p)$ to one"; see Section 2.3.1. An odds ratio of 3 means that success is 3 times as likely as failure. Thus, according

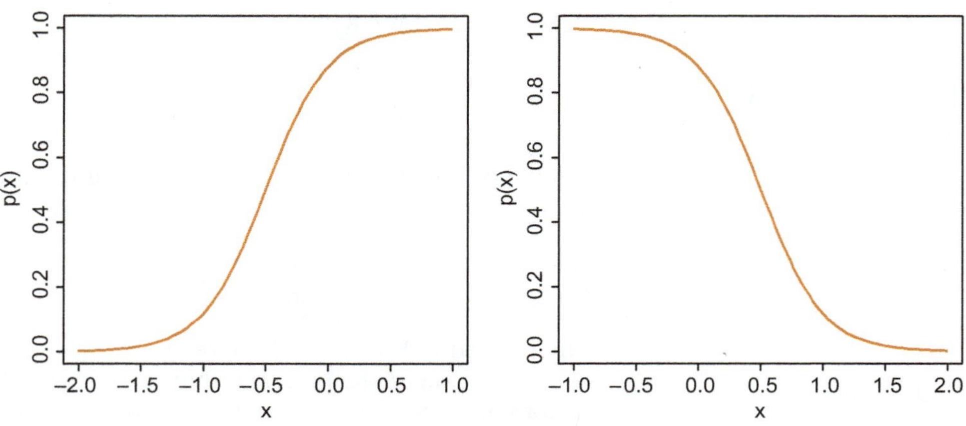

Figure 12-13 Examples of the logistic regression function (12.4.27): $\beta_0 = 2$ (both), $\beta_1 = 4$ (left), and $\beta_1 = -4$ (right).

[10] For example, F. van Der Meulen, T. Vermaat, and P. Willems (2011). Case study: An application of logistic regression in a six-sigma project in health care, *Quality Engineering*, 23: 113–124.

[11] In mathematical terminology, the logistic function is the inverse of the logit function.

to the logistic regression model, the odds ratio changes by a (multiplicative) factor of e^{β_1} as x changes by 1 unit.

Fitting the model (12.4.26), and its version with multiple covariates, to data, that is, estimating the model parameters, is done by the method of maximum likelihood (see Section 6.3.3). The details are involved and will not be given. Instead we will demonstrate the use of R commands for obtaining the maximum likelihood estimators. The R function for fitting logistic regression models is *glm*. The syntax of *glm* is similar to that of *lm* except we must also specify what is called the *family*. This is illustrated in the context of the following example.

Example 12.4-4

The file *FluShotData.txt* has simulated flu shot data from a study on the effect of age and a health awareness index on a person's decision to get a flu shot. Use R to complete the following.

(a) Fit a two-predictor logistic regression model to the data, and test the significance of the two predictors at level of significance 0.05.

(b) Fit the logistic regression model that includes the interaction of the two predictors to the model of part (a). Construct 95% CIs for the regression coefficients.

(c) Using the model with the interaction term, give an estimate of the probability that a 35-year-old person with health awareness index 50 will get the flu shot. Do the same for a 45-year-old person with index 50.

Solution

(a) With the data having been read into the data frame *fd*, the R commands

```
fit1=glm(Shot~Age+Health.Aware, family=binomial( ),
  data=fd); summary(fit1)
```

generate output, part of which is

```
Coefficients:
               Estimate   Std. Error   z value   Pr(> |z|)
(Intercept)     -21.585        6.418    -3.363     0.0008
Age               0.222        0.074     2.983     0.0029
Health.Aware      0.204        0.063     3.244     0.0012
AIC: 38.416
```

Thus, both age and health awareness are significant at the 0.05 level (p-values of 0.003 and 0.001, respectively).

(b) The command *fit2=glm(Shot~Age*Health.Aware, family=binomial( ), data=fd); summary(fit2)* generates output, part of which is

```
Coefficients:
                    Estimate   Std. Error   z value   Pr(> |z|)
(Intercept)           26.759       23.437     1.142       0.254
Age                   -0.882        0.545    -1.618       0.106
Health.Aware          -0.822        0.499    -1.647       0.099
Age:Health.Aware       0.024        0.012     1.990       0.047
AIC: 32.283
```

Note that now age and health awareness are not significant at $\alpha = 0.05$ but their interaction is. As advised in the context of an MLR model, if the interaction of two variables is significant, both variables should be included in the model. Note also the drop in the AIC value incurred by the inclusion of the interaction term. The additional command

```
confint(fit2)
```

produces the following 95% CIs for the regression parameters.

	2.5%	97.5%
(Intercept)	-11.80	84.08
Age	-2.25	-0.038
Health.Aware	-2.066	-0.047
Age:Health.Aware	0.006	0.054

c) The R command

```
predict(fit2, list(Age=c(35, 45), Health.Aware=c(50, 50)),
  type="response")
```

generates the output

```
      1       2
0.02145 0.30809
```

Thus, the probability that a 35-year-old with health awareness index 50 will get the shot is 0.021, while the probability for a 45-year-old with the same index is 0.308. ■

For additional information on the topic of logistic regression see *Applied Logistic Regression* by Hosmer, Lemeshow, and Sturdivant.[12]

Exercises

1. A computer activity. This exercise uses computer simulations to demonstrate that, for heteroscedastic data, unweighted/ordinary least squares (OLS) analysis underestimates the variability of the estimated slope, while weighted least squares (WLS) analysis estimates it correctly. The simulated regression model has regression function $\mu_{Y|X}(x) = 3 - 2x$ and error variance function $\text{Var}(Y|X = x) = (1 + 0.5x^2)^2$.

(a) Generate a set of 100 (*x*, *y*) values by *x=rnorm(100, 0, 3); y=3-2*x+rnorm(100, 0, sapply(x, function (x){1+0.5*x**2}))*. Then use *fit=lm(y~x); summary (fit)*, and *fitw=lm(y~x, weights=(1+0.5*x**2)** (-2)); summary(fitw)* and report the standard error of the slope estimate obtained from the OLS and WLS analyses.

(b) The following commands generate 1000 sets of y-values. Like in part (a), obtain the OLS and WLS slope estimates for each set of 100 (x, y) values, and compute the standard deviation of the 1000 OLS slope estimates and of the 1000 WLS slope estimates. (The x-values stay the same for all 1000 sets of 100 (x, y) values.) These standard deviations are the simulation-based approximations of the true variability of the slope estimates. Compare them with the standard errors or the slope estimates reported in part (a), and comment. The needed commands for this simulation are:

```
beta=rep(0, 1000); betaw=rep(0, 1000);
  for(i in 1:1000) { y=3-2*x+rnorm(100, 0,
  sapply(x, function(x) {1+0.5*x**2}));
```

[12] D. H. Hosmer, Jr., S. Lemeshow, and R. Sturdivant (2013). *Applied Logistic Regression*, 3rd Edition, Hoboken: Wiley.

```
beta[i]=lm(y~x)$coefficients[2];
betaw[i]=lm(y~x, weights=(1+
0.5*x**2)**(-2))$coefficients[2]};
sd(beta); sd(betaw)
```

2. Perform the following analysis for the stackloss data described in Exercise 2 in Section 12.3.

(a) Using the residuals from fitting an MLR model on the three covariates with no polynomial or interaction terms, check the homoscedasticity assumption using the plot and formal test described in Example 12.3-3, and state your conclusion.

(b) Use the steps outlined in Section 12.4.1 to fit the MLR model by weighted LS. Report the p-value of the model utility test.

(c) Use *confint(fit); confint(fitw)* to obtain the 95% OLS and WLS CIs for the regression parameters. Which method gives shorter cIs?

3. Use *edu=read.table("EducationData.txt", header=T)* to import the 1975 education expenditure by state data set into the R data frame *edu*.[13] In this data set, the response Y is per capita education expenditure, $X1$ is per capita income, $X2$ is proportion of population under 18, and $X3$ is proportion in urban areas. (The covariate "Region" will not be used in this exercise.)

(a) Use *edu.fit=lm(Y~X1+X2+X3, data=edu); plot(edu.fit, which=1)* to fit the MLR model by ordinary LS and to plot the residuals versus the predicted values. On the basis of this plot would you suspect the homoscedasticity assumption? Check the homoscedasticity assumption using the formal test described in Example 12.3-3, and state your conlusion.

(b) Use the steps outlined in Section 12.4.1 to fit the MLR model by weighted LS.

(c) Use *confint(edu.fit); confint(fitw)* to obtain the 95% OLS and WLS CIs for the regression parameters.

4. For the regression model formulation, (12.4.11), of the additive two-factor design, which is $Y_{ijk} = \mu + \alpha_i + \beta_j + \epsilon_{ijk}$, $\sum_i \alpha_i = \sum_j \beta_j = 0$, show that

$$\alpha_i = \beta_i^A, \quad \text{for } i = 1, \ldots, a-1,$$

$$\beta_j = \beta_j^B, \quad \text{for } j = 1, \ldots, b-1,$$

$$\alpha_a = -\beta_1^A - \cdots - \beta_{a-1}^A, \qquad \text{and}$$

$$\beta_b = -\beta_1^B - \cdots - \beta_{b-1}^B.$$

5. A company is investigating whether a new driving route reduces the time to deliver goods from its factory to a nearby port for export. Data on 48 delivery times with the standard route and 34 delivery times with the new route are given in *DriveDurat.txt*.

(a) The regression model formulation of the two-sample problem requires only one indicator variable. Specialize the indicator variable formula (12.4.3) to the two-sample case.

(b) If x denotes the indicator variable of part (a), the regression model formulation of the two samples is written as $Y_i = \beta_0 + \beta_1 x_i + \epsilon_i$, $i = 1, \ldots, n_1 + n_2$. How are the regression parameters β_0, β_1 related to the population means μ_1, μ_2?

(c) With data imported into the R data frame *dd*, use the commands *y=dd$duration; x=rep(0, length(y)); x[which(dd$route==1)]=1; x[which(dd$route==2)] =-1* to set the duration times in the object y and to define the indicator variable. Compare the p-value from the model utility test using OLS analysis of y on x with the p-value from the two-sample T test that assumes equal variances. (*Hint.* Use *summary(lm(y~x))* and *t.test(y~dd$route, var.equal=T)*.)

(d) Test for heteroscedasticity using (i) Levene's test and (ii) the regression type test. Report the two p-values. (*Hint.* Use *library(lawstat); levene.test(y, x)* for Levene's test, and *dd.fit=lm(y~x); summary(lm(abs(rstandard(dd.fit))~poly(fitted(dd.fit),2)))* for the regression type test (report the p-value for the model utility test).)

(e) Are the two population means significantly different at $\alpha = 0.05$? Test using WLS analysis and the T test without the equal variances assumption. (*Hint.* For the WLS analysis use *efit=lm(abs(resid(dd.fit))~poly(fitted(dd.fit), 2)); w=1/fitted(efit)**2; summary(lm(y~x, weights= w))*, and use the p-value for the model utility test.)

6. With the data frame *edu* containing the education data of Exercise 3, define indicator variables R1, R2, and R3 to represent the four levels of the variable "Region" according to the contrast coding of (12.4.3). For example, the following commands define R1: *R1=rep(0, length(edu$R)); R1[which(edu$R==1)]=1; R1[which(edu$R==4)]=-1.* (Note that *edu$R* is an accepted abbreviation of *edu$Region*.) Use WLS analysis to test the significance of the variable "Region" at level of significance 0.1. (*Hint.* Use the steps outlined in Section 12.4.1 to estimate the weights for the WLS analysis, using covariates X1, X2, X3, R1, R2, and R3. Let *fitFw* and *fitRw* be the output objects from fitting the full and the reduced (i.e., without R1, R2, and R3) models; for example *fitFw = lm(Y~X1+X2+X3+R1+R2+R3, weights=w, data=edu)*. Then use *anova(fitFw, fitRw)* to test the joint significance of R1, R2, and R3.)

7. Using the *cpus* data from Example 12.4-3 and the R commands given in (12.4.20), perform forward

[13] S. Chatterjee and B. Price (1977). *Regression Analysis by Example*. New York: Wiley, 108. This data set is also available in R in the package *robustbase*.

selection with the function *step*. Obtain the statistical analysis for the final model, and report the p-value for each predictor variable and of the model utility test. (*Hint.* Set *fit=step(lm(perf~1, data=cpus), list(upper=~1+syct+mmin+mmax+cach+chmin+chmax), direction="forward")* and use *summary(step(fit, direction="forward")).*)

8. In this exercise we will apply backward elimination, at p-to-remove 0.1, for variable selection on the "state" data set described in Exercise 3 in Section 12.3. Stepwise variable selection with p-to-remove (or p-to-enter) of your choice is not automatic in R, but the process is greatly facilitated by the *update* function described in Exercise 3 in Section 12.3. Use *st=read.table("State.txt", header=T)* to import the data into the R data frame *st* and complete the following.

(a) Use *h=lm(Life.Exp~ . , data=st); summary(h)* to fit the full model. (Here "." means "use all predictors in the data set.") Because the largest p-value, which is 0.965, corresponds to "Area" and is > 0.1, continue with *h=update(h, . ~ . -Area); summary(h)* to fit the model without the predictor "Area."

(b) Continue removing the predictor with the largest p-value until all p-values are smaller than 0.1. Give the R^2 for the final model and compare it with that of the full model.

9. Using the "state" data of Exercise 8, perform criterion-based variable selection with each of the criteria C_p, adjusted R^2, and BIC. (*Hint.* Invoke the command *library(leaps)*, create the outcome object *vs.out* as in (12.4.14) (use the abbreviated *lm* statement given in part (a) of Exercise 8), and then use (12.4.15) and (12.4.16) to create plots like that in Figure 12-10 in order to select the best model.)

10. *UScereal* is a built-in data frame in R's package *MASS* with $n = 65$ rows on 11 variables. Three of the variables ("mfr" for manufacturer, "shelf" for the display shelf with three categories counting from the floor, and "vitamins") are categorical, and the others are quantitative. Type *library(MASS); ?UScereal* for a full description. In this exercise we will apply variable selection methods to determine the best model for predicting "calories," the number of calories in one portion, using the seven quantitative predictors. Use *uscer=UScereal[, -c(1, 9, 11)]* to generate the data frame *uscer* without the three categorical variables.

(a) Perform criterion-based variable selection with each of the criteria C_p, adjusted R^2, and BIC, and give the model selected by each criterion. (*Hint.* Use *library(leaps); vs.out=regsubsets(calories~ . , nbest=3, data=uscer)*, and then use (12.4.15) and (12.4.16) to create plots like that in Figure 12-10 in order to select the best model according to each criterion.)

(b) Use Cook's D to determine if there are any influential observations. (*Hint.* Use *cer.out=lm(calories~ . , data=uscer); plot(cer.out, which =4)* to construct a plot like Figure 12-12.)

(c) Create a new data frame, *usc*, by removing the influential observations; perform criterion-based variable selection with each of the criteria C_p, adjusted R^2 and BIC; and give the model selected by each criterion. Does the final model seem reasonable? (*Hint.* See (12.4.22).)

11. Researchers interested in understanding how the composition of the cement affects the heat evolved during the hardening of cement, measured the heat in calories per gram (y), the percent of tricalcium aluminate (x1), the percent of tricalcium silicate (x2), the percent of tetracalcium alumino ferrite (x3), and the percent of dicalcium silicate (x4) for 13 batches of cement. The data are in *CementVS.txt*. Import the data into the R data frame *hc*, fit the MLR model with *hc.out=lm (y~ . , data=hc); summary(hc.out)*, and complete the following.

(a) Are any of the variables significant at level $\alpha = 0.05$? Is your answer compatible with the R^2 value and the p-value for the model utility test? If not, what is a possible explanation?

(b) Compute the variance inflation factors for each variable. Is multicollinearity an issue with this data?

(c) Remove the variable with the highest variance inflation factor and fit the reduced MLR model. Are any of the variables significant at level 0.05? Is there much loss in terms of reduction in R^2 or adjusted R^2? (*Hint.* The data frame *hc1* created by *hc1=hc [, -5]* does not include x4. Alternatively, you can use *hc1.out=update(hc.out, .~. -x4); summary(hc1.out).*)

(d) Starting with the full model, apply backward elimination with p-to-remove 0.15; see Exercise 8. Give the variables retained in the final model. Is there much loss in terms of the adjusted R^2 of the final model?

12. The SAT data in *SatData.txt* are extracted from the *1997 Digest of Education Statistics*, an annual publication of the US Department of Education. Its columns correspond to the variables name of state, current expenditure per pupil, pupil/teacher ratio, estimated average annual teacher salary (in thousands of dollars), percent of all eligible students taking the SAT, average verbal SAT score, average math SAT score, and average total score on the SAT.

(a) Read the data into the R data frame *sat* and construct a scatterplot matrix with *pairs(sat)*. Does the Salary vs Total scatterplot suggest that increasing teachers salary will have a positive or negative effect on student SAT score? Confirm your impression by fitting a least squares line through the scatterplot and reporting the value of the slope. (Use *summary(lm(Total~Salary, data=sat)).*)

(b) Fit the MLR model for predicting the total SAT score in terms of all available covariates: *summary(lm(Total~Salary+ExpendPP+PupTeachR+PercentEll, data=sat))*. Comment on whether increasing teacher salary, while keeping all other predictor variables the same, appears to have a positive or negative effect on student SAT score. Is your comment compatible with your answer in part (a)? If not, suggest a possible reason for the incompatibility.

(c) Which, if any, of the predictor variables appear to be significant predictors of the student SAT score? Base your answer on each variable's p-value at level of significance 0.05.

(d) Report R^2, adjusted R^2, and the p-value for the model utility test. Are these values compatible with your comments in part (c) regarding the significance of each individual predictor? If not, suggest a possible reason for the incompatibility.

(e) Compute the variance inflation factor for each predictor variable and comment on whether or not they suggest multicollinearity of the predictors.

13. The data set in *LaysanFinchWt.txt* has measurements of weight and five other physical characteristics of 43 female Laysan finches.[14] Import the data set into the R data frame *lf*, use *lf.out=lm(wt~ . , data=lf); summary(lf.out)* to fit the MLR model explaining weight in terms of the other variables, and complete the following.

(a) Report the R^2 and the adjusted R^2. Is the model utility test significant at $\alpha = 0.05$? Are any of the predictor variables significant at $\alpha = 0.05$?

(b) Compute the variance inflation factors for each variable. Is multicollinearity an issue with this data? What is a typical side effect of this degree of multicollinearity?

(c) Apply a criterion-based variable selection method using the R command *libaray (leaps); vs.out=regsubsets(wt~ . , nbest=3, data=lf)*. You may use either C_p, BIC, or adjusted R^2 to construct a plot like Figure 12-10 and to identify the best model according to the criterion chosen.

(d) Fit the model identified by the variable selection procedure of part (c), and report the R^2, the adjusted R^2, the p-value for the model utility test, and the p-values for each of the predictors in the model. Compare these with the results obtained in part (a). Finally, compute and report the variance inflation factors for the variables in the model. Is multicollinearity an issue now?

14. A reliability study investigated the probability of failure within 170 hours of operation of a product under accelerated life testing conditions. A random sample of 38 products was used in the study. The data set *FailStress.txt* contains simulated results (1 for failure, 0 for no failure) together with a variable quantifying the stress level for each product. Import the data into the data frame *fs*, and set *x=fs$stress; y=fs$fail.*

(a) Assume the logit transformation of $p(x)$ is a linear function of x, and fit a logistic regression model to the data. Test the significance of the stress variable at level $\alpha = 0.05$.

(b) Use the fitted model to estimate the probability of failure at stress level 3.5.

(c) Fit the logistic regression model that includes a quadratic term of the stress variable to model the logit transformation of $p(x)$. Construct 95% CIs for the regression parameters. Use the CIs to test the hypothesis at $\alpha = 0.05$ that the coefficient of the quadratic component is zero. (*Hint.* Use *fit=glm(y~x+I(x**2), family=binomial( ), data=ld); summary(fit)* to fit the quadratic logistic regression model.)

[14] Original data collected by Dr. S. Conant, University of Hawaii, from 1982 to 1992 for conservation-related research.

Chapter 13

Statistical Process Control

13.1 Introduction and Overview

Product quality and cost are the two major factors affecting purchasing decisions, both by individual consumers and businesses. Industry has come to realize that high quality and cost efficiency are not incompatible goals. The active use of statistics plays an important role in raising quality levels and decreasing production costs.

Product quality is not assured by thorough inspection and evaluation of every single product item produced. In fact, 100% inspection schemes are inefficient in most practical settings. It is much more efficient to build quality into the manufacturing process. Statistical *process control* represents a breakthrough in the strategy for achieving cost-efficient, high-quality manufacturing. Careful (re-)examination of the production process, which also includes the design and assembly states, may result in lower costs and increased productivity. Indeed, the production process serves as the common platform for implementing both quality control and quality improvement programs. Quality improvement programs, which have sprung from the idea that a product's quality should be continuously improved, rely mainly on Pareto charts (see Section 1.5.3) and designed factorial, or fractional factorial, experiments (see Section 11.4). This chapter discusses the use of *control charts*, which are the main tools for monitoring the production process in order to maintain product quality at desired levels.

The quality of a product is represented by certain quality characteristic(s). For example, comfort and safety features, gas mileage, and acceleration are some quality characteristics of cars; while air permeability is an important quality characteristic of woven fabrics used in parachute sails and air bags. The number of scratches in optical lenses and the proportion of defective integrated circuits made in a day are also important quality characteristics of their respective production processes. If the average (i.e., population mean) of the quality characteristic is what it is desired to be, that is, equal to a *target value*, the production process is called **in control** (with respect to the mean). If the average characteristic deviates from its target value, the process is said to be **out of control**.

High-quality manufacturing processes are also characterized by product uniformity, that is, limited variation in the quality characteristic(s) among the product items produced. For example, uniformity of cars (of a certain make and model) with respect to, say, gas mileage, is indicative of high-quality car manufacturing. The degree of uniformity is inversely proportional to the intrinsic variability (see Section 1.2) of the quality characteristic. If the variability, measured by the range or standard deviation, of the quality characteristic stays at a specified level, which is the naturally occurring (also called *uncontrolled*, or *common cause*) intrinsic variability, the process is in control (with respect to uniformity). If the variability increases, the process is out of control.

To put the above discussion in statistical notation, let Y_t denote the quality characteristic of a product item selected randomly at time t, and set μ_t and σ_t^2 for the population mean and variance of the product's quality characteristic at time t. Also, let μ_0 and σ_0^2 denote the target values for the mean and variance of the quality characteristic. For as long as $\mu_t = \mu_0$ and $\sigma_t^2 = \sigma_0^2$ the process is in control. Imagine that, over time, a change in the processing conditions occurs and, as a consequence, the mean and/or the variance deviate from their target values. Thus, if t_* denotes the (unknown to us) time when a change occurs, we have that, for $t \geq t_*$, $\mu_t \neq \mu_0$ and/or $\sigma_t^2 > \sigma_0^2$. Similarly, a change in the production process at time t_* can increase the probability of a defective product, or the rate at which unwanted features, such as scratches and other defects/abnormalities, occur on the finished product. In words, starting from time t_*, the process is out of control, which means that the quality of the products produced do not meet the desired standards. The purpose of control charts is to indicate when the production process drifts out of control. When an out-of-control state is indicated, an investigation is launched to determine what caused the out-of-control drift. Once causes, hereafter referred to as *assignable causes*, have been identified, corrective actions are taken.

To monitor the state of the production process, control charts rely on a series of samples taken at selected time points. Such a series of samples is shown in Figure 13-1, where the unknown time t_* when the process drifts out of control is also indicated. A statistic, such as the sample mean or sample variance, is computed for each sample in turn, as it becomes available. Control charts incorporate the sampling distribution of the computed statistic and ideas similar to those of hypothesis testing and CIs in order to indicate, at each time point, whether or not the new sample suggests that the process is out of control.

When the mean and/or variance of the quality characteristic (also referred to as *process mean* and *process variability*) drift away from their target values, it is of interest to detect the change as quickly as possible. On the other hand, a false alarm causes an unnecessary disruption in the production process, due to the investigation to identify assignable causes. It is thus desirable that, at each time point when a sample is selected and its conformance to the target mean and/or variance is tested, an actual out-of-control state be detected with high probability. At the same time, it is desirable to have a low probability for false alarms. These notions, which are

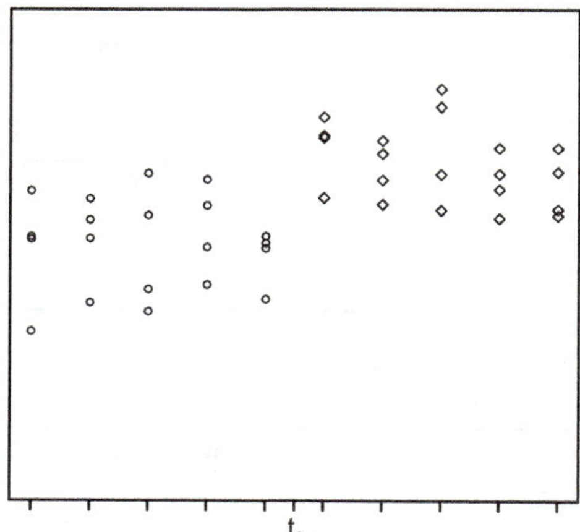

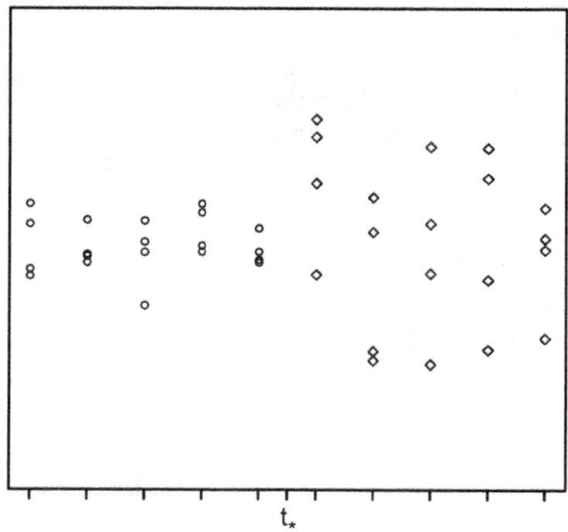

Figure 13-1 Series of samples of size 4. Changes in the mean (left panel) and the variance (right panel) occur at time t_*.

similar to type I and type II error probabilities in hypothesis testing (see Section 8.4.1), form the basis for evaluating the efficiency of a control chart.

Process Capability, Process Yield, and the Six Sigma Ethic Very often, the dual objectives of keeping the process mean and process variability at specified target levels, are replaced by an alternative objective that is expressed in terms of the proportion of items whose quality characteristic lies within desirable *specification limits*. For example, in aluminum cold rolling processes, the coolant viscosity, which has a major impact on surface quality, should be between 2.7 ± 0.2 centistokes. In this example, $2.7 - 0.2 = 2.5$ is the *lower specification limit*, or LSL, and $2.7 + 0.2$ is the *upper specification limit*, or USL for the coolant production process. Since batches of coolant with viscosity outside the specification limits result in inferior surface quality, the coolant manufacturing process should produce such batches with low probability. Because the proportion of product items whose quality characteristic lies outside the specification limits depends on both the mean value and variance of the quality characteristic (see Figure 13-2), controlling the proportion of product items whose quality characteristic lies outside the specification limits is an indirect way of controlling the mean value and the variance.

Process capability refers to the ability of a process to produce output within the specification limits. A quantification for process capability is given by the *capability index*:

$$C = \frac{\text{USL} - \text{LSL}}{6\sigma}, \tag{13.1.1}$$

where σ is the standard deviation of the quality characteristic, assuming that the mean value is the midpoint, μ_0, between USL and LSL.[1] Note that the capability index makes sense only for processes that are in statistical control. The higher a process's capability index, the smaller the probability that it will produce an item with quality characteristic outside the specification limits. A process is said to be at *six sigma (6σ)* quality level if its capability index is 2. The so-called *process yield* of a 6σ process is 99.9999998%, meaning the probability (under normality) of producing

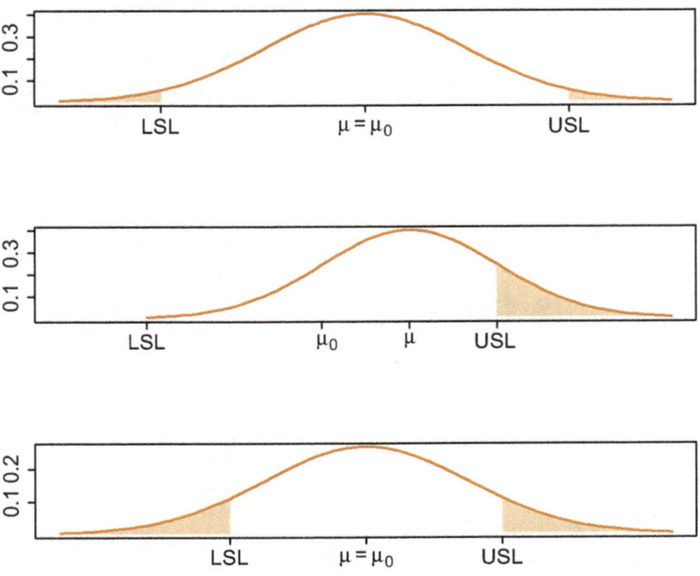

Figure 13-2 Probability of producing an item with quality characteristic outside the specification limits (shaded area) when mean and variance are equal to target values (top panel), only the mean is larger than the target value (middle panel), and only the variance is larger than the target value (bottom panel).

[1] See http://en.wikipedia.org/wiki/Process_capability_index for variations of this capability index.

an item outside the specification limits is 2×10^{-9}. Starting with Motorola in 1985 and General Electric in 1995, 6σ quality has become the standard in high-quality manufacturing.

The Assumption of Rational Subgroups *Rational subgroups* is a technical term used to describe a sequence of samples, such as shown in Figure 13-1, provided the observations taken at each time point are independent and identically distributed (iid) and different samples are independent. Each sample is a rational subgroup and offers a snapshot of the state of the process at the time it is taken. Observations from different samples need not be identically distributed as the state of the process may have changed.

The control charts we will discuss rely on the assumption that the samples taken at the different inspection time points form rational subgroups. When observations within a rational subgroup are not independent, in which case they are said to be *autocorrelated* or *serially correlated*, their sample variance tends to underestimate the population variance. Unless the autocorrelation is properly taken into account, the control charts may produce frequent false alarms.

Historically, the idea of a control chart was first proposed by Dr. Walter A. Shewhart in 1924 while he was working at Bell Labs. Shewhart's thinking and use of the control chart were further popularized by W. Edwards Deming, who is also credited with their wide adoption by the Japanese manufacturing industry throughout the 1950s and 1960s.

The next section discusses the use of the $\overline{X}$ *chart*, which is the most common chart for monitoring the process mean. Charts for controlling the variability are discussed in Section 13.3. Charts for controlling the proportion of defective items and charts for controlling the number of defects/abnormalities per item are presented in Section 13.4. The charts in Sections 13.2 through 13.4 are referred to as *Shewhart-type* control charts. Finally, the *cumulative sum*, or *CUSUM*, chart and the *exponentially weighted moving average*, or *EWMA*, chart, both of which have better efficiency properties than the $\overline{X}$ chart for small deviations from the target value of the process mean, are discussed in Section 13.5.

13.2 The $\overline{X}$ Chart

In addition to the assumption of rational subgroups, the $\overline{X}$ chart, which will be described in this section, requires that the normality assumption is, at least approximately, satisfied. In addition, construction of the $\overline{X}$ chart relies on the tacit assumption that the variance of the quality characteristic is in control. For this reason, charts for controlling the process variation (see Section 13.3) are typically the first to be employed. On the other hand, the $\overline{X}$ chart is more suitable for explaining the underlying ideas of control charts, and thus it is presented first.

There are two versions of the $\overline{X}$ chart. One assumes known target values for the mean value and variance, while the other uses estimates of these target values. The latter is mainly used in the early stages of a process control program when, more often than not, the true target values are unknown. In addition to describing the two versions of the $\overline{X}$ chart, this section introduces the concept of *average run length*, which is used for evaluating the performance of control charts.

13.2.1 $\overline{X}$ CHART WITH KNOWN TARGET VALUES

Let μ_0, σ_0 denote the target values for the mean and standard deviation of a quality characteristic, which is assumed to have the normal distribution. Then, if $\overline{X}$ denotes the mean of a sample of size n, taken when the process is in control, that is, when the true mean and standard deviation of the quality characteristic are equal to their target values, properties of the normal distribution imply that it lies within $\mu_0 \pm 3\sigma_0/\sqrt{n}$ with high probability. In particular,

$$P\left(\mu_0 - 3\frac{\sigma_0}{\sqrt{n}} \le \overline{X} \le \mu_0 + 3\frac{\sigma_0}{\sqrt{n}}\right) = 0.9973. \qquad \textbf{(13.2.1)}$$

This fact underlies the construction of the $\overline{X}$ chart.

Let $\overline{x}_1, \overline{x}_2, \overline{x}_3, \ldots$ denote the sample means of the samples (rational subgroups) of size n taken at inspection time points $1, 2, 3, \ldots$. The $\overline{X}$ chart consists of a scatterplot of the points $(1, \overline{x}_1), (2, \overline{x}_2), (3, \overline{x}_3), \ldots$ together with horizontal lines drawn at the *lower control limit*, or (LCL), and the *upper control limit*, or (UCL), where

$$\text{LCL} = \mu_0 - 3\frac{\sigma_0}{\sqrt{n}} \quad \text{and} \quad \text{UCL} = \mu_0 + 3\frac{\sigma_0}{\sqrt{n}}. \qquad \textbf{(13.2.2)}$$

If a point $(i, \overline{x}_i)$ falls above the horizontal line at UCL or below the horizontal line at LCL the process is declared out of control, and an investigation to identify assignable causes is launched. If all points are within the control limits, then there is no reason to suspect that the process is out of control.

The upper and lower control limits given in (13.2.2) are called 3σ control limits. They are the most frequently used and the default control limits in statistical software.

Example 13.2-1

The file *CcShaftDM.txt* contains 20 samples (rational subgroups) of $n = 5$ shaft diameter measurements used in a process control study.[2] Each sample was taken on a different day. The target mean value for the shaft diameter is 0.407 and the historic standard deviation is 0.0003.

(a) Construct an $\overline{X}$ chart with 3σ control limits.

(b) If the specification limits for these machine shafts is 0.407 ± 0.00025, use the data to estimate the process yield.

Solution

(a) Application of the formulas for the control limits in (13.2.2) gives

$$\text{LCL} = 0.407 - 3\frac{0.0003}{\sqrt{5}} = 0.4065975 \quad \text{and}$$

$$\text{UCL} = 0.407 + 3\frac{0.0003}{\sqrt{5}} = 0.4074025.$$

Importing the data into the R data frame *ShaftDiam*, the R commands *plot(rowMeans(ShaftDiam), ylim=c(0.4065, 0.4076)); abline(h=c(0.4065975, 0.407, 0.4074025))* plot the subgroup means and draw horizontal lines at the control limits and the target value. Alternatively, the $\overline{X}$ chart can be constructed with a customized R function available in the R package *qcc*. With the package installed and loaded to the current R session by *install.packages("qcc")* and *library(qcc)*, respectively, the R command for the 3σ $\overline{X}$ chart with known target values is

[2] Dale G. Sauers (1999). Using the Taguchi loss function to reduce common-cause variation. *Quality Engineering*, 12(2): 245–252.

> **R Command for a 3σ $\overline{X}$ Chart with Specified Target Mean Value and Standard Deviation**
>
> ```
> qcc(ShaftDiam, type="xbar", center=0.407,
>   std.dev=0.0003)
> ```
> **(13.2.3)**

The output of this command is the $\overline{X}$ chart shown in Figure 13-3. The chart shows two points above the upper control limit, suggesting that the process is out of control. The number of observations beyond the control limits is also given at the bottom of the chart, together with additional information.

Assuming normality, relation (13.2.1) gives that, when the process is in control, the probability that each individual subsample mean falls within the 3σ control limits is 0.9973. The options *nsigmas* and *confidence.level* can be used to construct alternative $\overline{X}$ charts. For example, *qcc(ShaftDiam, type="xbar", center=0.407, std.dev=0.0003, nsigmas=2)* constructs a 2σ $\overline{X}$ chart, while *qcc(ShaftDiam, type="xbar", center=0.407, std.dev=0.0003, confidence.level=0.99)* constructs an $\overline{X}$ chart with probability 0.99 for each individual subsample mean to fall within the control limits.

A final comment on the use of the R function *qcc* has to do with the fact that data files often present the data in two columns, one corresponding to the actual measurements and a second column specifying the sample each measurement comes from. For illustration purposes, the shaft diameter data is given in this alternative form in the file *CcShaftD.txt*. With the data imported into the R data frame *SD*, the $\overline{X}$ chart of Figure 13-3 can be constructed with the R commands *Diam=qcc.groups(SD$ShaftDiam, SD$Day); qcc(Diam, type="xbar", center=0.407, std.dev=0.0003)*. The first of these commands converts the data into the form suitable for use in the *qcc* function.

Figure 13-3 3σ $\overline{X}$ chart for the shaft diameter data: Known target values.

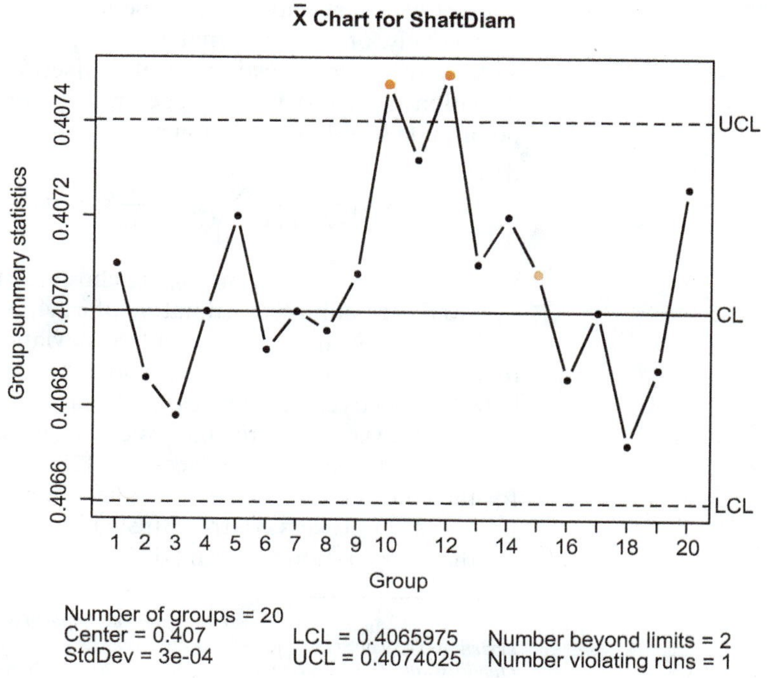

(b) The number of shafts within the specification limits, obtained with the R command *sum(ShaftDiam>=0.407-0.00025&ShaftDiam<=0.407+0.00025)*, is 56. Since there is a total of 100 diameter measurements, the process yield is 56%.

13.2.2 $\overline{X}$ CHART WITH ESTIMATED TARGET VALUES

As already mentioned, in the early stages of a process control program, the true target values are often unknown. For example, the target mean value and variance of characteristics such as the length where a cap is electronically placed on a syringe, or the stress resistance of aluminum sheets, might be unknown when monitoring of the respective production processes begins. In such cases, the initial purpose of process monitoring is to maintain the product quality at current levels. To do so, the current levels of the mean and standard deviation must first be estimated by collecting samples from time periods when the process is assumed to be in control.

Let $\overline{x}_1, \ldots, \overline{x}_k$, be the sample means of k samples (rational subgroups) taken when the process is believed to be in control. It is recommended that the subgroup sample size be at least 3 and the total number of observations in the k samples be at least 60; for example, at least $k = 20$ samples each of size at least 3. Then the mean value is estimated by averaging the k sample means:

$$\widehat{\mu} = \frac{1}{k}\sum_{i=1}^{k} \overline{x}_i. \tag{13.2.4}$$

There are two commonly used estimators of the standard deviation. One is based on the subgroup standard deviations, and the other on the subgroup ranges, where the range of a sample is defined as the difference between the largest and smallest sample values. Thus, if $x_{i1}, \ldots, x_{in}$ are the n observations in the ith subgroup, its range is computed as

$$r_i = \max\{x_{i1}, \ldots, x_{in}\} - \min\{x_{i1}, \ldots, x_{in}\}.$$

Let $s_1, \ldots, s_k$ and $r_1, \ldots, r_k$ be sample standard deviations and sample ranges, respectively, of the k rational subgroups. (Note that, even though the sample variance is an unbiased estimator of σ^2 [see Proposition 6.2-1], the sample standard deviation is biased for σ.) The two commonly used unbiased estimators of the population standard deviation are

$$\widehat{\sigma}_1 = \frac{1}{A_n}\frac{1}{k}\sum_{i=1}^{k} s_i = \frac{1}{A_n}\overline{s}, \quad \text{and} \quad \widehat{\sigma}_2 = \frac{1}{B_n}\frac{1}{k}\sum_{i=1}^{k} r_i = \frac{1}{B_n}\overline{r}, \tag{13.2.5}$$

where the constants A_n and B_n are chosen so that, if the normality assumption holds, $\widehat{\sigma}_1$ and $\widehat{\sigma}_2$ are unbiased estimators of σ. A_n values are less than one, indicating that the average of the sample standard deviations underestimates σ; for example, for $n = 3, 4, 5$, the value of A_n, rounded to three decimal places, is 0.886, 0.921, and 0.940, respectively. B_n values are greater than one, indicating that the average of the sample ranges overestimates σ; for example, for $n = 2, 3, 4, 5$, the value of B_n, rounded to three decimal places, is 1.128, 1.693, 2.058, and 2.325, respectively. Tables for the constants A_n and B_n are available in many textbooks and training materials on statistical process control.[3] Instead of reproducing the tables here, we rely on R for the computation of $\widehat{\sigma}_1$ and $\widehat{\sigma}_2$.

---

[3] A table of A_n values is also available at *http://en.wikipedia.org/wiki/Unbiased_estimation_of_standard_deviation*. The values of B_n for n from 2 to 25 can be found in D. C. Montgomery (1996). *Introduction to Statistical Quality Control*, 3rd Edition, New York: John Wiley & Sons, Inc.

The 3σ control limits for the $\overline{X}$ chart with estimated parameters are

$$\mathrm{LCL} = \widehat{\mu} - 3\frac{\widehat{\sigma}}{\sqrt{n}} \quad \text{and} \quad \mathrm{UCL} = \widehat{\mu} + 3\frac{\widehat{\sigma}}{\sqrt{n}}, \tag{13.2.6}$$

where $\widehat{\mu}$ is given in (13.2.4) and $\widehat{\sigma}$ is either one of the two estimators of σ given in (13.2.5).

Example 13.2-2

The R data frame *pistonrings*,[4] available in the R package *qcc*, consists of 40 samples, each of size 5, of the inside diameter of piston rings for an automotive engine. The first column of the data frame gives the diameter measurements, and the second column is a sample (rational subgroup) indicator. The first 25 samples were taken when the process was thought to be in control; this is indicated by "TRUE" on the third column of the data frame. Use the data to estimate the in-control diameter mean and standard deviation, construct a 3σ $\overline{X}$ chart, and check if any of the remaining subgroups indicate an out-of-control state.

Solution

Use *library(qcc); data(pistonrings); attach(pistonrings)* to import the *pistonrings* data frame to the current R session, and *piston=qcc.groups(diameter, sample)* to arrange the 40 samples into 40 rows (each of size 5) in the object *piston*. The R command for estimating the in-control diameter mean and standard deviation from the first 25 subgroups and constructing a 3σ $\overline{X}$ chart is

R Command for a 3σ $\overline{X}$ Chart with Estimated Mean Value and Standard Deviation

```
qcc(piston[1:25,], type="xbar", newdata=piston[26:40, ])
```
(13.2.7)

The output of this command is the $\overline{X}$ chart shown in Figure 13-4. The vertical line at the 25.5 mark on the x-axis separates the first 25 samples, called *calibration data*, from the remaining samples that are to be tested using estimates of the mean and standard deviation obtained from the calibration data. The chart shows three points, corresponding to samples numbered 37, 38, and 39, above the upper control limit, suggesting that the process is out of control.

The $\overline{X}$ chart shown in Figure 13-4 uses the estimator of σ based on the sample ranges, that is, $\widehat{\sigma}_2$, given in (13.2.5). A longer version of the command in (13.2.7) is:

```
qcc(piston[1:25,], type="xbar", newdata=piston[26:40, ],
  std.dev="UWAVE-R").
```

To use the estimator $\widehat{\sigma}_1$, given in (13.2.5), simply replace *std.dev="UWAVE-R"* by *std.dev="UWAVE-SD"*. With the pistonrings data, the choice of $\widehat{\sigma}$ does not make a noticeable difference in the resulting $\overline{X}$ chart. In fact, for this data set, $\widehat{\sigma}_1 = 0.009829977$, while $\widehat{\sigma}_2 = 0.009785039$ (also displayed at the bottom of Figure 13-4). Finally, the options *nsigmas* and *confidence.level*, described in connection with (13.2.3), can be used to construct alternative $\overline{X}$ charts. ■

[4] D. C. Montgomery (1991). *Introduction to Statistical Quality Control*, 2nd Edition, New York: John Wiley & Sons, 206–213.

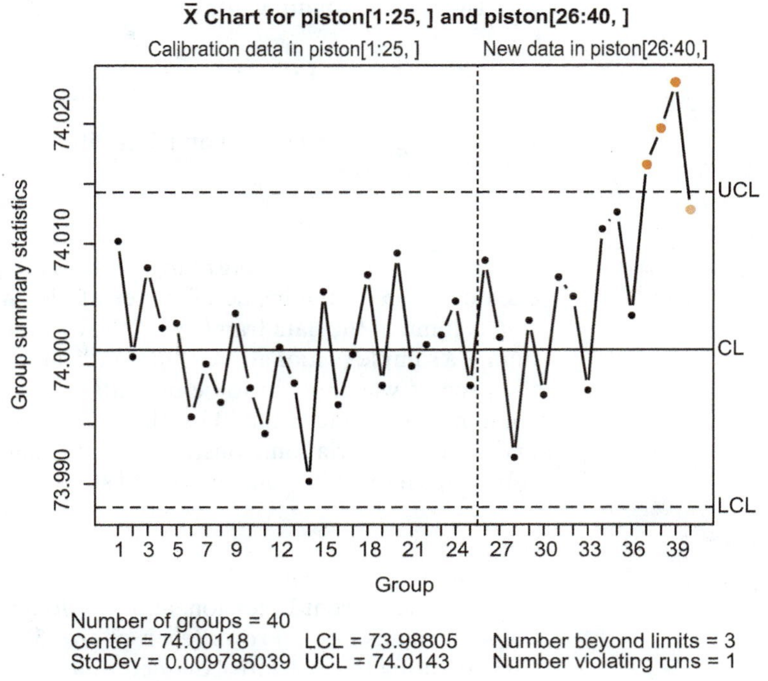

Figure 13-4 3σ $\overline{X}$ chart for the pistonrings data: Estimated μ and σ.

Recomputing the Estimated Target Values It is possible that the sample mean of one or more of the calibration subgroups falls outside the control limits. This is an indication that, contrary to previous belief, the process might not have been in control during the collection of the calibration data. If an assignable cause can be identified, a judgement must be made if the assignable cause affected only the subgroups whose sample means fall outside the control limits. If so, these subgroups must be removed from the calibration data, and the target values be recomputed using the remaining subgroups in the calibration data. For example, had Figure 13-4 shown the means $\overline{x}_4$, $\overline{x}_9$, $\overline{x}_{18}$ to lie outside the control limits, the elimination of subgroups 4, 9, and 18 from the calibration data set and the construction of the 3σ $\overline{X}$ chart using estimated target values from the reduced calibration data would have been done with the R commands

R Commands for Deleting Subgroups from the Calibration Data

```
qcc(piston[c(1:25)[-c(4, 9, 18)], ], type="xbar",
  newdata=piston[26:40, ])
```
(13.2.8)

The resulting chart shows the same three points above the upper control limit.

13.2.3 THE *X* CHART

When only $n = 1$ observation is collected at each inspection time point, the $\overline{X}$ chart is called the X chart. Because computation of either the sample range or standard deviation requires a sample of size at least 2, the methods described for estimating the in-control standard deviation do not apply if $n = 1$. In such cases, the most commonly used estimator uses the *average moving range*,

$$\overline{\text{MR}} = \frac{1}{k-1}\sum_{i=1}^{k-1} \text{MR}_i, \quad \text{where } \text{MR}_i = |X_{i+1} - X_i|,$$

where k is either the total number of inspection time points or the number of inspection time points when the process is believed to be in control. In particular, σ is estimated by

$$\widehat{\sigma} = \overline{\text{MR}}/1.128, \tag{13.2.9}$$

and the resulting control limits for the X chart are

$$\text{LCL} = \widehat{\mu} - 3\frac{\overline{\text{MR}}}{1.128} \quad \text{and} \quad \text{UCL} = \widehat{\mu} + 3\frac{\overline{\text{MR}}}{1.128}, \tag{13.2.10}$$

where, in this case, $\widehat{\mu}$ is simply the average $\overline{x}$ of all observations or of those taken when the process is believed to be in control. If the target value, μ_0, of the process mean is known, μ_0 can be used instead of $\widehat{\mu}$ in (13.2.10).

With the data in the R object x, the R commands for the X chart are

R Commands for a 3σ X Chart

```
qcc(x, type="xbar.one")
qcc(x[1:k], type="xbar.one", newdata=x[k+1:length(x)])
```
(13.2.11)

In the first of the commands in (13.2.11), $\overline{x}$ and the average moving range (and hence the estimator of σ) are computed from the entire data set, while in the second command they are computed from the first k observations, where k is the number of inspection time points when the process is believed to be in control. If the target value, μ_0, of the process mean is known, the option *center=*μ_0 can be used with either of the two commands (e.g., *qcc(x, type="xbar.one", center=0.0)*, if $\mu_0 = 0.0$).

To illustrate the construction of X charts, 50 observations were randomly generated, the first 30 from the $N(0, 1)$ distribution and the last 20 from $N(1, 1)$, and saved in the file *SqcSimDatXone.txt*. Read the data into the data frame *simd*, and copy the 50 values into the R object x by *x=simd$x*. Then the commands in (13.2.11) with k and *k+1* replaced by 30 and 31, respectively, produce the X charts shown in Figure 13-5. Note that, up to three decimal places, the estimated standard deviation in the two charts is the same. This is due to the fact that the process variability remains in control, so only one of 49 moving ranges has the potential of being larger (though in this case $|x_{31} - x_{30}|$ happens to be relatively small). What really makes a difference in the two charts is the estimated mean, which is larger in the left panel chart. This difference in the estimated center results in three points being in color in the right panel chart. The significance of colored points is explained in the following section.

13.2.4 AVERAGE RUN LENGTH AND SUPPLEMENTAL RULES

As already mentioned, it is desirable that false alarms, that is, out-of-control signals when the process is in control, happen with low probability. If the quality characteristic is normally distributed and the target variance is known to be σ_0^2, the probability that a 3σ $\overline{X}$ chart issues a false alarm is

$$P_{\mu_0,\sigma_0}\left(\overline{X} < \mu_0 - 3\frac{\sigma_0}{\sqrt{n}}\right) + P_{\mu_0,\,\sigma_0}\left(\overline{X} > \mu_0 + 3\frac{\sigma_0}{\sqrt{n}}\right) = 2\Phi(-3) = 0.0027. \tag{13.2.12}$$

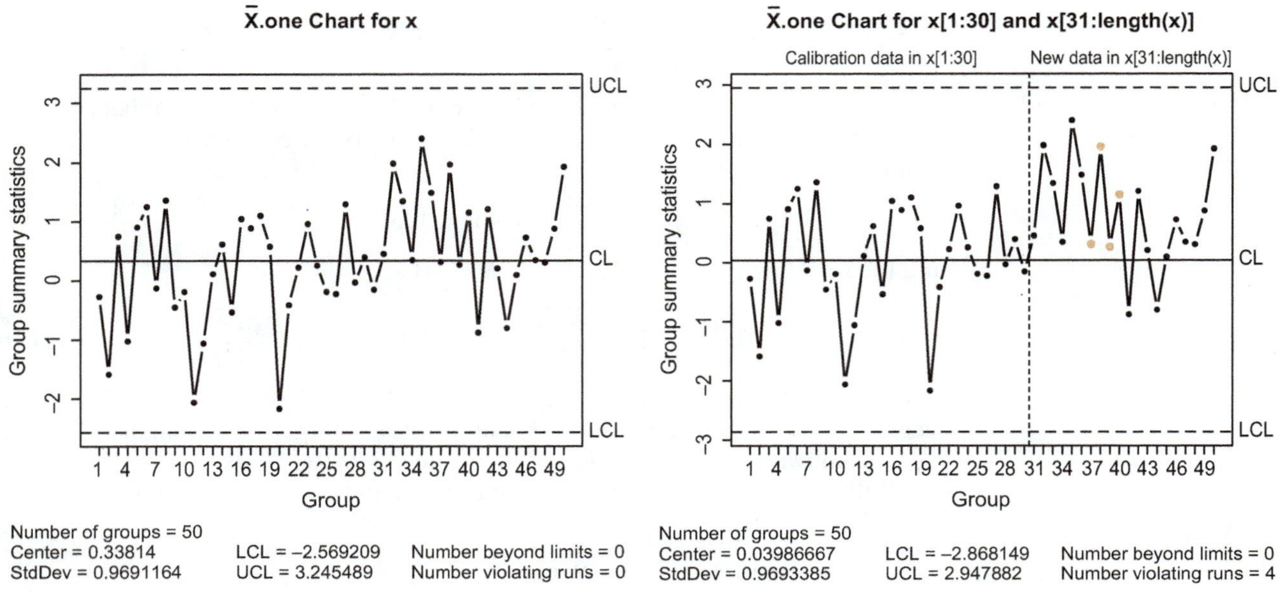

Figure 13-5 X charts: μ and σ estimated from all data (left panel) and from the first 30 observations (right panel).

(The subscript μ_0, σ_0 in P_{μ_0,σ_0} is used to indicate that the probability is computed under the assumption that the true mean of $\overline{X}$ is μ_0 and its true variance is σ_0^2/n, i.e., that the process mean is in control.) On the other hand, if the process shifts out of control, it is desirable to have a high probability of an out-of-control signal. For the $\overline{X}$ chart, such probabilities are calculated below.

The number of inspection time points until the first out-of-control signal is called a *run length*. Intuitively, long run lengths are expected when the probability of an out-of-control signal is low, and short run lengths are expected when this probability is high. Under the assumption of rational subgroups, this connection between run lengths and the probability of an out-of-control signal can be made precise.

Consider a sequence of independent Bernoulli trials whose probability of "success" is p. In the context of statistical process control, each inspection time point corresponds to a Bernoulli trial, where "success" is the event that the corresponding sample mean lies outside the control limits. For example, according to (13.2.12), if the process is in control, and assuming normality and known σ, $p = 0.0027$ for a 3σ $\overline{X}$ chart. The number of Bernoulli trials until the first success is a random variable having the geometric distribution (see Section 3.4.3). Thus, under the assumption of rational subgroups, run lengths are geometric random variables. The mean value of a run length is called *average run length*, or *ARL*. In terms of ARL, the desirable properties of any control chart can be restated as follows: *long ARL when the process is in control, and short ARL when the process is out of control.*

From relation (3.4.13) we have that the mean value of a geometric random variable is $1/p$. Thus, (13.2.12) implies that, under the assumptions of rational subgroups, normality, and σ known to be σ_0, the in-control ARL for a 3σ $\overline{X}$ chart is $1/(2\Phi(-3)) = 1/0.0027 = 370.4$. Thus, on average, false alarms happen about every 370 inspections of an in-control process. Similarly, the in-control ARL for a 2.7σ $\overline{X}$ chart and that for a 3.1σ $\overline{X}$ chart are found to be 144.2 and 516.7, respectively (Exercise 1). These ARLs are approximately correct with estimated σ and with the normality assumption being approximately satisfied.

The probability of an out-of-control signal when (only) the process mean has shifted out of control depends on the size of the shift, that is, $|\mu - \mu_0|$, where μ denotes the current value of the process mean and μ_0 is the target value. If $\mu = \mu_0 + \Delta\sigma$, so $\mu - \mu_0 = \Delta\sigma$, where Δ can be positive or negative (i.e., $|\mu - \mu_0| = |\Delta|\sigma$),

$$
\begin{aligned}
&P_{\mu,\sigma}\left(\overline{X} < \mu_0 - 3\frac{\sigma}{\sqrt{n}}\right) + P_{\mu,\sigma}\left(\overline{X} > \mu_0 + 3\frac{\sigma}{\sqrt{n}}\right)\\
&= P_{\mu,\sigma}\left(\frac{\overline{X}}{\sigma/\sqrt{n}} < \frac{\mu_0}{\sigma/\sqrt{n}} - 3\right) + P_{\mu,\sigma}\left(\frac{\overline{X}}{\sigma/\sqrt{n}} > \frac{\mu_0}{\sigma/\sqrt{n}} + 3\right)\\
&= P_{\mu,\sigma}\left(\frac{\overline{X}-\mu}{\sigma/\sqrt{n}} < \frac{\mu_0-\mu}{\sigma/\sqrt{n}} - 3\right) + P_{\mu,\sigma}\left(\frac{\overline{X}-\mu}{\sigma/\sqrt{n}} > \frac{\mu_0-\mu}{\sigma/\sqrt{n}} + 3\right)\\
&= P_{\mu,\sigma}\left(\frac{\overline{X}-\mu}{\sigma/\sqrt{n}} < -\sqrt{n}\Delta - 3\right) + P_{\mu,\sigma}\left(\frac{\overline{X}-\mu}{\sigma/\sqrt{n}} > -\sqrt{n}\Delta + 3\right)\\
&= \Phi(-3-\sqrt{n}\Delta) + 1 - \Phi(3-\sqrt{n}\Delta).
\end{aligned}
\tag{13.2.13}
$$

Note that, for a fixed value of n, the value of (13.2.13) depends only on the absolute value of Δ, not its sign (Exercise 1). Keeping n fixed, differentiation of (13.2.13) reveals that the probability of an out-of-control signal increases as $|\Delta|$ increases. For example, if $n = 5$, the R commands

```
Delta=seq(.2, 1.4, .2); p=1+pnorm(-3-sqrt(5)*Delta)-
  pnorm(3-sqrt(5)*Delta); p; 1/p
```

give the values shown in Table 13-1 for the probability of an out-of-control signal (P(signal) in the table), and the corresponding ARL for Δ values from 0.2 to 1.4 in increments of 0.2. (The table also shows the in-control ARL, i.e., $\Delta = 0$, which was derived before.)

Similarly, the probability of an out-of-control signal increases with the sample size for each fixed $|\Delta|$; see Exercise 1 for an example.

From Table 13-1, it is clear that a 3σ $\overline{X}$ chart is slow to detect small shifts in the process mean. This prompted investigators at the Western Electric Company[5] to develop supplemental stopping rules. According to these rules, called *Western Electric rules*, the process is also declared out of control if any of the following is violated:

1. Two out of three consecutive points exceed the 2σ limits on the same side of the center line.
2. Four out of five consecutive points exceed the 1σ limits on the same side of the center line.
3. Eight consecutive points fall on one side of the center line.

Table 13-1 Probability of out-of-control signal and ARL for shifts of $\Delta\sigma$, with $n = 5$

Δ	0	0.2	0.4	0.6	0.8	1.0	1.2	1.4
P(signal)	0.0027	0.0056	0.0177	0.0486	0.1129	0.2225	0.3757	0.5519
ARL	370.4	177.7	56.6	20.6	8.9	4.5	2.7	1.8

[5] Western Electric Company (1956). *Statistical Quality Control Handbook*. Indianapolis: Western Electric Company.

Rules 1 and 2 are referred to as *scan rules*, while rule 3 is referred to as the *runs rule*. In control charts produced by R, the points violating the runs rule are marked in color (see Figures 13-3, 13-4, and 13-5). (Note that, according to R's default settings, a violation of the runs rule occurs at the seventh consecutive point on the same side of the center line. In general, there is some variation of the rules industry.)

The computation of the in-control and out-of-control ARLs of $\overline{X}$ charts with the supplemental rules is considerably more complicated and beyond the scope of this book.[6]

Exercises

1. Complete the following:

(a) Verify that the ARLs for a 2.7σ $\overline{X}$ chart and a 3.1σ $\overline{X}$ chart, under the assumptions of rational subgroups, normality, and known σ_0, are 144.2 and 516.7, respectively.

(b) Show that, keeping n fixed, the value of (13.2.13) for $\Delta = |\Delta|$ is the same as its value for $\Delta = -|\Delta|$.

(c) Use (13.2.13) to compute the probability of an out-of-control signal and the corresponding ARL for $n = 3, \ldots, 7$ when $\Delta = 1$.

2. A case study deals with the length at which a cap is electronically placed on syringes manufactured in a Midwestern pharmaceutical plant.[7] The cap should be tacked at an approximate length of 4.95 in. If the length is less than 4.92 in. or larger than 4.98 in., the syringe must be scrapped. In total, 47 samples of size 5 were taken, with the first 15 samples used for calibration. After sample 32, the maintenance technician was called to adjust the machine. The data (transformed by subtracting 4.9 and multiplying times 100) can be found in *SqcSyringeL.txt*.

(a) Use commands similar to those given in Example 13.2-2 to construct a 3σ $\overline{X}$ chart, using the estimator $\widehat{\sigma}_2$ given in (13.2.5). What does the chart suggest regarding the effect of the adjustment made after the 32nd sample? Has the adjustment brought the process back in control?

(b) Use a command similar to (13.2.8) to delete the second subgroup and re-construct the 3σ $\overline{X}$ chart with the new control limits. Have your conclusions in part (a) changed?

(c) Repeat parts (a) and (b) using the estimator $\widehat{\sigma}_1$ given in (13.2.5). Have the conclusions from parts (a) and (b) changed?

(d) Use commands similar to those given in Example 13.2-1(b), and upper and lower specification limits of 4.92 and 4.98, to estimate the process yield after the adjustment made. (*Hint.* Import the data into the R data frame *syr*, and transform it back to the original scale by *x=syr$x/100+4.9*. Use *sum(x[161:235]>=4.92&x[161:235]<=4.98)*.)

3. A case study reports on controlling the viscosity of a coolant used in an aluminum cold rolling process, as the viscosity has a major impact on the surface quality of the aluminum produced.[8] The data can be found in *SqcCoolVisc.txt*. Use commands similar to those of (13.2.11) to construct the following charts.

(a) A 3σ X chart with the center and standard deviation computed from the entire data set.

(b) A 3σ X chart with the center and standard deviation computed from the first 25 observations, taken when the process is believed to be in control.

For each of the charts, identify any points that fall outside the control limits, as well as any points suggesting an out-of-control state according to the General Electric supplemental rules.

4. A study involves 48 daily unit rate construction labor productivity data.[9] The first 30 daily measurements correspond to an unimpacted period and can be used for calibration. The data are given in *SqcLaborProd.txt*. Use commands similar to those of (13.2.11) to construct a 3σ X chart with the center and standard deviation computed from the first 30 observations.

(a) Is the process in control during the calibration period or after the calibration period?

(b) Remove the 12th and 13th daily measurements and answer again the questions in part (a).

---

[6] See J. Glaz, J. Naus, and S. Wallenstein (2001). *Scan Statistics.* New York: Springer-Verlag, and N. Balakrishnan and M. V. Koutras (2002). *Runs and Scans with Applications*, New York: John Wiley.

[7] LeRoy A. Franklin and Samar N. Mukherjee (1999-2000). An SPC case study on stabilizing syringe lengths. *Quality Engineering*, 12: 65–71.

[8] Bryan Dodson (1995). Control charting dependent data: A case study. *Quality Engineering*, 7: 757–768.

[9] Ronald Gulezian and Frederic Samelian (2003). Baseline determination in construction labor productivity-loss claims. *Journal of Management in Engineering*, 19: 160–165.

13.3 The S and R Charts

The commonly used charts for monitoring process variability are the S chart and the R chart. As mentioned earlier, one of the underlying assumptions for the $\overline{X}$ chart is that the process is in control with respect to variability. Thus, S and R charts should always be performed first, and, if the process is declared out of control, there is no need to proceed with the $\overline{X}$ chart.

Let $s_1, \ldots, s_k$ and $r_1, \ldots, r_k$ be the sample standard deviations and sample ranges, respectively, of k rational subgroups, and let $\bar{s}$, $\bar{r}$ denote their respective averages. The expressions for the lower and upper control limits of the S and R charts use the constants A_n and B_n, which are introduced in (13.2.5), and the additional constant C_n, which is the standard deviation of the range of a sample of size n from the standard normal distribution. These expressions are given in Table 13-2, keeping in mind that the LCL is not allowed to take a negative value. In particular, LCL is set to zero whenever its expression results in a negative value. This happens if $n \leq 5$ for the S chart and if $n \leq 6$ for the R chart. Tables for C_n exist in many textbooks;[10] for example, for $n = 3, 4, 5$, the value of C_n, rounded to three decimal places, is 0.888, 0.880, 0.864. However, we will rely on R commands for the construction of S and R charts. These are described in the following example.

Example 13.3-1

Use the R data frame *pistonrings*, which was also used in Example 13.2-2, to construct S and R charts by computing $\bar{s}$ and $\bar{r}$ from all available subgroups. Repeat this using only the first 25 subgroups for computing $\bar{s}$ and $\bar{r}$.

Solution

As in Example 13.2-2, use *library(qcc); data(pistonrings); attach(pistonrings); piston= qcc.groups(diameter, sample)* to import the *pistonrings* data frame to the current R session and to create the object *piston* suitable for use in the *qcc* function. The R command for constructing S charts with $\bar{s}$ computed using all available subgroups and then using only the first 25 subgroups, respectively, are

R Commands for S Charts

```
qcc(piston, type="S")
qcc(piston[1:25,], type="S", newdata=piston[26:40, ])
```
(13.3.1)

The charts are shown in Figure 13-6. Neither of these charts suggests that the process variability is out of control. Thus, there is no reason to suspect the validity of the in-control process variability assumption, which is needed for the $\overline{X}$ charts of Example 13.2-2.

Table 13-2 Lower and upper control limits of S and R charts

	LCL	UCL
S Chart	$\bar{s} - 3\bar{s}\sqrt{1 - A_n^2}/A_n$	$\bar{s} + 3\bar{s}\sqrt{1 - A_n^2}/A_n$
R Chart	$\bar{r} - 3C_n\bar{r}/B_n$	$\bar{r} + 3C_n\bar{r}/B_n$

[10] D. C. Montgomery (1996). *Introduction to Statistical Quality Control*, 3rd Edition, New York: John Wiley & Sons, A-15, constant d_3.

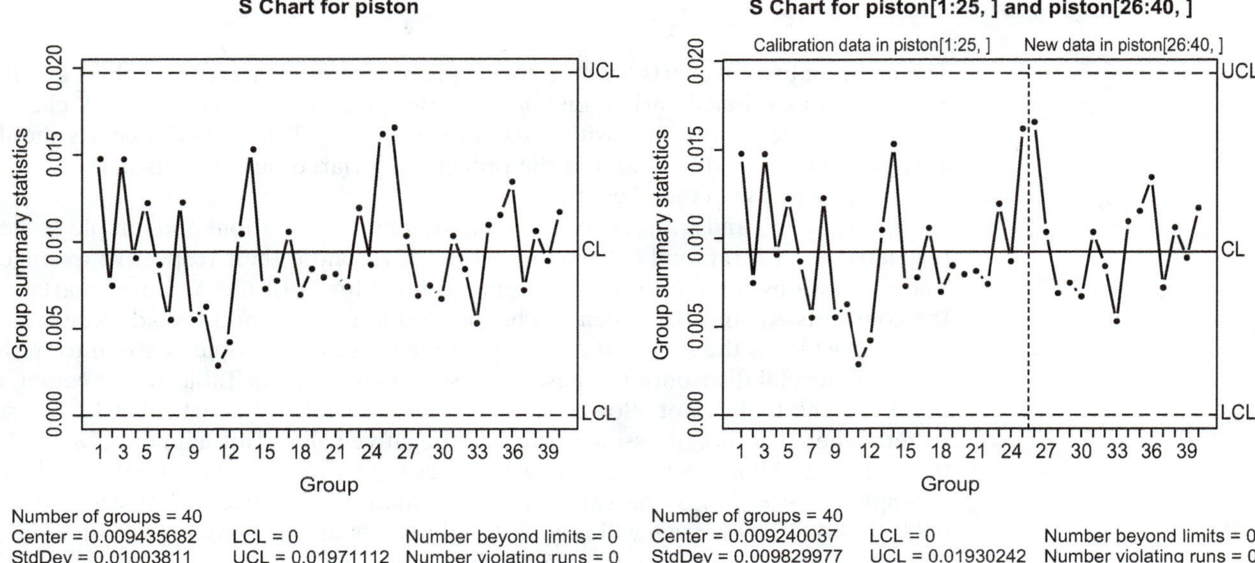

Figure 13-6 *S* charts: $\bar{s}$ computed from all available subgroups (left panel) and from the first 25 subgroups (right panel).

Replacing *type="S"* by *type="R"* in (13.3.1) produces the corresponding *R* charts. For the *pistonrings* data, these charts are similar to the *S* charts of Figure 13-6 and are not shown.

REMARK 13.3-1

(a) For the relatively small subgroup sizes typically used in quality control studies, the sample standard deviation and range are not (even approximately) normally distributed. Thus, probability calculations like those of relations (13.2.1) and (13.2.13) cannot be made for the *S* and *R* charts of Table 13-2.

(b) Because the 3σ-type control limits used in the *S* and *R* charts are not based on sound probabilistic properties, it is recommended that both charts be constructed, especially if $n \leq 7$ or 10. For larger n values, the *S* chart is preferable.

(c) Under the assumption that the underlying distribution of the quality characteristic is normal, a different *S* chart, based on the χ^2 CI for σ given in (7.3.20), is possible; see Exercise 3. ◁

Exercises

1. Use the data of Exercise 2 in Section 13.2 and commands similar to those given in Example 13.3-1 to construct an *S* and an *R* chart. What do these charts suggest regarding the effect on the process variability of the adjustment made after the 32nd sample?

2. A case study deals with semiconductor wafer diameter.[11] The data set, consisting of samples of size 2 that are taken from 20 lots, can be found in *SqcSemicondDiam.txt*.

(a) Use commands similar to those given in Example 13.3-1 to construct an *S* and an *R* chart. Comment on what these charts suggest about the process variability.

(b) Use commands similar to those given in Example 13.2-2 to construct a 3σ $\overline{X}$ chart, using the estimator $\widehat{\sigma}_2$ given in (13.2.5). What does the chart suggest regarding the process mean?

(c) Repeat part (b) using the estimator $\widehat{\sigma}_1$ given in (13.2.5). Has the conclusion from part (a) changed?

[11] Charles R. Jensen (2002). Variance component calculations: Common methods and misapplications in the semiconductor industry, *Quality Engineering*, 14: 645–657.

3. The χ^2 CI for σ given in (7.3.20) suggests a χ^2-based S chart. The 99.7% control limits of this chart, using the square root of the average of the k sample variances, $\widetilde{s} = \sqrt{(1/k)\sum_{i=1}^{k} s_i^2}$, as estimator of the in-control standard deviation, are given by

$$\text{LCL} = \widetilde{s}\sqrt{\frac{\chi^2_{n-1,0.9985}}{n-1}} \quad \text{and} \quad \text{UCL} = \widetilde{s}\sqrt{\frac{\chi^2_{n-1,0.0015}}{n-1}}.$$

Implementation of the χ^2-based S chart in R is not automatic, but its construction for the *pistonrings* data frame is demonstrated next. Let *piston* be the R object crated in Example 13.2-2. The R commands

```
stilde=sqrt(mean(apply(piston[1:25,  ], 1, var)))
n=5; LCL=stilde*sqrt(qchisq(0.0015, n-1)/(n-1))
UCL=stilde*sqrt(qchisq(0.9985, n-1)/(n-1))
```

compute the lower and upper control limits, and the further commands

```
sdv=apply(piston, 1, sd)
plot(1:40, sdv, ylim=c(0.0015, 0.0208), pch=4,
  main="Chi square S chart")
axis(4, at=c(LCL, UCL), lab=c("LCL", "UCL"))
abline(h=LCL); abline(h=UCL)
```

plot the standard deviations from the 40 subgroups and the control limits. Use the above commands to construct the χ^2-based S chart for the *pistonrings* data.

13.4 The p and c Charts

The p chart is used for controlling the binomial probability p which, in quality control settings, typically refers to the proportion of defective/nonconforming items produced. The c chart (c for *count*) is used for controlling the Poisson parameter λ, which typically refers to the average number of nonconformances per item or per specified area of surface, or of the number of accidents in an industrial plant or outbreaks of a certain disease, etc. In the quality control literature, count data used for the p and c charts are jointly referred to as *attribute data*.

13.4.1 THE p CHART

Let D_i be the number of defective items in a sample of size n taken at inspection times $i = 1, 2, \ldots$, and set $\widehat{p}_i = D_i/n$. In this context, the assumption of rational subgroups means that the D_i, and hence the $\widehat{p}_i$, are independent. Also assume that at each inspection time point, different items are defective or non-defective independently of each other. If the process is in control, and the in-control probability that an item is defective is p, then, by (4.4.4) and (4.4.9),

$$E(\widehat{p}_i) = p \quad \text{and} \quad \text{Var}(\widehat{p}_i) = \frac{p(1-p)}{n}.$$

Moreover, if $np \geq 5$ and $n(1-p) \geq 5$, the DeMoivre-Laplace Theorem (Theorem 5.4-2) implies that $\widehat{p}$ has approximately a normal distribution. Thus, the 3σ control limits are

$$\text{LCL} = p - 3\sqrt{\frac{p(1-p)}{n}} \quad \text{and} \quad \text{UCL} = p + 3\sqrt{\frac{p(1-p)}{n}}. \qquad \textbf{(13.4.1)}$$

If LCL is negative, it is replaced by zero. When p is unknown, the control limits in (13.4.1) are computed with p replaced by

$$\overline{p} = \frac{1}{k}\sum_{i=1}^{k} \widehat{p}_i,$$

where $\widehat{p}_1, \ldots, \widehat{p}_k$ are obtained when the process is believed to be in control. Note that because the probability of a defective item is typically quite small, the sample size n is much larger here than for the $\overline{X}$ chart.

The implementation of the p chart in R is illustrated in the following example.

Example 13.4-1

The R data frame *orangejuice*,[12] available in the R package *qcc*, consists of 54 samples (rational subgroups), each of size $n = 50$, of orange juice cans collected at half-hour intervals. Each can was inspected after filling to determine whether the liquid could leak either at the side seam or around the bottom joint, and the number of defective (leaking) cans for each sample was recorded. The first 30 samples were taken when the machine was in continuous operation, but the last 24 samples were taken after an adjustment was made. It is believed that the process was in control before the adjustment. Construct a 3σ p chart using the first 30 samples as the calibration data set.

Solution

Use *library(qcc)* to load the package *qcc* to the current R session, and *data(orangejuice); attach(orangejuice)* so the columns of the data frame *orangejuice* can be referred to by their names. The three columns of this data frame are D for the number of defectives in each sample, *size* for the sample size of each sample, and *trial*, which is a logical variable taking the value *TRUE* and *FALSE* for the first 30 and last 24 samples, respectively. With this information, the R command for the desired p chart is:

R Command for *p* Charts

```
qcc(D[trial], sizes=size[trial], type="p",
  newdata=D[!trial], newsizes=size[!trial])
```
(13.4.2)

Note that *c(1:30)* and *c(31:54)* could have been used instead of *trial* and *!trial*, respectively, in (13.4.2). The p chart produced by this command, shown in the left panel of Figure 13-7, reveals that (a) the sample proportions $\widehat{p}_{15}$ and $\widehat{p}_{23}$, both of which are used to obtain $\overline{p}$, are above the 3σ UCL, and (b) the adjustment that happened after the 30th subgroup appears to have decreased the proportion of defectives produced (which is a good thing!).

Upon investigation, it turns out that sample 15 used a different batch of cardboard, while sample 23 was obtained when an inexperienced operator was temporarily assigned to the machine. Because assignable causes for these two points have been found, samples 15 and 23 need to be removed from the calibration set and the control limits must be recomputed (see the discussion on recomputing the estimated target values in Section 13.2.2). The R command for doing so is

R Command for Recomputing Control Limits in *p* Charts

```
qcc(D[c(1:30)[-c(15,23)]], sizes=size[c(1:30)[-c(15,23)]],
type="p", newdata=D[!trial], newsizes=size[!trial])
```
(13.4.3)

[12] D. C. Montgomery (1991). *Introduction to Statistical Quality Control*, 2nd Edition, New York: John Wiley & Sons, 152–155.

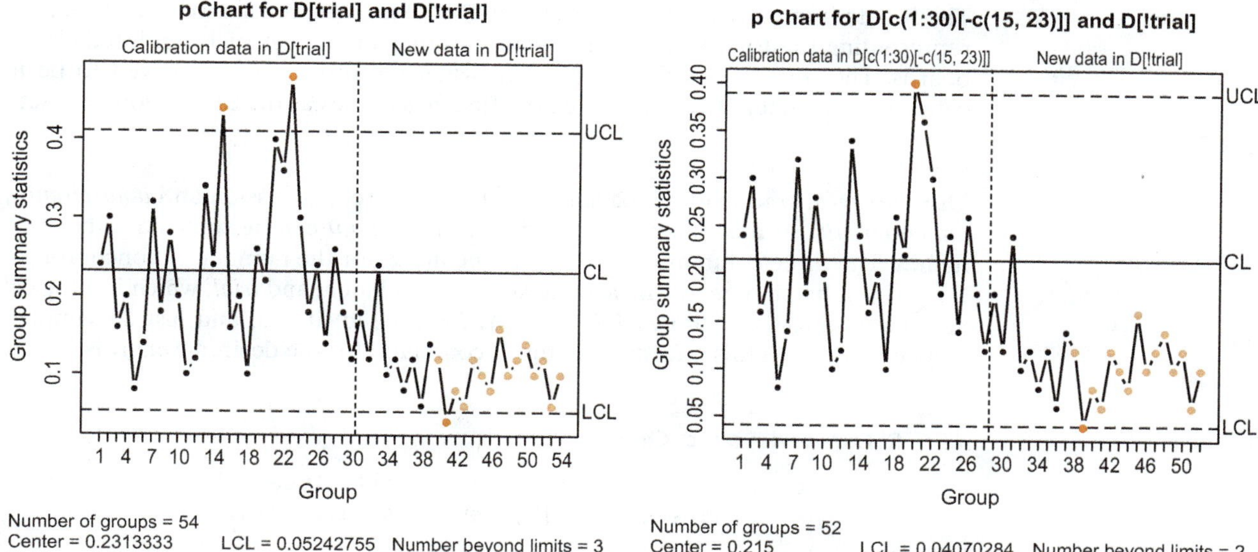

Figure 13-7 *p* Charts for the *orangejuice* data: The first 30 of 54 subgroups used for calibration (left panel), and after removing two subgroups from the calibration set (right panel).

The p chart produced by this command is shown in the right panel of Figure 13-7. Now a different sample proportion from the (reduced) calibration data set is above the 3σ UCL. However, since no assignable cause has been found, it is not removed. This chart also suggests that the adjustment, which happened after the 30th subgroup, reduced the proportion of defective items produced. ■

The *np* Chart Related to the p chart is the np chart, which plots the D_i with corresponding 3σ limits $np \pm 3\sqrt{np(1-p)}$. The np chart is constructed by replacing *type="p"* by *type="np"* in (13.4.2) and (13.4.3).

13.4.2 THE *c* CHART

Let C_i be the observed count of some type (nonconformances per item or occurrences of some event per specified area or time) from the ith subgroup, $i = 1, 2, \ldots$. It is assumed that the counts follow a Poisson distribution. Let λ be the in-control parameter value of the Poisson distribution. By (3.4.18), the in-control mean and variance of each C_i are

$$E(C_i) = \lambda \quad \text{and} \quad \text{Var}(C_i) = \lambda.$$

Moreover, if $\lambda \geq 15$, then C_i has approximately a normal distribution. Thus, the 3σ control limits are

$$\text{LCL} = \lambda - 3\sqrt{\lambda} \quad \text{and} \quad \text{UCL} = \lambda + 3\sqrt{\lambda}. \tag{13.4.4}$$

If LCL is negative, it is replaced by zero. When λ is unknown, the control limits in (13.4.4) are computed with λ replaced by

$$\overline{\lambda} = \frac{1}{k}\sum_{i=1}^{k} C_i,$$

where $C_1, \ldots, C_k$ are obtained when the process is believed to be in control.

The implementation of the c chart in R is illustrated in the following example.

Example 13.4-2

The R data frame *circuit*,[13] available in the R package *qcc*, consists of 46 counts, each representing the total number of nonconformities in batches of 100 printed circuit boards. The first 26 samples were taken when the process was believed to be in control. Construct a 3σ c chart using the first 26 samples as the calibration data set.

Solution

Use *library(qcc)* to load the package *qcc* to the current R session and *data(circuit); attach(circuit)* so the columns of the data frame *circuit* can be referred to by their names. The three columns of this data frame are x for the number of nonconformities in each batch, *size* for the sample size of each batch, and *trial*, which is a logical variable taking the value *TRUE* and *FALSE* for the first 26 and last 20 samples, respectively. With this information, the R command for the desired c chart is:

R Command for c Charts

```
qcc(x[trial], sizes=size[trial], type="c",
  newdata=x[!trial], newsizes=size[!trial])
```
(13.4.5)

The c chart produced by this command, shown in the left panel of Figure 13-8, reveals that (a) counts C_6 and C_{20}, both of which are used to obtain $\overline{\lambda}$, are outside the 3σ limits, and (b) there is a violation of the runs rule. However, the run of points below the center line involves points from the calibration set and thus can be discounted. (The c chart for the last 20 points, generated by the command *qcc(x[!trial], sizes=size[!trial], center=19.84615, std.dev=4.454902, type="c")* does not show any violations.)

Upon investigation, it turns out that sample 6 was examined by a new inspector who was not trained to recognize several types of nonconformities that could have

Figure 13-8 c Charts for the *circuit* data: The first 26 of 46 counts used for calibration (left panel), and after removing two counts from the calibration set (right panel).

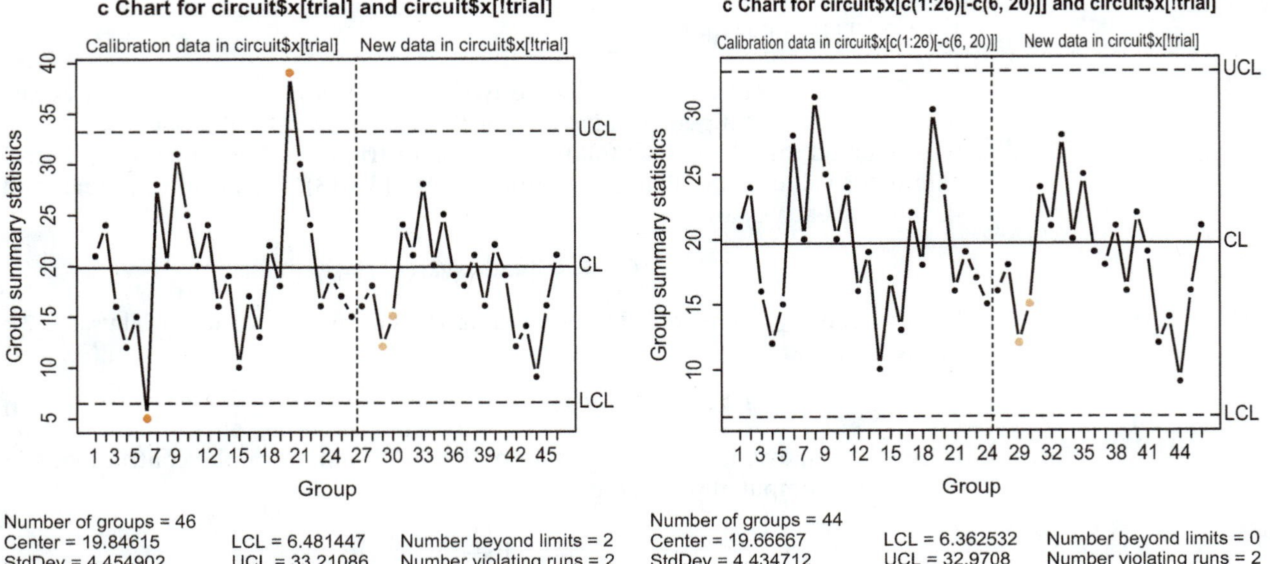

[13] D. C. Montgomery (1991). *Introduction to Statistical Quality Control*, 2nd Edition, New York: John Wiley & Sons, 173–175.

been present. Furthermore, the unusually large number of nonconformities in sample 20 resulted from a temperature control problem in the wave soldering machine, which was subsequently repaired. Because assignable causes for these two points have been found, samples 6 and 20 need to be removed from the calibration set and the control limits must be recomputed (see the discussion on recomputing the estimated target values in Section 13.2.2). The R command for doing so is

R Command for Recomputing Control Limits in c Charts

```
qcc(x[c(1:26)[-c(6, 20)]], sizes=size[c(1:26)[-c(6, 20)]],
  type="c", newdata=x[!trial], newsizes=size[!trial])
```
(13.4.6)

The c chart produced by this command is shown in the right panel of Figure 13-8. This chart shows violations of the runs rule, which can be discounted for the same reasons discussed above.

The u Chart In cases where the Poisson count pertains to the total number of nonconformities in batches of items, as is the case in Example 13.4-2, a related chart, called a u chart (u for *unit*), plots the average number of defectives per unit, that is,

$$U_i = \frac{C_i}{n},$$

where n is the batch size, with corresponding 3σ control limits $\lambda_u \pm 3\sqrt{\lambda_u/n}$, where $\lambda_u = \lambda/n$ is the average number of nonconformities per unit. The u chart is constructed by replacing *type="c"* by *type="u"* in (13.4.5) and (13.4.6). The u chart can also be used if the batch sample sizes are not equal; see Exercise 3.

Exercises

1. The analysis of the R data set *orangejuice* in Example 13.4-1 revealed that the probability of a defective item decreased after the adjustment following the 30th inspection time point. To see if the smaller proportion of defectives continues, 40 additional samples, each of size $n = 50$, were taken. The data, made up of the last 24 samples in the *orangejuice* data frame and the additional new 40 samples, are available in the R data set *orangejuice2*. Using commands similar to those in Example 13.4-1, construct a 3σ p chart using the first 24 samples in the *orangejuice2* data frame as the calibration data set and state your conclusion.

2. The R data frame *pcmanufact*,[14] available in the R package *qcc*, consists of 20 counts, each representing the total number of nonconformities in batches of 5 computers each. Construct a 3σ c chart, with λ estimated from the entire data set. Explain the difference between a c chart and a u chart, and construct the corresponding 3σ u chart for these data.

3. Unequal sample sizes. For simplicity, all control limits presented in this chapter assume equal sample sizes. However, R commands allow also for unequal sample sizes. The R data frame *dyedcloth*,[15] available in the R package *qcc*, consists of 10 counts, each representing the total number of defects in batches of inspection units (50 square meters of dyed cloth). In this data set, the batch sample sizes are unequal. Use the R commands given at the end of Section 13.4.2 to construct a u chart.

---

[14] D. C. Montgomery (1991). *Introduction to Statistical Quality Control*, 2nd Edition, New York: John Wiley & Sons, 181–184.

[15] Ibid., 183–184.

13.5 CUSUM and EWMA Charts

The supplemental runs and scan rules were devised to shorten the fairly long ARLs of the 3σ $\overline{X}$ chart for small shifts in the process mean. Like the $\overline{X}$ chart, however, these rules base the stopping (i.e., out-of-control) decision only on the current subgroup mean.

An alternative approach for improving the performance of the $\overline{X}$ chart is to base the stopping decision on the current and past subgroup means. The CUSUM (for CUmulative SUM) chart and the EWMA (for Exponentially Weighted Moving Average) chart correspond to two different methods for incorporating information from past subgroups into the stopping decision. Both methods achieve shorter ARLs for small shifts of the process mean and, unlike the supplemental rules, do not decrease the in-control ARL. Moreover, these charts apply just as easily when the subgroup size is $n = 1$ as when $n > 1$.

13.5.1 THE CUSUM CHART

As the name suggests, the CUSUM chart is based on the *cumulative sums*, which are defined as

$$S_1 = z_1, \quad S_2 = S_1 + z_2, \ \ldots, \quad S_m = S_{m-1} + z_m = \sum_{i=1}^{m} z_i, \quad \ldots,$$

where, with μ_0 and σ_0 being the in-control mean and standard deviation,

$$z_i = \frac{\overline{x}_i - \mu_0}{\sigma_0/\sqrt{n}}. \tag{13.5.1}$$

If the subgroup size is $n = 1$, $\overline{x}_i$ in (13.5.1) is the single observation in subgroup i; if μ_0 and σ_0 are unknown, they are replaced by $\widehat{\mu}$ and $\widehat{\sigma}$, respectively, evaluated from the calibration data; see (13.2.4) and (13.2.5), or (13.2.9) if $n = 1$.

The original CUSUM monitoring scheme consists of plotting the points (i, S_i), $i = 1, 2, \ldots$, and using suitable control limits. If the process remains in control, the cumulative sums S_i are expected to fluctuate around zero, since (as random variables) the cumulative sums have mean value zero. If the mean shifts from its in-control value to $\mu_1 = \mu_0 + \Delta\sigma_0/\sqrt{n}$, the mean value of S_i is $i\Delta$. Thus, a trend develops in the plotted points, which is upward or downward, depending on whether Δ is > 0 or < 0. On the other hand, the standard deviation of S_i also increases with i ($\sigma_{S_i} = \sqrt{i}$), so the simple control limits of the Shewhart-type charts do not apply. The appropriate control limits resemble a V on its side; this is known as the *V-mask* CUSUM chart.

A different CUSUM monitoring scheme, called a *Decision Interval Scheme* or *DIS*, uses a chart with horizontal control limits whose interpretation is similar to the control limits in Shewhart charts. With the right choice of parameters, the two CUSUM monitoring schemes are equivalent. However, in addition to its easier interpretation, the DIS can be used for one-sided monitoring problems, and has a tabular form as well as some additional useful properties.[16] For this reason, we focus on the DIS and describe its tabular form and corresponding CUSUM chart.

The DIS CUSUM Procedure Let z_i be as defined in (13.5.1), set $S_0^L = 0$ and $S_0^H = 0$, and define S_i^L and S_i^H for $i = 1, 2, \ldots$ recursively as follows:

---

[16] See D. C. Montgomery (2005). *Introduction to Statistical Quality Control*, 5th Edition, New York: John Wiley & Sons.

$$S_i^L = \max\left(S_{i-1}^L - z_i - k,\ 0\right)$$
$$S_i^H = \max\left(S_{i-1}^H + z_i - k,\ 0\right) \qquad (13.5.2)$$

The constant k is called the *reference value*. The out-of-control decision uses a constant h called the *decision interval*. As soon as an $S_i^L \geq h$ or an $S_i^H \geq h$, an out-of-control signal occurs. If the out-of-control signal occurs because of $S_i^L \geq h$, there is indication that the process mean has shifted to a smaller value; if the out-of-control signal occurs because of $S_i^H \geq h$, there is indication that the process mean has shifted to a larger value.

The reference value k and the decision interval h can be chosen to achieve desired in-control and out-of-control (at specified departures from the target mean) ARLs; see the text in footnote 16 for details. In R, the default settings are $k = 0.5$ and $h = 5$. These choices yield good out-of-control ARLs for shifts of about $\Delta = \pm 1$ standard errors of $\overline{x}$ (i.e., $\mu_1 = \mu_0 \pm \sigma/\sqrt{n}$).

A display of the computations in (13.5.2) in a table yields the tabular form of the DIS CUSUM. The corresponding CUSUM chart consists of plotting the points (i, S_i^H) and $(i, -S_i^L)$ in the same figure, with horizontal lines at h and $-h$. If a point falls above the horizontal line at h or below the horizontal line at $-h$, the process is declared out of control, as discussed following (13.5.2).

Example 13.5-1

Let *piston* be the R object containing the 40 subgroups of the *pistonrings* data set arranged into 40 rows of size 5 (see Example 13.2-2).

(a) Construct the CUSUM chart for the first 25 subgroups and the CUSUM chart for all 40 subgroups using the first 25 subsamples as the calibration data. For both charts, use the default reference value and decision interval, and estimate σ from the subgroup standard deviations.

(b) Construct the tabular form of the DIS CUSUM procedure for the first 25 subgroups, using reference value $k = 0.5$ and decision interval $h = 5$.

Solution

(a) The R commands for the two CUSUM charts are

R Commands for CUSUM Charts

```
cusum(piston[1:25,], std.dev="UWAVE-SD")
cusum(piston[1:25,], newdata=piston[26:40,],        (13.5.3)
  std.dev="UWAVE-SD")
```

The charts produced by these commands are shown in Figure 13-9. According to the chart in the left panel, there is no reason to suspect that the process is out of control during the first 25 sampling periods. But the chart in the right panel shows the last four points above the decision interval, suggesting that the process mean has shifted to a larger value.

Omitting *std.dev="UWAVE-SD"* from the commands or instead using *std.dev="UWAVE-R"* results in using $\widehat{\sigma}_2$ of (13.2.5) as estimator of $\widehat{\sigma}$. To use a reference value k different from the default $k = 0.5$ and a decision interval different from the default $h = 5$, include, for example, *se.shift=2k* (the

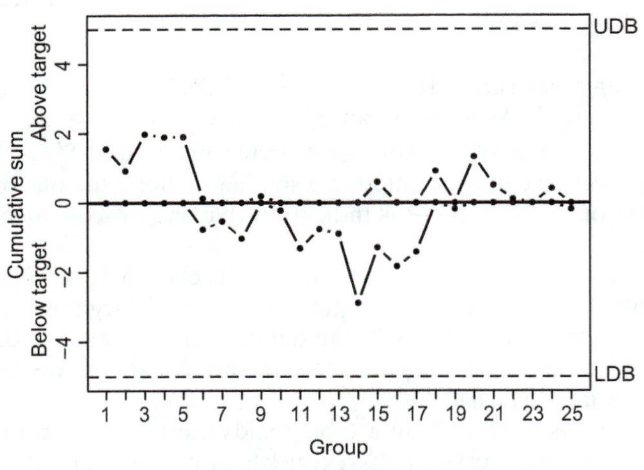

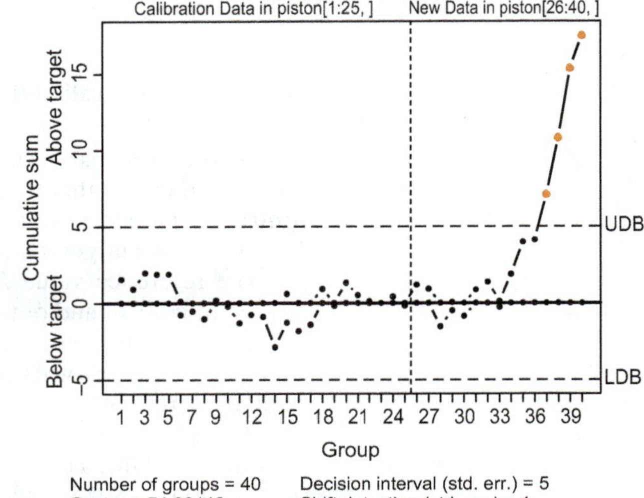

Figure 13-9 CUSUM charts for the *pistonrings* data: Only the first 25 subgroups (left panel), and all 40 subgroups with the first 25 serving as calibration data (right panel).

default value of $k = 0.5$ corresponds to *se.shift=1*) and *decision.interval=h*, respectively, in either of the two commands.

(b) The commands *mean(apply(piston[1:25,], 1, mean)); mean(apply(piston[1:25,], 1, sd))/0.94* yield $\widehat{\mu} = 74.00118$ (see (13.2.4)) and $\widehat{\sigma} = 0.00983$ (see $\widehat{\sigma}_1$ in (13.2.5)). (The more accurate value of 0.009829977 for $\widehat{\sigma}$, displayed in the CUSUM charts of Figure 13-9, is used in the computation that follows.) The standardized subgroup means z_i of (13.5.1) can be obtained with the command

```
z=(apply(piston[1:25, ], 1, mean)-74.00118)/
  (0.009829977/sqrt(5))
```

and the upper monitoring sums, S_i^H, are computed with the commands

```
SH=rep(0, 25); SH[1]=max(z[1]-0.5, 0)
for(i in 2:25){SH[i]=max(SH[i-1]+z[i]-0.5, 0)}
```

The lower monitoring sums, S_i^L, can be placed in the object *SL* using similar commands. Alternatively, both *SH* and *SL* can be obtained from the *cusum* function as

```
obj=cusum(piston[1:25, ], std.dev="UWAVE-SD")
SH=obj$pos; SL=-obj$neg
```

The results are displayed in Table 13-3, which is the tabular form of the CUSUM chart in the left panel of Figure 13-9. ■

Table 13-3 Tabular form of the CUSUM procedure

Sample	$\overline{x}_i$	z_i	S_i^H	S_i^L
1	74.0102	2.05272885	1.5527289	0.0000000
2	74.0006	−0.13102525	0.9217037	0.0000000
3	74.0080	1.55228521	1.9739889	0.0000000
4	74.0030	0.41491328	1.8889022	0.0000000
5	74.0034	0.50590303	1.8948052	0.0000000
6	73.9956	−1.26839717	0.1264080	0.7683972
7	74.0000	−0.26750988	0.0000000	0.5359071
8	73.9968	−0.99542791	0.0000000	1.0313350
9	74.0042	0.68788254	0.1878826	0.0000000
10	73.9980	−0.72245865	0.0000000	0.2224587
11	73.9942	−1.58686131	0.0000000	1.3093200
12	74.0014	0.05095426	0.0000000	0.7583658
13	73.9984	−0.63146889	0.0000000	0.8898347
14	73.9902	−2.49675886	0.0000000	2.8865936
15	74.0060	1.09733644	0.5973365	1.2892571
16	73.9966	−1.04092279	0.0000000	1.8301799
17	74.0008	−0.08553037	0.0000000	1.4157103
18	74.0074	1.41580058	0.9158006	0.0000000
19	73.9982	−0.67696377	0.0000000	0.1769638
20	74.0092	1.82525447	1.3252545	0.0000000
21	73.9998	−0.31300475	0.5122498	0.0000000
22	74.0016	0.09644914	0.1086989	0.0000000
23	74.0024	0.27842865	0.0000000	0.0000000
24	74.0052	0.91535693	0.4153570	0.0000000
25	73.9982	−0.67696377	0.0000000	0.1769638

13.5.2 THE EWMA CHART

A *weighted average* is an extension of the usual average where the contribution of each observation to the average depends on the weight it receives. A weighted average of $m+1$ numbers, $x_m, x_{m-1}, \ldots, x_1$, and μ is of the form

$$w_1 x_m + w_2 x_{m-1} + \cdots + w_m x_1 + w_{m+1}\mu, \qquad \textbf{(13.5.4)}$$

where the weights w_i are nonnegative and sum to 1. Setting $w_i = 1/(m+1)$ for all i yields the usual average of the $m+1$ numbers. For example, consider the four numbers $x_3 = 0.90, x_2 = 1.25, x_1 = -0.13$, and $\mu = 1.36$, and the set of weights $w_1 = 0.800, w_2 = 0.160, w_3 = 0.032$, and $w_4 = 0.008$; their average and weighted average are 0.845 and 0.92672, respectively. Because 96% of the weight goes to the first two numbers, the weighted average is determined, to a large extent, by the first two numbers, that is, by 0.9 and 1.25.

Weighted averages (13.5.4) with weights of the form

$$w_1 = \lambda, \quad w_2 = \lambda\overline{\lambda}, \quad w_3 = \lambda(\overline{\lambda})^2, \quad \ldots, \quad w_m = \lambda(\overline{\lambda})^{m-1}, \quad w_{m+1} = (\overline{\lambda})^m, \qquad \textbf{(13.5.5)}$$

where λ is some number between 0 and 1 and $\overline{\lambda} = 1 - \lambda$, are called *exponentially weighted* averages. Note that the first weight ($w_1 = \lambda$), which is assigned to x_m, is the largest, and subsequent weights $(w_2, \ldots, w_m)$ decrease exponentially fast. (However, if $\lambda < 0.5$, $w_{m+1} > w_m$.)

In statistical process control, the expression *exponentially weighted moving averages* (EWMA) refers to the sequence

$$z_1, z_2, z_3, \ldots, \tag{13.5.6}$$

where each z_m is an exponentially weighted average of the m subgroup means $\overline{x}_m, \ldots, \overline{x}_1$ and $\widehat{\mu}$ (or μ_0, if it is known). Thus, each z_m is given by (13.5.4), with $x_m, \ldots, x_1$ and μ replaced by $\overline{x}_m, \ldots, \overline{x}_1$ and $\widehat{\mu}$ (or μ_0, if known), respectively, and w_i given by (13.5.5). Setting $z_0 = \widehat{\mu}$ (or μ_0, if known), it can be shown that the EWMA can be computed recursively as

$$z_m = \lambda \overline{x}_m + (1 - \lambda) z_{m-1}, \quad m = 1, 2, \ldots . \tag{13.5.7}$$

The EWMA control chart plots the points (m, z_m), $m = 1, 2, \ldots$, and draws appropriate 3σ control limits for each z_m, that is, of the form $\mu_0 \pm 3\sigma_{z_m}$. More precisely, the control limits are computed as

Control Limits for the z_m EWMA

$$\begin{aligned} \text{LCL} &= \mu_0 - 3\frac{\sigma}{\sqrt{n}}\sqrt{\frac{\lambda}{2-\lambda}[1-(1-\lambda)^{2m}]} \\ \text{UCL} &= \mu_0 + 3\frac{\sigma}{\sqrt{n}}\sqrt{\frac{\lambda}{2-\lambda}[1-(1-\lambda)^{2m}]} \end{aligned} \tag{13.5.8}$$

If the in-control process mean and standard deviation are unknown, μ_0 and σ are replaced by estimators $\widehat{\mu}$ and $\widehat{\sigma}$; see (13.2.4) and (13.2.5), or (13.2.9) if $n = 1$.

The parameter λ in (13.5.8) is typically chosen between 0.1 and 0.5. Smaller values of λ yield better (shorter) ARLs for small shifts. The default value in R is $\lambda = 0.2$.

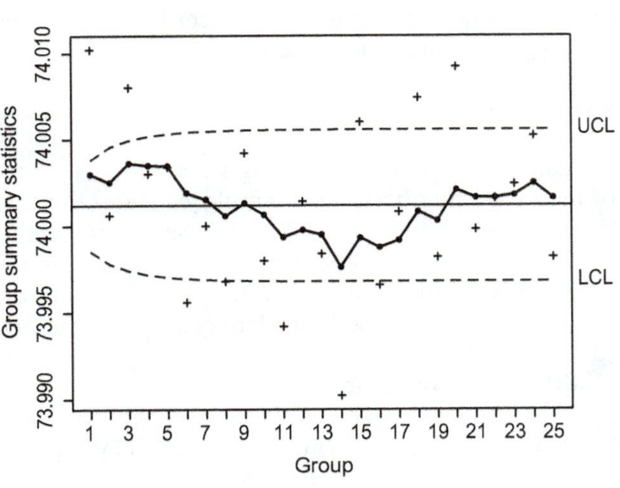

Figure 13-10 EWMA charts for the *pistonrings* data: Only the first 25 subgroups (left panel), and all 40 subgroups with the first 25 serving as calibration data (right panel).

The R commands that produced the EWMA charts shown in Figure 13-10 for the *pistonrings* data are

R Commands for EWMA Charts

```
ewma(piston[1:25,])
ewma(piston[1:25,], newdata=piston[26:40,])
```
(13.5.9)

Note that because the option *std.dev="UWAVE-SD"* is not used, the standard deviation shown at the bottom of the charts is computed as $\widehat{\sigma}_2$ in (13.2.5). To use values of λ different from the default value of 0.2, say 0.4, use the option *lambda=0.4*.

As Figure 13-10 makes clear, the control limits of the EWMA chart are not of equal width. However, the expression in the square root in (13.5.8) tends to $\lambda/(2-\lambda)$ exponentially fast, so that, after the first few inspection time points, the lines joining the control limits look horizontal. Finally, note that the charts also show the subgroup means (indicated with + signs).

Exercises

1. Completed the following:

(a) Use the data of Exercise 2 in Section 13.3 and commands similar to those in display (13.5.3) to construct a CUSUM chart for the semiconductor wafer diameter data. Comment on what the chart suggests regarding the process mean.

(b) Use commands similar to those in the display (13.5.9) to construct a EWMA chart for the semiconductor wafer diameter data. Comment on what the chart suggests regarding the process mean.

2. Redox potential or ORP (for oxidation-reduction potential) is a new method for assessing the efficacy of disinfectants used in water treatment plants. A case study reports 31 daily measurements of chlorine and sulfur dioxide used at a municipal wastewater treatment plant in California.[17] The chlorine measurement data are given in *SqcRedoxPotent.txt*. Use commands similar to those in the displays (13.5.3) and (13.5.9) to construct the CUSUM and EWMA charts for the chlorine data. Comment on what the charts suggest regarding the process mean.

[17] Y. H. Kim and R. Hensley (1997). Effective control of chlorination and dechlorination at wastewater treatment plants using redox potential, *Water Environment Research*, 69: 1008–1014.

TABLES

Table A.1 Cumulative Binomial Probabilities

		p								
n	*x*	0.1	0.2	0.3	0.4	0.5	0.6	0.7	0.8	0.9
5	0	0.591	0.328	0.168	0.078	0.031	0.010	0.002	0.000	0.000
	1	0.919	0.737	0.528	0.337	0.188	0.087	0.031	0.007	0.000
	2	0.991	0.942	0.837	0.683	0.500	0.317	0.163	0.058	0.009
	3	0.995	0.993	0.969	0.913	0.813	0.663	0.472	0.263	0.082
	4	1.000	1.000	0.998	0.990	0.699	0.922	0.832	0.672	0.410
10	0	0.349	0.107	0.028	0.006	0.001	0.000	0.000	0.000	0.000
	1	0.736	0.376	0.149	0.046	0.011	0.002	0.000	0.000	0.000
	2	0.930	0.678	0.383	0.167	0.055	0.012	0.002	0.000	0.000
	3	0.987	0.879	0.650	0.382	0.172	0.055	0.011	0.001	0.000
	4	0.988	0.967	0.850	0.633	0.377	0.166	0.047	0.006	0.000
	5	1.000	0.994	0.953	0.834	0.623	0.367	0.150	0.033	0.002
	6	1.000	0.999	0.989	0.945	0.828	0.618	0.350	0.121	0.013
	7	1.000	1.000	0.998	0.988	0.945	0.833	0.617	0.322	0.070
	8	1.000	1.000	1.000	0.998	0.989	0.954	0.851	0.624	0.264
	9	1.000	1.000	1.000	1.000	0.999	0.994	0.972	0.893	0.651
15	0	0.206	0.035	0.005	0.001	0.000	0.000	0.000	0.000	0.000
	1	0.549	0.167	0.035	0.005	0.001	0.000	0.000	0.000	0.000
	2	0.816	0.398	0.127	0.027	0.004	0.000	0.000	0.000	0.000
	3	0.944	0.648	0.297	0.091	0.018	0.002	0.000	0.000	0.000
	4	0.987	0.836	0.516	0.217	0.059	0.009	0.001	0.000	0.000
	5	0.998	0.939	0.722	0.403	0.151	0.034	0.004	0.000	0.000
	6	1.000	0.982	0.869	0.610	0.304	0.095	0.015	0.001	0.000
	7	1.000	0.996	0.950	0.787	0.500	0.213	0.050	0.004	0.000
	8	1.000	0.999	0.985	0.905	0.696	0.390	0.131	0.018	0.000
	9	1.000	1.000	0.996	0.966	0.849	0.597	0.278	0.061	0.002
	10	1.000	1.000	0.999	0.991	0.941	0.783	0.485	0.164	0.013
	11	1.000	1.000	1.000	0.998	0.982	0.909	0.703	0.352	0.056
	12	1.000	1.000	1.000	1.000	0.996	0.973	0.873	0.602	0.184
	13	1.000	1.000	1.000	1.000	1.000	0.995	0.965	0.833	0.451
	14	1.000	1.000	1.000	1.000	1.000	1.000	0.995	0.965	0.794
20	0	0.122	0.012	0.001	0.000	0.000	0.000	0.000	0.000	0.000
	1	0.392	0.069	0.008	0.001	0.000	0.000	0.000	0.000	0.000
	2	0.677	0.206	0.035	0.004	0.000	0.000	0.000	0.000	0.000
	3	0.867	0.411	0.107	0.016	0.001	0.000	0.000	0.000	0.000
	4	0.957	0.630	0.238	0.051	0.006	0.000	0.000	0.000	0.000
	5	0.989	0.804	0.416	0.126	0.021	0.002	0.000	0.000	0.000
	6	0.998	0.913	0.608	0.250	0.058	0.006	0.000	0.000	0.000
	7	1.000	0.968	0.772	0.416	0.132	0.021	0.001	0.000	0.000
	8	1.000	0.990	0.887	0.596	0.252	0.057	0.005	0.000	0.000
	9	1.000	0.997	0.952	0.755	0.412	0.128	0.017	0.001	0.000
	10	1.000	0.999	0.983	0.873	0.588	0.245	0.048	0.003	0.000
	11	1.000	1.000	0.995	0.944	0.748	0.404	0.113	0.010	0.000
	12	1.000	1.000	0.999	0.979	0.868	0.584	0.228	0.032	0.000
	13	1.000	1.000	1.000	0.994	0.942	0.750	0.392	0.087	0.002
	14	1.000	1.000	1.000	0.998	0.979	0.874	0.584	0.196	0.011
	15	1.000	1.000	1.000	1.000	0.994	0.949	0.762	0.370	0.043
	16	1.000	1.000	1.000	1.000	0.999	0.984	0.893	0.589	0.133
	17	1.000	1.000	1.000	1.000	1.000	0.996	0.965	0.794	0.323
	18	1.000	1.000	1.000	1.000	1.000	1.000	0.992	0.931	0.608
	19	1.000	1.000	1.000	1.000	1.000	1.000	0.999	0.988	0.878

Table A.2 Cumulative Poisson Probabilities

	λ									
x	0.1	0.2	0.3	0.4	0.5	0.6	0.7	0.8	0.9	1.0
0	0.905	0.819	0.741	0.670	0.607	0.549	0.497	0.449	0.407	0.368
1	0.995	0.982	0.963	0.938	0.910	0.878	0.844	0.809	0.772	0.736
2	1.000	0.999	0.996	0.992	0.986	0.977	0.966	0.953	0.937	0.920
3		1.000	1.000	0.999	0.998	0.997	0.994	0.991	0.987	0.981
4				1.000	1.000	1.000	0.999	0.999	0.998	0.996
5							1.000	1.000	1.000	0.999

	λ									
x	1.2	1.4	1.6	1.8	2.0	2.2	2.4	2.6	2.8	3.0
0	0.301	0.247	0.202	0.165	0.135	0.111	0.091	0.074	0.061	0.050
1	0.663	0.592	0.525	0.463	0.406	0.355	0.308	0.267	0.231	0.199
2	0.879	0.833	0.783	0.731	0.677	0.623	0.570	0.518	0.469	0.423
3	0.966	0.946	0.921	0.891	0.857	0.819	0.779	0.736	0.692	0.647
4	0.992	0.986	0.976	0.964	0.947	0.928	0.904	0.877	0.848	0.815
5	0.998	0.997	0.994	0.990	0.983	0.975	0.964	0.951	0.935	0.961
6	1.000	0.999	0.999	0.997	0.995	0.993	0.988	0.983	0.976	0.966
7		1.000	1.000	0.999	0.999	0.998	0.997	0.995	0.992	0.988
8				1.000	1.000	1.000	0.999	0.999	0.998	0.996
9							1.000	1.000	0.999	0.999

	λ									
x	3.2	3.4	3.6	3.8	4.0	4.2	4.4	4.6	4.8	5.0
0	0.041	0.033	0.027	0.022	0.018	0.015	0.012	0.010	0.008	0.007
1	0.171	0.147	0.126	0.107	0.092	0.078	0.066	0.056	0.048	0.040
2	0.380	0.340	0.303	0.269	0.238	0.210	0.185	0.163	0.143	0.125
3	0.603	0.558	0.515	0.473	0.433	0.395	0.359	0.326	0.294	0.265
4	0.781	0.744	0.706	0.668	0.629	0.590	0.551	0.513	0.476	0.440
5	0.895	0.871	0.844	0.816	0.785	0.753	0.720	0.686	0.651	0.616
6	0.955	0.942	0.927	0.909	0.889	0.867	0.844	0.818	0.791	0.762
7	0.983	0.977	0.969	0.960	0.949	0.936	0.921	0.905	0.887	0.867
8	0.994	0.992	0.998	0.984	0.979	0.972	0.964	0.955	0.944	0.932
9	0.998	0.997	0.996	0.994	0.992	0.989	0.985	0.980	0.975	0.968
10	1.000	0.999	0.999	0.998	0.997	0.996	0.994	0.992	0.990	0.986
11		1.000	1.000	0.999	0.999	0.999	0.998	0.997	0.996	0.995
12				1.000	1.000	1.000	0.999	0.999	0.999	0.998
13							1.000	1.000	1.000	0.999

Table A.3 The Cumulative Distribution Function for the Standard Normal Distribution: Values of $\Phi(z)$ for Nonnegative z

z	0.00	0.01	0.02	0.03	0.04	0.05	0.06	0.07	0.08	0.09
0.0	0.5000	0.5040	0.5080	0.5120	0.5160	0.5199	0.5239	0.5279	0.5319	0.5359
0.1	0.5398	0.5438	0.5478	0.5517	0.5557	0.5596	0.5636	0.5675	0.5714	0.5753
0.2	0.5793	0.5832	0.5871	0.5910	0.5948	0.5987	0.6026	0.6064	0.6103	0.6141
0.3	0.6179	0.6217	0.6255	0.6293	0.6331	0.6368	0.6406	0.6443	0.6480	0.6517
0.4	0.6554	0.6591	0.6628	0.6664	0.6700	0.6736	0.6772	0.6808	0.6844	0.6879
0.5	0.6915	0.6950	0.6985	0.7019	0.7054	0.7088	0.7123	0.7157	0.7190	0.7224
0.6	0.7257	0.7291	0.7324	0.7357	0.7389	0.7422	0.7454	0.7486	0.7517	0.7549
0.7	0.7580	0.7611	0.7642	0.7673	0.7704	0.7734	0.7764	0.7794	0.7823	0.7852
0.8	0.7881	0.7910	0.7939	0.7967	0.7995	0.8023	0.8051	0.8078	0.8106	0.8133
0.9	0.8159	0.8186	0.8212	0.8238	0.8264	0.8289	0.8315	0.8340	0.8365	0.8389
1.0	0.8413	0.8438	0.8461	0.8485	0.8508	0.8531	0.8554	0.8577	0.8599	0.8621
1.1	0.8643	0.8665	0.8686	0.8708	0.8729	0.8749	0.8770	0.8790	0.8810	0.8830
1.2	0.8849	0.8869	0.8888	0.8907	0.8925	0.8944	0.8962	0.8980	0.8997	0.9015
1.3	0.9032	0.9049	0.9066	0.9082	0.9099	0.9115	0.9131	0.9147	0.9162	0.9177
1.4	0.9192	0.9207	0.9222	0.9236	0.9251	0.9265	0.9279	0.9292	0.9306	0.9319
1.5	0.9332	0.9345	0.9357	0.9370	0.9382	0.9394	0.9406	0.9418	0.9429	0.9441
1.6	0.9452	0.9463	0.9474	0.9484	0.9495	0.9505	0.9515	0.9525	0.9535	0.9545
1.7	0.9554	0.9564	0.9573	0.9582	0.9591	0.9599	0.9608	0.9616	0.9625	0.9633
1.8	0.9641	0.9649	0.9656	0.9664	0.9671	0.9678	0.9686	0.9693	0.9699	0.9706
1.9	0.9713	0.9719	0.9726	0.9732	0.9738	0.9744	0.9750	0.9756	0.9761	0.9767
2.0	0.9772	0.9778	0.9783	0.9788	0.9793	0.9798	0.9803	0.9808	0.9812	0.9817
2.1	0.9821	0.9826	0.9830	0.9834	0.9838	0.9842	0.9846	0.9850	0.9854	0.9857
2.2	0.9861	0.9864	0.9868	0.9871	0.9875	0.9878	0.9881	0.9884	0.9887	0.9890
2.3	0.9893	0.9896	0.9898	0.9901	0.9904	0.9906	0.9909	0.9911	0.9913	0.9916
2.4	0.9918	0.9920	0.9922	0.9925	0.9927	0.9929	0.9931	0.9932	0.9934	0.9936
2.5	0.9938	0.9940	0.9941	0.9943	0.9945	0.9946	0.9948	0.9949	0.9951	0.9952
2.6	0.9953	0.9955	0.9956	0.9957	0.9959	0.9960	0.9961	0.9962	0.9963	0.9964
2.7	0.9965	0.9966	0.9967	0.9968	0.9969	0.9970	0.9971	0.9972	0.9973	0.9974
2.8	0.9974	0.9975	0.9976	0.9977	0.9977	0.9978	0.9979	0.9979	0.9980	0.9981
2.9	0.9981	0.9982	0.9982	0.9983	0.9984	0.9984	0.9985	0.9985	0.9986	0.9986
3.0	0.9987	0.9987	0.9987	0.9988	0.9988	0.9989	0.9989	0.9989	0.9990	0.9990

Table A.4 Percentiles of the T Distribution

df	90%	95%	97.5%	99%	99.5%	99.9%
1	3.078	6.314	12.706	31.821	63.657	318.309
2	1.886	2.920	4.303	6.965	9.925	22.327
3	1.638	2.353	3.183	4.541	5.841	10.215
4	1.533	2.132	2.777	3.747	4.604	7.173
5	1.476	2.015	2.571	3.365	4.032	5.893
6	1.440	1.943	2.447	3.143	3.708	5.208
7	1.415	1.895	2.365	2.998	3.500	4.785
8	1.397	1.860	2.306	2.897	3.355	4.501
9	1.383	1.833	2.262	2.822	3.250	4.297
10	1.372	1.812	2.228	2.764	3.169	4.144
11	1.363	1.796	2.201	2.718	3.106	4.025
12	1.356	1.782	2.179	2.681	3.055	3.930
13	1.350	1.771	2.160	2.650	3.012	3.852
14	1.345	1.761	2.145	2.625	2.977	3.787
15	1.341	1.753	2.132	2.603	2.947	3.733
16	1.337	1.746	2.120	2.584	2.921	3.686
17	1.333	1.740	2.110	2.567	2.898	3.646
18	1.330	1.734	2.101	2.552	2.879	3.611
19	1.328	1.729	2.093	2.540	2.861	3.580
20	1.325	1.725	2.086	2.528	2.845	3.552
21	1.323	1.721	2.080	2.518	2.831	3.527
22	1.321	1.717	2.074	2.508	2.819	3.505
23	1.319	1.714	2.069	2.500	2.807	3.485
24	1.318	1.711	2.064	2.492	2.797	3.467
25	1.316	1.708	2.060	2.485	2.788	3.450
26	1.315	1.706	2.056	2.479	2.779	3.435
27	1.314	1.703	2.052	2.473	2.771	3.421
28	1.313	1.701	2.048	2.467	2.763	3.408
29	1.311	1.699	2.045	2.462	2.756	3.396
30	1.310	1.697	2.042	2.457	2.750	3.385
40	1.303	1.684	2.021	2.423	2.705	3.307
80	1.292	1.664	1.990	2.374	2.639	3.195
∞	1.282	1.645	1.960	2.326	2.576	3.090

Table A.5 Percentiles of the Chi-square Distribution

df	0.5%	1%	2.5%	5%	10%	90%	95%	97.5%	99%	99.5%
1	0.000	0.000	0.001	0.004	0.016	2.706	3.841	5.024	6.635	7.879
2	0.010	0.020	0.051	0.103	0.211	4.605	5.991	7.378	9.210	10.597
3	0.072	0.115	0.216	0.352	0.584	6.251	7.815	9.348	11.345	12.838
4	0.207	0.297	0.484	0.711	1.064	7.779	9.488	11.143	13.277	14.860
5	0.412	0.554	0.831	1.145	1.610	9.236	11.070	12.833	15.086	16.750
6	0.676	0.872	1.237	1.635	2.204	10.645	12.592	14.449	16.812	18.548
7	0.989	1.239	1.690	2.167	2.833	12.017	14.067	16.013	18.475	20.278
8	1.344	1.646	2.180	2.733	3.490	13.362	15.507	17.535	20.090	21.955
9	1.735	2.088	2.700	3.325	4.168	14.684	16.919	19.023	21.666	23.589
10	2.156	2.558	3.247	3.940	4.865	15.987	18.307	20.483	23.209	25.188
11	2.603	3.053	3.816	4.575	5.578	17.275	19.675	21.920	24.725	26.757
12	3.074	3.571	4.404	5.226	6.304	18.549	21.026	23.337	26.217	28.300
13	3.565	4.107	5.009	5.892	7.042	19.812	22.362	24.736	27.688	29.819
14	4.075	4.660	5.629	6.571	7.790	21.064	23.685	26.119	29.141	31.319
15	4.601	5.229	6.262	7.261	8.547	22.307	24.996	27.488	30.578	32.801
16	5.142	5.812	6.908	7.962	9.312	23.542	26.296	28.845	32.000	34.267
17	5.697	6.408	7.564	8.672	10.085	24.769	27.587	30.191	33.409	35.718
18	6.265	7.015	8.231	9.390	10.865	25.989	28.869	31.526	34.805	37.156
19	6.844	7.633	8.907	10.117	11.651	27.204	30.144	32.852	36.191	38.582
20	7.434	8.260	9.591	10.851	12.443	28.412	31.410	34.170	37.566	39.997
21	8.034	8.897	10.283	11.591	13.240	29.615	32.671	35.479	38.932	41.401
22	8.643	9.542	10.982	12.338	14.041	30.813	33.924	36.781	40.289	42.796
23	9.260	10.196	11.689	13.091	14.848	32.007	35.172	38.076	41.638	44.181
24	9.886	10.856	12.401	13.848	15.659	33.196	36.415	39.364	42.980	45.559
25	10.520	11.524	13.120	14.611	16.473	34.382	37.652	40.646	44.314	46.928
26	11.160	12.198	13.844	15.379	17.292	35.563	38.885	41.923	45.642	48.290
27	11.808	12.879	14.573	16.151	18.114	36.741	40.113	43.195	46.963	49.645
28	12.461	13.565	15.308	16.928	18.939	37.916	41.337	44.461	48.278	50.993
29	13.121	14.256	16.047	17.708	19.768	39.087	42.557	45.722	49.588	52.336
30	13.787	14.953	16.791	18.493	20.599	40.256	43.773	46.979	50.892	53.672
40	20.707	22.164	24.433	26.509	29.051	51.805	55.758	59.342	63.691	66.766
60	35.534	37.485	40.482	43.188	46.459	74.397	79.082	83.298	88.379	91.952
80	51.172	53.540	57.153	60.391	64.278	96.578	101.879	106.629	112.329	116.321

Table A.6 Percentiles of the *F* Distribution (ν_1 = Numerator df; ν_2 = Denominator df)

		ν_1										
ν_2	α	1	2	3	4	5	6	7	8	12	24	1,000
1	0.10	39.86	49.50	53.59	55.83	57.24	58.20	58.91	59.44	60.71	62.00	63.30
	0.05	161.4	199.5	215.7	224.6	230.2	234.0	236.8	238.9	243.9	249.1	254.2
2	0.10	8.53	9.00	9.16	9.24	9.29	9.33	9.35	9.37	9.41	9.45	9.49
	0.05	18.51	19.00	19.16	19.25	19.30	19.33	19.35	19.37	19.41	19.45	19.49
3	0.10	5.54	5.46	5.39	5.34	5.31	5.28	5.27	5.25	5.22	5.18	5.13
	0.05	10.13	9.55	9.28	9.12	9.01	8.94	8.89	8.85	8.74	8.64	8.53
4	0.10	4.54	4.32	4.19	4.11	4.05	4.01	3.98	3.95	3.90	3.83	3.76
	0.05	7.71	6.94	6.59	6.39	6.26	6.16	6.09	6.04	5.91	5.77	5.63
5	0.10	4.06	3.78	3.62	3.52	3.45	3.40	3.37	3.34	3.27	3.19	3.11
	0.05	6.61	5.79	5.41	5.19	5.05	4.95	4.88	4.82	4.68	4.53	4.37
6	0.10	3.78	3.46	3.29	3.18	3.11	3.05	3.01	2.98	2.90	2.82	2.72
	0.05	5.99	5.14	4.76	4.53	4.39	4.28	4.21	4.15	4.00	3.84	3.67
7	0.10	3.59	3.26	3.07	2.96	2.88	2.83	2.78	2.75	2.67	2.58	2.47
	0.05	5.59	4.74	4.35	4.12	3.97	3.87	3.79	3.73	3.57	3.41	3.23
8	0.10	3.46	3.11	2.92	2.81	2.73	2.67	2.62	2.59	2.50	2.40	2.30
	0.05	5.32	4.46	4.07	3.84	3.69	3.58	3.50	3.44	3.28	3.12	2.93
10	0.10	3.29	2.92	2.73	2.61	2.52	2.46	2.41	2.38	2.28	2.18	2.06
	0.05	4.96	4.10	3.71	3.48	3.33	3.22	3.14	3.07	2.91	2.74	2.54
12	0.10	3.18	2.81	2.61	2.48	2.39	2.33	2.28	2.24	2.15	2.04	1.91
	0.05	4.75	3.89	3.49	3.26	3.11	3.00	2.91	2.85	2.69	2.51	2.30
14	0.10	3.10	2.73	2.52	2.39	2.31	2.24	2.19	2.15	2.05	1.94	1.80
	0.05	4.60	3.74	3.34	3.11	2.96	2.85	2.76	2.70	2.53	2.35	2.14
16	0.10	3.05	2.67	2.46	2.33	2.24	2.18	2.13	2.09	1.99	1.87	1.72
	0.05	4.49	3.63	3.24	3.01	2.85	2.74	2.66	2.59	2.42	2.24	2.02
20	0.10	2.97	2.59	2.38	2.25	2.16	2.09	2.04	2.00	1.89	1.77	1.61
	0.05	4.35	3.49	3.10	2.87	2.71	2.60	2.51	2.45	2.28	2.08	1.85
30	0.10	2.88	2.49	2.28	2.14	2.05	1.98	1.93	1.88	1.77	1.64	1.46
	0.05	4.17	3.32	2.92	2.69	2.53	2.42	2.33	2.27	2.09	1.89	1.63
50	0.10	2.81	2.41	2.20	2.06	1.97	1.90	1.84	1.80	1.68	1.54	1.33
	0.05	4.03	3.18	2.79	2.56	2.40	2.29	2.20	2.13	1.95	1.74	1.45
100	0.10	2.76	2.36	2.14	2.00	1.91	1.83	1.78	1.73	1.61	1.46	1.22
	0.05	3.94	3.09	2.70	2.46	2.31	2.19	2.10	2.03	1.85	1.63	1.30
1,000	0.10	2.71	2.31	2.09	1.95	1.85	1.78	1.72	1.68	1.55	1.39	1.08
	0.05	3.85	3.00	2.61	2.38	2.22	2.11	2.02	1.95	1.76	1.53	1.11

Table A.7 Percentiles of the Studentized Range Distribution ($Q_{\alpha,k,\nu}$ for $\nu = 0.10$ and $\alpha = 0.05$)

		k									
ν	α	2	3	4	5	6	7	8	9	10	11
5	0.10	2.85	3.72	4.26	4.66	4.98	5.24	5.46	5.65	5.82	5.96
	0.05	3.63	4.60	5.22	5.67	6.03	6.33	6.58	6.80	6.99	7.17
6	0.10	2.75	3.56	4.06	4.43	4.73	4.97	5.17	5.34	5.50	5.64
	0.05	3.46	4.34	4.90	5.30	5.63	5.89	6.12	6.32	6.49	6.65
7	0.10	2.68	3.45	3.93	4.28	4.55	4.78	4.97	5.14	5.28	5.41
	0.05	3.34	4.16	4.68	5.06	5.36	5.61	5.81	6.00	6.16	6.30
8	0.10	2.63	3.37	3.83	4.17	4.43	4.65	4.83	4.99	5.13	5.25
	0.05	3.26	4.04	4.53	4.89	5.17	5.40	5.60	5.77	5.92	6.05
10	0.10	2.56	3.27	3.70	4.02	4.26	4.46	4.64	4.78	4.91	5.03
	0.05	3.15	3.88	4.33	4.65	4.91	5.12	5.30	5.46	5.60	5.72
12	0.10	2.52	3.20	3.62	3.92	4.16	4.35	4.51	4.65	4.78	4.89
	0.05	3.08	3.77	4.20	4.51	4.75	4.95	5.12	5.26	5.39	5.51
13	0.10	2.50	3.18	3.59	3.88	4.12	4.30	4.46	4.60	4.72	4.83
	0.05	3.05	3.73	4.15	4.45	4.69	4.88	5.05	5.19	5.32	5.43
14	0.10	2.49	3.16	3.56	3.85	4.08	4.27	4.42	4.56	4.68	4.79
	0.05	3.03	3.70	4.11	4.41	4.64	4.83	4.99	5.13	5.25	5.36
16	0.10	2.47	3.12	3.52	3.80	4.03	4.21	4.36	4.49	4.61	4.71
	0.05	3.00	3.65	4.05	4.33	4.56	4.74	4.90	5.03	5.15	5.26
18	0.10	2.45	3.10	3.49	3.77	3.98	4.16	4.31	4.44	4.55	4.65
	0.05	2.97	3.61	4.00	4.28	4.49	4.67	4.82	4.95	5.07	5.17
20	0.10	2.44	3.08	3.46	3.74	3.95	4.12	4.27	4.40	4.51	4.61
	0.05	2.95	3.58	3.96	4.23	4.44	4.62	4.77	4.89	5.01	5.11
25	0.10	2.42	3.04	3.42	3.68	3.89	4.06	4.20	4.32	4.43	4.53
	0.05	2.91	3.52	3.89	4.15	4.36	4.53	4.67	4.79	4.90	4.99
30	0.10	2.40	3.02	3.39	3.65	3.85	4.02	4.15	4.27	4.38	4.47
	0.05	2.89	3.49	3.84	4.10	4.30	4.46	4.60	4.72	4.82	4.92
40	0.10	2.38	2.99	3.35	3.60	3.80	3.96	4.10	4.21	4.32	4.41
	0.05	2.86	3.44	3.79	4.04	4.23	4.39	4.52	4.63	4.73	4.82
60	0.10	2.36	2.96	3.31	3.56	3.75	3.91	4.04	4.15	4.25	4.34
	0.05	2.83	3.40	3.74	3.98	4.16	4.31	4.44	4.55	4.65	4.73
80	0.10	2.35	2.94	3.29	3.54	3.73	3.88	4.01	4.12	4.22	4.31
	0.05	2.81	3.38	3.71	3.95	4.13	4.28	4.40	4.51	4.60	4.69
∞	0.10	2.33	2.90	3.24	3.48	3.66	3.81	3.93	4.04	4.13	4.21
	0.05	2.77	3.31	3.63	3.86	4.03	4.17	4.29	4.39	4.47	4.55

Appendix

B

ANSWERS TO SELECTED EXERCISES

Chapter 1

Section 1.2

1. (a) The customers (of all dealerships of the car manufacturer) who bought a car the previous year.

(b) Not a hypothetical population.

3. (a) There are two populations, one for each shift. The cars that have and will be produced by each shift constitute the populations.

(b) Both populations are hypothetical.

(c) The number of nonconformances per car.

5. (a) There are two populations, one for each teaching method.

(b) The students that have and will take the course with each of the teaching methods.

(c) Both populations are hypothetical.

(d) The particular students whose test score will be recorded at the end of the semester.

Section 1.3

1. Choice (ii).

3. (a) All current drivers in his university town.

(b) No. **(c)** Convenience sample.

(d) Assuming the proportion of younger drivers who use their seat belt is smaller, it would underestimate.

5. Identify each pipe with a number from 1 to 90. Then write each of these numbers on 90 slips of paper, put them all in a box and, after mixing them thoroughly, select 5 slips, one at a time and without replacement. The R command *sample(seq(1, 90), size=5)* implements this process. A set of five numbers thus generated is 30, 62, 15, 54, 31.

7. One method is to take a simple random sample, of some size n, from the population of N customers (of all dealerships of that car manufacturer) who bought a car the previous year. Another method is to divide the population of the previous year's customers into three strata, according to the type of car each customer bought, and perform stratified sampling with proportionate allocation of sample sizes. That is, if N_1, N_2, N_3 denote the sizes of the three strata, take simple random samples of approximate (due to rounding) sizes $n_1 = n(N_1/N)$, $n_2 = n(N_2/N)$, $n_3 = n(N_3/N)$, respectively, from each of the three strata. Stratified sampling assures that the sample representation of the three strata equals their population representation.

9. No, because the method excludes samples consisting of n_1 cars from the first shift and $n_2 = 9 - n_1$ from the second shift for any (n_1, n_2) different from (6, 3).

Section 1.4

1. (a) The variable of interest is the number of scratches in each plate. The statistical population consists of 500 numbers, 190 zeros, 160 ones, and 150 twos.

(b) Quantitative. **(c)** Univariate.

3. (a) Univariate. **(b)** Quantitative.

(c) If N is the number cars of available for inspection, the statistical population consists of N numbers, $\{v_1, \ldots, v_N\}$, where v_i is the total number of engine and transmission non-conformances of the ith car.

(d) Bivariate.

Section 1.6

1. (a) $\overline{x}$. **(b)** S. **(c)** $\widehat{p}$.

3. $\widehat{p} = 4/14 = 0.286$. It estimates the proportion of time the ozone level was below 250.

5. (a) $\sigma^2 = 0.7691, \sigma = 0.877$.

(b) $S^2 = 0.9, S = 0.949$.

7. (a) $\mu = 0.92, \sigma^2 = 0.6736, \sigma = 0.8207$.

(b) $\overline{x} = 0.91, S^2 = 0.6686, S = 0.8177$.

10. (a) $\sigma^2_{\overline{X}} = 0.25$.

(b) $S_1^2 = 0, S_2^2 = 0.5, S_3^2 = 0.5, S_4^2 = 0$.

(c) $E(Y) = (0 + 0.5 + 0.5 + 0)/4 = 0.25$.

(d) $\sigma_X^2 = E(Y)$. If the sample variances in part (b) were computed according to a formula that divides by n instead of $n-1$, $E(Y)$ would have been 0.125.

11. **(a)** $\bar{x}_1 = 30, \bar{x}_2 = 30$. **(b)** $S_1^2 = 0.465, S_2^2 = 46.5$. **(c)** There is more uniformity among cars of type A (smaller variability in achieved gas mileage) so type A cars are of better quality.

13. **(a)** $\bar{y} = \frac{1}{n}\sum_{i=1}^n y_i = \frac{1}{n}\sum_{i=1}^n (c_1 + x_i) = c_1 + \frac{1}{n}\sum_{i=1}^n x_i = c_1 + \bar{x}$. Because $y_i - \bar{y} = x_i - \bar{x}$, $S_y^2 = \frac{1}{n-1}\sum_{i=1}^n (y_i - \bar{y})^2 = \frac{1}{n-1}\sum_{i=1}^n (x_i - \bar{x})^2 = S_x^2$. $S_y = \sqrt{S_y^2} = \sqrt{S_x^2} = S_x$.

(b) $\bar{y} = \frac{1}{n}\sum_{i=1}^n y_i = \frac{1}{n}\sum_{i=1}^n (c_2 x_i) = \frac{c_2}{n}\sum_{i=1}^n x_i = c_2\bar{x}$. Because $y_i - \bar{y} = c_2(x_i - \bar{x})$, $S_y^2 = \frac{1}{n-1}\sum_{i=1}^n (y_i - \bar{y})^2 = \frac{c_2^2}{n-1}\sum_{i=1}^n (x_i - \bar{x})^2 = c_2^2 S_x^2$. $S_y = \sqrt{S_y^2} = \sqrt{c_2^2 S_x^2} = |c_2|S_x$.

(c) Set $t_i = c_2 x_i$, so $y_i = c_1 + t_i$. From (a) and (b) we have $\bar{y} = c_1 + \bar{t} = c_1 + c_2\bar{x}$, $S_y^2 = S_t^2 = c_2^2 S_x^2$, and $S_y = S_t = |c_2|S_x$.

15. Because $x_i = 81.2997 + 10{,}000^{-1}y_i$, the results to Exercise 1.6.4-13 (with the roles of x_i and y_i reversed) give $S_X^2 = 10{,}000^{-2}S_Y^2 = 10{,}000^{-2}68.33 = 10^{-7}6.833$.

Section 1.7

1. **(a)** $\tilde{x} = 717$, $q_1 = (691 + 699)/2 = 695$, $q_3 = (734 + 734)/2 = 734$. **(b)** $734 - 695 = 39$.

(c) (100[19-0.5]/40)-th = 46.25th percentile.

3. **(a)** $x_{(1)} = 27.67$, $q_1 = 27.99$, $\tilde{x} = 28.64$, $q_3 = 29.52$, $x_{(n)} = 30.93$. **(b)** 29.768. **(c)** No.

Section 1.8

1. **(a)** The batches of cake.

(b) Baking time and temperature.

(c) 25 and 30 for baking time, and 275, 300, and 325 for temperature.

(d) (25, 275), (25, 300), (25, 325), (30, 275), (30, 300), (30, 325).

(e) Qualitative.

3. **(a)** $\alpha_1 = \mu_1 - \mu, \alpha_2 = \mu_2 - \mu, \alpha_3 = \mu_3 - \mu, \alpha_4 = \mu_4 - \mu, \alpha_5 = \mu_5 - \mu$, where $\mu = (\mu_1 + \mu_2 + \mu_3 + \mu_4 + \mu_5)/5$.

(b) $\frac{\mu_1+\mu_2}{2} - \frac{\mu_3+\mu_4+\mu_5}{3}$.

5. $\mu_1 - \mu_2, \mu_1 - \mu_3, \mu_1 - \mu_4$.

7. **(a)** Yes. (b) Paint type with levels T1, ..., T4, and location with levels L1, ..., L4. The treatments are (T1, L1), ..., (T1, L4), ..., (T4, L1), ..., (T4, L4).

12. The watering and location effects will be confounded. The three watering regimens should be employed in each location. The root systems in each location should be assigned randomly to a watering regimen.

15. **(a)** Of 2590 male applicants, about 1192 (1191.96, according to the major specific admission rates) were admitted. Similarly, of the 1835 female applicants, about 557 were admitted. Thus, the admission rates for men and women are 0.46 and 0.30, respectively.

(b) Yes. **(c)** No, because the major specific admission rates are higher for women for most majors.

16. **(a)** No, because the Pygmalion effect is stronger for female recruits.

(b) Here, $\overline{\mu} = (8 + 13 + 10 + 12)/4 = 10.75$. Thus, the main gender effects are $\alpha_F = (8 + 13)/2 - 10.75 = -0.25$, $\alpha_M = (10 + 12)/2 - 10.75 = 0.25$, and the main Pygmalion effects are $\beta_C = (8 + 10)/2 - 10.75 = -1.75$, $\beta_P = (13 + 12)/2 - 10.75 = 1.75$.

(c) $\gamma_{FC} = 8 - 10.75 + 0.25 + 1.75 = -0.75$, $\gamma_{FP} = 13 - 10.75 + 0.25 - 1.75 = 0.75$, $\gamma_{MC} = 10 - 10.75 - 0.25 + 1.75 = 0.75$, $\gamma_{MP} = 12 - 10.75 - 0.25 - 1.75 = -0.75$.

Chapter 2

Section 2.2

1. **(a)** $\{(1, 1), \ldots, (1, 6), \ldots, (6, 1), \ldots, (6, 6)\}$.

(b) $\{2, 3, \ldots, 12\}$. **(c)** $\{0, 1, \ldots, 6\}$. **(d)** $\{1, 2, 3, \ldots\}$.

3. **(a)** i. $T \cap M$. ii. $(T \cup M)^c$. iii. $(T \cap M^c) \cup (T^c \cap M)$.

5. **(a)** $A^c = \{x | x \geq 75\}$, the component lasts at least 75 time units.

(b) $A \cap B = \{x | 53 < x < 75\}$, the component lasts more than 53 but less than 75 time units.

(c) $A \cup B = \mathcal{S}$, the sample space.

(d) $(A - B) \cup (B - A) = \{x | 0 < x \leq 53 \text{ or } x \geq 75\}$, the component lasts either at most 53 or at least 75 time units.

8. **(a)** $e \in (A - B) \cup (B - A) \iff e \in A - B$ or $e \in B - A \iff e \in A \cup B$ and $e \notin A \cap B \iff e \in (A \cup B) - (A \cap B)$.

(b) $e \in (A \cap B)^c \iff e \in A - B$ or $e \in B - A$ or $e \in (A \cup B)^c \iff [e \in A - B$ or $e \in (A \cup B)^c]$ or $[e \in B - A$ or $e \in (A \cup B)^c] \iff e \in B^c$ or $e \in A^c \iff e \in A^c \cup B^c$.

(c) $e \in (A \cap B) \cup C \iff [e \in A$ and $e \in B]$ or $e \in C \iff [e \in A$ or $e \in C]$ and $[e \in B$ or $e \in C] \iff e \in (A \cup C) \cap (B \cup C)$.

9. **(a)** $\mathcal{S}_1 = \{(x_1, \ldots, x_5)\colon x_i = 5.3, 5.4, 5.5, 5.6, 5.7, i = 1, \ldots, 5\}$. $5^5 = 3125$.

(b) The collection of distinct averages, $(x_1 + \cdots + x_5)/5$, formed from the elements of $\mathcal{S}_1$. The commands *S1=expand.grid(x1=1:5, x2=1:5, x3=1:5, x4=1:5, x5=1:5); length(table(rowSums(S1)))* return 21 for the size of the sample space of the averages.

Section 2.3

1. $P(E_1) = P(E_2) = 0.5$, $P(E_1 \cap E_2) = 0.3$, $P(E_1 \cup E_2) = 0.7$, $P(E_1 - E_2) = 0.2$, $P((E_1 - E_2) \cup (E_2 - E_1)) = 0.4$.

3. $P(E_1) = P(E_2) = 4/5$, $P(E_1 \cap E_2) = 3/5$, $P(E_1 \cup E_2) = 1$, $P(E_1 - E_2) = 1/5$, $P((E_1 - E_2) \cup (E_2 - E_1)) = 2/5$.

6. **(a)** $2^5 = 32$. **(b)** $\mathcal{S} = \{0, 1, \ldots, 5\}$.

(c)

x	0	1	2	3	4	5
p(x)	0.031	0.156	0.313	0.313	0.156	0.031

8. $(26^2 \times 10^3)/(26^3 \times 10^4) = 0.0038$.

10. **(a)** $\binom{10}{5} = 252$. **(b)** $252/2 = 126$. **(c)** $\binom{12}{2} = 66$.

12. **(a)** $\binom{9}{5} = 126$. **(b)** $\binom{9}{5}/\binom{13}{5} = 0.098$.

14. **(a)** $\binom{30}{5} = 142{,}506$. **(b)** $\binom{6}{2}\binom{24}{3} = 30{,}360$.
(c) (i) $[\binom{6}{2}\binom{24}{3}]/\binom{30}{5} = 0.213$ (ii) $\binom{24}{5}/\binom{30}{5} = 0.298$.

16. **(a)** $\binom{10}{2,2,2,2,2} = 113{,}400$.

18. **(a)** $2^n = (1+1)^n = \sum_{k=0}^{n} \binom{n}{k} 1^k 1^{n-k} = \sum_{k=0}^{n} \binom{n}{k}$.
(b) $(a^2 + b)^4 = b^4 + 4a^2b^3 + 6a^4b^2 + 4a^6b + a^8$.

Section 2.4

1. $0.37 + 0.23 - 0.47 = 0.13$.

3. **(a)** The commands *attach(expand.grid(X1=50:53, X2= 50:53, X3=50:53)); table((X1+X2+X3)/3)/ length(X1)* generate a table of possible values for the average and corresponding probabilities.
(b) 0.3125. (This is found by summing the probabilities of 52, 52.33, 52.67 and 53, which are the values in the sample space of the average that are at least 52.)

5. **(a)** i. $E_1 = \{(>3, V), (<3, V)\}$, $P(E_1) = 0.25 + 0.30 = 0.55$. ii. $E_2 = \{(<3, V), (<3, D), (<3, F)\}$, $P(E_2) = 0.30 + 0.15 + 0.13 = 0.58$. iii. $E_3 = \{(>3, D), (<3, D)\}$, $P(E_3) = 0.10 + 0.15 = 0.25$. iv. $E_4 = E_1 \cup E_2 = \{(>3, V), (<3, V), (<3, D), (<3, F)\}$, $P(E_4) = 0.25 + 0.30 + 0.15 + 0.13 = 0.83$, and $E_5 = E_1 \cup E_2 \cup E_3 = \{(>3, V), (<3, V), (<3, D), (<3, F), (>3, D)\}$, $P(E_5) = 0.25 + 0.30 + 0.15 + 0.13 + 0.10 = 0.93$.
(b) $P(E_4) = 0.55 + 0.58 - 0.30 = 0.83$.
(c) $P(E_5) = 0.55 + 0.58 + 0.25 - 0.30 - 0 - 0.15 + 0 = 0.93$.

7. $P(E_1 \cup E_2 \cup E_3) = 0.95 + 0.92 + 0.9 - 0.88 - 0.87 - 0.85 + 0.82 = 0.99$.

9. Let E_4={at least two of the original four components work}, E_5= {at least three of the original four components work}∪ {two of the original four components work and the additional component works}. Then $E_4 \not\subset E_5$ because B = {exactly two of the original four components work and the additional component does not work}, which is part of E_4, is not in E_5. Thus, $E_4 \not\subset E_5$ and, hence, it is not necessarily true that $P(E_4) \le P(E_5)$.

11. **(a)** $A > B$={die A results in 4}, $B > C$={die C results in 2}, $C > D$={die C results in 6, or die C results in 2 and die D results in 1}, $D > A$={die D results in 5, or die D results in 1 and die A results in 0}.
(b) $P(A > B) = 4/6$, $P(B > C) = 4/6$, $P(C > D) = 4/6$, $P(D > A) = 4/6$.

Section 2.5

1. $P(>3 \mid >2) = P(>3)/P(>2) = (1+2)^2/(1+3)^2 = 9/16$.

3. **(a)** $P(A) = 0.2$. **(b)** $P(B|A) = 0.132/0.2 = 0.66$.
(c) $P(X = 1) = 0.2$, $P(X = 2) = 0.3$, $P(X = 3) = 0.5$.

5. **(a)** $P(\text{car} \cap (\text{import})) = P((\text{import})|\text{car})P(\text{car}) = 0.58 \times 0.36 = 0.209$.
(c) $P(\text{lease}) = 0.2 \times 0.42 \times 0.36 + 0.35 \times 0.58 \times 0.36 + 0.2 \times 0.7 \times 0.64 + 0.35 \times 0.3 \times 0.64 = 0.260$.

7. **(b)** $(0.98 - 0.96 \times 0.15)/0.85 = 0.984$.

9. **(a)** $0.9 \times 0.85 + 0.2 \times 0.15 = 0.795$.
(b) $0.9 \times 0.85/0.795 = 0.962$.

11. **(a)** $0.4 \times 0.2 + 0.3 \times 0.1 + 0.2 \times 0.5 + 0.3 \times 0.2 = 0.27$. **(b)** $0.3 \times 0.1/0.27 = 0.111$.

Section 2.6

1. No, because $P(E_2) = 2/10 \ne 2/9 = P(E_2|E_1)$.

3. **(a)** $0.9^{10} = 0.349$. **(b)** $0.1 \times 0.9^9 = 0.039$.
(c) $10 \times 0.1 \times 0.9^9 = 0.387$.

5. **(a)** $0.8^4 = 0.410$. **(b)** $0.9^3 = 0.729$.
(c) $0.8^4 \times 0.9^3 = 0.299$. It is assumed that cars have zero nonconformances independently of each other.

6. Yes. Because, by Proposition 2.6-1, the independence of T and M implies independence of T and $M^c = F$.

8. $P(E_1) = P(\{(1,6), (2,5), (3,4), (4,3), (5,2), (6,1)\}) = 1/6$, $P(E_2) = P(\{(3,1), \ldots, (3,6)\}) = 1/6$, $P(E_3) = P(\{(1,4), \ldots, (6,4)\}) = 1/6$. $P(E_1 \cap E_2) = P(\{(3,4)\}) = 1/36 = P(E_1)P(E_2)$, $P(E_1 \cap E_3) = P(\{(3,4)\}) = 1/36 = P(E_1)P(E_3)$, $P(E_2 \cap E_3) = P(\{(3,4)\}) = 1/36 = P(E_2)P(E_3)$. Finally, $P(E_1 \cap E_2 \cap E_3) = P(\{(3,4)\}) = 1/36 \ne P(E_1)P(E_2)P(E_3)$.

10. Let A, B, C, D be the events that components 1, 2, 3, 4, respectively, function. $P(\text{system functions}) = P(A \cap B) + P(C \cap D) - P(A \cap B \cap C \cap D) = 2 \times 0.9^2 - 0.9^4 = 0.9639$.

Chapter 3

Section 3.2

1. **(a)** No, yes. **(b)** $k = 1/1.1$.

3. **(a)** $1 - 0.7 = 0.3$.
(b) $p(x) = 0.2, 0.5, 0.2, 0.1$ for $x = 0, 1, 2, 3$, respectively.

5. **(a)** No, yes.
(b) (i) $k = 1/18$, $F(x) = 0$ for $x < 8$, $F(x) = (x^2 - 64)/36$ for $8 \le x \le 10$, $F(x) = 1$ for $x > 10$, $P(8.6 \le X \le 9.8) = 0.6133$. (ii) $P(X \le 9.8 | X \ge 8.6) = 0.6133/0.7233 = 0.8479$.

7. $\mathcal{S}_Y = (0, \infty)$. $F_Y(y) = P(Y \le y) = P(X \ge \exp(-y)) = 1 - \exp(-y)$. Thus, $f_Y(y) = \exp(-y)$.

9. **(a)** $P(X > 10) = P(D < 3) = 1/9$.

(b) Using Example 3.2-9, $F(x) = P(X \le x) = 1 - P(D \le 30/x) = 1 - 100(x^{-2}/9)$. Differentiating this we get $f_X(x) = 200(x^{-3}/9)$, for $x > 0$.

Section 3.3

2. **(a)** $E(X) = 2.1$ and $E(1/X) = 0.63333$.

(b) $1000/E(X) = 476.19 < E(1000/X) = 633.33$. Choose $1000/X$.

4. **(a)** $E(X) = 3.05$, $\mathrm{Var}(X) = 1.7475$.

(b) $E(15{,}000X) = 45{,}750$, $\mathrm{Var}(15{,}000X) = 393{,}187{,}500$.

7. **(a)** $\widetilde{\mu} = \sqrt{2}$, $\mathrm{IQR} = \sqrt{3} - 1 = 0.732$.

(b) $E(X) = 1.333$, $\mathrm{Var}(X) = 0.222$.

9. **(a)** $E(X) = \theta/(\theta + 1)$, $\mathrm{Var}(X) = \theta/[(\theta + 2)(\theta + 1)^2]$.

(b) $F_P(p) = 0$ for $p \le 0$, $F_P(p) = p^\theta$ for $0 < p < 1$, and $F_P(p) = 1$ for $p \ge 1$. **(c)** $\mathrm{IQR} = 0.75^{1/\theta} - 0.25^{1/\theta}$.

Section 3.4

1. **(a)** Binomial. **(b)** $\mathcal{S}_X = \{0, 1, \ldots, 5\}$, $p_X(x) = \binom{5}{x}0.3^x 0.7^{5-x}$ for $x \in \mathcal{S}_X$.

(c) $E(X) = 5 \times 0.3 = 1.5$, $\mathrm{Var}(X) = 5 \times 0.3 \times 0.7 = 1.05$.

(d) (i) 0.163 (ii) $E(9X) = 13.5$, $\mathrm{Var}(9X) = 85.05$.

3. **(a)** Binomial. **(b)** $n = 20$, $p = 0.01$. The command *1 − pbinom(1, 20, 0.01)* returns 0.0169 for the probability.

5. **(a)** Binomial. **(b)** $E(X) = 9$, $\mathrm{Var}(X) = 0.9$.

(c) 0.987. **(d)** 190, 90.

7. **(a)** Negative binomial. **(b)** $\mathcal{S} = \{1, 2, \ldots\}$. $P(X = x) = (1 - p)^{x-1}p$. **(c)** 3.333, 7.778.

9. **(a)** If X denotes the number of games until team A wins three games, we want $P(X \le 5)$. The command *pnbinom(2, 3, 0.6)* returns 0.6826. (b) It is larger. The more games they play, the more likely it is the better team will prevail.

11. **(a)** Negative binomial.

(b) $E(Y) = 5/0.01 = 500$, $\mathrm{Var}(Y) = 49500$.

13. **(a)** Hypergeometric.

(b) $\mathcal{S} = \{0, 1, 2, 3\}$, $p(x) = \binom{3}{x}\binom{17}{5-x}/\binom{20}{5}$ for $x \in \mathcal{S}$.

(c) 0.461. **(d)** 0.75, 0.5033.

15. **(a)** Hypergeometric.

(b) *phyper(3, 300, 9700, 50)* returns 0.9377.

(c) Binomial. **(d)** *pbinom(3, 50, .03)* returns 0.9372, quite close to that found in part (b).

17. 0.0144 = 1 − *ppois(2, 0.5)*.

19. **(a)** 2.6 and 3.8. **(b)** 0.0535.

(c) 0.167 = *(ppois(0, 3.8)*0.4)/(ppois(0, 2.6)*0.6 + ppois(0, 3.8)*0.4)*.

21. **(a)** hypergeometric(300, 9,700, 200); binomial(200, 0.03); Poisson(6).

(b) 0.9615, 0.9599, 0.9574. (Note that the Poisson approximation is quite good even though $p = 0.03$ is greater than 0.01.)

23. **(a)** Both say that an event occurred in [0, t] and no event occurred in (t, 1].

(b) 0.1624.

(c) (i) Both say that the event occurred before time t. (ii) $P(T \le t | X(1) = 1) = P([X(t) = 1] \cap [X(1) = 1])/P(X(1) = 1) = P([X(t) = 1] \cap [X(1) - X(t) = 0])/P(X(1) = 1) = e^{-\alpha t}(\alpha t)e^{-\alpha(1-t)}/\,(e^{-\alpha}\alpha) = t$.

Section 3.5

1. **(a)** $\lambda = 1/6$, $P(T > 4) = \exp(-\lambda 4) = 0.513$.

(b) $\sigma^2 = 36$, $x_{0.05} = 17.97$. **(c)** (i) 0.4346 (ii) six years.

3. $P(X \le s + t | X \ge s) = 1 - P(X > s + t | X \ge s) = 1 - P(X > t)$, by (3.5.3), and $P(X > t) = \exp\{-\lambda t\}$.

5. **(a)** 39.96. **(b)** 48.77. **(c)** 11.59.

(d) 0.0176 = *pbinom(3, 15, 0.5)*.

7. **(a)** *pnorm(9.8, 9, 0.4) - pnorm(8.6, 9, 0.4)* = 0.8186.

(b) 0.1323.

9. **(a)** *qnorm(0.1492, 10, 0.03)* = 9.97. **(b)** 0.9772.

10. **(a)** 0.147. **(b)** 8.95 mm.

Chapter 4

Section 4.2

1. **(a)** 0.20, 0.79, 0.09.

(b) $P_X(1) = 0.34$, $P_X(2) = 0.34$, $P_X(3) = 0.32$; $P_Y(1) = 0.34$, $P_Y(2) = 0.33$, $P_Y(3) = 0.33$.

3. **(a)** 0.705, 0.255.

(b) $p_X(8) = 0.42$, $p_X(10) = 0.31$, $p_X(12) = 0.27$, $p_Y(1.5) = 0.48$, $p_Y(2) = 0.405$, $p_Y(2.5) = 0.115$.

(c) 0.6296.

5. $p_{X_1}(0) = 0.3$, $p_{X_1}(1) = 0.3$, $p_{X_1}(2) = 0.4$, $p_{X_2}(0) = 0.27$, $p_{X_2}(1) = 0.38$, $p_{X_2}(2) = 0.35$, $p_{X_3}(0) = 0.29$, $p_{X_3}(1) = 0.34$, $p_{X_3}(2) = 0.37$.

7. **(a)** $k = 15.8667^{-1}$.

(b) $f_X(x) = k(27x - x^4)/3$, $0 \le x \le 2$, $f_Y(y) = ky^4/2$, if $0 \le y \le 2$, and $f_Y(y) = 2ky^2$, if $2 < y \le 3$.

Section 4.3

1. **(a)** $p_X(0) = 0.30$, $p_X(1) = 0.44$, $p_X(2) = 0.26$, $p_Y(0) = 0.24$, $p_Y(1) = 0.48$, $p_Y(2) = 0.28$. Since $0.30 \times 0.24 = 0.072 \neq 0.06$ they are not independent.

(b) $p_{Y|X=0}(0) = 0.06/0.30$, $p_{Y|X=0}(1) = 0.04/0.30$, $p_{Y|X=0}(2) = 0.20/0.30$, $p_{Y|X=1}(0) = 0.08/0.44$, $p_{Y|X=1}(1) = 0.30/0.44$, $p_{Y|X=1}(2) = 0.06/0.44$, $p_{Y|X=2}(0) = 0.10/0.26$, $p_{Y|X=2}(1) = 0.14/0.26$, $p_{Y|X=2}(2) = 0.02/0.26$. Since $p_{Y|X=0}(0) = 0.06/0.30 \neq p_{Y|X=1}(0) = 0.08/0.44$ they are not independent.

(c) 0.3161.

3. (a) $\mu_{Y|X}(8) = 1.64$, $\mu_{Y|X}(10) = 1.80$, $\mu_{Y|X}(12) = 2.11$.

(b) $1.64 \times 0.42 + 1.8 \times 0.31 + 2.11 \times 0.27 = 1.82$.

(c) Not independent because the regression function is not constant.

5. (a) Not independent because the conditional PMFs change with x.

(b)

y	0	1	$p_X(x)$
$p_{X,Y}(0, y)$	0.3726	0.1674	0.54
$p_{X,Y}(1, y)$	0.1445	0.0255	0.17
$p_{X,Y}(2, y)$	0.2436	0.0464	0.29
$p_Y(y)$	0.7607	0.2393	

Not independent because $0.7607 \times 0.54 = 0.4108 \neq 0.3726$.

8. (a) $\mu_{Y|X}(1) = 1.34$, $\mu_{Y|X}(2) = 1.2$, $\mu_{Y|X}(3) = 1.34$.

(b) 1.298.

10. (b) $\mu_{Y|X}(x) = 0.6x$.

(c) $E(Y) = 1.29$.

12. (a) Yes. **(b)** No. **(c)** No.

14. (a) 0.211.. **(b)** $\mu_{Y|X}(x) = 6.25 + x$.

16. (a) $\mu_{Y|X}(x) = 1/x$; 0.196.

(b) $(\log(6) - \log(5))^{-1}(1/5 - 1/6) = 0.183$.

18. (a) $f_{Y|X=x}(y) = (1 - 2x)^{-1}$ for $0 \leq y \leq 1 - 2x$, $E(Y|X = 0.3) = 0.5 - 0.3 = 0.2$.

(b) 0.25.

Section 4.4

1. $\mu = 132$, $\sigma^2 = 148.5$.

3. 0.81.

5. (a) 36, 1/12.

(b) 1080, 2.5.

(c) 0, 5.

7. $20 \times 4 + 10 \times 6 = 140$.

9. $\text{Cov}(X, Y) = 13.5$, $\text{Cov}(\varepsilon, Y) = 16$.

11. 18.08, 52.6336.

13. (a) $\text{Var}(X_1 + Y_1) = 11/12$, $\text{Var}(X_1 - Y_1) = 3/12$.

Section 4.5

1. −0.4059.

3. (a) $S_{X,Y} = 5.46$, $S_X^2 = 8.27$, $S_Y^2 = 3.91$, and $r_{X,Y} = 0.96$.

(b) $S_{X,Y}$, S_X^2, and S_Y^2 change by a factor of 12^2, but $r_{X,Y}$ remains unchanged.

6. (a) $X \sim \text{Bernoulli}(0.3)$.

(b) $Y|X = 1 \sim \text{Bernoulli}(2/9)$, $Y|X = 0 \sim \text{Bernoulli}(3/9)$.

(c) $p_{X,Y}(1, 1) = 0.3 \times 2/9$, $p_{X,Y}(1, 0) = 0.3 \times 7/9$, $p_{X,Y}(0, 1) = 0.7 \times 3/9$, $p_{X,Y}(0, 0) = 0.7 \times 6/9$.

(d) $p_Y(1) = 0.3 \times 2/9 + 0.7 \times 3/9 = 0.3$, so $Y \sim \text{Bernoulli}(0.3)$, which is the same as the distribution of X.

(e) −1/9.

9. (a) $\text{Cov}(X, Y) = E(X^3) - E(X)E(X^2) = 0$.

(b) $E(Y|X = x) = x^2$.

(c) Not a linear relationship, so not appropriate.

Section 4.6

1. (a) For $y = 0, \ldots, n$, $p_{P,Y}(0.6, y) = 0.2\binom{n}{y}0.6^y 0.4^{n-y}$, $p_{P,Y}(0.8, y) = 0.5\binom{n}{y}0.8^y 0.2^{n-y}$, $p_{P,Y}(0.9, y) = 0.3\binom{n}{y}0.9^y 0.1^{n-y}$.

(b) $p_Y(0) = 0.0171$, $p_Y(1) = 0.1137$, $p_Y(2) = 0.3513$, $p_Y(3) = 0.5179$.

3. (a) 45.879.

(b) 0.9076.

5. (a) It is the PDF of a bivariate normal with $\mu_X = 24$, $\mu_Y = 45.3$, $\sigma_X^2 = 9$, $\sigma_Y^2 = 36.25$, $\text{Cov}(X, Y) = 13.5$.

(b) 0.42612.

7. (a) 2.454×10^{-6}.

(b) $f_{X,Y}(x, y) = 0.25\lambda(x)\exp(-\lambda(x)y)$.

9. (a) 0.1304.

(b) 0.0130.

(c) 2.1165, −2.1165.

Chapter 5

Section 5.2

1. (a) $P(|X - \mu| > a\sigma) \leq \frac{\sigma^2}{(a\sigma)^2} = \frac{1}{a^2}$.

(b) 0.3173, 0.0455, 0.0027 compared to 1, 0.25, 0.1111; upper bounds are much larger.

3. (a) 0.6. **(b)** 0.922; lower bound is much smaller.

Section 5.3

3. (a) $N(180, 36)$; 0.202. **(b)** $N(5, 9)$; 0.159.

5. 139.

Section 5.4

2. (a) 0.8423, 0.8193.

(b) 0.8426. The approximation with continuity correction is more accurate.

4. (a) $N(4, 0.2222)$; $N(3, 0.2143)$, by the CLT; $N(1, 0.4365)$, by the independence of the two sample means.

(b) 0.9349.

6. 0.9945.

8. 33.

10. (a) 0.744.

(b) (i) Binomial(40, 0.744); 0.596. (ii) 0.607, 0.536; the approximation with continuity correction is closer to the true value.

12. 0.1185 (with continuity correction).

Chapter 6

Section 6.2

1. 286.36, 17.07.

3. $E(\widehat{\sigma}^2) = \frac{(n_1-1)E(S_1^2)+(n_2-1)E(S_2^2)}{n_1+n_2-2}$
$= \sigma^2 \frac{(n_1-1)+(n_2-1)}{n_1+n_2-2} = \sigma^2$.

5. $2\sigma/\sqrt{n}$. Yes.

7. **(a)** 0.5. **(b)** 0.298.

10. **(b)** 0.39, 45.20, 46.52.
(c) 0.375, 44.885, 46.42.

Section 6.3

1. $\widehat{\lambda} = 1/\overline{X}$. It is not unbiased.

3. $\widehat{\alpha} = 10.686$, $\widehat{\beta} = 10.6224$.

5. **(a)** $\widehat{p} = X/n$. It is unbiased.
(b) $\widehat{p} = 24/37$.
(c) $\widehat{p^2} = (24/37)^2$.
(d) No, because $E(\widehat{p}^2) = p(1-p)/n + p^2$.

7. **(a)** $\log\binom{X+4}{4} + 5\log p + (X)\log(1-p)$, $\widehat{p} = 5/(X+5)$.
(b) $\widehat{p} = 5/X$.
(c) 0.096, 0.106.

9. **(a)** $\widehat{\theta} = \overline{P}/(1-\overline{P})$.
(b) 0.202.

11. **(a)** $-78.7381 + 0.1952x$, 28.65.
(b) 24.82, 12.41.
(c) Fitted: 18.494, 23.961, 30.404, 41.142. Residuals: −2.494, 1.039, 3.596, −2.142.

Section 6.4

1. **(a)** $\text{Bias}(\widehat{\theta}_1) = 0$; $\widehat{\theta}_1$ is unbiased. $\text{Bias}(\widehat{\theta}_2) = -\theta/(n+1)$; $\widehat{\theta}_2$ is biased.
(b) $\text{MSE}(\widehat{\theta}_1) = \theta^2/(3n)$, $\text{MSE}(\widehat{\theta}_2) = 2\theta^2/[(n+1)(n+2)]$.
(c) $\text{MSE}(\widehat{\theta}_1) = 6.67$, $\text{MSE}(\widehat{\theta}_2) = 4.76$. Thus, $\widehat{\theta}_2$ is preferable.

Chapter 7

Section 7.3

1. **(a)** (37.47, 52.89). Normality.

3. **(a)** (206.69, 213.31). **(b)** Yes. **(c)** No.

5. **(b)** (256.12, 316.59). **(c)** (248.05, 291.16).

7. **(a)** (1.89, 2.59). **(b)** $(\sqrt{1.89}, \sqrt{2.59})$.

9. **(a)** (0.056, 0.104).
(b) The number who qualify and the number who do not qualify must be at least 8.

11. **(a)** (0.495, 0.802). **(b)** $(0.495^2, 0.802^2)$.

13. **(a)** $\widehat{\alpha}_1 = 193.9643$, $\widehat{\beta}_1 = 0.9338$, $S_\varepsilon^2 = 4118.563$.
(b) (0.7072, 1.1605); normality.
(c) For $X = 500$, (614.02, 707.75); for $X = 900$, (987.33, 1081.51); the CI at $X = 900$ is not appropriate because 900 is not in the range of X-values in the data set.

15. **(b)** For β_1: (−0.2237, −0.1172); for $\mu_{Y|X}(80)$: (9.089, 10.105).

17. *n=16; a=5; 1-2*(1-pbinom(n-a,n,0.5))* returns 0.9232; changing to *a=4* returns 0.9787.

19. (0.095, 0.153); yes.

Section 7.4

1. 195.

3. **(a)** 189. **(b)** 271.

Section 7.5

1. (2.56, 3.64). Normality.

3. **(a)** (41.17, 49.24).

4. **(a)** (7.66, 7.79).
(b) A distance of 12 feet is not in the range of X-values in the data set, so the desired PI would not be reliable.

Chapter 8

Section 8.2

1. **(a)** $H_0 : \mu \le 31$, $H_a : \mu > 31$.
(b) Adopt the method of coal dust cover.

3. **(a)** $H_a : \mu < \mu_0$.
(b) $H_a : \mu > \mu_0$.
(c) For (a), the new grille guard is not adopted; for (b), the new grille guard is adopted.

5. **(a)** $H_0 : p \ge 0.05$, $H_a : p < 0.05$.
(b) Not proceed.

7. **(a)** $\widehat{\mu}_{Y|X}(x) \ge C$.
(b) $\mu_{Y|X}(x)_0 + t_{n-2,0.05} S_{\widehat{\mu}_{Y|X}(x)}$.

9. **(a)** $|T_{H_0}| > t_{n-2,\alpha/2}$, where $T_{H_0} = (\widehat{\beta}_1 - \beta_{1,0})/S_{\widehat{\beta}_1}$.
(b) $T_{H_0} = (\widehat{\mu}_{Y|X}(x) - \mu_{Y|X}(x)_0)/S_{\widehat{\mu}_{Y|X}(x)}$.

Section 8.3

1. **(a)** $T_{H_0} = 1.94$, H_0 is rejected.
(b) No additional assumptions are needed.

3. **(a)** $H_0 : \mu \le 50$, $H_a : \mu > 50$.
(b) Reject H_0; normality.
(c) (i) Between 0.01 and 0.025 (ii) 0.02.

5. **(a)** H_0 is not rejected, so club should be established.
(b) 0.999.

7. **(a)** $H_0 : p \le 0.2$, $H_a : p > 0.2$.
(b) H_0 is not rejected; p-value= 0.41. Not enough evidence that the marketing would be profitable.

9. **(b)** R-squared = 0.975.
(c) $\widehat{\mu}_{Y|X}(x) = -0.402 + 1.020x$. 10.2.

(d) (i) $H_0 : \beta_1 \leq 1$, $H_a : \beta_1 > 1$ (ii) H_0 is not rejected; p-value $= 0.28$.

11. (a) $H_0 : \tilde{\mu} = 300$, $H_a : \tilde{\mu} \neq 300$; H_0 is not rejected, so not enough evidence to conclude that the claim is false.

(b) $H_0 : \tilde{\mu} \geq 300$, $H_a : \tilde{\mu} < 300$; H_0 is rejected, so there is enough evidence to conclude that the median increase is less than 300.

15. H_0 is rejected; p-value$= 0.036$; normality.

Section 8.4

1. (a) Type I.

(b) Type II.

3. (a) 0.102.

(b) 0.228.

(c) 0.098, 0.316; smaller probability of type I error and more power.

5. 79.

7. (a) 0.926. **(b)** 2319.

Chapter 9

Section 9.2

1. (a) $H_0 : \mu_1 = \mu_2$, $H_a : \mu_1 \neq \mu_2$. No, because $15{,}533^2/3{,}954^2 = 15.43$ is much larger than 2.

(b) $T_{H_0}^{SS} = 0.3275$; H_0 is not rejected; p-value $= 0.74$.

(c) $(-4{,}186.66, 5{,}814.66)$; since zero is included in the CI, H_0 is not rejected.

3. (a) $H_0 : \mu_1 \leq \mu_2$, $H_a : \mu_1 > \mu_2$. Yes, because $20.38^6/15.62^2 = 1.70 < 2$.

(b) $T_{H_0}^{EV} = 7.02$; H_0 is rejected; p-value$= 3.29 \times 10^{-10}$.

(c) (18.22, 40.18).

(d) *t.test(duration~route, data=dd, var.equal=T, alternative="greater")* gives the p-value in (b), and *t.test(duration~route, data=dd, var.equal=T, conf.level=0.99)* gives the CI in (c).

5. (a) $H_0 : \mu_1 - \mu_2 \leq 126$, $H_a : \mu_1 - \mu_2 > 126$. Yes, because $52.12/25.83 = 2.02 < 3$.

(b) $T_{H_0}^{EV} = 4.42$; H_0 is rejected; p-value$= 6.8 \times 10^{-5}$.

(c) (131.4, 140.74).

(d) Using $T_{H_0}^{EV}$, p-value $= 6.83 \times 10^{-5}$; using $T_{H_0}^{SS}$, p-value $= 8.38 \times 10^{-5}$.

7. (a) $H_0 : p_1 = p_2$, $H_a : p_1 \neq p_2$; H_0 is rejected; p-value$= 0.0027$.

(b) $(-0.0201, -0.0017)$.

(c) *prop.test(c(692, 1182), c(9396, 13985), correct=F, conf.level=0.99)* returns 0.0027 and $(-0.0201, -0.0017)$ for the p-value and 99% CI.

9. (a) $H_0 : p_1 - p_2 = 0$, $H_a : p_1 - p_2 \neq 0$; H_0 is not rejected; p-value$= 0.6836$.

(b) $(-0.148, 0.108)$.

Section 9.3

1. $H_0 : \tilde{\mu}_S - \tilde{\mu}_C \leq 0$, $H_a : \tilde{\mu}_S - \tilde{\mu}_C > 0$; p-value$= 0.007$, so H_0 is rejected at $\alpha = 0.05$; (14.10, 237.60).

3. (a) $H_0 : \tilde{\mu}_S - \tilde{\mu}_C = 0$, $H_a : \tilde{\mu}_S - \tilde{\mu}_C \neq 0$; $Z_{H_0} = 0.088$, H_0 is not rejected, p-value$= 0.93$.

(b) $(-33.00, 38.00)$.

5. *t.test(duration~route, alternative ="greater", data=dd); wilcox.test(duration~route, alternative="greater", data=dd)* produce p-values of 8.13×10^{-11} and 1.19×10^{-8}, respectively. *t.test(duration~route, conf.level=0.9, data=dd); wilcox.test(duration~route, conf.int=T, conf.level=0.9, data=dd)* produce 90% CIs of (22.58, 35.82) and (22.80, 37.70), respectively.

Section 9.4

1. p-value$= 0.79$, H_0 is not rejected.

3. p-value$= 0.11$, H_0 is not rejected. Normality.

Section 9.5

1. (a) p-value$= 0.78$, H_0 is not rejected. The differences should be normally distributed; this assumption is suspect due to an outlier.

(b) p-value$= 0.27$, H_0 is not rejected.

3. (a) p-value $= 0.037$, H_0 is rejected; $(-5.490, -0.185)$.

(b) p-value$= 0.037$, H_0 is rejected; $(-5.450, -0.100)$; quite similar to part (a).

(c) T test: p-value$= 0.215$, CI of $(-7.364, 1.690)$, H_0 is not rejected; Rank-sum test: p-value$= 0.340$, CI of $(-6.900, 2.300)$, H_0 is not rejected. Very different from (a) and (b).

5. $T_{H_0} = -2.505$, $MN = -2.5$, p-value$= 0.012$, H_0 is rejected at $\alpha = 0.05$.

Chapter 10

Section 10.2

1. (a) $H_0 : \mu_1 = \cdots = \mu_4$, $H_a : H_0$ is not true, p-value$= 0.1334$, H_0 is not rejected. Independent samples, normality, and homoscedasticity.

(b) (i) $\theta = (\mu_1 + \mu_2)/2 - (\mu_3 + \mu_4)/2$; $H_0 : \theta = 0$, $H_a : \theta \neq 0$. (ii) $T_{H_0} = -2.01$, p-value$= 0.056$, H_0 is rejected. The 90% CI is $(-1.17, -0.24)$. (After reading the data and attaching the data frame use: *sm=by(values, ind, mean); svar= by(values, ind, var); t = (sm[1] + sm[2])/2 – (sm[3] + sm[4])/2; st=sqrt(mean(svar)*2*(1/4)*(2/7)); t-qt(0.9, 24)*st; t+qt(0.9, 24)*st.)*) (iii) No, because $\theta \neq 0$ means that $H_0 : \mu_1 = \cdots = \mu_4$ is not true. The T test for a specialized contrast is more powerful than the F test for the equality of all means.

3. (a) $H_0 : \mu_1 = \cdots = \mu_4$, $H_a : H_0$ is not true, H_0 is rejected; independent samples, normality, homoscedasticity.

(b) p-value$= 8.32 \times 10^{-6}$, H_0 is rejected.

(c) After reading the data into the data frame df, the command *anova(aov(resid(aov(df$values~df$ind))**2~df$ind))* returns a p-value of 0.62, suggesting the homoscedasticity assumption is not contradicted by the data. The Shapiro-Wilk test returns a p-value of 0.13 suggest the normality assumption is not contradicted by the data.

5. (a) $H_0 : \mu_1 = \mu_2 = \mu_3$, H_a : H_0 is not true, $DF_{SSTr} = 2$, $DF_{SSE} = 24$, SSE = 24.839, MSTr = 0.0095, MSE = 1.035, F value = 0.009.

(b) H_0 is not rejected.

(c) p-value = 0.99, H_0 is not rejected.

7. (a) 14.72, $\chi^2_{3,0.05} = 7.81$, H_0 is rejected.

(b) 14.72, 0.002, H_0 is rejected.

9. (a) *fit=aov(values~ind, data=ff); anova(aov(resid(fit)^2~ff$ind))* return a p-value of 0.04; the assumption of homoscedasticity is suspect. *shapiro.test(resid(fit))* returns a p-value of 0.008; there is significant evidence that the assumption of normality is not valid.

(b) Kruskal-Wallis is recommended as its validity does not depend on these assumptions.

(c) 11.62, 0.02, H_0 is not rejected at level $\alpha = 0.01$.

11. $H_0 : p_1 = \cdots = p_5$, H_a : H_0 is not true. Chi-square test, 9.34, H_0 is not rejected at level $\alpha = 0.05$ (p-value= 0.053), independent samples.

Section 10.3

1. (a) $H_0 : \mu_S = \mu_C = \mu_G$, H_a : H_0 is not true.

(b) 0.32, H_0 is not rejected.

(c) No, because H_0 is not rejected.

3. (a) $H_0 : \mu_1 = \mu_2 = \mu_3$, H_a : H_0 is not true, $F_{H_0} = 7.744$, p-value= 0.0005, H_0 is rejected at level $\alpha = 0.05$.

(b) μ_1 and μ_2 are not significantly different, but μ_1 and μ_3 as well as μ_2 and μ_3 are.

5. (a) $H_0 : \mu_1 = \mu_2 = \mu_3$, H_a : H_0 is not true, H_0 is rejected at level $\alpha = 0.05$. Independent samples, normality, homoscedasticity.

(b) Tukey's 95% SCIs for $\mu_2 - \mu_1$, $\mu_3 - \mu_1$, and $\mu_3 - \mu_2$ are (−3.72, 14.97), (2.28, 20.97), and (−3.35, 15.35), respectively. Only teaching methods 1 and 3 are significantly different.

(c) The p-values for the F test on the squared residuals and the Shapiro test are 0.897 and 0.847, respectively. The procedures in (a) and (b) are valid.

7. Bonferroni's 95% SCIs for $p_1 - p_2$, $p_1 - p_3$, and $p_2 - p_3$, are (−0.158, 0.166), (−0.022, 0.268), and (−0.037, 0.276). respectively. None of the contrasts are significantly different from zero.

Section 10.4

1. (a) The assumption of independent samples, required for the ANOVA F procedure of Section 10.2.1, does not hold for this data. The appropriate model is $X_{ij} = \mu + \alpha_i + b_j + \varepsilon_{ij}$, with $\sum_i \alpha_i = 0$ and $\mathrm{Var}(b_j) = \sigma_b^2$.

(b) $H_0 : \alpha_1 = \alpha_2 = \alpha_3$, H_a : H_0 is not true.

(c) With the data read in *cr*, the commands *st=stack(cr); MF=st$ind; temp=as.factor(rep(1:length(cr$MF1), 3)); summary(aov(st$values~MF+temp))* generate the ANOVA table, which gives a p-value of 1.44×10^{-10} for testing H_0. H_0 is rejected.

(d) Ignoring the block effect, the command *summary(aov(st$values~MF))* returns a p-value of 0.991, suggesting no mole fraction effect. This analysis is inappropriate because the samples are not independent.

3. (a) $X_{ij} = \mu + \alpha_i + b_j + \varepsilon_{ij}$, with $\sum_i \alpha_i = 0$ and $\mathrm{Var}(b_j) = \sigma_b^2$; the parameters α_i specify the treatment effects and b_j represent the random block (pilot) effects.

(b) $H_0 : \alpha_1 = \cdots = \alpha_4$, H_a : H_0 is not true.

(c) *fit=aov(times~design+pilot, data=pr); anova(aov(resid(fit)**2~pr$design+pr$pilot))* return p-values of 0.231 and 0.098 for the design and pilot effects on the residual variance, suggesting there is no strong evidence against the homoscedasticity assumption. *shapiro.test(resid(fit))* returns a p-value of 0.80, suggesting the normality assumption is reasonable for this data.

(d) The further command *anova(fit)* returns p-values of 0.00044 and 1.495×10^{-6} for the design and pilot effects on the response time. The null hypothesis in part (b) is rejected.

5. (a) "fabric".

(b) $DF_{SSTr} = 3$, $DF_{SSB} = 4$, $DF_{SSE} = 12$, SSTr = 2.4815, MSTr = 0.8272, MSB = 1.3632, MSE = 0.0426, $F^{Tr}_{H_0} = 19.425$. Because $19.425 > F_{3,12,0.05} = 3.49$, the hypothesis that the four chemicals do not differ is rejected at level 0.05. The p-value, found by *1-pf(19.425, 3, 12)*, is 6.72×10^{-5}.

7. (a) Bonferroni's 95% SCIs for $\mu_A - \mu_B$, $\mu_A - \mu_C$, and $\mu_B - \mu_C$ are (−10.136, 8.479), (−9.702, 2.188), and (−8.301, 2.444), respectively. None of the differences is significantly different at experiment-wise significance level 0.05.

(b) Bonferroni's 95% SCIs for $\widetilde{\mu}_A - \widetilde{\mu}_B$, $\widetilde{\mu}_A - \widetilde{\mu}_C$, and $\widetilde{\mu}_B - \widetilde{\mu}_C$ are (−8.30, 3.75), (−13.9, −0.1), and (−11.0, 3.3), respectively. The difference $\widetilde{\mu}_A - \widetilde{\mu}_C$ is significantly different from zero at experiment-wise error rate of 0.05, but the other differences are not.

Chapter 11

Section 11.2

1. (a) $F^{GS}_{H_0} = 0.6339$ with p-value of 0.4286; the hypothesis of no interaction effects is not rejected. $F^{G}_{H_0} = 141.78$ with p-value of less than 2.2×10^{-16}; the hypothesis of no main growth hormone effects is rejected. $F_{H_0}^S = 18.96$ with p-value of 4.46×10^{-5}; the hypothesis of no main sex steroid effects is rejected.

(d) The p-values for testing the hypotheses of no main growth effects, no main sex steroid effects, and no interaction effects on the residual variance are, respectively, 0.346, 0.427, and 0.299; none of these hypotheses is rejected. The p-value for the normality test is 0.199, so the normality assumption appears to be reasonable.

3. (a) The hypothesis of no interaction between gender and dose.

(b) $F_{H_0}^{GD} = 0.0024$ with p-value of 0.9617; the hypothesis of no interaction effects is retained.

5. (b) The p-values for testing the hypotheses of no main day effects, no main section effects, and no interaction effects on the residual variance are, respectively, 0.3634, 0.8096, and 0.6280; none of these hypotheses is rejected. The p-value for the normality test is 0.3147, so the normality assumption appears to be reasonable.

(c) The pairs of days, except for (M, T), (M, W) and (T, W), are significantly different at experiment-wise error rate $\alpha = 0.01$. The pairs of newspaper sections (Sports, Business) and (Sports, News) are also significantly different.

7. The new SSE is 157,565 with 14 degrees of freedom. The values of the F statistics for "tree species" and "flake size" are 7.876 and 0.294, with corresponding p-values of 0.005 and 0.596. H_0^A is rejected at level 0.05, but H_0^B is not rejected.

9. (a) With "period" being the row factor and "trap" being the column factor, the p-values are 0.0661 and 0.0005. Both null hypotheses are rejected at level 0.05.

(b) The p-value for Tukey's one degree of freedom test for additivity is 0.0012, suggesting that the factors interact.

(c) Yes.

11. (a) With "Auxin" being the row factor and "Kinetin" being the column factor, the p-values are 2.347×10^{-11} and 4.612×10^{-11}. Both null hypotheses are rejected at level 0.05.

(b) All pairs of levels of the "Auxin" factor, except for (0.1, 0.5), (0.1, 2.5), and (0.5, 2.5), are significantly different at experiment-wise error rate of 0.01.

(c) The p-value for Tukey's one degree of freedom test for additivity is 1.752×10^{-6}, suggesting that the factors interact.

Section 11.3

1. (a) $X_{ijk\ell} = \alpha_i + \beta_j + \gamma_k + (\alpha\beta)_{ij} + (\alpha\gamma)_{ik} + (\beta\gamma)_{jk} + (\alpha\beta\gamma)_{ijk} + \varepsilon_{ijk\ell}$. All main effects and interactions, except for the main factor B effect (p-value 0.081) and the three-way interaction effects (p-value 0.996), are significant at level 0.05.

(b) $X_{ijk\ell} = \alpha_i + \beta_j + \gamma_k + (\alpha\beta)_{ij} + (\alpha\gamma)_{ik} + (\beta\gamma)_{jk} + \varepsilon_{ijk\ell}$. All main effects and interactions, except for the main factor B effect (p-value 0.073), are significant at level 0.05.

(c) With the data read in *h2o*, and *fit1* defined by *fit1=aov(y~MS*SH*MH, data=h2o)*, the command *h2o$res=resid(fit1); anova(aov(res**2~MS*SH*MH, data=h2o))* yields the p-value of 0.0042 for the main factor C effect on the residual variance. The two other p-values are less than 0.05, suggesting the homoscedasticity assumption does not hold. The p-value of the Shapiro-Wilk test for normality is 0.07, though in the presence of heteroscedasticity this is not easily interpretable.

(d) After the square root arcsine transformation on the response variable, only the main factor C effect on the residual variance has a p-value less than 0.05 (0.012). The p-value of the Shapiro-Wilk test is 0.64, suggesting the normality assumption is tenable.

3. (a) $x_{221} = ab = 16$, $x_{112} = c = 12$, and $x_{122} = bc = 18$.

(c) $\alpha_1 = -1.375$, $\beta_1 = -1.625$, $\gamma_1 = -2.375$, $(\alpha\beta)_{11} = 0.875$, $(\alpha\gamma)_{11} = -0.375$, $(\beta\gamma)_{11} = 0.875$, $(\alpha\beta\gamma)_{111} = 1.375$.

(d) $SSA = 16 \times 1.375^2 = 30.25$, $SSB = 16 \times 1.625^2 = 42.25$, $SSC = 16 \times 2.375^2 = 90.25$, $SSAB = 16 \times 0.875^2 = 12.25$, $SSAC = 16 \times 0.375^2 = 2.25$, $SSBC = 16 \times 0.875^2 = 12.25$, $SSABC = 16 \times 1.375^2 = 30.25$.

(e) $F_{H_0}^A = 30.25/((12.25 + 2.25 + 12.25 + 30.25)/4) = 2.12$, $F_{H_0}^B = 42.25/((12.25 + 2.25 + 12.25 + 30.25)/4) = 2.96$, $F_{H_0}^C = 90.25/((12.25 + 2.25 + 12.25 + 30.25)/4) = 6.33$; these test statistics are all less than $F_{1,4,0.05}$, so none of the main effects is significantly different from zero.

5. (a) True. **(b)** True.

(c) $F_{H_0}^A = 11.27$, $F_{H_0}^B = 1.01$, $F_{H_0}^C = 2.92$; $F_{H_0}^A > F_{1,4,0.05} = 7.709$ so the hypothesis of no main factor A effects is rejected. The other test statistics are all less than $F_{1,4,0.05}$, so the main effects of factors B and C are not significantly different from zero.

Section 11.4

1. If θ_1 is the effect of block 1, and $\theta_2 = -\theta_1$ is the effect of block 2, then the mean μ_{ijk} of X_{ijk} is $\mu_{ijk} = \mu + \alpha_1 + \beta_j + \gamma_k + (\alpha\beta)_{ij} + (\alpha\gamma)_{ik} + (\beta\gamma)_{jk} + (\alpha\beta\gamma)_{ijk} + \theta_1$ if (i, j, k) is one of $(1, 1, 1), (2, 2, 1), (2, 1, 2), (1, 2, 2)$, and the same expression but with θ_1 replaced by θ_2 if (i, j, k) is one of the other four sets of indices. Now use the expressions for the contrasts estimating the different effects, which are given in Example 11.3.1, to verify that θ_1 and θ_2 cancel each other in all contrasts except the one for the three-factor interaction.

3. (a) *ABDE*.

(b) *BE*.

(c) *G=rbind(c(1, 1, 1, 0, 0), c(0, 0, 1, 1, 1)); conf.design(G, p=2)* for part (a), and *G=rbind(c(0, 1, 1, 1, 0), c(0, 0, 1, 1, 1)); conf.design(G, p=2)* for part (b).

5. (a) *block=c(rep(2,16)); for(i in c(1, 4, 6, 7, 9, 12, 14, 15))block[i]=1; sr$block=block* generate the "block" variable in the data frame *sr*, and *anova(aov(y~block+A*B*C, data=sr))* returns the ANOVA table.

(b) The three main effects and the AB interaction effect are significantly different from zero.

7. (a) *G=rbind(c(1, 1, 0, 0), c(0, 0, 1, 1)); conf.design(G, p=2)* give the allocation of runs into the four blocks.

(b) *x1=rep(4, 16); for(i in c(1, 4, 13, 16))x1[i]=1; for(i in c(5, 8, 9, 12))x1[i]=2; for(i in c(2, 3, 14, 15))x1[i]=3; block=c(x1, x1); aw$block=as.factor(block)* generate the "block" as part of the data frame *aw*.

(c) *anova(aov(y~block+A*B*C*D, data=aw))* returns the ANOVA table. Only the main effects for factors A and B are significant (p-values of 3.66×10^{-5} and 4.046×10^{-11}).

9. The contrasts $(-\bar{x}_{211}+\bar{x}_{121}+\bar{x}_{112}-\bar{x}_{222})/4$, $(\bar{x}_{211}-\bar{x}_{121}+\bar{x}_{112}-\bar{x}_{222})/4$, and $(\bar{x}_{211}+\bar{x}_{121}-\bar{x}_{112}-\bar{x}_{222})/4$ estimate $\alpha_1-(\beta\gamma)_{11}$, $\beta_1-(\alpha\gamma)_{11}$, and $\gamma_1-(\alpha\beta)_{11}$, respectively. Thus, the alias pairs are $[A, BC]$, $[B, AC]$, $[C, AB]$.

11. (a) $[A, BCDE]$, $[B, ACDE]$, $[C, ABDE]$, $[D, ABCE]$, $[E, ABCD]$, $[AB, CDE]$, $[AC, BDE]$, $[AD, BCE]$, $[AE, BCD]$, $[BC, ADE]$, $[BD, ACE]$, $[BE, ACD]$, $[CD, ABE]$, $[CE, ABD]$, $[DE, ABC]$.

(b) With the data read into the data frame *df*, the sums of squares of the classes of aliased effects can be obtained by the command *anova(aov(y~A*B*C*D*E, data=df))*. It is not possible to test for the significance of the effects.

Chapter 12

Section 12.2

1. (a) 58.5.

(b) 46.8.

(c) Increases by 2.7.

(d) $\beta_0 = 46.8$, $\beta_1 = 2.7$, $\beta_2 = 0.9$.

3. (a) $-3.2, -0.4, 1.0$.

(b) $\beta_2 = 0.7$, $\beta_1 = -0.4$, $\beta_0 = -12.1$.

Section 12.3

1. (a) 10.4577 – 0.00023GS – 4.7198WS + 0.0033T; $R^2 = 0.8931$; p-value$= 0.021$; yes.

(b) 5.9367; −1.0367.

(c) *r1=lm(abs(rstandard(fit))~poly(fitted(fit),2)); summary(r1)* returns p-values of 0.60 and 0.61 for the two regression slope parameters, suggesting the homoscedasticity assumption is not contradicted; *shapiro.test(rstandard(fit))* returns a p-value of 0.93, suggesting the normality assumption is not contradicted.

(d) Only the coefficient of "Wind Speed" is significantly different from zero at level 0.05 (p-value of 0.026).

(e) (1.30, 19.62), (−0.0022, 0.0017), (−8.51, −0.92), (−0.171, 0.177).

3. (a) $70.94 + 10^{-5}5.18x_1 - 10^{-5}2.18x_2 + 10^{-2}3.382x_3 - 10^{-1}3.011x_4 + 10^{-2}4.893x_5 - 10^{-3}5.735x_6 - 10^{-8}7.383x_7$; R^2adj: 0.6922; p-value: $10^{-10}2.534$. The model is useful for predicting life expectancy.

(b) The variables "Income," "Illiteracy," and "Area" are not significant at level 0.05 (p-value of 0.9993).

(c) R^2 is almost the same for both models; this is consistent with the p-value in part (b) as they both suggest that "Income," "Illiteracy," and "Area" are not significant predictors; R^2adj is bigger for the reduced model, since the (almost identical) R^2 is adjusted for fewer predictors.

(d) *r1=lm(abs(rstandard(h2))~poly(fitted(h2), 2)); summary(r1)* returns p-values of 0.67 and 0.48 for the two regression slope parameters, suggesting the homoscedasticity assumption is not contradicted; *shapiro.test(rstandard(h2))* returns a p-value of 0.56, suggesting the normality assumption is not contradicted.

(e) 71.796; $71.796 + (-0.3001)(5 - 10.3) = 73.386$; *predict(h2, data.frame(Population=21,198, Murder=5, HS.Grad=62.6, Frost=20), interval="prediction")* returns the fitted value (73.386) and a 95% prediction interval of (71.630, 75.143).

5. (a) R^2: 0.962; adjusted R^2: 0.946; significant (p-value $= 2.4 \times 10^{-5}$).

(b) The quadratic and cubic terms are jointly significant at level 0.01 (p-value$= 9.433 \times 10^{-5}$).

(c) Polynomial terms of order 4-8 are not jointly significant at level 0.01 (p-value 0.79); R^2: 0.9824, adjusted R^2: 0.912; compared to those in (a), R^2 is somewhat bigger, but R^2adj is somewhat smaller, consistent with the non-significance of the higher order polynomial terms.

Section 12.4

2. (a) *r1=lm(abs(rstandard(fit))~poly(fitted(fit),2)); summary(r1)* returns a p-value of 0.038 for the model utility test, suggesting violation of the homoscedasticity assumption.

(b) p-value$= 10^{-10}1.373$.

(c) WLS.

5. **(a)** $X = 1$ or -1 depending on whether the observation comes from population 1 or 2.

   (b) $\mu_1 = \beta_0 + \beta_1$, $\mu_2 = \beta_0 - \beta_1$.

   (c) Both p-values are $10^{-10}6.609$.

   (d) 0.0898, 0.0775.

   (e) $< 10^{-16}2.2$, $10^{-10}1.626$.

7. mmax: $10^{-15}1.18$; cach: $10^{-6}5.11$; mmin: $10^{-15}4.34$; chmax: $10^{-11}3.05$; syct: 0.00539; model utility test: $< 10^{-16}2.2$.

11. **(a)** No variable is significant at level 0.05. The R^2 value of 0.9824 and the p-value of $10^{-7}4.756$ for the model utility test suggest that at least some of the variables should be significant. This is probably due to multicollinearity.

    (b) 38.50, 254.42, 46.87, 282.51. Yes.

    (c) Variables *x1* and *x2* are highly significant. No.

    (d) *x1* and *x2*. No.

13. **(a)** R^2: 0.3045, R^2adj: 0.2105. Yes (p-value= 0.016). No.

    (b) 13.95, 14.19, 1.92, 1.33, 1.20. Yes. Non-significance of predictors, significant model utility test.

    (c) According to the C_p criterion, the best model retains only "lowid" and "tarsus."

    (d) R^2: 0.2703, R^2adj: 0.2338, model utility test: 0.0018, lowid: 0.0059, tarsus: 0.0320. Somewhat smaller R^2, somewhat larger R^2adj, smaller p-value for the model utility test. The new variance inflation factors are both equal to 1.0098. Multicollinearity is not an issue now.

Chapter 13

Section 13.2

2. **(a)** It brought the subgroup means within the control limits.

   (b) No.

   (c) No.

   (d) 0.987.

4. **(a)** No, no.

   (b) Yes, no.

Section 13.3

3. Both charts show most points after the calibration period to be below the center line. It appears that the adjustment had no effect on the process variability.

Section 13.4

1. The chart produced by the commands *data(orangejuice2); attach(orangejuice2); qcc(D[trial], sizes=size[trial], type="p", newdata=D[!trial], newsizes=size[!trial])* suggests that the process remains in control.

Section 13.5

1. **(a)** The process is out of control (5 points outside the control limits).

   (b) One point shown out of control.

INDEX